# Telomeres and Telomerase

A subject collection from *Cold Spring Harbor Perspectives in Biology*

# Telomeres and Telomerase

A subject collection from *Cold Spring Harbor Perspectives in Biology*

EDITED BY

**Julia Promisel Cooper**
*University of Colorado Anschutz Medical Campus*

**Eros Lazzerini Denchi**
*National Cancer Institute, Bethesda*

**Joachim Lingner**
*Ecole Polytechnique Fédérale de Lausanne*

**Hilda A. Pickett**
*Children's Medical Research Institute*

**COLD SPRING HARBOR LABORATORY PRESS**
Cold Spring Harbor, New York • www.cshlpress.org

# Telomeres and Telomerase

A subject collection from *Cold Spring Harbor Perspectives in Biology*
Articles online at www.cshperspectives.org

| | |
|---|---|
| Executive Editor | Richard Sever |
| Project Supervisor | Barbara Acosta |
| Editorial Assistant | Danett Gil |
| Permissions Administrator | Carol Brown |
| Production Editor | Diane Schubach |
| Production Manager/Cover Designer | Denise Weiss |
| Publisher | John Inglis |

*Front cover artwork:* Micrograph of a mitotic HT1080 6TG human fibrosarcoma cell depleted of the telomere-protective factor TRF2. The cell was cytocentrifuged onto a glass slide and stained for DNA (DAPI, blue), telomeres by fluorescence in situ hybridization (green), and $\gamma$-H2AX by immunofluorescence (red). Deprotected telomeres are identified by colocalization of telomeric and $\gamma$-H2AX signals. Image credit: Anthony J. Cesare, Head of the Genome Integrity Unit, Children's Medical Research Institute, The University of Sydney, Australia.

*Library of Congress Cataloging-in-Publication Data*

Names: Cooper, Julia P., 1961- editor | Lazzerini Denchi, Eros editor | Lingner, Joachim editor | Pickett, Hilda A. editor
Title: Telomeres and telomerase : a subject collection from Cold Spring Harbor Perspectives in biology / edited by Julia Promisel Cooper, University of Colorado Anschutz Medical Campus, Eros Lazzerini Denchi, National Cancer Institute, Joachim Lingner, Ecole Polytechnique Fédérale de Lausanne and Hilda A. Pickett, Children's Medical Research Institute.
Other titles: Telomeres and telomerase (Cold Spring Harbor Laboratory Press)
Description: Cold Spring Harbor, New York : Cold Spring Harbor Laboratory Press, [2026] | "A subject collection from Cold Spring Harbor Perspectives in Biology"--title page. | Includes bibliographical references and index. | Summary: "The ends of chromosomes comprise specific DNA sequences called telomeres. These are bound to by specialist proteins and may play a role in aging. This volume summarizes recent progress in our understanding of the molecular details of the mechanisms that maintain and replicate telomeres"-- Provided by publisher.
Identifiers: LCCN 2025011683 (print) | ISBN 9781621824923 hardcover | ISBN 9781621824930 epub
Subjects: LCSH: Telomere | Telomerase
Classification: LCC QH600.3 .T4554 2025 (print) | LCC QH600.3 (ebook)
LC record available at https://lccn.loc.gov/2025011683
LC ebook record available at https://lccn.loc.gov/2025011684

For a complete catalog of all Cold Spring Harbor Laboratory Press publications, visit our website at www.cshlpress.org.

# Contents

Contents

# Preface

THE PAST TWO DECADES HAVE BEEN TRANSFORMATIVE for telomere biology and telomerase research. Since the publication of *Telomeres, Second Edition*, 2006, the field has experienced an exponential increase in publications, reflecting both the depth and breadth of discovery. In 2009, the award of the Nobel Prize in Physiology or Medicine to Elizabeth Blackburn, Carol Greider, and Jack Szostak recognized the pioneering work that first brought telomeres and telomerase to the forefront of modern biology. This milestone underscored the central importance of telomeres to genome stability, cellular aging, and cancer.

In the years since, the pace of progress has only accelerated. Powerful new technologies, from single-molecule imaging to large-scale genomics, proteomics, and artificial intelligence, have enabled ever more detailed insight into telomere structure, regulation, and function. The scope of the literature has expanded accordingly, as has the translational impact of the field, with advances in diagnostics, therapeutic strategies, and clinical applications.

This volume is intended as a compendium, presenting an opportunity to lay out the striking progress made in this rapidly advancing field. As with the literature itself, the chapters in this volume have expanded, covering topics that range from the molecular architecture of telomeres and telomerase to the mechanisms underlying the Alternative Lengthening of Telomeres (ALT) Pathway, and the growing interface between telomere biology, human health, and therapeutic innovation.

We are well aware that this research topic is vast and have endeavored to assemble a representative slice that captures the diversity of perspectives and the dynamism of ongoing discovery. The contributions in this volume are authored by many of the leading scientists in the field, whose insights collectively offer both an authoritative reference and an inspiration for future research.

In bringing together this collection, our aim has been to showcase how far the field has advanced since 2006, and to provide a foundation for the next generation of breakthroughs. Telomere biology continues to exemplify the interplay of fundamental curiosity and translational promise, ensuring its place at the heart of biomedical research for years to come.

We thank Danett Gil, Barbara Acosta, Richard Sever, and their colleagues at Cold Spring Harbor Laboratory Press for their considerable effort in bringing this volume to completion, and for their patience in coordinating chapters and accommodating the inevitable delays that come with assembling a work of this scope. Their support has been invaluable in keeping this project on track. We feel fortunate to work in such a vibrant and rapidly evolving field, and we hope that this new edition of an old favorite will not only serve as a trusted reference but also spark thought, provoke investigation, and motivate new directions for the years ahead.

JULIA PROMISEL COOPER
EROS LAZZERINI DENCHI
JOACHIM LINGNER
HILDA A. PICKETT

# How Shelterin Orchestrates the Replication and Protection of Telomeres

Titia de Lange

Laboratory for Cell Biology and Genetics, Rockefeller University, New York, New York 10065, USA

*Correspondence:* delange@rockefeller.edu

Efforts to determine how telomeres solve the end-protection problem led to the discovery of shelterin, a conserved six-subunit protein complex that specifically binds to the long arrays of telomeric TTAGGG repeats at vertebrate chromosome ends. The mechanisms by which shelterin prevents telomeres from being detected as sites of DNA damage and how shelterin prevents inappropriate DNA repair pathways are now largely known. More recently, shelterin has emerged as a central player in solving the second major problem at telomeres: how to complete the duplication of telomeric DNA. This end-replication problem results from the inability of the canonical DNA replication machinery to maintain the DNA at chromosome ends. Shelterin solves this problem by recruiting two enzymes that can replenish the lost telomeric repeats: telomerase and CST-Polα/primase. How shelterin accomplishes these critical tasks is reviewed here.

## SHELTERIN COMPOSITION

Given the multitude of its functions, human shelterin is a remarkably simple complex comprising just six distinct subunits: TRF1, TRF2, Rap1, TIN2, TPP1, and POT1 (Fig. 1A). The TRF proteins are named Telomeric Repeat binding Factors for their ability to recognize the double-strand (ds) TTAGGG repeats that make up metazoan telomeres (Zhong et al. 1992). Their specificity for telomeric DNA derives from carboxy-terminal Myb/SANT domains, simple three-helix bundles that stably engage telomeric DNA when two TRF proteins form a dimer through the homotypic interactions of their TRF homology (TRFH) domains (Fig. 1B; Bianchi et al. 1997; Broccoli et al. 1997). In mammals, TRF1 and TRF2 evolved short, dis-ordered amino-terminal domains with distinct charge features—a positively charged basic domain in TRF2 and a negatively charged acidic domain in TRF1 (Poulet et al. 2012; Myler et al. 2021). TRF1, the most recent addition in the evolution of the human complex, emerged through a duplication of the metazoan TRF2-like TRF gene and diverged from TRF2 in structure and function (Fig. 1C; Myler et al. 2021). In both proteins, the Myb domain is separated from the TRFH by a Hinge region that serves as a flexible linker (Fig. 1A). The Hinge of TRF2 is important for protein interactions, containing a Rap1-binding motif (RBM), a TIN2-binding motif (TBM), and the inhibitor of DNA damage response domain, which binds and affects the Mre11-Rad50-Nbs1 (MRN) complex (see below). Like TRF2, TRF1 binds to TIN2, but this

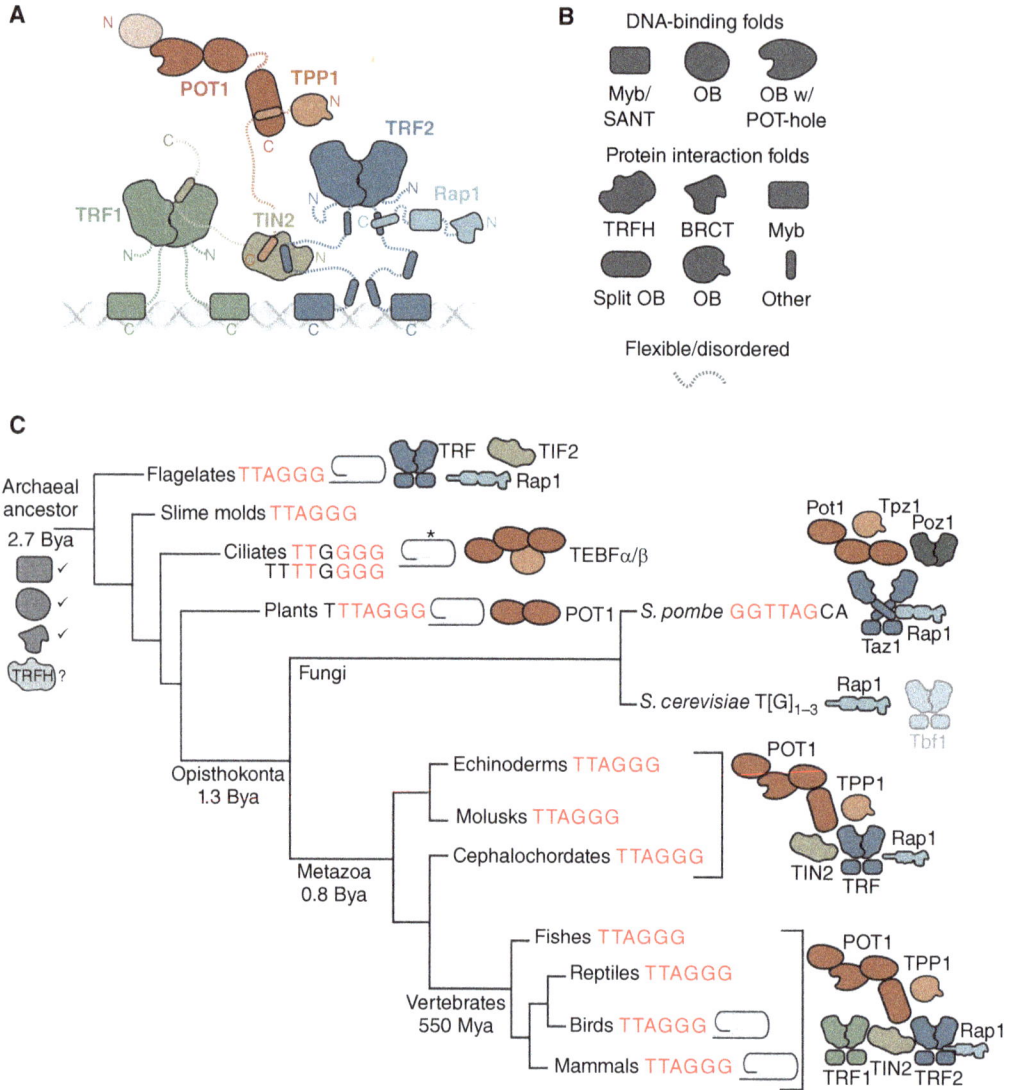

Figure 1. The structure and evolution of shelterin. (*A*) Schematic of the six subunits of mammalian shelterin and their interactions. For simplicity, shelterin is not drawn as the fully dimeric complex that can be reconstituted in vitro (Zinder et al. 2022). The fourth oligosaccharide/oligonucleotide-binding (OB) fold at the amino terminus of POT1 is absent from primate and rodent POT1s (Myler et al. 2021). (*B*) Summary of the protein folds that are found in shelterin components. (*C*) Schematic highlighting the conservation of shelterin components, telomeric repeats, and t-loops during eukaryotic evolution.

interaction is mediated by a site that includes F142 in the TRFH domain rather than in its Hinge (Fig. 1A; Chen et al. 2008). The analogous TRFH site in TRF2 (situated around F120 in the short form of TRF2) does not bind to TIN2 but functions to bring shelterin accessory factors to telomeres (e.g., the Apollo exonuclease) (Chen et al. 2008). The dual binding of TIN2 to TRF1 and TRF2 stabilizes both proteins on telomeres, creating a shelterin "core" complex composed of TRF1-TIN2-TRF2-Rap1 (Liu et al. 2004a; Ye et al. 2004a). This core of four shelterin subunits

Cite this article as *Cold Spring Harb Perspect Biol* doi: 10.1101/cshperspect.a041685

is abundant at telomeres, possibly numbering in the 100s–1000s at each chromosome end (Takai et al. 2010).

A subset of shelterin cores (perhaps 10%) (Takai et al. 2010) are associated with TPP1/POT1, a heterodimer that is largely composed of oligosaccharide/oligonucleotide-binding (OB) folds. Two OB folds in POT1 bind to single-stranded (ss) telomeric DNA and its third (split) OB fold interacts with the recruitment domain (RD) of TPP1. TPP1 has a single amino-terminal OB fold that interacts with hTERT (Wan et al. 2009; Abreu et al. 2010; Nandakumar et al. 2012; Zhong et al. 2012) (see below) and a long unstructured region at its carboxy terminus, linking the RD to its TBM (Fig. 1A, B). It is estimated that there may be ~50 copies of TPP1/POT1 per telomere (Takai et al. 2010), which would be an excess over the available ssDNA-binding sites in the 3′ overhang (~10 per telomere based on the recent estimate of ~100 nt of ssDNA per end) (Takai et al. 2024). The telomeric accumulation of TPP1/POT1 is dependent on the interaction of TPP1 with the monomeric TRFH domain of TIN2 in the shelterin core (Houghtaling et al. 2004; Liu et al. 2004b; Ye et al. 2004b). Given that there is more TIN2 than TPP1 at telomeres, it will be interesting to determine what limits the accumulation of TPP1/POT1.

## SHELTERIN STRUCTURE

Shelterin carries three types of DNA-binding folds (ssDNA-binding OB folds; dsDNA-binding Myb/SANT domains; and the specialized amino-terminal OB fold of POT1 that binds both the ssDNA as well as 5′ end of the C-rich strand with its "POT-hole") and four distinct protein interaction folds (the dimeric TRFH domains of TRF1 and TRF2; the monomeric TRFH of TIN2; the Myb domain of Rap1; the OB fold of TPP1; and the BRCT domain of Rap1). Structures of each of these folds have been obtained from either crystallographic analysis or high-resolution cryogenic-electron microscopy (cryo-EM) (for review, see Hu et al. 2024). In addition, structures are available of short peptides that function as interaction inter-

faces, such as the binding of the carboxy terminus of Rap1 to TRF2, a peptide of Apollo bound to the F120 region of TRF2, peptides of TRF2 and TPP1 bound to the TIN2 TRFH, and the RD region of TPP1 bound to the carboxy terminus of POT1 (for review, see Hu et al. 2024). However, attempts to determine the structure of full-length shelterin subunits, let alone the whole complex, have been stymied by flexible linkers (Fig. 1A; Zinder et al. 2022). The only exception is POT1 whose cryo-EM structure was recently resolved in complex with its binding partner CST (Cai et al. 2024). Negative-stain EM revealed the extensive conformational flexibility of shelterin, and mass photometry showed that the whole complex can form a dimer with two copies of each of the six subunits (Zinder et al. 2022). It was speculated that the conformational heterogeneity of shelterin may have functional significance since the complex likely has to bind to DNA in a variety of conformations (e.g., branched DNA, nucleosomal DNA, linker DNA, etc.) while also interacting with distinct shelterin accessory proteins.

## SHELTERIN EVOLUTION

In most eukaryotes, telomeres contain protein complexes with one or more of the four major structural folds (OB, Myb, TRFH, and BRCT) observed in mammalian shelterin (Fig. 1C). For instance, trypanosome telomeres have a TRF ortholog containing Myb and TRFH domains (Li et al. 2005), a Rap1 ortholog with Myb and BRCT domains (Li et al. 2005), and a TIN2 ortholog with a TRFH domain (Fig. 1C; Jehi et al. 2014). Most eukaryotes also carry OB fold–containing ssDNA-binding proteins at their telomeres (Fig. 1C; for review, see Lewis and Wuttke 2012). A nearly complete shelterin complex, containing orthologs of Rap1, POT1, TPP1 (Tpz1), and a TRF (Taz1) is found at the telomeres of *Schizosaccharomyces pombe* (Fig. 1C). However, unlike most TRFs, *S. pombe* Taz1 does not use its TRFH-like domain for dimerization (Deng et al. 2015; Xue et al. 2017; T Germe and J Cooper, pers. comm.).

The conservation of shelterin suggests an ancient origin of the complex, perhaps dating back to the earliest eukaryotes. Compellingly, three of the four protein folds typically observed in shelterin (OB, Myb, and BRCT) are found in nontelomeric proteins in Archaea. Whether Archaea also have a TRFH-containing protein is difficult to discern due to the low level of conservation of this fold. Nonetheless, it seems likely that Archaea had most of the components to form a shelterin-like complex before the first linear chromosomes emerged. They may have obtained an alphaproteobacterial mobile element that carried the reverse transcriptase precursor of telomerase (Nakamura and Cech 1998), and they likely contained a form of RPA related to the CST complex (Cai and de Lange 2023). Therefore, the building blocks for a complete telomere protection and maintenance system may have preceded the emergence of linear chromosomes (for speculations on how linear chromosomes emerged, see de Lange 2015).

## THE END-PROTECTION PROBLEM: THE THREAT OF DNA DAMAGE SIGNALING

The first cell harboring linear chromosomes must have had a mechanism to avoid activation of the DNA damage response (DDR) at its chromosome ends. In most present-day eukaryotes, including mammals, DNA ends can be detected by two related kinases, ATM and ATR (Fig. 2A–D). The ATM kinase responds to DNA ends through the agency of the MRN complex, which can slide along DNA until a free end is reached (Fig. 2A; for reviews, see Stracker and Petrini 2011; Lee and Paull 2021; Hopfner 2023). At a free end, MRN adopts a closed conformation that can bind and activate the ATM kinase (Fig. 2A). This mechanism for DNA end detection by Mre11 and Rad50 is ancient, preceding the evolution of eukaryotes (for review, see Hopfner 2023). The mechanism of ATM activation by MRN is still not fully understood but it is known to involve the conversion of inactive ATM dimers into active monomers (Fig. 2A; Bakkenist and Kastan 2003). When ATM is activated, it phosphorylates the H2A variant

H2AX in the nearby nucleosomes, initiating a cascade of protein modifications and protein-binding events that create cytologically observable DNA damage foci. Importantly, ATM can enforce cell cycle arrest through phosphorylation of Chk2 and p53 (for review, see Ciccia and Elledge 2010).

Activation of the ATR kinase at DNA ends requires the presence of ssDNA, generated at sites of replication stress or resulting from 5′ end resection at DNA double-strand breaks (DSBs) (for review, see Ciccia and Elledge 2010). The ssDNA is first detected by RPA, an abundant ssDNA-binding protein, which recruits ATR through a direct interaction with its binding partner ATRIP (Fig. 2D). Locally bound ATR is activated when it interacts with TOPBP1, a multifunctional protein that bears the ATR-activating domain (Fig. 2D). TOPBP1 is bound to the 9-1-1 clamp, which is loaded by Rad17/RFC positioned at the 5′ end adjacent to where ATR/ATRIP resides on the ssDNA (Fig. 2D). As a result, ATR is activated when there is a 5′ ds–ss junction close to the ssDNA. Once activated, ATR, like ATM, phosphorylates an effector kinase (Chk1) that blocks cell cycle progression and phosphorylates H2AX (creating γ-H2AX), initiating the formation of DNA damage foci. Alternatively, ATR can be activated by ETAA1 bound to RPA (Bass et al. 2016; Feng et al. 2016; Haahr et al. 2016; Lee et al. 2016).

As telomeres carry sufficient ssDNA next to a 5′ ds–ss transition to allow ATR activation and represent a free DNA where MRN can activate ATM, they are threatened by both pathways. Without shelterin repressing both kinases, the ATM/ATR signaling cascades will induce cell cycle arrest and senescence or apoptosis.

Telomeres also contain DNA structures that can be recognized by the poly(ADP-ribose) polymerase PARP1. PARP1 needs to be controlled at telomeres because persistent activation of PARP1 could lead to inactivation of shelterin proteins through their PARsylation, PARsylation of telomeric DNA (Wondisford et al. 2024), and depletion of the cellular pool of $NAD^+$, the precursor for poly(ADP-ribose) chains. Because the mechanism of PARP1 regulation at telomeres is not well understood, this

Cite this article as *Cold Spring Harb Perspect Biol* doi: 10.1101/cshperspect.a041685

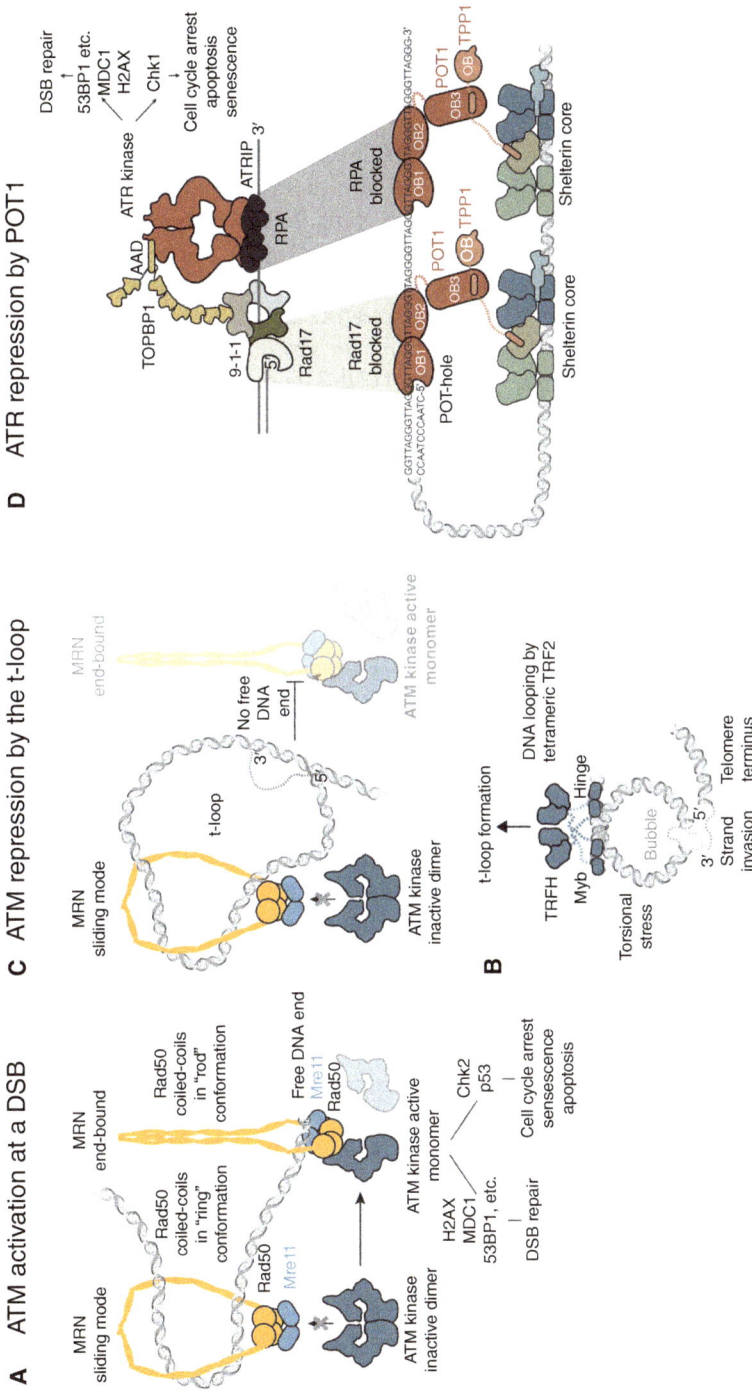

**Figure 2.** How shelterin prevents that activation of ATM and ATR signaling at telomeres. (*A*) Speculative model of the mechanism of ATM activation by the Mre11–Rad50–Nbs1 (MRN) complex in which the "closed" state of MRN at a free DNA end and allows the conversion of inactive ATM dimers into active monomers. (*B*) Proposed mechanism for t-loop formation by tetrameric TRF2. (*C*) Proposed model for how the constrained DNA end in the t-loop structure prevents MRN from attaining the "closed" conformation and prevents MRN-mediated activation of ATM at telomeres. (*D*) How POT1 blocks ATR signaling through preventing binding of RPA and Rad17 to telomere ends. (*Top*) Schematic of the activation of ATR kinase signaling at a resected DNA end. (*Bottom*) Cartoon depicting POT1 blocking RPA accumulation and thus ATR binding at telomeres through its ssDNA-binding activity. In addition, the POT-hole in the amino-terminal OB fold of POT1 can prevent ATR activation by blocking Rad17/RFC from gaining access to the 5′ end of the C-rich strand.

review will not delve into this aspect of the end-protection problem.

## BLOCKING ATM: RESTRAINING THE END IN THE t-LOOP

The primary role of shelterin is to avoid the activation of the ATM and ATR kinases at chromosome ends. Failure at this task, even at just a few of the 92–184 telomeres in a diploid human cell, could induce cell cycle arrest and cell death. Shelterin accomplishes this task using TRF2 and POT1, which block ATM and ATR signaling, respectively, by interfering with an early step in their activation pathways.

When TRF2 is removed from telomeres, the ATM signaling cascade becomes locally activated, leading to Chk2- and p53-mediated cell cycle arrest. Despite the absence of TRF2, POT1 can continue to repress ATR signaling at telomeres because it remains bound to TPP1-TIN2-TRF1. TRF2 prevents ATM signaling by hiding the chromosome end in the t-loop structure (Doksani et al. 2013). In the t-loop, the 3′ overhang invades the telomeric dsDNA creating a lariat structure in which the telomere end is locked down (Griffith et al. 1999). The protective feature of the t-loop is not dependent on the DNA loop or the sequestration of the 3′ overhang per se. Rather, t-loops protect telomeres by imposing a topological constraint on the DNA end. A free DNA end that can pass through the Rad50 coiled-coiled arms is required for the activation of the ATM kinase by MRN (Fig. 2A). When the DNA end is restrained, as in the t-loop structure, MRN will remain in its "sliding" conformation and ATM is not activated (Fig. 2B). It is likely that the first eukaryotes already used some version of the t-loop strategy to prevent the Mre11 complex from engaging telomere ends since t-loops are found throughout eukaryotes (Fig. 1C).

TRF2 is the only subunit in shelterin that is required for t-loop formation (Doksani et al. 2013). TRF2 does not require the presence of other shelterin components, including its binding partners Rap1 and TIN2, for t-loop formation and does not rely on the Rad51 recombinase for this feat (Timashev and De Lange

2020). Recent data has provided insights into how TRF2 creates t-loops (Goldfarb et al. 2024). Unlike TRF1, which binds DNA as a dimer, TRF2 forms a tetramer on telomeric DNA. Tetramerization by TRF2 depends on its TRFH domain as well as a second dimerization interface that involves the Hinge and the Myb domains (Fig. 2C). A tetrameric TRF2 is proposed to have the ability to induce torsional stress by looping the DNA between two TRF2-bound segments (Fig. 2C). The resulting torsional stress is expected to unwind the DNA in the loop, creating a bubble that can capture the 3′ overhang, thereby forming a t-loop (Fig. 2C). In support of this model, an engineered tetramer form of TRF1 formed t-loops and repressed ATM signaling (Goldfarb et al. 2024). Conversely, a TRF2 mutant lacking one of its two dimerization domains localized to telomeres but failed to repress ATM signaling, and this defect was partially restored by addition of an orthogonal dimerization module.

Work with a temperature-sensitive (ts) mutant of TRF2 showed that ATM is activated at telomeres in both $G_1$ and in S phase within hours of shifting cells to the nonpermissive temperature (Konishi and de Lange 2008). Conversely, ATM signaling is rapidly dampened by a shift to the permissive temperature of this TRF2 mutant. The simplest interpretation of these results is that t-loops are short-lived in absence of TRF2 and/or need to be constantly reformed by TRF2. Such dynamic behavior echoes the transient nature of other DNA looping reactions in the nucleus (Bartman et al. 2016; Gabriele et al. 2022).

Interestingly, the presence of t-loops creates a problem for the replisome when it moves outward to replicate the telomeric DNA. It is thought that mounting topological stress between the fork and the base of the t-loop overwhelms topoisomerases and results in fork arrest. This problem is solved by the TRF2-bound RTEL1 helicase, which can remove t-loops, presumably through its D-loop resolution activity (Vannier et al. 2012; Sarek et al. 2015, 2019). CDK-mediated phosphorylation of a site in the TRF2 Hinge controls the binding of RTEL1 such that the enzyme only resolves t-loops in S phase (Sarek et al. 2019). The finding that t-loops disappear, and that ATM becomes activated, when RTEL1 is permanently bound to TRF2—a situation created by a phosphomimic mutation of the TRF2 CDK site—further argues that ATM signaling is repressed through t-loop formation (Sarek et al. 2019).

## BLOCKING ATR: A COVER-UP BY POT1

POT1 helps to solve the end-protection problem by preventing the activation of ATR signaling at telomeres. The t-loop structure that is so effective in the case of ATM repression does not solve the problem posed by ATR signaling. This was demonstrated by deletion of the two mouse POT1 proteins (POT1a and POT1b), which leads to an ATR-dependent DNA damage response at most (if not all) telomeres despite the presence of t-loops (Hockemeyer et al. 2006; Denchi and de Lange 2007; Doksani et al. 2013). The failure of the t-loop to avert ATR activation is likely due to the structure at the base of the t-loop, where ssDNA is positioned close to the 5′ end of the telomere, allowing RPA to load ATR/ATRIP and 9-1-1-bound TOPBP1 to activate the kinase (see Fig. 2D). A series of experiments showed that the ability of POT1 to prevent activation of ATR signaling is dependent on its ability to bind to ssDNA with its two amino-terminal OB folds and its interaction with TPP1 and the rest of shelterin (Palm et al. 2009; Flynn et al. 2011; Frescas and de Lange 2014; Pinzaru et al. 2016; Gu et al. 2017; Kim et al. 2021). When POT1 is removed or when it lacks DNA-binding activity, RPA can be detected at telomeres (Gong and de Lange 2010; Flynn et al. 2011). The competition between POT1 and RPA for the ssDNA is stacked in favor of RPA, which is 200-fold more abundant in the nucleus and has the same affinity for ss TTAGGG repeats as POT1 (Takai et al. 2011). Yet, POT1 can prevent RPA from gaining a foothold on the telomeric ssDNA by virtue of its link to the shelterin core anchored on the ds telomeric DNA. When this connection is severed, POT1 loses the ability to accumulate at telomeres and RPA takes its place on the ssDNA (Takai et al. 2011). The importance of POT1-

tethering to the shelterin core was revealed by replacing the POT1 DNA-binding domain with that of RPA70 that was mutated to prevent its interaction with ATRIP (Kratz and de Lange 2018). This RPA-POT1 chimera localized to telomeres by binding to TPP1 and conferred significant protection from ATR signaling. However, this chimera was not as effective as the wild-type POT1. The reason for this partial function has now been revealed through a careful reappraisal of the DNA-binding features of POT1, which showed that it can use its most amino-terminal OB fold to cap over the 5′ end of the telomere (Fig. 2D; Tesmer et al. 2023). The structure of this OB fold bound to the 5′ end revealed a cavity, referred to as the POT-hole, that recognizes the 5′-phosphate at the telomere end. When this OB fold is engaged at the telomere 5′ end, Rad17/RFC is presumably prevented from loading 9-1-1/TOPBP1 and the activation of ATR is averted. Thus, POT1 deals with the threat of the ATR kinase by covering up the binding sites of the two initiators of this pathway, RPA and Rad17/RFC (Fig. 2D).

As expected, the POT-hole is present in mouse POT1a, the primary repressor of ATR signaling in mouse cells (Tesmer et al. 2023). Unexpectedly, POT1b, which lacks the POT1-hole, can also repress ATR signaling although it is less effective than POT1a. This difference is due to the binding of POT1b, but not POT1a, to CST (Kratz and de Lange 2018). How CST interferes with the ability of POT1b to block ATR is not understood. Interestingly, a POT1b mutant that cannot bind to CST is as effective at repressing ATR as POT1a (Kratz and de Lange 2018). This implies that the POT1-hole is not required for blocking ATR. A second question concerns the evolution of POT1b. POT1a and POT1b arose from duplication of the ancestral POT1 gene in one rodent lineage and their sequence and function diverged (Hockemeyer et al. 2006). While POT1a took on the role of ATR repression, POT1b became dedicated to the control of 5′ end resection (through its ability to prevent excessive resection by TRF2-bound Apollo) and the postreplicative fill-in at telomeres (through the recruitment of CST-Polα/primase) (Wu et al. 2012). Why then was the POT-hole lost from POT1b? Could the POT-hole interfere with the regulation of resection and/or fill-in? If so, how is such interference circumvented at telomeres that have a single POT-hole-bearing POT1?

## THE END-PROTECTION PROBLEM: THE THREAT OF INAPPROPRIATE DSB REPAIR

In addition to creating the threat of cell death resulting from ATM or ATR kinase signaling, the ends of linear chromosomes can be subject to deleterious DSB repair. The principal problem is the joining of a telomere of one chromatid to that of another (in another chromosome or the sister chromatid), which results in a product with two centromeres. Such dicentric chromosomes are inherently unstable and can break during cell division, initiating the cycles of breakage-fusion-bridge events that originally led McClintock to speculate on the protective nature of telomeres (McClintock 1939, 1941). There are two pathways that can fuse one telomere to another—classical nonhomologous end joining (c-NHEJ) and alternative end joining (alt-EJ, also referred to as a-EJ, MMEJ, and TMEJ) (Billing and Sfeir 2025). In mammalian cells, alt-EJ is not a major threat to genome integrity because it is repressed at telomeres (and at DSBs) by Ku70/80 and it is largely confined to mitosis (Brambati et al. 2023).

In contrast to alt-EJ, telomeres are at constant risk of becoming fused by c-NHEJ, which is initiated when the ring-shaped Ku70/80 complex loads on a free DNA end (Fig. 3A; for review, see Stinson and Loparo 2021). Ku70/80 then binds to DNA-PKcs, which, like ATM and ATR, is a large PI3-kinase related kinase. DNA-PKcs promotes c-NHEJ through protein phosphorylation, including an autophophorylation step that ultimately removes the kinase from the DNA ends so that they can be ligated (Fig. 3A). The ligation is executed by DNA Ligase IV (lig4) in the context of a short-range complex formed by XRCC4 and XLF (Fig. 3A).

Shelterin prevents c-NHEJ at telomeres primarily by hiding the telomere end in the t-loop (Fig. 3B). Since Ku70/80 require a free DNA for the initiation of this pathway, the t-loop strategy is effective for the avoidance of c-NHEJ as it is for the repression of ATM signaling (Fig. 2B). The importance of t-loops for the repression of c-NHEJ was demonstrated with an engineered tetrameric form TRF1 (Goldfarb et al. 2024). This tetrameric TRF1 also blocks ATM signaling (see above), which in itself is needed for the cNHEJ of telomeres, confounding the interpretation of the lack of telomere fusions in this setting (why ATM [or ATR] signaling is required for the cNHEJ of telomeres is explained below). To determine whether the t-loops formed by tetrameric TRF1 block c-NHEJ, it was necessary to activate ATR activation at the telomeres. Even in that context, no or very little c-NHEJ occurred at telomeres that lacked TRF2 but had t-loops formed by tetrameric TRF1 (Goldfarb et al. 2024). Thus, under these conditions, t-loops are the primary ploy by which shelterin prevents c-NHEJ at telomeres.

At telomeres that lack t-loops, such as the telomeres synthesized by leading-strand DNA synthesis, c-NHEJ can be inhibited by Rap1 (Eickhoff et al. 2025) as was suggested by early biochemical work and experiments in which Rap1 was artificially positioned at telomeres (Bae and Baumann 2007; Sarthy et al. 2009). The mechanism involves the binding of the BRCT domain of Rap1 to Ku70 at the same site where it binds to Lig4 (Fig. 3C). Presumably, when this site is occupied by Rap1, Lig4 cannot interact with Ku70 and no ligation can occur. This ploy is particularly important in the context of leading-end telomeres where DNA-PK plays an important role in activating the Apollo nuclease (Sonmez et al. 2024), making the loading of DNA-PK at telomeres a requirement for the generation of the 3′ overhang that ultimately will protect telomeres by forming the t-loop. The ability of Rap1 to prevent the bound DNA-PK from mediating fusions at the blunt leading-end telomeres appears to have evolved to solve this problem. The inhibition of c-NHEJ by Rap1 also explains why telomeres that have activated ATM signaling and are therefore inferred to be in a linear state, often resist c-NHEJ (Cesare et al. 2013; Van Ly et al. 2018).

Telomere fusions formed through c-NHEJ occur primarily in $G_1$ and are slow to accumu-

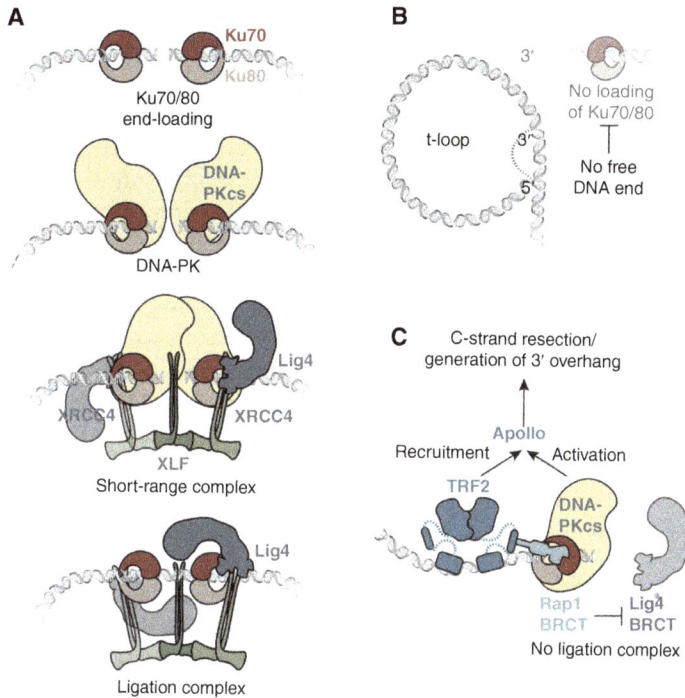

**Figure 3.** How shelterin blocks classical nonhomologous end joining (c-NHEJ). (*A*) Simplified schematic of c-NHEJ. (*B*) Cartoon depicting why Ku70/80 cannot load onto the telomere end when the telomere is in the t-loop configuration. (*C*) Cartoon depicting how Rap1 can block c-NHEJ at leading-strand telomeres where binding of DNA-PK is required to activate the Apollo nuclease.

late. A metaphase spread in which most telomeres are joined is usually the result of several cell divisions in absence of TRF2 (Celli and de Lange 2005). This timeline contrasts with the rapid repair of ionizing radiation (IR)-induced DSBs by c-NHEJ, which takes a few hours. The major difference is that telomeres are dispersed in the nucleus, whereas the two ends of a DSB remain closely positioned. Telomere–telomere fusions require that the unprotected telomeres become more mobile and sample greater subnuclear volumes that increase the chance of an encounter with another unprotected telomere. Increased mobility was first demonstrated at unprotected telomeres and then also shown for DSBs (Dimitrova et al. 2008; Dion et al. 2012; Miné-Hattab and Rothstein 2012; Lottersberger et al. 2015). At dysfunctional telomeres, mobility is increased through 53BP1. The mechanism by which 53BP1 promotes the mobility of sites of DNA damage is still not fully worked out but

has been shown to involve the LINC complex, kinesins, and microtubules that form invaginations in the nuclear envelope (Lottersberger et al. 2015; Shokrollahi et al. 2024). Because chromatin mobility is critical for efficient fusion of telomeres, this form of c-NHEJ is dependent on ATM kinase signaling, whereas c-NHEJ of DSBs is not.

Homology-directed repair (HDR) involving telomeres can also be detrimental. Although HDR at telomeres will not create dicentric chromosomes, it can change the length of telomeres through an unequal exchange. Since the daughter cell that ends up with the shortened telomere may have lost replicative potential, HDR needs to be repressed. HDR repression is also important to avoid telomerase-independent telomere maintenance, which could thwart the telomere tumor suppressor pathway.

At DSBs, HDR starts with the generation of a 3′ overhang through resection (for reviews, see

Ciccia and Elledge 2010; Cejka and Symington 2021). This 3′ overhang is then used by the Rad51 recombinase to find a stretch of homology (usually in the sister chromatid), creating a 3′ end that can be extended by DNA polymerase δ (Maloisel et al. 2008). This 3′ end extension underlies both break-induced replication, the alternative telomere maintenance pathway called ALT, and synthesis-dependent strand annealing (Llorente et al. 2008; Dilley et al. 2016; Verma and Greenberg 2016). However, if the D-loop captures the second DNA end, extension of both 3′ ends can result in a dHJ. Such dHJs can be resolved by HJ resolvases, such as SLX4/SLX1, Mus81/EMI1, and Gen1, or dissolved by the BTR complex, composed of the Bloom's syndrome (BLM) helicase, topoisomerase 3α, and RMI1/RMI2.

The protection of telomeres from HDR requires the presence of both Rap1 and a POT1 protein (either POT1a or POT1b in the mouse) (Palm et al. 2009; Sfeir et al. 2010; Glousker et al. 2020). In mouse cells, HDR is also repressed by Ku70/80, which similarly counteracts HDR at DSBs (Celli et al. 2006). Therefore, detection of HDR at mouse telomeres requires removal of Ku70/80 as well as either Rap1 or POT1. By contrast, loss of POT1 from human telomeres unleashes HDR despite the presence of Ku70/80 (Glousker et al. 2020). The outcome of HDR at telomeres is readily monitored based on telomere sister chromatid exchanges (T-SCEs) using a technique called chromosome orientation fluorescence in situ hybridization wherein ultraviolet (UV)/Hoechst-treated metaphases of BrdU-treated cells allow detection of remaining parental telomeric strands with differentially labeled C- and G-strand probes (Bailey et al. 2001). Evidence in human cells indicates that POT1 functions by preventing loading of Rad51 on the ss telomeric DNA (Glousker et al. 2020). Rap1 has been reported to interact with SLX4 (Rai et al. 2016), a scaffold protein that binds to several nucleases (XPF-ERCC1, Mus81, and SLX1) that can mediate HJ resolution but how this interaction limits T-SCEs is not clear. Rap1 has also been reported to act together with the Basic domain of TRF2 to limit PARP1 activation (Rai et al. 2016), thereby po-

tentially limiting the recruitment of HJ resolvase activities. However, while TRF2 and TIN2 were previously shown to independently prevent PARP1 accumulation at telomeres, Rap1 did not have this effect (Schmutz et al. 2017). Thus, the mechanism by which Rap1 limits T-SCEs requires further analysis. It is also still unclear what the origin is of the DSBs in telomeres that initiate HDR.

HJ resolvases also threaten telomere integrity when branch-migration at the base of the t-loop generates a dHJ (Wang et al. 2004; Poulet et al. 2009; Saint-Léger et al. 2014; Schmutz et al. 2017). Cleavage of the dHJ can lead to loss of the looped part of the t-loop and severe telomere shortening. This process is blocked by the Basic domain of TRF2, which binds the three-way branched structure at the base of the t-loop, thereby preventing branch migration. Replacement of the Basic domain by other branched DNA-binding proteins (RuvC or RuvA) is sufficient to guard telomeres against t-loop cleavage (Schmutz et al. 2017). In addition, the Basic domain prevents the activation of PARP1, which promotes the recruitment of HJ resolvases (Rai et al. 2016). Finally, telomeres are protected from t-loop cleavage by the BLM RecQ helicase, which can remove dHJ formed at the base of the t-loop using its ability to branch-migrate these structures (Schmutz et al. 2017).

Like DSBs, telomeres are at risk of processing of the 5′ ended strand by exonucleases. There are two main pathways for 5′ end resection (Cejka and Symington 2021). One pathway relies on Exo1, which is loaded at DSBs by MRN. The other pathway is mediated by the endonuclease DNA2, which can digest ssDNA formed by one of two RecQ helicases (BLM or WRN). Uncontrolled resection through either pathway is a potential threat to telomere integrity. At DSBs, 5′ end resection is counteracted by fill-in synthesis by CST-Polα/primase (Mirman et al. 2018, 2022; Mirman and de Lange 2020), the same enzyme that is responsible for the maintenance of the telomeric C-strand (see below). Similarly, CST-Polα/primase counteracts resection at telomeres. Since the role of CST-Polα/primase in mitigating resection has been reviewed recently (Lim and Cech 2021; Cai and de Lange 2023; Mirman et al. 2023), this issue will not be discussed here.

Exonucleolytic shortening of the C-rich strand is an important aspect of telomere husbandry since it is required to regenerate the 3′ overhang at blunt telomere ends formed by leading-strand DNA synthesis. This process is regulated by TRF2, which recruits the Apollo 5′ exonuclease and DNA protein kinase (PK), and when bound to the leading ends activates the nuclease activity of Apollo (Sonmez et al. 2024). In addition to Apollo, Exo1 has been implicated in the 5′ end resection of the C-rich strand (Wu et al. 2012). Whether DNA2/WRN (or BLM) are also involved in the regeneration of the 3′ overhang is not known.

Upon loss of either TRF2 or POT1, telomeres become prone to DNA resection by the nucleases that normally act at DSBs. This resection is largely counteracted by CST-Polα/primase, which is either brought to the telomere by POT1 (Cai et al. 2024) or (in the case of POT1 loss) by the 53BP1/Rif1/shieldin complex (Mirman et al. 2018). When TRF2 (and/or POT1a/b) are deleted from 53BP1-deficient cells, telomeres accumulate extremely long regions of ss G-rich DNA (Sfeir and de Lange 2012; Kibe et al. 2016).

## HOW SHELTERIN HELPS TO SOLVE THE TWO END-REPLICATION PROBLEMS

One of the two telomere end-replication problems originates from incomplete copying of the G-rich telomeric DNA strand by the replisome, a deficiency that can be counteracted by telomerase (Fig. 4A,B; Lingner et al. 1995). The 3′ overhang is a critical feature of telomeres that is lost during leading-strand DNA synthesis, which creates a (near) blunt end (Takai et al. 2024). G-strand loss is readily mitigated by telomerase, but since telomerase does not act on a blunt end, initial 5′ end resection is needed (Fig. 4A). Shelterin promotes resection of the 5′ (C-rich) strand (for review, see de Lange 2018), creating a 3′ overhang that is used as primer by telomerase and can form the protective t-loop. When telomerase is absent, C-strand resection also re-

creates the 3′ overhang but now both G-rich and C-rich sequences are lost and telomeres shorten.

In addition to generating the primer for telomerase, shelterin also facilitates the recruitment of the enzyme (reviewed in Hockemeyer and Collins 2015; Martin and Hockemeyer 2025). Telomerase is brought to the telomeric chromatin through a direct interaction between the TEL-patch in the OB fold of TPP1 and the hTERT component of telomerase (Xin et al. 2007; Abreu et al. 2010; Nandakumar et al. 2012; Zhong et al. 2012; Zhang et al. 2013; Sekne et al. 2022). When this interaction is severed, telomeres shorten and, consistently, a mutation in the TEL patch of TPP1 leads to dyskeratosis congenita (DC), a disease caused by defective telomere maintenance. Interestingly, DC can also be caused by mutations in a small region in the carboxy-terminal half of TIN2 (Walne et al. 2008; Savage 2022). How this so-called DC patch in TIN2 contributes to telomere maintenance is still unclear, although some studies suggested that the TIN2 DC patch contributes to telomerase recruitment (Yang et al. 2011; Frank et al. 2015).

In addition to the critical role of shelterin in the recruitment of telomerase, shelterin controls telomerase-mediated telomere elongation. A poorly understood negative feedback loop is thought to enforce telomere length homeostasis, acting in *cis* to prevent telomerase from further elongating telomeres that are too long. The main shelterin components that limit telomerase in this manner are TRF1, TIN2, and POT1 (for reviews, see Hockemeyer and Collins 2015; Martin and Hockemeyer 2025). Mutations in TIN2 or POT1 can lead to excessively long telomeres that increase the risk of cancer later in life (Schmutz et al. 2020; Kim et al. 2021). It is possible that TRF1, TIN2, and POT1 act through recruitment of CST to telomeres since deletion of CST also leads to uncontrolled telomerase-mediated telomere elongation (Chen et al. 2012) but this epistatic relationship has not been tested. Furthermore, it is unclear how and at what step telomerase is blocked, although in vitro experiments indicate that both POT1 and CST can directly interfere with telomerase activity (Kelleher et al. 2005; Chen et al. 2012;

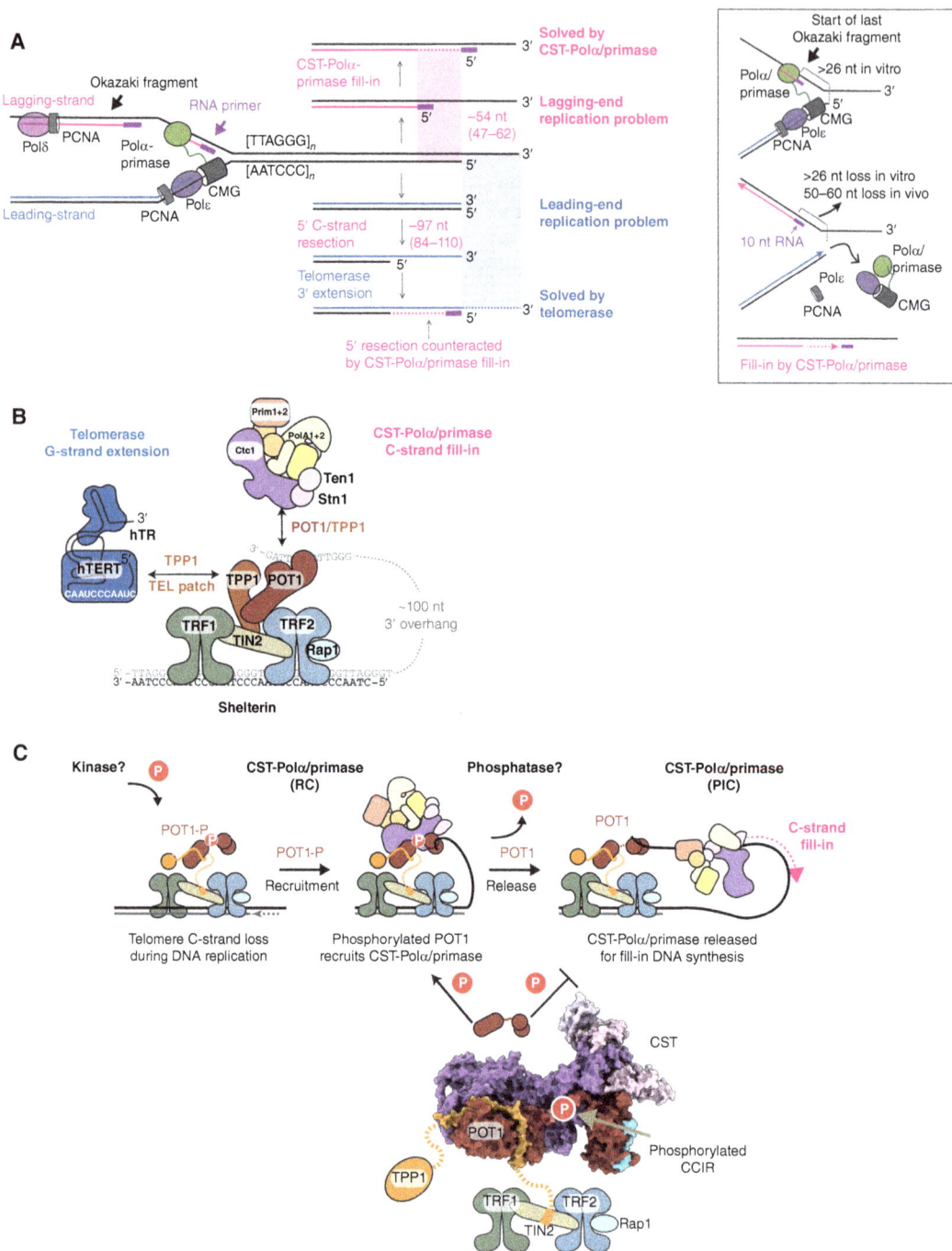

**Figure 4.** How shelterin governs the duplication of telomeric DNA. (*A*) Cartoon depicting the two end-replication problems. One problem involves loss of the G-rich 3′ overhang and is solved by telomerase. The second problem involves the loss of C-rich sequences, which is solved by CST-Polα/primase. (*Inset*) The loss of C-strand sequences during lagging-strand replication is due to the inability of the replisome to allow Polα/primase to act at (or beyond) the end of the duplex. (*Inset*: modified from Takai et al. 2024, Copyright © 2024, The Authors). (*B*) Cartoon showing how shelterin recruits both telomerase and CST-Polα/primase. (*C*) Model for how phosphorylation of POT1 controls the recruitment and activation of CST-Polα/primase. (Panel modified, with permission, from Cai et al. 2024, © Elsevier.)

Zaug et al. 2021). Finally, addition of POT1 and TPP1 can increase the processivity of telomerase in vitro (Wang et al. 2007) but whether processivity is relevant to telomere length regulation in vivo is not known because telomerase appears to be able to extend human telomeres in the absence of POT1 (Martin et al. 2025).

It was long assumed that there would be no sequence loss at the lagging-end telomere because it was assumed that the last Okazaki fragments could start anywhere along the 3′ overhang (Chow et al. 2012). It was recently shown that this assumption was erroneous. According to data obtained with an in vitro reconstituted eukaryotic replication system, the replisome does not allow Polα/primase to initiate Okazaki fragment synthesis beyond the 5′ ds–ss transition (Fig. 4A, inset; Takai et al. 2024). In fact, the last Okazaki fragment is initiated in a broad zone preceding the ds–ss transition by 26–200 nt. Similarly, in vivo this second end-replication problem results in loss of ∼50–60 nt of C-strand sequences from the lagging ends with each cell division. This C-strand loss is mitigated by CST-Polα/primase fill-in, not telomerase (Takai et al. 2024). CST-Polα/primase fill-in is also important to counteract excessive resection of the C-rich strand during the processing of the leading-strand end. Consistent with a critical role for CST in telomere maintenance, mutations in this pathway lead to a premature aging syndrome, Coats plus, and, in rare cases, DC (Anderson et al. 2012; Cai and de Lange 2023). The involvement of CST-Polα/primase in telomere maintenance was not recognized earlier because, unlike telomerase, CST is essential for cell viability barring analysis of the long-term effects of CST loss.

Whereas telomerase is recruited by TPP1, CST-Polα/primase is recruited and regulated by POT1 (Fig. 4C; Cai et al. 2024). Structural analysis indicates that POT1 can bind to CST-Polα/primase in a catalytically inactive state where the active site of POLA1 is blocked. This recruitment requires phosphorylation of POT1 in its Ctc1 cleft interaction region (CCIR). The CCIR phosphorylation has emerged as an important regulatory switch that determines whether CST-Polα/primase can attain a state competent for the fill-in reaction. It is predicted that a kinase dictates the binding of catalytically inactive CST-Polα/primase to POT1, whereas a phosphatase promotes the release of the enzyme from shelterin in an active state that can execute the fill-in synthesis.

## OUTLOOK

Shelterin has now emerged as a protein complex that governs most (if not all) of the important events at telomeres: repression of the DDR, avoidance of DSB repair, and maintenance of both the C- and G-rich telomeric DNA strands. Some of the strategies used by shelterin, including many that involve shelterin accessory factors (e.g., tankyrase) had to be ignored here due to length limitations of this review. Similarly, the important role of TRF1 in the duplication of the telomeric DNA by the replisome had to be left out (Martínez et al. 2009; Sfeir et al. 2009). While inherently incomplete, this review aims to highlight the central role shelterin plays in all aspects of telomere function. Compared to telomerase and some telomeric proteins in unicellular organisms, the work on shelterin has started relatively recently. We can therefore anticipate many new insights into the function of this remarkable complex, including future findings that contradict the views expressed here.

## ACKNOWLEDGMENTS

I am grateful to Drs. Jamie Phipps, Giordano Reginato, and Alex Stuart for comments on the manuscript. In the last 10 years, research in my laboratory has been supported by grants from the National Institutes of Health, the Breast Cancer Research Foundation, the Melanoma Research Alliance, and the STARR cancer consortium.

## REFERENCES

*Reference is also in this subject collection.

Abreu E, Aritonovska E, Reichenbach P, Cristofari G, Culp B, Terns RM, Lingner J, Terns MP. 2010. TIN2-tethered TPP1 recruits human telomerase to telomeres in vivo. *Mol Cell Biol* **30:** 2971–2982. doi:10.1128/MCB.00240-10

Anderson BH, Kasher PR, Mayer J, Szynkiewicz M, Jenkinson EM, Bhaskar SS, Urquhart JE, Daly SB, Dickerson JE, O'Sullivan J, et al. 2012. Mutations in CTC1, encoding conserved telomere maintenance component 1, cause Coats plus. *Nat Genet* **44**: 338–342. doi:10.1038/ng.1084

Bae NS, Baumann P. 2007. A RAP1/TRF2 complex inhibits nonhomologous end-joining at human telomeric DNA ends. *Mol Cell* **26**: 323–334. doi:10.1016/j.molcel.2007.03.023

Bailey SM, Cornforth MN, Kurimasa A, Chen DJ, Goodwin EH. 2001. Strand-specific postreplicative processing of mammalian telomeres. *Science* **293**: 2462–2465. doi:10.1126/science.1062560

Bakkenist CJ, Kastan MB. 2003. DNA damage activates ATM through intermolecular autophosphorylation and dimer dissociation. *Nature* **421**: 499–506. doi:10.1038/nature01368

Bartman CR, Hsu SC, Hsiung CC, Raj A, Blobel GA. 2016. Enhancer regulation of transcriptional bursting parameters revealed by forced chromatin looping. *Mol Cell* **62**: 237–247. doi:10.1016/j.molcel.2016.03.007

Bass TE, Luzwick JW, Kavanaugh G, Carroll C, Dungrawala H, Glick GG, Feldkamp MD, Putney R, Chazin WJ, Cortez D. 2016. ETAA1 acts at stalled replication forks to maintain genome integrity. *Nat Cell Biol* **18**: 1185–1195. doi:10.1038/ncb3415

Bianchi A, Smith S, Chong L, Elias P, de Lange T. 1997. TRF1 is a dimer and bends telomeric DNA. *EMBO J* **16**: 1785–1794. doi:10.1093/emboj/16.7.1785

*Billing D, Sfeir A. 2025. The role of microhomology-mediated end joining (MMEJ) at dysfunctional telomeres. *Cold Spring Harb Perspect Biol* **17**: a041687. doi:10.1101/cshperspect.a041687

Brambati A, Sacco O, Porcella S, Heyza J, Kareh M, Schmidt JC, Sfeir A. 2023. RHINO directs MMEJ to repair DNA breaks in mitosis. *Science* **381**: 653–660. doi:10.1126/science.adh3694

Broccoli D, Smogorzewska A, Chong L, de Lange T. 1997. Human telomeres contain two distinct Myb-related proteins, TRF1 and TRF2. *Nat Genet* **17**: 231–235. doi:10.1038/ng1097-231

Cai SW, de Lange T. 2023. CST-Polα/primase: the second telomere maintenance machine. *Genes Dev* **37**: 555–569. doi:10.1101/gad.350479.123

Cai SW, Zinder JC, Svetlov V, Bush MW, Nudler E, Walz T, de Lange T. 2022. Cryo-EM structure of the human CST-Polα/primase complex in a recruitment state. *Nat Struct Mol Biol* **29**: 813–819. doi:10.1038/s41594-022-00766-y

Cai SW, Takai H, Zaug AJ, Dilgen TC, Cech TR, Walz T, de Lange T. 2024. POT1 recruits and regulates CST-Polα/primase at human telomeres. *Cell* **187**: 3638–3651.e18. doi:10.1016/j.cell.2024.05.002

Cejka P, Symington LS. 2021. DNA end resection: mechanism and control. *Annu Rev Genet* **55**: 285–307. doi:10.1146/annurev-genet-071719-020312

Celli GB, de Lange T. 2005. DNA processing is not required for ATM-mediated telomere damage response after TRF2 deletion. *Nat Cell Biol* **7**: 712–718. doi:10.1038/ncb1275

Celli GB, Denchi EL, de Lange T. 2006. Ku70 stimulates fusion of dysfunctional telomeres yet protects chromosome ends from homologous recombination. *Nat Cell Biol* **8**: 885–890. doi:10.1038/ncb1444

Cesare AJ, Hayashi MT, Crabbe L, Karlseder J. 2013. The telomere deprotection response is functionally distinct from the genomic DNA damage response. *Mol Cell* **51**: 141–155. doi:10.1016/j.molcel.2013.06.006

Chen Y, Yang Y, van Overbeek M, Donigian JR, Baciu P, de Lange T, Lei M. 2008. A shared docking motif in TRF1 and TRF2 used for differential recruitment of telomeric proteins. *Science* **319**: 1092–1096. doi:10.1126/science.1151804

Chen LY, Redon S, Lingner J. 2012. The human CST complex is a terminator of telomerase activity. *Nature* **488**: 540–544. doi:10.1038/nature11269

Chow TT, Zhao Y, Mak SS, Shay JW, Wright WE. 2012. Early and late steps in telomere overhang processing in normal human cells: the position of the final RNA primer drives telomere shortening. *Genes Dev* **26**: 1167–1178. doi:10.1101/gad.187211.112

Ciccia A, Elledge SJ. 2010. The DNA damage response: making it safe to play with knives. *Mol Cell* **40**: 179–204. doi:10.1016/j.molcel.2010.09.019

de Lange T. 2015. A loopy view of telomere evolution. *Front Genet* **6**: 321. doi:10.3389/fgene.2015.00321

de Lange T. 2018. Shelterin-mediated telomere protection. *Annu Rev Genet* **52**: 223–247. doi:10.1146/annurev-genet-032918-021921

Denchi EL, de Lange T. 2007. Protection of telomeres through independent control of ATM and ATR by TRF2 and POT1. *Nature* **448**: 1068–1071. doi:10.1038/nature06065

Deng W, Wu J, Wang F, Kanoh J, Dehe PM, Inoue H, Chen J, Lei M. 2015. Fission yeast telomere-binding protein Taz1 is a functional but not a structural counterpart of human TRF1 and TRF2. *Cell Res* **25**: 881–884. doi:10.1038/cr.2015.76

Dilley RL, Verma P, Cho NW, Winters HD, Wondisford AR, Greenberg RA. 2016. Break-induced telomere synthesis underlies alternative telomere maintenance. *Nature* **539**: 54–58. doi:10.1038/nature20099

Dimitrova N, Chen YC, Spector DL, de Lange T. 2008. 53BP1 promotes non-homologous end joining of telomeres by increasing chromatin mobility. *Nature* **456**: 524–528. doi:10.1038/nature07433

Dion V, Kalck V, Horigome C, Towbin BD, Gasser SM. 2012. Increased mobility of double-strand breaks requires Mec1, Rad9 and the homologous recombination machinery. *Nat Cell Biol* **14**: 502–509. doi:10.1038/ncb2465

Doksani Y, Wu JY, de Lange T, Zhuang X. 2013. Super-resolution fluorescence imaging of telomeres reveals TRF2-dependent T-loop formation. *Cell* **155**: 345–356. doi:10.1016/j.cell.2013.09.048

Eickhoff P, Sonmez C, Fisher CEL, Inian O, Roumeliotis TI, Dello Stritto A, Mansfeld J, Choudhary JS, Guettler S, Lottersberger F, et al. 2025. Chromosome end protection by RAP1-mediated inhibition of DNA-PK. *Nature* doi:10.1038/s41586-025-08896-1

Feng S, Zhao Y, Xu Y, Ning S, Huo W, Hou M, Gao G, Ji J, Guo R, Xu D. 2016. Ewing tumor-associated antigen 1 interacts with replication protein A to promote restart of stalled replication forks. *J Biol Chem* **291:** 21956–21962. doi:10.1074/jbc.C116.747758

Flynn RL, Centore RC, O'Sullivan RJ, Rai R, Tse A, Songyang Z, Chang S, Karlseder J, Zou L. 2011. TERRA and hnRNPA1 orchestrate an RPA-to-POT1 switch on telomeric single-stranded DNA. *Nature* **471:** 532–536. doi:10.1038/nature09772

Frank AK, Tran DC, Qu RW, Stohr BA, Segal DJ, Xu L. 2015. The shelterin TIN2 subunit mediates recruitment of telomerase to telomeres. *PLoS Genet* **11:** e1005410. doi:10.1371/journal.pgen.1005410

Frescas D, de Lange T. 2014. Binding of TPP1 protein to TIN2 protein is required for POT1a,b protein-mediated telomere protection. *J Biol Chem* **289:** 24180–24187. doi:10.1074/jbc.M114.592592

Gabriele M, Brandão HB, Grosse-Holz S, Jha A, Dailey GM, Cattoglio C, Hsieh TS, Mirny L, Zechner C, Hansen AS. 2022. Dynamics of CTCF- and cohesin-mediated chromatin looping revealed by live-cell imaging. *Science* **376:** 496–501. doi:10.1126/science.abn6583

Glousker G, Briod AS, Quadroni M, Lingner J. 2020. Human shelterin protein POT1 prevents severe telomere instability induced by homology-directed DNA repair. *EMBO J* **39:** e104500. doi:10.15252/embj.2020104500

Goldfarb AM, Sasi NK, Dilgen TC, Cai SW, Myler LW, de Lange T. 2024. Tetrameric TRF2 forms t-loop to protect telomeres from ATM signaling and cNHEJ. bioRxiv doi:10.1101/2024.06.27.600884

Gong Y, de Lange T. 2010. A Shld1-controlled POT1a provides support for repression of ATR signaling at telomeres through RPA exclusion. *Mol Cell* **40:** 377–387. doi:10.1016/j.molcel.2010.10.016

Griffith JD, Comeau L, Rosenfield S, Stansel RM, Bianchi A, Moss H, de Lange T. 1999. Mammalian telomeres end in a large duplex loop. *Cell* **97:** 503–514. doi:10.1016/S0092-8674(00)80760-6

Gu P, Wang Y, Bisht KK, Wu L, Kukova L, Smith EM, Xiao Y, Bailey SM, Lei M, Nandakumar J, et al. 2017. Pot1 OB-fold mutations unleash telomere instability to initiate tumorigenesis. *Oncogene* **36:** 1939–1951. doi:10.1038/onc.2016.405

Haahr P, Hoffmann S, Tollenaere MA, Ho T, Toledo LI, Mann M, Bekker-Jensen S, Räschle M, Mailand N. 2016. Activation of the ATR kinase by the RPA-binding protein ETAA1. *Nat Cell Biol* **18:** 1196–1207. doi:10.1038/ncb3422

Hockemeyer D, Collins K. 2015. Control of telomerase action at human telomeres. *Nat Struct Mol Biol* **22:** 848–852. doi:10.1038/nsmb.3083

Hockemeyer D, Daniels JP, Takai H, de Lange T. 2006. Recent expansion of the telomeric complex in rodents: two distinct POT1 proteins protect mouse telomeres. *Cell* **126:** 63–77. doi:10.1016/j.cell.2006.04.044

Hopfner KP. 2023. Mre11-Rad50: the DNA end game. *Biochem Soc Trans* **51:** 527–538. doi:10.1042/BST20220754

Houghtaling BR, Cuttonaro L, Chang W, Smith S. 2004. A dynamic molecular link between the telomere length regulator TRF1 and the chromosome end protector TRF2. *Curr Biol* **14:** 1621–1631. doi:10.1016/j.cub.2004.08.052

Hu H, Yan HL, Nguyen THD. 2024. Structural biology of shelterin and telomeric chromatin: the pieces and an un-finished puzzle. *Biochem Soc Trans* **52:** 1551–1564. doi:10.1042/BST20230300

Jehi SE, Wu F, Li B. 2014. *Trypanosoma brucei* TIF2 suppresses VSG switching by maintaining subtelomere integrity. *Cell Res* **24:** 870–885. doi:10.1038/cr.2014.60

Kelleher C, Kurth I, Lingner J. 2005. Human protection of telomeres 1 (POT1) is a negative regulator of telomerase activity in vitro. *Mol Cell Biol* **25:** 808–818. doi:10.1128/MCB.25.2.808-818.2005

Kibe T, Zimmermann M, de Lange T. 2016. TPP1 blocks an ATR-mediated resection mechanism at telomeres. *Mol Cell* **61:** 236–246. doi:10.1016/j.molcel.2015.12.016

Kim WT, Hennick K, Johnson J, Finnerty B, Choo S, Short SB, Drubin C, Forster R, McMaster ML, Hockemeyer D. 2021. Cancer-associated POT1 mutations lead to telomere elongation without induction of a DNA damage response. *EMBO J* **40:** e107346. doi:10.15252/embj.2020107346

Konishi A, de Lange T. 2008. Cell cycle control of telomere protection and NHEJ revealed by a ts mutation in the DNA-binding domain of TRF2. *Genes Dev* **22:** 1221–1230. doi:10.1101/gad.1634008

Kratz K, de Lange T. 2018. Protection of telomeres 1 proteins POT1a and POT1b can repress ATR signaling by RPA exclusion, but binding to CST limits ATR repression by POT1b. *J Biol Chem* **293:** 14384–14392. doi:10.1074/jbc.RA118.004598

Lee JH, Paull TT. 2021. Cellular functions of the protein kinase ATM and their relevance to human disease. *Nat Rev Mol Cell Biol* **22:** 796–814. doi:10.1038/s41580-021-00394-2

Lee YC, Zhou Q, Chen J, Yuan J. 2016. RPA-binding protein ETAA1 Is an ATR activator involved in DNA replication stress response. *Curr Biol* **26:** 3257–3268. doi:10.1016/j.cub.2016.10.030

Lewis KA, Wuttke DS. 2012. Telomerase and telomere-associated proteins: structural insights into mechanism and evolution. *Structure* **20:** 28–39. doi:10.1016/j.str.2011.10.017

Li B, Espinal A, Cross GA. 2005. Trypanosome telomeres are protected by a homologue of mammalian TRF2. *Mol Cell Biol* **25:** 5011–5021. doi:10.1128/MCB.25.12.5011-5021.2005

Lim CJ, Cech TR. 2021. Shaping human telomeres: from shelterin and CST complexes to telomeric chromatin organization. *Nat Rev Mol Cell Biol* **22:** 283–298. doi:10.1038/s41580-021-00328-y

Lingner J, Cooper JP, Cech TR. 1995. Telomerase and DNA end replication: no longer a lagging strand problem? *Science* **269:** 1533–1534. doi:10.1126/science.7545310

Liu D, O'Connor MS, Qin J, Songyang Z. 2004a. Telosome, a mammalian telomere-associated complex formed by multiple telomeric proteins. *J Biol Chem* **279:** 51338–51342. doi:10.1074/jbc.M409293200

Liu D, Safari A, O'Connor MS, Chan DW, Laegeler A, Qin J, Songyang Z. 2004b. PTOP interacts with POT1 and regulates its localization to telomeres. *Nat Cell Biol* **6:** 673–680. doi:10.1038/ncb1142

Llorente B, Smith CE, Symington LS. 2008. Break-induced replication: what is it and what is it for? *Cell Cycle* **7:** 859–864. doi:10.4161/cc.7.7.5613

Lottersberger F, Karssemeijer RA, Dimitrova N, de Lange T. 2015. 53BP1 and the LINC complex promote microtubule-dependent DSB mobility and DNA repair. *Cell* **163:** 880–893. doi:10.1016/j.cell.2015.09.057

Maloisel L, Fabre F, Gangloff S. 2008. DNA polymerase δ is preferentially recruited during homologous recombination to promote heteroduplex DNA extension. *Mol Cell Biol* **28:** 1373–1382. doi:10.1128/MCB.01651-07

* Martin A, Hockemeyer D. 2025. Regulation of human telomerase: from molecular interactions to population genetics. *Cold Spring Harb Perspect Biol* doi:10.1101/cshperspect.a041693

Martin A, Schabort J, Bartke-Croughan R, Tran S, Preetham A, Lu R, Ho R, Gao J, Jenkins S, et al. 2025. *Genes Dev* **39:** 445–462. doi:10.1101/gad.352492.124

Martínez P, Thanasoula M, Muñoz P, Liao C, Tejera A, McNees C, Flores JM, Fernández-Capetillo O, Tarsounas M, Blasco MA. 2009. Increased telomere fragility and fusions resulting from TRF1 deficiency lead to degenerative pathologies and increased cancer in mice. *Genes Dev* **23:** 2060–2075. doi:10.1101/gad.543509

McClintock B. 1939. The behavior in successive nuclear divisions of a chromosome broken at meiosis. *Proc Natl Acad Sci* **25:** 405–416. doi:10.1073/pnas.25.8.405

McClintock B. 1941. The stability of broken ends of chromosomes in *Zea mays*. *Genetics* **26:** 234–282. doi:10.1093/genetics/26.2.234

Miné-Hattab J, Rothstein R. 2012. Increased chromosome mobility facilitates homology search during recombination. *Nat Cell Biol* **14:** 510–517. doi:10.1038/ncb2472

Mirman Z, de Lange T. 2020. 53BP1: a DSB escort. *Genes Dev* **34:** 7–23. doi:10.1101/gad.333237.119

Mirman Z, Lottersberger F, Takai H, Kibe T, Gong Y, Takai K, Bianchi A, Zimmermann M, Durocher D, de Lange T. 2018. 53BP1-RIF1-shieldin counteracts DSB resection through CST- and Polα-dependent fill-in. *Nature* **560:** 112–116. doi:10.1038/s41586-018-0324-7

Mirman Z, Sasi NK, King A, Chapman JR, de Lange T. 2022. 53BP1-shieldin-dependent DSB processing in BRCA1-deficient cells requires CST-Polα-primase fill-in synthesis. *Nat Cell Biol* **24:** 51–61. doi:10.1038/s41556-021-00812-9

Mirman Z, Cai S, de Lange T. 2023. CST/Polα/primase-mediated fill-in synthesis at DSBs. *Cell Cycle* **22:** 379–389. doi:10.1080/15384101.2022.2123886

Myler LR, Kinzig CG, Sasi NK, Zakusilo G, Cai SW, de Lange T. 2021. The evolution of metazoan shelterin. *Genes Dev* **35:** 1625–1641. doi:10.1101/gad.348835.121

Nakamura TM, Cech TR. 1998. Reversing time: origin of telomerase. *Cell* **92:** 587–590. doi:10.1016/S0092-8674(00)81123-X

Nandakumar J, Bell CF, Weidenfeld I, Zaug AJ, Leinwand LA, Cech TR. 2012. The TEL patch of telomere protein TPP1 mediates telomerase recruitment and processivity. *Nature* **492:** 285–289. doi:10.1038/nature11648

Palm W, Hockemeyer D, Kibe T, de Lange T. 2009. Functional dissection of human and mouse POT1 proteins. *Mol Cell Biol* **29:** 471–482. doi:10.1128/MCB.01352-08

Pinzaru AM, Hom RA, Beal A, Phillips AF, Ni E, Cardozo T, Nair N, Choi J, Wuttke DS, Sfeir A, et al. 2016. Telomere replication stress induced by POT1 inactivation accelerates tumorigenesis. *Cell Rep* **15:** 2170–2184. doi:10.1016/j.celrep.2016.05.008

Poulet A, Buisson R, Faivre-Moskalenko C, Koelblen M, Amiard S, Montel F, Cuesta-Lopez S, Bornet O, Guerlesquin F, Godet T, et al. 2009. TRF2 promotes, remodels and protects telomeric Holliday junctions. *EMBO J* **28:** 641–651. doi:10.1038/emboj.2009.11

Poulet A, Pisano S, Faivre-Moskalenko C, Pei B, Tauran Y, Haftek-Terreau Z, Brunet F, Le Bihan YV, Ledu MH, Montel F, et al. 2012. The N-terminal domains of TRF1 and TRF2 regulate their ability to condense telomeric DNA. *Nucleic Acids Res* **40:** 2566–2576. doi:10.1093/nar/gkr1116

Rai R, Chen Y, Lei M, Chang S. 2016. TRF2-RAP1 is required to protect telomeres from engaging in homologous recombination-mediated deletions and fusions. *Nat Commun* **7:** 10881. doi:10.1038/ncomms10881

Saint-Léger A, Koelblen M, Civitelli L, Bah A, Djerbi N, Giraud-Panis MJ, Londoño-Vallejo A, Ascenzioni F, Gilson E. 2014. The basic N-terminal domain of TRF2 limits recombination endonuclease action at human telomeres. *Cell Cycle* **13:** 2469–2474. doi:10.4161/cc.29422

Sarek G, Vannier JB, Panier S, Petrini JHJ, Boulton SJ. 2015. TRF2 recruits RTEL1 to telomeres in S phase to promote t-loop unwinding. *Mol Cell* **57:** 622–635. doi:10.1016/j.molcel.2014.12.024

Sarek G, Kotsantis P, Ruis P, Van Ly D, Margalef P, Borel V, Zheng XF, Flynn HR, Snijders AP, Chowdhury D, et al. 2019. CDK phosphorylation of TRF2 controls t-loop dynamics during the cell cycle. *Nature* **575:** 523–527. doi:10.1038/s41586-019-1744-8

Sarthy J, Bae NS, Scrafford J, Baumann P. 2009. Human RAP1 inhibits non-homologous end joining at telomeres. *EMBO J* **28:** 3390–3399. doi:10.1038/emboj.2009.275

Savage SA. 2022. Dyskeratosis congenita and telomere biology disorders. *Hematology Am Soc Hematol Educ Program* **2022:** 637–648. doi:10.1182/hematology.2022000394

Schmutz I, Timashev L, Xie W, Patel DJ, de Lange T. 2017. TRF2 binds branched DNA to safeguard telomere integrity. *Nat Struct Mol Biol* **24:** 734–742. doi:10.1038/nsmb.3451

Schmutz I, Mensenkamp AR, Takai KK, Haadsma M, Spruijt L, de Voer RM, Choo SS, Lorbeer FK, van Grinsven EJ, Hockemeyer D, et al. 2020. TINF2 is a haploinsufficient tumor suppressor that limits telomere length. *eLife* **9:** e61235. doi:10.7554/eLife.61235

Sekne Z, Ghanim GE, van Roon AM, Nguyen THD. 2022. Structural basis of human telomerase recruitment by TPP1-POT1. *Science* **375:** 1173–1176. doi:10.1126/science.abn6840

Sfeir A, de Lange T. 2012. Removal of shelterin reveals the telomere end-protection problem. *Science* **336:** 593–597. doi:10.1126/science.1218498

Sfeir A, Kosiyatrakul ST, Hockemeyer D, MacRae SL, Karlseder J, Schildkraut CL, de Lange T. 2009. Mammalian telomeres resemble fragile sites and require TRF1 for efficient replication. *Cell* **138:** 90–103. doi:10.1016/j.cell.2009.06.021

Sfeir A, Kabir S, van Overbeek M, Celli GB, de Lange T. 2010. Loss of Rap1 induces telomere recombination in the absence of NHEJ or a DNA damage signal. *Science* **327**: 1657–1661. doi:10.1126/science.1185100

Shokrollahi M, Stanic M, Hundal A, Chan JNY, Urman D, Jordan CA, Hakem A, Espin R, Hao J, Krishnan R, et al. 2024. DNA double-strand break-capturing nuclear envelope tubules drive DNA repair. *Nat Struct Mol Biol* **31**: 1319–1330. doi:10.1038/s41594-024-01286-7

Sonmez C, Toia B, Eickhoff P, Matei AM, El Beyrouthy M, Wallner B, Douglas ME, de Lange T, Lottersberger F. 2024. DNA-PK controls Apollo's access to leading-end telomeres. *Nucleic Acids Res* **52**: 4313–4327. doi:10.1093/nar/gkae105

Stinson BM, Loparo JJ. 2021. Repair of DNA double-strand breaks by the nonhomologous end joining pathway. *Annu Rev Biochem* **90**: 137–164. doi:10.1146/annurev-biochem-080320-110356

Stracker TH, Petrini JH. 2011. The MRE11 complex: starting from the ends. *Nat Rev Mol Cell Biol* **12**: 90–103. doi:10.1038/nrm3047

Takai KK, Hooper S, Blackwood S, Gandhi R, de Lange T. 2010. In vivo stoichiometry of shelterin components. *J Biol Chem* **285**: 1457–1467. doi:10.1074/jbc.M109.038026

Takai KK, Kibe T, Donigian JR, Frescas D, de Lange T. 2011. Telomere protection by TPP1/POT1 requires tethering to TIN2. *Mol Cell* **44**: 647–659. doi:10.1016/j.molcel.2011.08.043

Takai H, Aria V, Borges P, Yeeles JTP, de Lange T. 2024. CST-polymerase α-primase solves a second telomere end-replication problem. *Nature* **627**: 664–670. doi:10.1038/s41586-024-07137-1

Tesmer VM, Brenner KA, Nandakumar J. 2023. Human POT1 protects the telomeric ds-ss DNA junction by capping the 5′ end of the chromosome. *Science* **381**: 771–778. doi:10.1126/science.adi2436

Timashev LA, De Lange T. 2020. Characterization of t-loop formation by TRF2. *Nucleus* **11**: 164–177. doi:10.1080/19491034.2020.1783782

Van Ly D, Low RRJ, Frölich S, Bartolec TK, Kafer GR, Pickett HA, Gaus K, Cesare AJ. 2018. Telomere loop dynamics in chromosome end protection. *Mol Cell* **71**: 510–525.e6. doi:10.1016/j.molcel.2018.06.025

Vannier JB, Pavicic-Kaltenbrunner V, Petalcorin MI, Ding H, Boulton SJ. 2012. RTEL1 dismantles T loops and counteracts telomeric G4-DNA to maintain telomere integrity. *Cell* **149**: 795–806. doi:10.1016/j.cell.2012.03.030

Verma P, Greenberg RA. 2016. Noncanonical views of homology-directed DNA repair. *Genes Dev* **30**: 1138–1154. doi:10.1101/gad.280545.116

Walne AJ, Vulliamy T, Beswick R, Kirwan M, Dokal I. 2008. TINF2 mutations result in very short telomeres: analysis of a large cohort of patients with dyskeratosis congenita and related bone marrow failure syndromes. *Blood* **112**: 3594–3600. doi:10.1182/blood-2008-05-153445

Wan M, Qin J, Songyang Z, Liu D. 2009. OB fold-containing protein 1 (OBFC1), a human homolog of yeast Stn1, associates with TPP1 and is implicated in telomere length regulation. *J Biol Chem* **284**: 26725–26731. doi:10.1074/jbc.M109.021105

Wang RC, Smogorzewska A, de Lange T. 2004. Homologous recombination generates T-loop-sized deletions at human telomeres. *Cell* **119**: 355–368. doi:10.1016/j.cell.2004.10.011

Wang F, Podell ER, Zaug AJ, Yang Y, Baciu P, Cech TR, Lei M. 2007. The POT1-TPP1 telomere complex is a telomerase processivity factor. *Nature* **445**: 506–510. doi:10.1038/nature05454

Wondisford AR, Lee J, Lu R, Schuller M, Groslambert J, Bhargava R, Schamus-Haynes S, Cespedes LC, Opresko PL, Pickett HA, et al. 2024. Deregulated DNA ADP-ribosylation impairs telomere replication. *Nat Struct Mol Biol* **31**: 791–800. doi:10.1038/s41594-024-01279-6

Wu P, Takai H, de Lange T. 2012. Telomeric 3′ overhangs derive from resection by Exo1 and Apollo and fill-in by POT1b-associated CST. *Cell* **150**: 39–52. doi:10.1016/j.cell.2012.05.026

Xin H, Liu D, Wan M, Safari A, Kim H, Sun W, O'Connor MS, Songyang Z. 2007. TPP1 is a homologue of ciliate TEBP-β and interacts with POT1 to recruit telomerase. *Nature* **445**: 559–562. doi:10.1038/nature05469

Xue J, Chen H, Wu J, Takeuchi M, Inoue H, Liu Y, Sun H, Chen Y, Kanoh J, Lei M. 2017. Structure of the fission yeast *S. pombe* telomeric Tpz1-Poz1-Rap1 complex. *Cell Res* **27**: 1503–1520. doi:10.1038/cr.2017.145

Yang D, He Q, Kim H, Ma W, Songyang Z. 2011. TIN2 protein dyskeratosis congenita missense mutants are defective in association with telomeres. *J Biol Chem* **286**: 23022–23030. doi:10.1074/jbc.M111.225870

Ye JZ, Donigian JR, van Overbeek M, Loayza D, Luo Y, Krutchinsky AN, Chait BT, de Lange T. 2004a. TIN2 binds TRF1 and TRF2 simultaneously and stabilizes the TRF2 complex on telomeres. *J Biol Chem* **279**: 47264–47271. doi:10.1074/jbc.M409047200

Ye JZ, Hockemeyer D, Krutchinsky AN, Loayza D, Hooper SM, Chait BT, de Lange T. 2004b. POT1-interacting protein PIP1: a telomere length regulator that recruits POT1 to the TIN2/TRF1 complex. *Genes Dev* **18**: 1649–1654. doi:10.1101/gad.1215404

Zaug AJ, Lim CJ, Olson CL, Carilli MT, Goodrich KJ, Wuttke DS, Cech TR. 2021. CST does not evict elongating telomerase but prevents initiation by ssDNA binding. *Nucleic Acids Res* **49**: 11653–11665. doi:10.1093/nar/gkab942

Zhang Y, Chen LY, Han X, Xie W, Kim H, Yang D, Liu D, Songyang Z. 2013. Phosphorylation of TPP1 regulates cell cycle-dependent telomerase recruitment. *Proc Natl Acad Sci* **110**: 5457–5462. doi:10.1073/pnas.1217733110

Zhong Z, Shiue L, Kaplan S, de Lange T. 1992. A mammalian factor that binds telomeric TTAGGG repeats in vitro. *Mol Cell Biol* **12**: 4834–4843. doi:10.1128/mcb.12.11.4834-4843.1992

Zhong FL, Batista LF, Freund A, Pech MF, Venteicher AS, Artandi SE. 2012. TPP1 OB-fold domain controls telomere maintenance by recruiting telomerase to chromosome ends. *Cell* **150**: 481–494. doi:10.1016/j.cell.2012.07.012

Zinder JC, Olinares PDB, Svetlov V, Bush MW, Nudler E, Chait BT, Walz T, de Lange T. 2022. Shelterin is a dimeric complex with extensive structural heterogeneity. *Proc Natl Acad Sci* **119**: e2201662119. doi:10.1073/pnas.2201662119

# Telomeric Repeat-Containing RNA: Biogenesis, Regulation, and Functions

Patricia L. Abreu,[1,5] Valentina Riva,[1,3,5] Luca Zardoni,[1,4,5] and Claus M. Azzalin[1,2]

[1]GIMM - Gulbenkian Institute for Molecular Medicine, 1649-035 Lisbon, Portugal

[2]Faculty of Medicine, University of Lisbon, 1649-028 Lisbon, Portugal

*Correspondence:* claus.azzalin@gimm.pt

Telomeric repeat-containing RNA (TERRA) molecules are transcripts comprising extended stretches of telomeric G-rich repeats, which are generated from telomeres or intrachromosomal loci. TERRA production is an evolutionarily conserved process observed across all eukaryotic kingdoms. While originally thought to localize and function only at telomeres, it is now clear that TERRA is involved in numerous cellular pathways beyond telomere maintenance, including gene expression regulation and signaling of dysfunctional telomeres to the cytoplasm and the extracellular environment. In this work, we will review key aspects of TERRA biogenesis, regulation, and functional relevance and propose models to reconcile the multiple and sometimes contradictory functions ascribed to TERRA. Based on TERRA interaction with proteins involved in disparate cellular processes, we also suggest that the full spectrum of TERRA-associated functions is still far from being completely unveiled. We anticipate that further study of this complex and fascinating RNA will reveal additional surprises in the future.

For decades, telomeres were considered transcriptionally silent due to their heterochromatic nature and ability to silence proximal reporter genes through the so-called telomere position effect (TPE) (Gottschling et al. 1990; Baur et al. 2001; Pedram et al. 2006). The first evidence arguing against this dogma dates back to 1989, when Rudenko and van der Ploeg reported the existence of RNA transcripts containing telomeric repeats in the six protozoa *Trypanosoma brucei, Trypanosoma equiperdum, Trypanosoma lewisi, Leptomonas seymourii, Crithidia fasciculata,* and *Leishmania major* (Rudenko and Van der Ploeg 1989). A few years later, telomeric transcripts originating from DNA loop structures located at the ends of lampbrush chromosomes were observed in chicken (*Gallus gallus domesticus*), turkey (*Meleagris gallopavo*), and pigeon (*Columba livia*) (Solovei et al. 1994). However, it took 13 more years to obtain definitive proof that a conserved feature of eukaryotic cells is the production of a nuclear RNA dubbed telomeric repeat-containing RNA or telomeric repeat-containing RNA (TERRA) (Azzalin et al. 2007).

TERRA has been identified in a variety of eukaryotes with linear chromosomes, including

---

[3]Current address: Institute of Science and Technology Austria, Am Campus 1, 3400 Klosterneuburg, Austria.
[4]Current address: Istituto di Genetica Molecolare Luigi Luca Cavalli-Sforza, CNR, 27100 Pavia, Italy.
[5]These authors contributed equally to this work.

Cite this article as *Cold Spring Harb Perspect Biol* doi: 10.1101/cshperspect.a041683

humans, the rodents *Mus musculus* and *Cricetulus griseus*, the yeasts *Saccharomyces cerevisiae* and *Schizosaccharomyces pombe*, the plants *Arabidopsis thaliana*, *Nicotiana tabacum*, and *Ballantinia antipoda*, the fish *Danio rerio*, and the nematode *Caenorhabditis elegans* (Azzalin et al. 2007; Luke et al. 2008; Schoeftner and Blasco 2008; Vrbsky et al. 2010; Bah et al. 2012; Greenwood and Cooper 2012; Majerová et al. 2014; Idilli et al. 2020; Manzato et al. 2023). In all cases, TERRA derives from the transcription of telomeric DNA repeat sequences with the C-rich strand functioning as a template (Fig. 1). RNA polymerase (RNAP)II is the main enzyme-producing TERRA across species (Azzalin et al. 2007; Luke et al. 2008; Schoeftner and Blasco 2008; Bah et al. 2012); however, a few exceptions have been reported, such as in *Trypanosoma brucei* and *Trypanosoma equiperdum*, where TERRA is a product of RNAPI (Rudenko and Van der Ploeg 1989; Saha et al. 2021).

In this paper, we focus on the main features and functions of TERRA in human, mouse, and yeast cells, as the majority of information is derived from studies on those organisms. For the main features of TERRA transcripts across different eukaryotes, refer to Table 1. Additional transcripts generated from chromosome ends were reported to exist in yeasts and other organisms and include an RNA containing telomeric C-rich repeats only (dubbed ARIA) (Fig. 1) and two antiparallel RNAs comprising subtelomeric sequences and apparently devoid of telomeric repeats (ARRET and αARRET) (Fig. 1; Luke et al. 2008; Vrbsky et al. 2010; Bah et al. 2012; Greenwood and Cooper 2012). The molecular features and the functions of these RNAs remain largely unexplored.

## GENERAL FEATURES OF TERRA AND TELOMERE TRANSCRIPTION

In human cells, TERRA transcripts originating from nearly all chromosome ends were identified using molecular biology-based techniques such as northern blotting and reverse transcription quantitative-PCR (RT-qPCR), as well as

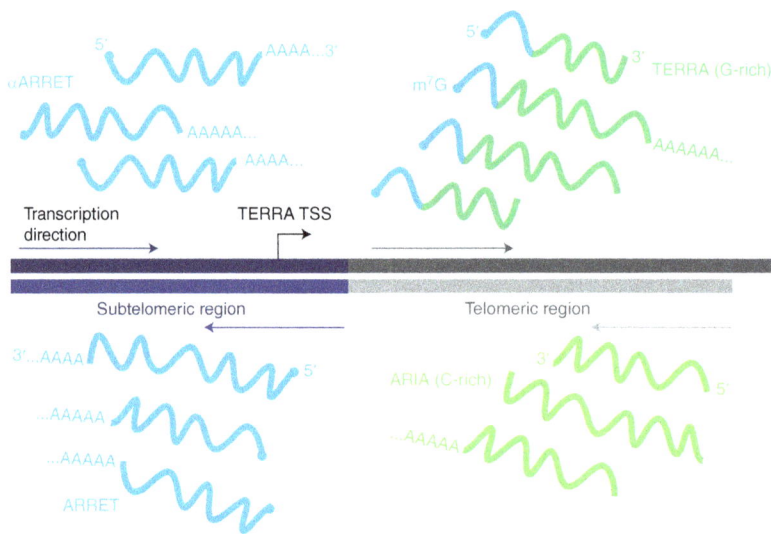

**Figure 1.** Schematic representation of transcripts emanating from natural chromosome ends in eukaryotes. Telomeric repeat-containing RNA (TERRA) and ARIA contain telomeric G-rich and C-rich repeats (green), respectively. TERRA also contains a subtelomeric tract (blue) that precedes the telomeric repeats. ARRET and αARRET comprise only subtelomeric sequences. A fraction of TERRA and ARIA molecules carry poly(A) tails at their 3′ ends, while ARRET and αARRET are fully polyadenylated. TERRA, ARRET, and αARRET contain 7-methylguanosine (m$^7$G) caps at their 5′ ends, while for ARIA this has not been tested. The direction of transcription for each species is indicated by arrows.

**Table 1.** Telomeric repeat-containing RNA (TERRA) across different eukaryotes

| Organism | Polymerase | Chromosomal origin | 5′ modification | 3′ modification | Transcript size | References |
|---|---|---|---|---|---|---|
| *Homo sapiens* | II | All subtelomeres | 7-methylguanosine-cap | Poly(A) (~10%) | 100 b to 9 kb | Azzalin et al. (2007), Schoeftner and Blasco (2008), Porro et al. (2010), and Rodrigues et al. (2024) |
| *Mus musculus* | II | Pseudoautosomal region of Xq/Yq; minimal contribution from other genomic loci | ND | Poly(A) (~10%) | 100 b to 9 kb | Schoeftner and Blasco (2008), Chu et al. (2017a,b), and Viceconte et al. (2021) |
| *Saccharomyces cerevisiae* | II | All subtelomeres | 7-methylguanosine-cap | Poly(A) (~90% in Rat1 mutants; ~10% in wt) | 100–1200 b | Luke et al. (2008), Pfeiffer and Lingner (2012), and Rodrigues and Lydall (2018) |
| *Schizosaccharomyces pombe* | II | All subtelomeres | 7-methylguanosine-cap | Poly(A) (~10%) | 100–2000 b | Bah et al. (2012) and Greenwood and Cooper (2012) |
| *Trypanosoma* | I (*Trypanosoma brucei*, *Trypanosoma equiperdum*); II (*Trypanosoma lewisi*) | Single telomere downstream from the active VSG subtelomeric gene | ND | Poly(A) (~50% in *Trypanosoma brucei*) | 800 b to 10 kb (*T. brucei*) | Rudenko and Van der Ploeg (1989), Damasceno et al. (2017), and Saha et al. (2021) |
| *Leishmania* | II (*Leishmania major*) | Multiple subtelomeres (*L. major*) | ND | Poly(A) (*L. major*, *Leishmania braziliensis*) | 100 b to 10 kb (*L. major*, *Leishmania amazonesis*, *L. braziliensis*) | Rudenko and Van der Ploeg (1989), Damasceno et al. (2017), and Morea et al. (2021) |
| Birds[a] | ND | Lampbrush chromosomes | ND | ND | 1.5–45 kb | Solovei et al. (1994) |
| *Danio rerio* | ND | ND | ND | ND | 100 b to 6 kb | Schoeftner and Blasco (2008) |
| Plants[b] | ND | Multiple subtelomeres | ND | ND | Few b to several kb | Vrbsky et al. (2010) and Majerová et al. (2014) |
| *Caenorhabditis elegans* | ND | Multiple subtelomeres | ND | ND | 100 b to 6 kb | Manzato et al. (2023) |

[a]Chicken (*Gallus gallus domesticus*), turkey (*Meleagris gallopavo*), domestic pigeon (*Columba livia*).
[b]*Arabidopsis thaliana*, *Nicotiana tabacum*, *Ballantinia antipoda*.
(ND) Not determined.

short- and long-read-based RNA sequencing (Azzalin et al. 2007; Nergadze et al. 2009; Porro et al. 2010, 2014a; Arnoult et al. 2012; Arora et al. 2014; Feretzaki et al. 2019; Rodrigues et al. 2024). Human TERRA transcripts range in size from 100 bases (b) to more than 9 kilobases (kb) and contain a unique subtelomeric sequence specific to the chromosome of origin followed by $(UUAGGG)_n$ tracts of varying lengths. TERRA from different chromosome ends differs in abundance, with TERRA produced from the long arm of chromosome 7 being among the most represented in several cell lines (Feretzaki et al. 2019; Rodrigues et al. 2024). Like other RNAs generated by RNAPII, human TERRA molecules possess a $5'$ 7-methylguanosine $(m^7G)$ cap, and although canonical polyadenylation sites are not present in the telomeric repeats, ~10% of TERRA transcripts are $3'$ end polyadenylated in a telomere-specific manner (Fig. 1; Azzalin and Lingner 2008; Porro et al. 2010; Savoca et al. 2023). In cervical carcinoma HeLa cells, TERRA poly(A)$^-$ and poly(A)$^+$ transcripts have half-lives of about three and more than eight hours, respectively, indicating that polyadenylation stabilizes TERRA (Porro et al. 2010). Total TERRA levels are not constant during cell cycle progression. In telomerase-positive cell lines, including HeLa and fibrosarcoma HT1080 cells, TERRA levels are highest in the $G_1$ phase and decrease in the S and $G_2$ phases (Porro et al. 2010; Flynn et al. 2011; Arnoult et al. 2012). An exception to this is found in cancer cells that elongate telomeres through the so-called alternative lengthening of telomeres (for more on ALT, see O'Sullivan and Greenberg 2025) mechanism and in cells from immunodeficiency, centromeric region instability, facial anomalies syndrome (ICF) patients; in both cases, TERRA remains constant throughout the cell cycle phases, likely due to the impairment of one or more TERRA regulatory mechanisms (Flynn et al. 2015; Sagie et al. 2017).

In RNA fluorescence in situ hybridization (FISH) experiments with probes detecting UUAGGG repeats, human TERRA is found in numerous nuclear foci, the majority of which colocalize with telomeres (Azzalin et al. 2007; Schoeftner and Blasco 2008; Arnoult et al.

2012). Notably, the protocols usually employed for TERRA RNA-FISH involve pre-extraction of soluble cellular material and only detect chromatin-bound TERRA. However, cellular fractionation and molecular biology-based experiments revealed that a fraction of TERRA is soluble in the nucleoplasm and that TERRA transcript localization is regulated by polyadenylation. Approximately 60% of poly(A)$^-$ TERRA is in the nucleoplasm while the remaining 40% is chromatin-bound. Conversely, ~80% of poly(A)$^+$ TERRA is in the nucleoplasm (Porro et al. 2010). The mechanisms retaining TERRA at telomeric chromatin are not fully understood, although they appear to involve interactions between TERRA and shelterin proteins, such as TRF1 and TRF2 (Deng et al. 2009; Lee et al. 2018; Abreu et al. 2022), or the formation of RNA:DNA hybrids (telomeric R-loops or telR-loops) between TERRA UUAGGG sequences and the C-rich, template telomeric DNA (Fig. 2). TelR-loops are expected to form cotranscriptionally when RNAPII traverses the telomeric tract, as it was shown that in vitro transcription of human telomeric repeats is associated with robust formation of R-loops (Arora et al. 2014). TelR-loops can also form in *trans* upon TERRA invasion of a telomeric tract aided by proteins like TRF2, RAD51, and RAD51AP1 (Lee et al. 2018; Feretzaki et al. 2020; Kaminski et al. 2022; Yadav et al. 2022). Several factors restricting telR-loop levels were identified in different human cell types, including the endoribonuclease RNaseH1, the ATPase/translocase FANCM, the shelterin protein TRF1, and the multiprotein THO complex (Arora et al. 2014; Lee et al. 2018; Pan et al. 2019; Silva et al. 2019; Fernandes and Lingner 2023).

The promoter regions of human TERRA were first proposed to be CpG dinucleotide-rich tandem repeats of 29 bp, thought to exist at only half of the human subtelomeres and located ~1 kb from the telomeric tract (Nergadze et al. 2009). This implied that the length heterogeneity of TERRA transcripts derives mostly from their $3'$ ends, a notion supported by the observation that experimentally elongated telomeres generate longer TERRA species (Arnoult et al. 2012). It was also proposed that a fraction

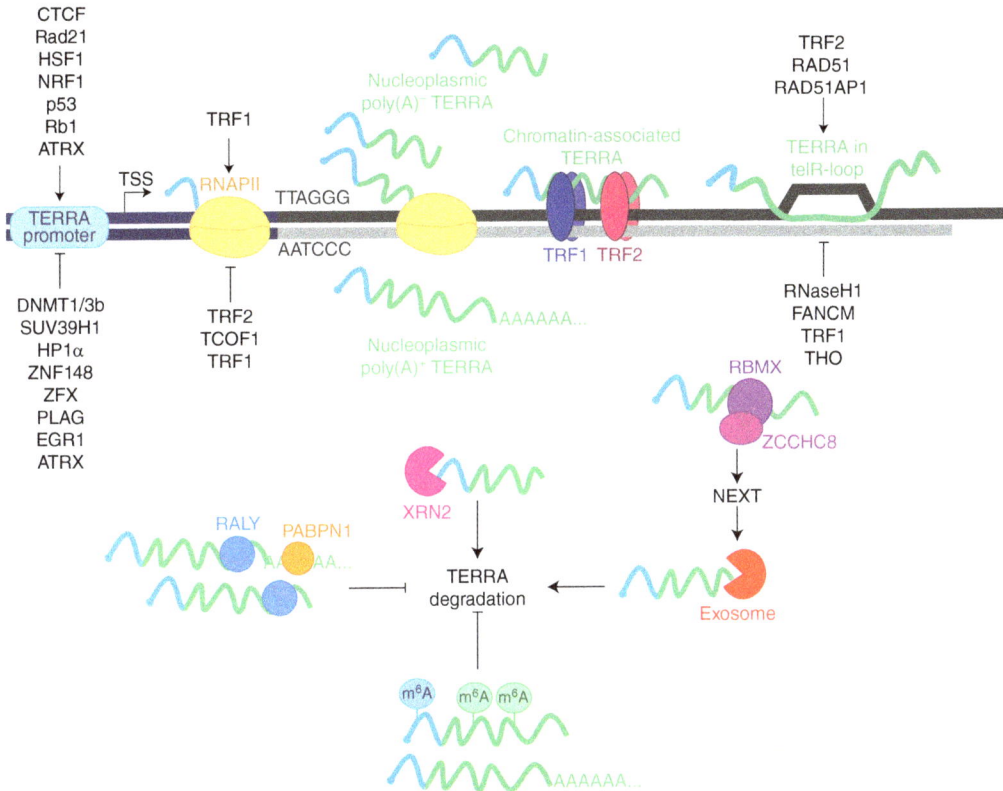

**Figure 2.** Telomeric repeat-containing RNA (TERRA) in human cells. TERRA is produced by RNA polymerase (RNAP)II starting from subtelomeric promoters often composed of CpG dinucleotide-rich repeat sequences (29 bp repeat, see text). RNAPII proceeds toward the end of the chromosome and uses the C-rich strand as a template. TERRA promoters and RNAPII are positively or negatively regulated by several factors including chromatin modifiers, transcription factors, and shelterin proteins. ATRX and TRF1 appear to both promote and suppress TERRA transcription according to the cell type. Once transcribed, TERRA diffuses in the nucleoplasm or remains associated with telomeres likely through interactions with TRF1 and TRF2 or by hybridizing with the C-rich strand and establishing telomeric R-loops (telR-loops). TelR-loops can form in *trans*, aided by TRF2, RAD51, and RAD51AP1, and are negatively regulated by RNaseH1, FANCM, TRF1, and the THO complex. TERRA in the nucleoplasm is degraded through pathways involving members of the nuclear exosome (3′-to-5′ degradation) and XRN2 (5′-to-3′ degradation). TERRA-binding proteins, including RALY and PABPN1, and m⁶A-modified nucleotides stabilize TERRA transcripts.

of human TERRA originates from promoters of unknown sequence located 5–10 kb from the telomere, and that TERRA UUAGGG tracts only extend for a few hundreds of bases (Porro et al. 2014a). Recently, the combination of long-read-based TERRA sequencing with the availability of T2T human reference genomes confirmed that CpG-rich repeats of sequences that are identical or highly similar to the originally

identified 29 bp consensus motif (herein globally referred to as 29 bp repeats) are indeed the main promoter sequences of human TERRA and that they are more common than originally estimated, being present at 36 chromosome ends (Rodrigues et al. 2024). TERRA transcription start sites (TSSs) were identified at 35 chromosome ends and they are all located not more than 1500 bp from the first telomeric repeat, confirming that TERRA heterogeneity largely stems from its $3'$ end, at least in the cell lines that were analyzed in the study (Rodrigues et al. 2024). It remains possible that some TERRA TSSs, more distant from the telomeric tract and corresponding to less abundant transcripts, might have escaped detection.

In mouse cells, TERRA is substantially more abundant than in human cells and comprises long $(UUAGGG)_n$ sequences, produced by RNAPII, a fraction of which are polyadenylated (Azzalin et al. 2007; Schoeftner and Blasco 2008). Contrarily to humans, where TERRA primarily originates from telomeres, more than 90% of mouse TERRA transcripts originate from the pseudoautosomal region of the Xq/Yq subtelomeres, which contain long stretches of interstitial TTAGGG repeats (Chu et al. 2017a,b; Vicenconte et al. 2021). TERRA molecules from the telomeres of chromosome 18q and an internal region of chromosome 2 also contribute to the cellular TERRA pool although at very low levels (López de Silanes et al. 2014; Vicenconte et al. 2021). The chromosomal location of mouse TERRA TSSs and promoters has not been identified yet. TERRA localization in differentiated mouse cells differs from that in humans, as it is found in one to three prominent foci located in the vicinity of the inactive sex chromosomes. Additionally, smaller foci can also be observed and partly colocalize with telomeres (Schoeftner and Blasco 2008; López de Silanes et al. 2014; Vicenconte et al. 2021). Interestingly, in mouse embryonic stem cells (mESCs), TERRA marks both sex chromosomes, while it associates only with the distal telomeric end of the heterochromatinized sex chromosomes upon differentiation, pointing to the involvement of TERRA in X chromosome inactivation (Zhang et al. 2009; Chu et al. 2017b).

A large amount of information on TERRA comes from work in yeasts. In the budding yeast *S. cerevisiae*, the first yeast where telomere transcription was studied, TERRA is transcribed from all telomeres and ranges in size from 100 to 1200 bases, as shown by northern blot experiments using RNA from cells with inactivated Rat1, a $5'$-to-$3'$ RNA exonuclease responsible

for TERRA degradation in the nucleus. In the same strains, ~90% of TERRA transcripts are polyadenylated mainly through the action of the poly(A) polymerase Pap1 (Luke et al. 2008; Iglesias et al. 2011; Balk et al. 2013; Pfeiffer et al. 2013). However, in wild-type cells, only ~10% of TERRA transcripts appear to be polyadenylated (Rodrigues and Lydall 2018). Budding yeast TERRA molecules contain $m^7G$ caps and, although promoter sequences have not yet been identified, TERRA TSSs were mapped in a few X-only and Y′ chromosome ends <400 bp away from the first telomeric repeat (Fig. 3; Pfeiffer and Lingner 2012; Bauer et al. 2022; Guintini et al. 2022). *S. cerevisiae* TERRA levels fluctuate during cell-cycle progression, being the highest at the $G_1$/S transition and decreasing during the late S phase (Graf et al. 2017). Moreover, TERRA can form telR-loops, which are dismantled by RNaseH1, RNaseH2, and the THO complex, and are stabilized by the RNA-binding protein Npl3 (Fig. 3; Pfeiffer et al. 2013; Graf et al. 2017; Pérez-Martinez et al. 2020).

In the fission yeast *S. pombe*, TERRA transcripts are between 200 and 5000 bases in length, at least in mutants with long telomeres (Bah et al. 2012; Greenwood and Cooper 2012). *S. pombe* TERRA transcripts have a $m^7G$ cap and ~10% of them are $3'$ end polyadenylated. TERRA TSSs were mapped 211 bp upstream of the telomeric sequence of chromosomes I and II, though TERRA promoter sequences remain unknown. A fraction of TERRA associates with chromatin and forms discrete nuclear foci, which might colocalize with telomeres. The mechanisms supporting TERRA retention on chromatin in fission yeast are unclear; however, it is known that the majority of chromatin-bound TERRA is not polyadenylated and that TERRA is able to form telR-loops, which are degraded by RNaseH2 (Fig. 3; Bah et al. 2012; Moravec et al. 2016; Hu et al. 2019).

## TERRA TRANSCRIPTION REGULATION

TERRA production is regulated by multiple factors, including heterochromatin, transcription factors, telomere length, and shelterin. In humans, the 29 bp promoter repeats are methyl-

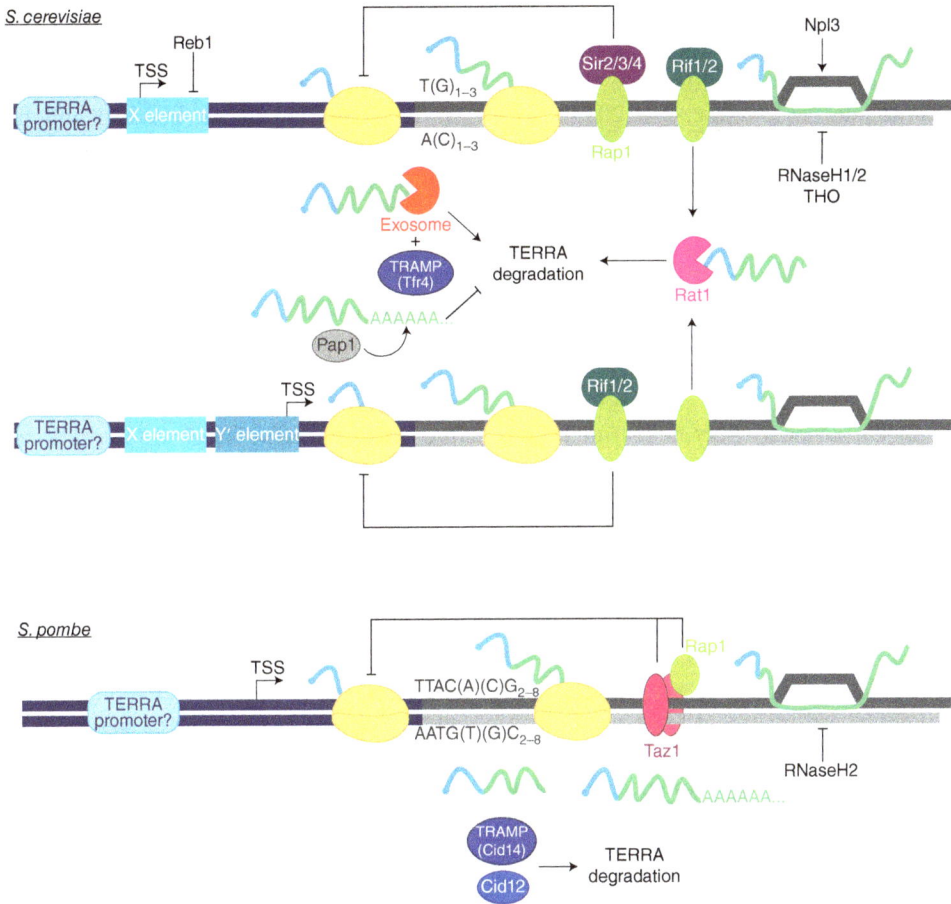

**Figure 3.** Telomeric repeat-containing RNA (TERRA) in yeast cells. In *Saccharomyces cerevisiae* (*upper* panel), TERRA is produced by RNA polymerase (RNAP)II starting from promoters of unclear sequences, although they are likely to be located within the X and the Y′ elements. Rap1 and Sir2, 3, and 4 histone deacetylases suppress RNAPII-mediated TERRA transcription from the X-only chromosome ends, while Rap1 and Rif1 and 2 suppress TERRA transcription from Y′ chromosome ends. A fraction of TERRA forms telomeric R-loops (telR-loops), which are restricted by RNaseH1 and 2 and the THO complex, and stabilized by Npl3. TERRA is degraded through pathways involving members of the nuclear exosome and the Trf4/Air2/Mtr4p polyadenylation (TRAMP) complex (3′-to-5′ degradation) and the Rat1 exonuclease (5′-to-3′ degradation). Pap1-mediated polyadenylation stabilizes TERRA transcripts. In *Schizosaccharomyces pombe* (*lower* panel), TERRA is produced by RNAPII starting from promoters of unknown sequence. The shelterin proteins Taz1 and Rap1 suppress TERRA transcription. TERRA is able to form telR-loops, which are restricted by RNaseH2. A fraction of TERRA is polyadenylated and soluble in the nucleoplasm. The noncanonical poly(A) polymerases Cid14, part of the TRAMP complex, and Cid12 promote TERRA degradation.

ated at CpG dinucleotides by the DNA methyltransferases DNMT1 and DNMT3b, resulting in TERRA transcription repression (Fig. 2). Colorectal carcinoma HCT116 cells lacking both DNMTs show increased RNAPII binding to 29 bp and telomeric repeats and strongly increased TERRA levels compared to parental cells (Nergadze et al. 2009; Feretzaki et al. 2019). Twenty-nine base pairs repeat hypomethylation is associated with increased TERRA also in ICF patient cells carrying mutations in the *DNMT3B* gene, and in ALT cells (Yehezkel et al. 2008, 2013; Nergadze et al. 2009; Ng et al. 2009; Arora et al. 2014). In mice, however, the inactivation of DNMT3b does not increase TERRA levels, possibly because mouse TERRA promoters do not contain CpG dinucleotides (Toubiana et al. 2020).

Histone modifications play a crucial role in regulating TERRA transcription. TERRA levels are increased in mESCs or mouse embryonic fibroblasts (MEFs) knocked-out for the histone methyltransferases SUV39H or SUV420H, which catalyze the deposition of the heterochromatin marks histone 3 trimethylated at lysine 9 (H3K9me3) and histone 4 trimethylated at lysine 20 (H4K20me3), respectively (Schoeftner and Blasco 2008). Likewise, SUV39H1 and HP1α, which create and bind the heterochromatic mark H3K9me3, respectively, and can in turn recruit SUV420H, also repress TERRA transcription in telomerase-positive human cells (Fig. 2; Arnoult et al. 2012). H3K4 methylation, which is typically associated with active chromatin, is present at human telomeric repeats and subtelomeric TERRA promoters and correlates with transcriptional activation and increased TERRA levels (Caslini et al. 2009; Negishi et al. 2015). Similarly, in human cells, treatment with trichostatin A, a histone deacetylase inhibitor, increased TERRA levels (Azzalin and Lingner 2008; Farnung et al. 2012). In *S. cerevisiae*, the histone deacetylases Sir2, Sir3, and Sir4, which support deacetylation of H4K16, H3K9, and H3K14 and are recruited to telomeres by Rap1, repress TERRA transcription from X-only telomeres (Fig. 3; Iglesias et al. 2011). Heterochromatin does not seem to play a major role in the regulation of TERRA transcription in fission yeast, where no changes in TERRA levels were observed upon loss of the heterochromatic factor Swi6, the *pombe* ortholog of HP1α, or the histone H3K9 methyltransferase Clr4 (Greenwood and Cooper 2012).

There is also evidence that in some organisms, TERRA expression negatively correlates with telomere length. As mentioned above, increased TERRA transcription was reported in human ICF cells, which carry short telomeres, when compared to healthy control cells, and in oncogene-transformed human fibroblasts with critically short telomeres (Yehezkel et al. 2008; Sagie et al. 2017; Nassour et al. 2023). Furthermore, expression of the human telomerase cat-

alytic subunit hTERT induced telomere elongation and repressed TERRA transcription in human fibroblasts and cancer cells (Arnoult et al. 2012). Finally, in budding and fission yeasts, TERRA transcription increased upon telomere shortening induced by telomerase inactivation (Cusanelli et al. 2013; Moravec et al. 2016; Graf et al. 2017). It remains to be established what mechanistic events stand behind the communication between telomere length and TERRA transcription. Because shorter telomeres in mouse cells lose marks of heterochromatin and telomere elongation in human cells increases telomeric H3K9me3 and HP1α (Benetti et al. 2007; Arnoult et al. 2012), it is very likely that changes in chromatin compaction mediate telomere length and TERRA cross talk, at least in mammalian cells.

Several transcription regulatory proteins have been implicated in TERRA production. Human CTCF binds to several human subtelomeres, at sites frequently overlapping with those of the cohesin subunits SMC1 and Rad21. Depletion of CTCF or Rad21 impaired RNAPII recruitment and decreased TERRA levels (Deng et al. 2012; Beishline et al. 2017). The transcription factors HSF1, NRF1, and p53 promote TERRA transcription in human cells in response to heat shock, oxidative stress, and etoposide treatment or serum starvation, respectively (Diman et al. 2016; Tutton et al. 2016; Le Berre et al. 2019). The transcription factor Rb1 was also proposed to positively regulate TERRA transcription in human and mouse cells (Fig. 2; Gonzalez-Vasconcellos et al. 2017). On the other hand, transcription factors such as ZNF148, ZFX, PLAG1, and EGR1 in human cells, and Reb1 in budding yeast, were shown to inhibit TERRA transcription (Figs. 2 and 3; Feretzaki et al. 2019; Bauer et al. 2022). Finally, both in human and mouse cells, the chromatin remodeler ATRX regulates TERRA transcription, although differently according to the model system used (Fig. 2). In human telomerase-positive glioma cell lines 8MGBA and HS683 and the immortalized fibroblast SW39 line, ATRX depletion decreased TERRA levels (Episkopou et al. 2014; Eid et al. 2015). On the other hand, no changes in TERRA levels were ob-

Cite this article as *Cold Spring Harb Perspect Biol* doi: 10.1101/cshperspect.a041683

served in mouse forebrain tissue or MEFs lacking ATRX, while increased TERRA levels were reported in mESCs and HeLa cells with inactivated ATRX (Goldberg et al. 2010; Flynn et al. 2015; Levy et al. 2015; Feretzaki et al. 2019).

Members of the shelterin complex regulate TERRA transcription across diverse organisms. Increased TERRA was induced by depletion of TRF1 or TRF2 in human and mouse cell lines (Fig. 2; Caslini et al. 2009; Porro et al. 2014a,b; Zhang et al. 2017; Sadhukhan et al. 2018; Marión et al. 2019; Nassour et al. 2019; Cao et al. 2020; Porreca et al. 2020; Nie et al. 2021). Similarly, TERRA levels are elevated in *S. pombe* mutants ablated of Taz1, the TRF1 and TRF2 ortholog (Fig. 3; Greenwood and Cooper 2012). However, TRF1 and possibly TRF2 might suppress TERRA expression in a cell-type-dependent manner; indeed, TERRA down-regulation was observed upon TRF1 depletion in human Hela Kyoto cells, the human hepatocarcinoma cell lines SNU-368 and SNU-739, and the immortalized mouse myoblast cell line C2C12 (Schoeftner and Blasco 2008; Scheibe et al. 2013; Cao et al. 2020), whereas no change in TERRA levels was caused by inactivation of TRF1 in MEFs (Sfeir et al. 2009). How TRF1 regulates TERRA expression remains to be clarified.

The ability of TRF2 to suppress TERRA transcription depends on its TRFH domain, which is able to compact chromatin, likely reducing RNA-PII accessibility (Poulet et al. 2012; Porro et al. 2014a). Additionally, TRF2 promotes the recruitment of the nucleolar protein TCOF1 to telomeres during S phase, also contributing to decreased TERRA transcription, as TCOF1 suppresses RNAPII activation (Fig. 2; Nie et al. 2021). The shelterin component Rap1 suppresses TERRA expression in budding and fission yeasts (Iglesias et al. 2011; Bah et al. 2012; Lorenzi et al. 2015). While it is not clear how fission yeast Rap1 regulates TERRA, budding yeast Rap1 suppresses TERRA transcription by recruiting Sir2/3/4 proteins at X-only telomeres and Rif1 and Rif2 at Y′ telomeres (Fig. 3; Iglesias et al. 2011). Rap1 functions in regulating TERRA expression might be poorly conserved across species, as Rap1 depletion in human HeLa cells did not alter TERRA levels (Porro et al. 2014a).

## TERRA DEGRADATION REGULATION

TERRA cellular levels are also regulated posttranscriptionally through RNA degradation. In human cancer cells, TERRA molecules are bound by the RNA-binding proteins RALY and PABPN1, the latter being implicated in targeting RNAs for degradation by the nuclear exosome complex (Fig. 2). RALY inactivation reduced the levels of poly(A)$^-$ TERRA transcripts, while PABPN1 depletion in cells lacking RALY reduced the levels of both poly(A)$^-$ and poly(A)$^+$ TERRA (Savoca et al. 2023). Interestingly, PABPN1 depletion alone or depletion of the nuclear exosome component EXOSC3 led to an increase in TERRA levels, suggesting that the exosome is involved in TERRA degradation (Savoca et al. 2023). Consistently, EXOSC9, another exosome component, was reported to promote TERRA degradation in MCF10-2A human mammary epithelial cells (Quttina et al. 2023). Moreover, TERRA binds to the ribonucleoprotein RBMX, and the interaction of RBMX with ZCCHC8, a main component of the nuclear exosome targeting (NEXT) complex, was proposed to promote exosome-mediated degradation of TERRA in HeLa and osteosarcoma U2OS cells (Fig. 2; Liu et al. 2023). Only some subcomplexes of the exosome might be involved in TERRA degradation because inactivation of the exosome subunits DIS3 and EXOSC10 did not alter TERRA levels in HCT116 cells (Reiss et al. 2023). Of note, members of the exosome complex were found to be enriched in a screen for TERRA interactors in human but not in mouse cells, suggesting different TERRA degradation mechanisms in the two species (Scheibe et al. 2013; Viceconte et al. 2021).

Another factor involved in the degradation of TERRA transcripts in human and budding yeast cells is the exonuclease XRN2/Rat1 (Figs. 2 and 3). In human cells, degron-induced inactivation of XRN2 led to an increase in TERRA and telR-loop levels (Reiss et al. 2023). In budding yeast cells, Rat1 was shown to degrade TERRA through pathways dependent on Rap1 and the Rif1/2 protein complex (Iglesias et al. 2011). Moreover, in Rat1 yeast mutants, Pap1-mediated polyadenylation stabilizes TERRA transcripts, while the Trf4/Air2/Mtr4p polyadenylation (TRAMP) complex, in concert with the exosome, support TERRA degradation (Luke et al. 2008). The TRAMP complex is also involved in TERRA degradation in fission yeast, since strains expressing a mutant Cid14, the fission yeast ortholog of Trf4, showed increased levels of TERRA (Bah et al. 2012). Finally, RNA modifications can also influence TERRA stability. Recent work in different human ALT cell lines revealed that m$^6$A-modified nucleotides, which can be found both in the subtelomeric and telomeric part of TERRA transcripts, stabilize TERRA (Fig. 2; Chen et al. 2022; Vaid et al. 2024).

## TERRA FUNCTIONS

The functions associated with TERRA and telomere transcription are diverse and often linked to the ability of TERRA to interact with proteins and recruit them to telomeres or other genomic regions. The molecular mechanisms regulating TERRA–protein interactions are far from being fully elucidated; however, studies performed both in vitro and in cells point to TERRA folding into RNA G-quadruplex (rG4) structures as a crucial event (Collie et al. 2010; Xu et al. 2010; Biffi et al. 2012; Liu et al. 2017; Ghosh and Singh 2020; Roach et al. 2020; Mei et al. 2021; Abreu et al. 2022). While we will review here the most well-characterized functions of TERRA, its interaction with factors involved in disparate cellular processes suggests that TERRA participates in so-far unexplored pathways. Table 2 summarizes the known TERRA interacting proteins in human and mouse cells and the proposed functions for these interactions.

### TERRA and Telomere Integrity Maintenance

TERRA supports telomere homeostasis throughout several processes, including DNA replication and heterochromatin deposition. TERRA directly binds the origin recognition complex (ORC) subunit ORC1 and stimulates its interaction with the amino-terminal domain of TRF2, forming a ternary complex. This stabilized complex promotes the efficient recruitment of other ORC proteins, including ORC2, thus facilitating the initiation of telomeric DNA replication. Supporting this model, transfection of cells with short interfering RNAs (siRNAs) targeting UUAGGG sequences led to a decrease in ORC2 association with telomeres and telomere instability (Deng et al. 2009). However, it should be noted that TERRA is mostly nuclear, while the siRNA degradation machinery is cytoplasmic; hence, the effects exerted by TERRA siRNAs on ORC recruitment and telomere stability might be indirect. Nonetheless, other lines of evidence support a direct role for TERRA in telomeric DNA replication. For example, regulated TERRA transcription induced by CTCF promotes proper telomere replication (Beishline et al. 2017). Moreover, in HeLa cells, TERRA is found at active replication forks and remains associated with newly synthesized DNA for several hours after DNA duplication (Gylling et al. 2020).

Notably, deregulated TERRA transcription and/or mislocalization to telomeres can cause telomeric replication defects. Depletion of members of the nonsense-mediated mRNA decay (NMD) machinery including UPF1, SMG1, and SMG6 caused aberrant accumulation of TERRA at telomeres and telomere fragility in human cancer cells (Azzalin et al. 2007; Chawla et al. 2011). Similarly, the depletion of TCOF1, which abolishes the suppression of TERRA transcription in the S phase, led to the appearance of fragile telomeres (Nie et al. 2021). These results are consistent with early observations in budding yeast, where ectopically increasing TERRA transcription from the subtelomeric region of chromosome 7L caused replication defects in *cis* (Maicher et al. 2012). Intriguingly, similar experiments were performed in HeLa cells by introducing an inducible cytomegalovirus (CMV) promoter upstream of a telomeric tract. In that case, increasing TERRA transcription did not cause detectable telomeric defects, likely because the increase in TERRA was modest when compared to that induced at yeast 7L telomeres (Arora et al. 2012; Maicher et al. 2012). Hence, TERRA and telomere transcription might induce replication defects only if brought to exceptionally high and nonphysiological levels or if combined with lesions altering the telomeric chromatin.

Cite this article as *Cold Spring Harb Perspect Biol* doi: 10.1101/cshperspect.a041683

**Table 2.** Human and mouse telomeric repeat-containing RNA (TERRA) interactors

| Family/complex | Protein name | Experimental system | Validation | Cellular process | References |
|---|---|---|---|---|---|
| Shelterin | TRF1 | Raji cells (nuclear extracts) | RNA affinity purification followed by LC/MS/MS and WB | ND | Deng et al. (2009) |
| | | HCT116 cells | RIP with endogenous or FLAG-tagged TRF1 | | |
| | | U2OS cells | RIP with endogenous TRF1 | | Scheibe et al. (2013) |
| | | Hela S3 cells | SILAC-based quantitative MS and siRNA depletion | TERRA expression and localization | |
| | | Human embryonic stem cells (hESCs) | RIP with endogenous TRF1 | ND | Zeng et al. (2017) |
| | | Recombinant protein | EMSA | TelR-loop formation | Lee et al. (2018) |
| | | Recombinant protein | ELISA | TRF1 association with telomeric DNA | Kang et al. (2021) |
| | | HEK293T cells | RIP with FLAG-tagged TRF1 | ND | Ghisays et al. (2021) |
| | | Hela cells (chromatin fraction) | RNA pull-down and WB | ND | Vohhodina et al. (2021) |
| | | Recombinant protein | EMSA | TERRA–TRF2 association | Abreu et al. (2022) |
| | TRF2 | Raji cells (nuclear extracts) | RNA affinity purification followed by LC/MS/MS and WB | TERRA localization to telomeres; heterochromatin formation | Deng et al. (2009) |
| | | HCT 116 cells | RIP with endogenous or FLAG-tagged TRF2 | | |
| | | U2OS cells | RIP with endogenous TRF2 | | |
| | | Recombinant protein | EMSA | ND | Biffi et al. (2012) |
| | | Recombinant protein | ELISA | | |
| | | Recombinant protein | EMSA | TRF2-mediated DNA compaction | Poulet et al. (2012) |
| | | Hela S3 cells | SILAC-based quantitative MS and siRNA depletion | TERRA levels and localization | Scheibe et al. (2013) |
| | | Recombinant protein | EMSA | TelR-loop formation | Lee et al. (2018) |

*Continued*

**Table 2.** *Continued*

| Family/complex | Protein name | Experimental system | Validation | Cellular process | References |
|---|---|---|---|---|---|
| | | Mouse embryonic fibroblasts (MEFs) | RIP with endogenous TRF2 | ND | Viceconte et al. (2021) |
| | | Recombinant protein | Fluorescence polarization assay | TERRA expression and localization; telomere length and stability maintenance | Mei et al. (2021) |
| | | LOX cells | RIP with endogenous TRF2 | | |
| | | Hela cells (chromatin fraction) | RNA pull-down and WB | ND | Vohhodina et al. (2021) |
| | | Recombinant protein | EMSA | TelR-loop formation | Abreu et al. (2022) |
| | | U2OS cells | RIP with endogenous TRF2 | ND | Wang et al. (2023) |
| | | Recombinant protein | EMSA | ND | |
| | RAP1 | Raji cells (nuclear extracts) | RNA affinity purification followed by LC/MS/MS and WB | ND | Deng et al. (2009) |
| | | Hela cells (chromatin fraction) | RNA pull-down and WB | ND | Vohhodina et al. (2021) |
| | TIN2 | Hela cells (chromatin fraction) | RNA pull-down and WB | ND | Vohhodina et al. (2021) |
| Telomerase and telomerase-associated proteins | Telomerase | HEK 293T cells (nuclear extracts) | IP with endogenous or Myc-tagged hTERT | Telomerase activation | Redon et al. (2010) |
| | | Purified protein | IP with FLAG or Myc-tagged hTERT | | |
| | Dyskerin | Raji cells (nuclear extracts) | RNA affinity purification followed by LC/MS/MS and WB | ND | Deng et al. (2009) |
| Heterochromatin proteins | HP1α | HCT 116 cells | RIP with endogenous or Flag-tagged HP1α | Heterochromatin formation | Deng et al. (2009) |
| | | Recombinant protein | Biolayer interferometry and EMSA | ND | Roach et al. (2020) |
| | HP1β | HCT 116 cells | RIP with FLAG-tagged HP1β | Heterochromatin formation | Deng et al. (2009) |
| Histone marks | H3K9me3 | HCT 116 cells | RIP with endogenous H3K9me3 | Heterochromatin formation | Deng et al. (2009) |
| | | Hela cells | RIP with endogenous H3K9me3 | Heterochromatin formation | Porro et al. (2014a) |
| | | hESCs | RIP with endogenous H3K9me3 | Heterochromatin formation | Zeng et al. (2017) |
| Histone acetyltransferases | MORF4L2 (part of NuA4 complex) | Hela S3 cells | SILAC-based quantitative MS and siRNA depletion | TERRA expression and localization | Scheibe et al. (2013) |

*Continued*

Cite this article as *Cold Spring Harb Perspect Biol* doi: 10.1101/cshperspect.a041683

Table 2. Continued

| Family/complex | Protein name | Experimental system | Validation | Cellular process | References |
|---|---|---|---|---|---|
| Histone methyltransferases | PRC2 complex | Recombinant proteins (complex) | EMSA | ND | Wang et al. (2017) |
| | SUZ12 (part of PRC2) | U2OS cells (nuclear extracts) | RNA pull-down and WB | Heterochromatin formation; recruitment of SUZ12 to telomeres | Montero et al. (2018) |
| | | Mouse Trp53$^{-/-}$ 2i-grown induced pluripotent stem cells (iPSCs) | RIP with endogenous SUZ12 | Expression of pluripotency and differentiation genes | Marión et al. (2019) |
| | | Mouse R1/E ESCs | SILAC-based quantitative MS | ND | Viceconte et al. (2021) |
| | | MEFs | RIP with endogenous SUZ12 | | |
| | EZH2 (part of PCR2) | U2OS cells (nuclear extracts) | RNA pull-down and WB | Heterochromatin formation | Montero et al. (2018) |
| | SUV39H1 | Hela cells | RIP with endogenous SUV39H1 | Heterochromatin formation | Porro et al. (2014a) |
| | | Recombinant protein | EMSA | Heterochromatin formation | Zeng et al. (2017) |
| | | hESCs | RIP with endogenous H3K9me3 | | |
| Histone demethylases | LSD1 | Hela cells | RIP with endogenous LSD1 RNA pull-down in nuclear extracts and WB | LSD1 interaction with MRE11; processing of uncapped telomeres | Porro et al. (2014b) |
| | | HEK293T cells | RIP with endogenous or HA-tagged LSD1 | | |
| | | Recombinant protein | RNA pull-down and WB EMSA | | |
| | | U2OS cells | RIP with endogenous LSD1 RNA pull-down with Flag-LSD1 in cell lysates and WB | TERRA localization; phase separation; telR-loop formation | Xu et al. (2024) |
| | | Recombinant proteins (LSD1 alone or in complex with coREST) | EMSA | | |

Continued

**Table 2.** *Continued*

| Family/complex | Protein name | Experimental system | Validation | Cellular process | References |
|---|---|---|---|---|---|
| | | Recombinant LSD1-coREST complex (individual proteins or in complex) | EMSA | LSD1 enzymatic activity | Hirschi et al. (2016) |
| ORCs | ORC1 | Raji cells (nuclear extracts) | RNA affinity purification followed by LC/MS/MS and WB | Heterochromatin formation | Deng et al. (2009) |
| | | HCT 116 cells | RIP with endogenous ORC1 | | |
| | | Recombinant protein | EMSA | | |
| | ORC2 | Raji cells (nuclear extracts) | RNA affinity purification followed by LC/MS/MS and WB | Heterochromatin formation | Deng et al. (2009) |
| | ORC4 | Raji cells (nuclear extracts) | RNA affinity purification followed by LC/MS/MS and WB | Heterochromatin formation | Deng et al. (2009) |
| hnRNPs | hnRNPA1 | Primary MEFs | Biotin pull-down assay followed by LC-MALDI TOF/TOF and WB validation; siRNA depletion; RIP with endogenous hnRNPA1 | Telomere capping and telomere length maintenance | López de Silanes et al. (2010) |
| | | HEK 293T cells (nuclear extracts) | IP with endogenous hnRNPA1 | ND | Redon et al. (2010) |
| | | HEK 293T cells | IP with Myc-tagged hnRNPA1 (in nuclear extracts) | Telomerase activation | Redon et al. (2013) |
| | | | RIP with endogenous hnRNPA1 | | |
| | | HT1080 cells | RIP with ProteinA-tagged hnRNPA1 | | |
| | | Recombinant protein | EMSA | | |
| | | Hela S3 cells | SILAC-based quantitative MS; siRNA depletion | TERRA expression | Scheibe et al. (2013) |
| | | Hela cells (nuclear extracts) | RNA pull-down and WB | ND | Porro et al. (2014b) |
| | | HEK293T cells | RIP with endogenous hnRNPA1 | | |
| | | Recombinant protein | EMSA | ND | Liu et al. (2017) and Liu and Xu (2018) |

*Continued*

**Table 2.** *Continued*

| Family/complex | Protein name | Experimental system | Validation | Cellular process | References |
|---|---|---|---|---|---|
| | | Recombinant protein (UP1 or UP1+RGG domains) | Isothermal titration calorimetry | ND | Ghosh and Singh (2020) |
| | | Hela cells with long or short telomeres | RIP with endogenous hnRNPA1; siRNA depletion | ND | Fernandes and Lingner (2023) |
| | hnRNPA2/B1 | Primary MEFs | Biotin pull-down assay followed by LC-MALDI TOF/TOF and WB; siRNA depletion; IP with endogenous hnRNPA2/B1 (in nuclear extracts) | TERRA localization | López de Silanes et al. (2010) |
| | | U2OS cells | RIP with endogenous hnRNPA2/B1 | TERRA localization | Vaid et al. (2024) |
| | hnRNPCL2/RALY | Hela cells | RNA pull-down in cell lysates and WB | TERRA expression and localization | Savoca et al. (2023) |
| | hnRNPD/AUF1 | Primary MEFs | Biotin pull-down assay followed by LC-MALDI TOF/TOF and WB; siRNA depletion; IP with endogenous hnRNPD (in nuclear extracts) | TERRA localization; telomere length maintenance | López de Silanes et al. (2010) |
| | hnRNPF | Primary MEFs | Biotin pull-down assay followed by LC-MALDI TOF/TOF (and WB validation); siRNA depletion; IP with endogenous hnRNPF (in nuclear extracts) | TERRA expression and localization; telomere capping and telomere length maintenance | López de Silanes et al. (2010) |
| | hnRNPG/RBMX | U2OS cells (nuclear extracts) | RNA pull-down and WB | TERRA stability; telR-loop levels | Liu et al. (2023) |
| | hnRNPM | Primary MEFs | Biotin pull-down assay followed by LC-MALDI TOF/TOF and WB; siRNA depletion; IP with endogenous hnRNPM (in nuclear extracts) | TERRA localization; telomere length maintenance | López de Silanes et al. (2010) |
| | | Hela S3 cells | SILAC-based quantitative MS and siRNA depletion | TERRA localization | Scheibe et al. (2013) |

*Continued*

Table 2. *Continued*

| Family/complex | Protein name | Experimental system | Validation | Cellular process | References |
|---|---|---|---|---|---|
| | hnRNPP2/TLS/FUS | Recombinant protein or TLS RGG3 motif | EMSA | Heterochromatin formation; telomere length | Takahama et al. (2013) |
| | | Hela cells | RIP with Flag-tagged TLS | | |
| | | Recombinant TLS RGG3 motif | EMSA | Heterochromatin formation | Takahama et al. (2015) |
| | | Hela cells | RIP with Flag-tagged TLS or Flag-tagged RGG3 | | |
| | | Recombinant TLS RGG3 motif | EMSA | ND | Takahama and Oyoshi (2013) and Kondo et al. (2018) |
| | | Mouse embryonic stem cells (mESCs) (nuclear extracts) H1299 cells (nuclear extracts) | RNA pull-down and WB | ND | Petti et al. (2019) |
| Other RNA-binding motif containing proteins | GRSF1 | Hela S3 cells | SILAC-based quantitative MS and siRNA depletion | ND | Scheibe et al. (2013) |
| | NONO | mESCs (nuclear extracts) H1299 cells (nuclear extracts) | RNA pull-down and WB | TERRA localization; telR-loop formation; telomere stability maintenance | Petti et al. (2019) |
| | | C57Bl/6J mice male and female PGCs | iDRIP and IF-TERRA FISH | ND | Brieño-Enríquez et al. (2019) |
| | SFPQ | mESCs (nuclear extracts) H1299 cells (nuclear extracts) | RNA pull-down and WB | TERRA localization; telR-loop formation; telomere stability maintenance; homologous recombination | Petti et al. (2019) |
| | | C57Bl/6J mice male and female gonadal somatic cells | iDRIP and IF-TERRA FISH | ND | Brieño-Enríquez et al. (2019) |

*Continued*

Cite this article as *Cold Spring Harb Perspect Biol* doi: 10.1101/cshperspect.a041683

**Table 2.** *Continued*

| Family/complex | Protein name | Experimental system | Validation | Cellular process | References |
|---|---|---|---|---|---|
| | RBM14 | U2OS cells | RIP with endogenous RBM14 | TERRA expression; telR-loop formation | Wang et al. (2023) |
| | | Recombinant protein | EMSA | | |
| | Nucleolin | Recombinant protein | EMSA and Microscale Thermophoresis | ND | Khan et al. (2023) |
| | PABPN1 | Hela cells | RIP with endogenous PABPN1 | TERRA expression | Savoca et al. (2023) |
| Recombinases and associated proteins | Rad51 | Recombinant protein | EMSA | TERRA localization; telR-loop formation | Feretzaki et al. (2020) |
| | | Hela cells | RIP with endogenous RAD51 and siRNA depletion | | |
| | | U2OS, LM216J, Hela LT cells | RIP with endogenous RAD51 | ND | Kaminski et al. (2022) |
| | RAD51AP1 | Recombinant protein | EMSA | TelR-loop formation | Yadav et al. (2022) |
| | | Recombinant protein | EMSA | TelR-loop formation | Kaminski et al. (2022) |
| | | U2OS, LM216J, and Hela LT cells | RIP with endogenous RAD51AP1 | | |
| Helicases | BLM | Raji cells (nuclear extracts) | RNA affinity purification followed by LC/MS/MS and WB | ND | Deng et al. (2009) |
| | | Mouse R1/E ESCs | SILAC-based quantitative MS | ND | Viceconte et al. (2021) |
| | | MEFs | RIP with endogenous BLM | ND | |
| | DDX21 | Recombinant protein (carboxy-terminal G4-binding domain) | Microscale Thermophoresis Isothermal titration calorimetry | ND | McRae et al. (2018) |
| | RTEL1 | Recombinant protein (RTEL1Δ762) | Fluorescence anisotropy and EMSA | TERRA expression and localization | Ghisays et al. (2021) |
| | | HEK293T cells | RIP with Flag-tagged RTEL1 | | |
| | SETX | Hela cells (chromatin fraction) | RNA pull-down and WB | ND | Vohhodina et al. (2021) |

*Continued*

**Table 2.** *Continued*

| Family/complex | Protein name | Experimental system | Validation | Cellular process | References |
|---|---|---|---|---|---|
| SNF/SWI family | ARID1A | Hela S3 cells | SILAC-based quantitative MS and siRNA depletion | TERRA expression | Scheibe et al. (2013) |
| | ATRX | Mouse ESCs | iDRIP-MS | Gene expression at nontelomeric sites; ATRX localization to telomeres | Chu et al. (2017a) |
| | | Recombinant protein | EMSA | | |
| BRCA1 B complex | BRCA1 | Hela cells (chromatin fraction) | RNA pull-down and WB | TERRA and telR-loop expression | Vohhodina et al. (2021) |
| | | U2OS, T98G, and HME cells (chromatin fraction) | RIP with endogenous BRCA1 | | |
| | BARD1 | Hela cells (chromatin fraction) | RNA pull-down and WB | ND | Vohhodina et al. (2021) |
| THO complex | THOC1 and THOC2 | Hela cells with long or short telomeres | RIP with endogenous or HA-tagged THOC1 and with endogenous THOC2; siRNA depletion | TERRA localization; telR-loop formation; telomere stability maintenance | Fernandes and Lingner (2023) |
| | | HEK293E cells | RIP with endogenous or HA-tagged THOC1 and with endogenous THOC2 | ND | |
| XPF nuclease family | ERCC4/XPF | U2OS cells | iDRIP-MS and RIP with endogenous XPF | Break-induced telomere synthesis activation | Guh et al. (2022) |
| High-mobility group proteins | HMGA1 | Hela S3 cells | SILAC-based quantitative MS and siRNA depletion | TERRA localization | Scheibe et al. (2013) |
| | HMGB1 | Hela S3 cells | SILAC-based quantitative MS and siRNA depletion | TERRA localization | Scheibe et al. (2013) |
| | HMGB2 | Hela S3 cells | SILAC-based quantitative MS and siRNA depletion | TERRA expression | Scheibe et al. (2013) |
| YTH domain family | YTHDC1 | U2OS cells | RIP with HA-tagged YTHDC1 | TERRA expression and stability | Chen et al. (2022) |
| Innate immune sensors | ZBP1 | $IMR90^{E6E7}$ cells | RIP with Flag-tagged ZBP1 | MAVS-mediated immune response and autophagy | Nassour et al. (2023) |

*Continued*

**Table 2.** *Continued*

| Family/complex | Protein name | Experimental system | Validation | Cellular process | References |
|---|---|---|---|---|---|
| Exoribonucleases | XRN2 | Hela cells (chromatin fraction) | RNA pull-down and WB | ND | Vohhodina et al. (2021) |
| | C19orf43 | HEK293T cells | RIP with Flag-tagged C19orf43 | TERRA and telR-loop expression; protection of sister telomeres in mitosis | Sze et al. (2023) |
| Topoisomerases | TOPO I | Raji cells (nuclear extracts) | RNA affinity purification followed by LC/MS/MS and WB | ND | Deng et al. (2009) |
| | TOPO II | Raji cells (nuclear extracts) | RNA affinity purification followed by LC/MS/MS and WB | ND | Deng et al. (2009) |
| DNA damage factors | 53BP1 | Hela cells (chromatin fraction) | RNA pull-down and WB | ND | Vohhodina et al. (2021) |
| | MDC1 | Hela cells (chromatin fraction) | RNA pull-down and WB | ND | Vohhodina et al. (2021) |
| Others | PSIP1 | Hela S3 cells | SILAC-based quantitative MS and siRNA depletion | TERRA expression | Scheibe et al. (2013) |
| | RCC2 | Hela S3 cells | SILAC-based quantitative MS and siRNA depletion | TERRA expression | Scheibe et al. (2013) |
| | USP39 | Hela S3 cells | SILAC-based quantitative MS and siRNA depletion | TERRA expression | Scheibe et al. (2013) |
| | SRRT | Hela S3 cells | SILAC-based quantitative MS and siRNA depletion | TERRA expression and localization | Scheibe et al. (2013) |
| | ZNF691 | Hela S3 cells | SILAC-based quantitative MS and siRNA depletion | TERRA localization | Scheibe et al. (2013) |
| | DBT | Hela S3 cells | SILAC-based quantitative MS and siRNA depletion | TERRA expression | Scheibe et al. (2013) |
| | PC | Hela S3 cells | SILAC-based quantitative MS and siRNA depletion | TERRA expression | Scheibe et al. (2013) |
| | EBNA1 | Raji cells (nuclear extracts) | RNA affinity purification followed by LC/MS/MS and WB | ND | Deng et al. (2009) |
| | PARP1 | Raji cells (nuclear extracts) | RNA affinity purification followed by LC/MS/MS and WB | ND | Deng et al. (2009) |

*Continued*

Table 2. *Continued*

| Family/complex | Protein name | Experimental system | Validation | Cellular process | References |
|---|---|---|---|---|---|
| | MecP2 | Raji cells (nuclear extracts) | RNA affinity purification followed by LC/MS/MS and WB | ND | Deng et al. (2009) |
| | DNA-PK/PRKDC | Raji cells (nuclear extracts) | RNA affinity purification followed by LC/MS/MS and WB | ND | Deng et al. (2009) |
| | HuR | Primary MEFs | Biotin pull-down assay followed by LC-MALDI TOF/TOF and WB; siRNA depletion; IP with endogenous HuR (in nuclear extracts) | ND | López de Silanes et al. (2010) |
| | INIP | Hela S3 cells | SILAC-based quantitative MS and siRNA depletion | ND | Scheibe et al. (2013) |
| | TJP2 | Hela S3 cells | SILAC-based quantitative MS and siRNA depletion | ND | Scheibe et al. (2013) |

(ELISA) enzyme-linked immunosorbent assay, (EMSA) electrophoretic mobility shift assay, (FISH) fluorescence in situ hybridization, (iDRIP) identification of direct RNA interacting proteins, (IF) indirect immunofluorescence, (IP) immunoprecipitation, (LC-MALDI TOF/TOF) liquid chromatography with matrix-assisted laser desorption/ionization time of flight mass spectrometry/time of flight mass spectrometry, (LC/MS/MS) liquid chromatography with tandem mass spectrometry, (ND) not determined, (SILAC) stable isotope labeling by amino acids in cell culture, (RIP) RNA immunoprecipitation, (WB) western blot.

Cite this article as *Cold Spring Harb Perspect Biol* doi: 10.1101/cshperspect.a041683

TERRA transcripts regulate telomeric heterochromatin formation by recruiting chromatin-modifying complexes to telomeres. TERRA directly interacts with SUV39H1 and promotes H3K9me3 deposition and recruitment of HP1α. Additionally, TERRA rG4s interact directly with HP1α, further facilitating its recruitment to telomeres (Arnoult et al. 2012; Porro et al. 2014a; Zeng et al. 2017; Roach et al. 2020). TERRA binds to and recruits the transcription repressor Polycomb repressive complex 2 (PRC2), thus facilitating H3K27 trimethylation. This event is followed by the establishment of H3K9me3, H4K20me3, and HP1α at telomeres (Wang et al. 2017; Montero et al. 2018; Marión et al. 2019; Viceconte et al. 2021). The RNA-binding protein TLS/FUS binds to TERRA rG4s and telomeric DNA, forming a ternary complex that deposits H3K9me3 and H4K20me3 at telomeres through the recruitment of SUV39H and SUV420H (Takahama et al. 2013, 2015; Kondo et al. 2018). Importantly, as mentioned above, HP1α and H3K9me3 suppress TERRA transcription. This unveils the existence of a regulated negative feedback loop, similar to TPE, in which TERRA suppresses its own production. This TERRA-mediated TERRA silencing likely occurs in a telomere-length-dependent manner and only once telomeres have been properly heterochromatinized (Fig. 4).

## TERRA and Telomerase

Shortly after its discovery, human TERRA was proposed to function as a telomerase inhibitor due to the complementarity of the G-rich TERRA sequence to the telomerase RNA template. Indeed, human TERRA inhibits telomerase activity in vitro and physically interacts with hTERT in vitro and in cell extracts (Schoeftner and Blasco 2008; Redon et al. 2010). Moreover, TERRA knockdown in mouse cells using locked nucleic acids (LNAs) increased total telomerase activity (Chu et al. 2017a). However, TERRA role in regulating telomerase is more complex than originally believed. Under conditions where human TERRA transcription was enhanced, for example when a CMV promoter was placed in front of a telomere or when the *DNMT1* and *DNMT3* genes were deleted, telomerase-mediated telo-

mere elongation remained unaffected (Farnung et al. 2012). Hence, TERRA transcription alone does not consistently inhibit telomerase in cells and might require the activity of TERRA interacting proteins. For instance, the nuclear ribonucleoprotein hnRNPA1, which binds TERRA rG4s, alleviates TERRA-mediated telomerase inhibition in vitro, suggesting a mechanism of TERRA-mediated inhibition of telomerase possibly during specific cell-cycle phases or at particular telomeres only (Redon et al. 2010, 2013; Scheibe et al. 2013; Porro et al. 2014b; Liu et al. 2017; Liu and Xu 2018; Ghosh and Singh 2020; Fernandes and Lingner 2023).

Further confounding the roles of TERRA in regulating telomerase, TERRA and telomere transcription can stimulate telomerase activity at telomeres in yeasts. In *S. cerevisiae*, live imaging studies showed that TERRA produced from shortened telomeres colocalizes with the telomerase RNA component TLC1 in the nucleoplasm of S phase cells, forming clusters that ultimately relocate to the telomeres that produced TERRA (Cusanelli et al. 2013). In *S. pombe*, polyadenylated TERRA, comprising very short telomeric repeat sequences, interacts with the telomerase catalytic subunit Trt1. Moreover, increased telomere transcription, achieved by placing inducible promoters upstream of telomeric tracts in cells, stimulated telomerase recruitment and telomerase-mediated elongation in *cis* (Moravec et al. 2016).

Thus, depending on the system and circumstance, TERRA and telomere transcription can both activate and inhibit telomerase at chromosome ends. Given that, as mentioned above, TERRA transcription increases when telomeres shorten, an intriguing model posits that TERRA produced from short telomeres and carrying short telomeric RNA tracts could diffuse in the nucleoplasm, interact with telomerase, and direct it back to the telomeres of origin to facilitate re-elongation. Once elongated, telomeres begin producing longer TERRA molecules containing extended telomeric repeat tracts, which remain associated with telomeric chromatin and function as a telomerase inhibitor to prevent excessive or deregulated telomere elongation in *cis* (Fig. 4). Supporting and further extending this

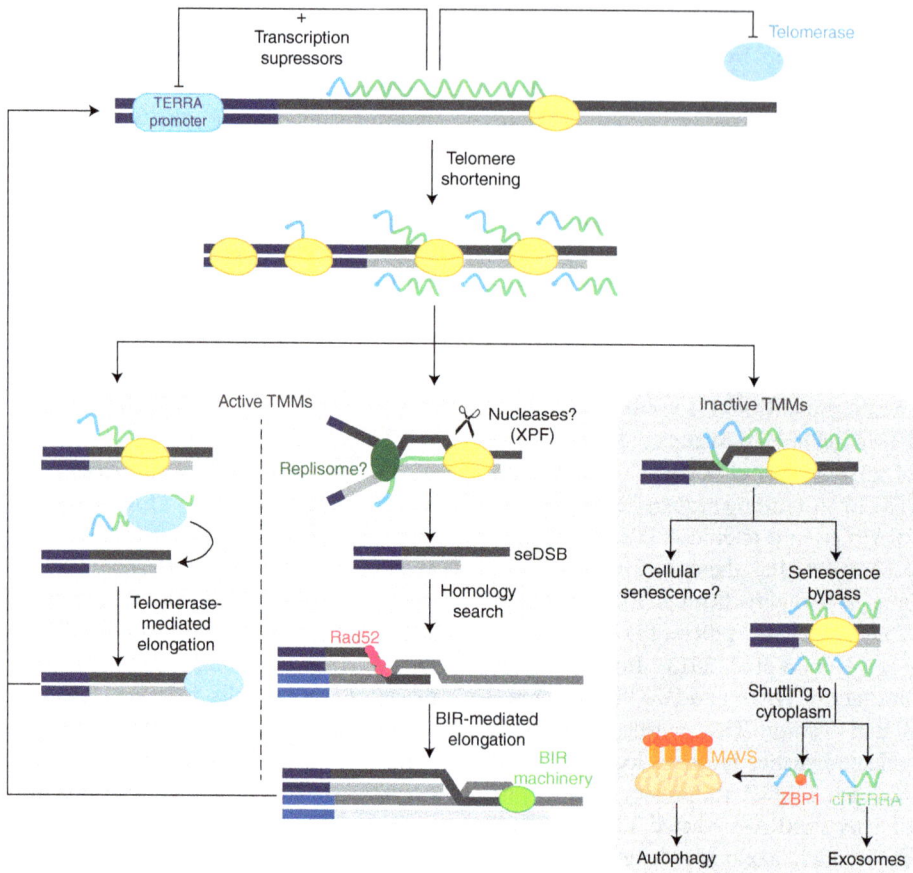

**Figure 4.** Speculative model for telomeric repeat-containing RNA (TERRA)-mediated functions upon telomere shortening. At long and intact telomeres, TERRA transcription is kept low and TERRA molecules containing long G-rich repeat tracts remain associated with the telomeric chromatin, where they maintain TERRA promoter silencing and avert activation of telomerase, if the latter is present. Upon telomere shortening, TERRA transcription increases and TERRA engages in different pathways according to whether telomere maintenance mechanisms (TMMs) are active in cells or not. In telomerase-positive cells (*left*), nucleoplasmic TERRA associates with telomerase and brings it back to the telomere of origin to promote elongation. In ALT cells (*middle*), a single-ended double-stranded break (seDSB) is generated at the short telomere either through collisions between the replication and the transcription machineries or by nuclease-mediated processing of telomeric R-loops (telR-loops). The seDSB uses a longer telomere as a template to initiate a break-induced replication (BIR) reaction and become elongated. In the absence of an active TMM (*right*), TERRA might induce cellular senescence, through pathways still to be clarified. If senescence is bypassed, TERRA produced from ultra-short telomeres shuttles to the cytoplasm where it interacts with ZBP1 and activates autophagy or becomes part of exosomes and is transported to the extracellular environment as cell-free (cf) TERRA.

model, live imaging and confocal microscopy in human cells demonstrated that TERRA molecules can associate in *trans* with long telomeres, counteracting telomerase localization (Bettin et al. 2024). Hence, TERRA from the same short telomere could not only recruit telomerase to the telomere of origin but also localize in *trans*

to longer telomeres and prevent telomerase from elongating them.

## TERRA and ALT

ALT cancer cells are telomerase-negative and elongate telomeres through a specialized break-induced replication (BIR) machinery (for more on ALT, see O'Sullivan and Greenberg 2025). TERRA is highly transcribed in ALT cells and localizes to ALT-associated PML bodies (APBs). In addition, ALT cells contain elevated telomeric telR-loops (Ng et al. 2009; Lovejoy et al. 2012; Arora et al. 2014). While these observations suggested an involvement of TERRA and telomere transcription in ALT, the first direct proof was achieved by targeting an RNAPII repressor domain to TERRA 29 bp promoters using transcription activator-like effectors (TERRA TALEs) (Silva et al. 2021). Reducing telomere transcription in ALT cells diminished the levels of TERRA and telR-loops, as expected, and alleviated ALT hallmarks such as telomeric replication stress, APBs, and de novo telomeric DNA synthesis in the $G_2/M$ phase. Furthermore, prolonged inhibition of TERRA transcription resulted in the accumulation of telomere-free chromosome ends, consistent with ALT inactivation (Silva et al. 2021). Similar conclusions were also obtained using a Cas9-based system degrading TERRA UUAGGG repeats in ALT cells, further attesting to crucial roles for TERRA in triggering and/or supporting ALT (Guh et al. 2022).

How do TERRA and telR-loops support ALT? While transcribing TERRA, RNAPII may codirectionally collide with the replisome at telomeres, thus inducing replication stress, fork collapse, and single-ended DNA double-stranded breaks (seDSBs). Alternatively, specific nucleases able to recognize and process telR-loops, such as XPF, may directly convert them into seDSBs (Guh et al. 2022). SeDSBs could then enter a BIR reaction, starting with RAD52-mediated homology search and invasion of longer donor telomeres (Fig. 4). As proposed for telomerase activation, short telomeres may produce more TERRA and accumulate telR-loops, ensuring that only short telomeres serve as BIR substrates. Supporting this hypothesis, recent studies in budding yeast models for ALT have shown that homology-directed repair is activated only at critically short telomeres, and that TERRA levels increase as telomeres shorten and decrease again following elongation (Misino et al. 2022). To add to the complexity of TERRA and ALT, telR-loops might also promote ALT BIR by forming at donor telomeres, possibly in *trans*, and successively being converted into D-loop intermediates that can serve as landing platforms for seDSBs (Yadav et al. 2022).

TERRA TALEs were also fused to an RNAPII activator and used to study the effects of increasing TERRA transcription in ALT cells (Silva et al. 2022). As expected, induced TERRA transcription increased telomeric replication and ALT activity, confirming that TERRA sustains ALT. However, despite heightened ALT activity, cells rapidly accumulated telomere-free ends through a reaction dependent on the activity of the structure-specific endonuclease Mus81 (Silva et al. 2022). These observations indicate that, while telomere transcription, TERRA, and telR-loops are essential for ALT, they must be precisely controlled to avoid excessive telomere dysfunction. Consistently, depletion of telR-loop suppressors, including RNAseH1 and FANCM, induced telomeric replication stress and telomere instability in ALT cells (Arora et al. 2014; Pan et al. 2017, 2019; Lu et al. 2019; Silva et al. 2019).

## TERRA and Senescence

In the absence of a telomere maintenance mechanism such as telomerase and ALT, cells undergoing telomere shortening eventually accumulate critically short telomeres that trigger an irreversible proliferation arrest state known as replicative senescence (d'Adda di Fagagna et al. 2003). TERRA is involved in senescence establishment, although the available information is somewhat contradictory and begs further investigation. In human cells, the link between telomere transcription and replicative senescence has not been extensively explored. However, insights come from studies of syndromes characterized by premature replicative senescence. Fibroblasts from patients with ICF or Hutchinson–Gilford progeria syndrome (HGPS) exhibit premature senescence (Gonzalo et al. 2017; Gagliardi et al. 2018) and are characterized by elevated levels of TERRA and telR-loops (Yehezkel et al. 2008; Decker et al. 2009; Sagie et al. 2017; and unpubl. results). In addition, a decrease in markers of facultative heterochromatin and an increase in transcription were observed at the 11q TERRA promoter in healthy human fibroblasts entering senescence (Thijssen et al. 2013). Based on this correlative evidence, one can speculate that TERRA, produced from short telomeres, could induce senescence initiation, perhaps through telR-loop formation (Fig. 4). Supporting this idea, inactivation of the THO complex subunits Hpr1 and Tho2 in telomerase-deficient budding yeast cells led to telR-loop accumulation and accelerated senescence, and this acceleration was partially suppressed by RNaseH1 overexpression (Yu et al. 2014). Similarly, the expression of an antisense RNA molecule that interacts with, and likely inhibits, TERRA delayed senescence in a manner that was epistatic with inactivation of the Dot1 histone H3K79 methyltransferase (Wanat et al. 2018).

However, other studies in budding yeast showed that telomere transcription can delay senescence onset. In presenescent telomerase-negative cells, TERRA and telR-loops accumulate at critically short telomeres due to the inhibition of RNA degradation. Rat1 and RNaseH2 are present at long telomeres where they degrade TERRA and telR-loops throughout S phase. In contrast, the recruitment of those enzymes is diminished at short telomeres, and TERRA and telR-loops accumulate. Because telomere transcription and telR-loops can trigger BIR-mediated telomere elongation, it has been proposed that this regulatory mechanism allows cells to re-elongate the first critically short telomeres, thereby delaying senescence rather than inducing it. In addition, destabilization of telR-loops via ablation of the nucleoprotein Npl3 or overexpression of RNaseH1 reduced telR-loops and accelerated senescence (Maicher et al. 2012; Balk et al. 2013; Graf et al. 2017; Pérez-Martinez et al. 2020; Misino et al. 2022).

Overall, while it is clear that telomere transcription, TERRA and telR-loops, have a role during senescence establishment, they might regulate senescence onset differently according to, for example, the number of critically short telomeres simultaneously accumulating in cells or the genetic background. Because cellular senescence is directly involved in suppressing cancer development and in promoting organ and tissue aging, this aspect of TERRA biology is extremely exciting and has implications for aging and disease biology.

## Extra-Telomeric Functions of TERRA

TERRA does not act only at telomeric chromatin but can also migrate to other genomic loci or to the cytoplasm and exert additional functions. In the nucleus, TERRA regulates gene expression. Comprehensive identification of RNA-binding proteins and capture hybridization analysis of RNA targets (ChIRP-CHART) experiments in mESCs revealed that TERRA binds not only to telomeres but also to thousands of sites across the mouse genome, which may or may not contain telomeric repeat sequences (Chu et al. 2017a). Many of those TERRA-binding sites are located within genic units, a fraction of which are also bound by ATRX. TERRA and ATRX exert opposing roles at those loci, with TERRA activating gene expression and ATRX repressing it (Chu et al. 2017a). In another study, TRF1 inactivation in mESCs was shown to increase TERRA levels and drive TERRA accumulation at Polycomb and pluripotency genes. This was concomitant with alteration in the expressions of those genes and induction of cell-fate programs and the loss of the naive state, suggesting that TERRA-mediated gene regulation is involved in cell differentiation (Marión et al. 2019).

In human cells, replicative senescence can be bypassed by the presence of oncogenic lesions such as p53 inactivation. After senescence bypass and additional cell divisions, telomeres become extremely short and trigger the so-called replicative crisis, a state characterized by telomeric fusions, genome rearrangements, and extensive death of the cell population through autophagy activation (Nassour et al. 2019). In oncogene-expressing human fibroblasts entering crisis, TERRA is produced at very high levels and it shuttles from the nucleus to the cytoplasm, where it associates with the cytosolic innate immunity sensor protein ZBP1. In turn, ZBP1 activates the mitochondrial signaling protein MAVS and induces autophagic cell death (Fig. 4; Nassour et al. 2023). A different study reported that transfection of synthetic TERRA and rG4-mimicking oligonucleotides in human cancer cells down-regulated the expression of innate immune genes including STAT1, ISG15, and OAS3, which are commonly found up-regulated in various cancers (Hirashima and Seimiya 2015). Altogether, these data show that TERRA exerts tumor suppressor roles in the cytoplasm, including eliminating precancerous cells by triggering crisis-associated cell death and down-regulating innate immune gene expression in cancer cells, possibly to counteract their malignancy.

TERRA molecules are also part of extracellular inflammatory exosomes. Cell-free TERRA molecules (cfTERRA) could be isolated from the exosome fractions of healthy and cancerous mouse tissues, human plasma, and cell culture medium (Fig. 4). CfTERRA increased in exosomes produced by human cultured cells expressing a dominant negative variant of TRF2 that induces telomere dysfunction (Wang et al. 2015; Wang and Lieberman 2016). Exosome vesicles containing cfTERRA can enter the bloodstream and trigger an inflammatory response by stimulating the production of cytokines such as IL-6 and TNF-α in peripheral blood mononuclear cells. Strikingly, cfTERRA was able to trigger the inflammatory response more robustly than telomeric DNA (Wang et al. 2015; Wang and Lieberman 2016). In light of these results, cfTERRA was proposed to be an "alarmin" able to induce immune cells to produce inflammatory cytokines and eliminate cells with dysfunctional telomeres (Wang et al. 2015; Wang and Lieberman 2016).

Finally, while TERRA was always considered to be a long noncoding RNA (lncRNA), it was recently proposed that human TERRA transcripts can undergo non-ATG translation and produce two repeat dipeptide proteins, one containing a valine–arginine (VR) motif and the other a glycine–leucine (GL) motif (Al-Turki and Griffith 2023). Synthetically produced VR dipeptide proteins bind nucleic acids and localize to replication forks in vitro and both VR and GL dipeptide proteins form long filaments with amyloid properties. Also, immunofluorescence studies using an antibody raised against VR repeats showed that VR proteins may localize both to the nucleus and cytoplasm and be more abundant in cells with elevated TERRA levels, including ALT and ICF cells. Induction of telomere dysfunction via knockdown of TRF2 increased the amounts of cytoplasmic VR proteins, and transient depletion of TERRA using LNAs generated large nuclear VR aggregates (Al-Turki and Griffith 2023). TERRA-derived proteins could pinpoint an additional role of TERRA as a sensor of telomere dysfunction, and execute one or more of the many functions associated with amyloid proteins including translation regulation or hormone storage. While TERRA should possibly not be considered a lncRNA any longer, additional studies must be performed to understand when exactly TERRA is translated and the physiological and pathological relevance of TERRA-derived proteins.

## CONCLUSIONS

Starting from a lncRNA molecule once suspected by many to be merely transcriptional noise, TERRA has come a long way. This multifaceted RNA is integral not only to telomere dynamics but also to broader cellular processes such as gene expression regulation, inter- and intracellular signaling, cell proliferation, and cell death. TERRA is involved in processes that can affect organismal fitness, including organ and tissue development and aging, disease etiology and progression, and immune response. The physical interactions of TERRA with proteins involved in diverse cellular pathways hold promise for discovering additional functions of this RNA. Moreover, the newly discovered translation of TERRA into polypeptides with amyloid properties is bound to open completely new lines of investigation, aiming at understanding not only the physiological functions of those proteins but also their potential involvement in neurodegeneration, as is the case for other amyloid proteins.

Many questions in TERRA biology are still open, including how exactly short telomeres activate TERRA transcription and, in turn, how this event modulates telomerase activity, ALT, and senescence establishment. In addition, while TERRA has been studied mostly in unicellular organisms or cultured mammalian cells, it is now more necessary than ever to shift toward multicellular organisms, including vertebrates. The scarcity of methods for efficient and amenable TERRA depletion remains one of the main challenges in the field and could represent a major drawback for the generation of model vertebrates to study TERRA at the organismal level. Nonetheless, the continuous and rapid development of novel genome editing approaches and small molecule design pipelines is expected to help overcome this impasse, allowing us to address a large number of so far unanswered yet exciting questions.

Finally, TERRA and its regulatory circuits might become attractive targets for novel therapeutic strategies to combat age-associated diseases, such as cancer and neurodegeneration, and to improve the quality of life for patients affected by genetic syndromes like ICF and progeria. The potential of TERRA in advancing our understanding and treatment of those conditions marks an exciting frontier in biomedical research.

## COMPETING INTEREST STATEMENT

Claus M. Azzalin is a founder and shareholder of TessellateBIO.

## ACKNOWLEDGMENTS

Due to space restrictions, we were unable to cite all the significant work on TERRA. We apologize to the authors whose work we could not include. The Azzalin laboratory is supported by Portuguese national funds through FCT—Fundação para a Ciência e a Tecnologia, I.P. (project 2021.00143.CEECIND to C.M.A. and project PTDC/MED-ONC/7864/2020 to P.L.A. and C.M.A.), by LaCaixa Foundation (project LCF/PR/HP21/52310016 to C.M.A.), and by TessellateBIO.

## REFERENCES

*Reference is also in this subject collection.

Abreu PL, Lee YW, Azzalin CM. 2022. In vitro characterization of the physical interactions between the long non-coding RNA TERRA and the telomeric proteins TRF1 and TRF2. *Int J Mol Sci* **23**: 10463. doi:10.3390/ijms231810463

Al-Turki TM, Griffith JD. 2023. Mammalian telomeric RNA (TERRA) can be translated to produce valine-arginine and glycine-leucine dipeptide repeat proteins. *Proc Natl Acad Sci* **120**: e2221529120. doi:10.1073/pnas.2221529120

Arnoult N, Van Beneden A, Decottignies A. 2012. Telomere length regulates TERRA levels through increased trimethylation of telomeric H3K9 and HP1α. *Nat Struct Mol Biol* **19**: 948–956. doi:10.1038/nsmb.2364

Arora R, Brun CM, Azzalin CM. 2012. Transcription regulates telomere dynamics in human cancer cells. *RNA* **18**: 684–693. doi:10.1261/rna.029587.111

Arora R, Lee Y, Wischnewski H, Brun CM, Schwarz T, Azzalin CM. 2014. RNaseh1 regulates TERRA-telomeric DNA hybrids and telomere maintenance in ALT tumour cells. *Nat Commun* **5**: 5220. doi:10.1038/ncomms6220

Azzalin CM, Lingner J. 2008. Telomeres: the silence is broken. *Cell Cycle* **7**: 1161–1165. doi:10.4161/cc.7.9.5836

Azzalin CM, Reichenbach P, Khoriauli L, Giulotto E, Lingner J. 2007. Telomeric repeat containing RNA and RNA surveillance factors at mammalian chromosome ends. *Science* **318**: 798–801. doi:10.1126/science.1147182

Bah A, Wischnewski H, Shchepachev V, Azzalin CM. 2012. The telomeric transcriptome of *Schizosaccharomyces pombe*. *Nucleic Acids Res* **40**: 2995–3005. doi:10.1093/nar/gkr1153

Balk B, Maicher A, Dees M, Klermund J, Luke-Glaser S, Bender K, Luke B. 2013. Telomeric RNA-DNA hybrids affect telomere-length dynamics and senescence. *Nat Struct Mol Biol* **20**: 1199–1205. doi:10.1038/nsmb.2662

Bauer SL, Grochalski TNT, Smialowska A, Åström SU. 2022. Sir2 and Reb1 antagonistically regulate nucleosome occupancy in subtelomeric X-elements and repress TERRAs by distinct mechanisms. *PLoS Genet* **18**: e1010419. doi:10.1371/journal.pgen.1010419

Baur JA, Zou Y, Shay JW, Wright WE. 2001. Telomere position effect in human cells. *Science* **292**: 2075–2077. doi:10.1126/science.1062329

Beishline K, Vladimirova O, Tutton S, Wang Z, Deng Z, Lieberman PM. 2017. CTCF driven TERRA transcription facilitates completion of telomere DNA replication. *Nat Commun* **8**: 2114. doi:10.1038/s41467-017-02212-w

Benetti R, García-Cao M, Blasco MA. 2007. Telomere length regulates the epigenetic status of mammalian telomeres and subtelomeres. *Nat Genet* **39**: 243–250. doi:10.1038/ng1952

Bettin N, Querido E, Gialdini I, Grupelli GP, Goretti E, Cantarelli M, Andolfato M, Soror E, Sontacchi A, Jurikova K, et al. 2024. TERRA transcripts localize at long telomeres to regulate telomerase access to chromosome ends. *Sci Adv* **10**: eadk4387. doi:10.1126/sciadv.adk4387

Biffi G, Tannahill D, Balasubramanian S. 2012. An intramolecular G-quadruplex structure is required for binding of telomeric repeat-containing RNA to the telomeric protein TRF2. *J Am Chem Soc* **134**: 11974–11976. doi:10.1021/ja305734x

Brieño-Enríquez MA, Moak SL, Abud-Flores A, Cohen PE. 2019. Characterization of telomeric repeat-containing RNA (TERRA) localization and protein interactions in primordial germ cells of the mouse. *Biol Reprod* **100:** 950–962. doi:10.1093/biolre/ioy243

Cao H, Zhai Y, Ji X, Wang Y, Zhao J, Xing J, An J, Ren T. 2020. Noncoding telomeric repeat-containing RNA inhibits the progression of hepatocellular carcinoma by regulating telomerase-mediated telomere length. *Cancer Sci* **111:** 2789–2802. doi:10.1111/cas.14442

Caslini C, Connelly JA, Serna A, Broccoli D, Hess JL. 2009. MLL associates with telomeres and regulates telomeric repeat-containing RNA transcription. *Mol Cell Biol* **29:** 4519–4526. doi:10.1128/MCB.00195-09

Chawla R, Redon S, Raftopoulou C, Wischnewski H, Gagos S, Azzalin CM. 2011. Human UPF1 interacts with TPP1 and telomerase and sustains telomere leading-strand replication. *EMBO J* **30:** 4047–4058. doi:10.1038/emboj.2011.280

Chen L, Zhang C, Ma W, Huang J, Zhao Y, Liu H. 2022. METTL3-mediated m6A modification stabilizes TERRA and maintains telomere stability. *Nucleic Acids Res* **50:** 11619–11634. doi:10.1093/nar/gkac1027

Chu HP, Cifuentes-Rojas C, Kesner B, Aeby E, Lee HG, Wei C, Oh HJ, Boukhali M, Haas W, Lee JT. 2017a. TERRA RNA antagonizes ATRX and protects telomeres. *Cell* **170:** 86–101.e16. doi:10.1016/j.cell.2017.06.017

Chu HP, Froberg JE, Kesner B, Oh HJ, Ji F, Sadreyev R, Pinter SF, Lee JT. 2017b. PAR-TERRA directs homologous sex chromosome pairing. *Nat Struct Mol Biol* **24:** 620–631. doi:10.1038/nsmb.3432

Collie GW, Haider SM, Neidle S, Parkinson GN. 2010. A crystallographic and modelling study of a human telomeric RNA (TERRA) quadruplex. *Nucleic Acids Res* **38:** 5569–5580. doi:10.1093/nar/gkq259

Cusanelli E, Romero CA, Chartrand P. 2013. Telomeric noncoding RNA TERRA is induced by telomere shortening to nucleate telomerase molecules at short telomeres. *Mol Cell* **51:** 780–791. doi:10.1016/j.molcel.2013.08.029

d'Adda di Fagagna F, Reaper PM, Clay-Farrace L, Fiegler H, Carr P, von Zglinicki T, Saretzki G, Carter NP, Jackson SP. 2003. A DNA damage checkpoint response in telomere-initiated senescence. *Nature* **426:** 194–198. doi:10.1038/nature02118

Damasceno JD, Silva G, Tschudi C, Tosi LR. 2017. Evidence for regulated expression of telomeric repeat-containing RNAs (TERRA) in parasitic trypanosomatids. *Mem Inst Oswaldo Cruz* **112:** 572–576. doi:10.1590/0074-02760170054

Decker ML, Chavez E, Vulto I, Lansdorp PM. 2009. Telomere length in Hutchinson–Gilford progeria syndrome. *Mech Ageing Dev* **130:** 377–383. doi:10.1016/j.mad.2009.03.001

Deng Z, Norseen J, Wiedmer A, Riethman H, Lieberman PM. 2009. TERRA RNA binding to TRF2 facilitates heterochromatin formation and ORC recruitment at telomeres. *Mol Cell* **35:** 403–413. doi:10.1016/j.molcel.2009.06.025

Deng Z, Wang Z, Stong N, Plasschaert R, Moczan A, Chen HS, Hu S, Wikramasinghe P, Davuluri RV, Bartolomei MS, et al. 2012. A role for CTCF and cohesin in subtelomere chromatin organization, TERRA transcription, and telomere end protection. *EMBO J* **31:** 4165–4178. doi:10.1038/emboj.2012.266

Diman A, Boros J, Poulain F, Rodriguez J, Purnelle M, Episkopou H, Bertrand L, Francaux M, Deldicque L, Decottignies A. 2016. Nuclear respiratory factor 1 and endurance exercise promote human telomere transcription. *Sci Adv* **2:** e1600031. doi:10.1126/sciadv.1600031

Eid R, Demattei MV, Episkopou H, Augé-Gouillou C, Decottignies A, Grandin N, Charbonneau M. 2015. Genetic inactivation of *ATRX* leads to a decrease in the amount of telomeric cohesin and level of telomere transcription in human glioma cells. *Mol Cell Biol* **35:** 2818–2830. doi:10.1128/MCB.01317-14

Episkopou H, Draskovic I, Van Beneden A, Tilman G, Mattiussi M, Gobin M, Arnoult N, Londoño-Vallejo A, Decottignies A. 2014. Alternative lengthening of telomeres is characterized by reduced compaction of telomeric chromatin. *Nucleic Acids Res* **42:** 4391–4405. doi:10.1093/nar/gku114

Farnung BO, Brun CM, Arora R, Lorenzi LE, Azzalin CM. 2012. Telomerase efficiently elongates highly transcribing telomeres in human cancer cells. *PLoS ONE* **7:** e35714. doi:10.1371/journal.pone.0035714

Feretzaki M, Renck Nunes P, Lingner J. 2019. Expression and differential regulation of human TERRA at several chromosome ends. *RNA* **25:** 1470–1480. doi:10.1261/rna.072322.119

Feretzaki M, Pospisilova M, Valador Fernandes R, Lunardi T, Krejci L, Lingner J. 2020. RAD51-dependent recruitment of TERRA lncRNA to telomeres through R-loops. *Nature* **587:** 303–308. doi:10.1038/s41586-020-2815-6

Fernandes RV, Lingner J. 2023. The THO complex counteracts TERRA R-loop-mediated telomere fragility in telomerase$^+$ cells and telomeric recombination in ALT$^+$ cells. *Nucleic Acids Res* **51:** 6702–6722. doi:10.1093/nar/gkad448

Flynn RL, Centore RC, O'Sullivan RJ, Rai R, Tse A, Songyang Z, Chang S, Karlseder J, Zou L. 2011. TERRA and hnRNPA1 orchestrate an RPA-to-POT1 switch on telomeric single-stranded DNA. *Nature* **471:** 532–536. doi:10.1038/nature09772

Flynn RL, Cox KE, Jeitany M, Wakimoto H, Bryll AR, Ganem NJ, Bersani F, Pineda JR, Suvà ML, Benes CH, et al. 2015. Alternative lengthening of telomeres renders cancer cells hypersensitive to ATR inhibitors. *Science* **347:** 273–277. doi:10.1126/science.1257216

Gagliardi M, Strazzullo M, Matarazzo MR. 2018. DNMT3B functions: novel insights from human disease. *Front Cell Dev Biol* **6:** 140. doi:10.3389/fcell.2018.00140

Ghisays F, Garzia A, Wang H, Canasto-Chibuque C, Hohl M, Savage SA, Tuschl T, Petrini JHJ. 2021. RTEL1 influences the abundance and localization of TERRA RNA. *Nat Commun* **12:** 3016. doi:10.1038/s41467-021-23299-2

Ghosh M, Singh M. 2020. Structure specific recognition of telomeric repeats containing RNA by the RGG-box of hnRNPA1. *Nucleic Acids Res* **48:** 4492–4506. doi:10.1093/nar/gkaa134

Goldberg AD, Banaszynski LA, Noh KM, Lewis PW, Elsaesser SJ, Stadler S, Dewell S, Law M, Guo X, Li X, et al. 2010. Distinct factors control histone variant H3.3 localization at specific genomic regions. *Cell* **140:** 678–691. doi:10.1016/j.cell.2010.01.003

Gonzalez-Vasconcellos I, Schneider R, Anastasov N, Alonso-Rodriguez S, Sanli-Bonazzi B, Fernández JL,

Atkinson MJ. 2017. The Rb1 tumour suppressor gene modifies telomeric chromatin architecture by regulating TERRA expression. *Sci Rep* **7**: 42056. doi:10.1038/srep 42056

Gonzalo S, Kreienkamp R, Askjaer P. 2017. Hutchinson-Gilford Progeria syndrome: a premature aging disease caused by LMNA gene mutations. *Ageing Res Rev* **33**: 18–29. doi:10.1016/j.arr.2016.06.007

Gottschling DE, Aparicio OM, Billington BL, Zakian VA. 1990. Position effect at *S. cerevisiae* telomeres: reversible repression of Pol II transcription. *Cell* **63**: 751–762. doi:10.1016/0092-8674(90)90141-Z

Graf M, Bonetti D, Lockhart A, Serhal K, Kellner V, Maicher A, Jolivet P, Teixeira MT, Luke B. 2017. Telomere length determines TERRA and R-loop regulation through the cell cycle. *Cell* **170**: 72–85.e14. doi:10.1016/j.cell.2017.06.006

Greenwood J, Cooper JP. 2012. Non-coding telomeric and subtelomeric transcripts are differentially regulated by telomeric and heterochromatin assembly factors in fission yeast. *Nucleic Acids Res* **40**: 2956–2963. doi:10.1093/nar/gkr1155

Guh CY, Shen HJ, Chen LW, Chiu PC, Liao IH, Lo CC, Chen Y, Hsieh YH, Chang TC, Yen CP, et al. 2022. XPF activates break-induced telomere synthesis. *Nat Commun* **13**: 5781. doi:10.1038/s41467-022-33428-0

Guintini L, Paillé A, Graf M, Luke B, Wellinger RJ, Conconi A. 2022. Transcription of ncRNAs promotes repair of UV induced DNA lesions in *Saccharomyces cerevisiae* subtelomeres. *PLoS Genet* **18**: e1010167. doi:10.1371/journal.pgen.1010167

Gylling HM, Gonzalez-Aguilera C, Smith MA, Kaczorowski DC, Groth A, Lund AH. 2020. Repeat RNAs associate with replication forks and post-replicative DNA. *RNA* **26**: 1104–1117. doi:10.1261/rna.074757.120

Hirashima K, Seimiya H. 2015. Telomeric repeat-containing RNA/G-quadruplex-forming sequences cause genome-wide alteration of gene expression in human cancer cells in vivo. *Nucleic Acids Res* **43**: 2022–2032. doi:10.1093/nar/gkv063

Hirschi A, Martin WJ, Luka Z, Loukachevitch LV, Reiter NJ. 2016. G-quadruplex RNA binding and recognition by the lysine-specific histone demethylase-1 enzyme. *RNA* **22**: 1250–1260. doi:10.1261/rna.057265.116

Hu Y, Bennett HW, Liu N, Moravec M, Williams JF, Azzalin CM, King MC. 2019. RNA-DNA hybrids support recombination-based telomere maintenance in fission yeast. *Genetics* **213**: 431–447. doi:10.1534/genetics.119.302606

Idilli AI, Cusanelli E, Pagani F, Berardinelli F, Bernabé M, Cayuela ML, Poliani PL, Mione MC. 2020. Expression of tert prevents ALT in zebrafish brain tumors. *Front Cell Dev Biol* **8**: 65. doi:10.3389/fcell.2020.00065

Iglesias N, Redon S, Pfeiffer V, Dees M, Lingner J, Luke B. 2011. Subtelomeric repetitive elements determine TERRA regulation by Rap1/Rif and Rap1/Sir complexes in yeast. *EMBO Rep* **12**: 587–593. doi:10.1038/embor.2011.73

Kaminski N, Wondisford AR, Kwon Y, Lynskey ML, Bhargava R, Barroso-González J, García-Expósito L, He B, Xu M, Mellacheruvu D, et al. 2022. RAD51AP1 regulates ALT-HDR through chromatin-directed homeostasis of TERRA. *Mol Cell* **82**: 4001–4017.e7. doi:10.1016/j.molcel.2022.09.025

Kang S, Cao J, Zhang M, Li X, Guo QL, Zeng H, Wei Z, Gong X, Wang J, Liu B, et al. 2021. Transcriptional regulation of telomeric repeat-containing RNA by acridine derivatives. *RNA Biol* **18**: 2261–2277. doi:10.1080/15476286.2021.1899652

Khan Y, Azam T, Sundar JS, Maiti S, Ekka MK. 2023. Biophysical characterization of nucleolin domains crucial for interaction with telomeric and TERRA G-quadruplexes. *Biochemistry* **62**: 1249–1261. doi:10.1021/acs.biochem.2c00641

Kondo K, Mashima T, Oyoshi T, Yagi R, Kurokawa R, Kobayashi N, Nagata T, Katahira M. 2018. Plastic roles of phenylalanine and tyrosine residues of TLS/FUS in complex formation with the G-quadruplexes of telomeric DNA and TERRA. *Sci Rep* **8**: 2864. doi:10.1038/s41598-018-21142-1

Le Berre G, Hossard V, Riou JF, Guieysse-Peugeot AL. 2019. Repression of TERRA expression by subtelomeric DNA methylation is dependent on NRF1 binding. *Int J Mol Sci* **20**: 2791. doi:10.3390/ijms20112791

Lee YW, Arora R, Wischnewski H, Azzalin CM. 2018. TRF1 participates in chromosome end protection by averting TRF2-dependent telomeric R loops. *Nat Struct Mol Biol* **25**: 147–153. doi:10.1038/s41594-017-0021-5

Levy MA, Kernohan KD, Jiang Y, Bérubé NG. 2015. ATRX promotes gene expression by facilitating transcriptional elongation through guanine-rich coding regions. *Hum Mol Genet* **24**: 1824–1835. doi:10.1093/hmg/ddu596

Liu X, Xu Y. 2018. HnRNPA1 specifically recognizes the base of nucleotide at the loop of RNA G-quadruplex. *Molecules* **23**: 237. doi:10.3390/molecules23010237

Liu X, Ishizuka T, Bao HL, Wada K, Takeda Y, Iida K, Nagasawa K, Yang D, Xu Y. 2017. Structure-dependent binding of hnRNPA1 to telomere RNA. *J Am Chem Soc* **139**: 7533–7539. doi:10.1021/jacs.7b01599

Liu J, Zheng T, Chen D, Huang J, Zhao Y, Ma W, Liu H. 2023. RBMX involves in telomere stability maintenance by regulating TERRA expression. *PLoS Genet* **19**: e1010937. doi:10.1371/journal.pgen.1010937

López de Silanes I, Stagno d'Alcontres M, Blasco MA. 2010. TERRA transcripts are bound by a complex array of RNA-binding proteins. *Nat Commun* **1**: 33. doi:10.1038/ncomms1032

López de Silanes I, Graña O, De Bonis ML, Dominguez O, Pisano DG, Blasco MA. 2014. Identification of TERRA locus unveils a telomere protection role through association to nearly all chromosomes. *Nat Commun* **5**: 4723. doi:10.1038/ncomms5723

Lorenzi LE, Bah A, Wischnewski H, Shchepachev V, Soneson C, Santagostino M, Azzalin CM. 2015. Fission yeast Cactin restricts telomere transcription and elongation by controlling Rap1 levels. *EMBO J* **34**: 115–129. doi:10.15252/embj.201489559

Lovejoy CA, Li W, Reisenweber S, Thongthip S, Bruno J, de Lange T, De S, Petrini JH, Sung PA, Jasin M, et al. 2012. Loss of ATRX, genome instability, and an altered DNA damage response are hallmarks of the alternative lengthening of telomeres pathway. *PLoS Genet* **8**: e1002772. doi:10.1371/journal.pgen.1002772

Lu R, O'Rourke JJ, Sobinoff AP, Allen JAM, Nelson CB, Tomlinson CG, Lee M, Reddel RR, Deans AJ, Pickett HA. 2019. The FANCM-BLM-TOP3A-RMI complex suppresses alternative lengthening of telomeres (ALT). *Nat Commun* **10:** 2252. doi:10.1038/s41467-019-10180-6

Luke B, Panza A, Redon S, Iglesias N, Li Z, Lingner J. 2008. The Rat1p 5′ to 3′ exonuclease degrades telomeric repeat-containing RNA and promotes telomere elongation in *Saccharomyces cerevisiae*. *Mol Cell* **32:** 465–477. doi:10.1016/j.molcel.2008.10.019

Maicher A, Kastner L, Dees M, Luke B. 2012. Deregulated telomere transcription causes replication-dependent telomere shortening and promotes cellular senescence. *Nucleic Acids Res* **40:** 6649–6659. doi:10.1093/nar/gks358

Majerová E, Mandáková T, Vu GT, Fajkus J, Lysak MA, Fojtová M. 2014. Chromatin features of plant telomeric sequences at terminal vs. internal positions. *Front Plant Sci* **5:** 593. doi:10.3389/fpls.2014.00593

Manzato C, Larini L, Oss Pegorar C, Dello Stritto MR, Jurikova K, Jantsch V, Cusanelli E. 2023. TERRA expression is regulated by the telomere-binding proteins POT-1 and POT-2 in *Caenorhabditis elegans*. *Nucleic Acids Res* **51:** 10681–10699. doi:10.1093/nar/gkad742

Marión RM, Montero JJ, López de Silanes I, Graña-Castro O, Martínez P, Schoeftner S, Palacios-Fábrega JA, Blasco MA. 2019. TERRA regulate the transcriptional landscape of pluripotent cells through TRF1-dependent recruitment of PRC2. *eLife* **8:** e44656. doi:10.7554/eLife.44656

McRae EKS, Davidson DE, Dupas SJ, McKenna SA. 2018. Insights into the RNA quadruplex binding specificity of DDX21. *Biochim Biophys Acta Gen Subj* **1862:** 1973–1979. doi:10.1016/j.bbagen.2018.06.009

Mei Y, Deng Z, Vladimirova O, Gulve N, Johnson FB, Drosopoulos WC, Schildkraut CL, Lieberman PM. 2021. TERRA G-quadruplex RNA interaction with TRF2 GAR domain is required for telomere integrity. *Sci Rep* **11:** 3509. doi:10.1038/s41598-021-82406-x

Misino S, Busch A, Wagner CB, Bento F, Luke B. 2022. TERRA increases at short telomeres in yeast survivors and regulates survivor associated senescence (SAS). *Nucleic Acids Res* **50:** 12829–12843. doi:10.1093/nar/gkac1125

Montero JJ, López-Silanes I, Megías D, F. Fraga M, Castells-García A, Blasco MA. 2018. TERRA recruitment of polycomb to telomeres is essential for histone trymethylation marks at telomeric heterochromatin. *Nat Commun* **9:** 1548. doi:10.1038/s41467-018-03916-3

Moravec M, Wischnewski H, Bah A, Hu Y, Liu N, Lafranchi L, King MC, Azzalin CM. 2016. TERRA promotes telomerase-mediated telomere elongation in *Schizosaccharomyces pombe*. *EMBO Rep* **17:** 999–1012. doi:10.15252/embr.201541708

Morea EGO, Vasconcelos EJR, Alves CS, Giorgio S, Myler PJ, Langoni H, Azzalin CM, Cano MIN. 2021. Exploring TERRA during *Leishmania major* developmental cycle and continuous in vitro passages. *Int J Biol Macromol* **174:** 573–586. doi:10.1016/j.ijbiomac.2021.01.192

Nassour J, Radford R, Correia A, Fusté JM, Schoell B, Jauch A, Shaw RJ, Karlseder J. 2019. Autophagic cell death restricts chromosomal instability during replicative crisis. *Nature* **565:** 659–663. doi:10.1038/s41586-019-0885-0

Nassour J, Aguiar LG, Correia A, Schmidt TT, Mainz L, Przetocka S, Haggblom C, Tadepalle N, Williams A, Shokhirev MN, et al. 2023. Telomere-to-mitochondria signalling by ZBP1 mediates replicative crisis. *Nature* **614:** 767–773. doi:10.1038/s41586-023-05710-8

Negishi Y, Kawaji H, Minoda A, Usui K. 2015. Identification of chromatin marks at TERRA promoter and encoding region. *Biochem Biophys Res Commun* **467:** 1052–1057. doi:10.1016/j.bbrc.2015.09.176

Nergadze SG, Farnung BO, Wischnewski H, Khoriauli L, Vitelli V, Chawla R, Giulotto E, Azzalin CM. 2009. CpG-island promoters drive transcription of human telomeres. *RNA* **15:** 2186–2194. doi:10.1261/rna.1748309

Ng LJ, Cropley JE, Pickett HA, Reddel RR, Suter CM. 2009. Telomerase activity is associated with an increase in DNA methylation at the proximal subtelomere and a reduction in telomeric transcription. *Nucleic Acids Res* **37:** 1152–1159. doi:10.1093/nar/gkn1030

Nie X, Xiao D, Ge Y, Xie Y, Zhou H, Zheng T, Li X, Liu H, Huang H, Zhao Y. 2021. TRF2 recruits nucleolar protein TCOF1 to coordinate telomere transcription and replication. *Cell Death Differ* **28:** 1062–1075. doi:10.1038/s41418-020-00637-3

* O'Sullivan RJ, Greenberg RA. 2025. Mechanisms of alternative lengthening of telomeres. *Cold Spring Harb Perspect Biol* **17:** a041690. doi:10.1101/cshperspect.a041690

Pan X, Drosopoulos WC, Sethi L, Madireddy A, Schildkraut CL, Zhang D. 2017. FANCM, BRCA1, and BLM cooperatively resolve the replication stress at the ALT telomeres. *Proc Natl Acad Sci* **114:** e5940–e5949. doi:10.1073/pnas.1708065114

Pan X, Chen Y, Biju B, Ahmed N, Kong J, Goldenberg M, Huang J, Mohan N, Klosek S, Parsa K, et al. 2019. FANCM suppresses DNA replication stress at ALT telomeres by disrupting TERRA R-loops. *Sci Rep* **9:** 19110. doi:10.1038/s41598-019-55537-5

Pedram M, Sprung CN, Gao Q, Lo AW, Reynolds GE, Murnane JP. 2006. Telomere position effect and silencing of transgenes near telomeres in the mouse. *Mol Cell Biol* **26:** 1865–1878. doi:10.1128/MCB.26.5.1865-1878.2006

Pérez-Martinez L, Öztürk M, Butter F, Luke B. 2020. Npl3 stabilizes R-loops at telomeres to prevent accelerated replicative senescence. *EMBO Rep* **21:** e49087. doi:10.15252/embr.201949087

Petti E, Buemi V, Zappone A, Schillaci O, Broccia PV, Dinami R, Matteoni S, Benetti R, Schoeftner S. 2019. SFPQ and NONO suppress RNA:DNA-hybrid-related telomere instability. *Nat Commun* **10:** 1001. doi:10.1038/s41467-019-08863-1

Pfeiffer V, Lingner J. 2012. TERRA promotes telomere shortening through exonuclease 1-mediated resection of chromosome ends. *PLoS Genet* **8:** e1002747. doi:10.1371/journal.pgen.1002747

Pfeiffer V, Crittin J, Grolimund L, Lingner J. 2013. The THO complex component Thp2 counteracts telomeric R-loops and telomere shortening. *EMBO J* **32:** 2861–2871. doi:10.1038/emboj.2013.217

Porreca RM, Herrera-Moyano E, Skourti E, Law PP, Gonzalez Franco R, Montoya A, Faull P, Kramer H, Vannier JB. 2020. TRF1 averts chromatin remodelling, recombination and replication dependent-break induced replication at mouse telomeres. *eLife* **9**: e49817. doi:10.7554/eLife.49817

Porro A, Feuerhahn S, Reichenbach P, Lingner J. 2010. Molecular dissection of telomeric repeat-containing RNA biogenesis unveils the presence of distinct and multiple regulatory pathways. *Mol Cell Biol* **30**: 4808–4817. doi:10.1128/MCB.00460-10

Porro A, Feuerhahn S, Delafontaine J, Riethman H, Rougemont J, Lingner J. 2014a. Functional characterization of the TERRA transcriptome at damaged telomeres. *Nat Commun* **5**: 5379. doi:10.1038/ncomms6379

Porro A, Feuerhahn S, Lingner J. 2014b. TERRA-reinforced association of LSD1 with MRE11 promotes processing of uncapped telomeres. *Cell Rep* **6**: 765–776. doi:10.1016/j.celrep.2014.01.022

Poulet A, Pisano S, Faivre-Moskalenko C, Pei B, Tauran Y, Haftek-Terreau Z, Brunet F, Le Bihan YV, Ledu MH, Montel F, et al. 2012. The N-terminal domains of TRF1 and TRF2 regulate their ability to condense telomeric DNA. *Nucleic Acids Res* **40**: 2566–2576. doi:10.1093/nar/gkr1116

Quttina M, Waiters KD, Khan AF, Karami S, Peidl AS, Babajide MF, Pennington J, Merchant FA, Bawa-Khalfe T. 2023. Exosc9 initiates SUMO-dependent lncRNA TERRA degradation to impact telomeric integrity in endocrine therapy insensitive hormone receptor-positive breast cancer. *Cells* **12**: 2495. doi:10.3390/cells12202495

Redon S, Reichenbach P, Lingner J. 2010. The non-coding RNA TERRA is a natural ligand and direct inhibitor of human telomerase. *Nucleic Acids Res* **38**: 5797–5806. doi:10.1093/nar/gkq296

Redon S, Zemp I, Lingner J. 2013. A three-state model for the regulation of telomerase by TERRA and hnRNPA1. *Nucleic Acids Res* **41**: 9117–9128. doi:10.1093/nar/gkt695

Reiss M, Keegan J, Aldrich A, Lyons SM, Flynn RL. 2023. The exoribonuclease XRN2 mediates degradation of the long non-coding telomeric RNA TERRA. *FEBS Lett* **597**: 1818–1836. doi:10.1002/1873-3468.14639

Roach RJ, Garavís M, González C, Jameson GB, Filichev VV, Hale TK. 2020. Heterochromatin protein 1α interacts with parallel RNA and DNA G-quadruplexes. *Nucleic Acids Res* **48**: 682–693. doi:10.1093/nar/gkz1138

Rodrigues J, Lydall D. 2018. Paf1 and Ctr9, core components of the PAF1 complex, maintain low levels of telomeric repeat containing RNA. *Nucleic Acids Res* **46**: 621–634. doi:10.1093/nar/gkx1131

Rodrigues J, Alfieri R, Bione S, Azzalin C. 2024. TERRA ONTseq: a long read-based sequencing pipeline to study the human telomeric transcriptome. *RNA* **30**: 955–966. doi:10.1261/rna.079906.123

Rudenko G, Van der Ploeg LH. 1989. Transcription of telomere repeats in protozoa. *EMBO J* **8**: 2633–2638. doi:10.1002/j.1460-2075.1989.tb08403.x

Sadhukhan R, Chowdhury P, Ghosh S, Ghosh U. 2018. Expression of telomere-associated proteins is interdependent to stabilize native telomere structure and telomere dysfunction by G-quadruplex ligand causes TERRA up-regulation. *Cell Biochem Biophys* **76**: 311–319. doi:10.1007/s12013-017-0835-0

Sagie S, Toubiana S, Hartono SR, Katzir H, Tzur-Gilat A, Havazelet S, Francastel C, Velasco G, Chédin F, Selig S. 2017. Telomeres in ICF syndrome cells are vulnerable to DNA damage due to elevated DNA:RNA hybrids. *Nat Commun* **8**: 14015. doi:10.1038/ncomms14015

Saha A, Gaurav AK, Pandya UM, Afrin M, Sandhu R, Nanavaty V, Schnur B, Li B. 2021. *Tb*TRF suppresses the TERRA level and regulates the cell cycle-dependent TERRA foci number with a TERRA binding activity in its C-terminal Myb domain. *Nucleic Acids Res* **49**: 5637–5653. doi:10.1093/nar/gkab401

Savoca V, Rivosecchi J, Gaiatto A, Rossi A, Mosca R, Gialdini I, Zubovic L, Tebaldi T, Macchi P, Cusanelli E. 2023. TERRA stability is regulated by RALY and polyadenylation in a telomere-specific manner. *Cell Rep* **42**: 112406. doi:10.1016/j.celrep.2023.112406

Scheibe M, Arnoult N, Kappei D, Buchholz F, Decottignies A, Butter F, Mann M. 2013. Quantitative interaction screen of telomeric repeat-containing RNA reveals novel TERRA regulators. *Genome Res* **23**: 2149–2157. doi:10.1101/gr.151878.112

Schoeftner S, Blasco MA. 2008. Developmentally regulated transcription of mammalian telomeres by DNA-dependent RNA polymerase II. *Nat Cell Biol* **10**: 228–236. doi:10.1038/ncb1685

Sfeir A, Kosiyatrakul ST, Hockemeyer D, MacRae SL, Karlseder J, Schildkraut CL, de Lange T. 2009. Mammalian telomeres resemble fragile sites and require TRF1 for efficient replication. *Cell* **138**: 90–103. doi:10.1016/j.cell.2009.06.021

Silva B, Pentz R, Figueira AM, Arora R, Lee YW, Hodson C, Wischnewski H, Deans AJ, Azzalin CM. 2019. FANCM limits ALT activity by restricting telomeric replication stress induced by deregulated BLM and R-loops. *Nat Commun* **10**: 2253. doi:10.1038/s41467-019-10179-z

Silva B, Arora R, Bione S, Azzalin CM. 2021. TERRA transcription destabilizes telomere integrity to initiate break-induced replication in human ALT cells. *Nat Commun* **12**: 3760. doi:10.1038/s41467-021-24097-6

Silva B, Arora R, Azzalin CM. 2022. The alternative lengthening of telomeres mechanism jeopardizes telomere integrity if not properly restricted. *Proc Natl Acad Sci* **119**: e2208669119. doi:10.1073/pnas.2208669119

Solovei I, Gaginskaya ER, Macgregor HC. 1994. The arrangement and transcription of telomere DNA sequences at the ends of lampbrush chromosomes of birds. *Chromosome Res* **2**: 460–470. doi:10.1007/BF01552869

Sze S, Bhardwaj A, Fnu P, Azarm K, Mund R, Ring K, Smith S. 2023. TERRA R-loops connect and protect sister telomeres in mitosis. *Cell Rep* **42**: 113235. doi:10.1016/j.celrep.2023.113235

Takahama K, Oyoshi T. 2013. Specific binding of modified RGG domain in TLS/FUS to G-quadruplex RNA: tyrosines in RGG domain recognize 2′-OH of the riboses of loops in G-quadruplex. *J Am Chem Soc* **135**: 18016–18019. doi:10.1021/ja4086929

Takahama K, Takada A, Tada S, Shimizu M, Sayama K, Kurokawa R, Oyoshi T. 2013. Regulation of telomere length by G-quadruplex telomere DNA- and TERRA-binding protein TLS/FUS. *Chem Biol* **20**: 341–350. doi:10.1016/j.chembiol.2013.02.013

Takahama K, Miyawaki A, Shitara T, Mitsuya K, Morikawa M, Hagihara M, Kino K, Yamamoto A, Oyoshi T. 2015. G-Quadruplex DNA- and RNA-specific-binding proteins engineered from the RGG domain of TLS/FUS. *ACS*

Cite this article as *Cold Spring Harb Perspect Biol* doi: 10.1101/cshperspect.a041683

*Chem Biol* **10:** 2564–2569. doi:10.1021/acschembio.5b00 566

Thijssen PE, Tobi EW, Balog J, Schouten SG, Kremer D, El Bouazzaoui F, Henneman P, Putter H, Eline Slagboom P, Heijmans BT, et al. 2013. Chromatin remodeling of human subtelomeres and TERRA promoters upon cellular senescence: commonalities and differences between chromosomes. *Epigenetics* **8:** 512–521. doi:10.4161/epi.24450

Toubiana S, Larom G, Smoom R, Duszynski RJ, Godley LA, Francastel C, Velasco G, Selig S. 2020. Regulation of telomeric function by DNA methylation differs between humans and mice. *Hum Mol Genet* **29:** 3197–3210. doi:10 .1093/hmg/ddaa206

Tutton S, Azzam GA, Stong N, Vladimirova O, Wiedmer A, Monteith JA, Beishline K, Wang Z, Deng Z, Riethman H, et al. 2016. Subtelomeric p53 binding prevents accumulation of DNA damage at human telomeres. *EMBO J* **35:** 193–207. doi:10.15252/embj.201490880

Vaid R, Thombare K, Mendez A, Burgos-Panadero R, Djos A, Jachimowicz D, Lundberg KI, Bartenhagen C, Kumar N, Tümmler C, et al. 2024. METTL3 drives telomere targeting of TERRA lncRNA through m6A-dependent R-loop formation: a therapeutic target for ALT-positive neuroblastoma. *Nucleic Acids Res* **52:** 2648–2671. doi:10 .1093/nar/gkad1242

Viceconte N, Loriot A, Lona Abreu P, Scheibe M, Fradera Sola A, Butter F, De Smet C, Azzalin CM, Arnoult N, Decottignies A. 2021. PAR-TERRA is the main contributor to telomeric repeat-containing RNA transcripts in normal and cancer mouse cells. *RNA* **27:** 106–121. doi:10 .1261/rna.076281.120

Vohhodina J, Goehring LJ, Liu B, Kong Q, Botchkarev VV, Huynh M, Liu Z, Abderazzaq FO, Clark AP, Ficarro SB, et al. 2021. BRCA1 binds TERRA RNA and suppresses R-Loop-based telomeric DNA damage. *Nat Commun* **12:** 3542. doi:10.1038/s41467-021-23716-6

Vrbsky J, Akimcheva S, Watson JM, Turner TL, Daxinger L, Vyskot B, Aufsatz W, Riha K. 2010. siRNA-mediated methylation of Arabidopsis telomeres. *PLoS Genet* **6:** e1000986. doi:10.1371/journal.pgen.1000986

Wanat JJ, Logsdon GA, Driskill JH, Deng Z, Lieberman PM, Johnson FB. 2018. TERRA and the histone methyltransferase Dot1 cooperate to regulate senescence in budding yeast. *PLoS ONE* **13:** e0195698. doi:10.1371/journal.pone .0195698

Wang Z, Lieberman PM. 2016. The crosstalk of telomere dysfunction and inflammation through cell-free TERRA containing exosomes. *RNA Biol* **13:** 690–695. doi:10 .1080/15476286.2016.1203503

Wang Z, Deng Z, Dahmane N, Tsai K, Wang P, Williams DR, Kossenkov AV, Showe LC, Zhang R, Huang Q, et al. 2015. Telomeric repeat-containing RNA (TERRA) constitutes a nucleoprotein component of extracellular inflammatory exosomes. *Proc Natl Acad Sci* **112:** E6293–E6300. doi:10 .1073/pnas.1505962112

Wang X, Goodrich KJ, Gooding AR, Naeem H, Archer S, Paucek RD, Youmans DT, Cech TR, Davidovich C. 2017. Targeting of Polycomb repressive complex 2 to RNA by short repeats of consecutive guanines. *Mol Cell* **65:** 1056–1067.e5. doi:10.1016/j.molcel.2017.02.003

Wang Y, Zhu W, Jang Y, Sommers JA, Yi G, Puligilla C, Croteau DL, Yang Y, Kai M, Liu Y. 2023. The RNA-binding motif protein 14 regulates telomere integrity at the interface of TERRA and telomeric R-loops. *Nucleic Acids Res* **51:** 12242–12260. doi:10.1093/nar/gkad967

Xu Y, Suzuki Y, Ito K, Komiyama M. 2010. Telomeric repeat-containing RNA structure in living cells. *Proc Natl Acad Sci* **107:** 14579–14584. doi:10.1073/pnas.1001177107

Xu M, Senanayaka D, Zhao R, Chigumira T, Tripathi A, Tones J, Lackner RM, Wondisford AR, Moneysmith LN, Hirschi A, et al. 2024. TERRA-LSD1 phase separation promotes R-loop formation for telomere maintenance in ALT cancer cells. *Nat Commun* **15:** 2165. doi:10.1038/ s41467-024-46509-z

Yadav T, Zhang JM, Ouyang J, Leung W, Simoneau A, Zou L. 2022. TERRA and RAD51AP1 promote alternative lengthening of telomeres through an R- to D-loop switch. *Mol Cell* **82:** 3985–4000.e4. doi:10.1016/j.molcel.2022.09 .026

Yehezkel S, Segev Y, Viegas-Péquignot E, Skorecki K, Selig S. 2008. Hypomethylation of subtelomeric regions in ICF syndrome is associated with abnormally short telomeres and enhanced transcription from telomeric regions. *Hum Mol Genet* **17:** 2776–2789. doi:10.1093/hmg/ddn177

Yehezkel S, Shaked R, Sagie S, Berkovitz R, Shachar-Bener H, Segev Y, Selig S. 2013. Characterization and rescue of telomeric abnormalities in ICF syndrome type I fibroblasts. *Front Oncol* **3:** 35. doi:10.3389/fonc.2013.00035

Yu TY, Kao YW, Lin JJ. 2014. Telomeric transcripts stimulate telomere recombination to suppress senescence in cells lacking telomerase. *Proc Natl Acad Sci* **111:** 3377–3382. doi:10.1073/pnas.1307415111

Zeng S, Liu L, Sun Y, Lu G, Lin G. 2017. Role of telomeric repeat-containing RNA in telomeric chromatin remodeling during the early expansion of human embryonic stem cells. *FASEB J* **31:** 4783–4795. doi:10.1096/fj .201600939RR

Zhang LF, Ogawa Y, Ahn JY, Namekawa SH, Silva SS, Lee JT. 2009. Telomeric RNAs mark sex chromosomes in stem cells. *Genetics* **182:** 685–698. doi:10.1534/genetics.109 .103093

Zhang Y, Zeng D, Cao J, Wang M, Shu B, Kuang G, Ou TM, Tan JH, Gu LQ, Huang ZS, et al. 2017. Interaction of Quindoline derivative with telomeric repeat-containing RNA induces telomeric DNA-damage response in cancer cells through inhibition of telomeric repeat factor 2. *Biochim Biophys Acta Gen Subj* **1861:** 3246–3256. doi:10 .1016/j.bbagen.2017.09.015

# Epigenetics of Human Telomeres

Nicole Bettin,[1] Mélina Vaurs,[1] and Anabelle Decottignies

GEPI Research Group, de Duve Institute, UCLouvain, 1200 Woluwe-Saint-Lambert, Brussels, Belgium

*Correspondence:* anabelle.decottignies@uclouvain.be

Human telomeric heterochromatin is unusual in that it does not show the enrichment of canonical repressive histone marks H3K9me3 or H4K20me3 seen in constitutive heterochromatin. Instead, human telomeres exhibit both facultative heterochromatin and euchromatin marks, consistent with their epigenetically regulated transcription into TERRA noncoding RNA. Additionally, telomeric DNA is out of phase with the DNA helical repeat and has no nucleosome positioning signal. Yet, human telomeric DNA forms a columnar structure of tightly stacked nucleosomes, alternating with open states, and regulated by histone tails and shelterin protein binding. We discuss the proposed mechanisms regulating human telomeric chromatin and the consequences that telomeric chromatin properties have on various cellular processes, such as telomere transcription, the regulation of shelterin binding, and the activation of the alternative lengthening of telomeres mechanism. Together, we summarize current evidence on the combination of hetero- and euchromatic properties of human telomeres that may help explain their crucial protective functions and plasticity to regulate telomere maintenance pathways and damage signaling.

In 1942, Conrad Waddington first defined epigenetics as the study of "the mechanisms by which the genotype produces the phenotype in the context of development" (Waddington 1942; Cavalli and Heard 2019). Since then, epigenetics has been repeatedly redefined to account not only for transgenerational inheritance but also for mitotic inheritance of chromatin state (Cavalli and Heard 2019). Bird (2007) defines epigenetics as the study of "structural adaptations of chromosomal regions so as to register, signal or perpetuate altered activity states" (Cavalli and Heard 2019), and this definition probably better describes the epigenetic regulation of telomeres. The main carriers of epigenetic information include posttranslation-al modifications of histone tails and readers of these marks, such as heterochromatin protein 1 (HP1), Polycomb (PRC1 and PRC2) and Trithorax (COMPASS) complexes, noncoding RNAs, and DNA methylation (Cavalli and Heard 2019).

Histone-modifying complexes constantly remodel nucleosomes, the basic units of chromatin consisting of DNA wrapped around octamers of histone proteins made of H2A, H2B, H3, and H4 (Cavalli and Heard 2019). Post-translational modifications of histones, occurring primarily on lysine residues located on their amino-terminal tails, influence local chromatin structure and mark regions of either euchromatin or facultative or constitutive heterochroma-

---

Cite this article as *Cold Spring Harb Perspect Biol* doi: 10.1101/cshperspect.a041706

tin (Grewal 2023). Histone marks, such as H3K9me3, and DNA methylation are also highly interrelated and rely on each other for the formation of constitutive heterochromatin (Rose and Klose 2014). One of the main functions of constitutive heterochromatin at highly repeated DNA sequences is to suppress recombination and, thereby, promote genomic stability through chromatin compaction. Constitutive heterochromatin regions, such as pericentromeres, are characterized by a high density of H3K9me3 and H4K20me3 histone marks, and compaction is classically achieved through the oligomerization of H3K9me3-bound HP1α and β isoforms (HP1α/β) (Canzio et al. 2011). According to this definition, mouse telomeres and subtelomeres are classified as constitutive heterochromatin domains (García-Cao et al. 2004; Blasco 2007). However, in striking contrast to their mouse counterparts, although present, H3K9me3 and H4K20me3 marks do not show enrichment at human telomeres, as they do in pericentromeric regions of the human genome (Rosenfeld et al. 2009; O'Sullivan et al. 2010; Arnoult et al. 2012; Cubiles et al. 2018; Udroiu and Sgura 2020). Whether human telomeres can be classified as canonical constitutive heterochromatin regions thus stands as an open question.

Pioneering studies have revealed that telomeric DNA forms nucleosomes with an estimated repeat length of ∼157 bp in rat liver nuclei, a value much lower than the 197 bp average repeat of bulk rat liver chromatin (Makarov et al. 1993). This unusually short nucleosome repeat length appears to be a general feature that also applies to human telomeric chromatin (Makarov et al. 1993; Tommerup et al. 1994; Fajkus and Trifonov 2001; Mechelli et al. 2004; Episkopou et al. 2014). Together with the fact that telomeric DNA does not carry any nucleosome positioning signal, this led Fajkus and Trifonov (2001) to propose a columnar packing model of telomeric nucleosomes, with continuous winding of the DNA around the stacked histone cores. This model was recently confirmed by cryogenic electron microscopy (cryo-EM) (Soman et al. 2022a), suggesting that telomeric chromatin compaction occurs through mecha-

nisms distinct from the canonical mechanisms of constitutive heterochromatin formation.

Here, we will review current knowledge on the chromatin characteristics of human telomeres. We will then address the mechanisms regulating human telomeric chromatin, such as the binding of shelterin proteins, the role of histone tails, the noncoding RNA TERRA, and the interaction of telomeres with the nuclear lamina. Finally, we will delve deeper into the—sometimes interdependent—impacts of telomeric chromatin on various aspects of telomere biology, focusing on shelterin protein binding, telomere transcription, the so-called telomere position effect (TPE), the alternative lengthening of telomeres (ALT) mechanism, telomere movement, and the synthesis of telomeric DNA.

## HUMAN TELOMERES DISPLAY CHARACTERISTICS OF MULTIPLE CHROMATIN TYPES

Classically, heterochromatin is classified as either constitutive or facultative. Mouse telomeres, with enrichment in H3K9me3 and H4K20me3 marks, display the classical features of loci showing constitutive heterochromatin (García-Cao et al. 2004; Blasco 2007). Although H3K9me3 and H4K20me3 marks are present at human telomeres, they are not enriched as they are in constitutive heterochromatin regions such as pericentromeres (Rosenfeld et al. 2009; O'Sullivan et al. 2010; Arnoult et al. 2012; Cubiles et al. 2018; Udroiu and Sgura 2020). The reason why mouse and human telomeres show differences in their enrichment in constitutive heterochromatin marks could be related to the distinct length of their telomeres, as telomeres of laboratory mice are 5–10 times longer than human telomeres. Supporting this hypothesis, on the one hand, the density of H3K9me3 and H4K20me3 marks was decreased at the shortened telomeres of telomerase-deficient mice (Benetti et al. 2007) and, on the other hand, the experimental lengthening of human telomeres, via ectopic overexpression of telomerase, increased H3K9me3 telomeric density (Arnoult et al. 2012). The mechanisms underlying the increased density of H3K9me3 at elongated telomeres are not yet clear, but may

Cite this article as *Cold Spring Harb Perspect Biol* doi: 10.1101/cshperspect.a041706

involve various histone methyltransferases and, possibly, the chromatin remodeler ATRX. In support of this, long human telomeres exhibit an increased density of G-quadruplex (G4) structures (Yang et al. 2021), which were proposed to promote ATRX-regulated deposition of H3K9me3 (Teng et al. 2021). It has also been proposed that the epigenetic characteristics of telomeres are not spatially homogenous and that the proximal and distal parts of the telomeres might exhibit distinct chromatin properties (Udroiu and Sgura 2020). In this case, it is possible that the observed differences between human and mouse telomeres result from a shift toward the chromatin characteristics of the more distal telomeric repeats in cells with very long telomeres.

Facultative heterochromatin, originally attributed to developmentally regulated heterochromatinization, identifies genomic regions that can adopt open or compact conformations, being transcriptionally silent with the potential to transition to a euchromatic state allowing transcription (Trojer and Reinberg 2007). Known human genomic regions of facultative heterochromatin include the inactive X chromosome and autosomal imprinted genomic loci that typically display a panel of histone marks, including hypoacetylation, H3K27me3, H3K9me2, and H4K20me1 (Trojer and Reinberg 2007). Strikingly, human telomeres display strong enrichment of H4K20me1 (Cubiles et al. 2018) compared to pericentromeres and the presence of H3K27me3 and H3K9me2 marks, as well as low abundance of the H3K9ac mark (Arnoult et al. 2012; Porro et al. 2014; Cubiles et al. 2018). Facultative heterochromatin is further characterized by the presence of HP1γ isoform, which was also detected at human telomeres (Déjardin and Kingston 2009; Canudas et al. 2011; Arnoult et al. 2012), thus highlighting the fact that human telomeres exhibit some properties of facultative heterochromatin.

In addition to facultative heterochromatin marks, the H3K4me3 mark, associated with active transcription, was detected at human telomeric chromatin (Rosenfeld et al. 2009; Cubiles et al. 2018). This histone modification, when co-occurring with H3K27me3, marks the transcriptional start site region of CpG-rich bivalent promoters that set the genes in a poised state during the initial stages of organismal development (Voigt et al. 2013). Since a large fraction of human chromosome ends is characterized by CpG-rich subtelomeric promoters, located directly upstream of the *TTAGGG* repeats (Nergadze et al. 2009) and regulating the expression of the telomeric repeat-containing RNA, TERRA (Fig. 1A; Azzalin et al. 2007), telomeres may resemble bivalent genes. Interestingly, non-*CG* DNA methylation, previously linked to genomic imprinting and cellular reprogramming (Lister et al. 2013), was also detected at telomeres of human embryonic stem cells (Lister et al. 2011).

An overview of the telomeric chromatin marks mentioned above is given in Figure 1B. In summary, although we have not given an exhaustive list of histone marks detected at human telomeres, it appears that the picture of human telomeric chromatin is still rather unclear, and likely distinct from mouse telomeric chromatin.

## COLUMNAR PACKING OF HUMAN TELOMERIC NUCLEOSOMES AND THE ALTERNATIVE OPEN STATE

Despite the lack of H3K9me3 enrichment at human telomeres and the fact that, with its 6 bp *TTAGGG* repeats, telomeric DNA is out of phase with the 10.4 bp DNA helical repeat and lacks any nucleosome positioning signal, human telomeric nucleosomes are organized into regular chromatin structures. The crystal structure of a telomeric nucleosome core particle (NCP) reconstituted with recombinant human histone octamers revealed that its overall structure is similar to previously published NCP structures (Soman et al. 2020). In solution, however, the telomeric NCP exhibited increased dynamics and decreased stability compared to canonical nucleosomes (Soman et al. 2020). Going further, the Nordenskiöld group then characterized 3-kb-long telomeric chromatin fibers using negative stain electron microscopy and single-molecule magnetic tweezers and obtained the cryo-EM structure of the condensed telomeric tetranucleosome and its dinucleosome unit (Soman et al.

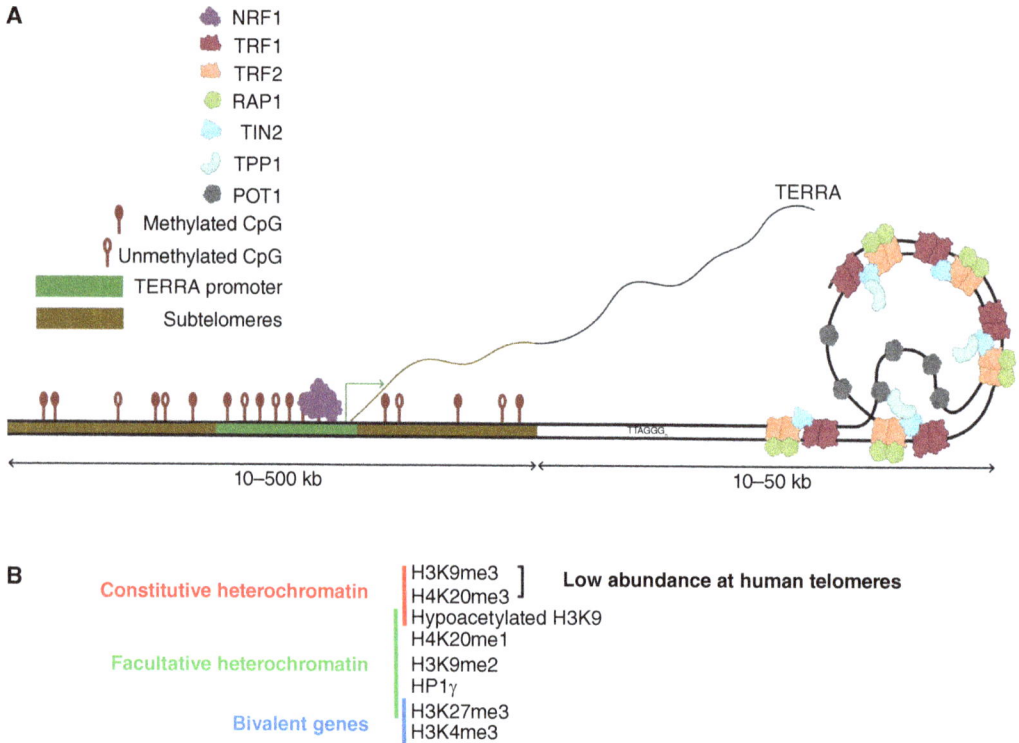

**Figure 1.** General organization of human chromosome ends. (*A*) Human subtelomeres, located directly upstream of telomeric repeats, are highly heterochromatic and exhibit CpG methylation. Subtelomeric promoters have been identified as drivers of the expression of TERRA noncoding RNA. A small subset of activators/repressors of TERRA transcription is shown. The ends of chromosomes fold into a T-loop structure stabilized by a complex of six telomere-specific proteins called shelterin (de Lange 2005). (TSS) Transcription start site, (DNMTs) DNA methyltransferases. (*B*) Selected chromatin marks detected at human telomeres (see text for details).

2022a). Consistent with the initial hypothesis of Fajkus and Trifonov (2001), the structure they obtained displayed close stacking of nucleosomes with a columnar arrangement, which was found to be primarily stabilized by the carboxy-terminal tail of H2A and the amino-terminal tails of other histones (Fig. 2A,B; Soman et al. 2022a). This work revealed that human telomere compaction occurs through mechanisms similar to archaeal histones (Mattiroli et al. 2017).

Further elucidation of the telomeric nucleosome structure revealed that an alternative open state coexists with the tightly stacked columnar telomeric fibers (Fig. 2A; Soman et al. 2022a). The alternative open state has been proposed to occur in the absence of heterochromatin decompaction and to result from unstacking and flip-ping out of one nucleosome to expose the H2A-H2B acidic patches (Soman et al. 2022a). This biphasic model of telomeric nucleosome structure provides an explanation for the access of factors required for telomere maintenance and the necessary DNA damage response (DDR) at telomeres, which is known to happen in the absence of chromatin decompaction (Timashev et al. 2017; Vancevska et al. 2017; Soman et al. 2022a). Several epigenetic modifications previously observed at telomeres, such as acetylation or methylation of lysine residues on the amino-terminal tails of H3 or H4, were predicted to be located at the interface between two stacked nucleosomes and to get exposed in the open state, indicating that these modifications may have an important structural role (Soman et al. 2022a).

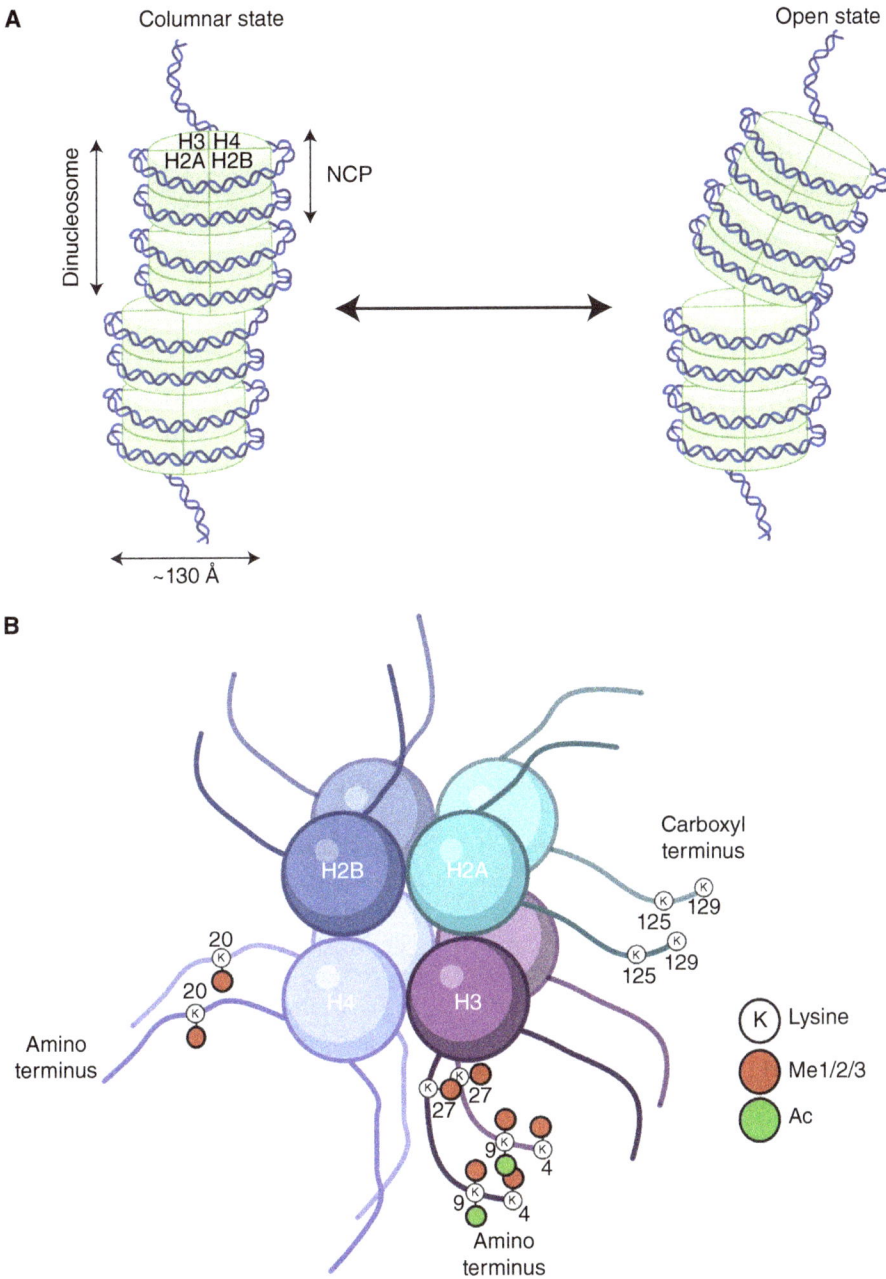

**Figure 2.** Columnar structure and alternative open state of human telomeric chromatin. (*A*) (*Left*) Structure of the human telomeric tetranucleosome. At telomeres, nucleosome core particles (NCPs) display close stacking with a columnar arrangement and comprise ~132 bp of DNA wrapped around the histone octamer, composed of two copies each of H2A, H2B, H3, and H4. (*Right*) The telomeric chromatin structure also exists in an alternative open state where one nucleosome is flipped out. (*B*) Schematic representation of the histone octamer. The carboxy-terminal tails of H2A, which comprise two positively charged lysine residues, K125 and K129, likely play an important role in mediating and stabilizing the columnar structure (Soman et al. 2022a). The posttranslational modifications on H3 and H4 amino-terminal tails that are mentioned in this review are highlighted. (Figure based on data from Soman et al. 2022a,b.)

Taken together, although obtained in the absence of shelterin proteins, this first cryo-EM structure of human telomeric chromatin supports the hypothesis of a distinct mode of chromatin compaction at telomeres compared to canonical constitutive heterochromatic loci and suggests that telomeres have a unique columnar/open conformation that likely ensures both protection and the highly specialized and dynamic functions of telomeres.

## REGULATION OF HUMAN TELOMERIC CHROMATIN

### By the Shelterin Complex

The interaction of telomeric DNA with the telomere-binding proteins TRF1 and TRF2 has long been recognized as an important regulator of chromatin compaction (Cacchione et al. 1997; Rossetti et al. 1998; de Lange 2005; Pisano et al. 2007, 2008; Ichikawa et al. 2014). However, although they share two important domains, the TRF homology (TRFH) dimerization domain and the Myb DNA-binding domain (Broccoli et al. 1997; Fairall et al. 2001), TRF1 and TRF2 have distinct effects on telomeric chromatin. The carboxy-terminal Myb DNA-binding domain of TRF2, by reducing the negative surface charge density of nucleosomal array fibers, mediates telomeric chromatin compaction (Baker et al. 2009), while its amino-terminal domain negatively regulates compaction (Poulet et al. 2012). Cumulatively, TRF2 functions to compact telomeric chromatin. Conversely, the acidic domain of TRF1 inhibits its ability to condense DNA (Poulet et al. 2012). The affinity of the TRF1-Myb domain for telomeric DNA is also higher than that of the TRF2-Myb domain (Hanaoka et al. 2005).

Recently, cryo-EM experiments confirmed distinct roles of TRF1 and TRF2 in telomeric chromatin compaction. Consistent with the compactor role of TRF2, the protein, whether full-length or lacking its amino-terminal part (TRF2$^{\Delta N}$), was found to induce and stabilize columnar fibers (Wong et al. 2024). The binding of TRF2/TRF2$^{\Delta N}$ to supergrooves on the surface of the columnar form of telomeric chromatin oc-

curred without eviction of histone proteins and increased fiber width, while reducing internucleosomal distance (Fig. 3A; Wong et al. 2024).

The impact of TRF1 binding on telomeric nucleosomes was reevaluated on a cryo-EM structure determined to 2.5 Å resolution using a reconstituted human telomeric NCP with a 145 bp telomeric DNA (Hu et al. 2023). The structural analysis suggested that TRF1 binds at the junction between the nucleosome and linker DNA (Fig. 3A; Hu et al. 2023), inducing a register shift of the nucleosomal DNA by 1 bp. This confirmed the previous findings that TRF1 recognizes telomeric repeats in a nucleosomal context and alters nucleosomal structure (Galati et al. 2006; Pisano et al. 2010), binding preferentially to the end of nucleosome (Galati et al. 2006) or to the linker DNA (Galati et al. 2015). The model suggests that the binding mode of TRF1 likely disfavors the binding of H1 to the linker DNA (Hu et al. 2023), and explains the underrepresentation of histone H1 at telomeres (Déjardin and Kingston 2009). The model, however, is not compatible with the columnar structure of human telomeres, in which the linker DNA is not accessible to protein binding (Soman et al. 2022a). To allow TRF1 binding, Hu et al. (2023) therefore proposed that the columnar structure would need to be disrupted. This hypothesis is consistent with the observation that TRF1 alters telomeric nucleosome spacing in vitro (Galati et al. 2015). The cryo-EM data further suggested that the interaction between the phosphorylated carboxy-terminal part of the TRF1-Myb domain and the amino-terminal tail of histone H3 plays a crucial role in stabilizing the complex (Hu et al. 2023). The phosphorylatable serine 435 residue of TRF1, which is involved in the interaction with histone H3, is highly conserved in mammals but absent in TRF2. This suggests that TRF2 binding may occur without the disruption of the columnar structure, nearby outer nucleosomal DNA gyres (Fig. 3A; Hu et al. 2023). Despite its inability to condense telomeric chromatin, it is, however, possible that, through its interaction with TIN2, which links TRF1 and TRF2, TRF1 nevertheless facilitates the compact state of telomeres by promoting intramolecular bonds at telomeres (Fig. 3B; Soman et al. 2022b).

Cite this article as *Cold Spring Harb Perspect Biol* doi: 10.1101/cshperspect.a041706

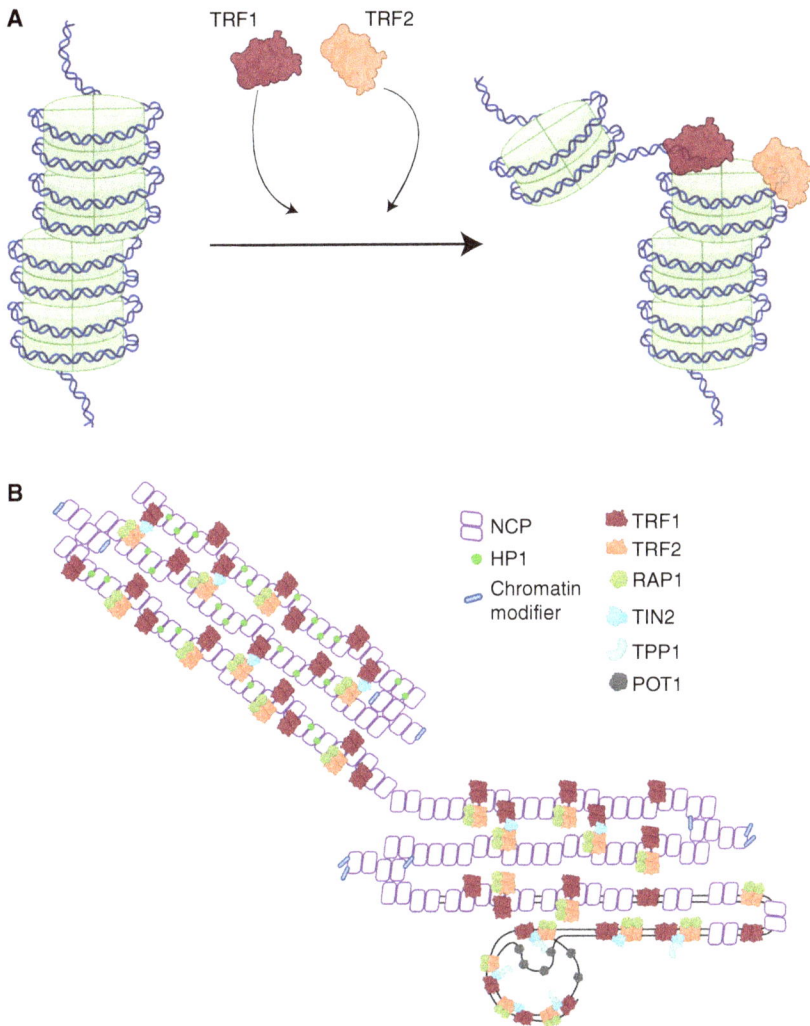

**Figure 3.** A model for the impact of the TRF1 and TRF2 shelterin proteins on human telomeric chromatin. (*A*) The binding of TRF2 on the surface of the telomeric columnar structure occurs without eviction of histones, resulting in no disruption of the packed state of telomeric chromatin. TRF1 binding, likely at the junction between the nucleosome and the linker DNA, instead, leads to a shift in the register of the nucleosomal DNA. The columnar structure of human telomeres needs to be disrupted to allow TRF1 binding. (Panel based on data from Hu et al. 2023.) (*B*) To account for the observed diameter of telomeric fibers of up to 35 nm, Soman et al. (2022b) proposed that TRF1, via its interaction with the shelterin protein TIN2, which can bridge TRF1 and TRF2, could contribute to higher-order formation of telomeric chromatin by promoting intramolecular bonds. (Panel based on data from Soman et al. 2022b.)

Finally, another study revealed that the role of TRF2 in establishing telomeric loops and protecting telomeres from ATM checkpoint activation depends on its ability to modify the topology of DNA by wrapping 90 bp of telomeric DNA around its TRFH domain (Benarroch-Popivker et al. 2016), suggesting a histone-like function for TRF2, possibly at the very ends of telomeres. Future investigations will be required to fully understand how TRF1, TRF2, and the other shelterin proteins regulate telomeric nucleosomal organization.

## By Amino- and Carboxy-Terminal Histone Tails

The carboxy-terminal tails of histone H2A have been proposed to stabilize the columnar structure of human telomeres (Soman et al. 2022a). Histone H2A, however, is not always present in nucleosomes because, upon activation of the DDR, it is replaced by the histone variant H2AX, which lacks the positively charged residues Lys125 and Lys129 of H2A and is phosphorylated on Ser139 by various DDR kinases (Fig. 4A). DDR activation could thus induce an overall shift from a positively charged H2A C-tail to a negatively charged γ-H2AX C-tail, which is expected to destabilize the telomeric chromatin structure (Fig. 4B; Soman et al. 2022a,b). From this point of view, DDR kinases, such as ATM or ATR, could thus alter the structure of telomeres.

In addition to the carboxy-terminal tail of H2A, the cryo-EM structure of the human telomeric NCP revealed that several posttranslational modifications on the amino-terminal histone tails, predicted to be located at the interface between two stacked nucleosomes, would likely modulate the open state and telomeric chromatin dynamics (Soman et al. 2022a). Several histone-modifying enzymes could therefore affect the structure of human telomeric chromatin, such as MLL1, the methyltransferase from the COMPASS complex responsible for H3K4me3

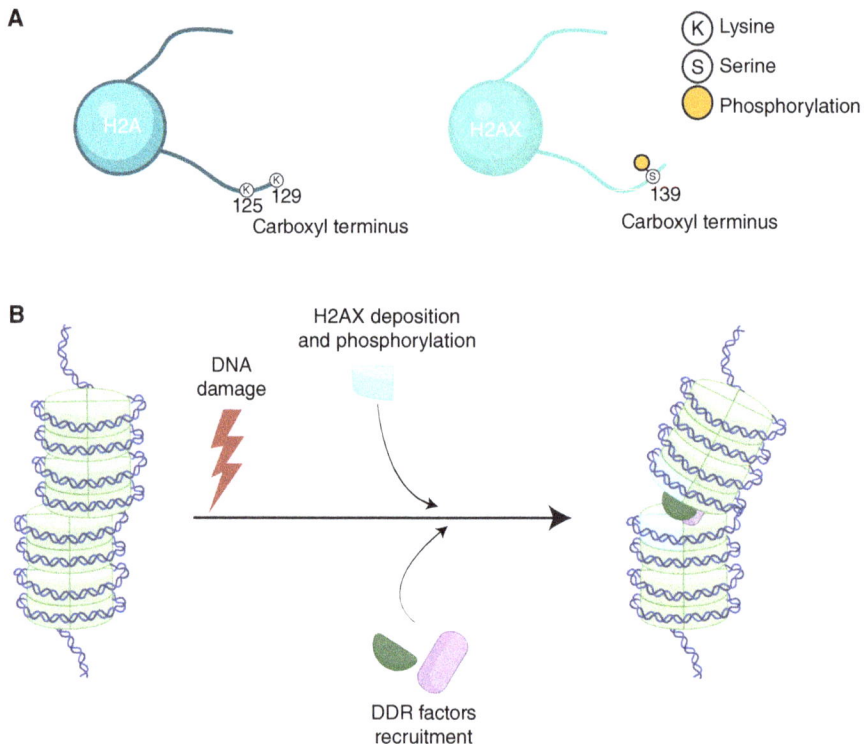

Figure 4. A model for the impact of DNA damage response (DDR) activation on the open state of human telomeric chromatin. (A) The carboxy-terminal tails of H2A and the histone variant H2AX are distinct and Lys125 and Lys129 residues are missing from the H2AX carboxy-terminal tail. (B) Activation of the DDR at telomeres can modulate the open state without massive decompaction of chromatin structure. Upon DNA damage, histone H2A is replaced by the H2AX variant and the latter is phosphorylated by DDR kinases, such as ATM and ATR, on Ser139 residue. This could favor the open state and allow the recruitment of additional DDR factors, which would ultimately lead to the repair of the damage. (Panel B based on data from Soman et al. 2022b.)

Cite this article as *Cold Spring Harb Perspect Biol* doi: 10.1101/cshperspect.a041706

(Caslini et al. 2009). Structural analysis demonstrated that the open state would be compatible with the recruitment of the SIRT6 NAD$^+$-dependent deacetylase of H3K9 to telomeres (Michishita et al. 2008; Soman et al. 2022a), suggesting that hypoacetylation of H3K9, by altering the charge of the amino-terminal tail of histone H3, may also regulate the human telomeric chromatin structure.

As discussed later, it is important to emphasize that some posttranslational modifications of amino-terminal histone tails may also indirectly impact telomeric chromatin properties through their impact on the expression of the TERRA telomeric long noncoding RNA.

## By TERRA Telomeric Noncoding RNA

The discovery that Xist noncoding RNA, through its recruitment to chromatin, silences an entire chromosome to equalize X chromosome expression in males and females propelled chromatin-associated RNAs to the rank of chromatin regulators (Johnson and Straight 2017). At telomeres, the RNA TERRA is also involved in chromatin regulation (Bettin et al. 2019). The first evidence was provided by the demonstration that TERRA, through its interaction with TRF2, facilitates the establishment of H3K9me3 and H3K9me2 marks and the recruitment of HP1 at human telomeres (Deng et al. 2009). Subsequently, correlative approaches unveiled that, in cells with highly elongated telomeres, telomeric establishment of H3K9me3 and recruitment of HP1α follow the same timing as TERRA expression throughout the cell cycle (Arnoult et al. 2012). An additional role of TERRA in shaping telomeric chromatin was provided by the identification of chromatin remodeling factors, such as ORC1 (origin of replication complex 1), NoRC (nuclear remodeling complex), and MORF4L2 (a component of the NuA histone acetyl transferase complex), among other proteins found to interact with the telomeric RNA (Deng et al. 2009; Postepska-Igielska et al. 2013; Scheibe et al. 2013; Chu et al. 2017a). TERRA has also been suggested to play a role in establishing heterochromatin at human subtelomeres, by influencing not only H3K9me3 density but also subtelomeric DNA

methylation (Deng et al. 2010). This, as we will discuss below, has the potential to modulate the transcriptional activity of the human subtelomeric promoters and places TERRA at the heart of a negative feedback mechanism controlling its own transcription.

Taken together, there is strong evidence that the telomeric RNA transcript TERRA modulates the deposition of both telomeric and subtelomeric heterochromatin marks. Whether TERRA is directly involved in the formation of the columnar structure of human telomeres and/or its alternative open state is however unknown.

## Through Interaction with the Nuclear Lamina

The perinuclear zone has been associated with the induction of heterochromatin, thanks to the demonstration that artificial tethering of genomic loci to the nuclear periphery favors gene silencing (Finlan et al. 2008; Reddy et al. 2008). Human telomeres were first reported to attach to the nuclear matrix more than 30 years ago (de Lange 1992), but it is not yet clear whether perinuclear localization of telomeres is involved in their heterochromatinization.

H3K9me2 appears to specifically mark the heterochromatin localized at the nuclear periphery (Poleshko et al. 2017), whereas euchromatin and H3K9me3-enriched heterochromatin are distributed in the nuclear interior (van Steensel and Belmont 2017; Poleshko et al. 2019). Consistent with the presence of the H3K9me2 mark at human telomeres, almost half the telomeres dynamically relocate to the nuclear periphery during nuclear assembly after mitosis (Crabbe et al. 2012). Several mechanisms of telomere anchoring to the nuclear lamina were proposed: (1) through interactions between the shelterin protein RAP1 and the inner nuclear membrane SUN1 (Crabbe et al. 2012), (2) through interactions between TRF2 or TRF1 and Lamin A or B1 (Burla et al. 2016; Pennarun et al. 2023), (3) through interactions between telomeres and Lamina-associated peptide 2α (LAP2α) (Dechat et al. 2004), or (4) through interactions between the subtelomeric insulator protein CTCF and A-type lamins (Ottaviani et al. 2009). Expression of progerin, a mutated form of Lamin A found in patients suffering from the Hutchinson–Gilford progeria syndrome (Arancio et al. 2014), reduces the association of telomeres with LAP2α and the global levels of H3K27me3 (Chojnowski et al. 2015). Whether the telomeric abundance of H3K27me3 was reduced, however, has not been evaluated and additional studies will be needed to clarify the role of the nuclear lamina in human telomere heterochromatinization.

## IMPACT OF TELOMERIC CHROMATIN ON TELOMERE BIOLOGY AND CELLULAR HOMEOSTASIS

### Shelterin Binding and Chromosome-End Protection

Resolution of cryo-EM structures revealed intimate contacts between shelterin proteins and histone tails, particularly between the TRF1-Myb domain and the amino-terminal tail of histone H3 (Hu et al. 2023). These observations suggest a reciprocal impact of the telomeric nucleosome on shelterin binding, consistent with the previous report that removal of histone amino-terminal tails significantly reduces the affinities of TRF1 for telomeric NCP (Galati et al. 2015). Posttranslational modifications of histone tails were proposed to affect the binding of shelterin proteins to telomeres and to promote the transition from a protected to a deprotected telomeric state (Galati et al. 2015). The same study further proposed that phosphorylation or acetylation of histone tails might regulate the binding affinity of TRF1 and TRF2 differentially, given their distinct amino-terminal domains, rich in acidic residues for TRF1 or in arginine residues for TRF2 (Galati et al. 2015). It will be of future interest to determine the impact of various posttranslational modifications of histone tails on the ability of shelterin proteins to bind telomeric chromatin.

Given the numerous functions played by the shelterin complex and, which are reviewed elsewhere, the interplay between the telomeric nucleosome and shelterin proteins may thus have a profound impact on telomere functions (Fig. 5).

## Telomere Transcription

Transcription of telomeres into TERRA is initiated from the heterochromatic subtelomeric regions of chromosomes (Fig. 1A; Nergadze et al. 2009). The discovery of transcriptionally permissive histone marks at telomeres, as well as the demonstration that H3K9me3, H3K9me2, and HP1γ are, in fact, associated with transcription elongation (Vakoc et al. 2005), were consistent with transcriptional activity at telomeres. There is clear evidence, however, that human telomere transcription is regulated by epigenetic modifications.

The first evidence in favor of an epigenetic regulation of telomere transcription came from the observation that, in cells derived from patients suffering from the immunodeficiency, centromeric region instability, facial anomalies (ICF) syndrome, caused by a mutation in the *DNMT3b* gene coding for DNA methyltransferase 3b, TERRA levels are strongly up-regulated (Yehezkel et al. 2008). Consistent with an impact of DNA methylation on telomere transcription, impaired activity of DNMT1 and DNMT3b in cancer cells increased TERRA expression (Nergadze et al. 2009), and TERRA levels were found to inversely correlate with subtelomeric DNA methylation in human cell lines (Ng et al. 2009; Arnoult et al. 2012; Diman et al. 2016). Both telomerase expression (Ng et al. 2009) and telomere length (Buxton et al. 2014) were also positively associated with methylation levels in subtelomeric DNA, although the underlying molecular mechanisms have not been elucidated. It was also proposed that the recruitment of DNMT3b at subtelomeres may imply specific telomeric histone marks as well as TERRA itself, in a negative feedback mechanism (Deng et al. 2010). The observation that the DNA methylation-sensitive transcription factor NRF1 regulates human telomere transcription, through its recruitment to bona fide binding sites located in the CpG-rich regions of human subtelomeres, also established a link between telomere transcriptional activity and DNA methylation (Fig. 1A; Diman et al. 2016; Le Berre et al. 2019).

In addition to subtelomeric DNA methylation, the repression of telomere transcription

Cite this article as *Cold Spring Harb Perspect Biol* doi: 10.1101/cshperspect.a041706

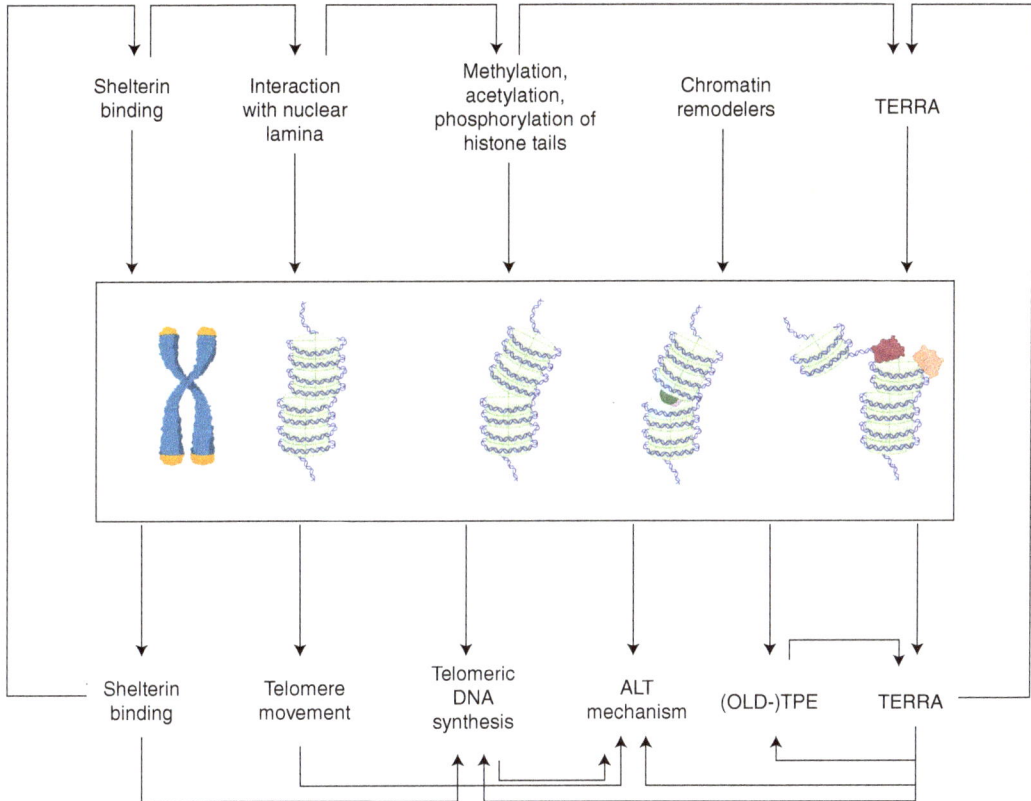

Figure 5. Upstream and downstream from human telomeric chromatin. See text for details.

may also result from increased telomeric density of H3K9me3 in human cells with overly elongated telomeres (Arnoult et al. 2012). The causative role for the H3K9me3 repressive mark in the down-regulation of telomere transcription was suggested by the observation that depletion of the SUV39H1 H3K9 methyltransferase abolished the observed telomere length-dependent repression of TERRA (Arnoult et al. 2012). SUV39H1 depletion, however, failed to increase TERRA expression in cells with nonelongated telomeres, suggesting that H3K9me3 may not regulate TERRA expression in human cells with physiological telomere length (Arnoult et al. 2012). In light of these data, the reported enrichment of the constitutive heterochromatin mark H3K9me3 at long mouse telomeres could possibly explain why mouse telomeres are silent. In fact, in mouse, TERRA molecules mainly come from intrachromosomal *TTAGGG*-rich

loci and not from telomeres (Chu et al. 2017b; Diman and Decottignies 2018; Viceconte et al. 2021). While DNA methylation and repressive histone marks negatively impact telomere transcription, recruitment of the MLL1 H3K4 methyltransferase to telomeres was found to positively correlate with TERRA transcription (Caslini et al. 2009).

From the above, it appears that TERRA is subject to strict epigenetic regulation in a process that likely involves various negative feedback mechanisms (Fig. 5), and that telomere transcription may be in a poised state, allowing rapid activation upon dedicated environmental stimuli or developmental cues (Voigt et al. 2013).

It is important to mention that, in light of the multiple roles that TERRA plays at telomeres (reviewed elsewhere), it is not always easy to distinguish a direct impact of telomeric chromatin on telomere functions and an indirect impact via regulation of TERRA expression.

## Telomere Position Effect

The spreading of the repressive chromatin environment of long telomeres was previously associated with a local transcriptional silencing mechanism dubbed the TPE. TPE was initially discovered in *Drosophila melanogaster*, where the insertion of genetic elements within telomeres was associated with lower expression levels mediated by an HP1- and H3K9me3-dependent mechanism (Hazelrigg et al. 1984; Perrini et al. 2004). Later, TPE was discovered in budding (Gottschling et al. 1990; Fourel et al. 1999; Pryde and Louis 1999) and fission yeast (Nimmo et al. 1994). While TPE has been well described in flies and yeast, its existence in mammalian cells has long remained controversial, likely due to the paucity of protein-encoding genes near human telomeres. Using an integrated telomeric transgene, Tennen et al. (2011) provided the first evidence for the existence of human TPE and showed that enhanced telomere silencing in response to telomere elongation required the SIRT6 H3K9 deacetylase. The discovery that human telomeres are transcribed (Azzalin et al. 2007) was next seized as an opportunity to assess the existence of TPE at endogenous human chromosomal loci. Isogenic cell lines with distinct telomere lengths revealed that long telomeres negatively regulate TERRA expression in a process involving the histone methyltransferase SUV39H1 and HP1α, providing the first demonstration that human telomere transcription is subject to TPE (Arnoult et al. 2012).

In addition to the phenomenon of TPE described above, evidences were later provided for the existence of a TPE mechanism which, in *trans*, induces gene silencing over long distances (TPE-OLD). The work by the laboratory of Shay and Wright demonstrated that, in the presence of long telomeres, chromosome looping can bring the telomere close to genes that are located up to 10 Mb away (Robin et al. 2014). Both the h*TERT* telomerase gene and the *PPP2R2C* tumor suppressor gene were found to be regulated by TPE-OLD (Kim et al. 2016; Jäger et al. 2022).

Interestingly, consistent with the demonstration that telomere shortening alleviates TPE-OLD (Robin et al. 2014), replicative senescence up-regulates the expression of *hTERT* and *PPP2R2C* genes (Kim et al. 2016; Jäger et al. 2022), providing new insights into how replicative aging changes the cell's susceptibility to cancer initiation. These observations go hand in hand with the demonstration of an age-dependent reprogramming of telomeric chromatin (O'Sullivan et al. 2010). The mechanisms underlying TPE-OLD are not entirely elucidated but likely rely on interactions between TRF2 and, possibly, TERRA, with interstitial *TTAGGG* repeats (Kim et al. 2016). The association of telomeres with the repressive environment of the nuclear lamina may further participate in TPE (Burla et al. 2016).

Thus, telomeric chromatin can regulate gene expression both locally—by affecting TERRA expression—and over long distances in a telomere length-dependent manner (Fig. 5).

## Activation of Alternative Lengthening of Telomeres

To acquire an indefinite replication potential, cells activate either telomerase or an alternative mechanism of telomere lengthening, dubbed ALT, which is based on homologous recombination (HR) events at telomeres (for review, see O'Sullivan and Greenberg 2025). How ALT is activated is still not fully understood, but it is clear that epigenetic alterations at telomeres alleviate the natural protection of telomeres against recombination to allow the access of HR factors. Increasing evidence has emerged over the past decade regarding the role of telomeric chromatin alterations on ALT activation (O'Sullivan and Almouzni 2014). One of the strongest evidences that chromatin dysfunction promotes ALT came from the observation that mutations in the *ATRX* or *DAXX* genes, which encode the subunits of a chromatin remodeling complex catalyzing H3.3 variant incorporation at telomeres (Goldberg et al. 2010; Wong et al. 2010; Clynes et al. 2015), correlate with ALT activation in pancreatic neuroendocrine tumors (Heaphy et al. 2011). Agreeing with the crucial role of H3.3 alterations in ALT activation, other

ALT tumors were subsequently found to harbor mutations in the *H3F3A* gene encoding H3.3 (Schwartzentruber et al. 2012; Wu et al. 2012). While several groups reported that the inactivation of the ATRX/DAXX/H3.3 pathway on its own is not enough to activate ALT (Lovejoy et al. 2012; O'Sullivan et al. 2014), the ectopic expression of ATRX is able to suppress ALT activity (Clynes et al. 2015; Napier et al. 2015).

ATRX depletion in human fibroblasts was associated with progressive telomeric chromatin decompaction and reduced telomeric density of H3 and H3.3 (Li et al. 2019). These observations were consistent with the reduced H3 and H4 histone density and the increased access to micrococcal nuclease (MNase) at telomeres of spontaneously immortalized ALT$^+$ATRX$^-$ cell lines (Episkopou et al. 2014). In support of a role for telomeric chromatin alterations in ALT induction, the codepletion of ASF1a/b H3.1/H3.3-H4 histone dimer chaperones was sufficient to activate ALT in vitro (O'Sullivan et al. 2014), even though the relevance for ALT activation in vivo remains to be established. Codepletion of ASF1a/b similarly increased the susceptibility of telomeric chromatin to MNase digestion (O'Sullivan et al. 2014). Together, these studies support a role for lack of compaction of telomeric chromatin in activation of the ALT mechanism (Fig. 5). It will be interesting to assess the contribution of various ALT telomere features, such as increased abundance of G4 structures (Yang et al. 2021), the presence of telomeric repeat variants, and binding of various nuclear receptors at ALT telomeres (Marzec et al. 2015) to this change in chromatin structure, and to test whether the columnar packing model of human telomeric nucleosomes is disrupted at ALT telomeres.

Although chromatin at ALT telomeres appears to be more relaxed, conflicting data have been reported regarding the abundance of repressive histone marks (Udroiu and Sgura 2020). In the immortalized human fibroblast lines used in Episkopou et al. (2014), the overall H3K9me3 density was reduced at ALT telomeres, but H3K9me3/H3 ratios were similar in ALT and telomerase-positive cells. These observations suggested that the methyltransferase activity on telomeric H3K9 was not reduced in ALT cells, in agreement with independent observations obtained in ALT and non-ALT cell lines (Conomos et al. 2014). Conversely, based on ChIP-seq data, ALT activity in childhood neuroblastoma was associated with an increased abundance of H3K9me3 marks at telomeric repeats (Hartlieb et al. 2021). In the latter study, however, telomeric repeat reads appeared to be mostly located outside of the terminal sequences of chromosomes and the amount of *TTAGGG* repeats obtained in non-ALT samples was very low (Hartlieb et al. 2021). A careful interpretation of the ChIP-seq data may therefore be needed to conclude on the abundance of these histone marks at the level of telomeres. In a separate study, the depletion of SETDB1 alone reduced some markers of ALT activity in human cancer cells (Gauchier et al. 2019). Whether this was the direct result of changes in H3K9me3 or H3K9me2 density at telomeres was however not evaluated. Interestingly, SETDB1 was proposed to primarily act on euchromatic regions of the genome (Schultz et al. 2002), further strengthening the idea that ALT telomeres do not show the canonical features of telomeric heterochromatin.

Whether ALT telomeres show increased or reduced abundance of repressive histone marks is therefore still unclear. A possible increased abundance of repressive histone marks at ALT telomeres, however, would still be in line with a reduction of telomere compaction which, as noted above, likely occurs through a mechanism independent of the canonical repressive histone marks of constitutive heterochromatin.

Although the impact of histone tail modifications on ALT telomeric chromatin compaction is still unclear, histone mark modifications are nevertheless likely to modulate ALT activity. The impact of histone marks on TERRA production emerged as an obvious candidate to promote ALT activity (reviewed elsewhere). Work from Pickett and Reddel's laboratories also demonstrated that the increased recruitment of the NuRD-ZNF827 complex at ALT telomeres not only causes histone hypoacetylation but also leads to (1) an increase in telomere–telomere interactions, (2) the recruitment of HR enzymes and the inhibition of shelterin protein binding, and (3) the enhancement of HR activity at ALT telomeres (Conomos et al. 2014; Pickett and Reddel 2015). Whether NuRD-ZNF827-dependent ALT telomere hypoacetylation impacts telomeric chromatin compaction has not been tested but would be important in supporting the hypothesis of a distinct impact of histone marks on ALT activity and telomeric chromatin compaction.

Therefore, similar to the lack of consensus on the epigenetic features of mammalian telomeres, the picture of epigenetic modifications at ALT telomeres is also unclear. In the future, it will be interesting to evaluate how histone modifications separately modulate ALT activity and chromatin compaction to gain a better overall picture of the impact of telomeric chromatin on ALT activity.

## Telomere Movement

Telomere movement is involved in various cellular processes such as telomere positioning in the nucleus (Pandita et al. 2007), end-to-end fusions (Dimitrova et al. 2008), and ALT activity (Jegou et al. 2009; Lamm et al. 2021) and is likely facilitated by telomeric chromatin decondensation (Fig. 5; Gao et al. 2018). Along these lines, the role of SIRT6 in facilitating the directional movement of damaged telomeres was related to its ability to recruit the ATP-dependent chromatin remodeling enzyme SNF2H (Gao et al. 2018), an enzyme involved in nucleosome repositioning (de La Serna et al. 2006). Consistent with the proposed lack of compaction of ALT telomeres and the link to telomere movement, a fraction of telomeres in U2OS ALT cells exhibit extended mobility, which has been proposed to promote ALT activity by facilitating telomere–telomere interactions and their association with PML bodies (Fig. 5; Jegou et al. 2009). The Smc5/6 complex, previously detected at ALT telomeres (Potts and Yu 2007), is required for the directed movement of heterochromatic double-strand breaks (Caridi et al. 2018), suggesting a similar role in the movement of ALT telomeres. Interestingly, telomere mobility is increased in Lamin A–depleted cells (de Vos et al. 2010; Bronshtein et al. 2015), reinforcing the idea that the interaction of telomeres with Lamin A could regulate telomeric chromatin.

## Telomeric DNA Synthesis

The low levels of telomeric H3K9 acetylation are necessary for the efficient replication of human telomeres and the prevention of structural abnormalities (Michishita et al. 2008). By creating a specialized chromatin state that allows the stable association of telomere-processing factors in S phase, such as WRN, the H3K9 histone deacetylase SIRT6 thereby protects telomeres from dysfunction (Fig. 5; Michishita et al. 2008). Further evidence supporting the role of histone hypoacetylation in telomeric DNA synthesis was provided by the demonstration that recruitment of the NuRD-ZNF827 complex to ALT telomeres, linked to the recruitment of histone deacetylases, enhanced the break-induced replication activity at telomeres, as suggested by the increased abundance of C-circles products (Conomos et al. 2014).

Telomerase also synthesizes telomeric DNA and its activity at telomeres may be modulated by telomeric chromatin, as the artificial tethering of HP1α, via its fusion to the TRF1 shelterin protein, attenuated the telomere extension mediated by telomerase (Chow et al. 2018). However, given that TERRA modulates human telomerase access to telomeres (Bettin et al. 2024), it remains to be verified whether the observed negative impact on telomerase-dependent extension was directly mediated by HP1α or whether it was a consequence of a possible deregulation of TERRA (Fig. 5). Similarly, TERRA transcripts have been implicated in various aspects of human telomere replication, and this, again, complicates the understanding of the direct role of histone modifications on telomeric DNA synthesis (Bettin et al. 2019).

Together, whether by direct or indirect mechanisms, telomeric chromatin thus appears to modulate various telomeric DNA synthesis pathways.

## CONCLUSION

In summary, we have emphasized that current knowledge calls for a reassessment of the traditional view of constitutive heterochromatin in human telomeric chromatin. We propose that the properties of human telomeric chromatin align more closely with those of facultative heterochromatin. Additionally, we have outlined the existence of tightly regulated and interconnected regulatory mechanisms that, together, likely account for the numerous important functions played by the telomeric chromatin structure (Fig. 5). The recent development of cryo-EM approaches will need to be implemented to study the properties of telomeric chromatin in experimental settings that better resemble physiological conditions by taking into account the diversity of histone variants/marks, telomere-binding proteins, or secondary structures, such as G-quadruplexes, as well as in the presence of TERRA noncoding RNA. Exciting times ahead!

## ACKNOWLEDGMENTS

We are grateful to Stefano Cacchione for critical comments on the manuscript. M.V. and A.D. are recipients of the Fonds National de la Recherche Scientifique (FNRS). N.B. is supported by the Fonds Maurange, managed by the King Baudouin Foundation (Brussels, Belgium). We are grateful to the de Duve Institute and the UCLouvain for constant support. We apologize to any authors whose work was not cited here due to space constraints.

## REFERENCES

*Reference is also in this subject collection.

Arancio W, Pizzolanti G, Genovese SI, Pitrone M, Giordano C. 2014. Epigenetic involvement in Hutchinson–Gilford progeria syndrome: a mini-review. *Gerontology* **60**: 197–203. doi:10.1159/000357206

Arnoult N, Van Beneden A, Decottignies A. 2012. Telomere length regulates TERRA levels through increased trimethylation of telomeric H3K9 and HP1α. *Nat Struct Mol Biol* **19**: 948–956. doi:10.1038/nsmb.2364

Azzalin CM, Reichenbach P, Khoriauli L, Giulotto E, Lingner J. 2007. Telomeric repeat containing RNA and RNA surveillance factors at mammalian chromosome ends. *Science* **318**: 798–801. doi:10.1126/science.1147182

Baker AM, Fu Q, Hayward W, Lindsay SM, Fletcher TM. 2009. The Myb/SANT domain of the telomere-binding protein TRF2 alters chromatin structure. *Nucleic Acids Res* **37**: 5019–5031. doi:10.1093/nar/gkp515

Benarroch-Popivker D, Pisano S, Mendez-Bermudez A, Lototska L, Kaur P, Bauwens S, Djerbi N, Latrick CM, Fraisier V, Pei B, et al. 2016. TRF2-mediated control of telomere DNA topology as a mechanism for chromosome-end protection. *Mol Cell* **61**: 274–286. doi:10.1016/j.molcel.2015.12.009

Benetti R, García-Cao M, Blasco MA. 2007. Telomere length regulates the epigenetic status of mammalian telomeres and subtelomeres. *Nat Genet* **39**: 243–250. doi:10.1038/ng1952

Bettin N, Oss Pegorar C, Cusanelli E. 2019. The emerging roles of TERRA in telomere maintenance and genome stability. *Cells* **8**: 246. doi:10.3390/cells8030246

Bettin N, Querido E, Gialdini I, Grupelli GP, Goretti E, Cantarelli M, Andolfato M, Soror E, Sontacchi A, Jurikova K, et al. 2024. TERRA transcripts localize at long telomeres to regulate telomerase access to chromosome ends. *Sci Adv* **10**: eadk4387. doi:10.1126/sciadv.adk4387

Bird A. 2007. Perceptions of epigenetics. *Nature* **447**: 396–398. doi:10.1038/nature05913

Blasco MA. 2007. The epigenetic regulation of mammalian telomeres. *Nat Rev Genet* **8**: 299–309. doi:10.1038/nrg2047

Broccoli D, Smogorzewska A, Chong L, de Lange T. 1997. Human telomeres contain two distinct Myb-related proteins, TRF1 and TRF2. *Nat Genet* **17**: 231–235. doi:10.1038/ng1097-231

Bronshtein I, Kepten E, Kanter I, Berezin S, Lindner M, Redwood AB, Mai S, Gonzalo S, Foisner R, Shav-Tal Y, et al. 2015. Loss of lamin A function increases chromatin dynamics in the nuclear interior. *Nat Commun* **6**: 8044. doi:10.1038/ncomms9044

Burla R, La Torre M, Saggio I. 2016. Mammalian telomeres and their partnership with lamins. *Nucleus* **7**: 187–202. doi:10.1080/19491034.2016.1179409

Buxton JL, Suderman M, Pappas JJ, Borghol N, McArdle W, Blakemore AI, Hertzman C, Power C, Szyf M, Pembrey M. 2014. Human leukocyte telomere length is associated with DNA methylation levels in multiple subtelomeric and imprinted loci. *Sci Rep* **4**: 4954. doi:10.1038/srep04954

Cacchione S, Cerone MA, Savino M. 1997. In vitro low propensity to form nucleosomes of four telomeric sequences. *FEBS Lett* **400**: 37–41. doi:10.1016/S0014-5793(96)01318-X

Canudas S, Houghtaling BR, Bhanot M, Sasa G, Savage SA, Bertuch AA, Smith S. 2011. A role for heterochromatin protein 1γ at human telomeres. *Genes Dev* **25**: 1807–1819. doi:10.1101/gad.17325211

Canzio D, Chang EY, Shankar S, Kuchenbecker KM, Simon MD, Madhani HD, Narlikar GJ, Al-Sady B. 2011. Chromodomain-mediated oligomerization of HP1 suggests a nucleosome-bridging mechanism for heterochromatin assembly. *Mol Cell* **41**: 67–81. doi:10.1016/j.molcel.2010.12.016

Caridi CP, D'Agostino C, Ryu T, Zapotoczny G, Delabaere L, Li X, Khodaverdian VY, Amaral N, Lin E, Rau AR, et al. 2018. Nuclear F-actin and myosins drive relocalization of heterochromatic breaks. *Nature* **559**: 54–60. doi:10.1038/s41586-018-0242-8

Caslini C, Connelly JA, Serna A, Broccoli D, Hess JL. 2009. MLL associates with telomeres and regulates telomeric repeat-containing RNA transcription. *Mol Cell Biol* **29:** 4519–4526. doi:10.1128/MCB.00195-09

Cavalli G, Heard E. 2019. Advances in epigenetics link genetics to the environment and disease. *Nature* **571:** 489–499. doi:10.1038/s41586-019-1411-0

Chojnowski A, Ong PF, Wong ES, Lim JS, Mutalif RA, Navasankari R, Dutta B, Yang H, Liow YY, Sze SK, et al. 2015. Progerin reduces LAP2α-telomere association in Hutchinson–Gilford progeria. *eLife* **4:** e07759. doi:10.7554/eLife.07759

Chow TT, Shi X, Wei JH, Guan J, Stadler G, Huang B, Blackburn EH. 2018. Local enrichment of HP1α at telomeres alters their structure and regulation of telomere protection. *Nat Commun* **9:** 3583. doi:10.1038/s41467-018-05840-y

Chu HP, Cifuentes-Rojas C, Kesner B, Aeby E, Lee HG, Wei C, Oh HJ, Boukhali M, Haas W, Lee JT. 2017a. TERRA RNA antagonizes ATRX and protects telomeres. *Cell* **170:** 86–101.e16. doi:10.1016/j.cell.2017.06.017

Chu HP, Froberg JE, Kesner B, Oh HJ, Ji F, Sadreyev R, Pinter SF, Lee JT. 2017b. PAR-TERRA directs homologous sex chromosome pairing. *Nat Struct Mol Biol* **24:** 620–631. doi:10.1038/nsmb.3432

Clynes D, Jelinska C, Xella B, Ayyub H, Scott C, Mitson M, Taylor S, Higgs DR, Gibbons RJ. 2015. Suppression of the alternative lengthening of telomere pathway by the chromatin remodelling factor ATRX. *Nat Commun* **6:** 7538. doi:10.1038/ncomms8538

Conomos D, Reddel RR, Pickett HA. 2014. NuRD-ZNF827 recruitment to telomeres creates a molecular scaffold for homologous recombination. *Nat Struct Mol Biol* **21:** 760–770. doi:10.1038/nsmb.2877

Crabbe L, Cesare AJ, Kasuboski JM, Fitzpatrick JA, Karlseder J. 2012. Human telomeres are tethered to the nuclear envelope during postmitotic nuclear assembly. *Cell Rep* **2:** 1521–1529. doi:10.1016/j.celrep.2012.11.019

Cubiles MD, Barroso S, Vaquero-Sedas MI, Enguix A, Aguilera A, Vega-Palas MA. 2018. Epigenetic features of human telomeres. *Nucleic Acids Res* **46:** 2347–2355. doi:10.1093/nar/gky006

Dechat T, Gajewski A, Korbei B, Gerlich D, Daigle N, Haraguchi T, Furukawa K, Ellenberg J, Foisner R. 2004. LAP2α and BAF transiently localize to telomeres and specific regions on chromatin during nuclear assembly. *J Cell Sci* **117:** 6117–6128. doi:10.1242/jcs.01529

Déjardin J, Kingston RE. 2009. Purification of proteins associated with specific genomic loci. *Cell* **136:** 175–186. doi:10.1016/j.cell.2008.11.045

de Lange T. 1992. Human telomeres are attached to the nuclear matrix. *EMBO J* **11:** 717–724. doi:10.1002/j.1460-2075.1992.tb05104.x

de Lange T. 2005. Shelterin: the protein complex that shapes and safeguards human telomeres. *Genes Dev* **19:** 2100–2110. doi:10.1101/gad.1346005

de la Serna IL, Ohkawa Y, Imbalzano AN. 2006. Chromatin remodelling in mammalian differentiation: lessons from ATP-dependent remodellers. *Nat Rev Genet* **7:** 461–473. doi:10.1038/nrg1882

Deng Z, Norseen J, Wiedmer A, Riethman H, Lieberman PM. 2009. TERRA RNA binding to TRF2 facilitates heterochromatin formation and ORC recruitment at telomeres. *Mol Cell* **35:** 403–413. doi:10.1016/j.molcel.2009.06.025

Deng Z, Campbell AE, Lieberman PM. 2010. TERRA, CpG methylation and telomere heterochromatin: lessons from ICF syndrome cells. *Cell Cycle* **9:** 69–74. doi:10.4161/cc.9.1.10358

De Vos WH, Houben F, Hoebe RA, Hennekam R, van Engelen B, Manders EM, Ramaekers FC, Broers JL, Van Oostveldt P. 2010. Increased plasticity of the nuclear envelope and hypermobility of telomeres due to the loss of A-type lamins. *Biochim Biophys Acta* **1800:** 448–458. doi:10.1016/j.bbagen.2010.01.002

Diman A, Decottignies A. 2018. Genomic origin and nuclear localization of TERRA telomeric repeat-containing RNA: from darkness to dawn. *FEBS J* **285:** 1389–1398. doi:10.1111/febs.14363

Diman A, Boros J, Poulain F, Rodriguez J, Purnelle M, Episkopou H, Bertrand L, Francaux M, Deldicque L, Decottignies A. 2016. Nuclear respiratory factor 1 and endurance exercise promote human telomere transcription. *Sci Adv* **2:** e1600031. doi:10.1126/sciadv.1600031

Dimitrova N, Chen YC, Spector DL, de Lange T. 2008. 53BP1 promotes non-homologous end joining of telomeres by increasing chromatin mobility. *Nature* **456:** 524–528. doi:10.1038/nature07433

Episkopou H, Draskovic I, Van Beneden A, Tilman G, Mattiussi M, Gobin M, Arnoult N, Londoño-Vallejo A, Decottignies A. 2014. Alternative lengthening of telomeres is characterized by reduced compaction of telomeric chromatin. *Nucleic Acids Res* **42:** 4391–4405. doi:10.1093/nar/gku114

Fairall L, Chapman L, Moss H, de Lange T, Rhodes D. 2001. Structure of the TRFH dimerization domain of the human telomeric proteins TRF1 and TRF2. *Mol Cell* **8:** 351–361. doi:10.1016/S1097-2765(01)00321-5

Fajkus J, Trifonov EN. 2001. Columnar packing of telomeric nucleosomes. *Biochem Biophys Res Commun* **280:** 961–963. doi:10.1006/bbrc.2000.4208

Finlan LE, Sproul D, Thomson I, Boyle S, Kerr E, Perry P, Ylstra B, Chubb JR, Bickmore WA. 2008. Recruitment to the nuclear periphery can alter expression of genes in human cells. *PLoS Genet* **4:** e1000039. doi:10.1371/journal.pgen.1000039

Fourel G, Revardel E, Koering CE, Gilson E. 1999. Cohabitation of insulators and silencing elements in yeast subtelomeric regions. *EMBO J* **18:** 2522–2537. doi:10.1093/emboj/18.9.2522

Galati A, Rossetti L, Pisano S, Chapman L, Rhodes D, Savino M, Cacchione S. 2006. The human telomeric protein TRF1 specifically recognizes nucleosomal binding sites and alters nucleosome structure. *J Mol Biol* **360:** 377–385. doi:10.1016/j.jmb.2006.04.071

Galati A, Micheli E, Alicata C, Ingegnere T, Cicconi A, Pusch MC, Giraud-Panis MJ, Gilson E, Cacchione S. 2015. TRF1 and TRF2 binding to telomeres is modulated by nucleosomal organization. *Nucleic Acids Res* **43:** 5824–5837. doi:10.1093/nar/gkv507

Gao Y, Tan J, Jin J, Ma H, Chen X, Leger B, Xu J, Spagnol ST, Dahl KN, Levine AS, et al. 2018. SIRT6 facilitates directional telomere movement upon oxidative damage. *Sci Rep* **8:** 5407. doi:10.1038/s41598-018-23602-0

García-Cao M, O'Sullivan R, Peters AH, Jenuwein T, Blasco MA. 2004. Epigenetic regulation of telomere length in mammalian cells by the Suv39h1 and Suv39h2 histone methyltransferases. *Nat Genet* **36:** 94–99. doi:10.1038/ng1278

Gauchier M, Kan S, Barral A, Sauzet S, Agirre E, Bonnell E, Saksouk N, Barth TK, Ide S, Urbach S, et al. 2019. SETDB1-dependent heterochromatin stimulates alternative lengthening of telomeres. *Sci Adv* **5:** eaav3673. doi:10.1126/sciadv.aav3673

Goldberg AD, Banaszynski LA, Noh KM, Lewis PW, Elsaesser SJ, Stadler S, Dewell S, Law M, Guo X, Li X, et al. 2010. Distinct factors control histone variant H3.3 localization at specific genomic regions. *Cell* **140:** 678–691. doi:10.1016/j.cell.2010.01.003

Gottschling DE, Aparicio OM, Billington BL, Zakian VA. 1990. Position effect at *S. cerevisiae* telomeres: reversible repression of Pol II transcription. *Cell* **63:** 751–762. doi:10.1016/0092-8674(90)90141-Z

Grewal SIS. 2023. The molecular basis of heterochromatin assembly and epigenetic inheritance. *Mol Cell* **83:** 1767–1785. doi:10.1016/j.molcel.2023.04.020

Hanaoka S, Nagadoi A, Nishimura Y. 2005. Comparison between TRF2 and TRF1 of their telomeric DNA-bound structures and DNA-binding activities. *Protein Sci* **14:** 119–130. doi:10.1110/ps.04983705

Hartlieb SA, Sieverling L, Nadler-Holly M, Ziehm M, Toprak UH, Herrmann C, Ishaque N, Okonechnikov K, Gartlgruber M, Park YG, et al. 2021. Alternative lengthening of telomeres in childhood neuroblastoma from genome to proteome. *Nat Commun* **12:** 1269. doi:10.1038/s41467-021-21247-8

Hazelrigg T, Levis R, Rubin GM. 1984. Transformation of white locus DNA in *Drosophila*: dosage compensation, zeste interaction, and position effects. *Cell* **36:** 469–481. doi:10.1016/0092-8674(84)90240-X

Heaphy CM, de Wilde RF, Jiao Y, Klein AP, Edil BH, Shi C, Bettegowda C, Rodriguez FJ, Eberhart CG, Hebbar S, et al. 2011. Altered telomeres in tumors with ATRX and DAXX mutations. *Science* **333:** 425. doi:10.1126/science.1207313

Hu H, van Roon AM, Ghanim GE, Ahsan B, Oluwole AO, Peak-Chew SY, Robinson CV, Nguyen THD. 2023. Structural basis of telomeric nucleosome recognition by shelterin factor TRF1. *Sci Adv* **9:** eadi4148. doi:10.1126/sciadv.adi4148

Ichikawa Y, Morohashi N, Nishimura Y, Kurumizaka H, Shimizu M. 2014. Telomeric repeats act as nucleosome-disfavouring sequences in vivo. *Nucleic Acids Res* **42:** 1541–1552. doi:10.1093/nar/gkt1006

Jäger K, Mensch J, Grimmig ME, Neuner B, Gorzelniak K, Türkmen S, Demuth I, Hartmann A, Hartmann C, Wittig F, et al. 2022. A conserved long-distance telomeric silencing mechanism suppresses mTOR signaling in aging human fibroblasts. *Sci Adv* **8:** eabk2814. doi:10.1126/sciadv.abk2814

Jegou T, Chung I, Heuvelman G, Wachsmuth M, Görisch SM, Greulich-Bode KM, Boukamp P, Lichter P, Rippe K. 2009. Dynamics of telomeres and promyelocytic leukemia nuclear bodies in a telomerase-negative human cell line. *Mol Biol Cell* **20:** 2070–2082. doi:10.1091/mbc.e08-02-0108

Johnson WL, Straight AF. 2017. RNA-mediated regulation of heterochromatin. *Curr Opin Cell Biol* **46:** 102–109. doi:10.1016/j.ceb.2017.05.004

Kim W, Ludlow AT, Min J, Robin JD, Stadler G, Mender I, Lai TP, Zhang N, Wright WE, Shay JW. 2016. Regulation of the human telomerase gene TERT by telomere position effect-over long distances (TPE-OLD): implications for aging and cancer. *PLoS Biol* **14:** e2000016. doi:10.1371/journal.pbio.2000016

Lamm N, Rogers S, Cesare AJ. 2021. Chromatin mobility and relocation in DNA repair. *Trends Cell Biol* **31:** 843–855. doi:10.1016/j.tcb.2021.06.002

Le Berre G, Hossard V, Riou JF, Guieysse-Peugeot AL. 2019. Repression of TERRA expression by subtelomeric DNA methylation is dependent on NRF1 binding. *Int J Mol Sci* **20:** 2791. doi:10.3390/ijms20112791

Li F, Deng Z, Zhang L, Wu C, Jin Y, Hwang I, Vladimirova O, Xu L, Yang L, Lu B, et al. 2019. ATRX loss induces telomere dysfunction and necessitates induction of alternative lengthening of telomeres during human cell immortalization. *EMBO J* **38:** e96659. doi:10.15252/embj.201796659

Lister R, Pelizzola M, Kida YS, Hawkins RD, Nery JR, Hon G, Antosiewicz-Bourget J, O'Malley R, Castanon R, Klugman S, et al. 2011. Hotspots of aberrant epigenomic reprogramming in human induced pluripotent stem cells. *Nature* **471:** 68–73. doi:10.1038/nature09798

Lister R, Mukamel EA, Nery JR, Urich M, Puddifoot CA, Johnson ND, Lucero J, Huang Y, Dwork AJ, Schultz MD, et al. 2013. Global epigenomic reconfiguration during mammalian brain development. *Science* **341:** 1237905. doi:10.1126/science.1237905

Lovejoy CA, Li W, Reisenweber S, Thongthip S, Bruno J, de Lange T, De S, Petrini JH, Sung PA, Jasin M, et al. 2012. Loss of ATRX, genome instability, and an altered DNA damage response are hallmarks of the alternative lengthening of telomeres pathway. *PLoS Genet* **8:** e1002772. doi:10.1371/journal.pgen.1002772

Makarov VL, Lejnine S, Bedoyan J, Langmore JP. 1993. Nucleosomal organization of telomere-specific chromatin in rat. *Cell* **73:** 775–787. doi:10.1016/0092-8674(93)90256-P

Marzec P, Armenise C, Pérot G, Roumelioti FM, Basyuk E, Gagos S, Chibon F, Déjardin J. 2015. Nuclear-receptor-mediated telomere insertion leads to genome instability in ALT cancers. *Cell* **160:** 913–927. doi:10.1016/j.cell.2015.01.044

Mattiroli F, Bhattacharyya S, Dyer PN, White AE, Sandman K, Burkhart BW, Byrne KR, Lee T, Ahn NG, Santangelo TJ, et al. 2017. Structure of histone-based chromatin in archaea. *Science* **357:** 609–612. doi:10.1126/science.aaj1849

Mechelli R, Anselmi C, Cacchione S, De Santis P, Savino M. 2004. Organization of telomeric nucleosomes: atomic force microscopy imaging and theoretical modeling. *FEBS Lett* **566:** 131–135. doi:10.1016/j.febslet.2004.04.032

Michishita E, McCord RA, Berber E, Kioi M, Padilla-Nash H, Damian M, Cheung P, Kusumoto R, Kawahara TL, Barrett JC, et al. 2008. SIRT6 is a histone H3 lysine 9 deacetylase that modulates telomeric chromatin. *Nature* **452:** 492–496. doi:10.1038/nature06736

Napier CE, Huschtscha LI, Harvey A, Bower K, Noble JR, Hendrickson EA, Reddel RR. 2015. ATRX represses alternative lengthening of telomeres. *Oncotarget* **6**: 16543–16558. doi:10.18632/oncotarget.3846

Nergadze SG, Farnung BO, Wischnewski H, Khoriauli L, Vitelli V, Chawla R, Giulotto E, Azzalin CM. 2009. CpG-island promoters drive transcription of human telomeres. *RNA* **15**: 2186–2194. doi:10.1261/rna.1748309

Ng LJ, Cropley JE, Pickett HA, Reddel RR, Suter CM. 2009. Telomerase activity is associated with an increase in DNA methylation at the proximal subtelomere and a reduction in telomeric transcription. *Nucleic Acids Res* **37**: 1152–1159. doi:10.1093/nar/gkn1030

Nimmo ER, Cranston G, Allshire RC. 1994. Telomere-associated chromosome breakage in fission yeast results in variegated expression of adjacent genes. *EMBO J* **13**: 3801–3811. doi:10.1002/j.1460-2075.1994.tb06691.x

O'Sullivan RJ, Almouzni G. 2014. Assembly of telomeric chromatin to create ALTernative endings. *Trends Cell Biol* **24**: 675–685. doi:10.1016/j.tcb.2014.07.007

* O'Sullivan RJ, Greenberg RA. 2025. Mechanisms of alternative lengthening of telomeres. *Cold Spring Harb Perspect Biol* **17**: a041690. doi:10.1101/cshperspect.a041690

O'Sullivan RJ, Kubicek S, Schreiber SL, Karlseder J. 2010. Reduced histone biosynthesis and chromatin changes arising from a damage signal at telomeres. *Nat Struct Mol Biol* **17**: 1218–1225. doi:10.1038/nsmb.1897

O'Sullivan RJ, Arnoult N, Lackner DH, Oganesian L, Haggblom C, Corpet A, Almouzni G, Karlseder J. 2014. Rapid induction of alternative lengthening of telomeres by depletion of the histone chaperone ASF1. *Nat Struct Mol Biol* **21**: 167–174. doi:10.1038/nsmb.2754

Ottaviani A, Schluth-Bolard C, Rival-Gervier S, Boussouar A, Rondier D, Foerster AM, Morere J, Bauwens S, Gazzo S, Callet-Bauchu E, et al. 2009. Identification of a perinuclear positioning element in human subtelomeres that requires A-type lamins and CTCF. *EMBO J* **28**: 2428–2436. doi:10.1038/emboj.2009.201

Pandita TK, Hunt CR, Sharma GG, Yang Q. 2007. Regulation of telomere movement by telomere chromatin structure. *Cell Mol Life Sci* **64**: 131–138. doi:10.1007/s00018-006-6465-0

Pennarun G, Picotto J, Bertrand P. 2023. Close ties between the nuclear envelope and mammalian telomeres: give me shelter. *Genes (Basel)* **14**: 775. doi:10.3390/genes14040775

Perrini B, Piacentini L, Fanti L, Altieri F, Chichiarelli S, Berloco M, Turano C, Ferraro A, Pimpinelli S. 2004. HP1 controls telomere capping, telomere elongation, and telomere silencing by two different mechanisms in *Drosophila*. *Mol Cell* **15**: 467–476. doi:10.1016/j.molcel.2004.06.036

Pickett HA, Reddel RR. 2015. Molecular mechanisms of activity and derepression of alternative lengthening of telomeres. *Nat Struct Mol Biol* **22**: 875–880. doi:10.1038/nsmb.3106

Pisano S, Marchioni E, Galati A, Mechelli R, Savino M, Cacchione S. 2007. Telomeric nucleosomes are intrinsically mobile. *J Mol Biol* **369**: 1153–1162. doi:10.1016/j.jmb.2007.04.027

Pisano S, Galati A, Cacchione S. 2008. Telomeric nucleosomes: forgotten players at chromosome ends. *Cell Mol Life Sci* **65**: 3553–3563. doi:10.1007/s00018-008-8307-8

Pisano S, Leoni D, Galati A, Rhodes D, Savino M, Cacchione S. 2010. The human telomeric protein hTRF1 induces telomere-specific nucleosome mobility. *Nucleic Acids Res* **38**: 2247–2255. doi:10.1093/nar/gkp1228

Poleshko A, Shah PP, Gupta M, Babu A, Morley MP, Manderfield LJ, Ifkovits JL, Calderon D, Aghajanian H, Sierra-Pagan JE, et al. 2017. Genome-nuclear lamina interactions regulate cardiac stem cell lineage restriction. *Cell* **171**: 573–587.e14. doi:10.1016/j.cell.2017.09.018

Poleshko A, Smith CL, Nguyen SC, Sivaramakrishnan P, Wong KG, Murray JI, Lakadamyali M, Joyce EF, Jain R, Epstein JA. 2019. H3k9me2 orchestrates inheritance of spatial positioning of peripheral heterochromatin through mitosis. *eLife* **8**: e49278. doi:10.7554/eLife.49278

Porro A, Feuerhahn S, Lingner J. 2014. TERRA-reinforced association of LSD1 with MRE11 promotes processing of uncapped telomeres. *Cell Rep* **6**: 765–776. doi:10.1016/j.celrep.2014.01.022

Postepska-Igielska A, Krunic D, Schmitt N, Greulich-Bode KM, Boukamp P, Grummt I. 2013. The chromatin remodelling complex NoRC safeguards genome stability by heterochromatin formation at telomeres and centromeres. *EMBO Rep* **14**: 704–710. doi:10.1038/embor.2013.87

Potts PR, Yu H. 2007. The SMC5/6 complex maintains telomere length in ALT cancer cells through SUMOylation of telomere-binding proteins. *Nat Struct Mol Biol* **14**: 581–590. doi:10.1038/nsmb1259

Poulet A, Pisano S, Faivre-Moskalenko C, Pei B, Tauran Y, Haftek-Terreau Z, Brunet F, Le Bihan YV, Ledu MH, Montel F, et al. 2012. The N-terminal domains of TRF1 and TRF2 regulate their ability to condense telomeric DNA. *Nucleic Acids Res* **40**: 2566–2576. doi:10.1093/nar/gkr1116

Pryde FE, Louis EJ. 1999. Limitations of silencing at native yeast telomeres. *EMBO J* **18**: 2538–2550. doi:10.1093/emboj/18.9.2538

Reddy L, Zullo JM, Bertolino E, Singh H. 2008. Transcriptional repression mediated by repositioning of genes to the nuclear lamina. *Nature* **452**: 243–247. doi:10.1038/nature06727

Robin JD, Ludlow AT, Batten K, Magdinier F, Stadler G, Wagner KR, Shay JW, Wright WE. 2014. Telomere position effect: regulation of gene expression with progressive telomere shortening over long distances. *Genes Dev* **28**: 2464–2476. doi:10.1101/gad.251041.114

Rose NR, Klose RJ. 2014. Understanding the relationship between DNA methylation and histone lysine methylation. *Biochim Biophys Acta* **1839**: 1362–1372. doi:10.1016/j.bbagrm.2014.02.007

Rosenfeld JA, Wang Z, Schones DE, Zhao K, DeSalle R, Zhang MQ. 2009. Determination of enriched histone modifications in non-genic portions of the human genome. *BMC Genomics* **10**: 143. doi:10.1186/1471-2164-10-143

Rossetti L, Cacchione S, Fuà M, Savino M. 1998. Nucleosome assembly on telomeric sequences. *Biochemistry* **37**: 6727–6737. doi:10.1021/bi9726180

Scheibe M, Arnoult N, Kappei D, Buchholz F, Decottignies A, Butter F, Mann M. 2013. Quantitative interaction screen of telomeric repeat-containing RNA reveals novel TERRA regulators. *Genome Res* **23**: 2149–2157. doi:10.1101/gr.151878.112

Schultz DC, Ayyanathan K, Negorev D, Maul GG, Rauscher FJ III. 2002. SETDB1: a novel KAP-1-associated histone H3, lysine 9-specific methyltransferase that contributes to HP1-mediated silencing of euchromatic genes by KRAB zinc-finger proteins. *Genes Dev* **16**: 919–932. doi:10.1101/gad.973302

Schwartzentruber J, Korshunov A, Liu XY, Jones DT, Pfaff E, Jacob K, Sturm D, Fontebasso AM, Quang DA, Tönjes M, et al. 2012. Driver mutations in histone H3.3 and chromatin remodelling genes in paediatric glioblastoma. *Nature* **482**: 226–231. doi:10.1038/nature10833

Soman A, Liew CW, Teo HL, Berezhnoy NV, Olieric V, Korolev N, Rhodes D, Nordenskiöld L. 2020. The human telomeric nucleosome displays distinct structural and dynamic properties. *Nucleic Acids Res* **48**: 5383–5396. doi:10.1093/nar/gkaa289

Soman A, Wong SY, Korolev N, Surya W, Lattmann S, Vogirala VK, Chen Q, Berezhnoy NV, van Noort J, Rhodes D, et al. 2022a. Columnar structure of human telomeric chromatin. *Nature* **609**: 1048–1055. doi:10.1038/s41586-022-05236-5

Soman A, Korolev N, Nordenskiöld L. 2022b. Telomeric chromatin structure. *Curr Opin Struct Biol* **77**: 102492. doi:10.1016/j.sbi.2022.102492

Teng YC, Sundaresan A, O'Hara R, Gant VU, Li M, Martire S, Warshaw JN, Basu A, Banaszynski LA. 2021. ATRX promotes heterochromatin formation to protect cells from G-quadruplex DNA-mediated stress. *Nat Commun* **12**: 3887. doi:10.1038/s41467-021-24206-5

Tennen RI, Bua DJ, Wright WE, Chua KF. 2011. SIRT6 is required for maintenance of telomere position effect in human cells. *Nat Commun* **2**: 433. doi:10.1038/ncomms1443

Timashev LA, Babcock H, Zhuang X, de Lange T. 2017. The DDR at telomeres lacking intact shelterin does not require substantial chromatin decompaction. *Genes Dev* **31**: 578–589. doi:10.1101/gad.294108.116

Tommerup H, Dousmanis A, de Lange T. 1994. Unusual chromatin in human telomeres. *Mol Cell Biol* **14**: 5777–5785. doi:10.1128/mcb.14.9.5777-5785.1994

Trojer P, Reinberg D. 2007. Facultative heterochromatin: is there a distinctive molecular signature? *Mol Cell* **28**: 1–13. doi:10.1016/j.molcel.2007.09.011

Udroiu I, Sgura A. 2020. Quantitative relationships between acentric fragments and micronuclei: new models and implications for curve fitting. *Int J Radiat Biol* **96**: 197–205. doi:10.1080/09553002.2020.1683638

Vakoc CR, Mandat SA, Olenchock BA, Blobel GA. 2005. Histone H3 lysine 9 methylation and HP1γ are associated with transcription elongation through mammalian chromatin. *Mol Cell* **19**: 381–391. doi:10.1016/j.molcel.2005.06.011

Vancevska A, Douglass KM, Pfeiffer V, Manley S, Lingner J. 2017. The telomeric DNA damage response occurs in the absence of chromatin decompaction. *Genes Dev* **31**: 567–577. doi:10.1101/gad.294082.116

van Steensel B, Belmont AS. 2017. Lamina-associated domains: links with chromosome architecture, heterochromatin, and gene repression. *Cell* **169**: 780–791. doi:10.1016/j.cell.2017.04.022

Viceconte N, Loriot A, Lona Abreu P, Scheibe M, Fradera Sola A, Butter F, De Smet C, Azzalin CM, Arnoult N, Decottignies A. 2021. PAR-TERRA is the main contributor to telomeric repeat-containing RNA transcripts in normal and cancer mouse cells. *RNA* **27**: 106–121. doi:10.1261/rna.076281.120

Voigt P, Tee WW, Reinberg D. 2013. A double take on bivalent promoters. *Genes Dev* **27**: 1318–1338. doi:10.1101/gad.219626.113

Waddington CH. 1942. The epigenotype. *Int J Endeavour* **41**: 10–13. doi:10.1093/ije/dyr184

Wong LH, McGhie JD, Sim M, Anderson MA, Ahn S, Hannan RD, George AJ, Morgan KA, Mann JR, Choo KH. 2010. ATRX interacts with H3.3 in maintaining telomere structural integrity in pluripotent embryonic stem cells. *Genome Res* **20**: 351–360. doi:10.1101/gr.101477.109

Wong SY, Soman A, Korolev N, Surya W, Chen Q, Shum W, van Noort J, Nordenskiöld L. 2024. The shelterin component TRF2 mediates columnar stacking of human telomeric chromatin. *EMBO J* **43**: 87–111. doi:10.1038/s44318-023-00002-3

Wu G, Broniscer A, McEachron TA, Lu C, Paugh BS, Becksfort J, Qu C, Ding L, Huether R, Parker M, et al. 2012. Somatic histone H3 alterations in pediatric diffuse intrinsic pontine gliomas and non-brainstem glioblastomas. *Nat Genet* **44**: 251–253. doi:10.1038/ng.1102

Yang SY, Chang EYC, Lim J, Kwan HH, Monchaud D, Yip S, Stirling PC, Wong JMY. 2021. G-quadruplexes mark alternative lengthening of telomeres. *NAR Cancer* **3**: zcab031. doi:10.1093/narcan/zcab031

Yehezkel S, Segev Y, Viegas-Péquignot E, Skorecki K, Selig S. 2008. Hypomethylation of subtelomeric regions in ICF syndrome is associated with abnormally short telomeres and enhanced transcription from telomeric regions. *Hum Mol Genet* **17**: 2776–2789. doi:10.1093/hmg/ddn177

# In the Loop: Unusual DNA Structures at Telomeric Repeats and Their Impact on Telomere Function

Elia Zanella and Ylli Doksani

IFOM ETS, The AIRC Institute of Molecular Oncology, 20139 Milan, Italy

*Correspondence:* ylli.doksani@ifom.eu

Telomeric repeats recruit the shelterin complex to prevent activation of the double-strand break response at chromosome ends. Thousands of TTAGGG repeats are present at each chromosome end to ensure telomere function. This abundance of G-rich repeats comes with the propensity to generate unusual DNA structures. The telomere loop (t-loop) structure, generated by strand invasion of the 3′ overhang in the internal repeats, contributes to telomere function. G4-DNA is promoted by the stretches of G-rich repeats in a single-stranded form and may affect telomere replication and elongation by telomerase. The intramolecular homology can lead to the formation of internal loops (i-loops) via intramolecular recombination at sites of telomeric damage, which can promote the excision of telomeric repeats as extrachromosomal circular DNA. Shelterin promotes t-loops, counteracting the accumulation of pathological structures either directly or via the recruitment of specialized helicases. Here, we will discuss the current evidence for the formation of unusual DNA structures at telomeres and possible implications for telomere function.

Linear chromosomes have evolved with telomeres as a solution to the intrinsic instability of free DNA ends. The solution consists of the capping of chromosomes with tandem repeats, bound by specialized proteins that protect the ends from being recognized and processed as DNA double-strand breaks. The maintenance of the terminal tandem repeats relies on telomerase, a specialized reverse transcriptase, which uses its associated RNA template to add new repeats to the 3′-end of each chromosome. This elongation mechanism is reminiscent of the reverse transcription of retrotransposons, which are considered the genetic ancestors of modern telomeres and telomerase (Cech et al. 1997; Eickbush 1997; Nakamura et al. 1997; de Lange 2004). The telomeric repeat sequence is encoded in the RNA template of telomerase and the most common motif in metazoans is TTAGGG (Moyzis et al. 1988; Greider and Blackburn 1989; Gomes et al. 2010). The G-rich filament is the one directly elongated by telomerase and is always found in the 5′ to 3′ orientation. This strand terminates in a 3′-single-stranded overhang, which is generated also in the absence of telomerase through 5′-end resection and appears to be essential for telomere function (Henderson and Blackburn 1989; Makarov et al. 1997). Due to

Cite this article as *Cold Spring Harb Perspect Biol* doi: 10.1101/cshperspect.a041694

the presence of the 3′-overhang and the inability of the DNA replication machinery to fully replicate the lagging strand, telomeric repeats are lost with each round of DNA replication (Watson 1972; Olovnikov 1973; Lingner et al. 1995; Takai et al. 2024). Apart from this gradual erosion, telomeric repeats are also lost as a consequence of DNA damage or replication stress (for a review, see Doksani 2019). Telomerase activity is generally sufficient to counterbalance telomere attrition, but in humans (and many other mammals), telomerase expression is switched off during development. Therefore, somatic cells can only perform a limited number of divisions before they "consume" their telomeres and undergo senescence (Hayflick 1965; d'Adda di Fagagna et al. 2003; Hockemeyer and Collins 2015). The proliferation limit imposed by telomeric erosion represents a double-edged sword for our physiology. It serves a tumor suppression function, particularly in early life, while also contributing to the aging process (Wright and Shay 1995; Campisi 2005). Short telomeres are also a genetic factor at the basis of a spectrum of diseases known as telomeropathies (Armanios and Blackburn 2012; Holohan et al. 2014).

Telomeric repeats are bound by specific factors, which are the executors of the end-protection function. Although this key molecular function is conserved, there is some degree of divergence and specialization in telomeric proteins across evolution (Wellinger and Zakian 2012; Sepsiova et al. 2016; Myler et al. 2021). Mammalian telomeres are associated with a 6-protein complex called shelterin (de Lange 2018). Three out of the six shelterin components bind telomeric repeats in a sequence-specific manner. TRF1 and TRF2 bind directly to double-stranded repeats, as homodimers through their MYB domains, while another component, POT1, binds the G-rich repeats in the single-stranded form through its OB-fold domains (Chong et al. 1995; Bilaud et al. 1996; Broccoli et al. 1997; Baumann and Cech 2001). TRF1 and TRF2 tether the rest of the complex to telomeres by recruiting in a redundant fashion TIN2, which interacts with TPP1, which interacts with POT1 (de Lange 2018). A sixth component of shelterin, Rap1, is recruited to telomeres via a direct inter-

action with TRF2 (Fig. 1; Li et al. 2000). Although shelterin is a complex, there is a significant separation of function among its components, also in the form of functional subcomplexes (Lim and Cech 2021). In broad strokes, TRF2 is fundamental in the repression of ATM signaling and classical nonhomologous end-joining, two key steps in the response to DNA double-strand breaks. POT1 is necessary to prevent ATR activation by the single-stranded DNA in the telomeric overhang and TRF1 is required for efficient replication of telomeric repeats. The recruitment of telomerase and some DNA repair activities (like alternative end-joining, 5′-end resection, and homologous recombination) is controlled in a more complex fashion by multiple shelterin components and other factors. The mechanisms by which shelterin protects telomeres from the double-strand break response and regulates telomere maintenance have been the subject of extensive studies (O'Sullivan and Karlseder 2010; Hockemeyer and Collins 2015; Lazzerini-Denchi and

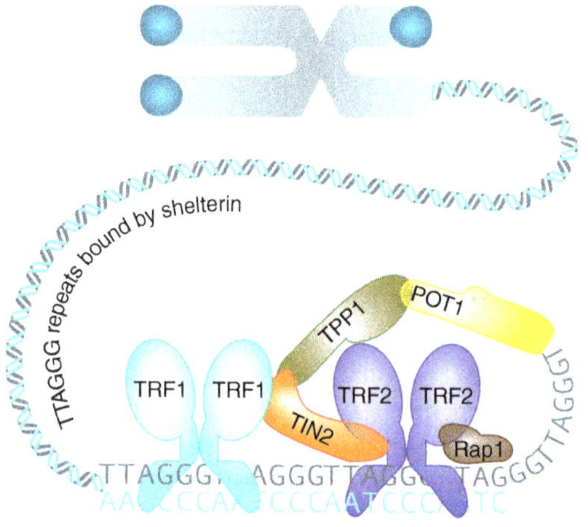

**Figure 1.** Schematic representation of the telomeric DNA and of the shelterin complex. The TTAGGG motif is the most common telomeric repeat in metazoans and terminates in a 3′-overhang, bound by POT1. The length of the TTAGGG repeats and of the 3′-overhang have been reduced for illustration purposes. For the same reason, only one shelterin complex is shown. Note that the visual rendering of the double helix, shown here and in the other figures, does not reproduce accurately the geometrical features of B-DNA.

Cite this article as *Cold Spring Harb Perspect Biol* doi: 10.1101/cshperspect.a041694

Sfeir 2016; de Lange 2018). It appears that shelterin employs both end-specific mechanisms, which prevent a DNA repair pathway from occurring at the actual end of the chromosome, as well as direct inhibition mechanisms that exclude a DNA repair pathway along the telomeric repeats altogether (Doksani 2019).

## UNUSUAL DNA STRUCTURES AT TELOMERIC REPEATS

Human telomeres range from 9 to 12 kb, while in mice, they can exceed 50 kb (Kipling and Cooke 1990; Alder et al. 2018). Therefore, there are thousands of copies of the TTAGGG motif repeated in tandem at the end each chromosome, which is about two orders of magnitude higher than at typical microsatellites (Richard et al. 2008). This abundance of repeats and the particular distribution of purines and pyrimidines in the two strands results in the propensity to undergo two types of structural transitions, that we will treat as two separate categories in this review.

The first is the formation of local secondary structures formed by alternative base-pairing like G4-DNA and, probably to a lesser extent, the i-motif and triplex DNA (see next paragraph). The second is the formation of large DNA structures mediated by conventional base-pairing or homology-mediated structures, like terminal telomere loops (t-loops) and internal loops (i-loops) (Griffith et al. 1999; Mazzucco et al. 2020). R-loops and reversed replication forks fall into this category as well, although they are more ubiquitous structures associated with transcription and replication stress genome-wide.

The propensity of telomeres to form unusual DNA structures has important implications both in telomere function and maintenance. t-loops are considered a functional feature of telomeres promoted by shelterin, involved in end protection, and invoked as a driver of telomere evolution (de Lange 2015; Tomáška et al. 2020). I-loops and G4-DNA, which may instead form at sites of telomeric damage or after prolonged G-strand exposure, could interfere with the replication and maintenance of telomeres.

Indeed, the fact that multiple specialized helicases (e.g., RTEL1, WRN, and BLM) are required for telomere maintenance and are actively recruited to telomeres by shelterin indicates that the latter has evolved the ability to harness key genome stability factors to keep unusual DNA structures in check at telomeres (Crabbe et al. 2004; Sfeir et al. 2009; Vannier et al. 2012; Zimmermann et al. 2014).

In the following paragraphs, we will provide a perspective view of unusual DNA structures that may form at telomeres and their potential consequences for telomere function.

## TELOMERIC STRUCTURES FORMED BY ALTERNATIVE BASE-PAIRING

Secondary DNA structures mediated by alternative base-pairing generally occur in a sequence-dependent manner and stretches of short repeats are the typical context that favors their formation (Khristich and Mirkin 2020). Telomeric repeats belong to this category and can form at least two types of structures with alternative base-pairing: (1) four-stranded DNA structures favored by stretches of Gs (G4-DNA) or Cs (i-motif) repeated in tandem, and (2) triplex DNA, favored by the asymmetric distribution of purines and pyrimidines (Veselkov et al. 1993; Neidle and Parkinson 2003).

Triplex DNA is formed by Hoogsteen base-pairing of a single-stranded DNA fragment with one of the two strands of a duplex, to form (in the case of telomeric DNA) a pyrimidine:purine–purine triplex structure (Fig. 2A; Frank-Kamenetskii and Mirkin 1995). Its formation is stimulated by homo-purine/pyrimidine stretches and has been extensively studied in the context of triplex-forming oligonucleotides (Knauert and Glazer 2001). Local transition of duplex DNA into intramolecular triplexes has been observed in vitro, but only in sequences containing mirror repeats under negative supercoiling (Mirkin et al. 1987; Kohwi and Kohwi-Shigematsu 1988). Therefore, this transition should not occur in double-stranded telomeric repeats. The single-stranded overhang, however, may pair with the double-

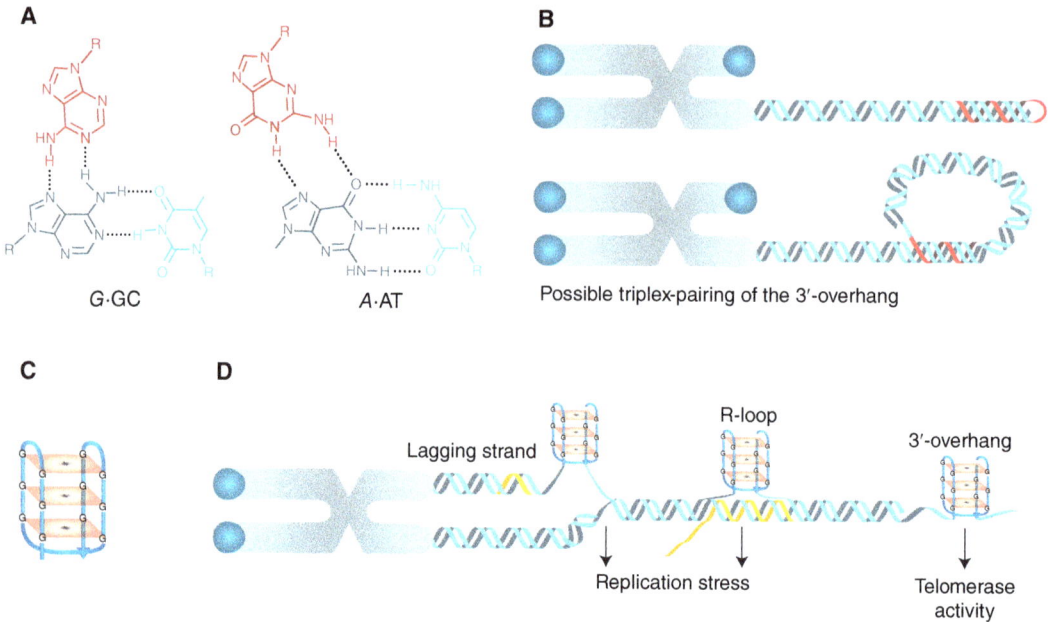

**Figure 2.** Telomeric structures are formed by alternative base-pairing. (*A*) Hoogsteen base-pairing that could mediate triplex DNA formation at telomeric repeats. (*B*) (*Top*) Schematic representation of the triplex pairing between the 3′-overhang and the internal telomeric repeats as proposed in Veselkov et al. (1993) for *Tetrahymena* telomeric repeats. (*Bottom*) A hypothetical parallel triplex pairing of the 3′-overhang with the internal repeats would generate a terminal loop that resembles the t-loop structure, although the physical properties of the loop junction would be different in this case. (*C*) Example of an antiparallel G4-DNA structure that can form at human telomeric repeats. Each planar arrangement of four guanines is stabilized by a monovalent cation at the center. While the planar arrangement of the guanines and their stacking is constant, strand orientation and loop extrusion may vary, generating a few different conformations of intramolecular G4-DNA (Neidle and Parkinson 2003). (*D*) Formation of G4-DNA at telomeric repeats in the context of the single-stranded G-rich filament. Depending on their location and half-life, G4 structures might interfere with DNA synthesis, while the presence of the G4-DNA on the telomeric overhang might affect telomerase binding and activity, depending on its strand orientation (Oganesian et al. 2006; Jansson et al. 2019).

stranded repeats, to form a triplex structure (Fig. 2B). Such structures have been observed in vitro, in physiological concentrations of salt and pH, with synthetic *Tetrahymena* telomeres made of TTGGGG repeats (Veselkov et al. 1993). The possibility that the telomeric overhang may pair with the internal repeats to form a stable triplex structure is rather intriguing, considering its potential implications for end protection. Triplex structures appear to form at GAA triplet repeats that become expanded in Friedrich ataxia (Follonier et al. 2013). However, studies with human TTAGGG repeats indicate a low probability of folding of the 3′-over-

hang in a triplex structure (Li et al. 2019). To date, there is no evidence (known to us) that such structures may be stabilized in vivo by factors involved in end protection.

Four-stranded DNA structures are generated by two or more planar arrangements of four guanines stabilized by hydrogen bonds and stacked on top of each other (G4-DNA) (Fig. 2C). Cytosine residues can also arrange in a similar structure, known as the i-motif (intercalated motif) and C4-DNA structures have been reported at the telomeric C-strand in vitro (Gehring et al. 1993; Ahmed et al. 1994; Kang et al. 1994). However, their formation requires

Cite this article as *Cold Spring Harb Perspect Biol* doi: 10.1101/cshperspect.a041694

hemi-protonation of cytidines, which occurs in nonphysiological acidic conditions, arguing against the spontaneous formation of the i-motif at telomeres in vivo (Phan et al. 2000). Four-stranded DNA structures with guanine residues do not have the same requirement and the formation of G4-DNA is heavily invoked as a relevant player in telomere biology (Bryan 2020).

## G4-DNA FORMATION AND CONSEQUENCES FOR TELOMERE FUNCTION

The observation of intramolecular G4 structures with *Tetrahymena* and *Oxytricha* telomeric oligonucleotides led to the hypothesis that G4-DNA might contribute to telomere function by stabilizing chromosome ends (Williamson et al. 1989). This idea was also inspired by the highly conserved G-rich nature of telomeres and by the existence of the 3′-overhang. Telomeric oligonucleotides at higher concentrations could also form intermolecular G4-DNA joining together two to four molecules, in a reaction stimulated by Tbpβ, a telomere-binding protein from *Oxytricha* (Sundquist and Klug 1989; Fang and Cech 1993). Intermolecular G4-DNA were suggested to mediate homolog chromosome pairing during meiosis, or telomere pairing and recombination; however, the lack of sequence specificity in G4-DNA formation and the potential to generate aggregates makes these hypotheses less appealing (Lipps 1980; Sen and Gilbert 1988; Sundquist and Klug 1989). While G4-DNA formation appears to be a universal feature of nearly all telomeric G-rich strands, the stability of the G4 structures depends on the number of uninterrupted guanines in the telomeric motif. Human (TTAGGG) telomeric repeats form less stable G4-DNA compared to those formed by *Tetrahymena* (TTGGGG) or *Oxytricha* (TTTTGGGG) telomeric repeats, with a difference in Tm of more than 15°C (Tran et al. 2011). In physiological conditions of pH and temperature, duplex DNA, stabilized by conventional base-pairing, prevails over G4-DNA (Phan and Mergny 2002). Therefore, the opportunities for G4-DNA formation at human telomeres are mostly limited to the single-stranded 3′-overhang, the lagging strand template during telomere replication, or the displaced G-rich strand at telomeric R-loops (Fig. 2D).

While the formation of G4-DNA at the telomeric overhang could be biochemically favored, the possibility that one or more of these structures contribute to end protection in human cells seems unlikely. First, it is not clear whether the presence of G4 structures at the telomeric overhang would indeed prevent the recruitment/activation of double-strand break response factors. Second, in multiple genetic settings where shelterin binding has been compromised, the telomeric repeats and the 3′-overhang per se do not offer much protection from double-strand break response activation (van Steensel et al. 1998; Denchi and de Lange 2007; Sfeir and de Lange 2012). Third, the shelterin component POT1, which binds directly to the single-stranded G-rich repeats, has been shown to resolve G4-DNA, arguing against a constant presence of this structure at the 3′-overhang (Zaug et al. 2005). Nevertheless, transient G4-DNA on the telomeric overhang may have specific regulatory functions during the recruitment of telomerase, contributing to telomere elongation and length homeostasis (Zahler et al. 1991; Paeschke et al. 2005; Oganesian et al. 2006; Moye et al. 2015; Jansson et al. 2019; Paudel et al. 2020).

On the contrary, the formation of G4-DNA during telomere replication is considered a major source of replication stress (Sfeir et al. 2009; Zimmermann et al. 2014; Drosopoulos et al. 2015; Yang et al. 2020). This view is supported by extensive studies with G4-stabilizing ligands that induce DNA damage and replication stress at telomeres and genome-wide (Gomez et al. 2006; Rizzo et al. 2009; Rodriguez et al. 2012; Vannier et al. 2012). In the absence of such ligands, however, it is not clear how frequently stable G4 structures form on the template DNA and interfere with telomere replication. In fact, events where the replicative helicase CMG encounters a G4 structure on the telomeric DNA should be very rare for two reasons. First, in a unidirectional telomere replication scenario, the replicative helicase moves on the C-rich telomeric strand and second, as mentioned above, G4-DNA should not form on double-stranded telomeric DNA (Phan and Mergny 2002). Furthermore, even if a G4 structure is somehow stabilized in front of the CMG helicase traveling on the leading strand, it should not represent an unsurmountable obstacle (Lerner and Sale 2019; Sparks et al. 2019). This does not necessarily mean that G4-DNA cannot affect leading strand synthesis. Uncoupling of the helicase and polymerase activities might allow the template to generate G4 structures that would interfere with DNA synthesis (Schiavone et al. 2014). In contrast, lagging strand replication would provide more opportunities for the template DNA to arrange in a G4 structure, due to the constant generation of single-stranded G-rich DNA at the fork. If unresolved, this structure could interfere with lagging strand synthesis by inhibiting both conventional and translesion DNA polymerases (Edwards et al. 2014; Zhang et al. 2019a). Consistent with the challenge it poses to replication, G4-DNA formation is counteracted by the shelterin component POT1 and by RPA, which compete for binding to the G-rich single-stranded DNA at telomeres (Zaug et al. 2005; Salas et al. 2006; Hwang et al. 2012; Chaires et al. 2020). In addition, several helicases and many other DNA-processing factors appear to be endowed with a G4 resolution activity (for a review, see Bochman et al. 2012). The list includes the helicases WRN, BLM, and RecQ1 (Sun et al. 1998; Fry and Loeb 1999; Mohaghegh et al. 2001; Huber et al. 2002; Popuri et al. 2008), FANCJ (London et al. 2008; Castillo Bosch et al. 2014), RTEL1 (Vannier et al. 2013; Kotsantis et al. 2020; Wu et al. 2020), DHX9, DDX11 (Chakraborty and Grosse 2011; Wu et al. 2012), DNA2 (Masuda-Sasa et al. 2008), Pif1 (Sanders 2010; Paeschke et al. 2013), the chromatin remodeler ATRX (Law et al. 2010), the DNA mismatch repair complex Mutsβ (Sakellariou et al. 2022), and the single-stranded DNA-binding complex CST (Zhang et al. 2019a). While it is possible that some of these factors are more specialized than others in dealing with G4-DNA, as is the case for Pif1 in yeast or FancJ in mammalian cells (Paeschke et al. 2011), the redundant nature of this activity warrants some caution while

invoking G4-DNA as the culprit of a telomere replication phenotype of a particular mutant.

Another important condition that can promote G4-DNA formation at telomeric repeats is accumulation of R-loops. These RNA–DNA hybrids form when the telomeric transcript, TERRA, displaces the G-rich strand in the duplex repeats (Azzalin et al. 2007; Pfeiffer et al. 2013; Arora et al. 2014; Graf et al. 2017). G4-DNA formation on the displaced strand of an R-loop is also supported by electron microscopy images of transcribing telomeric repeats (Duquette et al. 2004). This population of G4-DNA structures could be more long-lived than those forming at a replication fork, which can be resolved by replisome-associated helicases. In this view, R-loop-associated G4 structures might provide a reservoir of substrates detected by specific antibodies in immunofluorescence staining, or stabilized by treatment with G4 ligands (Müller et al. 2010; Biffi et al. 2013). Ultimately, these structures need to be resolved to prevent replication issues or induction of unscheduled recombination events (De Magis et al. 2019; Kumar et al. 2021; Kaminski et al. 2022; Yadav et al. 2022).

Although telomeres are hotspots for G4-DNA formation, it is important to consider that there are >300,000 G4-forming motifs in the human genome, enriched in promoters or other transcription regulatory regions (Balasubramanian et al. 2011; Hänsel-Hertsch et al. 2017). Hence, potential changes in relevant gene expression or in replication programs should be taken into consideration when studying the effect of G4-stabilizing ligands in telomere metabolism.

## TELOMERIC STRUCTURES FORMED BY CONVENTIONAL BASE-PAIRING

These homology-mediated structures are typically formed through an exchange of pairing partners among duplex DNA strands. Intermolecular strand exchange is a crucial step in homologous recombination and typically involves sister chromatids or homolog chromosomes. However, the abundance of homology along the telomere promotes intramolecular strand

Cite this article as Cold Spring Harb Perspect Biol doi: 10.1101/cshperspect.a041694 exchange events. As it will be described in detail later, these events are at the basis of the controlled formation of t-loops in the context of normal telomere function. However, in the context of telomere damage, unrestrained strand exchange may also result in the formation of i-loops, which can then lead to the excision of telomeric repeats as extrachromosomal circular DNA (Griffith et al. 1999; Mazzucco et al. 2020). Being direct tandem repeats, functional telomeres do not form cruciform/hairpin structures, which are typical of inverted repeats (Leach 1994). However, in instances of telomere–telomere fusions, long inverted repeats would be generated with the potential to extrude telomere-sized hairpin structures. Whether these structures form and their stability in the context of telomere fusions is not clear. The fact that some cells can undergo several DNA replication rounds, despite carrying multiple fused telomeres, argues against the stable presence of cruciforms at fused telomeres (Lazzerini Denchi et al. 2006). Other types of homology-mediated DNA structures that form at telomeres include R-loops, discussed above, and reversed forks associated with telomere replication problems.

## REPLICATION FORK REVERSAL AT TELOMERIC REPEATS AND IMPLICATIONS FOR TELOMERE MAINTENANCE

Reversed replication forks are Holliday junctions formed by reverse branch migration of a replication fork followed by pairing of the newly synthesized strands in a so-called chicken foot structure (Fig. 3A). Fork reversal is thought to occur in the presence of roadblocks associated with replication stress such as bulky DNA lesions, interstrand cross-links, accumulation of positive supercoiling, but also replication-transcription collisions and tightly bound proteins (for reviews, see Atkinson and McGlynn 2009; Neelsen and Lopes 2015). Telomeres are known to be sites of endogenous replication stress with multiple potential challenges to the replication forks like the presence of tightly bound proteins, i-loops, R-loops, G4-DNA, and t-loops (Ohki and Ishikawa 2004; Sarek et al. 2015; Lerner and Sale 2019; Mazzucco et al. 2020; Douglas and Diffley

2021; Silva et al. 2021; Radchenko et al. 2022). The fact that a highly conserved function of shelterin is to assist the replication of telomeric repeats emphasizes the biological relevance of telomere replication problems (Miller et al. 2006; Martinez et al. 2009; Sfeir et al. 2009). It seems that replication fork reversal is one structural manifestation of replication stress at telomeres. Indeed, we have recently shown by 2D gels and electron microscopy analysis an increased incidence of replication fork reversal both at ectopic telomeric repeats in an SV40 mini-chromosome and at endogenous mouse telomeres (Huda et al. 2023). Even telomeric replication fork models, similarly to other repetitive DNA, tend to undergo spontaneous fork reversal in vitro (Fouché et al. 2006). Reversed forks that maintain CMG association are thought to be competent for replication restart, which may occur either through a template-switching mechanism, or through branch migration assisted by RecQ1and WRN helicases (Quinet et al. 2017; Liu et al. 2023). However, a long-lived reversed fork might affect telomeres in multiple ways. In the absence of a protective structure, the reversed arm might initiate a double-strand break response during telomere replication or even engage in strand invasion events, initiating a recombination process known as break-induced replication. Indeed, break-induced replication is frequently associated with telomere replication stress (Dilley et al. 2016; Porreca et al. 2020; Yang et al. 2020). Similar pathways would be activated even if the telomeric reversed fork is cleaved by Holliday junction resolvases like MUS81 (Regairaz et al. 2011). Depending on the position of fork reversal, its cleavage might lead to loss of large fractions of telomeric repeats or even of whole telomeres.

One important consequence of fork reversal at telomeric repeats is recruitment of telomerase on the reversed arm. This scenario was first proposed as an explanation of the increased telomerase activity in the TRF1/TRF2 ortholog *taz1Δ* mutant in fission yeast, which is characterized by massive fork collapse at telomeres (Miller et al. 2006; Dehé et al. 2012). In mouse cells, an abnormal recruitment of telomerase at reversed forks was identified as the culprit of telomere replication failures in the absence of

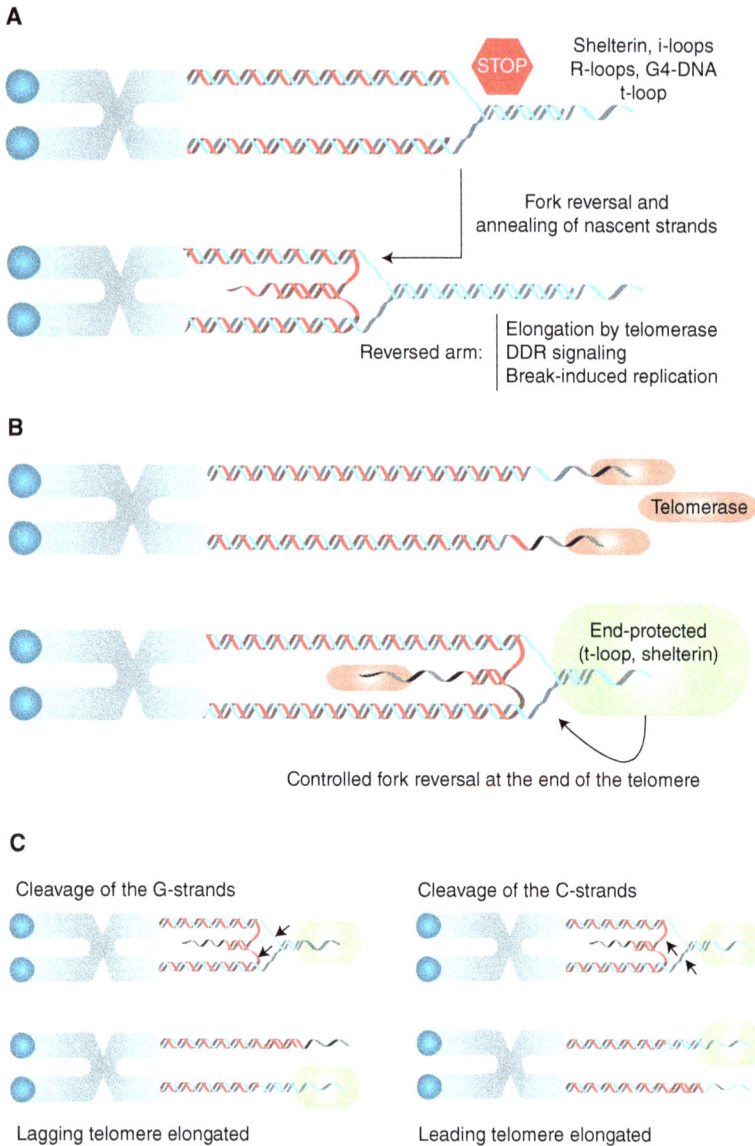

**Figure 3.** Consequences of replication fork reversal at telomeric repeats. (*A*) Schematic representation of replication fork reversal at telomeric repeats. Possible causes of fork reversal are listed on the *top* drawing ahead of the fork, while possible transactions of the reversed arm are listed on the *bottom*. (*B*) An alternative model for telomerase-mediated elongation on a telomeric reversed fork. (*Top*) The current model of telomerase acting on both leading and lagging telomeres in a postreplicative manner after replication termination. In this case, two unprotected telomere ends might be transiently generated in close proximity upon replication fork runoff. (*Bottom*) A proposed alternative model of telomerase action, on a reversed fork. Fork reversal would be actively induced at the very end of telomere replication, possibly by shelterin. Telomerase would be recruited to, and elongate the reversed arm, while the original end of the chromosome remains in a protected state. Resection of $5'$-end at the reversed arm, might be required to generate the $3'$-protruding end, necessary for telomerase activity. (*C*) Resolution of the reversed fork by a Holliday junction resolvase would lead to the generation of two telomeres. The newly added repeats would segregate either on the leading or on the lagging telomere, depending on the orientation of the Holliday junction cleavage. In this model, the original chromosome end remains in a protected state throughout the process.

Cite this article as *Cold Spring Harb Perspect Biol* doi: 10.1101/cshperspect.a041694

RTEL1 (Margalef et al. 2018). Finally, we recently provided direct evidence of telomerase-mediated elongation of reversed telomeric forks occurring at ectopic telomeric repeats in human cells (Huda et al. 2023). However, the absolute frequency of fork reversal and telomerase elongation events during unperturbed telomere replication is not known and therefore the consequences of this process in telomere metabolism remain to be investigated. In mouse cells with long telomeres, RTEL1 has been reported to prevent the accumulation of reversed forks and modulate telomerase activity on the reversed arm (Margalef et al. 2018). In contrast, in yeast cells, the elongation of reversed forks by telomerase has been proposed as a physiological mechanism that prevents telomere loss (Dehé et al. 2012; Matmati et al. 2020). In this view, it is tempting to speculate that fork reversal, at the very end of the telomere, might be a safer way to generate a substrate for telomerase elongation as it would allow the maintenance of end protection on the original chromosome terminus, instead of generating two unprotected DNA ends in close proximity (Fig. 3B,C). This scenario would require a controlled induction of fork reversal, in close proximity to the chromosome end, similar to what has been suggested for replication forks running into a double-strand break in yeast (Doksani et al. 2009).

Finally, an unscheduled recruitment of telomerase at nontelomeric reversed forks might contribute to the aberrant addition of telomeric repeats at genomic DNA breaks in checkpoint mutants (Zhang and Durocher 2010; Kinzig et al. 2024).

## T-LOOPS AND CHROMOSOME END PROTECTION

T-loops were identified by electron microscopy analysis of enriched telomeric DNA (Griffith et al. 1999). They are terminal structures, with a lasso-like appearance and a variable loop size (Griffith et al. 1999). First visualized in mouse and human cells, t-loops were subsequently reported at telomeres of protozoans, plants, and nematodes, suggesting that they represent a highly conserved feature of telomeres (for a review, see Tomáška et al. 2020). T-loops are thought to form by strand invasion of the 3′-overhang in the duplex telomeric repeats, generating a D-loop structure at the junction (Fig. 4A). Stabilization of this D-loop structure by interstrand psoralen cross-linking is thought to be required for t-loops to resist DNA isolation procedures. Indeed, upon deproteinization, the negative supercoiling that is normally accommodated within histone wrapping may exert a rotational force on the t-loop junction, thereby promoting its resolution. This is consistent with the fact that t-loop visualization in enriched telomeric chromatin does not require psoralen cross-linking (Nikitina and Woodcock 2004). The shelterin component TRF2 is necessary and sufficient to form t-loops in vitro and in vivo, while the other shelterin subunits or the recombinase Rad51 (which mediates strand invasion during homologous recombination) do not appear to contribute to the process (Griffith et al. 1999; Stansel et al. 2001; Doksani et al. 2013; Timashev and De Lange 2020). TRF2 could promote strand invasion by altering the writhe of telomeric DNA (i.e., the twisting of the double helix in space) through right-handed wrapping over its TRFH domain (Amiard et al. 2007; Benarroch-Popivker et al. 2016). The mechanism of t-loop formation by TRF2 is still an active area of research and recent results indicate that the ability of TRF2 to tetramerize on telomeric DNA is a key step in t-loop formation (Goldfarb et al. 2024). Furthermore, in vitro experiments with transcribing telomeric substrates indicate that R-loop-mediated strand openings could promote the formation of t-loops (Kar et al. 2016).

T-loop formation would prevent the detection of the chromosome end by the double-strand break response. For this reason, t-loops are thought to represent a structural solution to the end-protection problem. TRF2, the shelterin component necessary for t-loop formation, is also required for protection from ATM activation and classical nonhomologous end-joining (Doksani et al. 2013; Timashev and De Lange 2020). It is, however, important to point out that TRF2 seems to promote two redundant end-protection strategies: one through t-loop for-

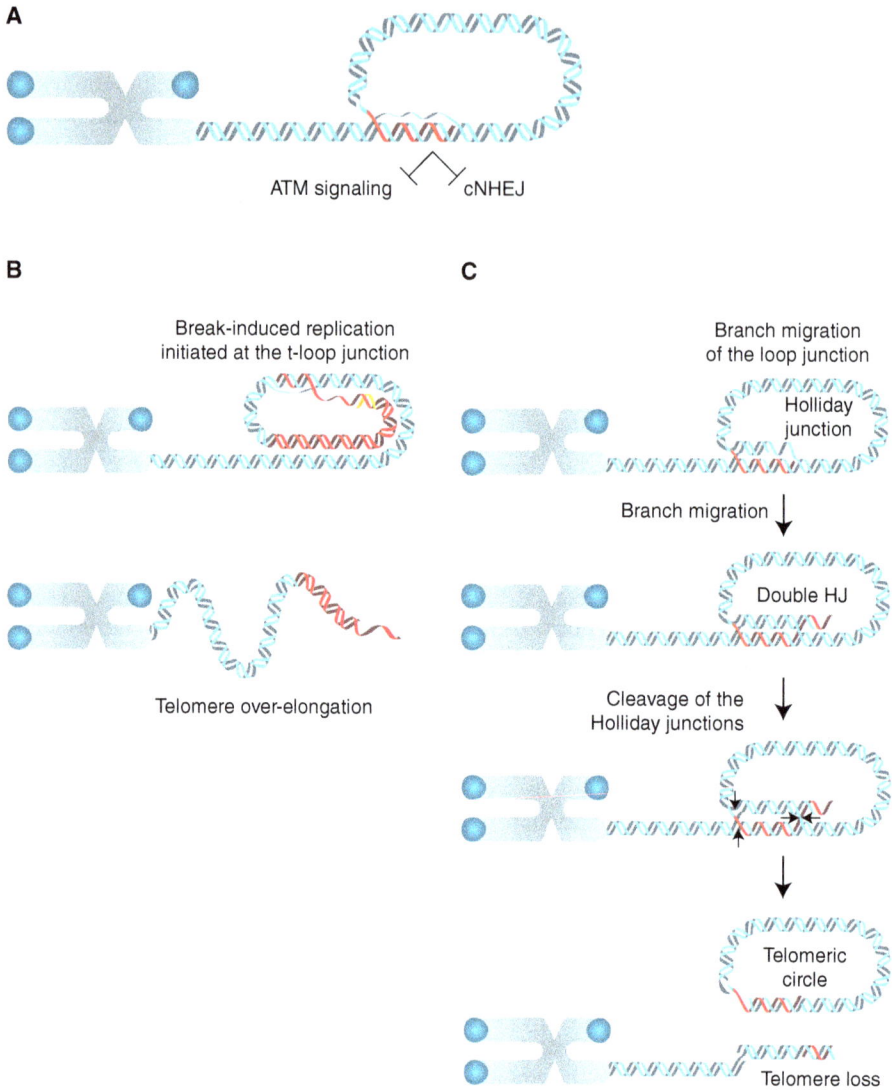

**Figure 4.** The t-loop structure. (*A*) Schematic representation of the t-loop structure. (*B*) The t-loop junction is identical to the strand invasion structure that initiates break-induced replication (BIR). BIR activation at the t-loop junction would lead to rapid, t-loop-sized telomere elongation events. (*C*) Reversed branch migration of the t-loop junction can lead to the pairing of the C-strand and formation of Holliday junctions at the base of the t-loop. Cleavage of the Holliday junction could lead to the generation of a telomeric circle, associated with loss of telomeric repeats.

mation, which makes the chromosome end inaccessible to nonhomologous end-joining (Goldfarb et al. 2024), and another through a direct inhibition of DDR factors (Karlseder et al. 2004; Lam et al. 2010; Wu et al. 2010; Okamoto et al. 2013; Ribes-Zamora et al. 2013; Doksani and de Lange 2016; Van Ly et al. 2018;

Myler et al. 2023). A relevant exception to the dependencies described above is represented by embryonic stem cells, where end protection is less reliant on TRF2. In this context, t-loops appear to form also in cells lacking TRF2, through unknown mechanisms (Markiewicz-Potoczny et al. 2021; Ruis et al. 2021).

Cite this article as *Cold Spring Harb Perspect Biol* doi: 10.1101/cshperspect.a041694

While being important for end protection, the t-loop junction needs to be protected from homologous recombination factors. The strand invasion step in t-loop formation is structurally identical to the one that initiates break-induced replication, which makes the t-loop a potential substrate of this pathway (Saini et al. 2013; Wilson et al. 2013). In this scenario, the 3′-end can recruit DNA polymerase δ and initiate conservative DNA synthesis through bubble migration, in a reaction that would lead to rapid, t-loop-sized, telomere elongation events (Fig. 4B). This activity has been proposed as an ancestral telomere maintenance mechanism, preceding the evolution of telomerase (de Lange 2015; Tomaska et al. 2019). However, this 3′-end elongation through intramolecular BIR would not be compatible with telomere length regulation, leading to uncontrolled, telomerase-independent telomere elongation. This scenario is reminiscent of the alternative telomere lengthening (ALT) mechanism in some cancer cells and is somehow avoided in normal cells (Bryan et al. 1997; Dilley et al. 2016). Another modification of the t-loop junction that can impact telomere maintenance is the reverse branch migration of the D-loop and pairing of the C-strand. This process would generate a single or double Holliday junction at the base of the loop (Fig. 4C). Cleavage of these structures by a resolvase and a subsequent nick sealing would generate an extrachromosomal telomeric circle with a reciprocal t-loop-sized telomere deletion event, a phenotype that has been reported in cells lacking the basic domain of TRF2 (Wang et al. 2004; Schmutz et al. 2017). Finally, the presence of the t-loop junction is invoked as an obstacle to replication fork progression at telomeric repeats (Vannier et al. 2012). Although the strand invasion structure per se should not block the CMG helicase, it is possible that the t-loop junction is bound more tightly by the shelterin complex and maybe other accessory factors, which could generate a replication fork barrier. Recruitment of the RTEL1 helicase by TRF2 has been proposed to dismantle the t-loop structure, allowing replication fork progression (Sarek et al. 2015). Importantly, RTEL1 recruitment requires an S-phase-specific dephosphor-

ylation of Ser365 of TRF2, which would ensure that t-loop resolution only occurs at the time of DNA replication (Sarek et al. 2019).

## I-LOOPS AND THE UNEXPECTED CONSEQUENCES OF TELOMERE DAMAGE

Tandem repeats are known to be hotspots for homologous recombination genome-wide, which in turn drives their instability (Harding et al. 1992). Repeat length variation can occur as a result of unequal exchanges between sister chromatids, strand slippage during replication, or double-strand break–initiated intramolecular recombination. These processes normally require canonical recombination factors, but early observations in *Escherichia coli* and *Saccharomyces cerevisiae* revealed the existence of recombination events leading to the deletion of repetitive elements that were independent of *recA* and *RAD52* genes, normally required for strand exchange (Lovett et al. 1993; Klein 1995; Bi and Liu 1996). A few mechanisms have been proposed to explain these noncanonical recombination-like events, but the experimental evidence in support has been scarce. Our recent work in visualizing the structure of purified telomeric repeats might provide insights on how some of these events initiate. In fact, electron microscopy images of purified telomeres from mouse and human cells revealed that telomeric DNA has a high propensity to form internal loop structures, which we named i-loops (Mazzucco et al. 2020). I-loops are a direct consequence of single-stranded DNA damage occurring in the context of tandem telomeric repeats. Since telomeres are made of repeats of a 6 bp motif, single-stranded DNA damage at these loci will almost certainly expose intramolecular homology. For example, two gaps occurring on opposite strands of the same molecule will be complementary and thus can simply anneal with each other, generating an i-loop in the process (Fig. 5). Upon branch migration and strand annealing, the structure at the basis of the i-loop could transform into a Holliday junction (Fig. 5). This model accommodates key observations: I-loops are strongly induced by DNA damage, they often occur in proximity to damage sites, and their formation/

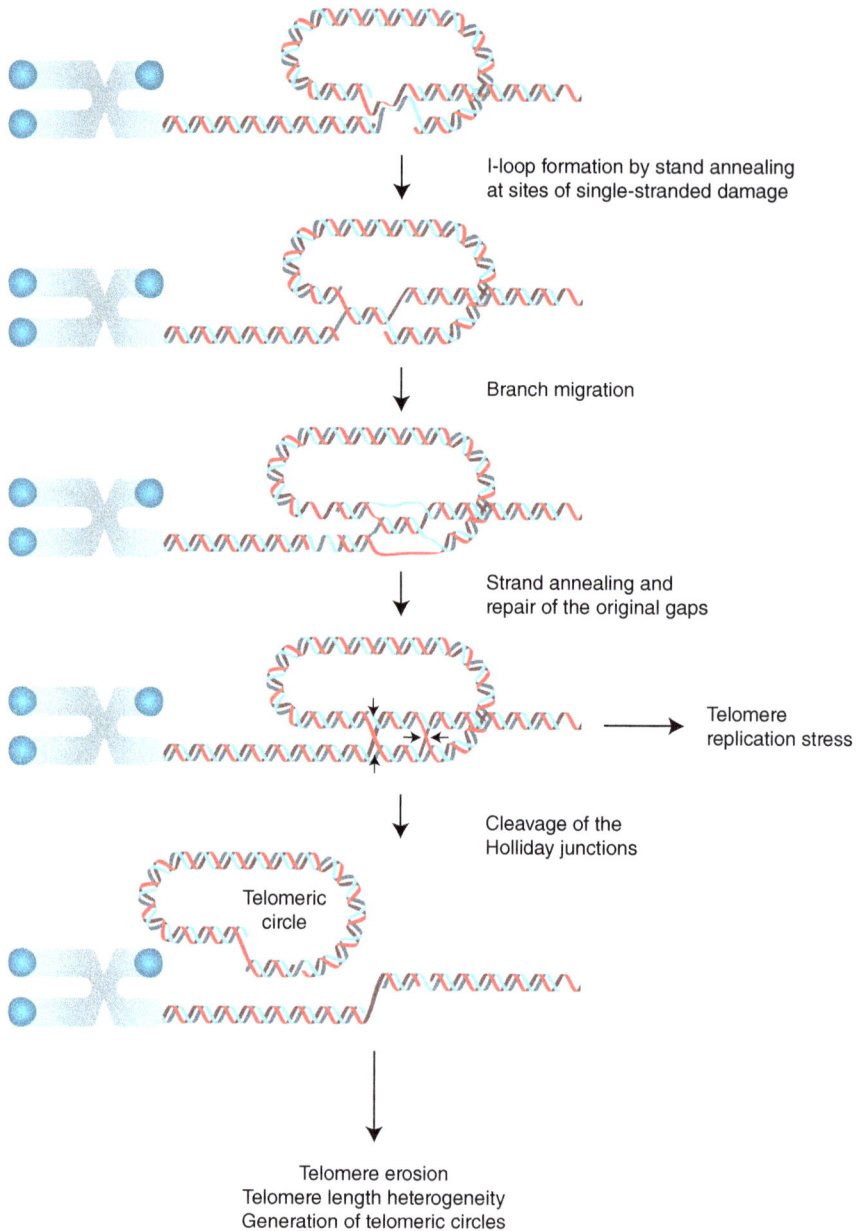

**Figure 5.** Formation of i-loops at damaged telomeric repeats and consequences for telomere maintenance. Two short single-stranded gaps on opposite strands, in the context of telomeric repeats, are complementary and therefore could pair with each other to generate an i-loop structure. Branch migration and pairing of the opposite strands would lead to the formation of an i-loop structure containing a double Holliday junction at the base. Upon repair of the single-stranded gaps that originated it, this i-loop structure could not be resolved by reversed branch migration and would therefore become trapped on the telomere. Such i-loop structures might hinder replication fork progression at telomeric repeats. Cleavage of the Holliday junctions at the base of the i-loop would lead to the excision of telomeric repeats in the form of extrachromosomal circular DNA, resulting in telomere erosion. While the scenario shown here represents the most spontaneous form of i-loop formation, similar transitions may be favored also in the presence of single-stranded gaps on the same telomeric strand, or in the presence of telomeric nicks.

Cite this article as *Cold Spring Harb Perspect Biol* doi: 10.1101/cshperspect.a041694

stabilization requires branch migration (Maz-zucco et al. 2020). Several features distinguish i-loops from the t-loop structure: (1) I-loops are internal to the telomeric molecule, while the t-loop is terminal; (2) one telomere can contain one or multiple i-loops, while it can only have one t-loop; (3) I-loops are on average smaller, with >75% of them being smaller than 4 kb, while t-loop size is randomly distributed over the length of the telomere; (4) I-loops can be visualized with or without psoralen cross-linking, both in 2D gels and electron micros-copy, while t-loop visualization in electron microscopy or superresolution microscopy re-quires psoralen cross-linking (Griffith et al. 1999; Doksani et al. 2013; Mazzucco et al. 2020). Probably the most important difference between t-loops and i-loops, however, is the fact that t-loops are considered physiological DNA structures, induced by shelterin with a function in end protection, while i-loops likely represent pathological structures that form as a conse-quence of DNA damage in the context of telo-meric repeats.

The first and probably most relevant conse-quence of i-loop formation is generation of ex-trachromosomal telomeric circles, likely due to the cleavage of the Holliday junction at the base of the i-loop (Mazzucco et al. 2020). I-loop for-mation and excision would explain the accumu-lation of telomeric circles in cells that maintain telomere length by a telomerase-independent mechanism known as alternative lengthening of telomeres (ALT cells), which are known to contain frequent nicks and gaps at telomeric repeats (Nabetani and Ishikawa 2009). Indeed, ALT telomeres contain frequent i-loops that mi-grate in a 2D-gel arc, commonly (and now in-correctly) referred to as the T-circle arc (Cesare and Griffith 2004; Wang et al. 2004). This arc, visualized in 2D gels, can be generated in the absence of circles and is largely made of i-loops (Mazzucco et al. 2020). Accumulation of DNA damage at telomeres and consequent formation of i-loops could be the common structural de-nominator of a long list of mutants in telomere metabolism, DNA damage, and DNA replica-tion that accumulate the i-loop arc in 2D gels (Deng et al. 2007; Tomaska et al. 2009; Gu et al.

2012; O'Sullivan et al. 2014; Li et al. 2017; Zhang et al. 2019b).

Importantly, excision of the i-loop as a telo-meric circle would be associated with an i-loop-sized deletion of telomeric repeats (Fig. 5). This process would be, by nature, stochastic, increas-ing in frequency with telomere damage and telomere length. A stochastic deletion of por-tions of telomeric repeats over time could ex-plain the inherent heterogeneity of telomeres and the phenomenon of telomere trimming (Hanish et al. 1994; Pickett et al. 2009; Wakai et al. 2014). I-loop formation and excision could contribute substantially to the stochastic telo-mere deletion events predicted by mathematical models of telomere shortening and the emer-gence of replicative senescence (Rubelj and Vondracek 1999; Proctor and Kirkwood 2003).

Apart from promoting extrachromosomal circular DNA formation and telomere loss, i-loops might represent yet another challenge to telomere replication. Indeed, the base of the i-loop structure would represent a qualitatively different challenge to a replication fork com-pared to the initial single-stranded gaps that in-duced i-loop formation. In particular, the CMG helicase alone might not be able to progress through double Holliday junction structures at the base of the i-loop. Recruitment of specialized helicases might be important in this setting to branch migrate or dissolve i-loop structures to allow replication of telomeric repeats. The BTR complex, involved in double Holliday junction resolution, is an obvious candidate for this job (Wu and Hickson 2003), but contribution by other helicases like RTEL1 or WRN cannot be excluded.

Finally, since i-loop formation is triggered upon the accumulation of DNA damage in the context of tandem telomeric repeats, it is possi-ble that the contribution of these structures in repeat instability is not limited to telomeres but might extend to other parts of the genome, 50% of which is made of repetitive elements.

## CONCLUDING REMARKS

Telomeric repeats have the potential to generate several unusual DNA structures that might contribute to telomere function or threaten telo-mere maintenance. These structures may repre-sent the ultimate molecular substrates of the multiple helicases and nucleases that the shel-terin complex actively recruits to telomeres. De-spite their relevance to telomere biology, our knowledge on the unusual telomeric structures is limited by our ability to probe the native struc-ture of DNA in vivo. In this view, each proposed structural model for telomeres ultimately repre-sents a framework that is continuously tested and validated against the available genetic and biochemical data on telomere function.

Our knowledge of the mechanisms of the DNA double-strand break response also derives in part from the study of dysfunctional telo-meres. Similarly, the mechanism at the basis of repeat instability throughout the genome might benefit from studies of unusual DNA structures and their consequences on the maintenance of telomeric repeats.

## ACKNOWLEDGMENTS

We apologize for the many relevant references that were not included due to space limita-tions. We are grateful to Titia de Lange, Fran-cisca Lottersberger, and members of the Y.D. laboratory for feedback and suggestions. Work in Y.D. laboratory is funded by the AIRC IG grant #28954, The Concern Foundation, and Worldwide Cancer Research.

## REFERENCES

Ahmed S, Kintanar A, Henderson E. 1994. Human telomeric C-strand tetraplexes. *Nat Struct Biol* **1**: 83–88. doi:10.1038/nsb0294-83

Alder JK, Hanumanthu VS, Strong MA, DeZern AE, Stanley SE, Takemoto CM, Danilova L, Applegate CD, Bolton SG, Mohr DW, et al. 2018. Diagnostic utility of telomere length testing in a hospital-based setting. *Proc Natl Acad Sci* **115**: E2358–E2365. doi:10.1073/pnas.1720427115

Amiard S, Doudeau M, Pinte S, Poulet A, Lenain C, Faivre-Moskalenko C, Angelov D, Hug N, Vindigni A, Bouvet P, et al. 2007. A topological mechanism for TRF2-enhanced strand invasion. *Nat Struct Mol Biol* **14**: 147–154. doi:10.1038/nsmb1192

Armanios M, Blackburn EH. 2012. The telomere syndromes. *Nat Rev Genet* **13**: 693–704. doi:10.1038/nrg3246

Arora R, Lee Y, Wischnewski H, Brun CM, Schwarz T, Az-zalin CM. 2014. RNaseh1 regulates TERRA-telomeric

DNA hybrids and telomere maintenance in ALT tumour cells. *Nat Commun* **5**: 5220. doi:10.1038/ncomms6220

Atkinson J, McGlynn P. 2009. Replication fork reversal and the maintenance of genome stability. *Nucleic Acids Res* **37**: 3475–3492. doi:10.1093/nar/gkp244

Azzalin CM, Reichenbach P, Khoriauli L, Giulotto E, Lingner J. 2007. Telomeric repeat–containing RNA and RNA surveillance factors at mammalian chromosome ends. *Science* **318**: 798–801. doi:10.1126/science.1147182

Balasubramanian S, Hurley LH, Neidle S. 2011. Targeting G-quadruplexes in gene promoters: a novel anticancer strategy. *Nat Rev Drug Discov* **10**: 261–275. doi:10.1038/nrd3428

Baumann P, Cech TR. 2001. Pot1, the putative telomere end-binding protein in fission yeast and humans. *Science* **292**: 1171–1175. doi:10.1126/science.1060036

Benarroch-Popivker D, Pisano S, Mendez-Bermudez A, Lo-totska L, Kaur P, Bauwens S, Djerbi N, Latrick CM, Frai-sier V, Pei B, et al. 2016. TRF2-mediated control of telo-mere DNA topology as a mechanism for chromosome-end protection. *Mol Cell* **61**: 274–286. doi:10.1016/j.molcel.2015.12.009

Bi X, Liu LF. 1996. recA-independent DNA recombination between repetitive sequences: mechanisms and implica-tions. *Prog Nucleic Acid Res Mol Biol* **54**: 253–292. doi:10.1016/S0079-6603(08)60365-7

Biffi G, Tannahill D, McCafferty J, Balasubramanian S. 2013. Quantitative visualization of DNA G-quadruplex struc-tures in human cells. *Nat Chem* **5**: 182–186. doi:10.1038/nchem.1548

Bilaud T, Koering CE, Binet-Brasselet E, Ancelin K, Pollice A, Gasser SM, Gilson E. 1996. The telobox, a Myb-related telomeric DNA binding motif found in proteins from yeast, plants and human. *Nucleic Acids Res* **24**: 1294–1303. doi:10.1093/nar/24.7.1294

Bochman ML, Paeschke K, Zakian VA. 2012. DNA second-ary structures: stability and function of G-quadruplex structures. *Nat Rev Genet* **13**: 770–780. doi:10.1038/nrg3296

Broccoli D, Smogorzewska A, Chong L, de Lange T. 1997. Human telomeres contain two distinct Myb-related pro-teins, TRF1 and TRF2. *Nat Genet* **17**: 231–235. doi:10.1038/ng1097-231

Bryan TM. 2020. G-quadruplexes at telomeres: friend or foe?. *Molecules* **25**: 3686. doi:10.3390/molecules25163686

Bryan TM, Englezou A, Dalla-Pozza L, Dunham MA, Reddel RR. 1997. Evidence for an alternative mechanism for maintaining telomere length in human tumors and tu-mor-derived cell lines. *Nat Med* **3**: 1271–1274. doi:10.1038/nm1197-1271

Campisi J. 2005. Senescent cells, tumor suppression, and organismal aging: good citizens, bad neighbors. *Cell* **120**: 513–522. doi:10.1016/j.cell.2005.02.003

Castillo Bosch P, Segura-Bayona S, Koole W, van Heteren JT, Dewar JM, Tijsterman M, Knipscheer P. 2014. FANCJ promotes DNA synthesis through G-quadruplex struc-tures. *EMBO J* **33**: 2521–2533. doi:10.15252/embj.201488663

Cech TR, Nakamura TM, Lingner J. 1997. Telomerase is a true reverse transcriptase. A review. *Biochemistry (Mosc)* **62**: 1202–1205.

Cesare AJ, Griffith JD. 2004. Telomeric DNA in ALT cells is characterized by free telomeric circles and heterogeneous t-loops. *Mol Cell Biol* **24:** 9948–9957. doi:10.1128/MCB.24.22.9948-9957.2004

Chaires JB, Gray RD, Dean WL, Monsen R, DeLeeuw LW, Stribinskis V, Trent JO. 2020. Human POT1 unfolds G-quadruplexes by conformational selection. *Nucleic Acids Res* **48:** 4976–4991. doi:10.1093/nar/gkaa202

Chakraborty P, Grosse F. 2011. Human DHX9 helicase preferentially unwinds RNA-containing displacement loops (R-loops) and G-quadruplexes. *DNA Repair (Amst)* **10:** 654–665. doi:10.1016/j.dnarep.2011.04.013

Chong L, van Steensel B, Broccoli D, Erdjument-Bromage H, Hanish J, Tempst P, de Lange T. 1995. A human telomeric protein. *Science* **270:** 1663–1667. doi:10.1126/science.270.5242.1663

Crabbe L, Verdun RE, Haggblom CI, Karlseder J. 2004. Defective telomere lagging strand synthesis in cells lacking WRN helicase activity. *Science* **306:** 1951–1953. doi:10.1126/science.1103619

d'Adda di Fagagna F, Reaper PM, Clay-Farrace L, Fiegler H, Carr P, Von Zglinicki T, Saretzki G, Carter NP, Jackson SP. 2003. A DNA damage checkpoint response in telomere-initiated senescence. *Nature* **426:** 194–198. doi:10.1038/nature02118

Dehé PM, Rog O, Ferreira MG, Greenwood J, Cooper JP. 2012. Taz1 enforces cell-cycle regulation of telomere synthesis. *Mol Cell* **46:** 797–808. doi:10.1016/j.molcel.2012.04.022

de Lange T. 2004. T-loops and the origin of telomeres. *Nat Rev Mol Cell Biol* **5:** 323–329. doi:10.1038/nrm1359

de Lange T. 2015. A loopy view of telomere evolution. *Front Genet* **6:** 321. doi:10.3389/fgene.2015.00321

de Lange T. 2018. Shelterin-mediated telomere protection. *Annu Rev Genet* **52:** 223–247. doi:10.1146/annurev-genet-032918-021921

De Magis A, Manzo SG, Russo M, Marinello J, Morigi R, Sordet O, Capranico G. 2019. DNA damage and genome instability by G-quadruplex ligands are mediated by R loops in human cancer cells. *Proc Natl Acad Sci* **116:** 816–825. doi:10.1073/pnas.1810409116

Denchi EL, de Lange T. 2007. Protection of telomeres through independent control of ATM and ATR by TRF2 and POT1. *Nature* **448:** 1068–1071. doi:10.1038/nature06065

Deng Z, Dheekollu J, Broccoli D, Dutta A, Lieberman PM. 2007. The origin recognition complex localizes to telomere repeats and prevents telomere-circle formation. *Curr Biol* **17:** 1989–1995. doi:10.1016/j.cub.2007.10.054

Dilley RL, Verma P, Cho NW, Winters HD, Wondisford AR, Greenberg RA. 2016. Break-induced telomere synthesis underlies alternative telomere maintenance. *Nature* **539:** 54–58. doi:10.1038/nature20099

Doksani Y. 2019. The response to DNA damage at telomeric repeats and its consequences for telomere function. *Genes (Basel)* **10:** 318. doi:10.3390/genes10040318

Doksani Y, de Lange T. 2016. Telomere-Internal double-strand breaks are repaired by homologous recombination and PARP1/Lig3-dependent end-joining. *Cell Rep* **17:** 1646–1656. doi:10.1016/j.celrep.2016.10.008

Doksani Y, Bermejo R, Fiorani S, Haber JE, Foiani M. 2009. Replicon dynamics, dormant origin firing, and terminal fork integrity after double-strand break formation. *Cell* **137:** 247–258. doi:10.1016/j.cell.2009.02.016

Doksani Y, Wu JY, de Lange T, Zhuang X. 2013. Super-resolution fluorescence imaging of telomeres reveals TRF2-dependent T-loop formation. *Cell* **155:** 345–356. doi:10.1016/j.cell.2013.09.048

Douglas ME, Diffley JFX. 2021. Budding yeast Rap1, but not telomeric DNA, is inhibitory for multiple stages of DNA replication in vitro. *Nucleic Acids Res* **49:** 5671–5683. doi:10.1093/nar/gkab416

Drosopoulos WC, Kosiyatrakul ST, Schildkraut CL. 2015. BLM helicase facilitates telomere replication during leading strand synthesis of telomeres. *J Cell Biol* **210:** 191–208. doi:10.1083/jcb.201410061

Duquette ML, Handa P, Vincent JA, Taylor AF, Maizels N. 2004. Intracellular transcription of G-rich DNAs induces formation of G-loops, novel structures containing G4 DNA. *Genes Dev* **18:** 1618–1629. doi:10.1101/gad.1200804

Edwards DN, Machwe A, Wang Z, Orren DK. 2014. Intramolecular telomeric G-quadruplexes dramatically inhibit DNA synthesis by replicative and translesion polymerases, revealing their potential to lead to genetic change. *PLoS ONE* **9:** e80664. doi:10.1371/journal.pone.0080664

Eickbush TH. 1997. Telomerase and retrotransposons: which came first? *Science* **277:** 911–912. doi:10.1126/science.277.5328.911

Fang G, Cech TR. 1993. The β subunit of *Oxytricha* telomere-binding protein promotes G-quartet formation by telomeric DNA. *Cell* **74:** 875–885. doi:10.1016/0092-8674(93)90467-5

Follonier C, Oehler J, Herrador R, Lopes M. 2013. Friedreich's ataxia-associated GAA repeats induce replication-fork reversal and unusual molecular junctions. *Nat Struct Mol Biol* **20:** 486–494. doi:10.1038/nsmb.2520

Fouché N, Özgür S, Roy D, Griffith JD. 2006. Replication fork regression in repetitive DNAs. *Nucleic Acids Res* **34:** 6044–6050. doi:10.1093/nar/gkl757

Frank-Kamenetskii MD, Mirkin SM. 1995. Triplex DNA structures. *Annu Rev Biochem* **64:** 65–95. doi:10.1146/annurev.bi.64.070195.000433

Fry M, Loeb LA. 1999. Human Werner syndrome DNA helicase unwinds tetrahelical structures of the fragile X syndrome repeat sequence $d(CGG)_n$. *J Biol Chem* **274:** 12797–12802. doi:10.1074/jbc.274.18.12797

Gehring K, Leroy JL, Guéron M. 1993. A tetrameric DNA structure with protonated cytosine·cytosine base pairs. *Nature* **363:** 561–565. doi:10.1038/363561a0

Goldfarb AM, Sasi NK, Dilgen TC, Cai SW, Myler LR, de Lange T. 2024. Tetrameric TRF2 forms t-loop to protect telomeres from ATM signaling and cNHEJ. bioRxiv doi:10.1101/2024.06.27.600884

Gomes NM, Shay JW, Wright WE. 2010. Telomere biology in Metazoa. *FEBS Lett* **584:** 3741–3751. doi:10.1016/j.febslet.2010.07.031

Gomez D, Wenner T, Brassart B, Douarre C, O'Donohue MF, El Khoury V, Shin-Ya K, Morjani H, Trentesaux C, Riou JF. 2006. Telomestatin-induced telomere uncapping is modulated by POT1 through G-overhang extension in

HT1080 human tumor cells. *J Biol Chem* **281:** 38721–38729. doi:10.1074/jbc.M605828200

Graf M, Bonetti D, Lockhart A, Serhal K, Kellner V, Maicher A, Jolivet P, Teixeira MT, Luke B. 2017. Telomere length determines TERRA and R-Loop regulation through the cell cycle. *Cell* **170:** 72–85.e14. doi:10.1016/j.cell.2017.06.006

Greider CW, Blackburn EH. 1989. A telomeric sequence in the RNA of *Tetrahymena* telomerase required for telomere repeat synthesis. *Nature* **337:** 331–337. doi:10.1038/337331a0

Griffith JD, Comeau L, Rosenfield S, Stansel RM, Bianchi A, Moss H, de Lange T. 1999. Mammalian telomeres end in a large duplex loop. *Cell* **97:** 503–514. doi:10.1016/S0092-8674(00)80760-6

Gu P, Min JN, Wang Y, Huang C, Peng T, Chai W, Chang S. 2012. CTC1 deletion results in defective telomere replication, leading to catastrophic telomere loss and stem cell exhaustion. *EMBO J* **31:** 2309–2321. doi:10.1038/emboj.2012.96

Hanish JP, Yanowitz JL, de Lange T. 1994. Stringent sequence requirements for the formation of human telomeres. *Proc Natl Acad Sci* **91:** 8861–8865. doi:10.1073/pnas.91.19.8861

Hänsel-Hertsch R, Di Antonio M, Balasubramanian S. 2017. DNA G-quadruplexes in the human genome: detection, functions and therapeutic potential. *Nat Rev Mol Cell Biol* **18:** 279–284. doi:10.1038/nrm.2017.3

Harding RM, Boyce AJ, Clegg JB. 1992. The evolution of tandemly repetitive DNA: recombination rules. *Genetics* **132:** 847–859. doi:10.1093/genetics/132.3.847

Hayflick L. 1965. The limited in vitro lifetime of human diploid cell strains. *Exp Cell Res* **37:** 614–636. doi:10.1016/0014-4827(65)90211-9

Henderson ER, Blackburn EH. 1989. An overhanging 3′ terminus is a conserved feature of telomeres. *Mol Cell Biol* **9:** 345–348.

Hockemeyer D, Collins K. 2015. Control of telomerase action at human telomeres. *Nat Struct Mol Biol* **22:** 848–852. doi:10.1038/nsmb.3083

Holohan B, Wright WE, Shay JW. 2014. Telomeropathies: an emerging spectrum disorder. *J Cell B* **205:** 289–299. doi:10.1083/jcb.201401012

Huber MD, Lee DC, Maizels N. 2002. G4 DNA unwinding by BLM and Sgs1p: substrate specificity and substrate-specific inhibition. *Nucleic Acids Res* **30:** 3954–3961. doi:10.1093/nar/gkf530

Huda A, Arakawa H, Mazzucco G, Galli M, Petrocelli V, Casola S, Chen L, Doksani Y. 2023. The telomerase reverse transcriptase elongates reversed replication forks at telomeric repeats. *Sci Adv* **9:** eadf2011. doi:10.1126/sciadv.adf2011

Hwang H, Buncher N, Opresko PL, Myong S. 2012. POT1-TPP1 regulates telomeric overhang structural dynamics. *Structure* **20:** 1872–1880. doi:10.1016/j.str.2012.08.018

Jansson LI, Hentschel J, Parks JW, Chang TR, Lu C, Baral R, Bagshaw CR, Stone MD. 2019. Telomere DNA G-quadruplex folding within actively extending human telomerase. *Proc Natl Acad Sci* **116:** 9350–9359. doi:10.1073/pnas.1814777116

Kaminski N, Wondisford AR, Kwon Y, Lynskey ML, Bhargava R, Barroso-González J, García-Expósito L, He B, Xu M, Mellacheruvu D, et al. 2022. RAD51AP1 regulates ALT-HDR through chromatin-directed homeostasis of TERRA. *Mol Cell* **82:** 4001–4017.e7. doi:10.1016/j.molcel.2022.09.025

Kang CH, Berger I, Lockshin C, Ratliff R, Moyzis R, Rich A. 1994. Crystal structure of intercalated four-stranded d(C3T) at 1.4 A resolution. *Proc Natl Acad Sci* **91:** 11636–11640. doi:10.1073/pnas.91.24.11636

Kar A, Willcox S, Griffith JD. 2016. Transcription of telomeric DNA leads to high levels of homologous recombination and t-loops. *Nucleic Acids Res* **44:** 9369–9380. doi:10.1093/nar/gkw779

Karlseder J, Hoke K, Mirzoeva OK, Bakkenist C, Kastan MB, Petrini JH, de Lange TL. 2004. The telomeric protein TRF2 binds the ATM kinase and can inhibit the ATM-dependent DNA damage response. *PLoS Biol* **2:** E240. doi:10.1371/journal.pbio.0020240

Khristich AN, Mirkin SM. 2020. On the wrong DNA track: molecular mechanisms of repeat-mediated genome instability. *J Biol Chem* **295:** 4134–4170. doi:10.1074/jbc.REV119.007678

Kinzig CG, Zakusilo G, Takai KK, Myler LR, de Lange T. 2024. ATR blocks telomerase from converting DNA breaks into telomeres. *Science* **383:** 763–770. doi:10.1126/science.adg3224

Kipling D, Cooke HJ. 1990. Hypervariable ultra-long telomeres in mice. *Nature* **347:** 400–402. doi:10.1038/347400a0

Klein HL. 1995. Genetic control of intrachromosomal recombination. *Bioessays* **17:** 147–159. doi:10.1002/bies.950170210

Knauert MP, Glazer PM. 2001. Triplex forming oligonucleotides: sequence-specific tools for gene targeting. *Hum Mol Genet* **10:** 2243–2251. doi:10.1093/hmg/10.20.2243

Kohwi Y, Kohwi-Shigematsu T. 1988. Magnesium ion-dependent triple-helix structure formed by homopurine-homopyrimidine sequences in supercoiled plasmid DNA. *Proc Natl Acad Sci* **85:** 3781–3785. doi:10.1073/pnas.85.11.3781

Kotsantis P, Segura-Bayona S, Margalef P, Marzec P, Ruis P, Hewitt G, Bellelli R, Patel H, Goldstone R, Poetsch AR. 2020. RTEL1 regulates G4/R-loops to avert replication-transcription collisions. *Cell Rep* **33:** 108546. doi:10.1016/j.celrep.2020.108546

Kumar C, Batra S, Griffith JD, Remus D. 2021. The interplay of RNA:DNA hybrid structure and G-quadruplexes determines the outcome of R-loop-replisome collisions. *eLife* **10:** e72286. doi:10.7554/eLife.72286

Lam YC, Akhter S, Gu P, Ye J, Poulet A, Giraud-Panis MJ, Bailey SM, Gilson E, Legerski RJ, Chang S. 2010. SNMIB/apollo protects leading-strand telomeres against NHEJ-mediated repair. *EMBO J* **29:** 2230–2241. doi:10.1038/emboj.2010.58

Law MJ, Lower KM, Voon HP, Hughes JR, Garrick D, Viprakasit V, Mitson M, De Gobbi M, Marra M, Morris A, et al. 2010. ATR-X syndrome protein targets tandem repeats and influences allele-specific expression in a size-dependent manner. *Cell* **143:** 367–378. doi:10.1016/j.cell.2010.09.023

Cite this article as *Cold Spring Harb Perspect Biol* doi: 10.1101/cshperspect.a041694

Lazzerini-Denchi E, Sfeir A. 2016. Stop pulling my strings—what telomeres taught us about the DNA damage response. *Nat Rev Mol Cell Biol* **17:** 364–378. doi:10.1038/nrm.2016.43

Lazzerini Denchi E, Celli G, de Lange T. 2006. Hepatocytes with extensive telomere deprotection and fusion remain viable and regenerate liver mass through endoreduplication. *Genes Dev* **20:** 2648–2653. doi:10.1101/gad.1453606

Leach DR. 1994. Long DNA palindromes, cruciform structures, genetic instability and secondary structure repair. *Bioessays* **16:** 893–900. doi:10.1002/bies.950161207

Lerner LK, Sale JE. 2019. Replication of G Quadruplex DNA. *Genes (Basel)* **10:** 95. doi:10.3390/genes10020095

Li B, Oestreich S, de Lange T. 2000. Identification of human Rap1: implications for telomere evolution. *Cell* **101:** 471–483. doi:10.1016/S0092-8674(00)80858-2

Li JS, Miralles Fusté J, Simavorian T, Bartocci C, Tsai J, Karlseder J, Lazzerini Denchi E. 2017. TZAP: a telomere-associated protein involved in telomere length control. *Science* **355:** 638–641. doi:10.1126/science.aah6752

Li N, Wang J, Ma K, Liang L, Mi L, Huang W, Ma X, Wang Z, Zheng W, Xu L, et al. 2019. The dynamics of forming a triplex in an artificial telomere inferred by DNA mechanics. *Nucleic Acids Res* **47:** e86. doi:10.1093/nar/gkz464

Lim CJ, Cech TR. 2021. Shaping human telomeres: from shelterin and CST complexes to telomeric chromatin organization. *Nat Rev Mol Cell Biol* **22:** 283–298. doi:10.1038/s41580-021-00328-y

Lingner J, Cooper JP, Cech TR. 1995. Telomerase and DNA end replication: no longer a lagging strand problem? *Science* **269:** 1533–1534. doi:10.1126/science.7545310

Lipps HJ. 1980. In vitro aggregation of the gene-sized DNA molecules of the ciliate *Stylonychia mytilus*. *Proc Natl Acad Sci* **77:** 4104–4107. doi:10.1073/pnas.77.7.4104

Liu W, Saito Y, Jackson J, Bhowmick R, Kanemaki MT, Vindigni A, Cortez D. 2023. RAD51 bypasses the CMG helicase to promote replication fork reversal. *Science* **380:** 382–387. doi:10.1126/science.add7328

London TB, Barber LJ, Mosedale G, Kelly GP, Balasubramanian S, Hickson ID, Boulton SJ, Hiom K. 2008. FANCJ is a structure-specific DNA helicase associated with the maintenance of genomic G/C tracts. *J Biol Chem* **283:** 36132–36139. doi:10.1074/jbc.M808152200

Lovett ST, Drapkin PT, Sutera VA Jr, Gluckman-Peskind TJ. 1993. A sister-strand exchange mechanism for recA-independent deletion of repeated DNA sequences in *Escherichia coli*. *Genetics* **135:** 631–642. doi:10.1093/genetics/135.3.631

Makarov VL, Hirose Y, Langmore JP. 1997. Long G tails at both ends of human chromosomes suggest a C strand degradation mechanism for telomere shortening. *Cell* **88:** 657–666. doi:10.1016/S0092-8674(00)81908-X

Margalef P, Kotsantis P, Borel V, Bellelli R, Panier S, Boulton SJ. 2018. Stabilization of reversed replication forks by telomerase drives telomere catastrophe. *Cell* **172:** 439–453.e14. doi:10.1016/j.cell.2017.11.047

Markiewicz-Potoczny M, Lobanova A, Loeb AM, Kirak O, Olbrich T, Ruiz S, Lazzerini Denchi E. 2021. TRF2-mediated telomere protection is dispensable in pluripotent stem cells. *Nature* **589:** 110–115. doi:10.1038/s41586-020-2959-4

Martínez P, Thanasoula M, Muñoz P, Liao C, Tejera A, McNees C, Flores JM, Fernández-Capetillo O, Tarsounas M, Blasco MA. 2009. Increased telomere fragility and fusions resulting from *TRF1* deficiency lead to degenerative pathologies and increased cancer in mice. *Genes Dev* **23:** 2060–2075. doi:10.1101/gad.543509

Masuda-Sasa T, Polaczek P, Peng XP, Chen L, Campbell JL. 2008. Processing of G4 DNA by Dna2 helicase/nuclease and replication protein A (RPA) provides insights into the mechanism of Dna2/RPA substrate recognition. *J Biol Chem* **283:** 24359–24373. doi:10.1074/jbc.M802244200

Matmati S, Lambert S, Géli V, Coulon S. 2020. Telomerase repairs collapsed replication forks at telomeres. *Cell Rep* **30:** 3312–3322.e3. doi:10.1016/j.celrep.2020.02.065

Mazzucco G, Huda A, Galli M, Piccini D, Giannattasio M, Pessina F, Doksani Y. 2020. Telomere damage induces internal loops that generate telomeric circles. *Nat Commun* **11:** 5297. doi:10.1038/s41467-020-19139-4

Miller KM, Rog O, Cooper JP. 2006. Semi-conservative DNA replication through telomeres requires Taz1. *Nature* **440:** 824–828. doi:10.1038/nature04638

Mirkin SM, Lyamichev VI, Drushlyak KN, Dobrynin VN, Filippov SA, Frank-Kamenetskii MD. 1987. DNA H form requires a homopurine-homopyrimidine mirror repeat. *Nature* **330:** 495–497. doi:10.1038/330495a0

Mohaghegh P, Karow JK, Brosh RM Jr, Bohr VA, Hickson ID. 2001. The Bloom's and Werner's syndrome proteins are DNA structure-specific helicases. *Nucleic Acids Res* **29:** 2843–2849. doi:10.1093/nar/29.13.2843

Moye AL, Porter KC, Cohen SB, Phan T, Zyner KG, Sasaki N, Lovrecz GO, Beck JL, Bryan TM. 2015. Telomeric G-quadruplexes are a substrate and site of localization for human telomerase. *Nat Commun* **6:** 7643. doi:10.1038/ncomms8643

Moyzis RK, Buckingham JM, Cram LS, Dani M, Deaven LL, Jones MD, Meyne J, Ratliff RL, Wu JR. 1988. A highly conserved repetitive DNA sequence, (TTAGGG)$_n$, present at the telomeres of human chromosomes. *Proc Natl Acad Sci* **85:** 6622–6626. doi:10.1073/pnas.85.18.6622

Müller S, Kumari S, Rodriguez R, Balasubramanian S. 2010. Small-molecule-mediated G-quadruplex isolation from human cells. *Nat Chem* **2:** 1095–1098. doi:10.1038/nchem.842

Myler LR, Kinzig CG, Sasi NK, Zakusilo G, Cai SW, de Lange T. 2021. The evolution of metazoan shelterin. *Genes Dev* **35:** 1625–1641. doi:10.1101/gad.348835.121

Myler LR, Toia B, Vaughan CK, Takai K, Matei AM, Wu P, Paull TT, de Lange T, Lottersberger F. 2023. DNA-PK and the TRF2 iDDR inhibit MRN-initiated resection at leading-end telomeres. *Nat Struct Mol Biol* **30:** 1346–1356. doi:10.1038/s41594-023-01072-x

Nabetani A, Ishikawa F. 2009. Unusual telomeric DNAs in human telomerase-negative immortalized cells. *Mol Cell Biol* **29:** 703–713. doi:10.1128/MCB.00603-08

Nakamura TM, Morin GB, Chapman KB, Weinrich SL, Andrews WH, Lingner J, Harley CB, Cech TR. 1997. Telomerase catalytic subunit homologs from fission yeast and human. *Science* **277:** 955–959. doi:10.1126/science.277.5328.955

Neelsen KJ, Lopes M. 2015. Replication fork reversal in eukaryotes: from dead end to dynamic response. *Nat Rev Mol Cell Biol* **16:** 207–220. doi:10.1038/nrm3935

Neidle S, Parkinson GN. 2003. The structure of telomeric DNA. *Curr Opin Struct Biol* **13:** 275–283. doi:10.1016/S0959-440X(03)00072-1

Nikitina T, Woodcock CL. 2004. Closed chromatin loops at the ends of chromosomes. *J Cell Biol* **166:** 161–165. doi:10.1083/jcb.200403118

Oganesian L, Moon IK, Bryan TM, Jarstfer MB. 2006. Extension of G-quadruplex DNA by ciliate telomerase. *EMBO J* **25:** 1148–1159. doi:10.1038/sj.emboj.7601006

Ohki R, Ishikawa F. 2004. Telomere-bound TRF1 and TRF2 stall the replication fork at telomeric repeats. *Nucleic Acids Res* **32:** 1627–1637. doi:10.1093/nar/gkh309

Okamoto K, Bartocci C, Ouzounov I, Diedrich JK, Yates JR III, Denchi EL. 2013. A two-step mechanism for TRF2-mediated chromosome-end protection. *Nature* **494:** 502–505. doi:10.1038/nature11873

Olovnikov AM. 1973. A theory of marginotomy. The incomplete copying of template margin in enzymic synthesis of polynucleotides and biological significance of the phenomenon. *J Theor Biol* **41:** 181–190. doi:10.1016/0022-5193(73)90198-7

O'Sullivan RJ, Karlseder J. 2010. Telomeres: protecting chromosomes against genome instability. *Nat Rev Mol Cell Biol* **11:** 171–181. doi:10.1038/nrm2848

O'Sullivan RJ, Arnoult N, Lackner DH, Oganesian L, Haggblom C, Corpet A, Almouzni G, Karlseder J. 2014. Rapid induction of alternative lengthening of telomeres by depletion of the histone chaperone ASF1. *Nat Struct Mol Biol* **21:** 167–174. doi:10.1038/nsmb.2754

Paeschke K, Simonsson T, Postberg J, Rhodes D, Lipps HJ. 2005. Telomere end-binding proteins control the formation of G-quadruplex DNA structures in vivo. *Nat Struct Mol Biol* **12:** 847–854. doi:10.1038/nsmb982

Paeschke K, Capra JA, Zakian VA. 2011. DNA replication through G-quadruplex motifs is promoted by the *Saccharomyces cerevisiae* Pif1 DNA helicase. *Cell* **145:** 678–691. doi:10.1016/j.cell.2011.04.015

Paeschke K, Bochman ML, Garcia PD, Cejka P, Friedman KL, Kowalczykowski SC, Zakian VA. 2013. Pif1 family helicases suppress genome instability at G-quadruplex motifs. *Nature* **497:** 458–462. doi:10.1038/nature12149

Paudel BP, Moye AL, Abou Assi H, El-Khoury R, Cohen SB, Holien JK, Birrento ML, Samosorn S, Intharapichai K, Tomlinson CG, et al. 2020. A mechanism for the extension and unfolding of parallel telomeric G-quadruplexes by human telomerase at single-molecule resolution. *eLife* **9:** e56428. doi:10.7554/eLife.56428

Pfeiffer V, Crittin J, Grolimund L, Lingner J. 2013. The THO complex component Thp2 counteracts telomeric R-loops and telomere shortening. *EMBO J* **32:** 2861–2871. doi:10.1038/emboj.2013.217

Phan AT, Mergny JL. 2002. Human telomeric DNA: G-quadruplex, I-motif and Watson–Crick double helix. *Nucleic Acids Res* **30:** 4618–4625. doi:10.1093/nar/gkf597

Phan AT, Guéron M, Leroy JL. 2000. The solution structure and internal motions of a fragment of the cytidine-rich strand of the human telomere. *J Mol Biol* **299:** 123–144. doi:10.1006/jmbi.2000.3613

Pickett HA, Cesare AJ, Johnston RL, Neumann AA, Reddel RR. 2009. Control of telomere length by a trimming mechanism that involves generation of t-circles. *EMBO J* **28:** 799–809. doi:10.1038/emboj.2009.42

Popuri V, Bachrati CZ, Muzzolini L, Mosedale G, Costantini S, Giacomini E, Hickson ID, Vindigni A. 2008. The human RecQ helicases, BLM and RECQ1, display distinct DNA substrate specificities. *J Biol Chem* **283:** 17766–17776. doi:10.1074/jbc.M709749200

Porreca RM, Herrera-Moyano E, Skourti E, Law PP, Gonzalez Franco R, Montoya A, Faull P, Kramer H, Vannier JB. 2020. TRF1 averts chromatin remodelling, recombination and replication dependent–break induced replication at mouse telomeres. *eLife* **9:** e49817. doi:10.7554/eLife.49817

Proctor CJ, Kirkwood TB. 2003. Modelling cellular senescence as a result of telomere state. *Aging Cell* **2:** 151–157. doi:10.1046/j.1474-9728.2003.00050.x

Quinet A, Lemaçon D, Vindigni A. 2017. Replication fork reversal: players and guardians. *Mol Cell* **68:** 830–833. doi:10.1016/j.molcel.2017.11.022

Radchenko EA, Aksenova AY, Volkov KV, Shishkin AA, Pavlov YI, Mirkin SM. 2022. Partners in crime: Tbf1 and Vid22 promote expansions of long human telomeric repeats at an interstitial chromosome position in yeast. *PNAS Nexus* **1:** pgac080. doi:10.1093/pnasnexus/pgac080

Regairaz M, Zhang YW, Fu H, Agama KK, Tata N, Agrawal S, Aladjem MI, Pommier Y. 2011. Mus81-mediated DNA cleavage resolves replication forks stalled by topoisomerase I-DNA complexes. *J Cell Biol* **195:** 739–749. doi:10.1083/jcb.201104003

Ribes-Zamora A, Indiviglio SM, Mihalek I, Williams CL, Bertuch AA. 2013. TRF2 interaction with Ku heterotetramerization interface gives insight into c-NHEJ prevention at human telomeres. *Cell Rep* **5:** 194–206. doi:10.1016/j.celrep.2013.08.040

Richard GF, Kerrest A, Dujon B. 2008. Comparative genomics and molecular dynamics of DNA repeats in eukaryotes. *Microbiol Mol Biol Rev* **72:** 686–727. doi:10.1128/MMBR.00011-08

Rizzo A, Salvati E, Porru M, D'Angelo C, Stevens MF, D'Incalci M, Leonetti C, Gilson E, Zupi G, Biroccio A. 2009. Stabilization of quadruplex DNA perturbs telomere replication leading to the activation of an ATR-dependent ATM signaling pathway. *Nucleic Acids Res* **37:** 5353–5364. doi:10.1093/nar/gkp582

Rodriguez R, Miller KM, Forment JV, Bradshaw CR, Nikan M, Britton S, Oelschlaegel T, Xhemalce B, Balasubramanian S, Jackson SP. 2012. Small-molecule-induced DNA damage identifies alternative DNA structures in human genes. *Nat Chem Biol* **8:** 301–310. doi:10.1038/nchembio.780

Rubelj I, Vondracek Z. 1999. Stochastic mechanism of cellular aging—abrupt telomere shortening as a model for stochastic nature of cellular aging. *J Theor Biol* **197:** 425–438. doi:10.1006/jtbi.1998.0886

Ruis P, Van Ly D, Borel V, Kafer GR, McCarthy A, Howell S, Blassberg R, Snijders AP, Briscoe J, Niakan KK, et al. 2021. TRF2-independent chromosome end protection during pluripotency. *Nature* **589:** 103–109. doi:10.1038/s41586-020-2960-y

Saini N, Ramakrishnan S, Elango R, Ayyar S, Zhang Y, Deem A, Ira G, Haber JE, Lobachev KS, Malkova A. 2013. Migrating bubble during break-induced replication drives conservative DNA synthesis. *Nature* **502:** 389–392. doi:10 .1038/nature12584

Sakellariou D, Bak ST, Isik E, Barroso SI, Porro A, Aguilera A, Bartek J, Janscak P, Peña-Diaz J. 2022. Mutsβ regulates G4-associated telomeric R-loops to maintain telomere integrity in ALT cancer cells. *Cell Rep* **39:** 110602. doi:10.1016/j.celrep.2022.110602

Salas TR, Petruseva I, Lavrik O, Bourdoncle A, Mergny JL, Favre A, Saintomé C. 2006. Human replication protein A unfolds telomeric G-quadruplexes. *Nucleic Acids Res* **34:** 4857–4865. doi:10.1093/nar/gkl564

Sanders CM. 2010. Human Pif1 helicase is a G-quadruplex DNA-binding protein with G-quadruplex DNA-unwinding activity. *Biochem J* **430:** 119–128. doi:10.1042/BJ20 100612

Sarek G, Vannier JB, Panier S, Petrini JHJ, Boulton SJ. 2015. TRF2 recruits RTEL1 to telomeres in S phase to promote t-loop unwinding. *Mol Cell* **57:** 622–635. doi:10.1016/j .molcel.2014.12.024

Sarek G, Kotsantis P, Ruis P, Van Ly D, Margalef P, Borel V, Zheng XF, Flynn HR, Snijders AP, Chowdhury D, et al. 2019. CDK phosphorylation of TRF2 controls t-loop dynamics during the cell cycle. *Nature* **575:** 523–527. doi:10 .1038/s41586-019-1744-8

Schiavone D, Guilbaud G, Murat P, Papadopoulou C, Sarkies P, Prioleau MN, Balasubramanian S, Sale JE. 2014. Determinants of G quadruplex-induced epigenetic instability in REV1-deficient cells. *EMBO J* **33:** 2507–2520. doi:10 .15252/embj.201488398

Schmutz I, Timashev L, Xie W, Patel DJ, de Lange T. 2017. TRF2 binds branched DNA to safeguard telomere integrity. *Nat Struct Mol Biol* **24:** 734–742. doi:10.1038/nsmb .3451

Sen D, Gilbert W. 1988. Formation of parallel four-stranded complexes by guanine-rich motifs in DNA and its implications for meiosis. *Nature* **334:** 364–366. doi:10.1038/ 334364a0

Sepsiova R, Necasova I, Willcox S, Prochazkova K, Gorilak P, Nosek J, Hofr C, Griffith JD, Tomaska L. 2016. Evolution of telomeres in Schizosaccharomyces pombe and its possible relationship to the diversification of telomere binding proteins. *PLoS ONE* **11:** e0154225. doi:10.1371/jour nal.pone.0154225

Sfeir A, de Lange T. 2012. Removal of shelterin reveals the telomere end-protection problem. *Science* **336:** 593–597. doi:10.1126/science.1218498

Sfeir A, Kosiyatrakul ST, Hockemeyer D, MacRae SL, Karlseder J, Schildkraut CL, de Lange T. 2009. Mammalian telomeres resemble fragile sites and require TRF1 for efficient replication. *Cell* **138:** 90–103. doi:10.1016/j.cell .2009.06.021

Silva B, Arora R, Bione S, Azzalin CM. 2021. TERRA transcription destabilizes telomere integrity to initiate break-induced replication in human ALT cells. *Nat Commun* **12:** 3760. doi:10.1038/s41467-021-24097-6

Sparks JL, Chistol G, Gao AO, Räschle M, Larsen NB, Mann M, Duxin JP, Walter JC. 2019. The CMG helicase bypasses DNA-Protein cross-links to facilitate their repair. *Cell* **176:** 167–181.e21. doi:10.1016/j.cell.2018.10.053

Stansel RM, de Lange T, Griffith JD. 2001. T-loop assembly in vitro involves binding of TRF2 near the 3′ telomeric

overhang. *EMBO J* **20:** 5532–5540. doi:10.1093/emboj/20 .19.5532

Sun H, Karow JK, Hickson ID, Maizels N. 1998. The Bloom's syndrome helicase unwinds G4 DNA. *J Biol Chem* **273:** 27587–27592. doi:10.1074/jbc.273.42.27587

Sundquist WI, Klug A. 1989. Telomeric DNA dimerizes by formation of guanine tetrads between hairpin loops. *Nature* **342:** 825–829. doi:10.1038/342825a0

Takai H, Aria V, Borges P, Yeeles JTP, de Lange T. 2024. CST-polymerase α-primase solves a second telomere end-replication problem. *Nature* **627:** 664–670. doi:10 .1038/s41586-024-07137-1

Timashev LA, De Lange T. 2020. Characterization of t-loop formation by TRF2. *Nucleus* **11:** 164–177. doi:10.1080/ 19491034.2020.1783782

Tomaska L, Nosek J, Kramara J, Griffith JD. 2009. Telomeric circles: universal players in telomere maintenance? *Nat Struct Mol Biol* **16:** 1010–1015. doi:10.1038/nsmb.1660

Tomaska L, Nosek J, Kar A, Willcox S, Griffith JD. 2019. A new view of the T-loop junction: implications for self-primed telomere extension, expansion of disease-related nucleotide repeat blocks, and telomere evolution. *Front Genet* **10:** 792. doi:10.3389/fgene.2019.00792

Tomáška L, Cesare AJ, AlTurki TM, Griffith JD. 2020. Twenty years of t-loops: a case study for the importance of collaboration in molecular biology. *DNA Repair (Amst)* **94:** 102901. doi:10.1016/j.dnarep.2020.102901

Tran PL, Mergny JL, Alberti P. 2011. Stability of telomeric G-quadruplexes. *Nucleic Acids Res* **39:** 3282–3294. doi:10 .1093/nar/gkq1292

Van Ly D, Low RRJ, Frölich S, Bartolec TK, Kafer GR, Pickett HA, Gaus K, Cesare AJ. 2018. Telomere loop dynamics in chromosome end protection. *Mol Cell* **71:** 510–525.e6. doi:10.1016/j.molcel.2018.06.025

Vannier JB, Pavicic-Kaltenbrunner V, Petalcorin MI, Ding H, Boulton SJ. 2012. RTEL1 dismantles T loops and counteracts telomeric G4-DNA to maintain telomere integrity. *Cell* **149:** 795–806. doi:10.1016/j.cell.2012.03.030

Vannier JB, Sandhu S, Petalcorin MI, Wu X, Nabi Z, Ding H, Boulton SJ. 2013. RTEL1 is a replisome-associated helicase that promotes telomere and genome-wide replication. *Science* **342:** 239–242. doi:10.1126/science.1241779

van Steensel B, Smogorzewska A, de Lange T. 1998. TRF2 protects human telomeres from end-to-end fusions. *Cell* **92:** 401–413. doi:10.1016/S0092-8674(00)80932-0

Veselkov AG, Malkov VA, Frank-Kamenetskll MD, Dobrynin VN. 1993. Triplex model of chromosome ends. *Nature* **364:** 496. doi:10.1038/364496a0

Wakai M, Abe S, Kazuki Y, Oshimura M, Ishikawa F. 2014. A human artificial chromosome recapitulates the metabolism of native telomeres in mammalian cells. *PLoS ONE* **9:** e88530. doi:10.1371/journal.pone.0088530

Wang RC, Smogorzewska A, de Lange T. 2004. Homologous recombination generates T-loop-sized deletions at human telomeres. *Cell* **119:** 355–368. doi:10.1016/j.cell .2004.10.011

Watson JD. 1972. Origin of concatemeric T7 DNA. *Nat New Biol* **239:** 197–201. doi:10.1038/newbio239197a0

Wellinger RJ, Zakian VA. 2012. Everything you ever wanted to know about *Saccharomyces cerevisiae* telomeres: beginning to end. *Genetics* **191**: 1073–1105. doi:10.1534/genet ics.111.137851

Williamson JR, Raghuraman MK, Cech TR. 1989. Monovalent cation-induced structure of telomeric DNA: the G-quartet model. *Cell* **59**: 871–880. doi:10.1016/0092-8674 (89)90610-7

Wilson MA, Kwon Y, Xu Y, Chung WH, Chi P, Niu H, Mayle R, Chen X, Malkova A, Sung P, et al. 2013. Pif1 helicase and Polδ promote recombination-coupled DNA synthesis via bubble migration. *Nature* **502**: 393–396. doi:10 .1038/nature12585

Wright WE, Shay JW. 1995. Time, telomeres and tumours: is cellular senescence more than an anticancer mechanism? *Trends Cell Biol* **5**: 293–297. doi:10.1016/S0962-8924(00) 89044-3

Wu L, Hickson ID. 2003. The Bloom's syndrome helicase suppresses crossing over during homologous recombination. *Nature* **426**: 870–874. doi:10.1038/nature02253

Wu P, van Overbeek M, Rooney S, de Lange T. 2010. Apollo contributes to G overhang maintenance and protects leading-end telomeres. *Mol Cell* **39**: 606–617. doi:10 .1016/j.molcel.2010.06.031

Wu Y, Sommers JA, Khan I, de Winter JP, Brosh RM Jr. 2012. Biochemical characterization of Warsaw breakage syndrome helicase. *J Biol Chem* **287**: 1007–1021. doi:10 .1074/jbc.M111.276022

Wu W, Bhowmick R, Vogel I, Özer Ö, Ghisays F, Thakur RS, Sanchez de Leon E, Richter PH, Ren L, Petrini JH, et al. 2020. RTEL1 suppresses G-quadruplex-associated R-loops at difficult-to-replicate loci in the human genome. *Nat Struct Mol Biol* **27**: 424–437. doi:10.1038/s41594-020-0408-6

Yadav T, Zhang JM, Ouyang J, Leung W, Simoneau A, Zou L. 2022. TERRA and RAD51AP1 promote alternative lengthening of telomeres through an R- to D-loop switch. *Mol Cell* **82**: 3985–4000.e4. doi:10.1016/j.molcel.2022.09 .026

Yang Z, Takai KK, Lovejoy CA, de Lange T. 2020. Break-induced replication promotes fragile telomere formation. *Genes Dev* **34**: 1392–1405. doi:10.1101/gad.328575.119

Zahler AM, Williamson JR, Cech TR, Prescott DM. 1991. Inhibition of telomerase by G-quartet DNA structures. *Nature* **350**: 718–720. doi:10.1038/350718a0

Zaug AJ, Podell ER, Cech TR. 2005. Human POT1 disrupts telomeric G-quadruplexes allowing telomerase extension in vitro. *Proc Natl Acad Sci* **102**: 10864–10869. doi:10 .1073/pnas.0504744102

Zhang W, Durocher D. 2010. De novo telomere formation is suppressed by the Mec1-dependent inhibition of Cdc13 accumulation at DNA breaks. *Genes Dev* **24**: 502–515. doi:10.1101/gad.1869110

Zhang M, Wang B, Li T, Liu R, Xiao Y, Geng X, Li G, Liu Q, Price CM, Liu Y, et al. 2019a. Mammalian CST averts replication failure by preventing G-quadruplex accumulation. *Nucleic Acids Res* **47**: 5243–5259. doi:10.1093/nar/ gkz264

Zhang T, Zhang Z, Shengzhao G, Li X, Liu H, Zhao Y. 2019b. Strand break-induced replication fork collapse leads to C-circles, C-overhangs and telomeric recombination. *PLoS Genet* **15**: e1007925. doi:10.1371/journal.pgen.1007925

Zimmermann M, Kibe T, Kabir S, de Lange T. 2014. TRF1 negotiates TTAGGG repeat-associated replication problems by recruiting the BLM helicase and the TPP1/POT1 repressor of ATR signaling. *Genes Dev* **28**: 2477–2491. doi:10.1101/gad.251611.114

# Oxidative Stress and DNA Damage at Telomeres

Patricia L. Opresko,[1,2,3] Samantha L. Sanford,[3] and Mariarosaria De Rosa[3]

[1]Department of Pharmacology and Chemical Biology, University of Pittsburgh School of Medicine, Pittsburgh, Pennsylvania 15261, USA

[2]Department of Environmental and Occupational Health, University of Pittsburgh School of Public Health, Pittsburgh, Pennsylvania 15261, USA

[3]UPMC Hillman Cancer Center at the University of Pittsburgh, Pittsburgh, Pennsylvania 15232, USA

*Correspondence:* plo4@pitt.edu

Oxidative stress is associated with increasing telomere shortening and telomere dysfunction, as well as with numerous pathologies in humans, including inflammatory diseases and cancer. Critically short and dysfunctional telomeres lose their ability to protect chromosome ends, which triggers irreversible growth arrest, termed senescence, or genomic instability. Telomeres are highly sensitive to damage from reactive oxygen species, which increase under conditions of oxidative stress. This work covers the evidence that oxidative damage to telomeric DNA alters telomere maintenance by various mechanisms and describes the DNA repair pathways important for preserving telomere function under oxidative stress conditions.

## OXIDATIVE STRESS AND TELOMERE MAINTENANCE

Oxidative stress, a condition caused by excess reactive oxygen species (ROS), is associated with DNA damage and accelerated telomere shortening and dysfunction. In aerobic organisms, oxygen ($O_2$) is vital for chemical reactions that fuel energy production and generate the metabolites required for life. However, $O_2$ consumption can give rise to ROS, which may damage cellular components, including DNA, because they are highly reactive. Oxidative stress arises from an imbalance between ROS production and the cellular antioxidant defense mechanisms that neutralize or detoxify ROS to prevent damage. Common ROS include superoxide radicals, hydroxyl radicals ($\cdot OH$), singlet oxygen ($^1O_2$), nitric oxide, and hypochlorous acid (Fig. 1; Cadet and Wagner 2013). Endogenous sources of ROS include inflammation, immune cells responding to infection or injury, and $O_2$ metabolism in mitochondria. Environmental sources of ROS include pollution, cigarette smoke, diet, and pesticides (Samet and Wages 2018; Sharifi-Rad et al. 2020). These exogenous exposures can also increase ROS through inflammation and by damaging mitochondria. Inflammatory conditions, and many environmental exposures linked to oxidative stress, are associated with increased telomere shortening or dysfunction (Fig. 1; Valdes et al. 2005; Martens and Nawrot 2016; Zhang et al. 2016; Reichert and Stier 2017; Assavanopakun et al. 2022; Armstrong and Boonekamp 2023; D'Angelo 2023; Zuo et al. 2024). ROS interaction with

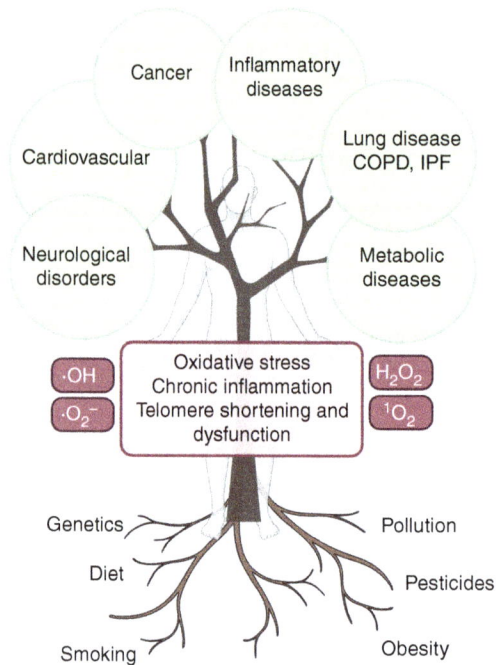

**Figure 1.** Schematic summarizing diseases, environmental exposures, and lifestyle factors associated with oxidative stress, chronic inflammation, and telomere shortening or dysfunction in humans. Oxidative stress is characterized by elevated reactive oxygen ($O_2$) species, including hydroxyl radical ($\cdot$OH), hydrogen peroxide ($H_2O_2$), superoxide ($\cdot O_2^-$), and singlet oxygen ($^1O_2$). Telomere lengths are measured either in white peripheral blood cells or in the affected/inflamed tissues, compared to unaffected individuals or tissues. Dysfunctional telomeres are detected by the localization of DNA damage response factors phosphorylated histone H2AX or 53BP1 at telomeres. (COPD) Chronic obstructive pulmonary disease, (IPF) idiopathic pulmonary fibrosis. (Figure generated with BioRender; www.biorender.com.)

DNA, including telomeres, produces chemical modifications (termed lesions) that drive genomic instability and contribute to disease (Markkanen 2017). Oxidative stress promotes numerous pathologies and diseases in humans, including inflammatory, neurological, pulmonary, cardiovascular, and metabolic (i.e., type II diabetes), as well as cancer (for reviews, see Lonkar and Dedon 2011; Malinin et al. 2011; Forman and Zhang 2021). Some studies show that these ROS-linked diseases are also associated with shortened telomeres or increased telomere dysfunction in humans (discussed below) (Fig. 1; Barnes et al. 2019; Rossiello et al. 2022). Therefore, environmental factors and diseases linked to oxidative stress are frequently associated with impaired telomere maintenance.

Telomeres shorten naturally with cell division due to the end-replication problem, but oxidative stress can influence the rate at which telomeres shorten and become dysfunctional (von Zglinicki 2002; Barnes et al. 2019). Changes in telomere length or structure can impair their function in protecting chromosome ends, causing the ends to be falsely recognized as chromosome breaks, which activates the DNA damage response (DDR). Dysfunctional telomeres are marked by the colocalization of DDR factors, termed DDR$^+$ telomeres or telomere dysfunction-induced foci (TIF). DDR activation signals exit from the cell cycle and triggers growth arrest, termed senescence, which can be irreversible (d'Adda di Fagagna et al. 2003; Takai et al. 2003). Evidence that human diseases characterized by oxidative stress, including those caused by chronic inflammation, are associated with impaired telomere maintenance derives from studies measuring average telomere lengths or DDR$^+$ telomeres in either peripheral blood mononuclear cells or affected tissues. For example, biopsies of inflamed colons from ulcerative colitis patients and precancerous lesions or tumors show shortened telomeres compared to normal adjacent tissues (O'Sullivan et al. 2002; Zhou et al. 2012; Ertunc et al. 2024). Inflamed livers from patients with chronic hepatitis and liver cirrhosis, as well as atherosclerotic lesions, also exhibit shorter telomeres, compared to unaffected tissues (Aikata et al. 2000; Nzietchueng et al. 2011; Rey et al. 2017). Affected tissues from patients with chronic obstructive pulmonary disease, fibrosis bronchiectasis, and metabolic liver diseases all show increased dysfunctional DDR$^+$ telomeres, indicated by phosphorylated histone H2AX ($\gamma$H2AX) at telomeres (Rossiello et al. 2022). Notably, an increase in DDR$^+$ telomeres is not always associated with increased telomere shortening.

Cite this article as *Cold Spring Harb Perspect Biol* doi: 10.1101/cshperspect.a041707

Studies in mice further support a correlation between oxidative stress and accelerated telomere shortening or dysfunction. Chronic depletion of the antioxidant glutathione in CAST/Ei mice accelerates telomere shortening in some tissues, including skin and testis (Cattan et al. 2008). Inflammation as a source of oxidative stress also contributes to telomere dysfunction. The loss of NF-κB, which regulates inflammatory genes, causes chronic inflammation, premature aging, and increased cell senescence in mice (Jurk et al. 2014). Affected tissues from these mice show increased dysfunctional DDR[+] telomeres, without changes in average telomere length. This and other studies reveal that oxidative stress can cause telomere dysfunction without causing shortening, resulting in cellular senescence.

Dysfunctional mitochondria and high $O_2$ consumption are important sources of ROS and can affect telomeres. The heart is among the organs consuming the most $O_2$. Dysfunctional DDR[+] telomeres increase with age in cardiomyocytes from mice without cell proliferation or telomere shortening, which correlates with increased oxidative DNA damage in hearts (Anderson et al. 2019). This study also shows a further increase in DDR[+] telomeres in cardiomyocytes from mouse models of oxidative stress caused by antioxidant enzyme deficiency or dysfunctional mitochondria, and that antioxidant treatments suppress this increase (Anderson et al. 2019). While these studies show no changes in telomere length, cardiomyocytes from mouse models of Duchenne muscular dystrophy or hypertensive heart failure do exhibit telomere shortening associated with mitochondria dysfunction or loss of telomeric antioxidant PRDX1, respectively (Chang et al. 2016; Brandt et al. 2022). These studies link oxidative stress and increased telomere dysfunction or shortening in both proliferating and nonproliferating cell types from mice.

Culturing human cells under oxidative stress conditions provides further evidence for impaired telomere maintenance. Early studies in the 1990s show mild oxidative stress correlates with accelerated telomere shortening in human fibroblasts and endothelial cells, whereas anti-oxidants and ROS scavengers correlate with decreased shortening and senescence (von Zglinicki 2002). Culturing cells at atmospheric 20% $O_2$ increases oxidative DNA damage and telomere shortening, compared to culturing at 3%–5% $O_2$ experienced by most cells in vivo (Parrinello et al. 2003; Richter and von Zglinicki 2007; Wang et al. 2010; Coluzzi et al. 2014). The link between mitochondria and telomere maintenance is also observed in cell culture. Oxidative stress due to mitochondrial dysfunction increases telomere shortening and dysfunction in cultured human cells, even when those cells express the telomere-lengthening enzyme telomerase (Saretzki et al. 2003; Passos et al. 2007). In addition, ROS produced by immune cells can affect telomeres in nonimmune cells. Neutrophils are an important source of ROS during inflammation and injury, and culturing neutrophils with human fibroblasts increases telomere shortening and premature senescence in the fibroblasts in a ROS-dependent manner (Lagnado et al. 2021). In summary, studies spanning human tissues, mouse models, and cell culture all reveal a general correlation between oxidative stress and increased telomere dysfunction, both with and without accelerating telomere shortening.

## TELOMERES ARE HOTSPOTS FOR OXIDATIVE DAMAGE

How does oxidative stress impair telomere maintenance? A widely accepted theory suggests this results from oxidative DNA damage at telomeres (von Zglinicki 2002). ROS generates nearly 100 different types of oxidatively damaged DNA bases in vitro; however, this number is much lower in cells, possibly due to some damaged bases being below detection limits or being unstable (Cadet and Wagner 2013). ROS from ionizing radiation in cells induces various damaged bases, including 8-oxo-7,8-dihydroguanine (8-oxoG), 2,6-diamino-4-hydroxy-5-formamidopyrimidine (FapyG), 8-oxo-7,8-dihydroadenine (8-oxoA), 2-hydroxyadenine, thymine glycol (Tg), as well as single-strand DNA breaks (SSBs) and abasic sites (Fig. 2; Cadet and Wagner 2013). Guanine has the lowest

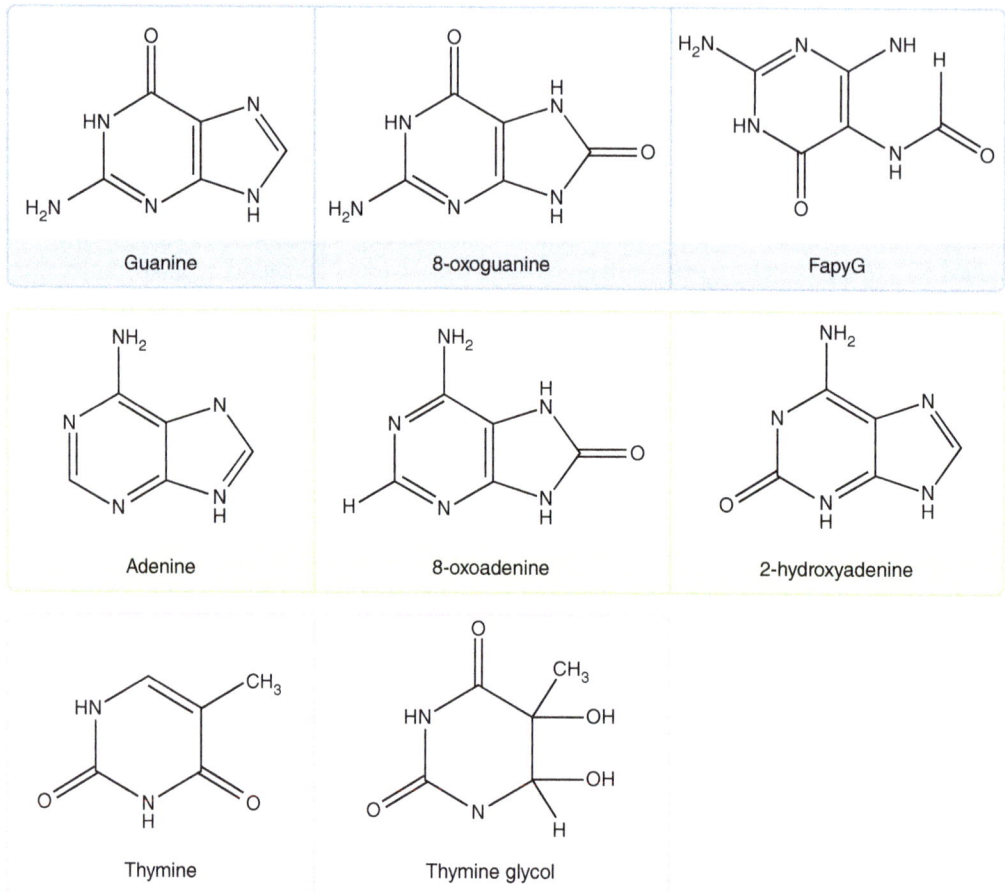

**Figure 2.** Common forms of oxidatively damaged bases are shown along with the natural base for reference. These include 8-oxo-7,8-dihydroguanine (8-oxoG), 2,6-diamino-4-hydroxy-5-formamidopyrimidine (FapyG), 8-oxo-7,8,-dihydroadenine (8-oxoA), 2-hydroxyadenine (2-OH-A), and thymine glycol (Tg). (Figure created with ChemDraw.)

reduction potential among the natural bases, making it the most susceptible to oxidation reactions (Crespo-Hernández et al. 2007). Guanine is even more susceptible to oxidation when present in G runs (Fukuzumi et al. 2005), as it exists in human telomeric TTAGGG repeats. 8-oxoG is among the most frequent oxidative lesion arising spontaneously an estimated 500–2800 lesions per cell per day, and higher under oxidative stress (Lindahl 1993; Tubbs and Nussenzweig 2017). For these reasons, 8-oxoG is the most well-studied type of oxidative DNA damage, especially within telomeres.

8-oxoG is commonly formed through ·OH addition to the eighth carbon in guanine (Fig. 2).

Highly reactive ·OH radicals arise in cells as products of Fenton reactions between iron ($Fe^{2+}$) and hydrogen peroxide ($H_2O_2$). Biochemical studies show TTAGGG sequences are preferred sites for iron binding, Fenton reactions, and 8-oxoG formation (Henle et al. 1999; Oikawa et al. 2001). Antioxidant peroxiredoxin 1 (PRDX1), which scavenges $H_2O_2$, is enriched at telomeres (Aeby et al. 2016), underscoring the importance of protecting telomeres from ·OH radicals. Other sources of 8-oxoG include $^1O_2$, a common product of UVA-radiation, and hypochlorous acid generated during inflammation (Agnez-Lima et al. 2012; Cadet and Wagner 2013). Quantifying 8-oxoG lesions

Cite this article as *Cold Spring Harb Perspect Biol* doi: 10.1101/cshperspect.a041707

at telomeres in cells is challenging because telomeres make up <0.025% of the total genome. Indirect methods for lesion detection in telomeres involve using repair enzymes to convert 8-oxoG to DNA breaks, which can be detected in telomere fragments by gel electrophoresis and Southern blotting, or as blocks to PCR amplification. These methods show more 8-oxoGs in telomeres compared to minisatellite sequences or the *36B4* gene locus in human and mouse cells after oxidant treatments (O'Callaghan et al. 2011; Rhee et al. 2011; Baquero et al. 2021). One study reported telomeres are more susceptible to 8-oxoG formation compared to the bulk genome in *Arabidopsis* plants (Castillo-González et al. 2022). More recent next-generation sequencing approaches to detect 8-oxoG show a high relative frequency of 8-oxoGs in telomeres and nucleosome-dense regions in *Saccharomyces cerevisiae* yeast (Wu et al. 2018). While the repetitive TTAGGG nature of human telomeres poses a challenge for sequencing, advances in sequencing technology hold promise for enabling facile 8-oxoG detection in the future.

## CONSEQUENCES OF OXIDATIVE DAMAGE AT TELOMERES

Are telomeres simply collateral damage from oxidative stress, or can oxidative damage at telomeres impact telomere function and cellular health? Studies have addressed this question by examining telomeres in cells that are unable to repair oxidative DNA damage, and by selectively targeting oxidative damage to the telomeres to assess the consequences.

### Base Excision Repair of Oxidative Damage at Telomeres

The impact of oxidative damage on telomere maintenance will depend on how the lesion is processed and repaired. Small, non-helix-distorting DNA lesions, such as 8-oxoG, are normally removed by base excision repair (BER) (Fig. 3). 8-oxoguanine glycosylase 1 (OGG1) recognizes 8-oxoG paired with C in duplex DNA with high specificity and hydrolyzes the

*N*-glycosidic bond to release the base, leaving an abasic residue (Roldán-Arjona et al. 1997; Rosenquist et al. 1997). OGG1 can also cleave the DNA backbone at the abasic residue, but its lyase activity is weak and cleavage normally occurs by AP endonuclease-1 (APE1), which stimulates OGG1 turnover (Hill et al. 2001). APE1 cleaves the DNA backbone 5′ of the abasic residue yielding an SSB with a 3′ hydroxyl and 5′ deoxyribose phosphate (dRP) (Krokan and Bjoras 2013). Poly(ADP-ribose) (PAR) polymerase 1 or 2 (PARP1 or PARP2) binds the SSB repair intermediate and synthesizes PAR chains, which help recruit downstream repair proteins (Schreiber et al. 2006). DNA polymerase (Pol) β cleaves the dRP and inserts 2′-deoxyguanosine 5′-triphosphate (dGTP) to fill the single-nucleotide gap, followed by DNA ligase III (LIG3) to seal the nick. Scaffold protein X-ray repair cross complementing 1 (XRCC1) interacts with and stabilizes Pol β and LIG3 and prevents PARP1 trapping at the SSB intermediate to promote BER completion (Demin et al. 2021). Typically, BER proceeds through this short-patch pathway in which one nucleotide is replaced. However, long-patch BER can occur if Pol β or another DNA polymerase inserts 2–10 nt, which produces a 5′ single-stranded flap that is cleaved by FEN1, followed by ligase 1 (LIG1) sealing the nick (Krokan and Bjoras 2013).

BER of 8-oxoG is essential for preventing mutations because the lesion can base pair incorrectly with adenine. Most DNA polymerases can insert C or A opposite 8-oxoG, but preferentially extend from the misinserted base causing G to T mutations (Markkanen 2017). To prevent mutations, DNA glycosylase MUTYH excises A opposite 8-oxoG, and then APE stimulates MUTYH turnover and cleaves the DNA backbone at the abasic site (Fig. 3; Yang et al. 2001). DNA polymerase λ (pol λ) inserts C opposite 8-oxoG, followed by long-patch BER, thereby allowing another opportunity for OGG1 excision of 8-oxoG (van Loon and Hübscher 2009; Burak et al. 2016). If Pol β reinserts dA opposite the 8-oxo-dG, this can cause futile MUTYH-initiated BER cycles that promote cell death (Hashimoto et al. 2004). Biallelic

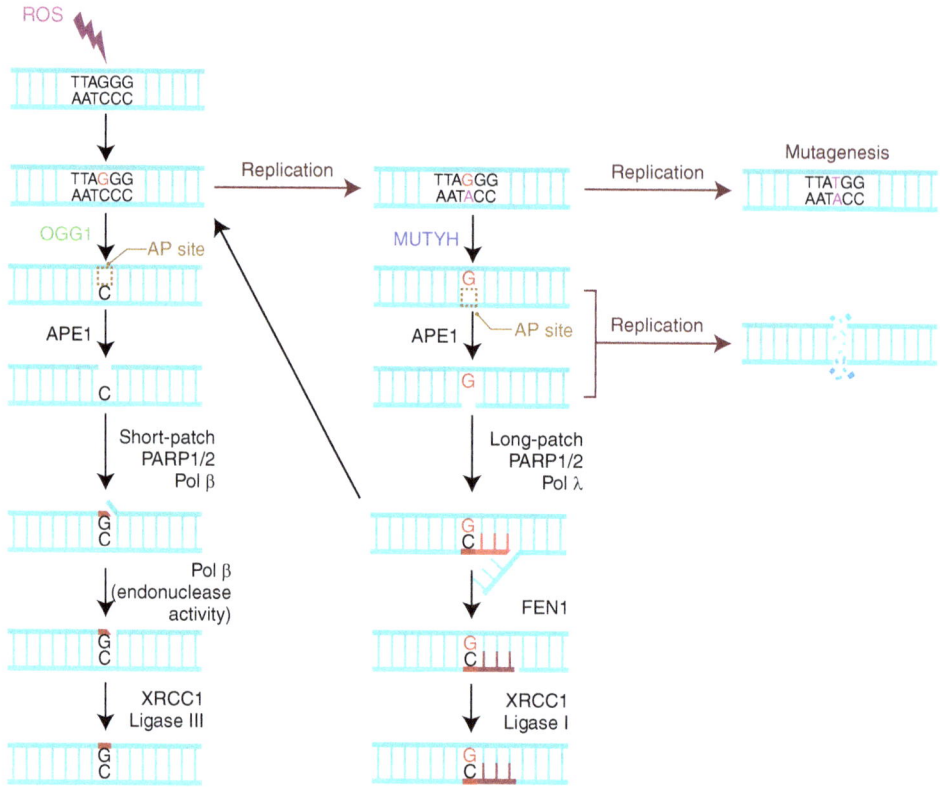

**Figure 3.** Base excision repair of 8-oxoguanine. Reactive oxygen species (ROS) interaction with telomeric TTAGGG repeats can convert G to 8-oxo-7,8-dihydroguanine (8-oxoG). 8-oxoguanine glycosylase 1 (OGG1) glycosylase excises 8-oxoG opposite C, leaving an apurinic (AP) site. AP endonuclease-1 (APE1) cleaves the DNA backbone 5′ of the AP site, producing single-strand DNA breaks (SSBs) with a 5′ deoxyribose phosphate (5′ dRP). Poly(ADP-ribose) polymerase 1 or 2 (PARP1 or PARP2) binds the SSB repair intermediate and synthesizes poly(ADP-ribose) (PAR) chains, which helps recruit downstream repair proteins. In short-patch base excision repair (BER) Pol β lyase activity cleaves the 5′ dRP and its polymerase activity inserts 2′-deoxyguanosine 5′-triphosphate (dGTP) to fill the single-nucleotide gap, and ligase III (LIG3) seals the SSB. Scaffold protein X-ray repair cross complementing 1 (XRCC1) stabilizes Pol β and LIG3. During DNA replication, if A is misinserted opposite 8-oxoG, this can cause G to T mutations. To prevent this, MUTYH excises A opposite 8-oxoG, and then APE1 cleaves at the AP. DNA polymerase λ (Pol λ) inserts C opposite 8-oxoG along with an additional 2–10 nt, followed by long-patch BER. FEN1 cleaves the 5′ single-stranded flap, and ligase 1 (LIG1) seals the SSB, thereby allowing another opportunity for OGG1 excision of 8-oxoG. If BER is aborted, the AP or SSB repair intermediate can cause replication fork collapse and a double-strand break. Black arrows depict canonical DNA repair steps. Brown arrows depict DNA replication leading to mutagenesis. Red G indicates 8-oxoG base and magenta bases indicate base mutation. (Figure generated with BioRender; www.biorender.com.)

germline MUTYH mutations cause the colorectal cancer predisposition syndrome MUTYH-associated polyposis (Al-Tassan et al. 2002), underscoring the importance of MUTYH in preventing mutagenesis and tumorigenesis in the face of oxidative stress and DNA damage.

Biochemical and cellular studies show that BER of 8-oxoG in telomeric sequences is influenced by the shelterin proteins, which bind telomeric DNA, and by secondary DNA structures in telomeres. Shelterin proteins TRF1, TRF2, and POT1 interact with BER proteins

APE1 and Pol β and stimulate the enzymatic steps in vitro (Muftuoglu et al. 2006; Miller et al. 2012), suggesting that shelterin proteins should not block BER or compete with BER proteins for binding to telomeres. Consistent with this, 8-oxoG lesions disrupt TRF1 and TRF2 binding, allowing repair proteins to gain access (Opresko et al. 2005). Regarding structural features, telomeric sequences can fold into secondary four-stranded structures termed G-quadruplexes formed by Gs base-pairing with other Gs via Hoogsteen bonds. OGG1 cannot remove 8-oxoG when the lesion resides within a G-quadruplex structure or single-stranded DNA (ssDNA) (Zhou et al. 2013, 2015). This is relevant because telomeres have long single-stranded overhangs that can fold into G-quadruplexes, and G-quadruplexes may form in ssDNA during telomere replication or transcription. How shelterin and G-quadruplexes influence the rate and efficiency of 8-oxoG repair, and BER in general, at telomeres in cells remains to be fully understood.

## 8-oxoG Lesions Alter Telomere Maintenance and Cellular Health

Studies in cells lacking the OGG1 repair enzyme reveal that unrepaired 8-oxoG lesions influence telomere length and function. A genetic screen in yeast *S. cerevisiae* found that strains lacking Ogg1 had longer telomeres compared to wild-type strains (Askree et al. 2004; Lu and Liu 2010). Similarly, *Ogg1*$^{-/-}$ mice have longer telomeres compared to wild-type mice (Wang et al. 2010). But when cells from *Ogg1*$^{-/-}$ mice are cultured at 20% $O_2$ or with oxidants, they show increased telomere shortening and losses, compared to cells from wild-type mice (Wang et al. 2010). Similarly, pharmacological disruption of BER with an OGG1 small molecule inhibitor increased telomere losses in human cancer cells treated with oxidants (Baquero et al. 2021). These studies suggest that while low 8-oxoG levels at telomeres may promote telomere lengthening, higher levels, which arise under oxidative stress conditions, impair telomere maintenance.

Repair-deficient cells provided the first evidence that unrepaired oxidative lesions can lead to telomere changes. However, in these cells, lesions also accumulate in the bulk genome, potentially affecting cellular replication and indirectly influencing telomere shortening rates. More recent approaches that target oxidative damage exclusively to telomeres allow the effects on telomeres and cellular health to be attributed directly to the telomere lesions. KillerRed fluorescent protein generates superoxide upon excitation with green light (550–580 nm) and localizes to telomeres when fused with TRF1. This allows selective superoxide production at telomeres, causing telomere shortening, loss, and fragility in HeLa cells, impaired growth in cancer cell lines and fibroblasts (Sun et al. 2015), and MUTYH recruitment to telomeres (Tan et al. 2020). Superoxide promotes ·OH production, which causes SSBs, oxidatively damaged purine and pyrimidine bases, including 8-oxoG and Tg, and abasic residues (Cadet and Wagner 2013).

Another system with fluorogen-activating peptide (FAP) fused to TRF1 is more specific for 8-oxoG production. When the photosensitizer dye di-iodinated malachite green (MG2I) binds to FAP, it can be excited with 660 nm light to produce $^1O_2$, which primarily reacts with G to produce 8-oxoG (Agnez-Lima et al. 2012; Barnes et al. 2022b). The generation of $^1O_2$ at telomeres causes telomere-specific 8-oxoG damage, and recruits OGG1 and downstream BER proteins XRCC1, PARP1, and PARP2 (Fouquerel et al. 2019; Barnes et al. 2022a; Kumar et al. 2022; Muoio et al. 2024). In HeLa cells, a single induction of telomeric 8-oxoG does not affect cell growth or telomere function, but chronic telomeric 8-oxoG production impairs cell growth and increases telomere shortening and losses (Fouquerel et al. 2019). The accumulation of telomeric 8-oxoG lesions in OGG1-deficient cells causes an even greater increase in telomere shortening and losses (Fouquerel et al. 2019). Furthermore, use of the FAP-TRF1 system revealed that Jurkat leukemia T cells are also sensitive to oxidative telomere damage (Wang et al. 2021). These studies provide direct evidence that oxidative damage at the telomere can promote telomere alterations and impair cell growth.

Studies in human cells from nondiseased tissues reveal that oxidative damage at telomeres can promote cellular aging. Human cancer cell lines are not suitable for investigating aging, because they have acquired mutations or changes in DDR signaling that enable bypass of senescence and cellular immortalization. Human skin fibroblasts (BJ-hTERT) and retinal epithelial cells (RPE1-hTERT) are widely used nondisease cell lines that have intact DDR pathways. They are derived from nondisease tissues but express telomerase to prevent replicative senescence. Targeted production of 8-oxoG in these nondisease cell lines reveals that they are more sensitive to oxidative telomere damage than cancer cell lines. A single induction of telomeric 8-oxoG in RPE1-hTERT and BJ-hTERT cells drives rapid, premature cell senescence within days after damage (Barnes et al. 2022a). Oxidative telomere damage fails to cause appreciable telomere shortening within this short recovery period; rather, the damage induces telomere fragility, a marker of impaired replication at telomeres. Fragile telomeres are dysfunctional and activate the DDR, thereby triggering p53-mediated cellular senescence (Sfeir et al. 2009). The production of 8-oxoG in nonreplicating cells does not significantly increase DDR⁺ telomeres or senescence (Barnes et al. 2022a). In summary, telomeric 8-oxoG damage triggers rapid senescence and telomere dysfunction in nondiseased human fibroblast and epithelial cells, even in the absence of telomere shortening.

## OXIDATIVE DAMAGE DISRUPTS TELOMERE REPLICATION

Oxidative damage is linked to replication stress at telomeres. Both ROS-induced oxidative base lesions and SSBs were predicted to cause telomere shortening in replicating cells by impairing DNA replication forks and causing replication stress (von Zglinicki 2002). Replication stress is defined as the slowing or stalling of DNA replication forks due to challenges such as DNA lesions and DNA polymerase inhibition, which activate response mechanisms to preserve replication

forks (Zeman and Cimprich 2014). Telomeres are difficult to replicate due to the repetitive sequence and secondary structures, including T-loops and G-quadruplexes (Lormand et al. 2013; Brenner and Nandakumar 2022). Collapsed replication forks that are not repaired can cause telomere loss, indicated by a chromatid end lacking staining with a telomere probe by fluorescent in situ hybridization (FISH) on metaphase chromosomes. Unrepaired 8-oxoG lesions in OGG1-deficient mouse cells exhibit loss specifically of telomeres replicated from the G-rich template strand, on which 8-oxoG lesions can arise (Wang et al. 2010). Similarly, the accumulation of unrepaired 8-oxoG lesions at telomeres promotes telomere losses in OGG1-deficient HeLa cells (Fouquerel et al. 2019). These studies provide evidence that 8-oxoG lesions produced at telomeres under oxidative stress can promote telomere losses in replicating cells, likely via replication stress.

As described above, replication stress at telomeres can also manifest as telomere fragility. Fragile telomeres appear as multiple telomeric DNA signals at a chromatid end on metaphase chromosomes stained by telomere FISH. These aberrations are called "fragile telomeres" because they increase with DNA replication inhibitors that cause breaks at common fragile sites in the genome (Sfeir et al. 2009). Culturing human cells under various pro-oxidant conditions increases telomere fragility (for review, see Barnes et al. 2023). Base damage may be partly responsible because targeted 8-oxoG formation at telomeres induces telomere fragility in various human cell lines (Barnes et al. 2022a). Fragile telomeres are thought to represent underreplicated single-strand gaps in telomeres, or byproducts of break-induced replication (BIR), a mechanism to recover DNA replication after a replication fork collapses into a one-ended DNA break (Yang et al. 2020). Mitotic DNA synthesis (MiDAS) is a form of BIR used to complete replication in late $G_2$ or M cell–cycle phases, of regions that failed to be replicated during S phase. MiDAS at telomeres can be visualized by the incorporation of detectable nucleotide analogs during mitosis at telomeres due to DNA synthesis. Telomere fragility is often

Cite this article as *Cold Spring Harb Perspect Biol* doi: 10.1101/cshperspect.a041707

associated with telomere MiDAS since both arise from replication stress (Barnes et al. 2023). The production of telomeric 8-oxoG in nondiseased human cells, and the accumulation of telomeric 8-oxoGs in HeLa cells, increases MiDAS at telomeres that resembles BIR DNA synthesis (Fouquerel et al. 2019; Barnes et al. 2022a). PARP2, a DNA-dependent ADP-ribosyl transferase enzyme, promotes a BIR form of MiDAS to prevent telomere losses arising from targeted telomeric 8-oxoG formation (Muoio et al. 2024). Collectively, these studies indicate that oxidative base damage at telomeres causes telomere fragility and dysfunction by interfering with telomere replication.

## OXIDATIVE DAMAGE TO NUCLEOTIDE POOLS IMPAIRS TELOMERE MAINTENANCE

The nucleotide precursors to DNA synthesis are more susceptible to oxidative damage than chromatin-protected DNA, especially free $2'$-deoxyguanosine $5'$-triphosphate (dGTP), which is converted to 8-oxo-dGTP upon reaction with ROS (Haghdoost et al. 2006). DNA polymerases can insert 8-oxo-dGTP opposite cytosine or adenine (for review, see Markkanen 2017). Therefore, 8-oxo-dGTP insertion can introduce damage into the genome and cause AT:CG transversion mutations if the A on the parental strand of the A:8-oxoG base pair is removed by MUTYH (Fig. 4; Hogg et al. 2007). AT:CG transversions have been reported in telomeric variant repeats (Lee et al. 2014), although whether they arise from 8-oxo-dGTP insertion and BER processing is unknown. Misinsertion of 8-oxo-dGTP during BER can also impair downstream ligation to seal the SSB, leading to an accumulation of toxic repair intermediates (Freudenthal et al. 2015). To prevent misinsertion of 8-oxo-dGTP during DNA replication or repair, MutT homolog 1 (MTH1), also called nudix hydrolase 1, hydrolyzes the damaged nucleotide triphosphate to 8-oxo-dGMP (Sakumi et al. 1993) as well as the less abundant oxidized dATPs, including 2-OH-dATP and 8-oxo-dATP (Fig. 4; Fujikawa et al. 1999). Therefore, the insertion of oxidized nucleotide triphosphates is another way by which oxidative damage can be introduced into telomeres. In support of this, MTH1 deficiency causes telomere shortening and losses under oxidative stress conditions, when oxidized dNTP levels are elevated (Fouquerel et al. 2016; Ahmed and Lingner 2018).

## OXIDATIVELY DAMAGED BASES AND dNTPs ALTER TELOMERASE ACTIVITY

8-oxoG impacts telomere maintenance by altering telomerase activity, either through this damage arising within the telomeric DNA or the dNTP pool. Telomerase reverse transcribes

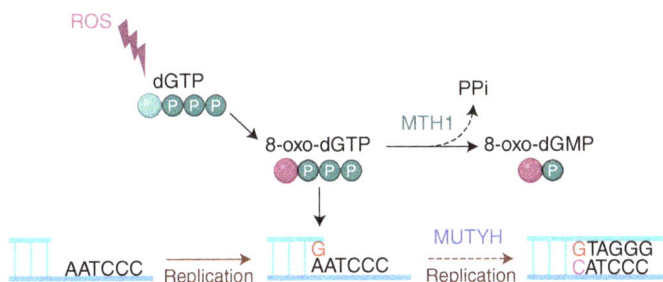

**Figure 4.** MutT homolog 1 (MTH1) hydrolyzes 8-oxo-dGTP to 8-oxo-dGMP and pyrophosphate (PPi) to prevent misincorporation of 8-oxo-dGTP opposite a template A during DNA replication or repair, which can cause mutations. If MUTYH removes the template A from the A:8-oxoG base pair, and polymerase λ (Pol λ) inserts C to allow for 8-oxoguanine glycosylase 1 (OGG1) removal of 8-oxo-7,8-dihydroguanine (8-oxoG) by base excision repair (BER), this will cause A:T to C:G transversions. Such mutagenic events can alter the TTAGGG repeats to GTAGGG (shown as an example), TGAGGG, or TTCGGG variants. Red G depicts 8-oxoG base and magenta C depicts base mutation. (Figure generated with BioRender; www.biorender.com.)

an integral RNA CCAAUC template to add GGTTAG repeats onto the $3'$ telomeric ssDNA overhang, then translocates along the product to add additional repeats to restore telomere lengths after replication (Fig. 5; Sanford et al. 2021). Like most DNA polymerases, telomerase can add 8-oxo-dGTP during DNA synthesis and prefers to misinsert 8-oxo-dGTP opposite rA, but unlike DNA polymerases, telomerase cannot extend the telomere and continue DNA synthesis after 8-oxo-dGTP addition (Aeby et al. 2016; Fouquerel et al. 2016; Sanford et al. 2020). However, telomerase can continue synthesis after inserting 2-OH-dATP, but the telomere products are shorter, indicating the damaged nucleotide impairs extension. Telomerase processivity factors POT1–TPP1 cannot overcome 8-oxo-dGTP or 2-OH-dATP inhibition of telomerase (Sanford et al. 2020). Depleting MTH1 to prevent the removal of oxidized dNTPs causes telomere losses and shortening and inhibits the addition of new telomeric repeats by telomerase (Fouquerel et al. 2016; Ahmed and Lingner 2018). In these studies, oxidized dNTPs were increased by culturing cells at 20% $O_2$ or by depleting antioxidant PRDX1, which is normally enriched at telomeres.

8-oxoG can also influence telomerase activity through its ability to impact telomere structure. Telomeric TTAGGG repeats in ssDNA, including the $3'$ overhang, can fold into G-quadruplex structures (Lee et al. 2005; Hwang et al. 2014). Four tandem TTAGGG repeats fold into very stable G-quadruplexes, which prevent telomerase from loading and extending the telomere (Fig. 5; Zahler et al. 1991; Fouquerel et al. 2016). Replacing G with 8-oxoG disrupts the Hoogsteen hydrogen bonding in the G-quadruplex and destabilizes the structure (Bielskute et al. 2019). Telomeric G-quadruplex folding in real time can be monitored by single-molecule Förster resonance energy transfer (smFRET), which detects when folding brings two strategically placed fluorescent dyes together in close proximity. Replacing a G with 8-oxoG does not completely unfold the G-quadruplex, but rather imparts dynamic fluctuations between partially unfolded and short-lived folded structures. G-quadruplex destabilization increases POT1

binding, as well as telomerase binding and extension activity (Fouquerel et al. 2016; Lee et al. 2017, 2020). Therefore, the ability of 8-oxoG to destabilize G-quadruplex structures and increase telomerase binding may partly explain why OGG1 loss leads to telomere lengthening in vivo under nonoxidative stress conditions when oxidative base damage is low (Lu and Liu 2010; Wang et al. 2010). Consequently, 8-oxoG can have both positive and negative effects on telomerase depending on the amount of ROS and oxidative damage. A preexisting 8-oxoG in the telomeric overhang can benefit telomere maintenance by disrupting the G-quadruplex structure to facilitate telomerase loading, but telomerase addition of 8-oxo-dGTP halts further elongation.

## OXIDATIVE DAMAGE STIMULATES ALT ACTIVITY

Cancers that lack telomerase maintain telomere lengths by activating a homology-directed repair (HDR) pathway termed alternative lengthening of telomeres (ALT). Telomere extension by ALT occurs in the $G_2$ cell–cycle phase and arises from telomeric DNA breaks that stimulate homologous recombination (HR) and telomere elongation by BIR (Dilley et al. 2016; Zhang et al. 2019; Sobinoff and Pickett 2020). Some studies show that oxidative stress increases hallmarks of ALT, including telomerase-independent telomere lengthening, ALT-associated PML bodies (APBs), and telomere sister chromatid exchanges indicative of HR (Coluzzi et al. 2017; De Vitis et al. 2019; Luxton et al. 2020). Links between oxidative stress and ALT may be due to the ability of oxidative damage to promote replication stress at telomeres. Replication stress drives and perpetuates ALT by promoting HDR mechanisms of telomere extension (Zhang and Zou 2020; Brenner and Nandakumar 2022; Lu and Pickett 2022). In yeast, $Ogg1$ deletion leads to telomere lengthening by an HDR mechanism (Lu and Liu 2010). In human cancer cells that use ALT, targeted 8-oxoG production at telomeres increases hallmarks of ALT activity, such as APBs and telomeric DNA synthesis during the $G_2$ phase when ALT is active

 Cite this article as *Cold Spring Harb Perspect Biol* doi: 10.1101/cshperspect.a041707

(Thosar et al. 2024). Supporting a link with replication stress, an acute induction of telomeric 8-oxoG also greatly increases telomere fragility, ATR kinase signaling, and telomere sister chromatid exchanges in ALT cancer cells (Thosar et al. 2024). Telomeres in ALT cells already experience high levels of replication stress, making them more sensitive to additional replication damage, including that caused by 8-oxoG damage (Sobinoff and Pickett 2020). Therefore, chronic telomeric 8-oxoG damage greatly inhibits the growth of ALT cancer cells, likely due to excessive ALT activity and replication stress, which can be detrimental (Thosar et al. 2024). But just as 8-oxoG can positively or negatively regulate telomerase elongation of telomeres, depending on where the lesion arises, 8-oxoG can also both promote and impair ALT. Specifically, when the DNA replication fork encounters damage at telomeres it may collapse into a break and trigger HDR mechanisms that promote ALT, but if 8-oxoG arises at the telomere in the $G_2$ cell–cycle phase it inhibits the ALT replisome and telomere extension (Thosar et al. 2024). However, in general, extensive oxidative damage at telomeres negatively impacts telomere maintenance.

## OTHER FORMS OF OXIDATIVE DAMAGE

Although 8-oxoG is the most well-studied oxidative lesion at telomeres, ROS induces other types of lesions that may also impact telomeres and therefore should be considered. SSBs arise spontaneously at an estimated 55,000 breaks per cell per day from endogenous sources (Tubbs and Nussenzweig 2017). An SSB can arise directly from oxidation of the deoxyribose in the DNA, or indirectly after glycosylase removal of an oxidatively damaged base and cleavage at the abasic residue (Caldecott 2024). Oxidant treatments induce SSBs at telomeres, particularly in PRDX1-deficient cells (von Zglinicki et al. 2000; Ahmed and Lingner 2020). PRDX1 is an antioxidant enriched at telomeres that protects against oxidant-induced SSBs, which leads to DSBs upon replication fork collapse (Ahmed and Lingner 2020). OGG1 depletion suppresses the increase in telomeric SSBs in PRDX1-defi-

cient cancer cells treated with $H_2O_2$, suggesting glycosylase-initiated BER at 8-oxoG lesions produces SSBs as repair intermediates. In addition, the loss of both OGG1 and MUTYH suppresses SSB repair intermediate formation at telomeres in BJ-hTERT fibroblasts after targeted production of 8-oxoG at telomeres, and partially rescues the damage-induced senescence (De Rosa et al. 2023). These studies show that SSBs arising directly or indirectly from ROS impact telomere maintenance and function.

As described earlier, oxidative stress can generate base damage other than 8-oxoG. Evidence that these lesions can impact telomeres is based mostly on studies in cells lacking various DNA glycosylases. The Nei endonuclease VII-family of DNA glycosylases (NEILs 1–3) can remove FapyG, FapyA, as well as various oxidatively damaged pyrimidines, including Tg and 5-hydroxyuracil, but have weak or no activity at 8-oxoG (Wallace 2013; Zhou et al. 2013). The NEILs can also remove spiroiminodihydantoin (Sp) and guanidinohydantoin (Gh) lesions, which are caused by further oxidation of 8-oxoG (Luo et al. 2001). These hydantoin lesions can block DNA replication and transcription because they distort the DNA double helix (Henderson et al. 2003; Kolbanovskiy et al. 2017). Biochemical studies show that NEIL glycosylases can remove hydantoin lesions from telomeric G-quadruplexes and ssDNA, and NEIL3 can remove Tg from telomeric G-quadruplexes and ssDNA (Zhou et al. 2013, 2015). NEIL1 and NEIL3 also show a preference for removing lesions from telomeric sequences in various contexts (Zhou et al. 2013). Whether they remove damaged bases from G-quadruplexes and ssDNA within telomeres in cells, and how this might impact telomere stability, are unknown. However, studies in NEIL-deficient cells suggest these glycosylases are important for telomere maintenance. NEIL3 acts in BER during DNA replication and localizes to telomeres during S and $G_2$ cell-cycle phases, which is further increased after oxidant treatment. Furthermore, NEIL3 depletion increases telomere losses and chromosome bridges that can arise from fusions of chromosome ends lacking telomeres (Zhou et al. 2017). NEIL2

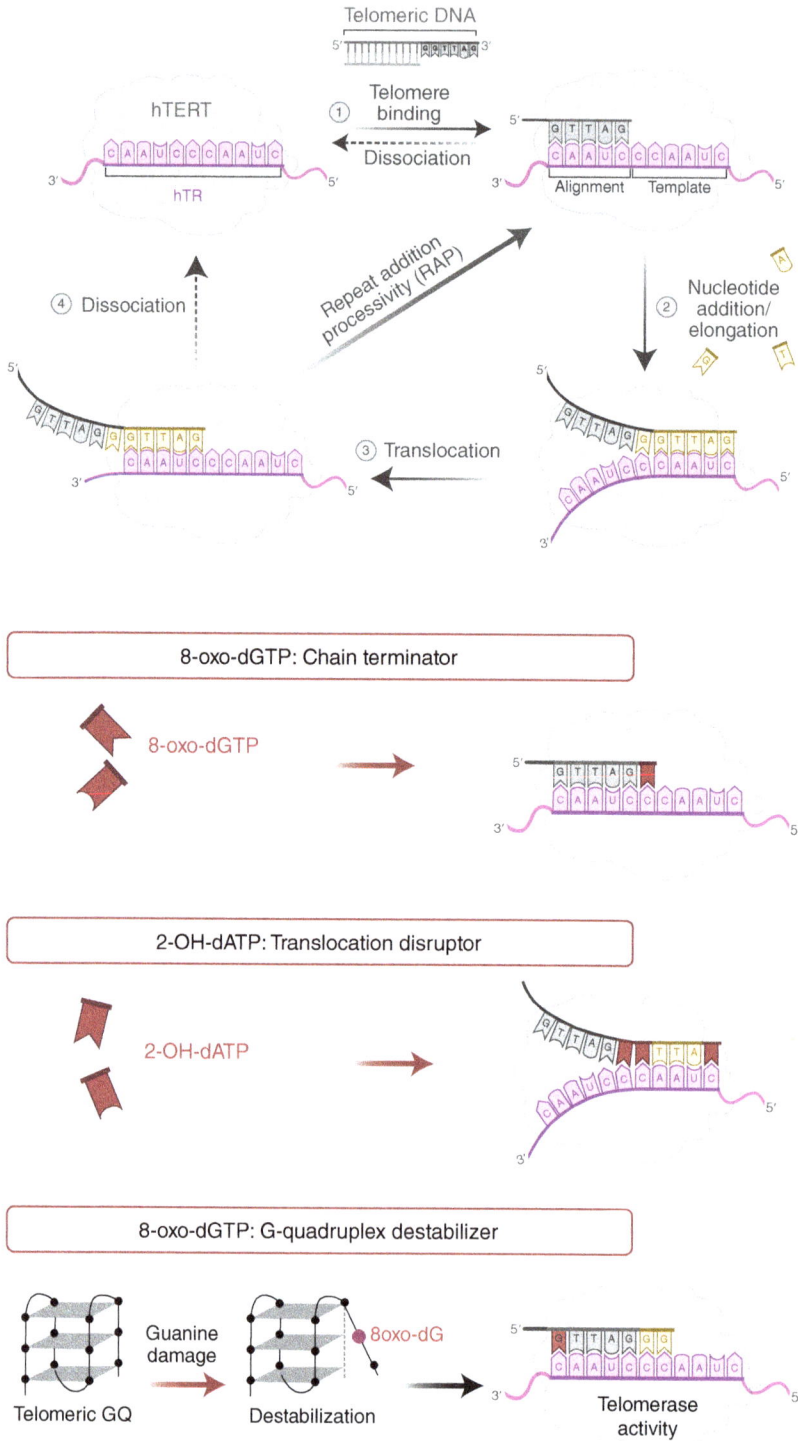

**Figure 5.** (*See following page for legend.*)

Cite this article as *Cold Spring Harb Perspect Biol* doi: 10.1101/cshperspect.a041707

acts in BER during transcription. Culturing cells from $Neil2^{-/-}$ mice at 20% $O_2$ increases telomere losses, compared to wild-type cells, similar to $Oggl^{-/-}$ cells cultured at 20% $O_2$ (see above) (Wang et al. 2010; Chakraborty et al. 2015). Whether unrepaired lesions interfere with telomere transcription in these cells is unknown. Since NEIL glycosylases remove various lesion types, it is difficult to know which lesions cause telomere defects in NEIL-deficient cells. Nevertheless, these studies suggest that oxidatively damaged bases that are normally removed by the NEIL glycosylases, may contribute to impaired telomere maintenance caused by oxidative stress.

Further evidence that Tg may disrupt telomere maintenance derives from studies in NTHL1 glycosylase-deficient cells. NTHL1 removes oxidized pyrimidine bases, including Tg, 5-hydroxycytosine, and 5-hydroxyuracil (Krokan and Bjoras 2013). $Nth1^{-/-}$ mice show increased telomere fragility in vivo, indicative of impaired telomere replication, and culturing cells from these mice at 20% $O_2$ increases telomere shortening, losses, and dysfunction more than in wild-type cells (Vallabhaneni et al. 2013), again similar to $Oggl^{-/-}$ cells (Wang et al. 2010). The precise lesion(s) that impair telomere maintenance in $Nth1^{-/-}$ cells is unknown. However, Tg is the most common oxidized thymine lesion and blocks DNA synthesis when present in the template strand, making it a likely candidate (McNulty et al. 1998). Furthermore, telomeres from $Nth1^{-/-}$ cells have higher levels of oxidized pyrimidines, compared to telomeres from wild-type cells, consistent with a repair deficiency (Vallabhaneni et al. 2013). Collectively, studies in NEIL- and NTH1-deficient cells provide evidence that oxidized pyrimidines may interfere with telomere replication.

## NUCLEOTIDE EXCISION REPAIR PROTEINS AND OXIDATIVE DAMAGE AT TELOMERES

While BER is the primary pathway for repairing oxidative base damage, proteins that function in nucleotide excision repair (NER) are also implicated in protecting telomeres from ROS-induced damage. NER primarily removes bulky lesions in the DNA that distort the DNA double helix, and involves more than 30 proteins that coordinate steps of lesion recognition, DNA duplex melting around the lesion, excision to release 24–32 nt that contain the lesion, DNA synthesis to fill the gap, and ligation to seal the nick (Spivak 2015). There is some evidence that NER proteins may also function in various aspects of BER. UV-DDB and XPC proteins, which are involved in lesion recognition during NER, stimulate OGG1 activity in vitro and localize to telomeres after targeted 8-oxoG production (D'Errico et al. 2006; Kumar et al. 2022). Additionally, XPG, an endonuclease involved in NER, stimulates the activity of NTH1 glycosylase (Klungland et al. 1999). This raises the possibility that XPG may cooperate with NTH1 to repair oxidized pyrimidines at telomeres, although this is unknown. However, cells deficient in some NER proteins show increased telomere defects under oxidative stress. Similar

---

**Figure 5.** Oxidatively damaged bases affect telomerase activity. (*Left* panel) The numbers represent each step in the telomerase catalytic cycle. Gray indicates the telomeric single-strand overhang, purple indicates the telomerase RNA template and hTERT, and yellow indicates the newly added nucleotides. Telomerase (1) binds the 3′ telomeric overhang, (2) reverse transcribes its RNA CCAAUC template to add a GGTTAG repeat, (3) translocates along the product to realign for another cycle of repeat addition, or (4) dissociates from the telomere, terminating further elongation. (*Right* panels; *top*) Telomerase can add 8-oxo-dGTP during telomere synthesis, but 8-oxo-dGTP acts as a chain terminator and halts further elongation. (*Middle*) Telomerase can continue DNA synthesis after inserting 2-OH-dATP, the damaged nucleotide impairs repeat addition processivity and the telomere products are shorter. (*Bottom*) A preexisting 8-oxo-7,8-dihydroguanine (8-oxoG) in the telomeric single-stranded DNA (ssDNA) can stimulate telomerase elongation of the telomere by disrupting the secondary G-quadruplex (GQ) structure that blocks telomerase loading. Whether 8-oxoG stimulates or impairs telomerase depends on whether it arises in the dNTP pool or the telomeric DNA overhang. (Figure generated with BioRender; www.biorender.com.)

to $Nth1^{-/-}$ cells, culturing cells from $Xpc^{-/-}$ mice at 20% $O_2$ increases telomere fragility more than wild-type cells (Stout and Blasco 2013; Vallabhaneni et al. 2013). In addition, $H_2O_2$ treatment of human cells deficient in XPB or XPD, the DNA helicases that melt the DNA duplex around the lesion, increases telomere losses as detected by telomere FISH on metaphase chromosomes (Gopalakrishnan et al. 2010; Low et al. 2022). These studies suggest that NER proteins may have a role in protecting telomeres after oxidative stress, either by stimulating BER and/or by facilitating the removal of oxidative lesions through NER.

## SUMMARY

This paper describes the evidence that oxidative stress correlates with accelerated telomere shortening and dysfunction provided from studies in human tissues, mouse models, cell culture, and biochemistry experiments. These studies show telomeres are not simply collateral damage from oxidative stress, due to their G-rich sequence, but that oxidative damage to telomeres impacts telomere function and maintenance. The results imply that oxidative damage to telomeres likely contributes to the pathogenesis of ROS and oxidative stress and may contribute to human diseases characterized by oxidative stress. Future advances in technologies that can accurately measure and quantify oxidative lesions and mutations arising from unrepaired lesions in the telomeres, will significantly advance the understanding of how oxidative stress impacts telomere maintenance in humans and other organisms. A better understanding of how the formation and processing of oxidative damage alters telomere maintenance and function will be valuable for developing strategies to protect telomeres in the face of oxidative stress to promote cellular health.

## ACKNOWLEDGMENTS

We thank Dr. Ryan Barnes, Ms. Libby Childs, and Ms. Theresa Heidenreich for critical reading of this paper and helpful feedback. We are grateful for support from National Institutes of Health (NIH) grants F32CA275287 (to S.L.S.), K99ES035871 (to M.D.R.), and R35ES030396 and R01CA207342 (to P.L.O.).

## REFERENCES

Aeby E, Ahmed W, Redon S, Simanis V, Lingner J. 2016. Peroxiredoxin 1 protects telomeres from oxidative damage and preserves telomeric DNA for extension by telomerase. *Cell Rep* **17**: 3107–3114. doi:10.1016/j.celrep.2016.11.071

Agnez-Lima LF, Melo JT, Silva AE, Oliveira AH, Timoteo AR, Lima-Bessa KM, Martinez GR, Medeiros MH, Di Mascio P, Galhardo RS, et al. 2012. DNA damage by singlet oxygen and cellular protective mechanisms. *Mutat Res Rev Mutat Res* **751**: 15–28. doi:10.1016/j.mrrev.2011.12.005

Ahmed W, Lingner J. 2018. PRDX1 and MTH1 cooperate to prevent ROS-mediated inhibition of telomerase. *Genes Dev* **32**: 658–669. doi:10.1101/gad.313460.118

Ahmed W, Lingner J. 2020. PRDX1 counteracts catastrophic telomeric cleavage events that are triggered by DNA repair activities post oxidative damage. *Cell Rep* **33**: 108347. doi:10.1016/j.celrep.2020.108347

Aikata H, Takaishi H, Kawakami Y, Takahashi S, Kitamoto M, Nakanishi T, Nakamura Y, Shimamoto F, Kajiyama G, Ide T. 2000. Telomere reduction in human liver tissues with age and chronic inflammation. *Exp Cell Res* **256**: 578–582. doi:10.1006/excr.2000.4862

Al-Tassan N, Chmiel NH, Maynard J, Fleming N, Livingston AL, Williams GT, Hodges AK, Davies DR, David SS, Sampson JR, et al. 2002. Inherited variants of MYH associated with somatic G:C→T:A mutations in colorectal tumors. *Nat Genet* **30**: 227–232. doi:10.1038/ng828

Anderson R, Lagnado A, Maggiorani D, Walaszczyk A, Dookun E, Chapman J, Birch J, Salmonowicz H, Ogrodnik M, Jurk D, et al. 2019. Length-independent telomere damage drives post-mitotic cardiomyocyte senescence. *EMBO J* **38**: e100492. doi:10.15252/embj.2018100492

Armstrong E, Boonekamp J. 2023. Does oxidative stress shorten telomeres in vivo? A meta-analysis. *Ageing Res Rev* **85**: 101854. doi:10.1016/j.arr.2023.101854

Askree SH, Yehuda T, Smolikov S, Gurevich R, Hawk J, Coker C, Krauskopf A, Kupiec M, McEachern MJ. 2004. A genome-wide screen for *Saccharomyces cerevisiae* deletion mutants that affect telomere length. *Proc Natl Acad Sci* **101**: 8658–8663. doi:10.1073/pnas.0401263101

Assavanopakun P, Sapbamrer R, Kumfu S, Chattipakorn N, Chattipakorn SC. 2022. Effects of air pollution on telomere length: evidence from in vitro to clinical studies. *Environ Pollut* **312**: 120096. doi:10.1016/j.envpol.2022.120096

Baquero JM, Benítez-Buelga C, Rajagopal V, Zhenjun Z, Torres-Ruiz R, Müller S, Hanna BMF, Loseva O, Wallner O, Michel M, et al. 2021. Small molecule inhibitor of OGG1 blocks oxidative DNA damage repair at telomeres and potentiates methotrexate anticancer effects. *Sci Rep* **11**: 3490. doi:10.1038/s41598-021-82917-7

Barnes RP, Fouquerel E, Opresko PL. 2019. The impact of oxidative DNA damage and stress on telomere homeostasis. *Mech Ageing Dev* **177**: 37–45. doi:10.1016/j.mad.2018.03.013

Barnes RP, de Rosa M, Thosar SA, Detwiler AC, Roginskaya V, Van Houten B, Bruchez MP, Stewart-Ornstein J, Opresko PL. 2022a. Telomeric 8-oxo-guanine drives rapid premature senescence in the absence of telomere shortening. *Nat Struct Mol Biol* **29**: 639–652. doi:10.1038/s41594-022-00790-y

Barnes RP, Thosar SA, Fouquerel E, Opresko PL. 2022b. Targeted formation of 8-oxoguanine in telomeres. *Methods Mol Biol* **2444**: 141–159. doi:10.1007/978-1-0716-2063-2_9

Barnes RP, Thosar SA, Opresko PL. 2023. Telomere fragility and MiDAS: managing the gaps at the end of the road. *Genes (Basel)* **14**: 348. doi:10.3390/genes14020348

Bielskute S, Plavec J, Podbevšek P. 2019. Impact of oxidative lesions on the human telomeric G-quadruplex. *J Am Chem Soc* **141**: 2594–2603. doi:10.1021/jacs.8b12748

Brandt M, Dörschmann H, Khraisat S, Knopp T, Ringen J, Kalinovic S, Garlapati V, Siemer S, Molitor M, Göbel S, et al. 2022. Telomere shortening in hypertensive heart disease depends on oxidative DNA damage and predicts impaired recovery of cardiac function in heart failure. *Hypertension* **79**: 2173–2184. doi:10.1161/HYPERTENSIONAHA.121.18935

Brenner KA, Nandakumar J. 2022. Consequences of telomere replication failure: the other end-replication problem. *Trends Biochem Sci* **47**: 506–517. doi:10.1016/j.tibs.2022.03.013

Burak MJ, Guja KE, Hambardjieva E, Derkunt B, Garcia-Diaz M. 2016. A fidelity mechanism in DNA polymerase lambda promotes error-free bypass of 8-oxo-dG. *EMBO J* **35**: 2045–2059. doi:10.15252/embj.201694332

Cadet J, Wagner JR. 2013. DNA base damage by reactive oxygen species, oxidizing agents, and UV radiation. *Cold Spring Harb Perspect Biol* **5**: a012559. doi:10.1101/cshperspect.a012559

Caldecott KW. 2024. Causes and consequences of DNA single-strand breaks. *Trends Biochem Sci* **49**: 68–78. doi:10.1016/j.tibs.2023.11.001

Castillo-González C, Barbero Barcenilla B, Young PG, Hall E, Shippen DE. 2022. Quantification of 8-oxoG in plant telomeres. *Int J Mol Sci* **23**: 4990. doi:10.3390/ijms23094990

Cattan V, Mercier N, Gardner JP, Regnault V, Labat C, Mäki-Jouppila J, Nzietchueng R, Benetos A, Kimura M, Aviv A, et al. 2008. Chronic oxidative stress induces a tissue-specific reduction in telomere length in CAST/Ei mice. *Free Radic Biol Med* **44**: 1592–1598. doi:10.1016/j.freeradbiomed.2008.01.007

Chakraborty A, Wakamiya M, Venkova-Canova T, Pandita RK, Aguilera-Aguirre L, Sarker AH, Singh DK, Hosoki K, Wood TG, Sharma G, et al. 2015. Neil2-null mice accumulate oxidized DNA bases in the transcriptionally active sequences of the genome and are susceptible to innate inflammation. *J Biol Chem* **290**: 24636–24648. doi:10.1074/jbc.M115.658146

Chang AC, Ong SG, LaGory EL, Kraft PE, Giaccia AJ, Wu JC, Blau HM. 2016. Telomere shortening and metabolic compromise underlie dystrophic cardiomyopathy. *Proc Natl Acad Sci* **113**: 13120–13125. doi:10.1073/pnas.1615340113

Coluzzi E, Colamartino M, Cozzi R, Leone S, Meneghini C, O'Callaghan N, Sgura A. 2014. Oxidative stress induces persistent telomeric DNA damage responsible for nuclear morphology change in mammalian cells. *PLoS ONE* **9**: e110963. doi:10.1371/journal.pone.0110963

Coluzzi E, Buonsante R, Leone S, Asmar AJ, Miller KL, Cimini D, Sgura A. 2017. Transient ALT activation protects human primary cells from chromosome instability induced by low chronic oxidative stress. *Sci Rep* **7**: 43309. doi:10.1038/srep43309

Crespo-Hernández CE, Close DM, Gorb L, Leszczynski J. 2007. Determination of redox potentials for the Watson–Crick base pairs, DNA nucleosides, and relevant nucleoside analogues. *J Phys Chem B* **111**: 5386–5395. doi:10.1021/jp0684224

d'Adda di Fagagna F, Reaper PM, Clay-Farrace L, Fiegler H, Carr P, Von Zglinicki T, Saretzki G, Carter NP, Jackson SP. 2003. A DNA damage checkpoint response in telomere-initiated senescence. *Nature* **426**: 194–198. doi:10.1038/nature02118

D'Angelo S. 2023. Diet and aging: the role of polyphenol-rich diets in slow down the shortening of telomeres: a review. *Antioxidants (Basel)* **12**: 2086. doi:10.3390/antiox12122086

Demin AA, Hirota K, Tsuda M, Adamowicz M, Hailstone R, Brazina J, Gittens W, Kalasova I, Shao Z, Zha S, et al. 2021. XRCC1 prevents toxic PARP1 trapping during DNA base excision repair. *Mol Cell* **81**: 3018–3030.e5. doi:10.1016/j.molcel.2021.05.009

D'Errico M, Parlanti E, Teson M, de Jesus BM, Degan P, Calcagnile A, Jaruga P, Bjørås M, Crescenzi M, Pedrini AM, et al. 2006. New functions of XPC in the protection of human skin cells from oxidative damage. *EMBO J* **25**: 4305–4315. doi:10.1038/sj.emboj.7601277

De Rosa M, Barnes RP, Detwiler AC, Nyalapatla PR, Wipf P, Opresko PL. 2023. OGG1 and MUTYH repair activity promote telomeric 8-oxoguanine induced senescence in human fibroblasts. bioRxiv doi:10.1101/2023.04.10.536247

De Vitis M, Berardinelli F, Coluzzi E, Marinaccio J, O'Sullivan RJ, Sgura A. 2019. X-rays activate telomeric homologous recombination mediated repair in primary cells. *Cells* **8**: 708. doi:10.3390/cells8070708

Dilley RL, Verma P, Cho NW, Winters HD, Wondisford AR, Greenberg RA. 2016. Break-induced telomere synthesis underlies alternative telomere maintenance. *Nature* **539**: 54–58. doi:10.1038/nature20099

Ertunc O, Smearman E, Zheng Q, Hicks JL, Brosnan-Cashman JA, Jones T, Gomes-Alexandre C, Trabzonlu L, Meeker AK, De Marzo AM, et al. 2024. Chromogenic detection of telomere lengths in situ aids the identification of precancerous lesions in the prostate. *Prostate* **84**: 148–157. doi:10.1002/pros.24633

Forman HJ, Zhang H. 2021. Targeting oxidative stress in disease: promise and limitations of antioxidant therapy. *Nat Rev Drug Discov* **20**: 689–709. doi:10.1038/s41573-021-00233-1

Fouquerel E, Lormand J, Bose A, Lee HT, Kim GS, Li J, Sobol RW, Freudenthal BD, Myong S, Opresko PL. 2016. Oxidative guanine base damage regulates human telomerase activity. *Nat Struct Mol Biol* **23**: 1092–1100. doi:10.1038/nsmb.3319

Fouquerel E, Barnes RP, Uttam S, Watkins SC, Bruchez MP, Opresko PL. 2019. Targeted and persistent 8-oxoguanine base damage at telomeres promotes telomere loss and crisis. *Mol Cell* 75: 117–130.e6. doi:10.1016/j.molcel .2019.04.024

Freudenthal BD, Beard WA, Perera L, Shock DD, Kim T, Schlick T, Wilson SH. 2015. Uncovering the polymerase-induced cytotoxicity of an oxidized nucleotide. *Nature* 517: 635–639. doi:10.1038/nature13886

Fujikawa K, Kamiya H, Yakushiji H, Fujii Y, Nakabeppu Y, Kasai H. 1999. The oxidized forms of dATP are substrates for the human MutT homologue, the hMTH1 protein. *J Biol Chem* 274: 18201–18205. doi:10.1074/jbc.274.26 .18201

Fukuzumi S, Miyao H, Ohkubo K, Suenobu T. 2005. Electron-transfer oxidation properties of DNA bases and DNA oligomers. *J Phys Chem A* 109: 3285–3294. doi:10 .1021/jp0459763

Gopalakrishnan K, Low GKM, Ting APL, Srikanth P, Slijepcevic P, Hande MP. 2010. Hydrogen peroxide induced genomic instability in nucleotide excision repair-deficient lymphoblastoid cells. *Genome Integr* 1: 16. doi:10.1186/ 2041-9414-1-16

Haghdoost S, Sjölander L, Czene S, Harms-Ringdahl M. 2006. The nucleotide pool is a significant target for oxidative stress. *Free Radic Biol Med* 41: 620–626. doi:10 .1016/j.freeradbiomed.2006.05.003

Hashimoto K, Tominaga Y, Nakabeppu Y, Moriya M. 2004. Futile short-patch DNA base excision repair of adenine: 8-oxoguanine mispair. *Nucleic Acids Res* 32: 5928–5934. doi:10.1093/nar/gkh909

Henderson PT, Delaney JC, Muller JG, Neeley WL, Tannenbaum SR, Burrows CJ, Essigmann JM. 2003. The hydantoin lesions formed from oxidation of 7,8-dihydro-8-oxoguanine are potent sources of replication errors in vivo. *Biochemistry* 42: 9257–9262. doi:10.1021/bi0347252

Henle ES, Han Z, Tang N, Rai P, Luo Y, Linn S. 1999. Sequence-specific DNA cleavage by Fe$^{2+}$-mediated fenton reactions has possible biological implications. *J Biol Chem* 274: 962–971. doi:10.1074/jbc.274.2.962

Hill JW, Hazra TK, Izumi T, Mitra S. 2001. Stimulation of human 8-oxoguanine-DNA glycosylase by AP-endonuclease: potential coordination of the initial steps in base excision repair. *Nucleic Acids Res* 29: 430–438. doi:10 .1093/nar/29.2.430

Hogg M, Aller P, Konigsberg W, Wallace SS, Doublié S. 2007. Structural and biochemical investigation of the role in proofreading of a β hairpin loop found in the exonuclease domain of a replicative DNA polymerase of the B family. *J Biol Chem* 282: 1432–1444. doi:10.1074/jbc.M605675200

Hwang H, Kreig A, Calvert J, Lormand J, Kwon Y, Daley JM, Sung P, Opresko PL, Myong S. 2014. Telomeric overhang length determines structural dynamics and accessibility to telomerase and ALT-associated proteins. *Structure* 22: 842–853. doi:10.1016/j.str.2014.03.013

Jurk D, Wilson C, Passos JF, Oakley F, Correia-Melo C, Greaves L, Saretzki G, Fox C, Lawless C, Anderson R, et al. 2014. Chronic inflammation induces telomere dysfunction and accelerates ageing in mice. *Nat Commun* 2: 4172. doi:10.1038/ncomms5172

Klungland A, Höss M, Gunz D, Constantinou A, Clarkson SG, Doetsch PW, Bolton PH, Wood RD, Lindahl T. 1999.

Base excision repair of oxidative DNA damage activated by XPG protein. *Mol Cell* 3: 33–42. doi:10.1016/S1097-2765(00)80172-0

Kolbanovskiy M, Chowdhury MA, Nadkarni A, Broyde S, Geacintov NE, Scicchitano DA, Shafirovich V. 2017. The nonbulky DNA lesions spiroiminodihydantoin and 5-guanidinohydantoin significantly block human RNA polymerase II elongation in vitro. *Biochemistry* 56: 3008–3018. doi:10.1021/acs.biochem.7b00295

Krokan HE, Bjoras M. 2013. Base excision repair. *Cold Spring Harb Perspect Biol* 5: a012583. doi:10.1101/cshper spect.a012583

Kumar N, Theil AF, Roginskaya V, Ali Y, Calderon M, Watkins SC, Barnes RP, Opresko PL, Pines A, Lans H, et al. 2022. Global and transcription-coupled repair of 8-oxoG is initiated by nucleotide excision repair proteins. *Nat Commun* 13: 974. doi:10.1038/s41467-022-28642-9

Lagnado A, Leslie J, Ruchaud-Sparagano MH, Victorelli S, Hirsova P, Ogrodnik M, Collins AL, Vizioli MG, Habiballa L, Saretzki G, et al. 2021. Neutrophils induce paracrine telomere dysfunction and senescence in ROS-dependent manner. *EMBO J* 40: e106048. doi:10.15252/ embj.2020106048

Lee JY, Okumus B, Kim DS, Ha T. 2005. Extreme conformational diversity in human telomeric DNA. *Proc Natl Acad Sci* 102: 18938–18943. doi:10.1073/pnas.0506144102

Lee M, Hills M, Conomos D, Stutz MD, Dagg RA, Lau LM, Reddel RR, Pickett HA. 2014. Telomere extension by telomerase and ALT generates variant repeats by mechanistically distinct processes. *Nucleic Acids Res* 42: 1733–1746. doi:10.1093/nar/gkt1117

Lee HT, Bose A, Lee CY, Opresko PL, Myong S. 2017. Molecular mechanisms by which oxidative DNA damage promotes telomerase activity. *Nucleic Acids Res* 45: 11752–11765. doi:10.1093/nar/gkx789

Lee HT, Sanford S, Paul T, Choe J, Bose A, Opresko PL, Myong S. 2020. Position-dependent effect of guanine base damage and mutations on telomeric G-quadruplex and telomerase extension. *Biochemistry* 59: 2627–2639. doi:10.1021/acs.biochem.0c00434

Lindahl T. 1993. Instability and decay of the primary structure of DNA. *Nature* 362: 709–715. doi:10.1038/36270 9a0

Lonkar P, Dedon PC. 2011. Reactive species and DNA damage in chronic inflammation: reconciling chemical mechanisms and biological fates. *Int J Cancer* 128: 1999–2009. doi:10.1002/ijc.25815

Lormand JD, Buncher N, Murphy CT, Kaur P, Lee MY, Burgers P, Wang H, Kunkel TA, Opresko PL. 2013. DNA polymerase δ stalls on telomeric lagging strand templates independently from G-quadruplex formation. *Nucleic Acids Res* 41: 10323–10333. doi:10.1093/nar/gkt813

Low GKM, Ting APL, Fok EDZ, Gopalakrishnan K, Zeegers D, Khaw AK, Jayapal M, Martinez-Lopez W, Hande MP. 2022. Role of xeroderma pigmentosum D (XPD) protein in genome maintenance in human cells under oxidative stress. *Mutat Res Genet Toxicol Environ Mutagen* 876–877: 503444. doi:10.1016/j.mrgentox.2022.503444

Lu J, Liu Y. 2010. Deletion of Ogg1 DNA glycosylase results in telomere base damage and length alteration in yeast. *EMBO J* 29: 398–409. doi:10.1038/emboj.2009.355

Cite this article as *Cold Spring Harb Perspect Biol* doi: 10.1101/cshperspect.a041707

Lu R, Pickett HA. 2022. Telomeric replication stress: the beginning and the end for alternative lengthening of telomeres cancers. *Open Biol* **12:** 220011. doi:10.1098/rsob.220011

Luo W, Muller JG, Rachlin EM, Burrows CJ. 2001. Characterization of hydantoin products from one-electron oxidation of 8-oxo-7,8-dihydroguanosine in a nucleoside model. *Chem Res Toxicol* **14:** 927–938. doi:10.1021/tx010072j

Luxton JJ, McKenna MJ, Taylor LE, George KA, Zwart SR, Crucian BE, Drel VR, Garrett-Bakelman FE, Mackay MJ, Butler D, et al. 2020. Temporal telomere and DNA damage responses in the space radiation environment. *Cell Rep* **33:** 108435. doi:10.1016/j.celrep.2020.108435

Malinin NL, West XZ, Byzova TV. 2011. Oxidation as "the stress of life." *Aging (Albany NY)* **3:** 906–910. doi:10.18632/aging.100385

Markkanen E. 2017. Not breathing is not an option: how to deal with oxidative DNA damage. *DNA Repair (Amst)* **59:** 82–105. doi:10.1016/j.dnarep.2017.09.007

Martens DS, Nawrot TS. 2016. Air pollution stress and the aging phenotype: the telomere connection. *Curr Environ Health Rep* **3:** 258–269. doi:10.1007/s40572-016-0098-8

McNulty JM, Jerkovic B, Bolton PH, Basu AK. 1998. Replication inhibition and miscoding properties of DNA templates containing a site-specific *cis*-thymine glycol or urea residue. *Chem Res Toxicol* **11:** 666–673. doi:10.1021/tx970225w

Miller AS, Balakrishnan L, Buncher NA, Opresko PL, Bambara RA. 2012. Telomere proteins POT1, TRF1 and TRF2 augment long-patch base excision repair in vitro. *Cell Cycle* **11:** 998–1007. doi:10.4161/cc.11.5.19483

Muftuoglu M, Wong HK, Imam SZ, Wilson DM III, Bohr VA, Opresko PL. 2006. Telomere repeat binding factor 2 interacts with base excision repair proteins and stimulates DNA synthesis by DNA polymerase β. *Cancer Res* **66:** 113–124. doi:10.1158/0008-5472.CAN-05-2742

Muoio D, Laspata N, Dannenberg RL, Curry C, Darkoa-Larbi S, Hedglin M, Uttam S, Fouquerel E. 2024. PARP2 promotes break induced replication-mediated telomere fragility in response to replication stress. *Nat Commun* **15:** 2857. doi:10.1038/s41467-024-47222-7

Nzietchueng R, Elfarra M, Nloga J, Labat C, Carteaux JP, Maureira P, Lacolley P, Villemot JP, Benetos A. 2011. Telomere length in vascular tissues from patients with atherosclerotic disease. *J Nutr Health Aging* **15:** 153–156. doi:10.1007/s12603-011-0029-1

O'Callaghan N, Baack N, Sharif R, Fenech M. 2011. A qPCR-based assay to quantify oxidized guanine and other FPG-sensitive base lesions within telomeric DNA. *BioTechniques* **51:** 403–411. doi:10.2144/000113788

Oikawa S, Tada-Oikawa S, Kawanishi S. 2001. Site-specific DNA damage at the GGG sequence by UVA involves acceleration of telomere shortening. *Biochemistry* **40:** 4763–4768. doi:10.1021/bi002721g

Opresko PL, Fan J, Danzy S, Wilson DM III, Bohr VA. 2005. Oxidative damage in telomeric DNA disrupts recognition by TRF1 and TRF2. *Nucleic Acids Res* **33:** 1230–1239. doi:10.1093/nar/gki273

O'Sullivan JN, Bronner MP, Brentnall TA, Finley JC, Shen WT, Emerson S, Emond MJ, Gollahon KA, Moskovitz AH, Crispin DA, et al. 2002. Chromosomal instability in ulcerative colitis is related to telomere shortening. *Nat Genet* **32:** 280–284. doi:10.1038/ng989

Parrinello S, Samper E, Krtolica A, Goldstein J, Melov S, Campisi J. 2003. Oxygen sensitivity severely limits the replicative lifespan of murine fibroblasts. *Nat Cell Biol* **5:** 741–747. doi:10.1038/ncb1024

Passos JF, Saretzki G, Ahmed S, Nelson G, Richter T, Peters H, Wappler I, Birket MJ, Harold G, Schaeuble K, et al. 2007. Mitochondrial dysfunction accounts for the stochastic heterogeneity in telomere-dependent senescence. *PLoS Biol* **5:** e110. doi:10.1371/journal.pbio.0050110

Reichert S, Stier A. 2017. Does oxidative stress shorten telomeres in vivo? A review. *Biol Lett* **13:** 20170463. doi:10.1098/rsbl.2017.0463

Rey S, Quintavalle C, Burmeister K, Calabrese D, Schlageter M, Quagliata L, Cathomas G, Diebold J, Molinolo A, Heim MH, et al. 2017. Liver damage and senescence increases in patients developing hepatocellular carcinoma. *J Gastroenterol Hepatol* **32:** 1480–1486. doi:10.1111/jgh.13717

Rhee DB, Ghosh A, Lu J, Bohr VA, Liu Y. 2011. Factors that influence telomeric oxidative base damage and repair by DNA glycosylase OGG1. *DNA Repair (Amst)* **10:** 34–44. doi:10.1016/j.dnarep.2010.09.008

Richter T, von Zglinicki T. 2007. A continuous correlation between oxidative stress and telomere shortening in fibroblasts. *Exp Gerontol* **42:** 1039–1042. doi:10.1016/j.exger.2007.08.005

Roldán-Arjona T, Wei YF, Carter KC, Klungland A, Anselmino C, Wang RP, Augustus M, Lindahl T. 1997. Molecular cloning and functional expression of a human cDNA encoding the antimutator enzyme 8-hydroxyguanine-DNA glycosylase. *Proc Natl Acad Sci* **94:** 8016–8020. doi:10.1073/pnas.94.15.8016

Rosenquist TA, Zharkov DO, Grollman AP. 1997. Cloning and characterization of a mammalian 8-oxoguanine DNA glycosylase. *Proc Natl Acad Sci* **94:** 7429–7434. doi:10.1073/pnas.94.14.7429

Rossiello F, Jurk D, Passos JF, d'Adda di Fagagna F. 2022. Telomere dysfunction in ageing and age-related diseases. *Nat Cell Biol* **24:** 135–147. doi:10.1038/s41556-022-00842-x

Sakumi K, Furuichi M, Tsuzuki T, Kakuma T, Kawabata S, Maki H, Sekiguchi M. 1993. Cloning and expression of cDNA for a human enzyme that hydrolyzes 8-oxo-dGTP, a mutagenic substrate for DNA synthesis. *J Biol Chem* **268:** 23524–23530. doi:10.1016/S0021-9258(19)49494-5

Samet JM, Wages PA. 2018. Oxidative stress from environmental exposures. *Curr Opin Toxicol* **7:** 60–66. doi:10.1016/j.cotox.2017.10.008

Sanford SL, Welfer GA, Freudenthal BD, Opresko PL. 2020. Mechanisms of telomerase inhibition by oxidized and therapeutic dNTPs. *Nat Commun* **11:** 5288. doi:10.1038/s41467-020-19115-y

Sanford SL, Welfer GA, Freudenthal BD, Opresko PL. 2021. How DNA damage and non-canonical nucleotides alter the telomerase catalytic cycle. *DNA Repair (Amst)* **107:** 103198. doi:10.1016/j.dnarep.2021.103198

Saretzki G, Murphy MP, von Zglinicki T. 2003. Mitoq counteracts telomere shortening and elongates lifespan of fibroblasts under mild oxidative stress. *Aging Cell* **2:** 141–143. doi:10.1046/j.1474-9728.2003.00040.x

Schreiber V, Dantzer F, Ame JC, de Murcia G. 2006. Poly (ADP-ribose): novel functions for an old molecule. *Nat Rev Mol Cell Biol* **7**: 517–528. doi:10.1038/nrm1963

Sfeir A, Kosiyatrakul ST, Hockemeyer D, MacRae SL, Karlseder J, Schildkraut CL, de Lange T. 2009. Mammalian telomeres resemble fragile sites and require TRF1 for efficient replication. *Cell* **138**: 90–103. doi:10.1016/j.cell.2009.06.021

Sharifi-Rad M, Anil Kumar NV, Zucca P, Varoni EM, Dini L, Panzarini E, Rajkovic J, Tsouh Fokou PV, Azzini E, Peluso I, et al. 2020. Lifestyle, oxidative stress, and antioxidants: back and forth in the pathophysiology of chronic diseases. *Front Physiol* **11**: 694. doi:10.3389/fphys.2020.00694

Sobinoff AP, Pickett HA. 2020. Mechanisms that drive telomere maintenance and recombination in human cancers. *Curr Opin Genet Dev* **60**: 25–30. doi:10.1016/j.gde.2020.02.006

Spivak G. 2015. Nucleotide excision repair in humans. *DNA Repair (Amst)* **36**: 13–18. doi:10.1016/j.dnarep.2015.09.003

Stout GJ, Blasco MA. 2013. Telomere length and telomerase activity impact the UV sensitivity syndrome xeroderma pigmentosum C. *Cancer Res* **73**: 1844–1854. doi:10.1158/0008-5472.CAN-12-3125

Sun L, Tan R, Xu J, LaFace J, Gao Y, Xiao Y, Attar M, Neumann C, Li GM, Su B, et al. 2015. Targeted DNA damage at individual telomeres disrupts their integrity and triggers cell death. *Nucleic Acids Res* **43**: 6334–6347. doi:10.1093/nar/gkv598

Takai H, Smogorzewska A, de Lange T. 2003. DNA damage foci at dysfunctional telomeres. *Curr Biol* **13**: 1549–1556. doi:10.1016/S0960-9822(03)00542-6

Tan J, Wang X, Hwang BJ, Gonzales R, Konen O, Lan L, Lu AL. 2020. An ordered assembly of MYH glycosylase, SIRT6 protein deacetylase, and Rad9-Rad1-Hus1 checkpoint clamp at oxidatively damaged telomeres. *Aging (Albany NY)* **12**: 17761–17785. doi:10.18632/aging.103934

Thosar SA, Barnes RP, Detwiler A, Bhargava R, Wondisford A, O'Sullivan RJ, Opresko PL. 2024. Oxidative guanine base damage plays a dual role in regulating productive ALT-associated homology-directed repair. *Cell Rep* **43**: 113656. doi:10.1016/j.celrep.2023.113656

Tubbs A, Nussenzweig A. 2017. Endogenous DNA damage as a source of genomic instability in cancer. *Cell* **168**: 644–656. doi:10.1016/j.cell.2017.01.002

Valdes AM, Andrew T, Gardner JP, Kimura M, Oelsner E, Cherkas LF, Aviv A, Spector TD. 2005. Obesity, cigarette smoking, and telomere length in women. *Lancet* **366**: 662–664. doi:10.1016/S0140-6736(05)66630-5

Vallabhaneni H, O'Callaghan N, Sidorova J, Liu Y. 2013. Defective repair of oxidative base lesions by the DNA glycosylase Nth1 associates with multiple telomere defects. *PLoS Genet* **9**: e1003639. doi:10.1371/journal.pgen.1003639

van Loon B, Hübscher U. 2009. An 8-oxo-guanine repair pathway coordinated by MUTYH glycosylase and DNA polymerase λ. *Proc Natl Acad Sci* **106**: 18201–18206. doi:10.1073/pnas.0907280106

von Zglinicki T. 2002. Oxidative stress shortens telomeres. *Trends Biochem Sci* **27**: 339–344. doi:10.1016/S0968-0004(02)02110-2

von Zglinicki T, Pilger R, Sitte N. 2000. Accumulation of single-strand breaks is the major cause of telomere shortening in human fibroblasts. *Free Radic Biol Med* **28**: 64–74. doi:10.1016/S0891-5849(99)00207-5

Wallace SS. 2013. DNA glycosylases search for and remove oxidized DNA bases. *Environ Mol Mutagen* **54**: 691–704. doi:10.1002/em.21820

Wang Z, Rhee DB, Lu J, Bohr CT, Zhou F, Vallabhaneni H, de Souza-Pinto NC, Liu Y. 2010. Characterization of oxidative guanine damage and repair in mammalian telomeres. *PLoS Genet* **6**: e1000951. doi:10.1371/journal.pgen.1000951

Wang L, Lu Z, Zhao J, Schank M, Cao D, Dang X, Nguyen LN, Nguyen LNT, Khanal S, Zhang J, et al. 2021. Selective oxidative stress induces dual damage to telomeres and mitochondria in human T cells. *Aging Cell* **20**: e13513. doi:10.1111/acel.13513

Wu J, McKeague M, Sturla SJ. 2018. Nucleotide-resolution genome-wide mapping of oxidative DNA damage by click-code-seq. *J Am Chem Soc* **140**: 9783–9787. doi:10.1021/jacs.8b03715

Yang H, Clendenin WM, Wong D, Demple B, Slupska MM, Chiang JH, Miller JH. 2001. Enhanced activity of adenine-DNA glycosylase (Myh) by apurinic/apyrimidinic endonuclease (APE1) in mammalian base excision repair of an A/GO mismatch. *Nucleic Acids Res* **29**: 743–752. doi:10.1093/nar/29.3.743

Yang Z, Takai KK, Lovejoy CA, de Lange T. 2020. Break-induced replication promotes fragile telomere formation. *Genes Dev* **34**: 1392–1405. doi:10.1101/gad.328575.119

Zahler AM, Williamson JR, Cech TR, Prescott DM. 1991. Inhibition of telomerase by G-quartet DNA structures. *Nature* **350**: 718–720. doi:10.1038/350718a0

Zeman MK, Cimprich KA. 2014. Causes and consequences of replication stress. *Nat Cell Biol* **16**: 2–9. doi:10.1038/ncb2897

Zhang JM, Zou L. 2020. Alternative lengthening of telomeres: from molecular mechanisms to therapeutic outlooks. *Cell Biosci* **10**: 30. doi:10.1186/s13578-020-00391-6

Zhang J, Rane G, Dai X, Shanmugam MK, Arfuso F, Samy RP, Lai MK, Kappei D, Kumar AP, Sethi G. 2016. Ageing and the telomere connection: an intimate relationship with inflammation. *Ageing Res Rev* **25**: 55–69. doi:10.1016/j.arr.2015.11.006

Zhang JM, Yadav T, Ouyang J, Lan L, Zou L. 2019. Alternative lengthening of telomeres through two distinct break-induced replication pathways. *Cell Rep* **26**: 955–968.e3. doi:10.1016/j.celrep.2018.12.102

Zhou X, Meeker AK, Makambi KH, Kosti O, Kallakury BV, Sidawy MK, Loffredo CA, Zheng YL. 2012. Telomere length variation in normal epithelial cells adjacent to tumor: potential biomarker for breast cancer local recurrence. *Carcinogenesis* **33**: 113–118. doi:10.1093/carcin/bgr248

Zhou J, Liu M, Fleming AM, Burrows CJ, Wallace SS. 2013. Neil3 and NEIL1 DNA glycosylases remove oxidative damages from quadruplex DNA and exhibit preferences for lesions in the telomeric sequence context. *J Biol Chem* **288**: 27263–27272. doi:10.1074/jbc.M113.479055

Zhou J, Fleming AM, Averill AM, Burrows CJ, Wallace SS. 2015. The NEIL glycosylases remove oxidized guanine lesions from telomeric and promoter quadruplex DNA structures. *Nucleic Acids Res* **43**: 4039–4054. doi:10.1093/nar/gkv252

Zhou J, Chan J, Lambelé M, Yusufzai T, Stumpff J, Opresko PL, Thali M, Wallace SS. 2017. NEIL3 repairs telomere damage during S phase to secure chromosome segregation at mitosis. *Cell Rep* **20:** 2044–2056. doi:10.1016/j.celrep.2017.08.020

Zuo S, Sasitharan V, Di Tanna GL, Vonk JM, De Vries M, Sherif M, Ádám B, Rivillas JC, Gallo V. 2024. Is exposure to pesticides associated with biological aging? A systematic review and meta-analysis. *Ageing Res Rev* **99:** 102390. doi:10.1016/j.arr.2024.102390

# Telomere Protection in Stem Cells

Marta Markiewicz-Potoczny and Eros Lazzerini Denchi

Laboratory of Genome Integrity, National Cancer Institute (NCI), National Institutes of Health (NIH), Bethesda, Maryland 20894, USA

*Correspondence:* eros.lazzerinidenchi@nih.gov

The natural ends of chromosomes resemble double-strand breaks (DSBs), which would activate the DNA damage response (DDR) pathway without the protection provided by a specialized protein complex called shelterin. Over the past decades, extensive research has uncovered the mechanism of action and the high degree of specialization provided by the shelterin complex to prevent aberrant activation of DNA repair machinery at chromosome ends in somatic cells. However, recent findings have revealed striking differences in the mechanisms of end protection in stem cells compared to somatic cells. In this review, we discuss what is known about the differences between stem cells and somatic cells regarding chromosome end protection.

Extensive research over the last few decades has dissected the mechanisms of end protection in mammalian somatic cells. This large body of work shows that telomere protection is provided by a core six-member protein complex termed shelterin (see de Lange 2005, 2025, and elsewhere in the literature). In somatic cells, a major role in end protection is played by the shelterin components TRF2, TRF1, and POT1. Specifically, TRF2 prevents chromosome ends from being recognized as double-stranded breaks (DSBs) by the ATM kinase, and becoming substrates of the nonhomologous end joining (NHEJ) DNA repair pathway (Celli and de Lange 2005; Denchi and de Lange 2007). POT1 binds the telomeric single-stranded overhang, prevents activation of the ATR-mediated DNA damage response (DDR), and suppresses homologous recombination at telomeres (Hockemeyer et al. 2006; Denchi and de Lange 2007;

Glousker et al. 2020). TRF1 facilitates replication of telomeric repeats, preventing DDR activation and, together with TRF2, suppresses activation of alt-NHEJ (Sfeir et al. 2009, 2010). These fundamental functions of the shelterin complex have been validated in different species as well as in various types of somatic cells, including immortalized fibroblasts, hepatocytes, neurons, epithelial cells, endothelial cells, and cardiomyocytes (van Steensel et al. 1998; Lazzerini Denchi et al. 2006; Martínez et al. 2014; Lobanova et al. 2017).

In contrast, the requirement and function of the shelterin complex in stem cells and during development have not been extensively studied. Stem cells are characterized by the ability to self-renew and differentiate and are essential for maintaining tissue homeostasis. Embryonic stem cells (ESCs) are found in the inner cell mass of embryos at the blastocyst stage and are unique in their ability to generate an entire organism through a hierarchical differentiation process that involves multiple cell divisions. Given the extensive number of divisions required to generate an entire organism, ESCs have a robust rate of telomere elongation, characterized by very high levels of telomerase activity, as well as evidence of alternative lengthening of telomeres (ALT)-like activity (Thomson et al. 1998; Liu et al. 2007; Zalzman et al. 2010; Le et al. 2021).

Indications that certain cells have a different mechanism of end protection compared to somatic cells can be observed during gametogenesis. Indeed, during meiosis, shelterin components are removed from telomeres in a process termed "telomere cap exchange" without compromising telomere integrity. Further evidence comes from the observation that in mouse ESCs (mESCs), TRF2 depletion does not affect telomere or cellular integrity (Markiewicz-Potoczny et al. 2021; Ruis et al. 2021). These data raise fundamental questions about telomere protection in pluripotent stem cells. In this review, we will explore the mechanisms of chromosome end protection exploited by stem cells.

## TELOMERE CAP EXCHANGE DURING MEIOSIS

Meiosis is the process by which four haploid gametes are generated through a single round of DNA replication followed by two cell divisions. During this process, homologous chromosomes are paired and undergo recombination. This is accomplished by the formation of a telomere cluster at the nuclear periphery, termed the "telomere bouquet" (Niwa et al. 2000; Scherthan et al. 2000; de La Roche Saint-André 2008; Moiseeva et al. 2017). The telomere bouquet allows cytoskeletal movements to be transmitted to chromosomes via the telomeres, facilitating homologous chromosome pairing (Scherthan et al. 2000; Li and Liu 2020).

In mammals, the formation of the telomere bouquet is coupled with a process known as "telomere cap exchange," wherein telomeric proteins are replaced to ensure the tethering of chromosome ends to the nuclear envelope (NE).

This process begins in early meiosis (prophase I) when telomeres attach to the inner nuclear membrane (INM) of the NE and move with the help of the cytoskeleton until they cluster in a "bouquet stage." In mammals, the telomere-binding site is formed by the assembly of the meiosis-specific linker of the nucleoskeleton and cytoskeleton complex (LINC), composed of SUN1 and KASH5 (Morimoto et al. 2012; Stewart and Burke 2014; Wang et al. 2020). The LINC complex spans the inner and outer nuclear membranes, with KASH5 interacting with the cytoskeleton and SUN1 tethering telomeres to the INM (Fig. 1; Ding et al. 2007; Morimoto et al. 2012; Horn et al. 2013; Burke 2018).

Another meiosis-specific complex, consisting of TERB1, TERB2, and MAJIN (collectively referred to as the TTM complex), tethers telomeres to the INM by interacting directly with the LINC complex and the double-stranded telomeric DNA-binding protein TRF1 (Fig. 1). The TTM complex binds to telomeric DNA, displaces the shelterin complex, and forms a link between the INM and chromosome ends (Shibuya et al. 2015; Wang et al. 2019). TERB1 is an MYB domain-containing protein required for telomere binding to the NE and for protecting telomeres from erosion during meiosis (Zhang et al. 2022). A second essential domain of TERB1, the TRFB domain, interacts with TRF1 (Shibuya et al. 2014). Highlighting the importance of the telomere cap exchange process, mutations in TERB1, TERB2, and MAJIN have been found in patients with a severe form of infertility known as nonobstructive azoospermia (Alhathal et al. 2020; Salas-Huetos et al. 2021).

In addition to the TTM complex, several other proteins have been shown to play a role in tethering telomeres to the NE during meiosis. Speedy/RINGO rapid inducer of $G_2/M$ progression in oocytes A (SPDYA) anchors telomeres to the INM and recruits CDK2 to telomeres (Fig. 1; Viera et al. 2015; Tu et al. 2017). SPDYA interacts with SUN1 of the LINC complex, and this interaction is crucial for assembling NE-attached telomeres (Chen et al. 2021). Recent research has shown that a meiosis-specific F-box protein, FBXO47, regulates the shelterin com-

Cite this article as *Cold Spring Harb Perspect Biol* doi: 10.1101/cshperspect.a041686

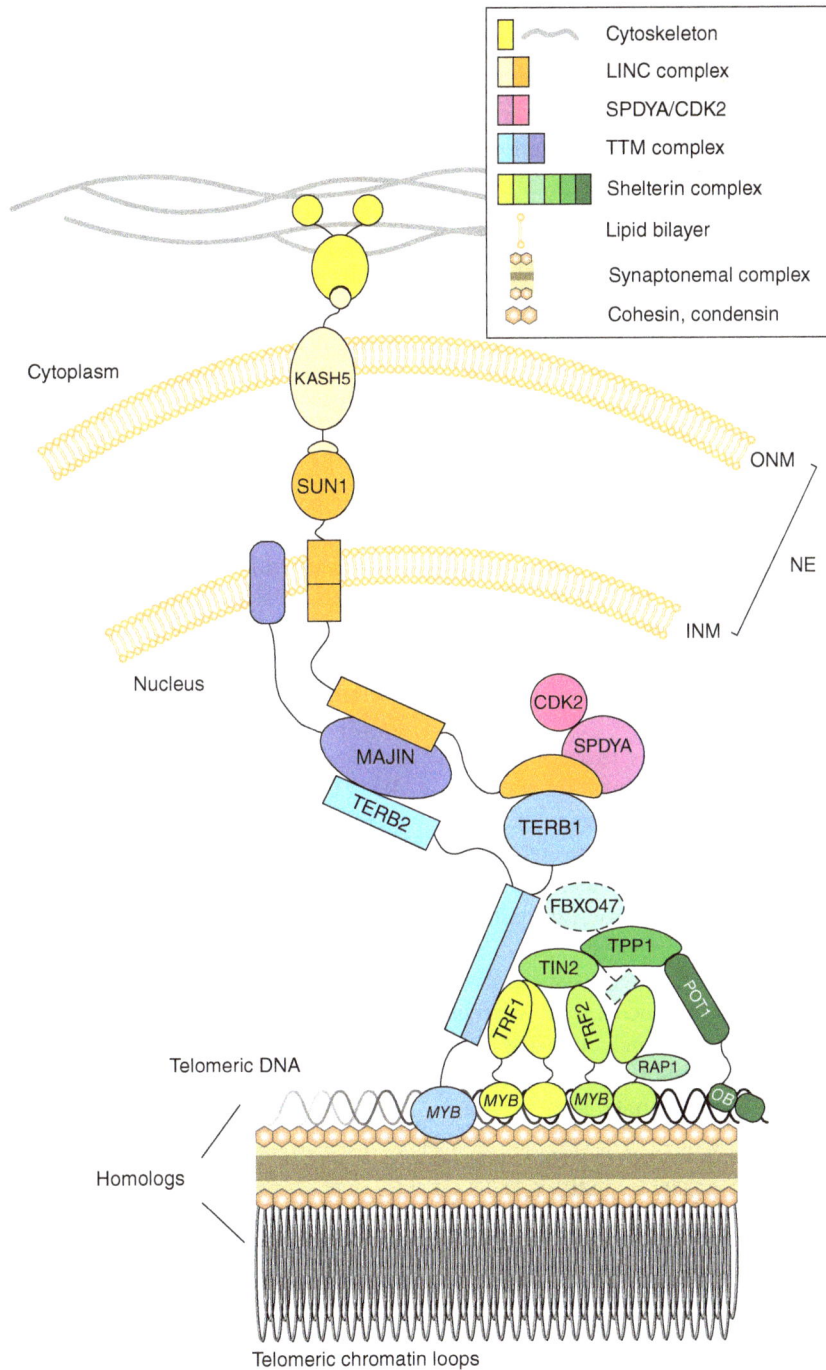

**Figure 1.** Meiosis-specific telomeric complex. Simplistic schematic representation of molecular interactions between telomeres and the nuclear membrane during meiosis. SUN1, a subunit of the linker of the nucleoskeleton and cytoskeleton complex (LINC)-complex, interacts with KASH5, anchoring the LINC complex to the outer nuclear membrane (ONM). Tethering of the telomeres to the nuclear envelope (NE) is promoted by the SUN1 interaction with the TMM complex (TERB1, TERB2, MAJIN) and the TMM–TRF1 interaction. Orange, LINC; blue, TMM; green, shelterin.

plex during meiosis in male mice (Hua et al. 2019; Li and Liu 2020). FBXO47 colocalizes with TRF2 at the nuclear periphery, and depletion of FBXO47 affects telomere bouquet formation, telomere-NE tethering, and chromosome movement (Hua et al. 2019).

Completion of telomere tethering to the NE would not be possible without cohesin, a protein complex conserved from yeast to humans, which is critical for keeping sister chromatids together from S-phase to anaphase during mitosis and meiosis (Fig. 1; Zhang et al. 2008). Although there are some structural differences between the meiotic and mitotic cohesin complexes, recent studies have shown that the meiosis-specific cohesin components also contribute to mitotic chromosome organization in ESCs (Choi et al. 2022). In particular, REC8, previously thought to be expressed only during early meiosis, has been shown to be involved in replication fork progression and mitotic chromosome morphogenesis in ESCs (Choi et al. 2022).

This body of work revealed that during meiosis, the protective shelterin complex is replaced by the TTM complex and accessory interactors, which provide novel functions to telomeres while maintaining telomere protection. These findings highlight the complexity and plasticity of telomere protection during development. The ability of ESCs to utilize the meiosis-specific cohesin complex raises the question of whether other meiotic complexes, such as TTM and LINC, might also have additional functions in ESCs beyond their conventional roles in meiosis.

## TELOMERIC CHROMATIN IN EMBRYONIC STEM CELLS

Epigenetic modifications are crucial for maintaining ESC pluripotency, which is defined as an ability to initiate all cell lineages of a mature organism (Meshorer and Misteli 2006; Pfaff et al. 2013; Kobayashi and Kikyo 2015; Ikeda et al. 2017). In general, stem cells display an open chromatin structure compared to somatic cells, which allows for dynamic and rapid changes in gene expression to drive differentiation (Schlesinger and Meshorer 2019). Both active (acety-

lated H3K9 and H3K4me3) and repressive (H3K9me3 and H3K27me3) chromatin marks undergo global changes during differentiation. This phenomenon also occurs at telomeres, where ESCs show fewer repressive telomeric chromatin marks compared to differentiated cells (Marion et al. 2009; Wong et al. 2009). Additionally, mESCs have higher levels of H3.3 at telomeres than somatic cells (Wong et al. 2009). As differentiation progresses, H3.3 levels at telomeres decrease, accompanied by an increase in repressive heterochromatin marks (Marion et al. 2009).

The ATRX/DAXX chaperone complex deposits H3.3 onto telomeres and pericentric heterochromatin (Lewis et al. 2010; Wong et al. 2010). In ESCs, the DAXX-ATRX-H3.3 pathway is essential for silencing hypomethylated repetitive DNA, whereas in adult somatic tissues, the silencing of repetitive DNA, including telomeres, strongly depends on DNA methylation (Elsässer et al. 2015; He and Ecker 2015). Transcription of telomeric repeats has been shown to generate a long noncoding RNA termed telomeric repeat-containing RNA (TERRA) (Azzalin et al. 2007, and reviewed in depth in Zanella and Doksani 2025). Expression of TERRA can induce telomere fragility and can be suppressed by DNA methylation (Feretzaki et al. 2020). DAXX, ATRX, H3.3, and PML all accumulate at telomeres in ESCs (Wong et al. 2009; Chang et al. 2013), unlike in most somatic cell types. Given ATRX/DAXX's role in protecting telomeres from replication defects (Li et al. 2019), it is plausible that these rapidly proliferating cells require extra protection to manage the replication challenges of telomeric repeats. Supporting this hypothesis, ATRX depletion in checkpoint-proficient human ESCs is lethal (Turkalo et al. 2023). Additionally, ATRX and H3.3 colocalize with telomeric DNA at PML bodies, which serve as platforms for maintaining telomeric chromatin integrity in ESCs (Chang et al. 2013). Interestingly, telomere-associated PML bodies have been linked to the pluripotency of mESCs (Yin et al. 2022).

In mESCs, DAXX and ATRX have been shown to be enriched at tandem repetitive elements, including telomeres and subtelomeres, and have been shown to play a critical role in transcriptional repression and the protection of repetitive elements (He and Ecker 2015). Strikingly, depletion of DAXX or ATRX led to an increase in fragile telomeres and dysregulated telomere length control in cells with reduced levels of DNA methylation, suggesting a critical role for this histone chaperone complex when methylation levels are low (He and Ecker 2015). Future research will likely clarify the connections between histone modifications and DNA methylation, TERRA expression, differentiation, and telomere function, with a particular emphasis on the role of DAXX and ATRX in these processes.

## ROLE OF THE SHELTERIN COMPLEX IN STEM CELLS

The six-subunit shelterin complex is comprised of two double-stranded DNA-binding proteins (TRF1 and TRF2), one single-stranded DNA-binding protein (POT1), and three proteins that do not bind to telomeric DNA directly (TPP1, TIN2, and RAP1). In somatic cells, binding of the shelterin complex to telomeric repeats is crucial for protecting chromosome ends, regulating telomerase recruitment, and recruiting additional factors to facilitate telomere replication and maintain telomere homeostasis. There is no evidence to suggest that the structure and composition of telomere-associated factors differ significantly between stem cells and somatic cells. However, differences in the expression levels of shelterin subunits do exist. For instance, TRF1 is expressed at higher levels in ESCs and induced pluripotent stem cells (iPSCs) compared to somatic cells (Boué et al. 2010). The significance of the increased TRF1 levels in pluripotent cells remains to be elucidated.

## TRF1

In somatic cells, TRF1 is an essential factor that facilitates DNA replication through telomeric repetitive sequences (Sfeir et al. 2009). TRF1 recruits BLM and the transcription and nucleotide excision repair (NER) factor TFIIH to enable telomeric replication to proceed (Martínez

et al. 2009; Sfeir et al. 2009; Yang et al. 2022). Additionally, TRF1 is able to unwrap telomeric nucleosomes, which further aids DNA replication (Hu et al. 2023). In agreement with its critical role in DNA replication at telomeres, TRF1 depletion in somatic cells triggers telomere fragility, a strong ATR-mediated DDR ultimately curbing cellular proliferation (Sfeir et al. 2009). In striking contrast, TRF1 depletion in pluripotent stem cells does not affect proliferation (Ruis et al. 2021). It is currently unclear whether TRF1 in mESCs is dispensable for telomere replication or whether mESCs can tolerate the defects associated with its depletion. Of note, TRF1 has also been implicated in the transcriptional regulation of pluripotency genes in mESCs through the recruitment of the Polycomb-repressive complex 2 (PRC2) (Marión et al. 2019), suggesting that TRF1 might additional, noncanonical roles in pluripotent stem cells.

## TRF2/RAP1

In somatic cells, TRF2 is an essential protein required to suppress ATM activation at telomeres (Denchi and de Lange 2007; Guo et al. 2007). Two nonmutually exclusive models for TRF2-mediated function have been proposed. One model suggests that TRF2 directly suppresses the DDR by inhibiting ATM kinase activation (Karlseder et al. 2004) or other components of the ATM signaling pathway (Okamoto et al. 2013; Myler et al. 2023; Khayat et al. 2024). Additionally, TRF2 facilitates the formation of a secondary protective structure at chromosome ends, known as the "T-loop," which prevents the detection of telomeres as sites of DNA damage (Doksani et al. 2013; Van Ly et al. 2018; Sarek et al. 2019). In line with its central role for end protection, TRF2 deletion causes dramatic phenotypes in somatic cells with a strong activation of the DDR and frequent end-to-end chromosome fusions. In striking contrast, mESCs can tolerate complete TRF2 depletion without significant telomeric defects (Markiewicz-Potoczny et al. 2021; Ruis et al. 2021). TRF2-null mESCs do not accumulate end-to-end chromosomal fusions and can proliferate indefinitely. However, upon differentiation, TRF2-

null mESCs activate a strong DDR, leading to telomere fusions and cell death (Markiewicz-Potoczny et al. 2021). These results suggest that TRF2-dependent end protection is only activated upon differentiation, suggesting a developmental switch that remains to be defined. Nevertheless, substantial evidence suggests that TRF2, while not essential in mESCs, still plays important functions in these cells in terms of telomere protection. Indeed, codepletion of TRF2 and additional shelterin components results in strong DDR activation and frequent NHEJ-dependent telomere fusions (Markiewicz-Potoczny et al. 2021; Ruis et al. 2021). Specifically, depletion of POT1B (or POT1A) in TRF2-null mESCs, but not in TRF2-proficient mESCs, leads to a complete loss of end protection (Markiewicz-Potoczny et al. 2021). Similarly, while TRF1 deletion alone does not cause telomere fusions in mESCs, codepletion with TRF2 results in significant telomere dysfunction and frequent fusions (Ruis et al. 2021). Moreover, the loss of both TRF2 and TPP1 from ESCs, but not the loss of either protein alone, induces telomere fusions and DDR activation (Ruis et al. 2021). These data strongly indicate that TRF2, while not essential for telomere protection on its own, contributes to end protection in conjunction with other shelterin components in mESCs.

TRF2 recruits the Repressor/Activator Protein 1 (RAP1), an evolutionarily conserved protein that provides major protective roles in yeast. In mammals, the precise role of RAP1 in telomere protection is less defined. RAP1-depleted mice are viable and do not display overt telomere defects (Sfeir et al. 2010), but RAP1 has been shown to protect telomeres in replicative senescence and when tethered to telomeres in the absence of TRF2 (Sarthy et al. 2009; Lototska et al. 2020). RAP1 also has nontelomeric functions, such as modulating the nuclear factor κB (NF-κB) signaling pathway, which regulates immune responses, inflammation, and cell survival (Teo et al. 2010). RAP1 can interact with NF-κB components, influencing its activity and downstream gene expression (Teo et al. 2010). Additionally, RAP1 can act as a transcriptional regulator, either activating or repressing target genes depending on the cellular context (Martinez et al. 2010, 2013). Recently, RAP1 was shown to modulate gene expression independently of TRF2-mediated binding to telomeres (Barry et al. 2022). In turn, RAP1 was shown to interact with the histone acetyltransferase activity of the TIP60/p400 complex. RAP1 enhances the repressive activity of TIP60/p400 across 2-cell (2C)-stage genes, including ZSCAN4 and the endogenous retrovirus MERVL (murine endogenous retrovirus-like element) (Barry et al. 2022). Unlike other shelterin components, RAP1 is not essential for cell viability, as its loss does not lead to cell death, suggesting it has distinct functions from other shelterin proteins, or that RAP1 functions can be compensated by other cellular mechanisms (Karlseder 2003; Celli and de Lange 2005; Barry et al. 2022). These observations indicate that ESCs and somatic cells respond differently to telomeric damage, and further research is needed to understand these differences.

## TIN2

TIN2 interacts with TRF1 and TRF2, serving as a bridge between telomeric repeat-binding factors and the rest of the shelterin complex. Complete loss of TIN2 is embryonic lethal, making it impossible to derive mESCs from $TIN2^{-/-}$ blastocysts (Chiang et al. 2004). However, mESCs with a homozygous mutation in TIN2 (Tin2S341X) that causes premature termination and reduced expression, are viable and exhibit ALT characteristics, as well as up-regulation of genes that are expressed at the 2C developmental stage, collectively known as 2C genes, such as ZSCAN4 (Yin et al. 2022). When induced to differentiate, these cells activate a strong DDR and lose ALT features, resulting in impaired differentiation potential. This further emphasizes that the differentiation status strongly influences the cellular response to telomeric defects.

## TPP1/POT1

TPP1 anchors the single-stranded DNA-binding protein POT1 to TIN2 and the rest of the shelterin complex. TPP1/POT1 play important

Cite this article as *Cold Spring Harb Perspect Biol* doi: 10.1101/cshperspect.a041686

roles in telomere length regulation as well as in telomere protection. A spontaneous point mutation that causes adrenocortical dysplasia in mice has been mapped to the gene encoding TPP1 in a mouse model termed *Acd/Tpp1* (Keegan et al. 2005). This mutation affects the splice donor site of the third intron of TPP1 resulting in the expression of a transcript that retains part of intron 3 and cannot produce a full-length protein (Else et al. 2007). Mice homozygous for the Tpp1/Acd allele are viable but exhibit a complex phenotype that includes growth retardation, hyperpigmentation, infertility, and malformations of the skeletal system (Else et al. 2007). In contrast, complete deletion of TPP1 is essential for mice and somatic cell proliferation (Kibe et al. 2010; Tejera et al. 2010; Takai et al. 2011). In the absence of TPP1, somatic cells initiate a DDR similar to that triggered by POT1 depletion, indicating that TPP1's main role is in recruiting POT1 (Hockemeyer et al. 2007; Kibe et al. 2010). Furthermore, checkpoint suppression is not sufficient to bypass the lethality associated with TPP1 depletion, further indicating the essential nature of TPP1 (Kibe et al. 2010). Strikingly, TPP1 depletion in human ESCs results in progressive telomere shortening (Boyle et al. 2020), consistent with its role in telomerase recruitment (Nandakumar et al. 2012; Zhong et al. 2012), yet cells with an inactive checkpoint can survive until they reach a critical telomere length.

POT1 binds the single-stranded telomeric 3′ overhang, and in doing so prevents activation of the ATR-mediated DDR (Hockemeyer et al. 2006; Denchi and de Lange 2007; Guo et al. 2007; Gong and de Lange 2010). Mice have two genes encoding POT1: POT1a and POT1b. Mice lacking POT1a are not viable, in contrast to viable POT1b-deficient mice (Hockemeyer et al. 2006). Depletion of POT1a or POT1b is tolerated in mESCs (Markiewicz-Potoczny et al. 2021), while the double deletion has not been reported. In human ESCs, POT1 mutations associated with different types of cancer are well-tolerated and result in telomere elongation (Kim et al. 2021). However, whether complete deletion of POT1 in human cells is compatible with proliferation has not been reported. Collectively, the

data reported so far suggest that the TPP1/POT1 complex plays a canonical role in telomere length regulation in pluripotent stem cells.

## RESPONSE TO SHORT TELOMERES IN PLURIPOTENT STEM CELLS

A functional telomere length maintenance pathway is critical for the proliferation and function of ESCs. For example, depletion of the telomerase RNA component in mESCs leads to severe differentiation defects, with several epigenetic defects, including genome-wide DNA hypomethylation and alteration in H3K27(me3) methylation levels (Pucci et al. 2013). Similarly, depletion of the catalytic component of telomerase resulted in severe phenotypes in human ESCs with the accumulation of single-stranded DNA, activation of ATR, and mitotic cell death caused by mitotic catastrophe (Fig. 2; Vessoni et al. 2021). Furthermore, several data show that reactivation of telomerase is critical in iPSCs (Agarwal et al. 2010; Batista et al. 2011), and that up-regulation of hTERT can significantly improve the proliferative ability of human ESCs (Yang et al. 2008). Collectively, these results show that maintaining telomeric repeats is crucial for the proliferation of pluripotent stem cells and suggest that telomerase is the major pathway contributing to telomere homeostasis in these cells.

## DIFFERENT RESPONSE TO TELOMERE DYSFUNCTION IN SOMATIC VERSUS PLURIPOTENT CELLS

How can the striking differences in the cellular response to depletion of the shelterin components (e.g., TRF2) between somatic cells and pluripotent stem cells be explained? Here, we highlight data that suggest divergent scenarios. Future research is needed to shed light on the precise differences between pluripotent stem cells and somatic cells, and to provide an explanation why pluripotent stem cells display so many distinctive features with regard to end protection. The following discussion points are highly speculative and are meant to be nonmu-

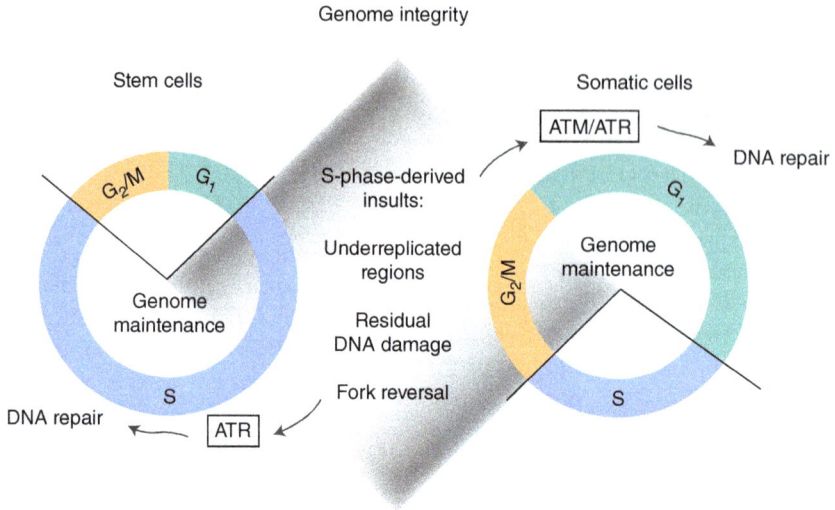

**Figure 2.** Simplistic model of differential control of genome integrity in stem and somatic cells. Underreplicated regions and potential DNA lesions can accumulate in embryonic stem cells and somatic cells at the end of S-phase. Somatic cells, having prolonged $G_1/G_2$ phases of the cell cycle, repair most of those lesions before entering the next S-phase through the ATM-mediated DNA damage response pathway. In contrast, due to the short $G_1/G_2$ phases of the cell cycle, the vast majority of DNA repair in ESCs occurs during S-phase in an ATR-dependent manner.

tually exclusive hypotheses and, as such, ready to be proven incorrect.

## DIRECT SUPPRESSION OF THE DDR

A potential explanation for the unique nature of end protection in pluripotent stem cells is that these cells have a unique tolerance to the type of damage that is exposed when telomeres are unprotected. This could be the result of an intrinsic property of the DDR pathway in these cells, or it could be selectively true at telomeres.

A major difference between ESCs and somatic cells is the cell cycle dynamics, with somatic cells spending most time in $G_1$ and ESCs spending most of their time in S-phase (Fig. 2; Ballabeni et al. 2011). Possibly due to the short cell-cycle gap phases ($G_1/G_2$), ESCs have been shown to accumulate single-stranded DNA gaps, replication fork reversal, replication stress and, as a result, display a basal activation of ATR-mediate DDR activation (Ahuja et al. 2016; Atashpaz et al. 2020). The data reported so far suggest that ESCs are fully capable of ac-

tivating ATR at telomeres similarly to what has been described in somatic cells. Indeed, depletion of POT1 in mESCs triggers ATR activation (Markiewicz-Potoczny et al. 2021). Similarly, critically short telomeres have also been shown to trigger ATR activation (Fig. 2; Vessoni et al. 2021). Moreover, induction of DNA lesions at telomeres using a TRF1-FOK1 nuclease triggers an ATR-dependent DDR in iPSCs, followed by HR-dependent DNA repair (Estep et al. 2024). These data suggest that telomeric lesions that trigger an ATR-mediated DDR elicit a similar response between pluripotent and somatic cells.

On the other hand, several data suggest that telomeric lesions that would normally elicit an ATM-mediated response are not recognized promptly in pluripotent stem cells. Indeed, TRF2 depletion, which triggers a strictly ATM-dependent DDR in somatic cells (Denchi and de Lange 2007), is well tolerated in mESCs (Markiewicz-Potoczny et al. 2021; Ruis et al. 2021). Furthermore, while TRF1-FOK1 expression in iPSCs triggers an ATR-dependent DDR, in mouse embryonic fibroblasts the same type of

Cite this article as *Cold Spring Harb Perspect Biol* doi: 10.1101/cshperspect.a041686

lesion activates ATM (Doksani and de Lange 2016). This suggests that in pluripotent stem cells, either ATM is unable to recognize deprotected telomeres, or additional factors are able to suppress this pathway in the absence of TRF2. Of note, TRF2-depleted mESCs do show a mild growth defect that could be rescued by ATM inhibition (Sarek et al. 2019), suggesting that ATM in this setting is induced possibly to a low level. Indeed, depletion of TRF2 results in a mild accumulation of γH2AX at TRF2-depleted telomeres (Markiewicz-Potoczny et al. 2021), consistent with mild activation of a DDR.

Finally, in TRF2-depleted mESCs, the depletion of other telomeric factors (e.g., POT1b/a, TRF1, TPP1) or the chromatin remodeling factor BRD2 unleashes the DDR, leading to 53BP1 localization and end-to-end chromosome fusions (Markiewicz-Potoczny et al. 2021; Ruis et al. 2021). Importantly, depletion of POT1a/b, TPP1, TRF1, or BRD2 in TRF2-proficient cells does not result in telomere fusions (Markiewicz-Potoczny et al. 2021; Ruis et al. 2021). These data demonstrate that while TRF2 is not essential for telomere protection in mESCs, it does play a role in protecting chromosome ends. However, TRF2 deletion in these cells does not fully activate the ATM-mediated DDR, possibly due to the presence of ESC-specific factors that suppress the DDR at telomeres, or because TRF2's functions are compensated by other proteins in ESCs. Supporting this, a loss-of-function genomic screen identified several genes that are synthetically lethal with TRF2 in mESCs (Markiewicz-Potoczny et al. 2021). Further research is needed to determine the roles of these factors in end protection. However, loss-of-function screens may not identify all essential genes involved in end protection, so complementary approaches such as proteomics or arrayed loss-of-function screens might be necessary to uncover additional ESC-specific factors.

## SECONDARY TELOMERIC STRUCTURE

An alternative explanation for the unique nature of end protection in pluripotent stem cells is related to the formation and stabilization of the protective secondary structure known as the "T-loop" (Doksani et al. 2013; Van Ly et al. 2018; Sarek et al. 2019). In somatic cells, T-loops are strictly dependent on the presence of TRF2 (Doksani and de Lange 2016; Van Ly et al. 2018; Sarek et al. 2019). Strikingly, in mESCs, TRF2 depletion does not impair the frequency at which T-loops can be detected, suggesting the existence of a TRF2-independent mechanism of T-loop formation (Ruis et al. 2021). It is currently unclear whether depletion of additional shelterin components (POT1a/b or TRF1) or the chromatin remodeling factor BRD2 is sufficient to disrupt T-loop formation. This would be the prediction based on the fact that codepletion of TRF2 with these factors leads to loss of end protection and telomere dysfunction. Future research is likely to uncover the mechanism of T-loop formation in pluripotent stem cells and define the role of the T-loop in telomere end protection. It is tempting to speculate that due to the high proliferation rate of ESCs, frequent T-loop resolution may require either a parallel mechanism to ensure T-loop formation or additional T-loop-independent layers of telomere protection.

## EXPRESSION OF PLURIPOTENT-SPECIFIC FACTORS: A POSSIBLE ROLE FOR 2C GENES IN TELOMERE PROTECTION

A simple model to explain the differences between pluripotent stem cells and other cell types is the expression of unique gene(s) involved in telomere protection. The transcriptional program of pluripotent stem cells is unique, and it is characterized by several factors that are not expressed later in development. Furthermore, depletion of shelterin components RAP1, TRF2, and TIN2 in mESCs has been associated with significant changes in gene expression (Teo et al. 2010; Yeung et al. 2013; Markiewicz-Potoczny et al. 2021; Barry et al. 2022; Yin et al. 2022; Liu et al. 2023). RAP1 has been shown to bind and modulate the activity of the TIP60/p400 complex, and, when depleted, resulted in the deregulation of this repressive complex and the induction of gene expression (Barry et al. 2022). This function of RAP1 is independent of its telomeric function, and it is currently unclear whether depletion of TRF2 and/or TIN2 results in gene deregulation in a RAP1/TIP60-dependent manner. Nevertheless, depletion of TRF2, TIN2, and RAP1 have all been shown to trigger up-regulation of a subset of common genes, which include ZSCAN4 and MERVL, as well as other 2C genes (Fig. 3). Murine ZSCAN4 consists of a cluster of highly homologous genes (ZSCAN4a–f) that has been implicated in multiple cellular activities including regulation of transcription, DNA methylation, genome stability, and telomere elongation (Zalzman et al. 2010; Macfarlan et al. 2012).

The function of 2C gene induction upon deletion of telomere-associated genes remains unclear. However, silencing of ZSCAN4 in TRF2-depleted mESCs is sufficient to trigger end-to-end chromosome fusions and induce cell death (Markiewicz-Potoczny et al. 2021). Furthermore, overexpression of ZSCAN4 in somatic cells is sufficient to provide partial suppression of telomere fusions (Markiewicz-Potoczny et al. 2021). These data suggest that

induction of 2C genes might represent a critical mechanism that allows pluripotent stem cells to tolerate depletion of factors that in somatic cells are required for end protection.

ZSCAN4 has also been implicated in recombination-mediated telomere elongation in a process that is similar to the ALT pathway and involves telomeric chromatin decompaction and increased telomeric H3K9me3, telomere elongation, the presence of APBs, and unstable chromosomal ends. In contrast to the canonical ALT pathway, meiosis-specific HR, as well as ATRX/DAXX, are present at ESCs telomeres (Zalzman et al. 2010; Dan et al. 2022; Yin et al. 2022). Indeed, depletion of ZSCAN4 in telomerase-deficient Terc$^{-/-}$ ESCs resulted in significant telomere shortening compared to ZSCAN4-proficient cell lines (Dan et al. 2022). Furthermore, upon depletion of either TIN2 or TRF2, mESCs showed increased levels of ZSCAN4 concomitant with telomere elongation (Markiewicz-Potoczny et al. 2021; Yin et al. 2022). While the mechanism of action of ZSCAN4 function remains to be fully defined, the current prevailing

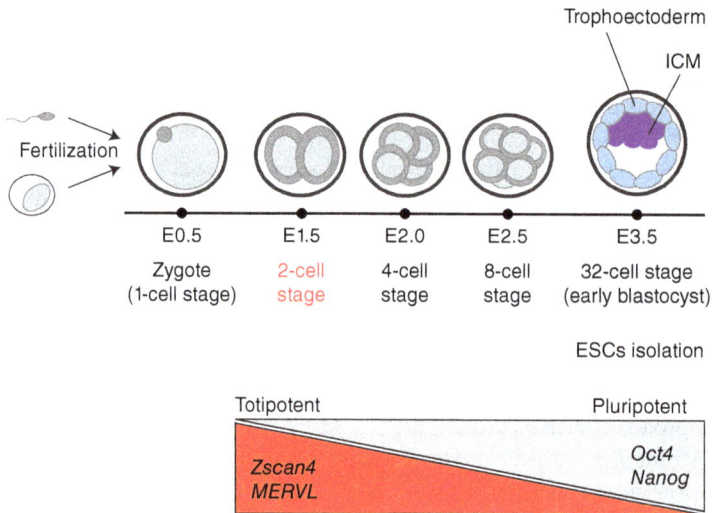

Figure 3. Potency regulation during preimplantation of the mouse embryo. Sequence of events throughout fertilization, zygote formation, and cell lineage generation. Embryonic stem cells (ESCs) are isolated from the E3.5 blastocyst. At the 2-cell (2C) stage, cells can give rise to any cell type or a complete embryo, including extraembryonic tissues. At this time, specific 2C genes, such as ZSCAN4 and MERVL, are up-regulated. After the 8-cell stage, the inner cell mass (ICM) with developing epiblast cells attains the state of pluripotency and competence to self-renew in vitro.

Cite this article as *Cold Spring Harb Perspect Biol* doi: 10.1101/cshperspect.a041686

hypothesis is that it promotes a homologous recombination (HR)-dependent pathway of telomere maintenance.

ZSCAN4 has been implicated in several additional functions beside telomere biology, including the protection of a subset of microsatellite repeat sequences that are prone to genomic instability (Srinivasan et al. 2020; Akiyama et al. 2024), controlling the protein stability of DNA methyltransferases (Dan et al. 2017), ensuring genome stability during reprogramming of iPSCs (Jiang et al. 2013; Cheng et al. 2020), as well as promoting Poly (ADP-ribose) polymerase 1 (PARP1)-dependent DNA repair (Tsai et al. 2023). Furthermore, activation of ZSCAN4 and other 2C genes has been observed not only in response to shelterin perturbations, but also in response to telomere shortening, exposure to genotoxic stress, and inhibition of the ATR signaling pathway (Zalzman et al. 2010; Nakai-Futatsugi and Niwa 2016; Atashpaz et al. 2020; Olbrich et al. 2021; Penev et al. 2022; Akiyama et al. 2024). These data suggest that induction of 2C genes might be a general response that pluripotent stem cells use to respond not only to telomeric damage, but in general to genotoxic lesions.

In striking contrast, somatic cells do not induce 2C gene expression following deletion of TRF2 or other shelterin components. For this reason, the role of 2C genes at telomeres seems to be specific and limited to this particular time point in development (Fig. 3). Nevertheless, ZSCAN4's function seems to extend beyond ESCs, as it was shown to be expressed in a small proportion of human cancers (Lee and Gollahon 2014; Portney et al. 2020; Dan et al. 2022; He et al. 2023). In these cells, ZSCAN4 expression correlates with DNA hypomethylation and facilitates the activation of an alternative mechanism of telomere elongation. These data suggest that pluripotent stem cells as well a subset of cancers may use 2C genes to provide an alternative mechanism of telomere protection and elongation. The exact mechanism of action of ZSCAN4 is still not well understood, and further work is necessary to determine the importance of ZSCAN4 in telomere maintenance during embryonic development and in cancer cells.

## CONCLUSIONS

Chromosome end protection is a fundamental process that enables genome stability while ensuring the cellular proliferation that is required during embryonic development and tissue and organism homeostasis (Fig. 4). The discovery that key factors required for end protection are not essential during early embryogenesis is surprising and raises the possibility that in these early stages of development, there are unique protective mechanism(s) at telomeres. In addition, the existence of alternative mechanisms to ensure telomere protection raises the possibility that cancer cells might hijack these processes to alleviate telomere dysfunction and enable proliferation in conditions of limited telomere maintenance mechanisms. Future work will likely shed light on the mechanism of telomere protection during early development and its potential involvement in telomere protection during other cellular stages.

Interestingly, depletion of telomere-associated factors in ESCs has a major impact on gene expression. These observations raise an intriguing question of how events occurring at chromosomal ends affect gene expression across the genome and whether similar effects are seen in other cell types. Thus far, several reports have shown that loss or displacement of telomere-associated proteins directly affect gene expression (Teo et al. 2010; Yeung et al. 2013; Markiewicz-Potoczny et al. 2021; Barry et al. 2022; Yin et al. 2022; Liu et al. 2023). Possibly, upon deletion of shelterin subunits, the remaining components might relocate and affect gene expression. For example, in the absence of TRF2, delocalization and/or destabilization of its binding partner RAP1 could affect gene expression based on its transcriptional function that is conserved from yeast to mammals (Shore 1994; Martinez et al. 2013; Platt et al. 2013; Yeung et al. 2013). In yeast, telomere shortening triggers the relocation of RAP1 to gene promoters (Platt et al. 2013). In mammals, depletion of RAP1 in mESCs, consistent with observations following TRF2 depletion, induces 2C gene expression (Barry et al. 2022). Thus, it is possible that telomere erosion, as well as the depletion of

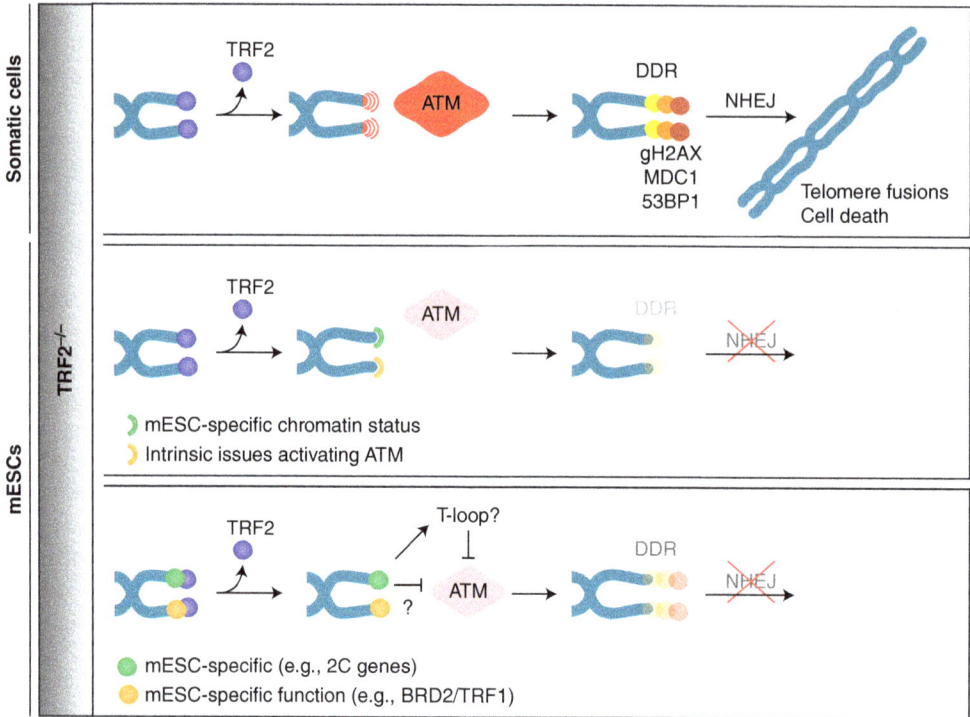

**Figure 4.** Model of cellular response to telomere deprotection. An overview of the different cellular responses to TRF2 depletion in somatic cells and pluripotent stem cells (mouse ESC [mESCs]). In somatic cells (*top* panel), TRF2-depleted telomeres are detected by ATM as sites of DNA damage, leading to the activation of the DNA damage response (DDR) pathway and nonhomologous end joining (NHEJ)-mediated chromosome fusions. In contrast, TRF2 deletion in mESCs does not activate a DDR response and does not result in NHEJ fusions. This could be explained by intrinsic properties of mESCs, such as unique telomeric chromatin status or defects in the ATM signaling pathway (*middle* panel). Alternatively, this could be explained by the expression of mESC-specific telomere protection factors (e.g., 2-cell genes) or by mESC-specific functions of factors that are also expressed in somatic cells (*bottom* panel). An example of the latter is represented by BRD2 and TRF1, factors that are expressed in various cell lines but are required only in ESCs to prevent end-to-end chromosome fusions in the absence of TRF2.

TRF2 or TIN2, could impact the transcriptional function of RAP1 by altering its stability and/or localization. The effects of shelterin removal on the "telomere cap exchange" during meiosis, the potential changes in gene expression, and the role of meiotic complexes at further stages of development, also remain to be verified. Given the role of ZSCAN4 in telomere protection and genome stability (Srinivasan et al. 2020; Markie-wicz-Potoczny et al. 2021; Akiyama et al. 2024), it will be interesting to define other cellular states in which telomere dysfunction could elicit ZSCAN4 induction. This novel finding opens several outstanding questions that integrate as-pects of DNA damage and repair, telomere biol-ogy, and cancer etiology.

Finally, a body of work presented here was performed on pluripotent cells. However, all the questions raised by recent findings are valid and stand for adult stem cells as well. Adult stem cells are noticeably different from ESCs in their pro-liferative capacity and differentiation potential. Telomeres of adult stem cells are shorter than those of pluripotent cells, and generally show low levels of telomerase activity. They also un-dergo replicative senescence, although not as rapidly as proliferating somatic cells (Lupatov and Yarygin 2022). It can be hypothesized that some of the telomere protection mechanisms utilized by ESCs may also be found in adult stem cells. Alternatively, adult stem cells may serve as an intermediate type of cell in which to study the switch between the novel mechanisms observed in ESCs and those known from somatic cells.

## ACKNOWLEDGMENTS

We apologize for the many relevant references that were not included due to space limitations. We are grateful to the members of the ELD laboratory for their feedback and suggestions. Work in the ELD laboratory is supported by the National Institutes of Health (NIH) Intramural Research Program of the NIH, National Cancer Institute (NCI), and Center for Cancer Research, projects 1-ZIA-BC011815 and 1-ZIA-BC 012010.

## REFERENCES

*Reference is also in this subject collection.

Agarwal S, Loh YH, McLoughlin EM, Huang J, Park IH, Miller JD, Huo H, Okuka M, Dos Reis RM, Loewer S, et al. 2010. Telomere elongation in induced pluripotent stem cells from dyskeratosis congenita patients. *Nature* **464**: 292–296. doi:10.1038/nature08792

Ahuja AK, Jodkowska K, Teloni F, Bizard AH, Zellweger R, Herrador R, Ortega S, Hickson ID, Altmeyer M, Mendez J, et al. 2016. A short $G_1$ phase imposes constitutive replication stress and fork remodelling in mouse embryonic stem cells. *Nat Commun* **7**: 10660. doi:10.1038/ncomms10660

Akiyama T, Ishiguro KI, Chikazawa N, Ko SBH, Yukawa M, Ko MSH. 2024. ZSCAN4-binding motif—TGCACAC is conserved and enriched in CA/TG microsatellites in both mouse and human genomes. *DNA Res* **31**: dsad029. doi:10.1093/dnares/dsad029

Alhathal N, Maddirevula S, Coskun S, Alali H, Assoum M, Morris T, Deek HA, Hamed SA, Alsuhaibani S, Mirdawi A, et al. 2020. A genomics approach to male infertility. *Genet Med* **22**: 1967–1975. doi:10.1038/s41436-020-0916-0

Atashpaz S, Samadi Shams S, Gonzalez JM, Sebestyen E, Arghavanifard N, Gnocchi A, Albers E, Minardi S, Faga G, Soffientini P, et al. 2020. ATR expands embryonic stem cell fate potential in response to replication stress. *eLife* **9**: e54756. doi:10.7554/eLife.54756

Azzalin CM, Reichenbach P, Khoriauli L, Giulotto E, Lingner J. 2007. Telomeric repeatcontaining RNA and RNA surveillance factors at mammalian chromosome ends. *Science* **318**: 798–801. doi:10.1126/science.1147182

Ballabeni A, Park IH, Zhao R, Wang W, Lerou PH, Daley GQ, Kirschner MW. 2011. Cell cycle adaptations of embryonic stem cells. *Proc Natl Acad Sci* **108**: 19252–19257. doi:10.1073/pnas.1116794108

Barry RM, Sacco O, Mameri A, Stojaspal M, Kartsonis W, Shah P, De Ioannes P, Hofr C, Côté J, Sfeir A. 2022. Rap1 regulates TIP60 function during fate transition between two-cell-like and pluripotent states. *Genes Dev* **36**: 313–330. doi:10.1101/gad.349039.121

Batista LF, Pech MF, Zhong FL, Nguyen HN, Xie KT, Zaug AJ, Crary SM, Choi J, Sebastiano V, Cherry A, et al. 2011. Telomere shortening and loss of self-renewal in dyskeratosis congenita induced pluripotent stem cells. *Nature* **474**: 399–402. doi:10.1038/nature10084

Boué S, Paramonov I, Barrero MJ, Izpisúa Belmonte JC. 2010. Analysis of human and mouse reprogramming of somatic cells to induced pluripotent stem cells. What is in the plate? *PLoS ONE* **5**: e12664. doi:10.1371/journal.pone.0012664

Boyle JM, Hennick KM, Regalado SG, Vogan JM, Zhang X, Collins K, Hockemeyer D. 2020. Telomere length set point regulation in human pluripotent stem cells critically depends on the shelterin protein TPP1. *Mol Biol Cell* **31**: 2583–2596. doi:10.1091/mbc.E19-08-0447

Burke B. 2018. LINC complexes as regulators of meiosis. *Curr Opin Cell Biol* **52**: 22–29. doi:10.1016/j.ceb.2018.01.005

Celli GB, de Lange T. 2005. DNA processing is not required for ATM-mediated telomere damage response after TRF2 deletion. *Nat Cell Biol* **7**: 712–718. doi:10.1038/ncb1275

Chang FT, McGhie JD, Chan FL, Tang MC, Anderson MA, Mann JR, Andy Choo KH, Wong LH. 2013. PML bodies provide an important platform for the maintenance of telomeric chromatin integrity in embryonic stem cells. *Nucleic Acids Res* **41**: 4447–4458. doi:10.1093/nar/gkt114

Chen Y, Wang Y, Chen J, Zuo W, Fan Y, Huang S, Liu Y, Chen G, Li Q, Li J, et al. 2021. The SUN1-SPDYA interaction plays an essential role in meiosis prophase I. *Nat Commun* **12**: 3176. doi:10.1038/s41467-021-23550-w

Cheng ZL, Zhang ML, Lin HP, Gao C, Song JB, Zheng Z, Li L, Zhang Y, Shen X, Zhang H, et al. 2020. The Zscan4-Tet2 transcription nexus regulates metabolic rewiring and enhances proteostasis to promote reprogramming. *Cell Rep* **32**: 107877. doi:10.1016/j.celrep.2020.107877

Chiang YJ, Kim SH, Tessarollo L, Campisi J, Hodes RJ. 2004. Telomere-associated protein TIN2 is essential for early embryonic development through a telomerase-independent pathway. *Mol Cell Biol* **24**: 6631–6634. doi:10.1128/MCB.24.15.6631-6634.2004

Choi EH, Yoon S, Koh YE, Hong TK, Do JT, Lee BK, Hahn Y, Kim KP. 2022. Meiosis-specific cohesin complexes display essential and distinct roles in mitotic embryonic stem cell chromosomes. *Genome Biol* **23**: 70. doi:10.1186/s13059-022-02632-y

Dan J, Rousseau P, Hardikar S, Veland N, Wong J, Autexier C, Chen T. 2017. Zscan4 inhibits maintenance DNA methylation to facilitate telomere elongation in mouse embryonic stem cells. *Cell Rep* **20**: 1936–1949. doi:10.1016/j.celrep.2017.07.070

Dan J, Zhou Z, Wang F, Wang H, Guo R, Keefe DL, Liu L. 2022. Zscan4 contributes to telomere maintenance in telomerase-deficient late generation mouse ESCs and human ALT cancer cells. *Cells* **11**: 456. doi:10.3390/cells 11030456

de Lange T. 2005. Shelterin: the protein complex that shapes and safeguards human telomeres. *Genes Dev* **19**: 2100–2110. doi:10.1101/gad.1346005

* de Lange T. 2025. Shelterin structure and function. *Cold Spring Harb Perspect Biol* doi:10.1101/cshperspect.a041685

de La Roche Saint-André C. 2008. Alternative ends: telomeres and meiosis. *Biochimie* **90**: 181–189. doi:10.1016/j.biochi.2007.08.010

Denchi EL, de Lange T. 2007. Protection of telomeres through independent control of ATM and ATR by TRF2 and POT1. *Nature* **448**: 1068–1071. doi:10.1038/nature06065

Ding X, Xu R, Yu J, Xu T, Zhuang Y, Han M. 2007. SUN1 is required for telomere attachment to nuclear envelope and gametogenesis in mice. *Dev Cell* **12**: 863–872. doi:10.1016/j.devcel.2007.03.018

Doksani Y, de Lange T. 2016. Telomere-internal double-strand breaks are repaired by homologous recombination and PARP1/Lig3-dependent end-joining. *Cell Rep* **17**: 1646–1656. doi:10.1016/j.celrep.2016.10.008

Doksani Y, Wu JY, de Lange T, Zhuang X. 2013. Super-resolution fluorescence imaging of telomeres reveals TRF2-dependent T-loop formation. *Cell* **155**: 345–356. doi:10.1016/j.cell.2013.09.048

Elsässer SJ, Noh KM, Diaz N, Allis CD, Banaszynski LA. 2015. Histone H3.3 is required for endogenous retroviral element silencing in embryonic stem cells. *Nature* **522**: 240–244. doi:10.1038/nature14345

Else T, Theisen BK, Wu Y, Hutz JE, Keegan CE, Hammer GD, Ferguson DO. 2007. Tpp1/Acd maintains genomic stability through a complex role in telomere protection. *Chromosome Res* **15**: 1001–1013. doi:10.1007/s10577-007-1175-5

Estep KN, Tobias JW, Fernandez RJ, Beveridge BM, Johnson FB. 2024. Telomeric DNA breaks in human induced pluripotent stem cells trigger ATR-mediated arrest and telomerase-independent telomere damage repair. *J Mol Cell Biol* **16**: mjad058. doi:10.1093/jmcb/mjad058

Feretzaki M, Pospisilova M, Valador Fernandes R, Lunardi T, Krejci L, Lingner J. 2020. RAD51-dependent recruitment of TERRA lncRNA to telomeres through R-loops. *Nature* **587**: 303–308. doi:10.1038/s41586-020-2815-6

Glousker G, Briod AS, Quadroni M, Lingner J. 2020. Human shelterin protein POT1 prevents severe telomere instability induced by homology-directed DNA repair. *EMBO J* **39**: e104500. doi:10.15252/embj.2020104500

Gong Y, de Lange T. 2010. A Shld1-controlled POT1a provides support for repression of ATR signaling at telomeres through RPA exclusion. *Mol Cell* **40**: 377–387. doi:10.1016/j.molcel.2010.10.016

Guo X, Deng Y, Lin Y, Cosme-Blanco W, Chan S, He H, Yuan G, Brown EJ, Chang S. 2007. Dysfunctional telomeres activate an ATM-ATR-dependent DNA damage response to suppress tumorigenesis. *EMBO J* **26**: 4709–4719. doi:10.1038/sj.emboj.7601893

He Y, Ecker JR. 2015. Non-CG methylation in the human genome. *Annu Rev Genomics Hum Genet* **16**: 55–77. doi:10.1146/annurev-genom-090413-025437

He HL, Lai HY, Chan TC, Hsing CH, Huang SK, Hsieh KL, Chen TJ, Li WS, Kuo YH, Shiue YL, et al. 2023. Low expression of ZSCAN4 predicts unfavorable outcome in urothelial carcinoma of upper urinary tract and urinary bladder. *World J Surg Oncol* **21**: 62. doi:10.1186/s12957-023-02948-4

Hockemeyer D, Daniels JP, Takai H, de Lange T. 2006. Recent expansion of the telomeric complex in rodents: two distinct POT1 proteins protect mouse telomeres. *Cell* **126**: 63–77. doi:10.1016/j.cell.2006.04.044

Hockemeyer D, Palm W, Else T, Daniels JP, Takai KK, Ye JZ, Keegan CE, de Lange T, Hammer GD. 2007. Telomere protection by mammalian Pot1 requires interaction with Tpp1. *Nat Struct Mol Biol* **14**: 754–761. doi:10.1038/nsmb1270

Horn HF, Kim DI, Wright GD, Wong ES, Stewart CL, Burke B, Roux KJ. 2013. A mammalian KASH domain protein coupling meiotic chromosomes to the cytoskeleton. *J Cell Biol* **202**: 1023–1039. doi:10.1083/jcb.201304004

Hu H, van Roon AM, Ghanim GE, Ahsan B, Oluwole AO, Peak-Chew SY, Robinson CV, Nguyen THD. 2023. Structural basis of telomeric nucleosome recognition by shelterin factor TRF1. *Sci Adv* **9**: eadi4148. doi:10.1126/sciadv.adi4148

Hua R, Wei H, Liu C, Zhang Y, Liu S, Guo Y, Cui Y, Zhang X, Guo X, Li W, et al. 2019. FBXO47 regulates telomere-inner nuclear envelope integration by stabilizing TRF2 during meiosis. *Nucleic Acids Res* **47**: 11755–11770. doi:10.1093/nar/gkz992

Ikeda K, Nagata S, Okitsu T, Takeuchi S. 2017. Cell fiber-based three-dimensional culture system for highly efficient expansion of human induced pluripotent stem cells. *Sci Rep* **7**: 2850. doi:10.1038/s41598-017-03246-2

Jiang J, Lv W, Ye X, Wang L, Zhang M, Yang H, Okuka M, Zhou C, Zhang X, Liu L, et al. 2013. Zscan4 promotes genomic stability during reprogramming and dramatically improves the quality of iPS cells as demonstrated by tetraploid complementation. *Cell Res* **23**: 92–106. doi:10.1038/cr.2012.157

Karlseder J. 2003. Telomere repeat binding factors: keeping the ends in check. *Cancer Lett* **194**: 189–197. doi:10.1016/S0304-3835(02)00706-1

Karlseder J, Hoke K, Mirzoeva OK, Bakkenist C, Kastan MB, Petrini JH, de Lange T. 2004. The telomeric protein TRF2 binds the ATM kinase and can inhibit the ATM-dependent DNA damage response. *PLoS Biol* **2**: E240. doi:10.1371/journal.pbio.0020240

Keegan CE, Hutz JE, Else T, Adamska M, Shah SP, Kent AE, Howes JM, Beamer WG, Hammer GD. 2005. Urogenital and caudal dysgenesis in adrenocortical dysplasia (ACD) mice is caused by a splicing mutation in a novel telomeric regulator. *Hum Mol Genet* **14**: 113–123. doi:10.1093/hmg/ddi011

Khayat F, Alshmery M, Pal M, Oliver AW, Bianchi A. 2024. Binding of the TRF2 iDDR motif to RAD50 highlights a convergent evolutionary strategy to inactivate MRN at telomeres. *Nucleic Acids Res* **52**: 7704–7719. doi:10.1093/nar/gkae509

Kibe T, Osawa GA, Keegan CE, de Lange T. 2010. Telomere protection by TPP1 is mediated by POT1a and POT1b. *Mol Cell Biol* **30**: 1059–1066. doi:10.1128/MCB.01498-09

Kim C, Sung S, Kim JS, Lee H, Jung Y, Shin S, Kim E, Seo JJ, Kim J, Kim D, et al. 2021. Telomeres reforged with non-telomeric sequences in mouse embryonic stem cells. *Nat Commun* **12:** 1097. doi:10.1038/s41467-021-21341-x

Kobayashi H, Kikyo N. 2015. Epigenetic regulation of open chromatin in pluripotent stem cells. *Transl Res* **165:** 18–27. doi:10.1016/j.trsl.2014.03.004

Lazzerini Denchi E, Celli G, de Lange T. 2006. Hepatocytes with extensive telomere deprotection and fusion remain viable and regenerate liver mass through endoreduplication. *Genes Dev* **20:** 2648–2653. doi:10.1101/gad.1453606

Le R, Huang Y, Zhang Y, Wang H, Lin J, Dong Y, Li Z, Guo M, Kou X, Zhao Y, et al. 2021. Dcaf11 activates Zscan4-mediated alternative telomere lengthening in early embryos and embryonic stem cells. *Cell Stem Cell* **28:** 732–747.e9. doi:10.1016/j.stem.2020.11.018

Lee K, Gollahon LS. 2014. Zscan4 interacts directly with human Rap1 in cancer cells regardless of telomerase status. *Cancer Biol Ther* **15:** 1094–1105. doi:10.4161/cbt.29220

Lewis PW, Elsaesser SJ, Noh KM, Stadler SC, Allis CD. 2010. Daxx is an H3.3-specific histone chaperone and cooperates with ATRX in replication-independent chromatin assembly at telomeres. *Proc Natl Acad Sci* **107:** 14075–14080. doi:10.1073/pnas.1008850107

Li M, Liu K. 2020. Protection of the shelterin complex is key for tethering telomeres to the nuclear envelope during meiotic prophase Idagger. *Biol Reprod* **102:** 771–772. doi:10.1093/biolre/ioz231

Li F, Deng Z, Zhang L, Wu C, Jin Y, Hwang I, Vladimirova O, Xu L, Yang L, Lu B, et al. 2019. ATRX loss induces telomere dysfunction and necessitates induction of alternative lengthening of telomeres during human cell immortalization. *EMBO J* **38:** e96659. doi:10.15252/embj.201796659

Liu L, Bailey SM, Okuka M, Muñoz P, Li C, Zhou L, Wu C, Czerwiec E, Sandler L, Seyfang A, et al. 2007. Telomere lengthening early in development. *Nat Cell Biol* **9:** 1436–1441. doi:10.1038/ncb1664

Liu M, Pan H, Kaur P, Wang LJ, Jin M, Detwiler AC, Opresko PL, Tao YJ, Wang H, Riehn R. 2023. Assembly path dependence of telomeric DNA compaction by TRF1, TIN2, and SA1. *Biophys J* **122:** 1822–1832. doi:10.1016/j.bpj.2023.04.014

Lobanova A, She R, Pieraut S, Clapp C, Maximov A, Denchi EL. 2017. Different requirements of functional telomeres in neural stem cells and terminally differentiated neurons. *Genes Dev* **31:** 639–647. doi:10.1101/gad.295402.116

Lototska L, Yue JX, Li J, Giraud-Panis MJ, Songyang Z, Royle NJ, Liti G, Ye J, Gilson E, Mendez-Bermudez A. 2020. Human RAP1 specifically protects telomeres of senescent cells from DNA damage. *EMBO Rep* **21:** e49076. doi:10.15252/embr.201949076

Lupatov AY, Yarygin KN. 2022. Telomeres and telomerase in the control of stem cells. *Biomedicines* **10:** 2335. doi:10.3390/biomedicines10102335

Macfarlan TS, Gifford WD, Driscoll S, Lettieri K, Rowe HM, Bonanomi D, Firth A, Singer O, Trono D, Pfaff SL. 2012. Embryonic stem cell potency fluctuates with endogenous retrovirus activity. *Nature* **487:** 57–63. doi:10.1038/nature11244

Marion RM, Strati K, Li H, Tejera A, Schoeftner S, Ortega S, Serrano M, Blasco MA. 2009. Telomeres acquire embryonic stem cell characteristics in induced pluripotent stem cells. *Cell Stem Cell* **4:** 141–154. doi:10.1016/j.stem.2008.12.010

Marión RM, Montero JJ, López de Silanes I, Graña-Castro O, Martínez P, Schoeftner S, Palacios-Fábrega JA, Blasco MA. 2019. TERRA regulate the transcriptional landscape of pluripotent cells through TRF1-dependent recruitment of PRC2. *eLife* **8:** e44656. doi:10.7554/eLife.44656

Markiewicz-Potoczny M, Lobanova A, Loeb AM, Kirak O, Olbrich T, Ruiz S, Lazzerini Denchi E. 2021. TRF2-mediated telomere protection is dispensable in pluripotent stem cells. *Nature* **589:** 110–115. doi:10.1038/s41586-020-2959-4

Martínez P, Thanasoula M, Muñoz P, Liao C, Tejera A, McNees C, Flores JM, Fernández-Capetillo O, Tarsounas M, Blasco MA. 2009. Increased telomere fragility and fusions resulting from TRF1 deficiency lead to degenerative pathologies and increased cancer in mice. *Genes Dev* **23:** 2060–2075. doi:10.1101/gad.543509

Martinez P, Thanasoula M, Carlos AR, Gómez-López G, Tejera AM, Schoeftner S, Dominguez O, Pisano DG, Tarsounas M, Blasco MA. 2010. Mammalian Rap1 controls telomere function and gene expression through binding to telomeric and extratelomeric sites. *Nat Cell Biol* **12:** 768–780. doi:10.1038/ncb2081

Martinez P, Gómez-López G, García F, Mercken E, Mitchell S, Flores JM, de Cabo R, Blasco MA. 2013. RAP1 protects from obesity through its extratelomeric role regulating gene expression. *Cell Rep* **3:** 2059–2074. doi:10.1016/j.celrep.2013.05.030

Martínez P, Ferrara-Romeo I, Flores JM, Blasco MA. 2014. Essential role for the TRF2 telomere protein in adult skin homeostasis. *Aging Cell* **13:** 656–668. doi:10.1111/acel.12221

Meshorer E, Misteli T. 2006. Chromatin in pluripotent embryonic stem cells and differentiation. *Nat Rev Mol Cell Biol* **7:** 540–546. doi:10.1038/nrm1938

Moiseeva V, Amelina H, Collopy LC, Armstrong CA, Pearson SR, Tomita K. 2017. The telomere bouquet facilitates meiotic prophase progression and exit in fission yeast. *Cell Discov* **3:** 17041. doi:10.1038/celldisc.2017.41

Morimoto A, Shibuya H, Zhu X, Kim J, Ishiguro K, Han M, Watanabe Y. 2012. A conserved KASH domain protein associates with telomeres, SUN1, and dynactin during mammalian meiosis. *J Cell Biol* **198:** 165–172. doi:10.1083/jcb.201204085

Myler LR, Toia B, Vaughan CK, Takai K, Matei AM, Wu P, Paull TT, de Lange T, Lottersberger F. 2023. DNA-PK and the TRF2 iDDR inhibit MRN-initiated resection at leading-end telomeres. *Nat Struct Mol Biol* **30:** 1346–1356. doi:10.1038/s41594-023-01072-x

Nakai-Futatsugi Y, Niwa H. 2016. Zscan4 is activated after telomere shortening in mouse embryonic stem cells. *Stem Cell Reports* **6:** 483–495. doi:10.1016/j.stemcr.2016.02.010

Nandakumar J, Bell CF, Weidenfeld I, Zaug AJ, Leinwand LA, Cech TR. 2012. The TEL patch of telomere protein TPP1 mediates telomerase recruitment and processivity. *Nature* **492:** 285–289. doi:10.1038/nature11648

Niwa O, Shimanuki M, Miki F. 2000. Telomere-led bouquet formation facilitates homologous chromosome pairing and restricts ectopic interaction in fission yeast meiosis. *EMBO J* **19**: 3831–3840. doi:10.1093/emboj/19.14.3831

Okamoto K, Bartocci C, Ouzounov I, Diedrich JK, Yates JR III, Denchi EL. 2013. A two-step mechanism for TRF2-mediated chromosome-end protection. *Nature* **494**: 502–505. doi:10.1038/nature11873

Olbrich T, Vega-Sendino M, Tillo D, Wu W, Zolnerowich N, Pavani R, Tran AD, Domingo CN, Franco M, Markiewicz-Potoczny M, et al. 2021. CTCF is a barrier for 2C-like reprogramming. *Nat Commun* **12**: 4856. doi:10.1038/s41467-021-25072-x

Penev A, Markiewicz-Potoczny M, Sfeir A, Lazzerini Denchi E. 2022. Stem cells at odds with telomere maintenance and protection. *Trends Cell Biol* **32**: 527–536. doi:10.1016/j.tcb.2021.12.007

Pfaff N, Lachmann N, Ackermann M, Kohlscheen S, Brendel C, Maetzig T, Niemann H, Antoniou MN, Grez M, Schambach A, et al. 2013. A ubiquitous chromatin opening element prevents transgene silencing in pluripotent stem cells and their differentiated progeny. *Stem Cells* **31**: 488–499. doi:10.1002/stem.1316

Platt JM, Ryvkin P, Wanat JJ, Donahue G, Ricketts MD, Barrett SP, Waters HJ, Song S, Chavez A, Abdallah KO, et al. 2013. Rap1 relocalization contributes to the chromatin-mediated gene expression profile and pace of cell senescence. *Genes Dev* **27**: 1406–1420. doi:10.1101/gad.218776.113

Portney BA, Arad M, Gupta A, Brown RA, Khatri R, Lin PN, Hebert AM, Angster KH, Silipino LE, Meltzer WA, et al. 2020. ZSCAN4 facilitates chromatin remodeling and promotes the cancer stem cell phenotype. *Oncogene* **39**: 4970–4982. doi:10.1038/s41388-020-1333-1

Pucci M, Rapino C, Di Francesco A, Dainese E, D'Addario C, Maccarrone M. 2013. Epigenetic control of skin differentiation genes by phytocannabinoids. *Br J Pharmacol* **170**: 581–591. doi:10.1111/bph.12309

Ruis P, Van Ly D, Borel V, Kafer GR, McCarthy A, Howell S, Blassberg R, Snijders AP, Briscoe J, Niakan KK, et al. 2021. TRF2-independent chromosome end protection during pluripotency. *Nature* **589**: 103–109. doi:10.1038/s41586-020-2960-y

Salas-Huetos A, Tüttelmann F, Wyrwoll MJ, Kliesch S, Lopes AM, Goncalves J, Boyden SE, Wöste M, Hotaling JM; GEMINI Consortium, et al. 2021. Disruption of human meiotic telomere complex genes TERB1, TERB2 and MAJIN in men with non-obstructive azoospermia. *Hum Genet* **140**: 217–227. doi:10.1007/s00439-020-02236-1

Sarek G, Kotsantis P, Ruis P, Van Ly D, Margalef P, Borel V, Zheng XF, Flynn HR, Snijders AP, Chowdhury D, et al. 2019. CDK phosphorylation of TRF2 controls T-loop dynamics during the cell cycle. *Nature* **575**: 523–527. doi:10.1038/s41586-019-1744-8

Sarthy J, Bae NS, Scrafford J, Baumann P. 2009. Human RAP1 inhibits non-homologous end joining at telomeres. *EMBO J* **28**: 3390–3399. doi:10.1038/emboj.2009.275

Scherthan H, Jerratsch M, Li B, Smith S, Hultén M, Lock T, de Lange T. 2000. Mammalian meiotic telomeres: protein composition and redistribution in relation to nuclear pores. *Mol Biol Cell* **11**: 4189–4203. doi:10.1091/mbc.11.12.4189

Schlesinger S, Meshorer E. 2019. Open chromatin, epigenetic plasticity, and nuclear organization in pluripotency. *Dev Cell* **48**: 135–150. doi:10.1016/j.devcel.2019.01.003

Sfeir A, Kosiyatrakul ST, Hockemeyer D, MacRae SL, Karlseder J, Schildkraut CL, de Lange T. 2009. Mammalian telomeres resemble fragile sites and require TRF1 for efficient replication. *Cell* **138**: 90–103. doi:10.1016/j.cell.2009.06.021

Sfeir A, Kabir S, van Overbeek M, Celli GB, de Lange T. 2010. Loss of Rap1 induces telomere recombination in the absence of NHEJ or a DNA damage signal. *Science* **327**: 1657–1661. doi:10.1126/science.1185100

Shibuya H, Ishiguro K, Watanabe Y. 2014. The TRF1-binding protein TERB1 promotes chromosome movement and telomere rigidity in meiosis. *Nat Cell Biol* **16**: 145–156. doi:10.1038/ncb2896

Shibuya H, Hernández-Hernández A, Morimoto A, Negishi L, Höög C, Watanabe Y. 2015. MAJIN links telomeric DNA to the nuclear membrane by exchanging telomere cap. *Cell* **163**: 1252–1266. doi:10.1016/j.cell.2015.10.030

Shore D. 1994. RAP1: a protean regulator in yeast. *Trends Genet* **10**: 408–412. doi:10.1016/0168-9525(94)90058-2

Srinivasan R, Nady N, Arora N, Hsieh LJ, Swigut T, Narlikar GJ, Wossidlo M, Wysocka J. 2020. Zscan4 binds nucleosomal microsatellite DNA and protects mouse two-cell embryos from DNA damage. *Sci Adv* **6**: eaaz9115. doi:10.1126/sciadv.aaz9115

Stewart CL, Burke B. 2014. The missing LINC: a mammalian KASH-domain protein coupling meiotic chromosomes to the cytoskeleton. *Nucleus* **5**: 3–10. doi:10.4161/nucl.27819

Takai KK, Kibe T, Donigian JR, Frescas D, de Lange T. 2011. Telomere protection by TPP1/POT1 requires tethering to TIN2. *Mol Cell* **44**: 647–659. doi:10.1016/j.molcel.2011.08.043

Tejera AM, Stagno d'Alcontres M, Thanasoula M, Marion RM, Martinez P, Liao C, Flores JM, Tarsounas M, Blasco MA. 2010. TPP1 is required for TERT recruitment, telomere elongation during nuclear reprogramming, and normal skin development in mice. *Dev Cell* **18**: 775–789. doi:10.1016/j.devcel.2010.03.011

Teo H, Ghosh S, Luesch H, Ghosh A, Wong ET, Malik N, Orth A, de Jesus P, Perry AS, Oliver JD, et al. 2010. Telomere-independent Rap1 is an IKK adaptor and regulates NF-κB-dependent gene expression. *Nat Cell Biol* **12**: 758–767. doi:10.1038/ncb2080

Thomson JA, Itskovitz-Eldor J, Shapiro SS, Waknitz MA, Swiergiel JJ, Marshall VS, Jones JM. 1998. Embryonic stem cell lines derived from human blastocysts. *Science* **282**: 1145–1147. doi:10.1126/science.282.5391.1145

Tsai LK, Peng M, Chang CC, Wen L, Liu L, Liang X, Chen YE, Xu J, Sung LY. 2023. ZSCAN4 interacts with PARP1 to promote DNA repair in mouse embryonic stem cells. *Cell Biosci* **13**: 193. doi:10.1186/s13578-023-01140-1

Tu Z, Bayazit MB, Liu H, Zhang J, Busayavalasa K, Risal S, Shao J, Satyanarayana A, Coppola V, Tessarollo L, et al. 2017. Speedy A-Cdk2 binding mediates initial telomere-nuclear envelope attachment during meiotic prophase I independent of Cdk2 activation. *Proc Natl Acad Sci* **114**: 592–597. doi:10.1073/pnas.1618465114

Turkalo TK, Maffia A, Schabort JJ, Regalado SG, Bhakta M, Blanchette M, Spierings DCJ, Lansdorp PM, Hockemeyer

D. 2023. A non-genetic switch triggers alternative telomere lengthening and cellular immortalization in ATRX deficient cells. *Nat Commun* **14:** 939. doi:10.1038/s41467-023-36294-6

Van Ly D, Low RRJ, Frölich S, Bartolec TK, Kafer GR, Pickett HA, Gaus K, Cesare AJ. 2018. Telomere loop dynamics in chromosome end protection. *Mol Cell* **71:** 510–525.e6. doi:10.1016/j.molcel.2018.06.025

van Steensel B, Smogorzewska A, de Lange T. 1998. TRF2 protects human telomeres from end-to-end fusions. *Cell* **92:** 401–413. doi:10.1016/S0092-8674(00)80932-0

Vessoni AT, Zhang T, Quinet A, Jeong HC, Munroe M, Wood M, Tedone E, Vindigni A, Shay JW, Greenberg RA, et al. 2021. Telomere erosion in human pluripotent stem cells leads to ATR-mediated mitotic catastrophe. *J Cell Biol* **220:** e202011014. doi:10.1083/jcb.202011014

Viera A, Alsheimer M, Gómez R, Berenguer I, Ortega S, Symonds CE, Santamaría D, Benavente R, Suja JA. 2015. CDK2 regulates nuclear envelope protein dynamics and telomere attachment in mouse meiotic prophase. *J Cell Sci* **128:** 88–99. doi:10.1242/jcs.154922

Wang Y, Chen Y, Chen J, Wang L, Nie L, Long J, Chang H, Wu J, Huang C, Lei M. 2019. The meiotic TERB1-TERB2-MAJIN complex tethers telomeres to the nuclear envelope. *Nat Commun* **10:** 564. doi:10.1038/s41467-019-08437-1

Wang G, Wu X, Zhou L, Gao S, Yun D, Liang A, Sun F. 2020. Tethering of telomeres to the nuclear envelope is mediated by SUN1-MAJIN and possibly promoted by SPDYA-CDK2 during meiosis. *Front Cell Dev Biol* **8:** 845. doi:10.3389/fcell.2020.00845

Wong LH, Ren H, Williams E, McGhie J, Ahn S, Sim M, Tam A, Earle E, Anderson MA, Mann J, et al. 2009. Histone H3.3 incorporation provides a unique and functionally essential telomeric chromatin in embryonic stem cells. *Genome Res* **19:** 404–414. doi:10.1101/gr.084947.108

Wong LH, McGhie JD, Sim M, Anderson MA, Ahn S, Hannan RD, George AJ, Morgan KA, Mann JR, Choo KH. 2010. ATRX interacts with H3.3 in maintaining telomere structural integrity in pluripotent embryonic stem cells. *Genome Res* **20:** 351–360. doi:10.1101/gr.101477.109

Yang C, Przyborski S, Cooke MJ, Zhang X, Stewart R, Anyfantis G, Atkinson SP, Saretzki G, Armstrong L, Lako M. 2008. A key role for telomerase reverse transcriptase unit in modulating human embryonic stem cell proliferation, cell cycle dynamics, and in vitro differentiation. *Stem Cells* **26:** 850–863. doi:10.1634/stemcells.2007-0677

Yang Z, Sharma K, de Lange T. 2022. TRF1 uses a noncanonical function of TFIIH to promote telomere replication. *Genes Dev* **36:** 956–969. doi:10.1101/gad.349975.122

Yeung F, Ramírez CM, Mateos-Gomez PA, Pinzaru A, Ceccarini G, Kabir S, Fernández-Hernando C, Sfeir A. 2013. Nontelomeric role for Rap1 in regulating metabolism and protecting against obesity. *Cell Rep* **3:** 1847–1856. doi:10.1016/j.celrep.2013.05.032

Yin S, Zhang F, Lin S, Chen W, Weng K, Liu D, Wang C, He Z, Chen Y, Ma W, et al. 2022. TIN2 deficiency leads to ALT-associated phenotypes and differentiation defects in embryonic stem cells. *Stem Cell Reports* **17:** 1183–1197. doi:10.1016/j.stemcr.2022.03.005

Zalzman M, Falco G, Sharova LV, Nishiyama A, Thomas M, Lee SL, Stagg CA, Hoang HG, Yang HT, Indig FE, et al. 2010. Zscan4 regulates telomere elongation and genomic stability in ES cells. *Nature* **464:** 858–863. doi:10.1038/nature08882

* Zanella E, Doksani Y. 2025. In the loop: unusual DNA structures at telomeric repeats and their impact on telomere function. *Cold Spring Harb Perspect Biol* doi:10.1101/cshperspect.a041694

Zhang N, Kuznetsov SG, Sharan SK, Li K, Rao PH, Pati D. 2008. A handcuff model for the cohesin complex. *J Cell Biol* **183:** 1019–1031. doi:10.1083/jcb.200801157

Zhang K, Tarczykowska A, Gupta DK, Pendlebury DF, Zuckerman C, Nandakumar J, Shibuya H. 2022. The TERB1 MYB domain suppresses telomere erosion in meiotic prophase I. *Cell Rep* **38:** 110289. doi:10.1016/j.celrep.2021.110289

Zhong FL, Batista LF, Freund A, Pech MF, Venteicher AS, Artandi SE. 2012. TPP1 OB-fold domain controls telomere maintenance by recruiting telomerase to chromosome ends. *Cell* **150:** 481–494. doi:10.1016/j.cell.2012.07.012

# Telomere Dynamics in Human Health and Disease

Duncan M. Baird

Division of Cancer and Genetics, School of Medicine, Cardiff University, Cardiff CF14 4XN, United Kingdom

*Correspondence:* bairddm@cardiff.ac.uk

Telomere function is critical for genomic stability; in the context of a functional TP53 response, telomere erosion leads to a $G_1/S$ cell-cycle arrest and the induction of replicative senescence, a process that is considered to underpin the ageing process in long-lived species. Abrogation of the TP53 pathway allows for continued cell division, telomere erosion, and the complete loss of telomere function; the ensuing genomic instability facilitates clonal evolution and malignant progression. Telomeres display extensive length heterogeneity in the population that is established at birth, and this affects the individual risk of a broad range of diseases, including cardiovascular disease and cancer. In this perspective, I discuss telomere length heterogeneity at the levels of the population, individual, and cell, and consider how the dynamics of these essential chromosomal structures contribute to human disease.

Telomeres are structures that cap the ends of eukaryotic chromosomes, preventing the recognition and repair of the natural chromosomal terminus by the cellular DNA damage response (DDR) apparatus (de Lange 2005). As a consequence of end-replication losses, telomeres shorten with ongoing cell division. Short telomeres can elicit a TP53-dependent cell-cycle arrest (d'Adda di Fagagna et al. 2003), referred to as replicative senescence, that provides a stringent tumor suppressive function in long-lived species (Deng et al. 2008). However, this natural limitation on proliferative capacity, together with the induction of replicative senescence, is considered a mechanism that may underlie age-related tissue deterioration and disease.

In the absence of a functional checkpoint response, ongoing cell division past the point of replicative senescence results in continued telomere erosion and ultimately the loss of the end-capping function (Counter et al. 1992). Telomeres can then be targeted for DNA repair, resulting in interchromosomal and intrachromosomal fusion of telomeres with other telomeres and nontelomeric loci (Capper et al. 2007). This can lead to the formation of dicentric chromosomes and the initiation of cycles of anaphase-bridging, breakage, and fusion resulting in large-scale genomic rearrangements, including nonreciprocal translocations (Artandi et al. 2000). Other repair processes also operate at short dysfunctional telomeres that can lead to more complex mutational outcomes including chromothripsis (Maciejowski et al. 2015; Cleal et al. 2019). This widespread telomere dysfunction, at the end of the replicative life span, leads to a replicative crisis and autophagic cell death activated via innate immune responses (Nassour et al. 2019, 2023). During crisis, there is a strong selection pressure for the up-regulation of telomere maintenance mechanisms, principally telomerase activity, that facilitates the escape from crisis. The process of crisis escape is considered a key mutational mechanism that drives genomic instability and clonal evolution during malignant progression (Artandi et al. 2000).

Consistent with the concept of antagonistic pleiotropy in long-lived species such as humans, telomeres are considered to play a role in the trade-offs between tumor suppression in early life and age-related disease in later life (Melzer et al. 2020). Consequently, there have been numerous studies associating telomere length with a broad range of human diseases, as well as psychosocial stresses and lifestyle factors, which are considered to be capable of modifying telomere length. From much of this work, causality has not been established and the use of error-prone methodologies together with small cohorts has led to potential biases toward the publication of positive associations (Pepper et al. 2018). However, recent very large-scale population studies and Mendelian randomization have started to bring some clarity to this area and demonstrate potential mechanistic links between telomere length and disease phenotypes (Telomeres Mendelian Randomization Collaboration et al. 2017; Codd et al. 2021). Considering the evolutionary trade-offs that determine species-specific telomere length settings, it is apparent that there is an optimal telomere length in humans that is not too long so to lose the limit on replicative capacity and tumor suppression, but not too short to drive age-related disease (Savage 2024a). This is exemplified by population studies demonstrating increased risks of cancer from constitutively long telomeres and other conditions, including atherosclerosis with shorter telomeres (Aviv 2012; Telomeres Mendelian Randomization Collaboration et al. 2017). Despite these constraints on telomere length, there is considerable telomere length heterogeneity in the human population. Variants within genes required for telomere maintenance can result in affected individuals born with short telomeres that can lead to a broad range of clinical manifestations that are consistent with telomeres limiting replicative life span and the induction of replicative senescence in specific cellular compartments (Armanios 2022). These extreme phenotypes are one end of a predominately genetically determined spectrum of telomere lengths also influenced by prenatal and early life exposures that exhibit a range of phenotypes in the human population.

In this perspective, I discuss telomere length dynamics and how these impact human health and disease.

## LIFELONG TELOMERE LENGTH HETEROGENEITY IS ESTABLISHED AT BIRTH

Telomere length exhibits considerable heterogeneity in the adult population. This is apparent from various cross-sectional surveys but was nicely exemplified in a study by Factor-Litvak et al. (2016), who analyzed telomere lengths in mother, father, and newborn trios using terminal restriction fragment (TRF) analysis. They showed that mean telomere length varied in mothers between 6.19 and 9.81 kb (mean 7.92 kb) and fathers between 5.83 and 9.88 kb (mean 7.70 kb). These ranges of 3.62 and 4.05 kb in mothers and fathers, respectively, represent ~50% of overall mean telomere lengths. However, it was also apparent from this study that the telomere length range in newborns was as great as that observed in the adults, 4.59 kb or 48% of overall mean telomere length (mean 9.50 kb, range 7.01–11.6 kb); indeed, there were no statistically significant differences in the variances between mothers, father, and newborns (Factor-Litvak et al. 2016). These observations were also consistent with other cross sectional studies of telomere length in the human population using different technologies that show a maintenance of telomere length heterogeneity across the age range (De Meyer et al. 2009; Weischer et al. 2013). This fundamentally important observation demonstrates that telomere length heterogeneity is established at birth and is maintained into adulthood. Moreover, these data indicate there is little requirement to invoke additional factors that create further interindividual telo-

Cite this article as *Cold Spring Harb Perspect Biol* doi: 10.1101/cshperspect.a041701

mere length heterogeneity during life. Consistent with these observations, longitudinal studies have shown that telomere length ranking, relative to the age-matched normal length range, changes little during life (Benetos et al. 2013). This is important because the conventional view was that stress and disease during life leads to chronic inflammation and an increase in the turnover of the hematopoietic stem cell (HSC) compartments (Zhang et al. 2016). As telomerase activity in HSCs is insufficient to counteract end-replication losses, increased turnover will be manifested as increased rates of telomere shortening. In this paradigm, telomere length is a biomarker of long-term chronic inflammation (Wong et al. 2014), that may in turn increase risk of disease, including cardiovascular disease (Pusceddu et al. 2020). Importantly also, this paradigm implies the possibility that individual rates of telomere erosion could be modulated by controlling rates of chronic inflammation. In contrast, the concept that telomere length, relative to the age-matched normal length range, is set early in life, implies it is not readily modifiable, that it precedes the onset of disease and may thus be causal. This was further exemplified in a large-scale analysis of leukocyte telomere length in participants of the UK Biobank using qPCR methodology. Analysis of 422,797 individuals showed that several modifiable behaviors and traits were associated with telomere length, but the effects were too small to modify the association of telomere length and disease (Bountziouka et al. 2022). In this study, healthy behaviors accounted for up to just 0.2% of the variation in telomere length; telomere length therefore appears to be set early in life and cannot be significantly changed by lifestyle factors to impact health and disease.

## DETERMINANTS OF TELOMERE LENGTH AT BIRTH

Telomere length at birth is underpinned by genetic and epigenetic factors, as well as maternal health during pregnancy and prenatal conditions. Prenatal adversity and intrauterine exposures can modulate telomere length at birth, as

well as in the placenta, which may functionally contribute to low birth weight (Davy et al. 2009; Biron-Shental et al. 2010a,b). These exposures include maternal psychosocial stress (Entringer et al. 2011), prenatal depression (Garcia-Martin et al. 2021), gestational diabetes mellitus (Xu et al. 2014; Hjort et al. 2018), and metformin treatment (Garcia-Martin et al. 2018). The underlying mechanisms have not been established, but these factors potentially contribute to telomere length heterogeneity at birth.

Telomere length in adult females is longer than in males (Gardner et al. 2014); this has previously been attributed to sex-specific differences in rates of telomere attrition with age (Bayne et al. 2007). However, it is now apparent that this is established at birth and thus may be genetically or prenatally determined (Factor-Litvak et al. 2016). In addition, the increase in paternal germline telomere length as a function of age (Allsopp et al. 1992; Baird et al. 2006) likely accounts for the correlation between paternal age at conception and telomere length in newborns, such that older fathers have children with longer telomeres (Unryn et al. 2005; De Meyer et al. 2007; Njajou et al. 2007; Arbeev et al. 2011).

Twin studies have been used to estimate the genetic contribution to telomere length heterogeneity in the population. These studies have consistently shown higher correlations in telomere length between monozygotic twins compared to dizygotic twins, indicating a substantial genetic influence on telomere length. Heritability estimates from these studies vary between 30% and 80% (Slagboom et al. 1994; Bischoff et al. 2005; Vasa-Nicotera et al. 2005; Andrew et al. 2006; Broer et al. 2013). This variation in heritability estimates across studies is likely to be largely due to methodological variability in study design, sample size, telomere length measurement techniques, and statistical analysis methods. Moreover, the dynamic nature of telomere length as a function of age, together with potential environmental interactions can contribute to differences in heritability estimates of telomere length.

Several genome-wide association studies (GWAS) have identified key loci that contribute to telomere length heterogeneity (Codd et al. 2010, 2013; Pooley et al. 2013; Li et al. 2020). One very large GWAS on telomere length in the human population using telomere length data from 472,174 participants of the UK Biobank, identified numerous genetic variants associated with telomere length variation at 138 genomic loci, shedding light on the molecular mechanisms underlying telomere biology and its implications for human health (Codd et al. 2021). GWAS have identified variants in genes directly involved in telomere maintenance pathways, such as those encoding components of the telomerase enzyme and the shelterin protein complex. For example, variants near the TERT and TERC genes, encoding the catalytic and RNA subunits of telomerase, have been associated with telomere length variation (Codd et al. 2013). Similarly, variants near genes encoding components of the shelterin complex proteins, including *ACD, TERF1, TERF2,* and *POT1,* and all three components of the CST complex *CTC1, SNT1,* and *TEN1,* have all been implicated in telomere length regulation (Codd et al. 2021), as have the *ATRX, PML,* and *SLX4* genes, whose encoded proteins are implicated in the establishment of the alternative lengthening of telomeres pathway (ALT), and the *UPS7* gene, the protein of which deubiquitinates ACD and POT1. In addition to known telomere maintenance genes, this GWAS uncovered 108 genomic loci not previously associated with telomere length variation. These loci include genes *RPA1* and *RPA2* with roles in DNA replication, including telomeres, genes involved in DNA repair activity, *SLX4, MCM4,* and *SAMHD1* at telomeres, but also genes encoding the translesion polymerases POLI and POLN not previously implicated in telomere metabolism (Codd et al. 2021).

The establishment of telomere length is therefore influenced by multiple genetic and environmental factors. It is clear from these studies that telomere length determination is genetically complex with numerous quantitative trait loci contributing to interindividual heterogeneity in telomere length at birth, which is modulated by prenatal conditions and parental age. Importantly, however, despite this considerable heterogeneity, numerous Mendelian randomization studies have demonstrated a causal relationship between heritable long or short telomeres and a wide range of diseases.

## TELOMERE DYNAMICS IN HUMAN CELLS

As outlined above, telomeres display considerable length heterogeneity in the human population, with the interindividual range in length representing nearly 50% of overall mean telomere length at birth. At the cellular level, telomere length can be considerably more heterogeneous. The full extent of telomere length heterogeneity was revealed using single telomere length analysis (STELA), a process that allows the full spectrum of telomere lengths to be determined from single chromosome ends, at the single-molecule level (Baird et al. 2003). For example, STELA of immune cell subsets revealed single telomeres in normal purified B cells obtained from healthy adults ranging from 1.6 to 22 kb, a range of 20.4 kb with a mean 9.9 kb over 200% of mean telomere length (Lin et al. 2010). Similar ranges were observed with STELA in T-cell subsets (Ahmed et al. 2016, 2020; Roger et al. 2023) and other normal tissues, including colorectal mucosa, gastric mucosa, and squamous esophageal epithelium (Roger et al. 2013; Letsolo et al. 2017). At each chromosome end, individuals inherit two telomeric alleles of unitary length; however, the natural dynamics of telomeres with ongoing cell division means that these individual alleles become progressively more heterogeneous during life.

In the absence of telomerase activity, the end-replication problem, coupled with DNA processing to create 3′ overhangs, leads to the gradual erosion of telomeres with ongoing cell division (Deng et al. 2008). Mathematical modeling of this process, with telomeres starting from a single telomere of unitary length, predicts a gradual decrease in the mean and increase in the variance of telomere length distributions with ongoing cell division (Levy et al. 1992). These models are consistent with the dynamics of telomeres observed in human cell cultures using TRF analysis of telomere length (Harley et al. 1990), but was most clearly observed using STELA where the analysis of single

 Cite this article as *Cold Spring Harb Perspect Biol* doi: 10.1101/cshperspect.a041701

telomeric alleles in clonal populations of cells revealed a clear decrease in mean with a commensurate increase in variance of the distributions consistent with telomere erosion occurring primarily by end-replication losses (Baird et al. 2003). Interestingly, however, additional rare outlying short telomeres were observed arising because of substantial telomere length changes. These events were detected in young proliferating cells, senescent cells, and cells expressing telomerase, as well as normal and malignant tissues and thus appear to be a normal aspect of telomere dynamics (Baird 2008). These apparently stochastic telomeric deletion events were not consistent with end-replication losses and created telomeres shorter than those observed in senescent cells. They did not appear to accumulate in culture with cell division, indicating that these shortened telomeres may have been processed, for example, by being re-lengthened or by being subjected to DNA repair activity to create a fused telomere; alternatively, the cells in which telomere deletion arose exited the cell cycle or were subjected to apoptosis. The underlying mechanism for these specific telomeric deletion events has not been formally established; however, they may relate to the phenomenon of telomere trimming, arising from the resolution of T-loops, in a process involving regulator of telomere elongation helicase 1 (Vannier et al. 2012).

In addition to end-replication losses, terminal processing, and stochastic deletion events, telomere dynamics are further complicated by the action of telomerase. Telomerase-mediated telomere elongation is dynamic and stochastic, such that different telomeres may receive distinct telomere elongation. During elongation, short telomeres are preferentially elongated by telomerase, but in equilibrium conditions, all telomeres are elongated equally adding ~60 nt at each telomere (Britt-Compton et al. 2009; Zhao et al. 2009, 2011). Higher-order chromatin structures may modify the accessibility of the chromosomal terminus to telomerase activity and different replication timings during S-phase may lead to differentials in telomeric elongation (Jády et al. 2006; Tomlinson et al. 2006; Chen et al. 2012; Redon et al. 2013).

Thus, replicative history, telomerase activity, and additional telomeric mutation contribute to the considerable intercellular telomere length heterogeneity observed in normal human somatic cells.

## TELOMERE DYSFUNCTION AND FUSION: THE CRISIS PARADIGM

In human cells, with functional DNA damage checkpoint responses, telomere shortening ultimately leads to the loss of the end-protective function and the triggering of a partial DDR referred to as replicative senescence, or mortality stage 1 (Wright and Shay 1992; d'Adda di Fagagna et al. 2003). In addition to a permanent cessation of cell division, cells undergoing replicative senescence typically exhibit characteristic morphological changes such as enlarged and flattened cell morphology, increased granularity, and changes to nuclear morphology (Sikora et al. 2016). Importantly, senescent cells exhibit changes in their secretory phenotype, known as the senescence-associated secretory phenotype (SASP) (Coppé et al. 2008). The SASP is characterized by the secretion of proinflammatory cytokines, chemokines, growth factors, and proteases that contribute to chronic inflammation, tissue remodeling, and the recruitment of immune cells to sites of senescent cell accumulation (Huang et al. 2022). While the SASP can have beneficial effects in certain contexts, such as promoting tissue repair and immune surveillance, chronic SASP activation can drive age-related pathologies, including cancer, cardiovascular disease, and neurodegenerative disorders (Wang et al. 2024a). Thus, telomere erosion limits its replicative life span, and this provides a stringent tumor-suppressive mechanism; however, in long-lived species, the accumulation of nonreplicating senescent cells, together with the SASP, may facilitate tissue deterioration and disease as a function of age (Campisi and Robert 2014; Schmitt et al. 2022).

Short telomeres in senescent cells are detected as DSBs leading to the activation of the key regulators of the DDR pathway ataxia telangiectasia mutated (ATM) and ATM and Rad3-related (ATR) kinases (d'Adda di Fagagna et al.

2003; Nassour et al. 2021). ATM and ATR phosphorylate downstream effector proteins involved in cell-cycle checkpoints, including TP53, which induces the expression of P21, a cyclin-dependent kinase inhibitor that inhibits the activity of cyclin-dependent kinases involved in cell-cycle progression (Shiloh and Ziv 2013). This leads to cell-cycle arrest at the $G_1/S$ checkpoint, preventing the proliferation of damaged cells. While senescent cells are subjected to a cell-cycle arrest, the telomeres are not repaired, and, thus, despite the presence of short telomeres, telomere fusion in senescent cells is no more common than that observed in young proliferating cells (Counter et al. 1992; Capper et al. 2007; Cesare et al. 2009; Kaul et al. 2012). The underling mechanism for the lack of repair of short telomeres in senescent cells is not clear; however, it is likely that these telomeres may retain sufficient TRF2 to inhibit DNA repair activity and fusions (Cesare and Karlseder 2012). Moreover, extensive chromatin remodeling occurs in senescent cells, which leads to the formation of senescence-associated heterochromatin foci (SAHF) (Narita et al. 2003). These densely packed heterochromatic regions contribute to the stable repression of cell-cycle genes and reinforce the irreversible growth arrest characteristic of senescence. SAHF formation can also lead to repression of genes involved in the DDR and may further prevent access of repair proteins to telomeres (Di Micco et al. 2011).

In the absence of functional TP53, the expression of P21 and other senescence-associated genes is reduced, and the presence of short telomeres does not illicit cell-cycle arrest. In this situation, cells continue to divide beyond their normal replicative capacity and telomeres continue to shorten (Counter et al. 1992). The extended life span of these cultures varies between cell types, but typically human fibroblast cultures exhibit additional 20–40 population doublings (PDs) before entering a state referred to as replicative crisis, or mortality stage 2 (Wright and Shay 1992). During a crisis, the expansion of the culture slows and stops; over time, cell death becomes greater than cell division and the culture comes to an end. Crisis is characterized by the presence of telomere fusion

events, widespread genomic instability, and autophagy-dependent cell death mediated via cGAS (cyclic GMP-AMP synthase)-STING (stimulator of interferon genes) (Counter et al. 1992; Capper et al. 2007; Nassour et al. 2019). Abrogation of the cGAS-STING pathway allows cells to continue to divide beyond M2, to a third cell growth plateau, M3 (Nassour et al. 2021). M2 and M3 crises represent the final proliferative life span barriers of human cells. Cells can only permanently escape crisis by the activation of a telomere maintenance mechanism to prevent telomere erosion, either via the establishment of telomerase activity, or following the induction of the alternative lengthening of telomeres pathway. The mechanism underlying the induction of crisis and escape is reviewed in Karlseder et al. (2024).

## THE MUTATIONAL IMPACT OF TELOMERE DYSFUNCTION AND FUSION

Telomere crisis represents the final proliferative life span barrier that cells must overcome to progress to malignancy. The widespread genome instability during crisis provides the genetic variation on which clonal selection can operate to drive clonal evolution. Thus, crisis is considered to be a critical event in tumorigenesis and is associated with the acquisition of genomic alterations that promote cancer development (Artandi and DePinho 2010).

Sensitive single-molecule PCR techniques allow for the detection and sequence characterization of single telomere fusion events in a background of $10^5$–$10^6$ cells. Telomere fusion events are rare in normal cells, occurring at similar frequencies irrespective of whether they are young and capable of proliferation or whether they are undergoing replicative senescence (Capper et al. 2007). The DNA sequence of rare telomere fusion events detected in normal cells revealed short telomeres at the fusion point that were considerably shorter than those observed in the bulk distribution, consistent with fusion between stochastic telomeric deletion events in normal cells (Capper et al. 2007). Thus, normal cells with otherwise intact and functional telomeres can be subjected to spora-

Cite this article as *Cold Spring Harb Perspect Biol* doi: 10.1101/cshperspect.a041701

dic telomeric deletion, and these telomeres can undergo telomere fusion, which can lead to the induction of large-scale chromosomal mutation.

Following the experimental abrogation of the TP53 pathway, cells continue to divide, and telomeres continue to erode to a point at which they are subjected to DNA repair activity that results in telomere fusion. Telomere fusions can be detected within five PDs from the PD point at which the culture would have undergone replicative senescence (Letsolo et al. 2010; Tankimanova et al. 2012). The frequency of telomere fusion, and the diversity of events generated, progressively increases as the telomeres continue to shorten and the cells enter replicative crisis (Capper et al. 2007; Tankimanova et al. 2012; Jones et al. 2014). Interestingly, as cells progress deeper into crisis, the diversity of telomere fusion events decreases and becomes dominated by a smaller number of clonal events (Capper et al. 2007; Letsolo et al. 2010). This is presumed to reflect the replicative dynamics of cells in culture, with the clonal fusion events being detected in the longest-lived clones. Importantly, this observation demonstrates that telomere fusion is not necessarily catastrophic for a cell undergoing crisis; instead, it can provide a temporary solution to the loss of telomere function and allows for additional replicative cycles.

Dysfunctional telomeres are susceptible to fusion through multiple DNA repair mechanisms. These repair pathways can contribute to genomic instability and chromosomal rearrangements. In the absence of the key shelterin component TRF2, chromosomes are subject to widespread fusions. This dramatic phenotype leads to the formation of long "trains" of chromosomes joined end to end at the telomeres. The formation of these structures is entirely dependent on LIG4, a key component of the classical nonhomologous end-joining (NHEJ) pathway (Smogorzewska et al. 2002). These observations inform the mechanism by which TRF2, and the shelterin complex, confer a key function of telomeres, that of distinguishing the natural end of the chromosome from nontelomeric DSBs and the prevention of aber-

rant DNA repair activity at the chromosomal termini. Telomeres rendered dysfunctional by the experimental loss of TRF2 still contain full lengthened telomere repeat arrays at the fusion points (Smogorzewska et al. 2002; Capper et al. 2007) and thus may not fully recapitulate the nature of naturally occurring dysfunctional telomeres. Indeed, it became apparent that telomeres rendered dysfunctional as a consequence of replicative erosion can be processed differently. The evidence for this came from observations in *Schizosaccharomyces pombe*, *Arabidopsis*, and mice, where fusion of short telomeres was observed in the absence of key components of the classical NHEJ pathway, including KU, DNA PKCS, and LIG4 (Baumann and Cech 2000; Heacock et al. 2004; Maser et al. 2007; Rai et al. 2010). Moreover, DNA sequence analysis of telomere fusion events obtained from human cells undergoing a crisis revealed a distinct mutational profile that was not consistent with classical NHEJ. Instead, fused telomeres were characterized by subtelomeric deletion of one, or both, of the participating telomeres and the presence of DNA sequence microhomology at the fusion point (Capper et al. 2007). This profile is consistent with the microhomology-mediated end-joining pathway (MMEJ) revealed experimentally in the absence of classical NHEJ. MMEJ requires microhomology and is error-prone, resulting in extensive DNA resection activity that creates deletions (Boulton and Jackson 1996; Göttlich et al. 1998; Ma et al. 2003; Yu and Gabriel 2003). MMEJ has been defined in physiological roles for mediating class switch recombination (Pan-Hammarström et al. 2005; Yan et al. 2007; Robert et al. 2009; Boboila et al. 2010a,b) and may be required for the processing of DSBs within repetitive DNA elements (Sfeir and Symington 2015).

Further work to establish the genetic requirements for the fusion of short dysfunctional telomeres revealed that LIG4-dependent classical NHEJ predominantly mediates interchromosomal fusion and MMEJ predominates with intrachromosomal telomere fusion (Jones et al. 2014; Liddiard et al. 2016). Furthermore, LIG1, the replicative ligase, was shown to have an essential and nonredundant role in mediating the fusion of sister chromatid telomeres that was decoupled from its engagement in DNA replication (Jones et al. 2014; Liddiard et al. 2019). DNA polymerase θ (POLQ) has emerged as a critical determinant of MMEJ, and this was first established in the context of telomere fusion in mouse embryonic fibroblasts following the depletion of both TRF1 and TRF2. Whole-genome sequence (WGS) analysis revealed nontelomeric insertions within the telomere–telomere fusion breakpoints that were dependent on POLQ (Mateos-Gomez et al. 2015). It is now apparent that POLQ is the key mediator of MMEJ, or what is now referred to as POLQ-mediated end joining (TMEJ), and this is the primary mechanism for end joining in M-phase (Ramsden et al. 2022; Brambati et al. 2023). Consistent with the roles of classical and MMEJ in mediating interchromosomal and intrachromosomal telomere fusion, POLQ appears to predominantly mediate intrachromosomal telomere fusions, including events involving centromeric satellite repeats (Liddiard et al. 2022).

Telomere dysfunction and fusion is a key mutational mechanism that can generate large-scale genomic rearrangements. The earliest observations of telomere dysfunction leading to genomic mutation were described by Barbara McClintock in the 1930s, who observed cycles of chromosomal breakage, fusion, and anaphase bridging (BFB) initiated following the loss of telomeres (McClintock 1941). In these cycles, dicentric chromosomes that arise from telomere fusion fail to properly segregate and instead form bridges between the dividing cells at anaphase. The bridges can break with the resulting daughter cells acquiring an unequal distribution of genetic material that will depend on whether the fusion is interchromosomal or intrachromosomal (Murnane 2006). Intrachromosomal fusion will lead to a terminal deletion in one daughter cell and inverted repeat in the other. As the broken chromosome ends lack telomeres, they can subsequently fuse with other broken ends; in the case of the daughter cell containing an inverted repeat, subsequent BFB cycles can lead to further amplification and additional deletion in the cell that acquired the initial deletion

(Murnane 2012). BFB cycles can only be stopped following the "healing" of chromosomes with the acquisition of a new telomere, either via the addition of telomere repeat sequences de novo at broken ends, or by translocation with a preexisting telomere (Sabatier et al. 2005).

The BFB paradigm has provided a framework for understanding how short dysfunctional telomeres can drive genomic mutation in cancer. However, the advent of genomic sequencing technologies has revealed the astonishing complexity of structural mutation in cancers, and subsequent data has implicated telomere dysfunction in the initiation of some of these types of events (Maciejowski et al. 2015; Cleal et al. 2019; Dewhurst et al. 2021). One of the first descriptions of these complex mutational phenomena was provided in 2011 following WGS of chronic lymphocytic leukemia (CLL) B cells from a single patient. This analysis revealed 42 rearrangements involving chromosome 4 and nine translocations involving three other chromosomes (Stephens et al. 2011). The breakpoints were clustered with an alternating copy number profile arising because of deletions between the breakpoints. The authors coined the term of chromothripsis to describe this mutational pattern, from the Greek *thripsis* to shatter, it was considered that the chromosome had "shattered" and had been religated in an apparently random order with missing sections of DNA. Subsequently, a large body of literature has described chromothriptic-like mutational patterns in the majority of cancer types, including glioblastoma, melanoma, and lung adenocarcinoma that exhibit chromothripsis in more than 50% of cases and up to 100% of liposarcoma cases (Cortés-Ciriano et al. 2020). Coincidently, chromothripsis is rare (≈1%) in CLL cases, the tumor type in which it was first discovered. Moreover, chromothripsis is not confined to cancer, as it has also been observed in some congenital defects (Kloosterman et al. 2011; Gamba et al. 2015). Other mutational processes have been characterized, including chromoanasynthesis and chromoplexy (Liu et al. 2011; Baca et al. 2013). These mutational phenomena are referred to under the collective term

 Cite this article as *Cold Spring Harb Perspect Biol* doi: 10.1101/cshperspect.a041701

of chromoanagenesis and may represent a continuum of large-scale mutational events, with distinct, but potentially overlapping, underlying mechanisms (Holland and Cleveland 2012). Chromothripsis is the most common and intensively studied; however, the underlying mechanistic basis of chromothripsis and how it can be initiated in the context of a replicative telomere crisis is yet to be fully established and it is possible that multiple mechanisms may have similar mutational outcomes (Liu et al. 2011; Maciejowski et al. 2015; Cleal et al. 2019; Cleal and Baird 2020; Umbreit et al. 2020; Dewhurst et al. 2021).

## TELOMERE DYNAMICS AND DISEASE ASSOCIATIONS

Telomere length heterogeneity in the human population, together with the decline in length as a function of age, has led to considerable interest in the potential association of telomere length and disease. The role that telomere biology plays in human disease is most obviously exemplified in the telomere biology disorders (TBDs). These are a group of rare genetic conditions characterized by abnormalities in telomere length maintenance and function. These disorders result from mutations in genes involved in telomere maintenance pathways, such as telomerase and shelterin complex components. TBDs can affect multiple organ systems and manifest with a wide range of clinical features. Dyskeratosis congenita (DC) is a TBD characterized by the triad of abnormal skin pigmentation, nail dystrophy, and leukoplakia in the mucous membranes. Individuals with DC are at increased risk of bone marrow failure, pulmonary fibrosis, liver disease, and certain cancers (Alter et al. 2009; Schratz and Armanios 2020). Aplastic anemia, in which bone marrow failure results in low blood cell counts, can also be caused by a TBD (Vulliamy et al. 2002). Idiopathic pulmonary fibrosis (IPF) is a progressive lung disease characterized by the formation of fibrotic scar tissue in the lungs, leading to impaired lung function and respiratory symptoms. While most cases of IPF are sporadic, a subset of individuals with familial IPF have mu-

tations in genes associated with telomere maintenance, such as TERT, TERC, and other shelterin complex components (Alder et al. 2008). While TBDs are caused by defined genetic variants, they can also display disease anticipation whereby the severity of the disease increases, and age of onset decreases, between generations (Vulliamy et al. 2004). This can be accounted for by the inheritance of ever shorter telomeres between the generations. In these situations, pedigrees may be observed where grandparental carriers are unaffected but exhibit telomere lengths less than the 50th percentile, a parental carrier with telomeres of around the first to the tenth percentile manifests symptoms in adulthood and their offspring with telomeres shorter than the first percentile exhibiting severe symptoms in early childhood. A detailed description of TBDs and their underlying biology is provided in more detail in the literature (see Savage 2024b).

TBDs provide direct evidence for a role of telomere length in disease but represent the extreme manifestations of telomere dysfunction. As telomere length is a continuous variable in the population, the presence of long and short telomeres and their association with disease has been extensively studied, of particular interest is the relationship between telomere length and cancer. Casual relationships between telomere length and health outcomes have been investigated with Mendelian randomization approaches using variants identified with GWAS that associate with telomere length. These studies have revealed decreased risk of cardiovascular disease, Alzheimer disease, interstitial lung disease, immunodeficiency, and celiac disease in individuals with genetic determined long telomeres (Haycock et al. 2017; Deng et al. 2022; Wang et al. 2024b), but increased risk of several cancers including glioma, kidney, serous low-malignancy-potential ovarian, bladder, neuroblastoma, melanoma, and lung (Walsh et al. 2015; Zhang et al. 2015; Haycock et al. 2017). The association between longer telomere lengths and cancer risk is not absolute as some studies have indicated an increased risk of pancreatic cancer in individuals with short telomeres (Campa et al. 2019), although this is controversial (Antwi et al. 2017). Interestingly, while short telomeres confer a decreased risk of cancer, they can also reduce survival from cancer (Weischer et al. 2013).

## TELOMERE LENGTH DYNAMICS AND DYSFUNCTION IN CANCER

Inherited short telomeres in normal somatic tissues confer a reduced risk of cancer. However, the original observations that telomere lengths in cancers tend to be shorter than that of patient-matched normal tissues implicated telomere dynamics and dysfunction in the progression to malignancy (de Lange et al. 1990; Hastie et al. 1990). It had been assumed that, in the context of limited telomerase activity, the telomere length differentials between normal and tumor tissue arise as a consequence of extensive cell division and replicative telomere erosion from the original cell to the clonal malignant tumor analyzed. However, analysis of early-stage lesions, including colorectal adenomatous polyps, reveals short telomeres and fusions occurring prior to disease progression (Roger et al. 2013). In the premalignant condition Barrett's esophagus, clonal patches of extreme telomere erosion have been observed (Maley et al. 2006; Letsolo et al. 2017). Telomere shortening was detected in colonocytes from patients with ulcerative colitis, a condition that increases the risk of colorectal cancer (Risques et al. 2008), and in early-stage cervical intraepithelial neoplasia (Maida et al. 2006). Telomere shortening was also detected in myelodysplasia, which predisposes to acute myeloid leukemia (AML) (Ohyashiki et al. 1999; Williams et al. 2017). CLL B-cell clones can exhibit extreme telomere erosion and fusion, consistent with these cells undergoing a telomere-driven crisis (Lin et al. 2010) and this exacerbated in the context of mutation in the ATM gene (Britt-Compton et al. 2012). Telomere erosion and dysfunction is a feature of late-stage CLL but was also detected in a subset of early-stage (Binet stage A) patients prior to clinical progression, where telomere fusion activity was observed together with large-scale genomic rearrangements that included telomeric regions (Lin et al. 2010).

Thus, short dysfunctional telomeres capable of fusion are detected in early-stage cancer prior to clinical progression and significant telomere shortening has been observed in premalignant lesions.

Detailed analysis of the telomere length spectrums in patient-matched normal tissues reveals that the telomere length distributions of normal cells overlap with those observed in cancers (Lin et al. 2010; Roger et al. 2013). In the context of colorectal cancer, these data imply that telomere erosion not only precedes the adenoma/carcinoma transition but is consistent with being preexistent in normal cells in which the initiating mutations occur (Roger et al. 2013). In this situation, normal cells with long telomeres give rise to adenomatous polyps that retain their long telomeres and stable genomes, whereas those with short telomeres produce adenomas with increased incidence of telomere fusion and chromosomal instability, and these clones may in turn have a greater probability of transition to carcinoma. Similarly, in CLL, the telomere length distributions observed in normal B cells overlapped with those observed in CLL B-cell clones from early- and late-stage disease (Lin et al. 2010). Thus, like colorectal adenomas, the telomere length distributions of early-stage CLL indicated an earlier origin for telomere erosion potentially within normal B cells. This situation was also observed when comparing patient-matched telomere length distributions in multiple myeloma and Barrett's esophagus (Hyatt et al. 2017; Letsolo et al. 2017).

Taken together, these observations point to an early origin for telomere erosion, with short telomeres being preexistent in subsets of normal cells, and short dysfunctional telomeres being present within premalignant and early-stage lesions.

The correlation between telomere length and genomic complexity implicates telomere dysfunction as a potential driver of tumor progression, a premise that is further strengthened by associations between short telomere length and a poorer clinical outcome in several tumor types, including prostate, myelodysplasia, chronic myeloid leukemia, CLL, breast, and colorectal (Bechter et al. 1998; Donaldson et al.

1999; Ohyashiki et al. 1999; Brümmendorf et al. 2000; Gertler et al. 2004; Fordyce et al. 2006). The use of telomere length was further refined in CLL by using high-resolution telomere length analysis, coupled with telomere fusion analysis, to functionally define the length below which telomere dysfunction was detected (Lin et al. 2014). Patients with CLL B cells with telomeres shorter than the telomere fusion threshold had significantly reduced overall survival that was even more prognostic in early-stage disease patients prior to clinical progression. Importantly also, the same telomere length threshold was predictive of outcome of the standard treatment for CLL, fludarabine, cyclophosphamide, rituximab (FCR) (Strefford et al. 2015; Norris et al. 2019), surpassing all other CLL disease markers currently employed. Combination of telomere length with IGHV mutation status and CD49d allowed the identification of long-term progression-free CLL patients treated with FCR (Pepper et al. 2022). These data imply that short telomeres and dysfunction occur early in CLL, thus facilitating clonal evolution and clinical progression, and may provide a clinically useful prognostic and predictive marker.

The nature of the disease and ease of sampling means some of the most detailed analysis of telomere dynamics has been undertaken in CLL; however, the telomere paradigm is not specific to CLL. Indeed, the application of high-resolution telomere length analysis together with the telomere fusion threshold defined in CLL, provided prognostic information for overall survival for myelodysplasia (Williams et al. 2017), multiple myeloma (Hyatt et al. 2017), and breast cancer (Simpson et al. 2015). Interestingly, while these parameters define prognosis in myelodysplasia, a precursor lesion for AML, no significant prognostic signature was identified in AML (Williams et al. 2017). Moreover, while short telomeres and dysfunction were observed in colorectal adenomatous polyps, there was no prognostic signature for telomere length in colorectal carcinoma (Roger et al. 2013). In these situations, it is possible that progression is accompanied by an up-regulation of telomerase activity that effectively homogenizes telomere length differentials, which were apparent in ear-

lier disease stages, thereby reducing the prognostic signature in these malignancies.

Taken together, the clinical data support the thesis that telomere erosion and dysfunction occur early in the progression to malignancy in several tumor types and may indeed be present in the initiating cell. Importantly, telomere-based markers may provide clinically useful tools in diverse cancers to inform the identification of those patients that require, and will benefit from, treatment. These data provide important evidence that the progression to malignancy is accompanied by the period of telomere dysfunction, consistent with a telomere crisis. Telomere crisis represents a situation of cellular vulnerability and as such may provide an opportunity for therapeutic intervention at the earliest stages of tumorigenesis. As the biological underpinnings of telomere crisis are dissected, new or existing agents may be used to target tumors and early-stage lesions exhibiting telomere dysfunction.

## ACKNOWLEDGMENTS

Work in the Baird laboratory is supported by Cancer Res UK (C17199/A29202), the Wales Cancer Res Centre, and Cancer Res Wales.

## REFERENCES

*Reference is also in this subject collection.

Ahmed R, Roger L, Costa Del Amo P, Miners KL, Jones RE, Boelen L, Fali T, Elemans M, Zhang Y, Appay V, et al. 2016. Human stem cell-like memory T cells are maintained in a state of dynamic flux. *Cell Rep* **17:** 2811–2818. doi:10.1016/j.celrep.2016.11.037

Ahmed R, Miners KL, Lahoz-Beneytez J, Jones RE, Roger L, Baboonian C, Zhang Y, Wang ECY, Hellerstein MK, McCune JM, et al. 2020. CD57+ memory T cells proliferate in vivo. *Cell Rep* **33:** 108501. doi:10.1016/j.celrep.2020.108501

Alder JK, Chen JJ, Lancaster L, Danoff S, Su SC, Cogan JD, Vulto I, Xie M, Qi X, Tuder RM, et al. 2008. Short telomeres are a risk factor for idiopathic pulmonary fibrosis. *Proc Natl Acad Sci* **105:** 13051–13056. doi:10.1073/pnas.0804280105

Allsopp RC, Vaziri H, Patterson C, Goldstein S, Younglai EV, Futcher AB, Greider CW, Harley CB. 1992. Telomere length predicts replicative capacity of human fibroblasts. *Proc Natl Acad Sci* **89:** 10114–10118. doi:10.1073/pnas.89.21.10114

Alter BP, Giri N, Savage SA, Rosenberg PS. 2009. Cancer in dyskeratosis congenita. *Blood* **113:** 6549–6557. doi:10.1182/blood-2008-12-192880

Andrew T, Aviv A, Falchi M, Surdulescu GL, Gardner JP, Lu X, Kimura M, Kato BS, Valdes AM, Spector TD. 2006. Mapping genetic loci that determine leukocyte telomere length in a large sample of unselected female sibling pairs. *Am J Hum Genet* **78:** 480–486. doi:10.1086/500052

Antwi SO, Bamlet WR, Broderick BT, Chaffee KG, Oberg A, Jatoi A, Boardman LA, Petersen GM. 2017. Genetically predicted telomere length is not associated with pancreatic cancer risk. *Cancer Epidemiol Biomarkers Prev* **26:** 971–974. doi:10.1158/1055-9965.EPI-17-0100

Arbeev KG, Hunt SC, Kimura M, Aviv A, Yashin AI. 2011. Leukocyte telomere length, breast cancer risk in the offspring: the relations with father's age at birth. *Mech Ageing Dev* **132:** 149–153. doi:10.1016/j.mad.2011.02.004

Armanios M. 2022. The role of telomeres in human disease. *Annu Rev Genomics Hum Genet* **23:** 363–381. doi:10.1146/annurev-genom-010422-091101

Artandi SE, DePinho RA. 2010. Telomeres and telomerase in cancer. *Carcinogenesis* **31:** 9–18. doi:10.1093/carcin/bgp268

Artandi SE, Chang S, Lee SL, Alson S, Gottlieb GJ, Chin L, DePinho RA. 2000. Telomere dysfunction promotes nonreciprocal translocations and epithelial cancers in mice. *Nature* **406:** 641–645. doi:10.1038/35020592

Aviv A. 2012. Genetics of leukocyte telomere length and its role in atherosclerosis. *Mutat Res* **730:** 68–74. doi:10.1016/j.mrfmmm.2011.05.001

Baca SC, Prandi D, Lawrence MS, Mosquera JM, Romanel A, Drier Y, Park K, Kitabayashi N, MacDonald TY, Ghandi M, et al. 2013. Punctuated evolution of prostate cancer genomes. *Cell* **153:** 666–677. doi:10.1016/j.cell.2013.03.021

Baird DM. 2008. Telomere dynamics in human cells. *Biochimie* **90:** 116–121. doi:10.1016/j.biochi.2007.08.003

Baird DM, Rowson J, Wynford-Thomas D, Kipling D. 2003. Extensive allelic variation and ultrashort telomeres in senescent human cells. *Nat Genet* **33:** 203–207. doi:10.1038/ng1084

Baird DM, Britt-Compton B, Rowson J, Amso NN, Gregory L, Kipling D. 2006. Telomere instability in the male germline. *Hum Mol Genet* **15:** 45–51. doi:10.1093/hmg/ddi424

Baumann P, Cech TR. 2000. Protection of telomeres by the Ku protein in fission yeast. *Mol Biol Cell* **11:** 3265–3275. doi:10.1091/mbc.11.10.3265

Bayne S, Jones ME, Li H, Liu JP. 2007. Potential roles for estrogen regulation of telomerase activity in aging. *Ann NY Acad Sci* **1114:** 48–55. doi:10.1196/annals.1396.023

Bechter OE, Eisterer W, Pall G, Hilbe W, Kuhr T, Thaler J. 1998. Telomere length and telomerase activity predict survival in patients with B cell chronic lymphocytic leukemia. *Cancer Res* **58:** 4918–4922.

Benetos A, Kark JD, Susser E, Kimura M, Sinnreich R, Chen W, Steenstrup T, Christensen K, Herbig U, von Bornemann Hjelmborg J, et al. 2013. Tracking and fixed ranking of leukocyte telomere length across the adult life course. *Aging Cell* **12:** 615–621. doi:10.1111/acel.12086

Biron-Shental T, Sukenik Halevy R, Goldberg-Bittman L, Kidron D, Fejgin MD, Amiel A. 2010a. Telomeres are shorter in placental trophoblasts of pregnancies complicated with intrauterine growth restriction (IUGR). *Early Hum Dev* **86:** 451–456. doi:10.1016/j.earlhumdev.2010.06.002

Biron-Shental T, Sukenik-Halevy R, Sharon Y, Goldberg-Bittman L, Kidron D, Fejgin MD, Amiel A. 2010b. Short telomeres may play a role in placental dysfunction in preeclampsia and intrauterine growth restriction. *Am J Obstet Gynecol* **202:** 381–387. doi:10.1016/j.ajog.2010.01.036

Bischoff C, Graakjaer J, Petersen HC, Hjelmborg JB, Vaupel JW, Bohr V, Koelvraa S, Christensen K. 2005. The heritability of telomere length among the elderly and oldest-old. *Twin Res Hum Genet* **8:** 433–439. doi:10.1375/twin.8.5.433

Boboila C, Jankovic M, Yan CT, Wang JH, Wesemann DR, Zhang T, Fazeli A, Feldman L, Nussenzweig A, Nussenzweig M, et al. 2010a. Alternative end-joining catalyzes robust IgH locus deletions and translocations in the combined absence of ligase 4 and Ku70. *Proc Natl Acad Sci* **107:** 3034–3039. doi:10.1073/pnas.0915067107

Boboila C, Yan C, Wesemann DR, Jankovic M, Wang JH, Manis J, Nussenzweig A, Nussenzweig M, Alt FW. 2010b. Alternative end-joining catalyzes class switch recombination in the absence of both Ku70 and DNA ligase 4. *J Exp Med* **207:** 417–427. doi:10.1084/jem.20092449

Boulton SJ, Jackson SP. 1996. *Saccharomyces cerevisiae* Ku70 potentiates illegitimate DNA double-strand break repair and serves as a barrier to error-prone DNA repair pathways. *EMBO J* **15:** 5093–5103. doi:10.1002/j.1460-2075.1996.tb00890.x

Bountziouka V, Musicha C, Allara E, Kaptoge S, Wang Q, Angelantonio ED, Butterworth AS, Thompson JR, Danesh JN, Wood AM, et al. 2022. Modifiable traits, healthy behaviours, and leukocyte telomere length: a population-based study in UK Biobank. *Lancet Healthy Longev* **3:** e321–e331. doi:10.1016/S2666-7568(22)00072-1

Brambati A, Sacco O, Porcella S, Heyza J, Kareh M, Schmidt JC, Sfeir A. 2023. RHINO directs MMEJ to repair DNA breaks in mitosis. *Science* **381:** 653–660. doi:10.1126/science.adh3694

Britt-Compton B, Capper R, Rowson J, Baird DM. 2009. Short telomeres are preferentially elongated by telomerase in human cells. *FEBS Lett* **583:** 3076–3080. doi:10.1016/j.febslet.2009.08.029

Britt-Compton B, Lin TT, Ahmed G, Weston V, Jones RE, Fegan C, Oscier DG, Stankovic T, Pepper C, Baird DM. 2012. Extreme telomere erosion in ATM-mutated and 11q-deleted CLL patients is independent of disease stage. *Leukemia* **26:** 826–830. doi:10.1038/leu.2011.281

Broer L, Codd V, Nyholt DR, Deelen J, Mangino M, Willemsen G, Albrecht E, Amin N, Beekman M, de Geus EJ, et al. 2013. Meta-analysis of telomere length in 19,713 subjects reveals high heritability, stronger maternal inheritance and a paternal age effect. *Eur J Hum Genet* **21:** 1163–1168. doi:10.1038/ejhg.2012.303

Brümmendorf TH, Holyoake TL, Rufer N, Barnett MJ, Schulzer M, Eaves CJ, Eaves AC, Lansdorp PM. 2000. Prognostic implications of differences in telomere length between normal and malignant cells from patients with chronic myeloid leukemia measured by flow cytometry. *Blood* **95:** 1883–1890. doi:10.1182/blood.V95.6.1883

Cite this article as *Cold Spring Harb Perspect Biol* doi: 10.1101/cshperspect.a041701

Campa D, Matarazzi M, Greenhalf W, Bijlsma M, Saum KU, Pasquali C, van Laarhoven H, Szentesi A, Federici F, Vodicka P, et al. 2019. Genetic determinants of telomere length and risk of pancreatic cancer: a PANDoRA study. *Int J Cancer* **144:** 1275–1283. doi:10.1002/ijc.31928

Campisi J, Robert L. 2014. Cell senescence: role in aging and age-related diseases. *Interdisc Top Gerontol* **39:** 45–61. doi:10.1159/000358899

Capper R, Britt-Compton B, Tankimanova M, Rowson J, Letsolo B, Man S, Haughton M, Baird DM. 2007. The nature of telomere fusion and a definition of the critical telomere length in human cells. *Genes Dev* **21:** 2495–2508. doi:10.1101/gad.439107

Cesare AJ, Karlseder J. 2012. A three-state model of telomere control over human proliferative boundaries. *Curr Opin Cell Biol* **24:** 731–738. doi:10.1016/j.ceb.2012.08.007

Cesare AJ, Kaul Z, Cohen SB, Napier CE, Pickett HA, Neumann AA, Reddel RR. 2009. Spontaneous occurrence of telomeric DNA damage response in the absence of chromosome fusions. *Nat Struct Mol Biol* **16:** 1244–1251. doi:10.1038/nsmb.1725

Chen LY, Redon S, Lingner J. 2012. The human CST complex is a terminator of telomerase activity. *Nature* **488:** 540–544. doi:10.1038/nature11269

Cleal K, Baird DM. 2020. Catastrophic endgames: emerging mechanisms of telomere-driven genomic instability. *Trends Genet* **36:** 347–359. doi:10.1016/j.tig.2020.02.001

Cleal K, Jones RE, Grimstead JW, Hendrickson EA, Baird DM. 2019. Chromothripsis during telomere crisis is independent of NHEJ, and consistent with a replicative origin. *Genome Res* **29:** 737–749. doi:10.1101/gr.240705.118

Codd V, Mangino M, van der Harst P, Braund PS, Kaiser M, Beveridge AJ, Rafelt S, Moore J, Nelson C, Soranzo N, et al. 2010. Common variants near TERC are associated with mean telomere length. *Nat Genet* **42:** 197–199. doi:10.1038/ng.532

Codd V, Nelson CP, Albrecht E, Mangino M, Deelen J, Buxton JL, Hottenga JJ, Fischer K, Esko T, Surakka I, et al. 2013. Identification of seven loci affecting mean telomere length and their association with disease. *Nat Genet* **45:** 422–427, 427e421-422. doi:10.1038/ng.2528

Codd V, Wang Q, Allara E, Musicha C, Kaptoge S, Stoma S, Jiang T, Hamby SE, Braund PS, Bountziouka V, et al. 2021. Polygenic basis and biomedical consequences of telomere length variation. *Nat Genet* **53:** 1425–1433. doi:10.1038/s41588-021-00944-6

Coppé JP, Patil CK, Rodier F, Sun Y, Muñoz DP, Goldstein J, Nelson PS, Desprez PY, Campisi J. 2008. Senescence-associated secretory phenotypes reveal cell-nonautonomous functions of oncogenic RAS and the p53 tumor suppressor. *PLoS Biol* **6:** 2853–2868. doi:10.1371/journal.pbio.0060301

Cortés-Ciriano I, Lee JJ, Xi R, Jain D, Jung YL, Yang L, Gordenin D, Klimczak LJ, Zhang CZ, Pellman DS, et al. 2020. Comprehensive analysis of chromothripsis in 2,658 human cancers using whole-genome sequencing. *Nat Genet* **52:** 331–341. doi:10.1038/s41588-019-0576-7

Counter CM, Avilion AA, LeFeuvre CE, Stewart NG, Greider CW, Harley CB, Bacchetti S. 1992. Telomere shortening associated with chromosome instability is arrested in immortal cells which express telomerase activity. *EMBO J* **11:** 1921–1929. doi:10.1002/j.1460-2075.1992.tb05245.x

d'Adda di Fagagna F, Reaper PM, Clay-Farrace L, Fiegler H, Carr P, Von Zglinicki T, Saretzki G, Carter NP, Jackson SP. 2003. A DNA damage checkpoint response in telomere-initiated senescence. *Nature* **426:** 194–198. doi:10.1038/nature02118

Davy P, Nagata M, Bullard P, Fogelson NS, Allsopp R. 2009. Fetal growth restriction is associated with accelerated telomere shortening and increased expression of cell senescence markers in the placenta. *Placenta* **30:** 539–542. doi:10.1016/j.placenta.2009.03.005

de Lange T. 2005. Shelterin: the protein complex that shapes and safeguards human telomeres. *Genes Dev* **19:** 2100–2110. doi:10.1101/gad.1346005

de Lange T, Shiue L, Myers RM, Cox DR, Naylor SL, Killery AM, Varmus HE. 1990. Structure and variability of human chromosome ends. *Mol Cell Biol* **10:** 518–527. doi:10.1128/mcb.10.2.518-527.1990

De Meyer T, Rietzschel ER, De Buyzere ML, De Bacquer D, Van Criekinge W, De Backer GG, Gillebert TC, Van Oostveldt P, Bekaert S, on behalf of the Asklepios Investigators. 2007. Paternal age at birth is an important determinant of offspring telomere length. *Hum Mol Genet* **16:** 3097–3102. doi:10.1093/hmg/ddm271

De Meyer T, Rietzschel ER, De Buyzere ML, Langlois MR, De Bacquer D, Segers P, Van Damme P, De Backer GG, Van Oostveldt P, Van Criekinge W, et al. 2009. Systemic telomere length and preclinical atherosclerosis: the Asklepios study. *Eur Heart J* **30:** 3074–3081. doi:10.1093/eurheartj/ehp324

Deng Y, Chan SS, Chang S. 2008. Telomere dysfunction and tumour suppression: the senescence connection. *Nat Rev Cancer* **8:** 450–458. doi:10.1038/nrc2393

Deng Y, Li Q, Zhou F, Li G, Liu J, Lv J, Li L, Chang D. 2022. Telomere length and the risk of cardiovascular diseases: a Mendelian randomization study. *Front Cardiovasc Med* **9:** 1012615. doi:10.3389/fcvm.2022.1012615

Dewhurst SM, Yao X, Rosiene J, Tian H, Behr J, Bosco N, Takai KK, de Lange T, Imieliński M. 2021. Structural variant evolution after telomere crisis. *Nat Commun* **12:** 2093. doi:10.1038/s41467-021-21933-7

Di Micco R, Sulli G, Dobreva M, Liontos M, Botrugno OA, Gargiulo G, dal Zuffo R, Matti V, d'Ario G, Montani E, et al. 2011. Interplay between oncogene-induced DNA damage response and heterochromatin in senescence and cancer. *Nat Cell Biol* **13:** 292–302. doi:10.1038/ncb2170

Donaldson L, Fordyce C, Gilliland F, Smith A, Feddersen R, Joste N, Moyzis R, Griffith J. 1999. Association between outcome and telomere DNA content in prostate cancer. *J Urol* **162:** 1788–1792. doi:10.1016/S0022-5347(05)68239-0

Entringer S, Epel ES, Kumsta R, Lin J, Hellhammer DH, Blackburn EH, Wüst S, Wadhwa PD. 2011. Stress exposure in intrauterine life is associated with shorter telomere length in young adulthood. *Proc Natl Acad Sci* **108:** E513–E518. doi:10.1073/pnas.1107759108

Factor-Litvak P, Susser E, Kezios K, McKeague I, Kark JD, Hoffman M, Kimura M, Wapner R, Aviv A. 2016. Leukocyte telomere length in newborns: implications for the role of telomeres in human disease. *Pediatrics* **137**: e20153927. doi:10.1542/peds.2015-3927

Fordyce CA, Heaphy CM, Bisoffi M, Wyaco JL, Joste NE, Mangalik A, Baumgartner KB, Baumgartner RN, Hunt WC, Griffith JK. 2006. Telomere content correlates with stage and prognosis in breast cancer. *Breast Cancer Res Treat* **99**: 193–202. doi:10.1007/s10549-006-9204-1

Gamba BF, Richieri-Costa A, Costa S, Rosenberg C, Ribeiro-Bicudo LA. 2015. Chromothripsis with at least 12 breaks at 1p36.33-p35.3 in a boy with multiple congenital anomalies. *Mol Genet Genomics* **290**: 2213–2216. doi:10.1007/s00438-015-1072-0

Garcia-Martin I, Penketh RJA, Janssen AB, Jones RE, Grimstead J, Baird DM, John RM. 2018. Metformin and insulin treatment prevent placental telomere attrition in boys exposed to maternal diabetes. *PLoS ONE* **13**: e0208533. doi:10.1371/journal.pone.0208533

Garcia-Martin I, Penketh RJA, Garay SM, Jones RE, Grimstead JW, Baird DM, John RM. 2021. Symptoms of prenatal depression associated with shorter telomeres in female placenta. *Int J Mol Sci* **22**: 7458. doi:10.3390/ijms22147458

Gardner M, Bann D, Wiley L, Cooper R, Hardy R, Nitsch D, Martin-Ruiz C, Shiels P, Sayer AA, Barbieri M, et al. 2014. Gender and telomere length: systematic review and meta-analysis. *Exp Gerontol* **51**: 15–27. doi:10.1016/j.exger.2013.12.004

Gertler R, Rosenberg R, Stricker D, Friederichs J, Hoos A, Werner M, Ulm K, Holzmann B, Nekarda H, Siewert JR. 2004. Telomere length and human telomerase reverse transcriptase expression as markers for progression and prognosis of colorectal carcinoma. *J Clin Oncol* **22**: 1807–1814. doi:10.1200/JCO.2004.09.160

Göttlich B, Reichenberger S, Feldmann E, Pfeiffer P. 1998. Rejoining of DNA double-strand breaks in vitro by single-strand annealing. *Eur J Biochem* **258**: 387–395. doi:10.1046/j.1432-1327.1998.2580387.x

Harley CB, Futcher AB, Greider CW. 1990. Telomeres shorten during ageing of human fibroblasts. *Nature* **345**: 458–460. doi:10.1038/345458a0

Hastie ND, Dempster M, Dunlop MG, Thompson AM, Green DK, Allshire RC. 1990. Telomere reduction in human colorectal carcinoma and with ageing. *Nature* **346**: 866–868. doi:10.1038/346866a0

Haycock PC, Hemani G, Aviv A. 2017. Telomere length and risk of cancer and non-neoplastic diseases: is survivin the Ariadne's thread?—reply. *JAMA Oncol* **3**: 1741–1742. doi:10.1001/jamaoncol.2017.2316

Heacock M, Spangler E, Riha K, Puizina J, Shippen DE. 2004. Molecular analysis of telomere fusions in Arabidopsis: multiple pathways for chromosome end-joining. *EMBO J* **23**: 2304–2313. doi:10.1038/sj.emboj.7600234

Hjort L, Vryer R, Grunnet LG, Burgner D, Olsen SF, Saffery R, Vaag A. 2018. Telomere length is reduced in 9- to 16-year-old girls exposed to gestational diabetes in utero. *Diabetologia* **61**: 870–880. doi:10.1007/s00125-018-4549-7

Holland AJ, Cleveland DW. 2012. Chromoanagenesis and cancer: mechanisms and consequences of localized, complex chromosomal rearrangements. *Nat Med* **18**: 1630–1638. doi:10.1038/nm.2988

Huang W, Hickson LJ, Eirin A, Kirkland JL, Lerman LO. 2022. Cellular senescence: the good, the bad and the unknown. *Nat Rev Nephrol* **18**: 611–627. doi:10.1038/s41581-022-00601-z

Hyatt S, Jones RE, Heppel NH, Grimstead JW, Fegan C, Jackson GH, Hills R, Allan JM, Pratt G, Pepper C, et al. 2017. Telomere length is a critical determinant for survival in multiple myeloma. *Br J Haematol* **178**: 94–98. doi:10.1111/bjh.14643

Jády BE, Richard P, Bertrand E, Kiss T. 2006. Cell cycle-dependent recruitment of telomerase RNA and Cajal bodies to human telomeres. *Mol Biol Cell* **17**: 944–954. doi:10.1091/mbc.e05-09-0904

Jones RE, Oh S, Grimstead JW, Zimbric J, Roger L, Heppel NH, Ashelford KE, Liddiard K, Hendrickson EA, Baird DM. 2014. Escape from telomere-driven crisis is DNA ligase III dependent. *Cell Rep* **8**: 1063–1076. doi:10.1016/j.celrep.2014.07.007

* Karlseder J. 2024. Crisis mechanisms and their escape. *Cold Spring Harb Perspect Biol* doi:10.1101/cshperspect.a041688

Kaul Z, Cesare AJ, Huschtscha LI, Neumann AA, Reddel RR. 2012. Five dysfunctional telomeres predict onset of senescence in human cells. *EMBO Rep* **13**: 52–59. doi:10.1038/embor.2011.227

Kloosterman WP, Guryev V, van Roosmalen M, Duran KJ, de Bruijn E, Bakker SC, Letteboer T, van Nesselrooij B, Hochstenbach R, Poot M, et al. 2011. Chromothripsis as a mechanism driving complex de novo structural rearrangements in the germline. *Hum Mol Genet* **20**: 1916–1924. doi:10.1093/hmg/ddr073

Letsolo BT, Rowson J, Baird DM. 2010. Fusion of short telomeres in human cells is characterized by extensive deletion and microhomology and can result in complex rearrangements. *Nucleic Acids Res* **38**: 1841–1852. doi:10.1093/nar/gkp1183

Letsolo BT, Jones RE, Rowson J, Grimstead JW, Keith WN, Jenkins GJ, Baird DM. 2017. Extensive telomere erosion is consistent with localised clonal expansions in Barrett's metaplasia. *PLoS ONE* **12**: e0174833. doi:10.1371/journal.pone.0174833

Levy MZ, Allsopp RC, Futcher AB, Greider CW, Harley CB. 1992. Telomere end-replication problem and cell aging. *J Mol Biol* **225**: 951–960. doi:10.1016/0022-2836(92)90096-3

Li C, Stoma S, Lotta LA, Warner S, Albrecht E, Allione A, Arp PP, Broer L, Buxton JL, Da Silva Couto Alves A, et al. 2020. Genome-wide association analysis in humans links nucleotide metabolism to leukocyte telomere length. *Am J Hum Genet* **106**: 389–404. doi:10.1016/j.ajhg.2020.02.006

Liddiard K, Ruis B, Takasugi T, Harvey A, Ashelford KE, Hendrickson EA, Baird DM. 2016. Sister chromatid telomere fusions, but not NHEJ-mediated inter-chromosomal telomere fusions, occur independently of DNA ligases 3 and 4. *Genome Res* **26**: 588–600. doi:10.1101/gr.200840.115

Liddiard K, Ruis B, Kan Y, Cleal K, Ashelford KE, Hendrickson EA, Baird DM. 2019. DNA ligase 1 is an essential mediator of sister chromatid telomere fusions in $G_2$ cell cycle phase. *Nucleic Acids Res* **47**: 2402–2424. doi:10.1093/nar/gky1279

Liddiard K, Aston-Evans AN, Cleal K, Hendrickson EA, Baird DM. 2022. POLQ suppresses genome instability and alterations in DNA repeat tract lengths. *NAR Cancer* **4:** zcac020. doi:10.1093/narcan/zcac020

Lin TT, Letsolo BT, Jones RE, Rowson J, Pratt G, Hewamana S, Fegan C, Pepper C, Baird DM. 2010. Telomere dysfunction and fusion during the progression of chronic lymphocytic leukemia: evidence for a telomere crisis. *Blood* **116:** 1899–1907. doi:10.1182/blood-2010-02-272104

Lin TT, Norris K, Heppel NH, Pratt G, Allan JM, Allsup DJ, Bailey J, Cawkwell L, Hills R, Grimstead JW, et al. 2014. Telomere dysfunction accurately predicts clinical outcome in chronic lymphocytic leukaemia, even in patients with early stage disease. *Br J Haematol* **167:** 214–223. doi:10.1111/bjh.13023

Liu P, Erez A, Nagamani SC, Dhar SU, Kolodziejska KE, Dharmadhikari AV, Cooper ML, Wiszniewska J, Zhang F, Withers MA, et al. 2011. Chromosome catastrophes involve replication mechanisms generating complex genomic rearrangements. *Cell* **146:** 889–903. doi:10.1016/j.cell.2011.07.042

Ma JL, Kim EM, Haber JE, Lee SE. 2003. Yeast Mre11 and Rad1 proteins define a Ku-independent mechanism to repair double-strand breaks lacking overlapping end sequences. *Mol Cell Biol* **23:** 8820–8828. doi:10.1128/MCB.23.23.8820-8828.2003

Maciejowski J, Li Y, Bosco N, Campbell PJ, de Lange T. 2015. Chromothripsis and Kataegis induced by telomere crisis. *Cell* **163:** 1641–1654. doi:10.1016/j.cell.2015.11.054

Maida Y, Kyo S, Forsyth NR, Takakura M, Sakaguchi J, Mizumoto Y, Hashimoto M, Nakamura M, Nakao S, Inoue M. 2006. Distinct telomere length regulation in premalignant cervical and endometrial lesions: implications for the roles of telomeres in uterine carcinogenesis. *J Pathol* **210:** 214–223. doi:10.1002/path.2038

Maley CC, Galipeau PC, Finley JC, Wongsurawat VJ, Li X, Sanchez CA, Paulson TG, Blount PL, Risques RA, Rabinovitch PS, et al. 2006. Genetic clonal diversity predicts progression to esophageal adenocarcinoma. *Nat Genet* **38:** 468–473. doi:10.1038/ng1768

Maser RS, Wong KK, Sahin E, Xia H, Naylor M, Hedberg HM, Artandi SE, DePinho RA. 2007. DNA-dependent protein kinase catalytic subunit is not required for dysfunctional telomere fusion and checkpoint response in the telomerase-deficient mouse. *Mol Cell Biol* **27:** 2253–2265. doi:10.1128/MCB.01354-06

Mateos-Gomez PA, Gong F, Nair N, Miller KM, Lazzerini-Denchi E, Sfeir A. 2015. Mammalian polymerase theta promotes alternative NHEJ and suppresses recombination. *Nature* **518:** 254–257. doi:10.1038/nature14157

McClintock B. 1941. The stability of broken ends of chromosomes in *Zea Mays*. *Genetics* **26:** 234–282. doi:10.1093/genetics/26.2.234

Melzer D, Pilling LC, Ferrucci L. 2020. The genetics of human ageing. *Nat Rev Genet* **21:** 88–101. doi:10.1038/s41576-019-0183-6

Murnane JP. 2006. Telomeres and chromosome instability. *DNA Repair (Amst)* **5:** 1082–1092. doi:10.1016/j.dnarep.2006.05.030

Murnane JP. 2012. Telomere dysfunction and chromosome instability. *Mutat Res* **730:** 28–36. doi:10.1016/j.mrfmmm.2011.04.008

Narita M, Nuñez S, Heard E, Lin AW, Hearn SA, Spector DL, Hannon GJ, Lowe SW. 2003. Rb-mediated heterochromatin formation and silencing of E2F target genes during cellular senescence. *Cell* **113:** 703–716. doi:10.1016/S0092-8674(03)00401-X

Nassour J, Radford R, Correia A, Fusté JM, Schoell B, Jauch A, Shaw RJ, Karlseder J. 2019. Autophagic cell death restricts chromosomal instability during replicative crisis. *Nature* **565:** 659–663. doi:10.1038/s41586-019-0885-0

Nassour J, Schmidt TT, Karlseder J. 2021. Telomeres and cancer: resolving the paradox. *Annu Rev Cancer Biol* **5:** 59–77. doi:10.1146/annurev-cancerbio-050420-023410

Nassour J, Aguiar LG, Correia A, Schmidt TT, Mainz L, Przetocka S, Haggblom C, Tadepalle N, Williams A, Shokhirev MN, et al. 2023. Telomere-to-mitochondria signalling by ZBP1 mediates replicative crisis. *Nature* **614:** 767–773. doi:10.1038/s41586-023-05710-8

Njajou OT, Cawthon RM, Damcott CM, Wu SH, Ott S, Garant MJ, Blackburn EH, Mitchell BD, Shuldiner AR, Hsueh WC. 2007. Telomere length is paternally inherited and is associated with parental lifespan. *Proc Natl Acad Sci* **104:** 12135–12139. doi:10.1073/pnas.0702703104

Norris K, Hillmen P, Rawstron A, Hills R, Baird DM, Fegan CD, Pepper C. 2019. Telomere length predicts for outcome to FCR chemotherapy in CLL. *Leukemia* **33:** 1953–1963. doi:10.1038/s41375-019-0389-9

Ohyashiki JH, Iwama H, Yahata N, Ando K, Hayashi S, Shay JW, Ohyashiki K. 1999. Telomere stability is frequently impaired in high-risk groups of patients with myelodysplastic syndromes. *Clin Cancer Res* **5:** 1155–1160.

Pan-Hammarström Q, Jones AM, Lähdesmäki A, Zhou W, Gatti RA, Hammarström L, Gennery AR, Ehrenstein MR. 2005. Impact of DNA ligase IV on nonhomologous end joining pathways during class switch recombination in human cells. *J Exp Med* **201:** 189–194. doi:10.1084/jem.20040772

Pepper GV, Bateson M, Nettle D. 2018. Telomeres as integrative markers of exposure to stress and adversity: a systematic review and meta-analysis. *R Soc Open Sci* **5:** 180744. doi:10.1098/rsos.180744

Pepper AGS, Zucchetto A, Norris K, Tissino E, Polesel J, Soe Z, Allsup D, Hockaday A, Ow PL, Hillmen P, et al. 2022. Combined analysis of IGHV mutations, telomere length and CD49d identifies long-term progression-free survivors in TP53 wild-type CLL treated with FCR-based therapies. *Leukemia* **36:** 271–274. doi:10.1038/s41375-021-01322-1

Pooley KA, Bojesen SE, Weischer M, Nielsen SF, Thompson D, Amin Al Olama A, Michailidou K, Tyrer JP, Benlloch S, Brown J, et al. 2013. A genome-wide association scan (GWAS) for mean telomere length within the COGS project: identified loci show little association with hormone-related cancer risk. *Hum Mol Genet* **22:** 5056–5064. doi:10.1093/hmg/ddt355

Pusceddu I, Herrmann W, Kleber ME, Scharnagl H, Hoffmann MM, Winklhofer-Roob BM, März W, Herrmann M. 2020. Subclinical inflammation, telomere shortening, homocysteine, vitamin B6, and mortality: the Ludwigshafen Risk and Cardiovascular Health Study. *Eur J Nutr* **59:** 1399–1411. doi:10.1007/s00394-019-01993-8

Rai R, Zheng H, He H, Luo Y, Multani A, Carpenter PB, Chang S. 2010. The function of classical and alternative non-homologous end-joining pathways in the fusion of dysfunctional telomeres. *EMBO J* **29:** 2598–2610. doi:10.1038/emboj.2010.142

Ramsden DA, Carvajal-Garcia J, Gupta GP. 2022. Mechanism, cellular functions and cancer roles of polymerase-theta-mediated DNA end joining. *Nat Rev Mol Cell Biol* **23:** 125–140. doi:10.1038/s41580-021-00405-2

Redon S, Zemp I, Lingner J. 2013. A three-state model for the regulation of telomerase by TERRA and hnRNPA1. *Nucleic Acids Res* **41:** 9117–9128. doi:10.1093/nar/gkt695

Risques RA, Lai LA, Brentnall TA, Li L, Feng Z, Gallaher J, Mandelson MT, Potter JD, Bronner MP, Rabinovitch PS. 2008. Ulcerative colitis is a disease of accelerated colon aging: evidence from telomere attrition and DNA damage. *Gastroenterology* **135:** 410–418. doi:10.1053/j.gastro.2008.04.008

Robert I, Dantzer F, Reina-San-Martin B. 2009. Parp1 facilitates alternative NHEJ, whereas Parp2 suppresses IgH/c-myc translocations during immunoglobulin class switch recombination. *J Exp Med* **206:** 1047–1056. doi:10.1084/jem.20082468

Roger L, Jones RE, Heppel NH, Williams GT, Sampson JR, Baird DM. 2013. Extensive telomere erosion in the initiation of colorectal adenomas and its association with chromosomal instability. *J Natl Cancer Inst* **105:** 1202–1211. doi:10.1093/jnci/djt191

Roger L, Miners KL, Leonard L, Grimstead JW, Price DA, Baird DM, Ladell K. 2023. T cell memory revisited using single telomere length analysis. *Front Immunol* **14:** 1100535. doi:10.3389/fimmu.2023.1100535

Sabatier L, Ricoul M, Pottier G, Murnane JP. 2005. The loss of a single telomere can result in instability of multiple chromosomes in a human tumor cell line. *Mol Cancer Res* **3:** 139–150. doi:10.1158/1541-7786.MCR-04-0194

Savage SA. 2024a. Telomere length and cancer risk: finding Goldilocks. *Biogerontology* **25:** 265–278. doi:10.1007/s10522-023-10080-9

* Savage SA. 2024b. Telomeres and human disease. *Cold Spring Harb Perspect Biol* doi:10.1101/cshperspect.a041684

Schmitt CA, Wang B, Demaria M. 2022. Senescence and cancer—role and therapeutic opportunities. *Nat Rev Clin Oncol* **19:** 619–636. doi:10.1038/s41571-022-00668-4

Schratz KE, Armanios M. 2020. Cancer and myeloid clonal evolution in the short telomere syndromes. *Curr Opin Genet Dev* **60:** 112–118. doi:10.1016/j.gde.2020.02.019

Sfeir A, Symington LS. 2015. Microhomology-mediated end joining: a back-up survival mechanism or dedicated pathway? *Trends Biochem Sci* **40:** 701–714. doi:10.1016/j.tibs.2015.08.006

Shiloh Y, Ziv Y. 2013. The ATM protein kinase: regulating the cellular response to genotoxic stress, and more. *Nat Rev Mol Cell Biol* **14:** 197–210. doi:10.1038/nrm3546

Sikora E, Mosieniak G, Sliwinska MA. 2016. Morphological and functional characteristic of senescent cancer cells. *Curr Drug Targets* **17:** 377–387. doi:10.2174/1389450116666151019094724

Simpson K, Jones RE, Grimstead JW, Hills R, Pepper C, Baird DM. 2015. Telomere fusion threshold identifies a poor prognostic subset of breast cancer patients. *Mol Oncol* **9:** 1186–1193. doi:10.1016/j.molonc.2015.02.003

Slagboom PE, Droog S, Boomsma DI. 1994. Genetic determination of telomere size in humans: a twin study of three age groups. *Am J Hum Genet* **55:** 876–882.

Smogorzewska A, Karlseder J, Holtgreve-Grez H, Jauch A, de Lange T. 2002. DNA ligase IV-dependent NHEJ of deprotected mammalian telomeres in G1 and G2. *Curr Biol* **12:** 1635–1644. doi:10.1016/S0960-9822(02)01179-X

Stephens PJ, Greenman CD, Fu B, Yang F, Bignell GR, Mudie LJ, Pleasance ED, Lau KW, Beare D, Stebbings LA, et al. 2011. Massive genomic rearrangement acquired in a single catastrophic event during cancer development. *Cell* **144:** 27–40. doi:10.1016/j.cell.2010.11.055

Strefford JC, Kadalayil L, Forster J, Rose-Zerilli MJ, Parker A, Lin TT, Heppel N, Norris K, Gardiner A, Davies Z, et al. 2015. Telomere length predicts progression and overall survival in chronic lymphocytic leukemia: data from the UK LRF CLL4 trial. *Leukemia* **29:** 2411–2414. doi:10.1038/leu.2015.217

Tankimanova M, Capper R, Letsolo BT, Rowson J, Jones RE, Britt-Compton B, Taylor AM, Baird DM. 2012. Mre11 modulates the fidelity of fusion between short telomeres in human cells. *Nucleic Acids Res* **40:** 2518–2526. doi:10.1093/nar/gkr1117

Telomeres Mendelian Randomization Collaboration; Haycock PC, Burgess S Nounu A, Zheng J, Okoli GN, Bowden J, Wade KH, Timpson NJ, Evans DM, et al. 2017. Association between telomere length and risk of cancer and non-neoplastic diseases: a Mendelian randomization study. *JAMA Oncol* **3:** 636–651. doi:10.1001/jamaoncol.2016.5945

Tomlinson RL, Ziegler TD, Supakorndej T, Terns RM, Terns MP. 2006. Cell cycle-regulated trafficking of human telomerase to telomeres. *Mol Biol Cell* **17:** 955–965. doi:10.1091/mbc.e05-09-0903

Umbreit NT, Zhang CZ, Lynch LD, Blaine LJ, Cheng AM, Tourdot R, Sun L, Almubarak HF, Judge K, Mitchell TJ, et al. 2020. Mechanisms generating cancer genome complexity from a single cell division error. *Science* **368:** eaba0712. doi:10.1126/science.aba0712

Unryn BM, Cook LS, Riabowol KT. 2005. Paternal age is positively linked to telomere length of children. *Aging Cell* **4:** 97–101. doi:10.1111/j.1474-9728.2005.00144.x

Vannier JB, Pavicic-Kaltenbrunner V, Petalcorin MI, Ding H, Boulton SJ. 2012. RTEL1 dismantles T loops and counteracts telomeric G4-DNA to maintain telomere integrity. *Cell* **149:** 795–806. doi:10.1016/j.cell.2012.03.030

Vasa-Nicotera M, Brouilette S, Mangino M, Thompson JR, Braund P, Clemitson JR, Mason A, Bodycote CL, Raleigh SM, Louis E, et al. 2005. Mapping of a major locus that determines telomere length in humans. *Am J Hum Genet* **76:** 147–151. doi:10.1086/426734

Vulliamy T, Marrone A, Dokal I, Mason PJ. 2002. Association between aplastic anaemia and mutations in telomerase RNA. *Lancet* **359:** 2168–2170. doi:10.1016/S0140-6736(02)09087-6

Vulliamy T, Marrone A, Szydlo R, Walne A, Mason PJ, Dokal I. 2004. Disease anticipation is associated with progressive telomere shortening in families with dyskeratosis congenita due to mutations in TERC. *Nat Genet* **36:** 447–449. doi:10.1038/ng1346

Walsh KM, Codd V, Rice T, Nelson CP, Smirnov IV, McCoy LS, Hansen HM, Elhauge E, Ojha J, Francis SS, et al. 2015. Longer genotypically estimated leukocyte telomere length is associated with increased adult glioma risk. *Oncotarget* **6:** 42468–42477. doi:10.18632/oncotarget.6468

Wang B, Han J, Elisseeff JH, Demaria M. 2024a. The senescence-associated secretory phenotype and its physiological and pathological implications. *Nat Rev Mol Cell Biol* doi:10.1038/s41580-024-00727-x

Wang B, Xiong Y, Li R, Zhang J, Zhang S. 2024b. Shorter telomere length increases the risk of lymphocyte immunodeficiency: a Mendelian randomization study. *Immun Inflamm Dis* **12:** e1251. doi:10.1002/iid3.1251

Weischer M, Nordestgaard BG, Cawthon RM, Freiberg JJ, Tybjaerg-Hansen A, Bojesen SE. 2013. Short telomere length, cancer survival, and cancer risk in 47102 individuals. *J Natl Cancer Inst* **105:** 459–468. doi:10.1093/jnci/djt016

Williams J, Heppel NH, Britt-Compton B, Grimstead JW, Jones RE, Tauro S, Bowen DT, Knapper S, Groves M, Hills RK, et al. 2017. Telomere length is an independent prognostic marker in MDS but not in de novo AML. *Br J Haematol* **178:** 240–249. doi:10.1111/bjh.14666

Wong JY, Vivo D, Lin I, Fang X, Christiani SC, C D. 2014. The relationship between inflammatory biomarkers and telomere length in an occupational prospective cohort study. *PLoS ONE* **9:** e87348. doi:10.1371/journal.pone.0087348

Wright WE, Shay JW. 1992. The two-stage mechanism controlling cellular senescence and immortalization. *Exp Gerontol* **27:** 383–389. doi:10.1016/0531-5565(92)90069-C

Xu J, Ye J, Wu Y, Zhang H, Luo Q, Han C, Ye X, Wang H, He J, Huang H, et al. 2014. Reduced fetal telomere length in gestational diabetes. *PLoS ONE* **9:** e86161. doi:10.1371/journal.pone.0086161

Yan CT, Boboila C, Souza EK, Franco S, Hickernell TR, Murphy M, Gumaste S, Geyer M, Zarrin AA, Manis JP, et al. 2007. Igh class switching and translocations use a robust non-classical end-joining pathway. *Nature* **449:** 478–482. doi:10.1038/nature06020

Yu X, Gabriel A. 2003. Ku-dependent and Ku-independent end-joining pathways lead to chromosomal rearrangements during double-strand break repair in *Saccharomyces cerevisiae*. *Genetics* **163:** 843–856. doi:10.1093/genetics/163.3.843

Zhang C, Doherty JA, Burgess S, Hung RJ, Lindström S, Kraft P, Gong J, Amos CI, Sellers TA, Monteiro AN, et al. 2015. Genetic determinants of telomere length and risk of common cancers: a Mendelian randomization study. *Hum Mol Genet* **24:** 5356–5366. doi:10.1093/hmg/ddv252

Zhang J, Rane G, Dai X, Shanmugam MK, Arfuso F, Samy RP, Lai MK, Kappei D, Kumar AP, Sethi G. 2016. Ageing and the telomere connection: an intimate relationship with inflammation. *Ageing Res Rev* **25:** 55–69. doi:10.1016/j.arr.2015.11.006

Zhao Y, Sfeir AJ, Zou Y, Buseman CM, Chow TT, Shay JW, Wright WE. 2009. Telomere extension occurs at most chromosome ends and is uncoupled from fill-in in human cancer cells. *Cell* **138:** 463–475. doi:10.1016/j.cell.2009.05.026

Zhao Y, Abreu E, Kim J, Stadler G, Eskiocak U, Terns MP, Terns RM, Shay JW, Wright WE. 2011. Processive and distributive extension of human telomeres by telomerase under homeostatic and nonequilibrium conditions. *Mol Cell* **42:** 297–307. doi:10.1016/j.molcel.2011.03.020

# Telomere Crisis Shapes Cancer Evolution

Joe Nassour[1] and Jan Karlseder[2]

[1]University of Colorado School of Medicine, Aurora, Colorado 80045, USA
[2]The Salk Institute for Biological Studies, La Jolla, California 92037, USA

*Correspondence:* joe.nassour@cuanschutz.edu; karlseder@salk.edu

Somatic mutations arise in normal tissues and precursor lesions, often targeting cancer-driver genes involved in cell cycle regulation. Most checkpoint-mutant clones, however, remain dormant throughout an individual's lifetime and seldom progress to malignancy, implying the presence of protective mechanisms that limit their expansion and malignant transformation. One such safeguard is telomere crisis—a potent tumor-suppressive barrier that eliminates cells lacking functional checkpoints and evading p53- and pRb-mediated surveillance. While the genomic instability unleashed during telomere crisis can drive clonal evolution, cell death is typically the dominant outcome, with only a rare subset of cells escaping elimination to initiate malignancy. Recognizing the dual role of telomere crisis—suppressing tumor initiation while enabling clonal evolution—is essential for understanding early cancer development and designing strategies to eliminate tumor-initiating cells.

Malignant transformation proceeds through the stepwise accumulation of driver mutations, typically arising every 20–30 cell divisions. Mutations in tumor-suppressor genes often remain latent until the loss of the remaining wild-type allele unmasks their tumorigenic potential. For cancer to develop, the initial mutant cell must proliferate sufficiently to inactivate the wild-type allele—commonly through loss of heterozygosity—and expand to a population size that increases the likelihood of acquiring additional driver mutations (Wright and Shay 2001). However, the replicative potential of most human somatic cells is inherently constrained, imposing a fundamental barrier to this multistep process (Hayflick 1965; Harley et al. 1992; Wright and Shay 1992). The concept of a finite replicative life span was first described in the early 1960s, when Leonard Hayflick and Paul Moorhead revealed that normal human cells undergo a limited number of divisions in culture (Hayflick and Moorhead 1961; Hayflick 1965). This proliferative restraint is governed by telomere attrition, resulting from incomplete DNA end replication, terminal processing, and stochastic deletions (Harley et al. 1990; Baird et al. 2003, 2006; Britt-Compton et al. 2006). Once telomeres erode beyond a certain threshold, their protective function collapses, triggering two distinct antiproliferative programs—replicative senescence (mortality stage 1, M1) and crisis (mortality stage 2, M2)—that together serve as tumor-suppressive barriers, preventing the propagation of altered cells that might otherwise progress toward malignancy (Fig. 1; Wright and Shay 1992).

Replicative senescence manifests as a stable cell cycle arrest that persists despite mitogenic signals and optimal growth conditions (Huang et al. 2022). This proliferative barrier is driven by the tumor-suppressive p53–p21$^{WAF1}$ and p16$^{INK4A}$–retinoblastoma protein (pRb) pathways, activated in response to DNA damage signals arising from dysfunctional telomeres (d'Adda di Fagagna et al. 2003; Takai et al.

2003; Herbig et al. 2004). Genetic or epigenetic alterations that disable senescence checkpoints permit continued proliferation and progressive telomere erosion until a second barrier is encountered—telomere crisis—characterized by severe telomere dysfunction and chromosomal instability that typically triggers widespread cell death via autophagy and innate immune pathways (Counter et al. 1992; Wright and Shay

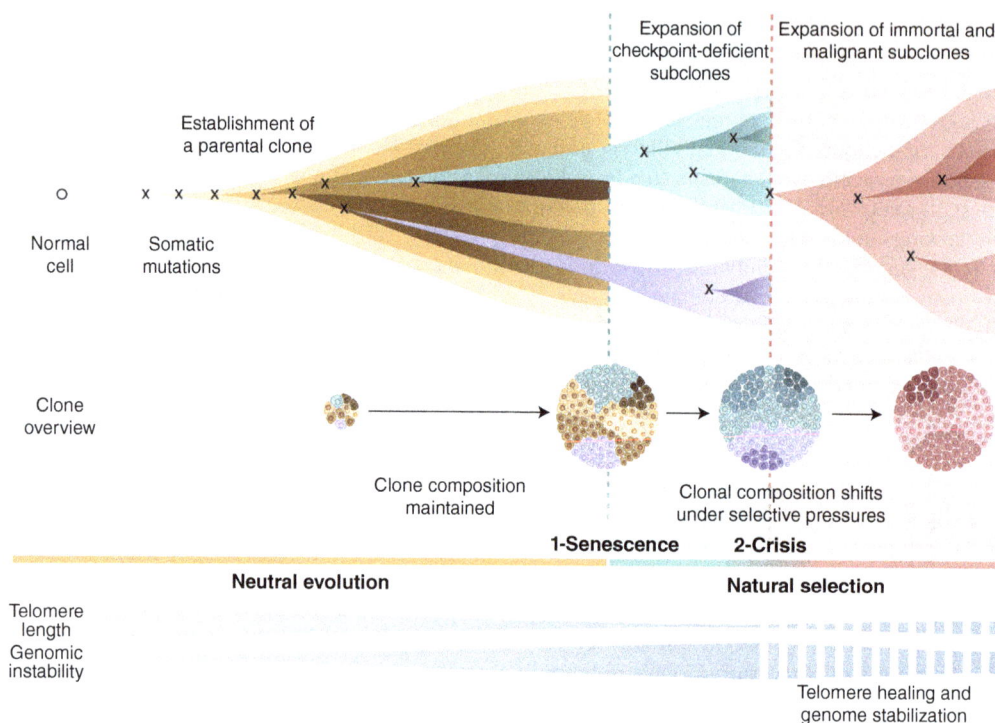

**Figure 1.** Telomere shortening imposes proliferative barriers and drives clonal evolution. Cancer arises through clonal evolution, with genetic variation providing the substrate for selection. Analyses of intratumoral heterogeneity (ITH) have yielded insights into how evolutionary forces govern tumor growth and progression. Central to this framework is cellular fitness—the ability of a tumor cell to survive, proliferate, and transmit its genotype. Increases in fitness promote clonal expansion and can lead to selective sweeps that favor dominant clones. In the early stages of tumorigenesis, telomere shortening imposes two distinct proliferative barriers: replicative senescence (M1) and crisis (M2). Senescence occurs when critically short telomeres trigger a DNA damage response (DDR), arresting cell division. Inactivation of p53 and retinoblastoma protein (pRb) allows cells to bypass this checkpoint, enabling continued proliferation and progressive telomere dysfunction. However, these checkpoint-deficient cells eventually reach crisis—a state marked by severe chromosomal instability (CIN) and widespread cell death. Although telomere crisis serves as a powerful tumor-suppressive barrier, the genomic instability it induces also accelerates clonal evolution by generating a broad spectrum of genetic alterations. Within this selective landscape, rare clones that resist cell death, suppress DDR signaling, or restore telomere maintenance —via telomerase reactivation or the alternative lengthening of telomeres (ALT) pathway—can escape crisis and drive malignant progression. Telomere crisis functions as an evolutionary bottleneck, eliminating most precancerous clones while selecting for resilient subpopulations that fuel tumor evolution. (M1) Mortality stage 1, (M2) mortality stage 2.

 Cite this article as *Cold Spring Harb Perspect Biol* doi: 10.1101/cshperspect.a041688

1992; Hayashi et al. 2015; Nassour et al. 2019, 2023). During crisis, cell populations typically stagnate or decline, as proliferative output is outweighed by cell death—positioning crisis as an intrinsic anticancer barrier targeting cells with acquired resistance to p53- and pRb-dependent checkpoints. However, this protective mechanism carries a cost: The genomic instability unleashed during crisis generates extensive genetic diversity upon which Darwinian selection can act, thereby fueling clonal evolution and malignant progression (Fig. 1; Maciejowski and de Lange 2017; Cleal and Baird 2020).

Early in tumorigenesis, nascent neoplastic clones undergo a brief period of telomere crisis, during which dysfunctional telomeres drive karyotypic complexity, preceding the acquisition of a telomere maintenance mechanism (TMM) and the emergence of invasive traits. Telomere-driven genomic instability stems from acute bursts of catastrophic events such as chromothripsis, as well as iterative breakage–fusion–bridge (BFB) cycles—each capable of substantially reconfiguring the genome (Riboni et al. 1997; Artandi et al. 2000; Artandi and DePinho 2010; Roger et al. 2013; Hermetz et al. 2014; Maciejowski et al. 2015; Cleal et al. 2018). While most checkpoint-deficient cells with altered genomes perish during crisis, a rare subset acquires adaptive traits—such as compromised DNA damage checkpoints, resistance to cell death pathways, or, most commonly, TMM activation—that restore telomere homeostasis and stabilize the genome. These adaptations enable individual cells to escape crisis and emerge as immortal cell populations poised for neoplastic transformation.

Viewed through this lens, telomere crisis serves as a double-edged sword in tumorigenesis—imposing a stringent proliferative barrier on p53- and pRb-deficient cells, while fostering the genetic and phenotypic diversity that provides fertile ground for cancer evolution. Defining the molecular mechanisms that govern telomere crisis—and understanding how rare cells circumvent this barrier during early tumorigenesis—is critical for developing strategies to intercept tumor-initiating cells before malignancy reaches an irreversible momentum. Here, we discuss how telomere crisis shapes early tumor evolution by eliminating checkpoint-deficient clones and selecting for a minority of resilient cells that acquire adaptive traits and progress toward malignant transformation.

## TELOMERE CRISIS AS A BARRIER TO MALIGNANT TRANSFORMATION

Cancer evolves through iterative cycles of mutation, competition, and clonal expansion, shaped by continuous selective pressures that refine its subclonal architecture (Fig. 1). Genetic and epigenetic instability generates a heterogeneous pool of subclones, each carrying alterations—ranging from single-nucleotide variants and copy-number changes to complex epigenetic modifications—that influence cellular fitness (Vendramin et al. 2021). Subclones with advantageous traits, such as enhanced proliferation, immune evasion, or survival signaling, undergo "clonal sweeps," outcompeting less fit counterparts. Meanwhile, negative selection eliminates subclones harboring deleterious mutations, further sculpting tumor composition (Reeves and Balmain 2024). Through repeated cycles of selection, tumors progressively acquire hallmark traits of malignancy, including unchecked proliferation, immune escape, and metastatic potential (Savy et al. 2025). Early models depicted tumor evolution as a predominantly linear sequence of oncogenic and tumor-suppressor mutations converging on a single dominant clone. However, advances in next-generation and single-cell sequencing have revealed extensive intratumor heterogeneity, where multiple subclones often coexist and evolve in parallel under distinct selective pressures. Modern perspectives now embrace a spectrum of evolutionary trajectories—linear, branched, neutral, and punctuated—driven by the dynamic interplay of selection, stochastic mutations, and punctuated genomic instability (Fig. 1; Davis et al. 2017).

A wide range of mutated genes—traditionally classified as driver or passenger—underlie human cancers. While passenger mutations often exert minimal influence on disease progression, driver mutations propel cells toward neoplasia (Ostroverkhova et al. 2023). For years, detecting somatic mutations in histologically normal tissues or early lesions was limited by technical challenges in identifying low-frequency variants and the scarcity of healthy donor material. Recent advances in deep mutational profiling—applied to small biopsies and microscopically defined clonal units—have revealed that ostensibly normal cells frequently carry numerous mutations. Many of these mutations affect canonical cancer-associated genes involved in cell growth and proliferation (Tsao et al. 2000; Baker et al. 2013; Sottoriva et al. 2015; Sun et al. 2017; Williams et al. 2018; Gerstung et al. 2020; Karlsson et al. 2023). For instance, driver mutations in core cell cycle regulators such as *TP53* (tumor protein p53), *RB1* (retinoblastoma protein), and *CDKN2A* (cyclin-dependent kinase inhibitor 2A) have been identified in histologically normal tissues and benign neoplasms at frequencies that sometimes match or surpass those in malignant tumors (Martincorena et al. 2015, 2018; Lee-Six et al. 2019; Lawson et al. 2020; Moore et al. 2020; Coorens et al. 2021; Machado et al. 2022; Mitchell et al. 2022; Wang et al. 2023). Despite their ability to expand and colonize tissue compartments, these initiated clones often remain clinically silent and rarely progress to full malignancy (Mustjoki and Young 2021; Wijewardhane et al. 2021). One striking example is the detection of *TP53* mutations in ∼80% of esophageal epithelial clones in older adults, despite an annual esophageal cancer incidence of only ∼1% (Martincorena et al. 2018; Yokoyama et al. 2019). Similarly, *TP53* mutations accumulate in sun-exposed keratinocytes, but only a small fraction of these mutant cells advance to cutaneous carcinomas (Nakazawa et al. 1994; Martincorena et al. 2015). Similar observations apply to other cell cycle regulators: *RB1* and *CDKN2A* mutations found in normal bronchial epithelia or pancreatic intraepithelial neoplasms seldom culminate in malignancy (Carrière et al. 2011; McWilliams et al. 2011; Tang et al. 2015), and *BRCA1*—strongly implicated in hereditary breast and ovarian cancer—can be detected in atypical ductal hyperplasia without necessarily progressing to invasive disease (King et al. 2005; Smith et al. 2007; Boyraz and Ly 2024).

Positively selected mutants can drive clonal expansion, yet in aging human tissues, such clones may persist for years or even decades without disrupting normal histology or progressing to malignancy (Dressler et al. 2022). This suggests that clonal expansion following loss of cell cycle checkpoints is transient, ultimately restrained by robust protective mechanisms that limit the proliferation of potentially malignant clones (Martincorena et al. 2015, 2018; Suda et al. 2018; Yokoyama et al. 2019; Tang et al. 2020; Kakiuchi and Ogawa 2021; Gomes 2022; Shah 2024). Given that the human body comprises ∼40 trillion cells across >200 specialized cell types, multiple layers of defense operate in concert to suppress oncogenic transformation. Among these layers is crisis—a telomere-driven checkpoint that induces cell death once primary p53 and pRb safeguards are circumvented—providing a stringent barrier against the unchecked expansion of mutant lineages on their path toward malignancy (Fig. 1).

Telomere crisis is viewed as the final proliferative barrier that must be overcome for malignancy to develop. However, the lack of specific biomarkers to identify and track crisis cells during cancer progression—as well as limited insight into the genetic mechanisms governing cell death during crisis—has made it challenging to demonstrate crisis as an anticancer barrier in vivo. These uncertainties have perpetuated the longstanding view that crisis is largely nonprogrammed, triggered by severe genomic instability beyond a threshold incompatible with cell viability. Recent findings, however, have revealed a dominant role for autophagy and innate immune signaling in regulating cell death during crisis. Disruption of these pathways allowed genetically unstable cells to evade clearance and bypass crisis, giving rise to populations harboring extensive chromosomal rearrangements (Nassour et al. 2019, 2023). Autophagy-deficient cells that survived crisis developed near-tetraploid karyotypes and exhibited numerous structural abnormalities—including nonreciprocal translocations, deletions, and duplications—reminiscent of those seen in epithelial cancers. These abnormalities arose at frequencies far exceeding those observed in cells under-

Cite this article as *Cold Spring Harb Perspect Biol* doi: 10.1101/cshperspect.a041688

going crisis (Nassour et al. 2019). Further studies are needed to determine whether impairing these programmed death mechanisms can indeed suppress crisis in vivo, thereby promoting cellular immortalization and fueling tumor progression.

## TELOMERE CRISIS AS A DRIVING FORCE IN MALIGNANT TRANSFORMATION

Dysfunctional telomeres activate DNA damage sensing and repair pathways, leading to inter- and intrachromosomal fusion events that give rise to chromosome circularization or, more commonly, dicentric chromosomes. These unstable chromosomal derivatives undergo BFB cycles, driving genomic alterations characteristic of early-stage cancer development, including aneuploidy, translocations, loss of heterozygosity, and focal gene amplifications (Riboni et al. 1997; Artandi et al. 2000; De Lange 2005; Roger et al. 2013; Cleal and Baird 2020). Such genomic instability can escalate into complex mutational phenomena, most notably chromothripsis and kataegis (Maciejowski et al. 2015; Cleal et al. 2019). The spectrum of telomere-driven genomic alterations now includes whole-genome duplication, reinforcing the view that cancer genome evolution is fundamentally rooted in telomere dysfunction (Davoli et al. 2010; Davoli and de Lange 2012; Maciejowski et al. 2015).

Telomere dysfunction and chromosomal aberrations often emerge early in epithelial carcinogenesis and tend to stabilize as malignancy progresses, coinciding with the activation of a TMM (Kim et al. 1994; Chadeneau et al. 1995; Meyerson et al. 1997; Mitelman et al. 1997; Shih et al. 2001). Studies of early premalignant lesions revealed that telomeres erode to critically short lengths—similar to those seen in crisis—often accompanied by telomere fusion, anaphase bridging, and chromosomal instability (Meeker et al. 2002, 2004; van Heek et al. 2002; Chin et al. 2004; Lin et al. 2010; Roger et al. 2013). Analysis of colorectal adenomatous polyps reveals short telomeres and fusions that precede the adenoma/carcinoma transition (Roger et al. 2013). In breast cancers, telomere-related genome instability—including anaphase and chromatin bridges—accumulates progressively as cells advance from usual ductal hyperplasia (UDH) to ductal carcinoma in situ (DCIS), along with a higher frequency of chromosomal aberrations in DCIS compared with UDH, evidence of telomere shortening in DCIS, and the activation of telomerase at this stage (Chin et al. 2004). In chronic lymphocytic leukemia (CLL), telomere erosion and dysfunction not only characterize late-stage disease but also appear in a subset of early-stage (Binet stage A) patients, where telomere fusions coincide with large-scale genomic rearrangements prior to clinical progression (Lin et al. 2010). These findings are further supported by modeling telomere crisis in late-generation telomerase-deficient mice lacking *Trp53*, where telomere dysfunction drives epithelial cancers displaying nonreciprocal translocations, as well as focal amplifications and deletions—mirroring the cytogenetic features of human carcinomas (Artandi et al. 2000; O'Hagan et al. 2002). Mouse models of telomerase reactivation following a period of telomere dysfunction facilitate the acquisition of copy number aberrations and aneuploidy, driving malignant phenotypes (Ding et al. 2012).

A transient phase of telomere dysfunction, consistent with crisis, is thought to arise during the early stages of human tumorigenesis, with escape from this state and acquisition of replicative immortality representing critical rate-limiting steps in malignant transformation (Li et al. 2014; Hadi et al. 2020). Once a TMM is established, telomere-driven genome instability is curtailed, allowing the genome to stabilize while retaining prior alterations. Most malignancies—as well as stem cells, germ cells, and many unicellular organisms—rely on the ribonucleoprotein complex telomerase as their primary TMM (Kim et al. 1994). However, ~10%–15% of malignancies—particularly those of mesenchymal origin (Henson and Reddel 2010)—lack telomerase expression and instead preserve their telomeres via the alternative lengthening of telomeres (ALT) pathway (Lundblad and Blackburn 1993; Bryan et al. 1995). Whether through telomerase or ALT, restoration of telomere stability allows genomic rearrangements acquired during crisis to be preserved and propagated, facilitating the emergence of malignant clones (Ding et al. 2012).

## THE TWO-STAGE MODEL OF REPLICATIVE AGING

The accumulation of driver mutations within a single somatic lineage is fundamentally limited by the intrinsic replicative capacity of human cells. An initial driver mutation may require ~20 population doublings (PDs) to achieve clonal expansion, with subsequent alterations demanding further rounds of replication. However, most normal human cells undergo replicative senescence after ~90 PDs and enter telomere crisis before reaching ~110 PDs, suggesting that many premalignant clones exhaust their proliferative capacity before acquiring the full complement of mutations necessary for malignant transformation. This concept underpins the two-stage model of replicative aging, originally proposed to explain how viral oncoproteins—such as simian virus 40 (SV40) T antigen or human papillomavirus type 16 (HPV16) E6/E7—can extend the life span of human fibroblasts without inducing cellular immortality (Wright et al. 1989; Wright and Shay 1992).

### Shelterin-Mediated Telomere Protection

Unlike circular chromosomes, linear chromosomes require a dedicated mechanism to prevent their DNA ends from being mistaken for DNA breaks—a concept identified by Hermann Muller and Barbara McClintock as the "end-protection problem" (McClintock 1941). Telomeres meet this need by forming specialized caps at chromosome ends, composed of repetitive DNA sequences and associated protein complexes that shield these termini from DNA damage surveillance and repair pathways (de Lange 2018). Understanding how telomeres fulfill this end-protection function is a fundamental question with profound implications for human health, as the loss of telomere integrity underlies telomere biology disorders, aging, and cancer.

Early studies in *Tetrahymena* revealed that linear eukaryotic chromosomes terminate with hexameric DNA repeats (TTGGGG), typically arranged in tandem arrays of ~20–70 repeats (Blackburn and Gall 1978). Subsequent research established that most eukaryotes possess similar telomeric architecture, characterized by short G-rich repeats oriented in the 5′–3′ direction, typically containing two to four consecutive guanines (Tomáška et al. 2020). These repeats exhibit remarkable evolutionary conservation across diverse taxa (Meyne et al. 1989). In humans, telomeres consist of multiple kilobases of TTAGGG repeats, a motif highly conserved among vertebrates, terminating in a single-stranded 3′ G overhang (Makarov et al. 1997; McElligott and Wellinger 1997). The double-stranded and single-stranded regions of telomeres are embedded by shelterin complexes, comprising six subunits: telomeric repeat-binding factor (TRF)1 (Bilaud et al. 1997; van Steensel and de Lange 1997) and TRF2 (Bilaud et al. 1997; Broccoli et al. 1997; van Steensel et al. 1998; Smogorzewska et al. 2000), protection of telomeres 1 (POT1) (Baumann and Cech 2001; Loayza and De Lange 2003), TRF1-interacting nuclear factor 2 (TIN2, also known as TINF2) (Kim et al. 1999), repressor/activator protein 1 (RAP1) (Li et al. 2000), and TIN2-interacting protein 1 (TPP1, also known as ACD [adreno-cortical dysplasia protein homolog]) (Fig. 2A; Ye et al. 2004). TRF1 and TRF2 homodimers bind directly to double-stranded telomeric DNA (Broccoli et al. 1997), whereas the POT1–TPP1 heterodimer interacts with the G overhang (Loayza and De Lange 2003). RAP1 associates with TRF2, and TIN2 functions as a central scaffold linking TRF1, TRF2, and TPP1 (Fig. 2A; Houghtaling et al. 2004; Ye et al. 2004; O'Connor et al. 2006; Takai et al. 2011).

Studies in mouse and human models lacking individual shelterin components have yielded key insights into how the shelterin complex prevents DNA damage signaling and repair pathways. Loss of shelterin subunits often activates ataxia-telangiectasia mutated (ATM)-, ataxia telangiectasia and Rad3-related (ATR)-, or poly-ADP-ribose polymerase 1 (PARP1)-dependent DNA damage responses (DDRs), rendering telomeres vulnerable to homologous recombination (HR), classical nonhomologous

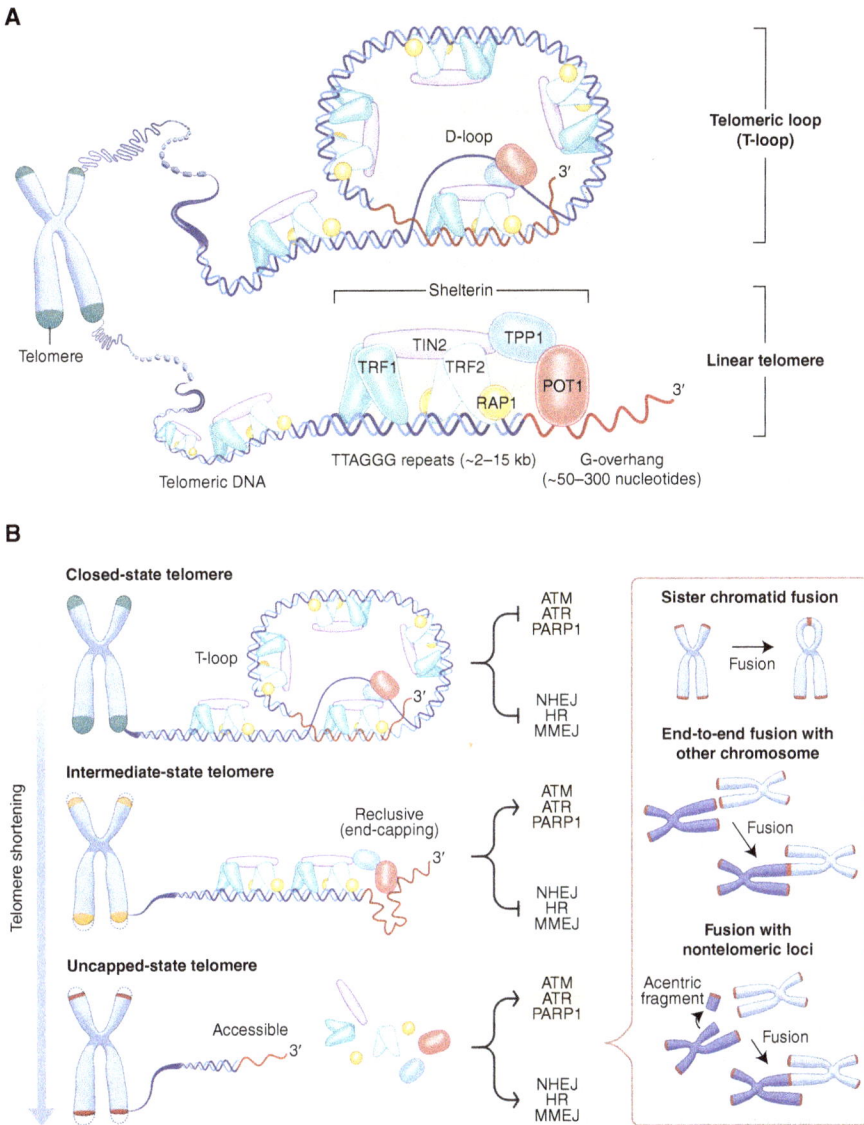

**Figure 2.** A three-state model of chromosome-end protection. (*A*) Graphical representation of human telomeres in linear and T-loop configurations. Telomeres distinguish naturally occurring chromosome ends from DNA double-strand breaks, preventing inappropriate repair events and the formation of dicentric chromosomes. They typically comprise several kilobases of repeated 5′-TTAGGG-3′ sequences, ending in a single-stranded G-rich 3′ overhang of 50–300 nt. These repeats are bound by the six-subunit shelterin complex (telomeric repeat-binding factor 1 [TRF1], TRF2, RAP1, TRF1-interacting nuclear factor 2 [TIN2], TPP1, and protection of telomeres 1 [POT1]). Telomeres can form a protective T-loop through invasion of the 3′ overhang into a double-stranded region of telomeric DNA, thereby avoiding recognition by DNA damage response (DDR) machinery. Shelterin represses key DDR enzymes (ataxia-telangiectasia mutated [ATM], ataxia telangiectasia and Rad3-related [ATR], and poly-ADP-ribose polymerase 1 [PARP1]) and blocks the double-strand break (DSB) repair pathways (classical nonhomologous end joining [c-NHEJ], alternative nonhomologous end joining [alt-NHEJ]/microhomology-mediated end joining [MMEJ], and homologous recombination [HR]). It also prevents excessive DNA end resection at telomeres, preserving genomic stability. (*B*) Graphical representation of human telomeres in closed-state, intermediate-state, and uncapped-state conformations, as suggested to occur during replicative aging or upon experimental disruption of telomeric repeat-binding factor 2 (TRF2). The intermediate state retains partial shelterin binding, limiting end joining despite DDR signaling; the uncapped state is DDR-positive and highly fusogenic due to insufficient shelterin coverage. Short, dysfunctional telomeres can fuse either with other short telomeres or with nontelomeric genomic loci, yielding dicentric chromosomes. Subsequent breakage of these dicentric chromosomes during cell division initiates breakage–fusion–bridge (BFB) cycles, leading to widespread genomic rearrangements. (RAP1) Repressor activator protein 1, (T-loop) telomere-loop.

end joining (NHEJ), and alternative end joining (alt-EJ; also known as microhomology-mediated end joining [MMEJ]) (Fig. 2B; de Lange 2018). These studies revealed the modular organization of shelterin, wherein distinct subunits block specific DNA damage signaling and repair pathways: TRF2 suppresses ATM signaling and NHEJ (van Steensel et al. 1998; Smogorzewska et al. 2002; Celli and de Lange 2005; Celli et al. 2006; Denchi and de Lange 2007; Okamoto et al. 2013; Ribes-Zamora et al. 2013; Badie et al. 2015); TRF1 and POT1 inhibit ATR activation (Veldman et al. 2004; Hockemeyer et al. 2005, 2006; Wu et al. 2006; Denchi and de Lange 2007; Guo et al. 2007; Martínez et al. 2009; Sfeir et al. 2009; Kibe et al. 2010; Takai et al. 2011; Glousker et al. 2020); RAP1 provides additional safeguards against NHEJ (Bae and Baumann 2007; Sarthy et al. 2009; Bombarde et al. 2010; Martinez et al. 2010; Sfeir et al. 2010; Kabir et al. 2014; Lototska et al. 2020; Eickhoff et al. 2025); and TRF2, TRF1, POT1, Ku70, and Ku80 collectively limit HR and MMEJ (Celli et al. 2006; Palm et al. 2009; Rai et al. 2010; Sfeir and de Lange 2012; Badie et al. 2015). Across eukaryotes, telomeres adopt a lariat-like T-loop (telomere loop) conformation, where the 3′ G-rich overhang invades the adjacent duplex telomeric DNA, thereby shielding chromosome ends from DNA damage signaling and repair (Fig. 2A; Griffith et al. 1999; Nikitina and Woodcock 2004; Raices et al. 2008). Initially identified by electron microscopy, T-loop dynamics have since been visualized at high resolution in vivo using superresolution techniques including STORM (stochastic optical reconstruction microscopy), STED (stimulated emission depletion microscopy), SIM (structured illumination microscopy), and Airyscan (superresolution detector module) (Doksani et al. 2013; Van Ly et al. 2018). The TRFH domain of TRF2 is required for the formation and stabilization of T-loops: partial TRF2 depletion, or expression of TRFH-domain mutants in $Trf2^{-/-}$ backgrounds, activates an ATM-dependent DDR that coincides with a transition from looped to linear telomeres (Doksani et al. 2013; Van Ly et al. 2018). A more detailed discussion of shelterin-mediated telomere protection can be found in de Lange (2025).

## Three States of Telomere End Protection

Unlike most cancer cell lines, which use TMMs to circumvent replicative limits, normal human somatic cells lack such pathways. As a result, telomeres progressively shorten with each division—by ~50–150 bp—due to incomplete end replication and telomere-specific processing (Harley et al. 1990). Beyond these predictable end-replication losses, telomeres can also undergo large-scale, seemingly stochastic deletion events, occasionally producing individual telomeres that are shorter than those observed in senescent or crisis-stage cultures (Baird et al. 2003, 2006; Britt-Compton et al. 2006; Capper et al. 2007). Accordingly, normal primary cells exhibit a finite replicative life span, characterized by an initial near-exponential growth phase followed by a plateau phase, during which proliferation ceases (Hayflick 1965; Counter et al. 1992; Harley et al. 1992; Wright and Shay 1992).

A conceptual model has been proposed in which progressive telomere shortening imposes proliferative limits through three distinct stages of telomere end protection, each marked by a unique macromolecular architecture (Fig. 2B; Cesare et al. 2009; Cesare and Karlseder 2012). In exponentially growing cells, telomeres reside in a protected "closed" state that conceals chromosome ends from DNA damage surveillance. As telomeres shorten and fall below a critical threshold, the T-loop structure collapses, exposing chromosome ends to DNA damage–sensing pathways while remaining refractory to end joining by NHEJ and MMEJ pathways (Cesare et al. 2009; Sarek et al. 2019; Romero-Zamora et al. 2025). This intermediate "linear" state underlies replicative senescence, wherein DDR-positive telomeres retain sufficient shelterin coverage to prevent catastrophic end-to-end fusions (d'Adda di Fagagna et al. 2003; Takai et al. 2003; Herbig et al. 2004; Kaul et al. 2012). Loss of cell cycle checkpoints allows cells to continue proliferating and drives further telomere erosion until a fully deprotected "uncapped" state is reached, triggering telomere crisis. In this uncapped state, reduced shelterin occupancy permits telomere ends to become vulnerable to aberrant repair through LIG4-dependent NHEJ and LIG1/LIG3-dependent MMEJ pathways (DNA ligase III and DNA ligase IV) (Fig. 2B; Jones et al. 2014; Hayashi et al. 2015; Maciejowski et al. 2015; Nassour et al. 2019).

## Replicative Senescence and Crisis

Replicative senescence is induced when one or a few critically short telomeres are perceived as DNA double-strand breaks and trigger a DDR that halts further cell division (d'Adda di Fagagna et al. 2003; Takai et al. 2003; Herbig et al. 2004; Kaul et al. 2012). While early studies associated senescence with the global loss of the 3′ G overhang (Stewart et al. 2003), subsequent work revealed that replicative senescence is induced by a change in the protective status of shortened telomeres rather than complete loss of telomeric DNA (Hemann et al. 2001; Karlseder et al. 2002). Senescent cells typically harbor telomeres in the 4–7 kb range but display DDR foci at only a few chromosome ends, consistent with the idea that a single critically short telomere can elicit a proliferative arrest (d'Adda di Fagagna et al. 2003; Deckbar et al. 2007; Löbrich and Jeggo 2007; Kaul et al. 2012). In this intermediate "linear" state, telomeres retain partial shelterin protection but become DDR-positive, activating ATM and ATR signaling pathways and forming telomere dysfunction-induced foci (TIF)—nuclear domains enriched in DNA repair factors including p53-binding protein 1 (53BP1), γ-H2AX, and MDC1 (d'Adda di Fagagna et al. 2003; Takai et al. 2003; Herbig et al. 2004; Kaul et al. 2012).

Intermediate-state deprotected telomeres permit mitotic progression and subsequently trigger a cell cycle arrest in $G_1$ phase through activation of the p53/p21$^{WAF1}$ and p16$^{INK4a}$/pRb tumor-suppressor pathways. These checkpoints inhibit cyclin-dependent kinases—CDK4 and CDK6, and in the case of p21$^{WAF1}$ also CDK2—which are essential for $G_1$/S transition and cell cycle progression. In human fibroblasts, bypassing senescence requires inactivating both p53 and pRb simultaneously,

whereas in murine cells, loss of either pathway alone often suffices—revealing a redundancy in regulating senescence (Shay et al. 1991a; Smogorzewska and de Lange 2002). Although telomere shortening correlates with increased p16$^{INK4a}$ levels, and telomerase-mediated telomere elongation reverses this increase, the precise role of p16$^{INK4a}$ in replicative senescence remains debated (Alcorta et al. 1996; Hara et al. 1996; Bodnar et al. 1998). Expression of a dominant-negative TRF2 allele (TRF2ΔBΔM) induces p16$^{INK4a}$-dependent arrest, but p16$^{INK4a}$ deficiency only partially rescues the proliferative block, implying that p16$^{INK4a}$ acts more as a reinforcing mechanism than as the central driver of replicative senescence (Jacobs and de Lange 2004).

Loss of p53 and pRb pathways renders cells insensitive to DDR signals emanating from short, dysfunctional telomeres and incapable of exiting the cell cycle. As a result, cells continue to divide, and their telomeres shorten beyond the senescence setpoint (Counter et al. 1992; Capper et al. 2007). These cultures, however, are not immortal, eventually succumbing to a second proliferative barrier often referred to as crisis, accompanied by a nearly complete loss of viable cells (Wright and Shay 1992). Unlike the cell cycle arrest typical of senescence, crisis is characterized by a growth plateau in which sustained proliferation is counterbalanced by extensive cell death, leading to an overall stagnation of the cell population (Wei and Sedivy 1999). The extended life span of these cultures varies between cell types, but typically human fibroblast cultures exhibit an additional 20–30 PDs before entering crisis. Mean telomere lengths during crisis often fall below 1 kb, suggesting that some chromosome ends erode into subtelomeric regions, thereby failing to suppress DNA damage signaling and repair. The molecular identity of this uncapped telomeric state remains undefined, but it has been proposed to involve telomeres completely devoid of shelterin components, rendering them accessible to DNA repair machinery. As a result, telomeres may undergo fusion with other telomeric ends or nontelomeric DSBs, generating chromosomal aberrations reminiscent of those observed in early-stage epithelial cancers (Capper et al. 2007; Tankimanova et al. 2012; Cleal et al. 2018).

## HALLMARKS OF TELOMERE CRISIS

Despite its established role as a tumor-suppressive barrier, the defining molecular and cellular features of telomere crisis remain incompletely understood. In this section, we provide a comprehensive overview of the emerging hallmarks of telomere crisis (Fig. 3).

### Telomere Fusions

Telomeres rendered dysfunctional by experimentally depleting TRF2 harbor full-length telomeric repeats at the fusion sites, raising questions about how accurately they mirror the naturally eroded telomeres encountered during crisis (van Steensel et al. 1998). Evidence from *Schizosaccharomyces pombe*, *Arabidopsis*, and mice indicates that fusions persist even when key NHEJ components—such as DNA-PKcs, LIG4, and 53BP1—are deleted, pointing at MMEJ as the predominant DNA repair pathway at short telomeres (Baumann and Cech 2000; Heacock et al. 2004; Capper et al. 2007; Maser et al. 2007). In human cells, telomere fusion breakpoints often retain only a few residual TTAGGG repeats, often accompanied by subtelomeric deletion of one, or both, of the participating telomeres and the presence of DNA sequence microhomology at the fusion point— hallmarks of error-prone MMEJ (Capper et al. 2007; Letsolo et al. 2010; Tankimanova et al. 2012).

Checkpoint-deficient human fibroblasts engineered to express HPV16 E6E7 oncoproteins (HCA2, IMR90, MRC5, and WI38) have provided a robust model to study telomere fusions during crisis. In MRC5 fibroblasts, single telomere length analysis (STELA) combined with XpYp-specific fusion PCR revealed that telomeres become fusogenic at an extremely short length, typically retaining fewer than five TTAGGG repeats. Approximately 47% of fusion junctions lacked detectable repeats, and the remainder averaged only five repeats per junction. Extensive deletions flanked by microhomology defined the predominant MMEJ repair signature, sharply contrasting with the blunt, minimal-deletion fusions characteristic of NHEJ after TRF2 depletion. Expanded analyses targeting subtelomeres at 17p and 21q in both MRC5 and IMR90 fibroblasts confirmed that MMEJ repair dominates at critically short telomeres across distinct chromosomal contexts (Capper et al. 2007; Letsolo et al. 2010).

Further mechanistic insight was gained through analysis of fibroblasts derived from patients with ataxia-telangiectasia-like disorder (ATLD), bearing hypomorphic MRE11 mutations (meiotic recombination 11 homolog) and expressing E6E7. Fusion PCR demonstrated that MRE11-deficient cells retained the capacity for telomere fusion but displayed increased microhomology usage and higher frequencies of insertional events, suggesting that MRE11 constrains resection and thereby suppresses MMEJ. Together, these findings establish that telomere crisis in human fibroblasts predominantly engages an MMEJ-driven mutagenic repair process distinct from the NHEJ pathway active after acute TRF2 loss (Tankimanova et al. 2012).

Extending earlier observations, single-molecule telomere fusion PCR combined with high-throughput sequencing in checkpoint-deficient human fibroblasts undergoing crisis revealed a striking enrichment of telomere fusions with actively transcribed genomic loci, particularly long genes prone to replication–transcription conflicts. Crisis progression was marked by widespread transcriptional reprogramming, the accumulation of copy number alterations, and a progressive shift toward MMEJ, reflected in longer microhomology tracts and increased templated insertions at fusion junctions. Functional disruption of DNA ligase 1 significantly reduced both the frequency and length of templated insertions, whereas loss of ligase 4 enhanced microhomology usage, underscoring distinct contributions of replication- and repair-associated pathways to the evolving telomere fusion landscape (Liddiard et al. 2021).

Additional mechanistic insights have emerged from HCT116 human colon carcinoma cells expressing a dominant-negative human telomerase reverse transcriptase (hTERT) allele,

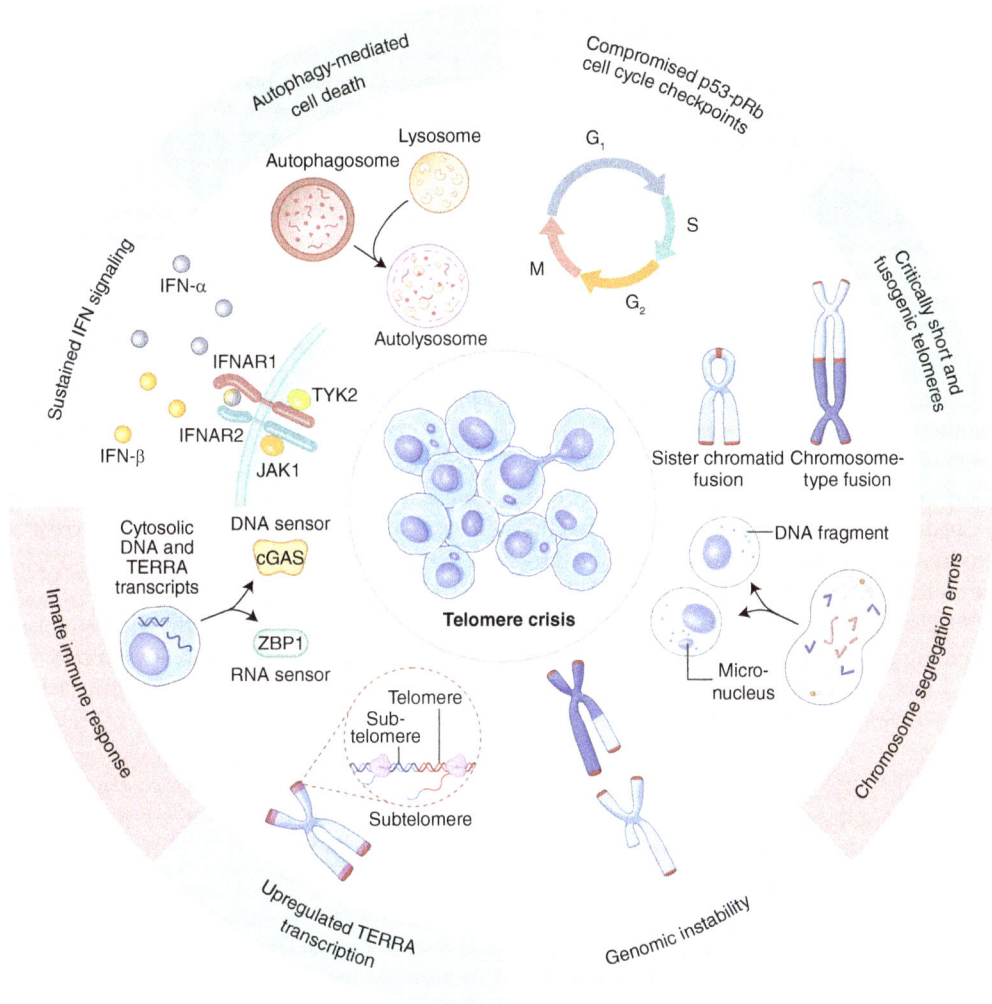

**Figure 3.** Hallmarks of telomere crisis. In the absence of functional cell cycle checkpoints, continued proliferation beyond the senescence barrier leads to telomere crisis, marked by the loss of end-capping function. Telomeres erode to the point where they are nearly devoid of repeats and can no longer be protected by the shelterin complex. These uncapped chromosome ends are recognized as DNA breaks and processed primarily through microhomology-mediated end joining (MMEJ), producing dicentric chromosomes and initiating breakage–fusion–bridge (BFB) cycles. This process drives widespread genomic instability, including amplifications, loss of heterozygosity, and translocations. The spectrum of telomere-induced genome alterations has recently expanded to include chromothripsis—a catastrophic form of chromosomal rearrangement observed in cancers and some congenital disorders. During crisis, cell death progressively increases until it surpasses proliferation, leading to culture collapse. This death response is triggered by the cytoplasmic accumulation of self-derived nucleic acids, which are sensed as damage-associated molecular patterns by two innate immune pathways. The cyclic GMP–AMP synthase (cGAS)– stimulator of interferon (IFN) genes (STING) axis is activated by cytoplasmic DNA, whereas the Z-DNA-binding protein 1 (ZBP1)–mitochondrial antiviral-signaling protein (MAVS) pathway responds to cytoplasmic telomeric repeat–containing RNA (TERRA) transcripts derived from damaged telomeres. These pathways converge to induce type I IFN signaling and sustained inflammation, culminating in a specialized form of autophagy that eliminates cells in crisis. (M2) Mortality stage 2, (DSB) double-strand break.

where telomere crisis appears to involve both NHEJ and MMEJ repair pathways. In this context, NHEJ-mediated fusions predominantly involve the telomeric repeat array, whereas MMEJ drives fusions within subtelomeric regions. LIG4-dependent NHEJ generates predominantly interchromosomal fusions with limited microhomology, while LIG3-dependent MMEJ promotes intrachromosomal fusions reliant on microhomology at junctions. Notably, LIG3—but not LIG4—is essential for crisis escape, highlighting a selective advantage conferred by MMEJ-mediated repair during telomere-driven genome remodeling (Jones et al. 2014).

Studying MMEJ poses inherent challenges, as its relatively low basal activity at DSBs is often overshadowed by the more dominant NHEJ pathway. Quantitative MMEJ assays remain limited, typically relying on computational analyses of next-generation sequencing data to detect microhomologies at repair junctions. Strategies that disable NHEJ or otherwise promote MMEJ—such as targeted depletion of shelterin components—can enhance MMEJ detection in mouse models. In Ku80-knockout MEFs lacking shelterin, abundant telomere fusions occur, most of which disappear upon inhibiting MMEJ factors, including PARP1 and LIG3. While NHEJ generally predominates at deprotected telomeres, MMEJ can serve as an alternative repair route in NHEJ-deficient contexts (Sfeir and de Lange 2012). Coupling shelterin depletion with Ku80 knockout provides a robust, quantitative readout of MMEJ through telomere fusion frequency, enabling the identification of critical MMEJ factors such as Polθ (Mateos-Gomez et al. 2015, 2017). Beyond Polθ, the nuclease APEX2 (encoding APE2 [apurinic/apyrimidinic endonuclease 2]) trims 3′ flaps formed during repair and was confirmed as an MMEJ factor using TRF2/Ku80-deficient telomere fusion assays (Fleury et al. 2023). APE2 depletion abrogated MMEJ-dependent telomere fusions, and Polθ knockout in APE2-deficient cells yielded no additive effect, placing Polθ and APE2 in the same epistatic pathway. More recently, CRISPR-Cas9-based genetic screens revealed RHNO1 and the 9-1-1 complex as putative MMEJ regulators during mitosis. Depleting RHNO1 or 9-1-1 components markedly reduced telomere fusions in TRF1/2 Ku80-deficient cells, confirming them as bona fide MMEJ factors (Brambati et al. 2023). Further investigation is required to define the regulatory mechanisms underlying telomere-driven MMEJ and to determine whether it represents a functionally distinct pathway from that operating at intrachromosomal DNA breaks.

## Genomic Instability

Single-cell DNA sequencing of MRC-5, IMR-90, HCA2, and TIG-3 fibroblasts—each stably expressing HPV16 E6E7 and undergoing telomere crisis—revealed extensive, heterogeneous copy number alterations, with recurrent involvement of chromosomes 9, 16, 17, 19, and, most prominently, chromosome 12. These alterations were associated with telomere fusion events, implicating telomere dysfunction and the subsequent formation of dicentric chromosomes as central drivers of genomic instability (Liddiard et al. 2021).

During cell division, dicentric chromosomes arising from telomere crisis are pulled toward opposing spindle poles in anaphase. Rather than breaking immediately, they often persist through mitosis as chromatin bridges (Fig. 4A; Maciejowski et al. 2015). These bridges typically rupture during the following interphase, leaving daughter cells with structurally altered chromosomes that carry terminal deletions or gains. Broken ends may then refuse, initiating BFB cycles (McClintock 1939). These cycles can conclude through ring chromosome formation, centromere inactivation, or the activation of TMMs that restore telomere capping at DNA breaks. Fusions involving different chromosomes frequently lead to nonreciprocal translocations and terminal deletions, while sister chromatid fusions can generate palindromic structures and terminal inverted repeats, promoting gene amplification in subsequent BFB cycles (Tanaka et al. 2005; Vukovic et al. 2007; Garsed et al. 2014; Hermetz et al. 2014; Li et al. 2014).

Loss of heterozygosity can result from the asymmetric breakage of dicentric chromosomes formed during telomere crisis, leading to the

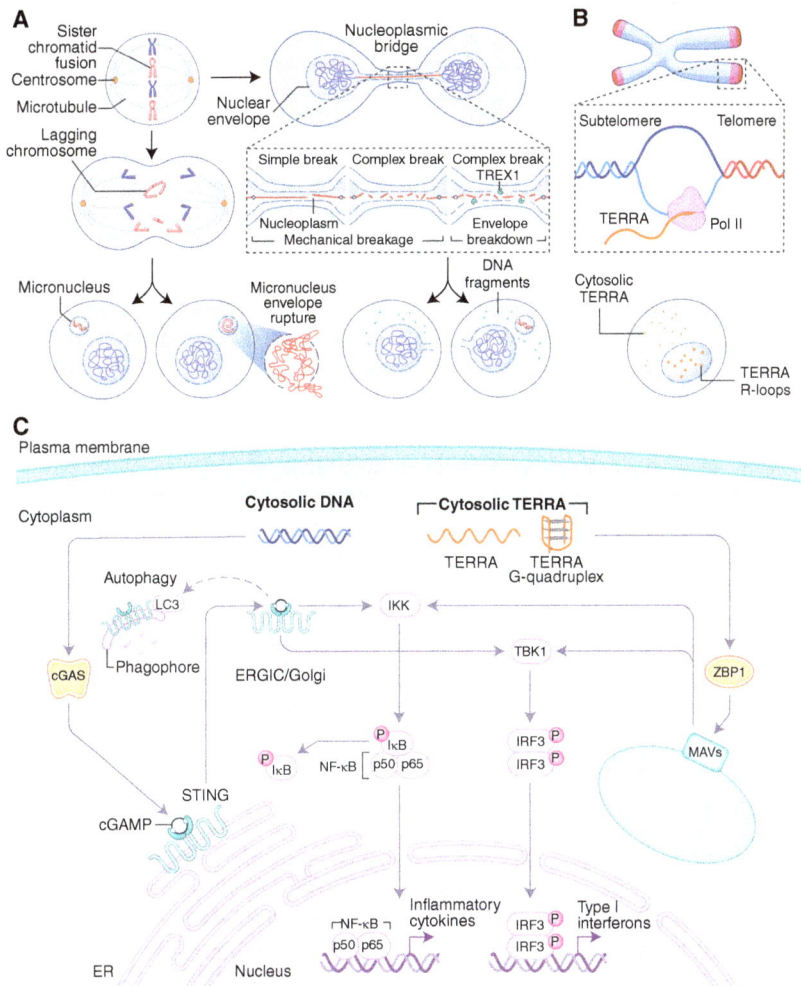

**Figure 4.** Telomere-associated molecular patterns and innate immunity. (*A*) Interchromosomal fusions typically produce dicentric chromosomes carrying two centromeres, whereas intrachromosomal fusions generate pseudo-dicentric chromosomes joined by fused ends. During anaphase, opposing spindle forces pull dicentric chromosomes apart, creating DNA bridges that can be resolved through multiple mechanisms. In one model, actomyosin-driven mechanical force alone can break these bridges, causing simple or extensive chromosome fragmentation. An alternative model posits that the 3′-to-5′ exonuclease three-prime repair exonuclease 1 (TREX1) mediates bridge resolution, requiring transient nuclear envelope breakdown. These processes often yield acentric DNA fragments, which either localize to micronuclei (MN) or are released into the cytosol. In some instances, dicentric chromosomes detach from spindle poles instead of breaking, resulting in lagging chromosomes that can be sequestered into MN. (*B*) TERRA is a long noncoding RNA transcribed by RNA polymerase II from subtelomeric regions on the C-rich telomeric strand. Polyadenylated telomeric repeat–containing RNA (TERRA) accumulates in the nucleoplasm, whereas nonpolyadenylated TERRA remains associated with telomeric chromatin. The 3′ terminus of TERRA facilitates interactions with telomere-bound proteins, driving the formation of R-loops. TERRA transcripts are also detected in the cytoplasm, where they may be packaged into exosomal vesicles. In the cytoplasm, TERRA can exist as single-stranded RNA, adopt G-quadruplex structures, or form RNA:DNA hybrids—each species capable of triggering innate immune signaling. (*C*) In the cytoplasm, telomere-derived double-stranded DNA (dsDNA) activates cyclic GMP–AMP synthase (cGAS), which in turn stimulates STING. Concurrently, cytosolic TERRA transcripts are recognized by Z-DNA-binding protein 1 (ZBP1), leading to mitochondrial antiviral-signaling protein (MAVS)-dependent signaling. Both pathways converge on the TANK-binding kinase 1 (TBK1)–interferon regulatory factor 3 (IRF3) and IκB kinase complex (IKK)–nuclear factor κB (NF-κB) axes, ultimately inducing the expression of interferon genes and proinflammatory cytokines. (BFB) Breakage–fusion–bridge, (c-NHEJ) classical nonhomologous end joining, (MMEJ) microhomology-mediated end joining, (STING) stimulator of interferon genes, (ER) endoplasmic reticulum.

loss of acentric fragments in daughter cells—a mechanism particularly detrimental when tumor-suppressor loci are affected (Cleal and Baird 2020). Nonreciprocal translocations may arise when broken chromosome ends invade ectopic loci via break-induced replication (BIR), a process documented in telomerase-deficient models and human tumors (Artandi et al. 2000; Shih et al. 2001; Llorente et al. 2008; Anand et al. 2013). Telomere fusion events further reveal promiscuous interactions with internal genomic sites, underscoring the genome-scarring potential of BFB cycles. Gene amplification frequently follows sister chromatid fusion, yielding dicentric isochromosomes that reenter BFB cycles upon breakage at internal fragile sites. Resulting amplicons, often arranged in inverted repeats, have been observed in multiple tumor types, including breast, pancreatic, esophageal, and hematological malignancies (Smith et al. 1990; Ma et al. 1993; Lo et al. 2002; Li et al. 2014).

Telomere dysfunction can trigger chromothripsis, a catastrophic event marked by the shattering and chaotic reassembly of chromosome arms (Stephens et al. 2011; Jones and Jallepalli 2012; Rausch et al. 2012). Chromothripsis is particularly common in *TP53*-deficient cancers and is strongly associated with unresolved chromatin bridges. In p53- and pRb-deficient epithelial cells, TRF2 inactivation triggers the formation of mitosis-surviving chromatin bridges that span daughter cells into $G_1$ (Maciejowski et al. 2015). These structures are enclosed in a contiguous nuclear envelope (NE) but often rupture, enabling nuclear–cytoplasmic mixing. Chromatin bridge resolution is mediated by TREX1, a cytoplasmic 3′ exonuclease that accesses DNA following envelope rupture and preferentially degrades nucleosome-poor regions (Maciejowski et al. 2015). This resection generates extensive single-stranded DNA decorated by replication protein A (RPA), ultimately leading to convergent degradation of both strands. Since only the bridge-contained portion of the chromosome is targeted, TREX1 activity provides a plausible source of localized DNA fragmentation characteristic of chromothripsis. Fragment religation within the primary nucleus

produces the clustered, yet regionally confined, rearrangements typical of this process (Maciejowski et al. 2015).

Building on earlier work, a recent study investigated the genomic consequences of telomere crisis using a model in which dominant-negative hTERT expression in HCT116 human colorectal carcinoma cells induced progressive telomere erosion and crisis. Whole-genome sequencing of post-crisis clones revealed extensive structural rearrangements characteristic of chromothripsis, including foldback inversions clustered around chromothriptic breakpoints, implicating dysfunctional telomeres as a key driver of genome instability. Notably, catastrophic genome restructuring was observed not in wild-type cells, but in cells deficient in core end-joining pathways: classical NHEJ ($LIG4^{-/-}$), MMEJ ($LIG3^{-/-}$: $TP53^{-/-}$), and the combined knockout ($LIG3^{-/-}$:$LIG4^{-/-}$). The resulting rearrangements frequently consisted of short DNA fragments with complex, nonrandom topologies, often involving local repair events. Enrichment of foldback inversions at rearrangement clusters supports a role for telomere dysfunction in initiating chromothripsis. These findings suggest that chromothriptic patterns arising from telomere crisis are driven by a replicative repair mechanism involving template switching (Cleal et al. 2019).

Alternatively, entire chromosomes mis-segregated into micronuclei (MN) can undergo DNA damage, fragmentation, and misrepair, leading to chromothripsis on a broader genomic scale (Crasta et al. 2012; Zhang et al. 2015; Ly et al. 2017). Recent studies reveal that fragmentation is exacerbated by pathological activation of the Fanconi anemia (FA) pathway, wherein FANCI-FANCD2 monoubiquitination promotes SLX4-XPF-ERCC1–mediated cleavage of underreplicated DNA during mitosis (excision repair cross-complementation group 1) (Engel et al. 2024). Subsequently, fragmented chromosomes are relegated predominantly via NHEJ, generating the complex rearrangement patterns characteristic of chromothripsis (Hu et al. 2023). Together, FA pathway-driven fragmentation and NHEJ-mediated reassembly cooperate to drive catastrophic genome restructuring following micronucleus formation.

Cite this article as *Cold Spring Harb Perspect Biol* doi: 10.1101/cshperspect.a041688

## Innate Immunity

During crisis, telomere dysfunction activates innate immune sensing and signaling pathways, driving the so-called "sterile inflammation" in the absence of pathogenic stimuli. The connection between telomere integrity and inflammatory responses is supported by findings in telomerase-deficient zebrafish and murine models (Blasco et al. 1997; Chin et al. 1999; Lex et al. 2020), which display accelerated telomere attrition, premature aging phenotypes, and elevated inflammation. In humans, short telomere syndromes are characterized by early-onset degenerative features and chronic inflammatory states (Armanios and Blackburn 2012; Armanios 2022; Revy et al. 2023), whereas individuals with persistent inflammation frequently exhibit compromised telomere maintenance and protection mechanisms (Kinouchi et al. 1998; O'Sullivan et al. 2002; Jonassaint et al. 2013). Innate immunity provides the first line of defense against invading pathogens and primes the subsequent adaptive immune response (Harapas et al. 2022). This pathway is mediated by pattern-recognition receptors (PRRs), which detect pathogen-associated molecular patterns (PAMPs) or damage-associated molecular patterns (DAMPs), including lipids, sugars, and nucleic acids (NAs) (Tomalka et al. 2022). Self-NAs resulting from replication stress (Coquel et al. 2018), genomic or mitochondrial DNA damage (Rongvaux et al. 2014; West et al. 2015), or endogenous retroelement (ERE) (Brégnard et al. 2016) activation have also been identified as molecular patterns capable of triggering immune surveillance sensors and contributing to sterile inflammation.

A central feature of innate immunity is the recognition of cytosolic DNA, derived from pathogens such as DNA viruses, retroviruses, bacteria, and parasites or from endogenous sources including mitochondrial and nuclear DNA (Miller et al. 2021). Under homeostatic conditions, the NE preserves nuclear compartmentalization and prevents aberrant immune activation by sequestering genomic DNA. However, when whole chromosomes or acentric fragments become entrapped in MN, their NE is inherently fragile and prone to spontaneous rupture or catastrophic collapse. Similar vulnerabilities are observed in nucleoplasmic bridges, where strands of DNA connecting daughter nuclei frequently experience NE rupture. Both MN and nucleoplasmic bridges are susceptible to NE rupture during interphase (NERDI), a phenomenon that exposes chromatin to the cytoplasm (Hatch et al. 2013; Maciejowski et al. 2015). Loss of NE integrity allows the cytosolic DNA sensor cyclic GMP–AMP synthase (cGAS) to access mislocalized genomic DNA, triggering STING-mediated innate immune signaling (stimulator of IFN genes [TMEM173]).

Rupture of the micronuclear envelope permits cGAS access to genomic DNA, establishing a mechanistic link between genome instability and type I IFN production, as well as proinflammatory cytokine secretion (Harding et al. 2017; Mackenzie et al. 2017; Mohr et al. 2021). Activation of this pathway drives the expression of IFN-stimulated genes (ISGs) with broad antiviral, antitumor, and immunomodulatory functions (Taffoni et al. 2021). While a fraction of cGAS resides in the nucleus, it is often inactivated when bound to nucleosomes. Thus, NE rupture alone may be insufficient for full cGAS activity; additional factors—including DNA damage inflicted by cytoplasmic nucleases—could remodel chromatin to facilitate cGAS oligomerization. In line with this, disruption of the histone pre-mRNA processing complex produces chromatin lacking linker histones, thereby releasing cGAS from nucleosomal constraints and inducing an IFN response (Mackenzie et al. 2017; Lahaye et al. 2018; Liu et al. 2018a; Jiang et al. 2019; Malireddi et al. 2019; Pathare et al. 2020).

During anaphase, dicentric chromatids formed in crisis often develop into extended chromatin bridges that break under spindle forces, initiating multiple BFB cycles and producing acentric DNA fragments that missegregate into MN (Fig. 4A). In some cases, dicentric chromosomes do not break but instead detach from both spindle poles and become sequestered in MN (McClintock 1939; Hoffelder et al. 2004; Terradas et al. 2010). Alternatively, DNA bridges may persist beyond cytokinesis—

reaching lengths of hundreds of microns in the subsequent $G_1$ phase (Fig. 4A; Maciejowski et al. 2015). Actomyosin-driven tension is essential for bridge rupture, as defects in myosin activation or actin assembly prolong their integrity (Umbreit et al. 2020). Moreover, the TREX1 exonuclease can sever chromatin bridges once the NE rupture exposes bridge DNA to the cytosol, generating single-stranded DNA fragments that facilitate breakage (Fig. 4A; Maciejowski et al. 2015). In both scenarios, the resulting chromosome fragments can be released into the cytosol, where they are detected by innate immune-sensing pathways. MN arising at mitotic exit, due to the incomplete partitioning of acentric chromosomal fragments or lagging chromosomes into daughter cells, frequently experience structural defects in the underlying lamina and nuclear pore complexes, collectively leading to the spontaneous collapse of the NE (Fig. 4A; Hatch et al. 2013). These disruptions not only impact crucial processes within MN, including DNA replication and repair, but also facilitate the access of cGAS to micronuclear DNA (Harding et al. 2017; Mackenzie et al. 2017; Liu et al. 2018b). Similarly, persistent interphase chromatin bridges induce transient NE ruptures, mixing nuclear and cytoplasmic contents and enabling cGAS detection of bridge DNA (Fig. 4A; Flynn et al. 2021).

Beyond cGAS–STING-mediated sensing of cytosolic DNA, telomere-driven innate immunity engages a distinct yet codependent pathway that selectively responds to telomeric stress. This involves the transcriptional upregulation of telomeric repeat–containing RNA (TERRA) at critically short or dysfunctional telomeres, where TERRA molecules function as immunogenic ligands that amplify innate immune signaling cascades (Fig. 4B; Nassour et al. 2023, 2024). TERRA is transcribed by RNA polymerase II from CpG-rich promoters in subtelomeric regions, using the C-rich telomeric strand as a template. As a result, TERRA transcripts comprise chromosome-specific subtelomeric sequences followed by variable numbers of telomeric UUAGGG repeats (Fig. 4B; Azzalin et al. 2007; Luke et al. 2008; Schoeftner and Blasco 2008; Bah et al. 2012; Greenwood and Cooper

2012; Diman and Decottignies 2018). TERRA expression is regulated by epigenetic mechanisms, including DNA methylation by DNA methyltransferase 1 (DNMT1) and DNA methyltransferase 3B (DNMT3B) (Nergadze et al. 2009), and by the homodimerization domain of TRF2 (Porro et al. 2014). Indeed, TRF2 depletion or expression of the dominant-negative TRF2ΔBΔM mutant disrupts telomere integrity and increases TERRA synthesis, mirroring the altered telomere state observed during replicative crisis (Porro et al. 2014; Wang and Lieberman 2016; Nassour et al. 2023). Polyadenylated TERRA transcripts predominantly localize to the nucleoplasm, whereas nonpolyadenylated forms associate with telomeric chromatin (Porro et al. 2010). The interaction of TERRA with telomeres is facilitated by its telomeric $3'$ end, allowing for binding to telomere-associated proteins and the formation of RNA:DNA hybrids (R-loops) (Feuerhahn et al. 2010; Porro et al. 2010; Balk et al. 2013; Pfeiffer et al. 2013; Silva et al. 2019). TERRA has additionally been detected in the cytosol and secreted in exosomes (Wang et al. 2015), although the mechanisms governing its nucleocytoplasmic and extracellular trafficking remain poorly understood. At telomeres, TERRA regulates replication (Beishline et al. 2017), promotes heterochromatin formation (Arnoult et al. 2012), recruits telomerase (Farnung et al. 2012; Cusanelli et al. 2013; Moravec et al. 2016; Lalonde and Chartrand 2020; Bettin et al. 2024), and supports HR in ALT cancer cells (Feuerhahn et al. 2010; Porro et al. 2010; Pfeiffer et al. 2013; Silva et al. 2019; Feretzaki et al. 2020).

Recent insights extend the spectrum of immunostimulatory RNA species beyond viral genomes and replication intermediates to include host-derived RNAs transcribed from EREs. Aberrant expression, mislocalization, or misprocessing of these self-derived RNAs can endow them with immunostimulatory properties, potentially driving sterile inflammation in autoimmune or autoinflammatory disorders (Atianand et al. 2017; Flores-Concha and Oñate 2020). TERRA serves as a prime example. By interacting with the NA sensor ZBP1 (Z-DNA-binding protein 1; also known as DAI or DLM-1),

TERRA triggers innate immune signaling and IFN responses (Fig. 4C). Originally identified as a cytosolic DNA receptor (Takaoka et al. 2007), ZBP1 is now recognized as a versatile sensor of Z-form NAs through its amino-terminal Zα1 and Zα2 domains, which bind NAs in a structure-specific manner independent of base sequence (Thapa et al. 2016; Maelfait et al. 2017; Sridharan et al. 2017; Jiao et al. 2020; Kesavardhana et al. 2020; Wang et al. 2020; Zhang et al. 2020; Karki et al. 2021). ZBP1 also harbors two RHIM (receptor-interacting protein homotypic interaction motif) domains (Kaiser et al. 2008; Rebsamen et al. 2009), enabling interactions with RIPK1 and RIPK3 that govern cell death, innate immunity, and inflammation, as demonstrated in mouse models challenged with influenza A virus, vaccinia virus, and select herpesviruses (Balachandran and Mocarski 2021).

Recent work has brought to light a novel role for the human-specific short isoform of ZBP1 (ZBP1-S) in initiating replicative crisis and restricting the expansion of p53- and pRb-deficient cells predisposed to neoplastic transformation (Nassour et al. 2023, 2024). During crisis, cGAS–STING signaling induces ZBP1-S as a canonical ISG, priming cells to sense telomeric stress through telomere-derived TERRA species, which act as messenger molecules linking telomere dysfunction to innate immune activation (Nassour et al. 2023). During crisis, ZBP1-S initiates cell death through an MAVS (mitochondrial antiviral-signaling protein)-dependent type I IFN response and autophagy, in the absence of detectable PANoptosis—a regulated inflammatory cell death program that integrates components of pyroptosis, apoptosis, and necroptosis (Malireddi et al. 2019).

MAVS, anchored to subcellular membranes via a carboxy-terminal transmembrane domain, orchestrates distinct antiviral signaling: Mitochondrial MAVS induces type I IFNs, whereas peroxisomal MAVS triggers type III IFN production (Seth et al. 2005; Dixit et al. 2010; Horner et al. 2011). It contains an amino-terminal caspase recruitment domain (CARD) that engages in homotypic interactions with the RNA sensors RIG-I (retinoic acid-inducible gene I) and MDA5 (melanoma differentiation-associ-

ated protein 5) (Gack et al. 2007; Oshiumi et al. 2010; Jiang et al. 2012; Oshiumi 2020). Upon binding these sensors, MAVS polymerizes on the mitochondrial membrane to assemble a "MAVS signalosome," which recruits TRAF effectors and concurrently activates two pathways: (1) TBK1/IKKε-mediated phosphorylation of IRF3 (interferon regulatory factor 3) and IRF7, inducing type I/III IFNs, and (2) IKKα/β/γ-mediated phosphorylation of IκB, leading to NF-κB-dependent inflammatory gene expression (nuclear factor κ light-chain enhancer of activated B cells) (Fig. 4C; Kawai et al. 2005; Meylan et al. 2005; Seth et al. 2005; Xu et al. 2005). Cells lacking ZBP1–MAVS–autophagy signaling resist death and proliferate beyond the crisis barrier, despite harboring critically short, fused telomeres—thereby highlighting a tumor-suppressive role for ZBP1-S in eliminating cells predisposed to neoplastic transformation (Nassour et al. 2023, 2024).

These findings reveal a novel telomere-driven tumor-suppressive mechanism, wherein dysfunctional telomeres activate innate immune responses through mitochondrial TERRA–ZBP1 complexes, facilitating the clearance of cells predisposed to malignant transformation. Yet, key questions remain—most notably, the molecular identity of the immunogenic TERRA ligands sensed by the short isoform of ZBP1. TERRA transcripts can adopt stable secondary structures, including hairpins and G-quadruplexes (RNA G-quadruplexes [RG4s]), which may mimic left-handed Z-form NAs—preferred ligands of ZBP1. Alternatively, TERRA may hybridize with the C-rich telomeric strand to form R-loops, promoting aberrant processing and cytoplasmic mislocalization, thereby triggering ZBP1 activation. A related mechanism has been described in which R-loops are converted into RNA–DNA hybrids that accumulate in the cytosol, eliciting robust innate immune signaling and type I IFN–driven cell death (Crossley et al. 2023).

## Autophagy-Dependent Cell Death

Cell death during telomere crisis was historically presumed to resemble apoptosis, based largely on indirect evidence from TUNEL (terminal deoxynucleotidyl transferase dUTP nick end labeling) and Annexin V staining—assays that mark general features of dying cells rather than apoptosis-specific mechanisms (Artandi et al. 2000; Karlseder et al. 2002). As a result, early interpretations likely overestimated the contribution of caspase-mediated apoptosis to the clearance of crisis cells. More recent studies, however, have redefined this view, revealing that macroautophagy (hereafter referred to as autophagy) is the dominant mechanism driving cell death during telomere crisis (Nassour et al. 2019). In human fibroblasts and epithelial cells deficient in both p53- and pRb-mediated cell cycle checkpoints, telomere crisis induces hallmark features of elevated autophagic flux—including accumulation of autophagosomes and autolysosomes, increased expression of autophagy-related proteins, and enhanced lysosomal activity—without evidence of PANoptosis. These findings identify autophagy as a regulated, primary effector of cell death during crisis (Nassour et al. 2019, 2021).

The role of autophagy during crisis fits within the broader concept of autophagy-dependent cell death (ADCD). ADCD refers to a form of regulated cell death that requires a functional autophagy machinery for execution and is mechanistically distinct from other programmed death pathways such as apoptosis or necroptosis. According to the Nomenclature Committee on Cell Death (NCCD), formal criteria for ADCD include the following: (1) enhanced autophagic flux during the death process, (2) prevention of death upon genetic or pharmacological inhibition of autophagy, and (3) absence of death via alternative pathways (Galluzzi et al. 2018). Crisis-induced death fulfills these criteria. Partial inhibition of autophagy confers resistance to telomere crisis, enabling checkpoint-deficient cells to evade clearance and accumulate widespread chromosomal abnormalities—including nonreciprocal translocations and deletions—while transitioning from a diploid to a near-tetraploid karyotype, indicative of impaired mitotic fidelity and cytokinesis. Together, these findings establish autophagy as a key effector of the crisis bar-

rier, essential for eliminating genomically unstable, checkpoint-defective cells and preventing malignant progression (Nassour et al. 2019, 2021). Beyond telomere crisis, ADCD has been observed in diverse biological contexts, including developmental models such as *Drosophila melanogaster*, where autophagy functions as a primary effector of programmed cell death (Denton et al. 2009). ADCD is also implicated in pathological settings, including neuronal death following ischemia (Shi et al. 2012) and the elimination of apoptosis-resistant cancer cells (Voss et al. 2010; Lian et al. 2011).

In early tumorigenesis, autophagy acts as a tumor-suppressive mechanism. By removing damaged mitochondria, limiting reactive oxygen species (ROS) production, and clearing misfolded proteins, autophagy preserves genomic integrity and prevents oncogenic transformation (Jalali et al. 2025). Mouse models support this view: Heterozygous deletion of *Beclin 1* or liver-specific knockout of *Atg5* or *Atg7* (autophagy-related gene 5 and autophagy-related gene 7) results in spontaneous tumor development and increased chromosomal instability (Qu et al. 2003; Zhang et al. 2007). In contrast, once malignancy is established, autophagy frequently adopts a tumor-promoting role. Tumor cells rely on autophagy to adapt to metabolic stress, hypoxia, and oxidative damage by recycling intracellular components to maintain bioenergetic balance (Degenhardt et al. 2006; Yang et al. 2011; Karsli-Uzunbas et al. 2014; Guo et al. 2016; Russell and Guan 2022). Understanding how autophagy switches from a protective to a destructive process—and identifying molecular checkpoints that regulate this transition—will be essential for the rational design of therapies that exploit lethal autophagy to eradicate cancer cells.

## CRISIS ESCAPE AS A GATEWAY TO IMMORTALIZATION AND TRANSFORMATION

Unlike rodent cells, which immortalize readily following oncogene expression, human somatic cells are intrinsically resistant to transformation. The only agents known to reproducibly—but in-

 Cite this article as *Cold Spring Harb Perspect Biol* doi: 10.1101/cshperspect.a041688

efficiently—extend the life span of human fibroblasts are DNA tumor viruses, including SV40, HPV, and adenovirus. Viral oncogenes such as SV40 large T antigen, HPV E6/E7, and adenoviral E1A/E1B promote life span extension by inactivating key tumor suppressors, notably p53 and pRb (Sack 1981; Shay et al. 1991b; De Silva et al. 1994). However, escape from crisis and subsequent immortalization remain rare events. Escape frequency was directly assessed in a comparative study of SV40 large T-antigen-transfected human lung fibroblasts and mammary epithelial cells (Shay et al. 1993). While nearly all fibroblast clones entered crisis and arrested, rare, immortalized derivatives emerged at a frequency of $\sim 3 \times 10^{-7}$. In contrast, mammary epithelial cells escaped crisis and became immortal at a substantially higher frequency of $\sim 10^{-5}$. These differences have been attributed to the greater chromosomal instability and ploidy complexity of fibroblasts, in contrast to the more permissive, pseudodiploid karyotype of epithelial cells. This intrinsic cellular susceptibility may underlie the clinical predominance of carcinomas—derived from epithelial tissues—over sarcomas, which originate from mesenchymal lineages (Shay and Wright 1989; Tsuyama et al. 1991; Shay et al. 1993; Montalto et al. 1999).

Insights from Li–Fraumeni syndrome (germline *TP53* mutation) suggest that telomerase is typically silent during crisis, becoming upregulated only in post-crisis immortal clones, coinciding with telomere stabilization and restored proliferation (Gollahon et al. 1998). While telomerase activation is the predominant outcome, a rare alternative route—ALT—has also been documented. In SV40-transformed JFCF-6 human fetal lung fibroblasts, escape from crisis led to the emergence of ALT-positive immortal clones, characterized by heterogeneous telomere lengths, C-circle accumulation, and ALT-associated PML bodies (Lovejoy et al. 2012; Dilley and Greenberg 2015).

While rare clones may escape crisis, immortality remains a distinct and infrequent outcome. Quantitative studies in SV40 large T-antigen-transfected human fibroblasts have shown that post-crisis outgrowth does not reliably lead to sustained proliferation. In IMR-90

human fetal lung fibroblasts transfected with SV40 LT, many clones resumed growth after crisis, yet the vast majority underwent a third mortality stage and failed to establish immortal lines. Only rare clones—approximately one in three million cells—achieved true immortalization, indicating that additional mutational events are required beyond crisis escape (Shay and Wright 1989). A follow-up study analyzing 35 post-crisis clones derived from SV40 LT-transfected IMR-90 and MRC-5 fibroblasts identified only two immortal lines—one maintained by telomerase activation and one by ALT—whereas the remaining 33 mortal clones failed to sustain telomere length and ceased dividing after a brief extension of life span past the crisis barrier (Montalto et al. 1999). A plausible scenario is that certain post-crisis clones transiently persist by reactivating low, subdetectable levels of telomerase. This limited activity may suffice to selectively heal the most critically short telomeres, while failing to prevent continued erosion of the remaining telomere pool. As proliferation proceeds, the cumulative burden of dysfunctional telomeres increases, and insufficient telomerase activity ultimately undermines genome integrity and compromises cellular fitness (Chiba et al. 2017).

Escape from telomere crisis and acquisition of an immortal phenotype do not inherently confer malignant potential. This principle has been delineated through studies tracking the transformation of human fibroblasts—primarily WI-38 and MRC-5 cells—using SV40 early region constructs, including SV40 LT alone, LT in combination with small T antigen, and replication-defective mutants. In MRC-5 fibroblasts transfected with replication-defective SV40 LT, most clones initially exhibited extended life spans but ultimately entered crisis. Only a small fraction—emerging at frequencies of $\sim 10^{-7}$—successfully escaped crisis and gave rise to immortalized lines (Imai et al. 1993). These post-crisis clones displayed variable phenotypes, with many acquiring features such as anchorage-independent growth, serum-independent proliferation, and altered morphology, yet the majority remained nontumorigenic (Imai et al. 1993; Ray and Kraemer 1993). Complete malignant transformation, however, required iterative cycles of chromosomal mutation and clonal selection, both in vitro and in vivo. WI-38 fibroblasts transfected with SV40 LT required extended serial passage and in vivo selection using gelatin sponge implantation in nude mice to achieve tumorigenicity. The resulting tumorigenic derivatives exhibited progressively shortened latency periods and increased tumor incidence, indicating that full transformation emerged through cumulative genomic alterations and selective expansion of malignant clones (Ray and Kraemer 1993).

Cytogenetic analysis revealed that persistent genomic instability, including anaphase bridges, dicentric chromosomes, and nonclonal structural abnormalities, was common to all postcrisis cells. However, tumorigenic derivatives exhibited markedly greater genomic complexity, suggesting that selective pressures within the microenvironment may shape the evolving genomic landscape, favoring the emergence of highly plastic and tumorigenic clones (Imai et al. 1993). Taken together, these findings demonstrate that while escape from crisis is necessary for continued proliferation and the attainment of replicative immortality, it is not sufficient for tumorigenicity. Instead, crisis escape marks an unstable intermediate state that remains reliant on ongoing genomic instability, clonal evolution, and microenvironmental selection to drive full neoplastic transformation.

## DNA Repair and Chromatin Remodeling in Crisis Escape

Using telomerase-inhibited HCT116 colorectal cancer cells as a model system, escape from telomere crisis was found to be strictly dependent on LIG3-mediated MMEJ. Although LIG4-deficient (NHEJ-deficient) cells could escape crisis and reengage telomerase activity, LIG3-null cells failed to do so—despite undergoing telomere shortening and fusion (Jones et al. 2014). Analysis of telomere fusions in LIG3- and LIG4-deficient HCT116 cells provides additional evidence for this mechanistic dichotomy. LIG3-deficient cells displayed a marked increase in interchromosomal telomere fusions, which gen-erate dicentric chromosomes. In contrast, cells capable of escaping crisis—including wild-type and LIG4-deficient lines—displayed a predominance of intrachromosomal (sister chromatid) fusions, a hallmark of MMEJ activity. These fusions are thought to drive BFB cycles and focal genomic rearrangements that enable telomerase reactivation and immortalization. MRC5 fibroblasts, which rarely escape crisis, exhibit similarly high interchromosomal fusion levels and low LIG3 expression, reinforcing the idea that fusion topology, governed by repair pathway choice, determines crisis outcome. These results suggest that while both NHEJ and MMEJ can fuse dysfunctional telomeres, only MMEJ permits a controlled, survivable genomic reshaping necessary for escape (Jones et al. 2014).

Further mechanistic insight comes from studies on DNA polymerase θ (POLQ), which facilitates MMEJ by annealing short homologous sequences and extending resected DNA ends. In POLQ-deficient HCT116 cells, high-throughput sequencing revealed a significant shift in telomere fusion profiles, with increased interchromosomal fusions. Notably, these interchromosomal fusions often involved centromeric α-satellite repeats, suggesting a predisposition for telomere–centromere interactions in the absence of POLQ. POLQ-deficient clones exhibited elevated genetic heterogeneity post-crisis, indicating that POLQ contributes to maintaining genomic stability during telomere crisis by promoting precise repair mechanisms and limiting aberrant recombination events (Liddiard et al. 2022). The role of PARP1, another MMEJ cofactor, was further elucidated in POLQ-deficient HCT116 and JJN-3 multiple myeloma cells. Treatment with PARP inhibitors (PARPi) such as olaparib and rucaparib selectively blocked clonal escape, triggering robust ATR–Chk1 activation, intrachromosomal fusions, and cell death—without affecting telomere erosion or TERT (telomerase reverse transcriptase) expression (Ngo et al. 2018).

ATRX (α-thalassemia/mental retardation X-linked), a chromatin remodeling factor, has emerged as a central regulator of crisis escape and immortalization. In telomerase-deficient, SV40-transformed IMR90 fibroblasts, ATRX loss induced premature proliferative arrest—many PDs earlier than control cells—reflecting an accelerated, crisis-like state characterized by acute replication stress and activation of DNA damage signaling at dysfunctional telomeres. With extended culture, immortal clones eventually arose from both ATRX-proficient and ATRX-deficient populations, but through distinct TMMs. ATRX-proficient clones reactivated telomerase, as evidenced by elevated TERT expression and absence of ALT-associated features. In contrast, all immortalized ATRX-deficient clones lacked TERT expression and exhibited hallmark indicators of ALT activation, including robust C-circle formation, extensive telomere length heterogeneity, and frequent ALT-associated PML bodies. Chromatin profiling revealed that ATRX loss promoted progressive telomeric chromatin decompaction and replication fork stalling. Consistent results were observed in primary human fibroblasts (HCA2 and MRC5) rendered permissive for crisis via HPV16 E6/E7 expression. In this model, only ATRX-deficient cells successfully escaped telomere crisis, doing so through stable engagement of the ALT pathway. Clones retaining ATRX expression failed to escape and ultimately perished, reinforcing the requirement for ATRX loss in facilitating ALT-mediated immortalization (Li et al. 2019; Geiller et al. 2022).

## Multiple Paths to Crisis Escape

Escape from crisis appears to be a dynamic, multistep process rather than a singular, rate-limiting mutational event traditionally attributed to the activation of a TMM. Instead, crisis resolution likely reflects a coordinated interplay among DNA repair pathway choice, telomeric chromatin remodeling, and the regulation of cell death programs. The relative contribution of these pathways may vary not only between cell types but also among individual cells within the same population, complicating efforts to define a universal or linear route to crisis escape. We propose a model in which the timing of TMM activation shapes the genomic architecture and malignant potential of post-crisis progeny. In some cases, early engagement of a TMM during crisis stabilizes telomeres before extensive genomic instability accumulates, resulting in post-crisis clones with relatively simple structural variants (SVs) and minimal chromothripsis. In contrast, loss of DDR components, chromatin remodelers, or cell death regulators may permit transient proliferation in a TMM-negative state, allowing cells to bypass crisis. During this permissive window, ongoing BFB cycles promote escalating genome rearrangement and increase the likelihood of catastrophic events such as chromothripsis. Once a TMM is eventually engaged, the genome becomes stabilized, yet often retains the structural imprints of prior instability. Although less frequent, this delayed TMM activation pathway likely gives rise to post-crisis clones with highly rearranged genomes and increased tumorigenic potential (Fig. 5).

We further propose that selective pressures inherent to the tissue microenvironment shape the evolutionary trajectory of post-crisis cells. Unlike in vitro models, where proliferative capacity is the dominant selective force, cells in vivo are subject to additional constraints including immune surveillance, metabolic stress, hypoxia, and interactions with the stroma and extracellular matrix. These pressures favor the emergence of clones with adaptations for survival, immune evasion, and invasiveness, thereby driving a broader and more aggressive spectrum of genomic diversity than typically observed in culture. Although in vitro systems remain indispensable for mechanistic dissection, they likely underestimate the complexity of genome evolution during crisis under physiological conditions. A deeper understanding of the trajectories underlying crisis escape will require high-resolution approaches capable of capturing cell-to-cell variability and dynamic transitions over time. Single-cell lineage tracing and genomic profiling will be essential to reconstruct the fates of individual cells as they progress through crisis toward immortalization. In parallel, ultrasensitive assays to detect nascent TMM activation will be needed to precisely define the timing and context of telomere stabilization. Finally, more physiologically relevant 3D or organoid culture systems will be essential to better recapitulate the in vivo

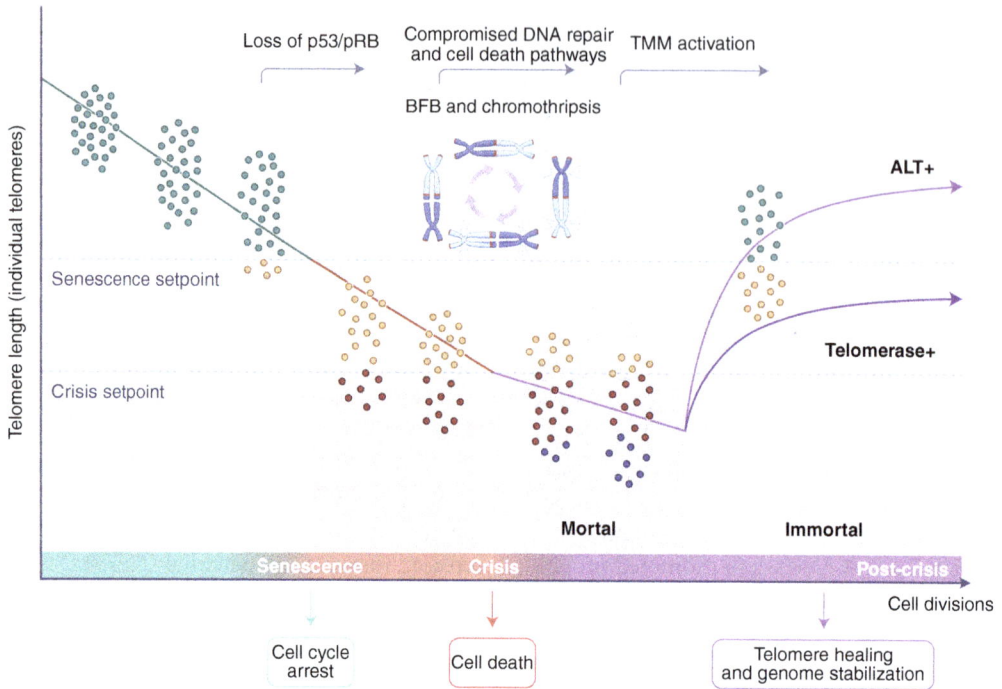

**Figure 5.** Multiple paths to crisis escape. Cancer arises through an evolutionary process as cells acquire genetic and epigenetic alterations that enable evasion of two distinct proliferative barriers: senescence and crisis. Senescence is the primary response to progressive telomere shortening, initiated when one or a few critically short telomeres are perceived as double-strand breaks (DSBs) and trigger a DNA damage response (DDR) that halts further cell division. Loss of p53 and retinoblastoma protein (pRb) renders cells insensitive to DDR signals from dysfunctional telomeres and incapable of exiting the cell cycle. As a result, checkpoint-deficient cells continue dividing, leading to further telomere attrition beyond the senescence threshold. However, these cells are not immortal and ultimately encounter crisis, characterized by chromosomal fusions and near-complete loss of viability. Individual cells may follow distinct trajectories through crisis. In some cases, early telomere maintenance mechanism (TMM) activation permits telomere stabilization and escape with minimal structural disruption. In other cases, TMM activation is delayed, often following loss of DDR components, chromatin regulators, or cell death pathways. These changes allow cells to bypass crisis and transiently expand in a TMM-negative state while remaining mortal. During this interval, ongoing breakage–fusion–bridge (BFB) cycles fuel cumulative genomic instability and increase the risk of catastrophic rearrangements, including chromothripsis. Once a TMM is eventually engaged, the genome stabilizes but often retains the structural hallmarks of earlier instability. These divergent paths through crisis yield post-crisis clones with varying degrees of genomic complexity and, in many cases, increased tumorigenic potential.

microenvironment and assess how external selective pressures influence not only the frequency of crisis escape, but also the genomic landscape and tumorigenic potential of post-crisis clones. Compared to conventional two-dimensional cultures, these models may more accurately capture the selective bottlenecks that shape clonal survival and genome evolution during and after crisis.

**ACKNOWLEDGMENTS**

J.N. is supported by the National Cancer Institute (K99CA252447, R00CA252447), the Cancer League of Colorado, the American Cancer Society (ACS IRG 22-154-59), and the Glenn Foundation for Medical Research Grant for Junior Faculty. J.K. is supported by the Salk Institute Cancer Center Core Grant (P30CA014195), the National Institutes of Health (RO1CA2 27934, RO1CA234047, RO1CA228211, RO1A G077324), the Donald and Darlene Shiley Chair, the Samuel Waxman Cancer Research Foundation, and the American Heart Association (19PABHI34610000).

## REFERENCES

*Reference is also in this subject collection.

Alcorta DA, Xiong Y, Phelps D, Hannon G, Beach D, Barrett JC. 1996. Involvement of the cyclin-dependent kinase inhibitor p16 (INK4a) in replicative senescence of normal human fibroblasts. *Proc Natl Acad Sci* **93**: 13742–13747. doi:10.1073/pnas.93.24.13742

Anand RP, Lovett ST, Haber JE. 2013. Break-induced DNA replication. *Cold Spring Harb Perspect Biol* **5**: a010397. doi:10.1101/cshperspect.a010397

Armanios M. 2022. The role of telomeres in human disease. *Annu Rev Genomics Hum Genet* **23**: 363–381. doi:10 .1146/annurev-genom-010422-091101

Armanios M, Blackburn EH. 2012. The telomere syndromes. *Nat Rev Genet* **13**: 693–704. doi:10.1038/nrg3246

Arnoult N, Van Beneden A, Decottignies A. 2012. Telomere length regulates TERRA levels through increased trimethylation of telomeric H3K9 and HP1α. *Nat Struct Mol Biol* **19**: 948–956. doi:10.1038/nsmb.2364

Artandi SE, DePinho RA. 2010. Telomeres and telomerase in cancer. *Carcinogenesis* **31**: 9–18. doi:10.1093/carcin/bgp268

Artandi SE, Chang S, Lee SL, Alson S, Gottlieb GJ, Chin L, DePinho RA. 2000. Telomere dysfunction promotes nonreciprocal translocations and epithelial cancers in mice. *Nature* **406**: 641–645. doi:10.1038/35020592

Atianand MK, Caffrey DR, Fitzgerald KA. 2017. Immunobiology of long noncoding RNAs. *Annu Rev Immunol* **35**: 177–198. doi:10.1146/annurev-immunol-041015-055459

Azzalin CM, Reichenbach P, Khoriauli L, Giulotto E, Lingner J. 2007. Telomeric repeat containing RNA and RNA surveillance factors at mammalian chromosome ends. *Science* **318**: 798–801. doi:10.1126/science.1147182

Badie S, Carlos AR, Folio C, Okamoto K, Bouwman P, Jonkers J, Tarsounas M. 2015. BRCA1 and CtIP promote alternative non-homologous end-joining at uncapped telomeres. *EMBO J* **34**: 410–424. doi:10.15252/embj .201488947

Bae NS, Baumann P. 2007. A RAP1/TRF2 complex inhibits nonhomologous end-joining at human telomeric DNA ends. *Mol Cell* **26**: 323–334. doi:10.1016/j.molcel.2007 .03.023

Bah A, Wischnewski H, Shchepachev V, Azzalin CM. 2012. The telomeric transcriptome of *Schizosaccharomyces pombe*. *Nucleic Acids Res* **40**: 2995–3005. doi:10.1093/nar/gkr1153

Baird DM, Rowson J, Wynford-Thomas D, Kipling D. 2003. Extensive allelic variation and ultrashort telomeres in senescent human cells. *Nat Genet* **33**: 203–207. doi:10.1038/ng1084

Baird DM, Britt-Compton B, Rowson J, Amso NN, Gregory L, Kipling D. 2006. Telomere instability in the male germline. *Hum Mol Genet* **15**: 45–51. doi:10.1093/hmg/ddi424

Baker AM, Graham TA, Wright NA. 2013. Pre-tumour clones, periodic selection and clonal interference in the origin and progression of gastrointestinal cancer: potential for biomarker development. *J Pathol* **229**: 502–514. doi:10.1002/path.4157

Balachandran S, Mocarski ES. 2021. Viral Z-RNA triggers ZBP1-dependent cell death. *Curr Opin Virol* **51**: 134–140. doi:10.1016/j.coviro.2021.10.004

Balk B, Maicher A, Dees M, Klermund J, Luke-Glaser S, Bender K, Luke B. 2013. Telomeric RNA-DNA hybrids affect telomere-length dynamics and senescence. *Nat Struct Mol Biol* **20**: 1199–1205. doi:10.1038/nsmb.2662

Baumann P, Cech TR. 2000. Protection of telomeres by the Ku protein in fission yeast. *Mol Biol Cell* **11**: 3265–3275. doi:10.1091/mbc.11.10.3265

Baumann P, Cech TR. 2001. Pot1, the putative telomere end-binding protein in fission yeast and humans. *Science* **292**: 1171–1175. doi:10.1126/science.1060036

Beishline K, Vladimirova O, Tutton S, Wang Z, Deng Z, Lieberman PM. 2017. CTCF driven TERRA transcription facilitates completion of telomere DNA replication. *Nat Commun* **8**: 2114. doi:10.1038/s41467-017-02212-w

Bettin N, Querido E, Gialdini I, Grupelli GP, Goretti E, Cantarelli M, Andolfato M, Soror E, Sontacchi A, Jurikova K, et al. 2024. TERRA transcripts localize at long telomeres to regulate telomerase access to chromosome ends. *Sci Adv* **10**: eadk4387. doi:10.1126/sciadv.adk4387

Bilaud T, Brun C, Ancelin K, Koering CE, Laroche T, Gilson E. 1997. Telomeric localization of TRF2, a novel human telobox protein. *Nat Genet* **17**: 236–239. doi:10.1038/ng1097-236

Blackburn EH, Gall JG. 1978. A tandemly repeated sequence at the termini of the extrachromosomal ribosomal RNA genes in *Tetrahymena*. *J Mol Biol* **120**: 33–53. doi:10.1016/0022-2836(78)90294-2

Blasco MA, Lee HW, Hande MP, Samper E, Lansdorp PM, DePinho RA, Greider CW. 1997. Telomere shortening and tumor formation by mouse cells lacking telomerase RNA. *Cell* **91**: 25–34. doi:10.1016/S0092-8674(01)80006-4

Bodnar AG, Ouellette M, Frolkis M, Holt SE, Chiu CP, Morin GB, Harley CB, Shay JW, Lichtsteiner S, Wright WE. 1998. Extension of life-span by introduction of telomerase into normal human cells. *Science* **279**: 349–352. doi:10.1126/science.279.5349.349

Bombarde O, Boby C, Gomez D, Frit P, Giraud-Panis MJ, Gilson E, Salles B, Calsou P. 2010. TRF2/RAP1 and DNA-PK mediate a double protection against joining at telomeric ends. *EMBO J* **29**: 1573–1584. doi:10.1038/emboj .2010.49

Boyraz B, Ly A. 2024. Spectrum of histopathologic findings in risk-reducing bilateral prophylactic mastectomy in patients with and without BRCA mutations. *Hum Pathol* **151**: 105534. doi:10.1016/j.humpath.2023.11.010

Brambati A, Sacco O, Porcella S, Heyza J, Kareh M, Schmidt JC, Sfeir A. 2023. RHINO directs MMEJ to repair DNA breaks in mitosis. *Science* **381**: 653–660. doi:10.1126/science.adh3694

Brégnard C, Guerra J, Déjardin S, Passalacqua F, Benkirane M, Laguette N. 2016. Upregulated LINE-1 activity in the Fanconi anemia cancer susceptibility syndrome leads to spontaneous pro-inflammatory cytokine production. *EBioMedicine* **8:** 184–194. doi:10.1016/j.ebiom.2016.05.005

Britt-Compton B, Rowson J, Locke M, Mackenzie I, Kipling D, Baird DM. 2006. Structural stability and chromosome-specific telomere length is governed by *cis*-acting determinants in humans. *Hum Mol Genet* **15:** 725–733. doi:10.1093/hmg/ddi486

Broccoli D, Smogorzewska A, Chong L, de Lange T. 1997. Human telomeres contain two distinct Myb-related proteins, TRF1 and TRF2. *Nat Genet* **17:** 231–235. doi:10.1038/ng1097-231

Bryan TM, Englezou A, Gupta J, Bacchetti S, Reddel RR. 1995. Telomere elongation in immortal human cells without detectable telomerase activity. *EMBO J* **14:** 4240–4248. doi:10.1002/j.1460-2075.1995.tb00098.x

Capper R, Britt-Compton B, Tankimanova M, Rowson J, Letsolo B, Man S, Haughton M, Baird DM. 2007. The nature of telomere fusion and a definition of the critical telomere length in human cells. *Genes Dev* **21:** 2495–2508. doi:10.1101/gad.439107

Carrière C, Gore AJ, Norris AM, Gunn JR, Young AL, Longnecker DS, Korc M. 2011. Deletion of Rb accelerates pancreatic carcinogenesis by oncogenic Kras and impairs senescence in premalignant lesions. *Gastroenterology* **141:** 1091–1101. doi:10.1053/j.gastro.2011.05.041

Celli GB, de Lange T. 2005. DNA processing is not required for ATM-mediated telomere damage response after TRF2 deletion. *Nat Cell Biol* **7:** 712–718. doi:10.1038/ncb1275

Celli GB, Denchi EL, de Lange T. 2006. Ku70 stimulates fusion of dysfunctional telomeres yet protects chromosome ends from homologous recombination. *Nat Cell Biol* **8:** 885–890. doi:10.1038/ncb1444

Cesare AJ, Karlseder J. 2012. A three-state model of telomere control over human proliferative boundaries. *Curr Opin Cell Biol* **24:** 731–738. doi:10.1016/j.ceb.2012.08.007

Cesare AJ, Kaul Z, Cohen SB, Napier CE, Pickett HA, Neumann AA, Reddel RR. 2009. Spontaneous occurrence of telomeric DNA damage response in the absence of chromosome fusions. *Nat Struct Mol Biol* **16:** 1244–1251. doi:10.1038/nsmb.1725

Chadeneau C, Siegel P, Harley CB, Muller WJ, Bacchetti S. 1995. Telomerase activity in normal and malignant murine tissues. *Oncogene* **11:** 893–898.

Chiba K, Lorbeer FK, Shain AH, McSwiggen DT, Schruf E, Oh A, Ryu J, Darzacq X, Bastian BC, Hockemeyer D. 2017. Mutations in the promoter of the telomerase gene *TERT* contribute to tumorigenesis by a two-step mechanism. *Science* **357:** 1416–1420. doi:10.1126/science.aao0535

Chin L, Artandi SE, Shen Q, Tam A, Lee SL, Gottlieb GJ, Greider CW, DePinho RA. 1999. P53 deficiency rescues the adverse effects of telomere loss and cooperates with telomere dysfunction to accelerate carcinogenesis. *Cell* **97:** 527–538. doi:10.1016/S0092-8674(00)80762-X

Chin K, de Solorzano CO, Knowles D, Jones A, Chou W, Rodriguez EG, Kuo WL, Ljung BM, Chew K, Myambo K, et al. 2004. In situ analyses of genome instability in breast cancer. *Nat Genet* **36:** 984–988. doi:10.1038/ng1409

Cleal K, Baird DM. 2020. Catastrophic endgames: emerging mechanisms of telomere-driven genomic instability. *Trends Genet* **36:** 347–359. doi:10.1016/j.tig.2020.02.001

Cleal K, Norris K, Baird D. 2018. Telomere length dynamics and the evolution of cancer genome architecture. *Int J Mol Sci* **19:** 482. doi:10.3390/ijms19020482

Cleal K, Jones RE, Grimstead JW, Hendrickson EA, Baird DM. 2019. Chromothripsis during telomere crisis is independent of NHEJ, and consistent with a replicative origin. *Genome Res* **29:** 737–749. doi:10.1101/gr.240705.118

Coorens THH, Oliver TRW, Sanghvi R, Sovio U, Cook E, Vento-Tormo R, Haniffa M, Young MD, Rahbari R, Sebire N, et al. 2021. Inherent mosaicism and extensive mutation of human placentas. *Nature* **592:** 80–85. doi:10.1038/s41586-021-03345-1

Coquel F, Silva MJ, Técher H, Zadorozhny K, Sharma S, Nieminuszczy J, Mettling C, Dardillac E, Barthe A, Schmitz AL, et al. 2018. SAMHD1 acts at stalled replication forks to prevent interferon induction. *Nature* **557:** 57–61. doi:10.1038/s41586-018-0050-1

Counter CM, Avilion AA, LeFeuvre CE, Stewart NG, Greider CW, Harley CB, Bacchetti S. 1992. Telomere shortening associated with chromosome instability is arrested in immortal cells which express telomerase activity. *EMBO J* **11:** 1921–1929. doi:10.1002/j.1460-2075.1992.tb05245.x

Crasta K, Ganem NJ, Dagher R, Lantermann AB, Ivanova EV, Pan Y, Nezi L, Protopopov A, Chowdhury D, Pellman D. 2012. DNA breaks and chromosome pulverization from errors in mitosis. *Nature* **482:** 53–58. doi:10.1038/nature10802

Crossley MP, Song C, Bocek MJ, Choi JH, Kousouros JN, Sathirachinda A, Lin C, Brickner JR, Bai G, Lans H, et al. 2023. R-loop-derived cytoplasmic RNA-DNA hybrids activate an immune response. *Nature* **613:** 187–194. doi:10.1038/s41586-022-05545-9

Cusanelli E, Romero CAP, Chartrand P. 2013. Telomeric noncoding RNA TERRA is induced by telomere shortening to nucleate telomerase molecules at short telomeres. *Mol Cell* **51:** 780–791. doi:10.1016/j.molcel.2013.08.029

d'Adda di Fagagna F, Reaper PM, Clay-Farrace L, Fiegler H, Carr P, Von Zglinicki T, Saretzki G, Carter NP, Jackson SP. 2003. A DNA damage checkpoint response in telomere-initiated senescence. *Nature* **426:** 194–198. doi:10.1038/nature02118

Davis A, Gao R, Navin N. 2017. Tumor evolution: linear, branching, neutral or punctuated? *Biochim Biophys Acta Rev Cancer* **1867:** 151–161. doi:10.1016/j.bbcan.2017.01.003

Davoli T, de Lange T. 2012. Telomere-driven tetraploidization occurs in human cells undergoing crisis and promotes transformation of mouse cells. *Cancer Cell* **21:** 765–776. doi:10.1016/j.ccr.2012.03.044

Davoli T, Denchi EL, de Lange T. 2010. Persistent telomere damage induces bypass of mitosis and tetraploidy. *Cell* **141:** 81–93. doi:10.1016/j.cell.2010.01.031

Deckbar D, Birraux J, Krempler A, Tchouandong L, Beucher A, Walker S, Stiff T, Jeggo P, Löbrich M. 2007. Chromosome breakage after G₂ checkpoint release. *J Cell Biol* **176:** 749–755. doi:10.1083/jcb.200612047

Degenhardt K, Mathew R, Beaudoin B, Bray K, Anderson D, Chen G, Mukherjee C, Shi Y, Gélinas C, Fan Y, et al. 2006.

Autophagy promotes tumor cell survival and restricts necrosis, inflammation, and tumorigenesis. *Cancer Cell* **10**: 51–64. doi:10.1016/j.ccr.2006.06.001

de Lange T. 2005. Telomere-related genome instability in cancer. *Cold Spring Harb Symp Quant Biol* **70**: 197–204. doi:10.1101/sqb.2005.70.032

de Lange T. 2018. Shelterin-mediated telomere protection. *Annu Rev Genet* **52**: 223–247. doi:10.1146/annurev-genet-032918-021921

* de Lange T. 2025. How shelterin orchestrates the replication and protection of telomeres. *Cold Spring Harb Perspect Biol* doi:10.1101/cshperspect.a041685

Denchi EL, de Lange T. 2007. Protection of telomeres through independent control of ATM and ATR by TRF2 and POT1. *Nature* **448**: 1068–1071. doi:10.1038/nature06065

Denton D, Shravage B, Simin R, Mills K, Berry DL, Baehrecke EH, Kumar S. 2009. Autophagy, not apoptosis, is essential for midgut cell death in *Drosophila*. *Curr Biol* **19**: 1741–1746. doi:10.1016/j.cub.2009.08.042

De Silva R, Whitaker NJ, Rogan EM, Reddel RR. 1994. HPV-16 E6 and E7 genes, like SV40 early region genes, are insufficient for immortalization of human mesothelial and bronchial epithelial cells. *Exp Cell Res* **213**: 418–427. doi:10.1006/excr.1994.1218

Dilley RL, Greenberg RA. 2015. ALTernative telomere maintenance and cancer. *Trends Cancer* **1**: 145–156. doi:10.1016/j.trecan.2015.07.007

Diman A, Decottignies A. 2018. Genomic origin and nuclear localization of TERRA telomeric repeat-containing RNA: from Darkness to Dawn. *FEBS J* **285**: 1389–1398. doi:10.1111/febs.14363

Ding Z, Wu CJ, Jaskelioff M, Ivanova E, Kost-Alimova M, Protopopov A, Chu GC, Wang G, Lu X, Labrot ES, et al. 2012. Telomerase reactivation following telomere dysfunction yields murine prostate tumors with bone metastases. *Cell* **148**: 896–907. doi:10.1016/j.cell.2012.01.039

Dixit E, Boulant S, Zhang Y, Lee ASY, Odendall C, Shum B, Hacohen N, Chen ZJ, Whelan SP, Fransen M, et al. 2010. Peroxisomes are signaling platforms for antiviral innate immunity. *Cell* **141**: 668–681. doi:10.1016/j.cell.2010.04.018

Doksani Y, Wu JY, de Lange T, Zhuang X. 2013. Super-resolution fluorescence imaging of telomeres reveals TRF2-dependent T-loop formation. *Cell* **155**: 345–356. doi:10.1016/j.cell.2013.09.048

Dressler L, Bortolomeazzi M, Keddar MR, Misetic H, Sartini G, Acha-Sagredo A, Montorsi L, Wijewardhane N, Repana D, Nulsen J, et al. 2022. Comparative assessment of genes driving cancer and somatic evolution in non-cancer tissues: an update of the Network of Cancer Genes (NCG) resource. *Genome Biol* **23**: 35. doi:10.1186/s13059-022-02607-z

Eickhoff P, Sonmez C, Fisher CEL, Inian O, Roumeliotis TI, dello Stritto A, Mansfeld J, Choudhary JS, Guettler S, Lottersberger F, et al. 2025. Chromosome end protection by RAP1-mediated inhibition of DNA-PK. *Nature* **642**: 1090–1096.

Engel JL, Zhang X, Wu M, Wang Y, Espejo Valle-Inclán J, Hu Q, Woldehawariat KS, Sanders MA, Smogorzewska A, Chen J, et al. 2024. The Fanconi anemia pathway induces chromothripsis and ecDNA-driven cancer drug resis-

tance. *Cell* **187**: 6055–6070.e22. doi:10.1016/j.cell.2024.08.001

Farnung BO, Brun CM, Arora R, Lorenzi LE, Azzalin CM. 2012. Telomerase efficiently elongates highly transcribing telomeres in human cancer cells. *PLoS One* **7**: e35714. doi:10.1371/journal.pone.0035714

Feretzaki M, Pospisilova M, Valador Fernandes R, Lunardi T, Krejci L, Lingner J. 2020. RAD51-dependent recruitment of TERRA lncRNA to telomeres through R-loops. *Nature* **587**: 303–308. doi:10.1038/s41586-020-2815-6

Feuerhahn S, Iglesias N, Panza A, Porro A, Lingner J. 2010. TERRA biogenesis, turnover and implications for function. *FEBS Lett* **584**: 3812–3818. doi:10.1016/j.febslet.2010.07.032

Fleury H, MacEachern MK, Stiefel CM, Anand R, Sempeck C, Nebenfuehr B, Maurer-Alcalá K, Ball K, Proctor B, Belan O, et al. 2023. The APE2 nuclease is essential for DNA double-strand break repair by microhomology-mediated end joining. *Mol Cell* **83**: 1429–1445.e8. doi:10.1016/j.molcel.2023.03.017

Flores-Concha M, Oñate ÁA. 2020. Long non-coding RNAs in the regulation of the immune response and trained immunity. *Front Genet* **11**: 718. doi:10.3389/fgene.2020.00718

Flynn PJ, Koch PD, Mitchison TJ. 2021. Chromatin bridges, not micronuclei, activate cGAS after drug-induced mitotic errors in human cells. *Proc Natl Acad Sci* **118**: e2103585118. doi:10.1073/pnas.2103585118

Gack MU, Shin YC, Joo CH, Urano T, Liang C, Sun L, Takeuchi O, Akira S, Chen Z, Inoue S, et al. 2007. TRIM25 RING-finger E3 ubiquitin ligase is essential for RIG-I-mediated antiviral activity. *Nature* **446**: 916–920. doi:10.1038/nature05732

Galluzzi L, Vitale I, Aaronson SA, Abrams JM, Adam D, Agostinis P, Alnemri ES, Altucci L, Amelio I, Andrews DW, et al. 2018. Molecular mechanisms of cell death: recommendations of the Nomenclature Committee on Cell Death 2018. *Cell Death Differ* **25**: 486–541. doi:10.1038/s41418-017-0012-4

Garsed DW, Marshall OJ, Corbin VDA, Hsu A, Di Stefano L, Schröder J, Li J, Feng ZP, Kim BW, Kowarsky M, et al. 2014. The architecture and evolution of cancer neochromosomes. *Cancer Cell* **26**: 653–667. doi:10.1016/j.ccell.2014.09.010

Geiller HEB, Harvey A, Jones RE, Grimstead JW, Cleal K, Hendrickson EA, Baird DM. 2022. ATRX modulates the escape from a telomere crisis. *PLoS Genet* **18**: e1010485. doi:10.1371/journal.pgen.1010485

Gerstung M, Jolly C, Leshchiner I, Dentro SC, Gonzalez S, Rosebrock D, Mitchell TJ, Rubanova Y, Anur P, Yu K, et al. 2020. The evolutionary history of 2,658 cancers. *Nature* **578**: 122–128. doi:10.1038/s41586-019-1907-7

Glousker G, Briod AS, Quadroni M, Lingner J. 2020. Human shelterin protein POT1 prevents severe telomere instability induced by homology-directed DNA repair. *EMBO J* **39**: e104500. doi:10.15252/embj.2020104500

Gollahon LS, Kraus E, Wu TA, Yim SO, Strong LC, Shay JW, Tainsky MA. 1998. Telomerase activity during spontaneous immortalization of Li–Fraumeni syndrome skin fibroblasts. *Oncogene* **17**: 709–717. doi:10.1038/sj.onc.1201987

Gomes CC. 2022. Recurrent driver mutations in benign tumors. *Mutat Res Rev Mutat Res* **789:** 108412. doi:10.1016/j.mrrev.2022.108412

Greenwood J, Cooper JP. 2012. Non-coding telomeric and subtelomeric transcripts are differentially regulated by telomeric and heterochromatin assembly factors in fission yeast. *Nucleic Acids Res* **40:** 2956–2963. doi:10.1093/nar/gkr1155

Griffith JD, Comeau L, Rosenfield S, Stansel RM, Bianchi A, Moss H, de Lange T. 1999. Mammalian telomeres end in a large duplex loop. *Cell* **97:** 503–514. doi:10.1016/S0092-8674(00)80760-6

Guo X, Deng Y, Lin Y, Cosme-Blanco W, Chan S, He H, Yuan G, Brown EJ, Chang S. 2007. Dysfunctional telomeres activate an ATM-ATR-dependent DNA damage response to suppress tumorigenesis. *EMBO J* **26:** 4709–4719. doi:10.1038/sj.emboj.7601893

Guo JY, Teng X, Laddha SV, Ma S, Van Nostrand SC, Yang Y, Khor S, Chan CS, Rabinowitz JD, White E. 2016. Autophagy provides metabolic substrates to maintain energy charge and nucleotide pools in Ras-driven lung cancer cells. *Genes Dev* **30:** 1704–1717. doi:10.1101/gad.283416.116

Hadi K, Yao X, Behr JM, Deshpande A, Xanthopoulakis C, Tian H, Kudman S, Rosiene J, Darmofal M, DeRose J, et al. 2020. Distinct classes of complex structural variation uncovered across thousands of cancer genome graphs. *Cell* **183:** 197–210.e32. doi:10.1016/j.cell.2020.08.006

Hara E, Smith R, Parry D, Tahara H, Stone S, Peters G. 1996. Regulation of p16CDKN2 expression and its implications for cell immortalization and senescence. *Mol Cell Biol* **16:** 859–867. doi:10.1128/MCB.16.3.859

Harapas CR, Idiiatullina E, Al-Azab M, Hrovat-Schaale K, Reygaerts T, Steiner A, Laohamonthonkul P, Davidson S, Yu CH, Booty L, et al. 2022. Organellar homeostasis and innate immune sensing. *Nat Rev Immunol* **22:** 535–549. doi:10.1038/s41577-022-00682-8

Harding SM, Benci JL, Irianto J, Discher DE, Minn AJ, Greenberg RA. 2017. Mitotic progression following DNA damage enables pattern recognition within micronuclei. *Nature* **548:** 466–470. doi:10.1038/nature23470

Harley CB, Futcher AB, Greider CW. 1990. Telomeres shorten during ageing of human fibroblasts. *Nature* **345:** 458–460. doi:10.1038/345458a0

Harley CB, Vaziri H, Counter CM, Allsopp RC. 1992. The telomere hypothesis of cellular aging. *Exp Gerontol* **27:** 375–382. doi:10.1016/0531-5565(92)90068-B

Hatch EM, Fischer AH, Deerinck TJ, Hetzer MW. 2013. Catastrophic nuclear envelope collapse in cancer cell micronuclei. *Cell* **154:** 47–60. doi:10.1016/j.cell.2013.06.007

Hayashi MT, Cesare AJ, Rivera T, Karlseder J. 2015. Cell death during crisis is mediated by mitotic telomere deprotection. *Nature* **522:** 492–496. doi:10.1038/nature14513

Hayflick L. 1965. The limited in vitro lifetime of human diploid cell strains. *Exp Cell Res* **37:** 614–636. doi:10.1016/0014-4827(65)90211-9

Hayflick L, Moorhead PS. 1961. The serial cultivation of human diploid cell strains. *Exp Cell Res* **25:** 585–621. doi:10.1016/0014-4827(61)90192-6

Heacock M, Spangler E, Riha K, Puizina J, Shippen DE. 2004. Molecular analysis of telomere fusions in *Arabidopsis*: multiple pathways for chromosome end-joining. *EMBO J* **23:** 2304–2313. doi:10.1038/sj.emboj.7600236

Hemann MT, Strong MA, Hao LY, Greider CW. 2001. The shortest telomere, not average telomere length, is critical for cell viability and chromosome stability. *Cell* **107:** 67–77. doi:10.1016/S0092-8674(01)00504-9

Henson JD, Reddel RR. 2010. Assaying and investigating alternative lengthening of telomeres activity in human cells and cancers. *FEBS Lett* **584:** 3800–3811. doi:10.1016/j.febslet.2010.06.009

Herbig U, Jobling WA, Chen BPC, Chen DJ, Sedivy JM. 2004. Telomere shortening triggers senescence of human cells through a pathway involving ATM, p53, and p21 (CIP1), but not p16(INK4a). *Mol Cell* **14:** 501–513. doi:10.1016/S1097-2765(04)00256-4

Hermetz KE, Newman S, Conneely KN, Martin CL, Ballif BC, Shaffer LG, Cody JD, Rudd MK. 2014. Large inverted duplications in the human genome form via a fold-back mechanism. *PLoS Genet* **10:** e1004139. doi:10.1371/journal.pgen.1004139

Hockemeyer D, Sfeir AJ, Shay JW, Wright WE, de Lange T. 2005. POT1 protects telomeres from a transient DNA damage response and determines how human chromosomes end. *EMBO J* **24:** 2667–2678. doi:10.1038/sj.emboj.7600733

Hockemeyer D, Daniels JP, Takai H, de Lange T. 2006. Recent expansion of the telomeric complex in rodents: two distinct POT1 proteins protect mouse telomeres. *Cell* **126:** 63–77. doi:10.1016/j.cell.2006.04.044

Hoffelder DR, Luo L, Burke NA, Watkins SC, Gollin SM, Saunders WS. 2004. Resolution of anaphase bridges in cancer cells. *Chromosoma* **112:** 389–397. doi:10.1007/s00412-004-0284-6

Horner SM, Liu HM, Park HS, Briley J, Gale M. 2011. Mitochondrial-associated endoplasmic reticulum membranes (MAM) form innate immune synapses and are targeted by hepatitis C virus. *Proc Natl Acad Sci* **108:** 14590–14595. doi:10.1073/pnas.1110133108

Houghtaling BR, Cuttonaro L, Chang W, Smith S. 2004. A dynamic molecular link between the telomere length regulator TRF1 and the chromosome end protector TRF2. *Curr Biol* **14:** 1621–1631. doi:10.1016/j.cub.2004.08.052

Huang W, Hickson LJ, Eirin A, Kirkland JL, Lerman LO. 2022. Cellular senescence: the good, the bad and the unknown. *Nat Rev Nephrol* **18:** 611–627. doi:10.1038/s41581-022-00601-z

Imai S, Saito F, Ikeuchi T, Segawa K, Takano T. 1993. Escape from in vitro aging in SV40 large T antigen-transformed human diploid cells: a key event responsible for immortalization occurs during crisis. *Mech Ageing Dev* **69:** 149–158. doi:10.1016/0047-6374(93)90079-7

Jacobs JJL, de Lange T. 2004. Significant role for p16INK4a in p53-independent telomere-directed senescence. *Curr Biol* **14:** 2302–2308. doi:10.1016/j.cub.2004.12.025

Jalali P, Shahmoradi A, Samii A, Mazloomnejad R, Hatamnejad MR, Saeed A, Namdar A, Salehi Z. 2025. The role of autophagy in cancer: from molecular mechanism to therapeutic window. *Front Immunol* **16:** 1528230. doi:10.3389/fimmu.2025.1528230

Jiang X, Kinch LN, Brautigam CA, Chen X, Du F, Grishin NV, Chen ZJ. 2012. Ubiquitin-induced oligomerization of the RNA sensors RIG-I and MDA5 activates antiviral innate immune response. *Immunity* **36:** 959–973. doi:10.1016/j.immuni.2012.03.022

Jiang H, Xue X, Panda S, Kawale A, Hooy RM, Liang F, Sohn J, Sung P, Gekara NO. 2019. Chromatin-bound cGAS is an inhibitor of DNA repair and hence accelerates genome destabilization and cell death. *EMBO J* **38:** e102718. doi:10.15252/embj.2019102718

Jiao H, Wachsmuth L, Kumari S, Schwarzer R, Lin J, Eren RO, Fisher A, Lane R, Young GR, Kassiotis G, et al. 2020. Z-nucleic-acid sensing triggers ZBP1-dependent necroptosis and inflammation. *Nature* **580:** 391–395. doi:10.1038/s41586-020-2129-8

Jonassaint NL, Guo N, Califano JA, Montgomery EA, Armanios M. 2013. The gastrointestinal manifestations of telomere-mediated disease. *Aging Cell* **12:** 319–323. doi:10.1111/acel.12041

Jones MJK, Jallepalli PV. 2012. Chromothripsis: chromosomes in crisis. *Dev Cell* **23:** 908–917. doi:10.1016/j.devcel.2012.10.010

Jones RE, Oh S, Grimstead JW, Zimbric J, Roger L, Heppel NH, Ashelford KE, Liddiard K, Hendrickson EA, Baird DM. 2014. Escape from telomere-driven crisis is DNA ligase III dependent. *Cell Rep* **8:** 1063–1076. doi:10.1016/j.celrep.2014.07.007

Kabir S, Hockemeyer D, de Lange T. 2014. TALEN gene knockouts reveal no requirement for the conserved human shelterin protein Rap1 in telomere protection and length regulation. *Cell Rep* **9:** 1273–1280. doi:10.1016/j.celrep.2014.10.014

Kaiser WJ, Upton JW, Mocarski ES. 2008. Receptor-interacting protein homotypic interaction motif-dependent control of NF-kappa B activation via the DNA-dependent activator of IFN regulatory factors. *J Immunol* **181:** 6427–6434. doi:10.4049/jimmunol.181.9.6427

Kakiuchi N, Ogawa S. 2021. Clonal expansion in non-cancer tissues. *Nat Rev Cancer* **21:** 239–256. doi:10.1038/s41568-021-00335-3

Karki R, Sundaram B, Sharma BR, Lee S, Malireddi RKS, Nguyen LN, Christgen S, Zheng M, Wang Y, Samir P, et al. 2021. ADAR1 restricts ZBP1-mediated immune response and PANoptosis to promote tumorigenesis. *Cell Rep* **37:** 109858. doi:10.1016/j.celrep.2021.109858

Karlseder J, Smogorzewska A, de Lange T. 2002. Senescence induced by altered telomere state, not telomere loss. *Science* **295:** 2446–2449. doi:10.1126/science.1069523

Karlsson K, Przybilla MJ, Kotler E, Khan A, Xu H, Karagyozova K, Sockell A, Wong WH, Liu K, Mah A, et al. 2023. Deterministic evolution and stringent selection during preneoplasia. *Nature* **618:** 383–393. doi:10.1038/s41586-023-06102-8

Karsli-Uzunbas G, Guo JY, Price S, Teng X, Laddha SV, Khor S, Kalaany NY, Jacks T, Chan CS, Rabinowitz JD, et al. 2014. Autophagy is required for glucose homeostasis and lung tumor maintenance. *Cancer Discov* **4:** 914–927. doi:10.1158/2159-8290.CD-14-0363

Kaul Z, Cesare AJ, Huschtscha LI, Neumann AA, Reddel RR. 2012. Five dysfunctional telomeres predict onset of senescence in human cells. *EMBO Rep* **13:** 52–59. doi:10.1038/embor.2011.227

Kawai T, Takahashi K, Sato S, Coban C, Kumar H, Kato H, Ishii KJ, Takeuchi O, Akira S. 2005. IPS-1, an adaptor triggering RIG-I- and Mda5-mediated type I interferon induction. *Nat Immunol* **6:** 981–988. doi:10.1038/ni1243

Kesavardhana S, Malireddi RKS, Burton AR, Porter SN, Vogel P, Pruett-Miller SM, Kanneganti TD. 2020. The Zα2 domain of ZBP1 is a molecular switch regulating influenza-induced PANoptosis and perinatal lethality during development. *J Biol Chem* **295:** 8325–8330. doi:10.1074/jbc.RA120.013752

Kibe T, Osawa GA, Keegan CE, de Lange T. 2010. Telomere protection by TPP1 is mediated by POT1a and POT1b. *Mol Cell Biol* **30:** 1059–1066. doi:10.1128/MCB.01498-09

Kim NW, Piatyszek MA, Prowse KR, Harley CB, West MD, Ho PL, Coviello GM, Wright WE, Weinrich SL, Shay JW. 1994. Specific association of human telomerase activity with immortal cells and cancer. *Science* **266:** 2011–2015. doi:10.1126/science.7605428

Kim SH, Kaminker P, Campisi J. 1999. TIN2, a new regulator of telomere length in human cells. *Nat Genet* **23:** 405–412. doi:10.1038/70508

King TA, Li W, Yee C, Gemignani ML, Olvera N, Brogi E, Robson ME, Offit K, Norton L, Borgen PI, et al. 2005. *BRCA* haploinsufficiency in human breast tumorigenesis. *J Clin Oncol* **23:** 9512–9512. doi:10.1200/jco.2005.23.16_suppl.9512

Kinouchi Y, Hiwatashi N, Chida M, Nagashima F, Takagi S, Maekawa H, Toyota T. 1998. Telomere shortening in the colonic mucosa of patients with ulcerative colitis. *J Gastroenterol* **33:** 343–348. doi:10.1007/s005350050094

Lahaye X, Gentili M, Silvin A, Conrad C, Picard L, Jouve M, Zueva E, Maurin M, Nadalin F, Knott GJ, et al. 2018. NONO detects the nuclear HIV capsid to promote cGAS-mediated innate immune activation. *Cell* **175:** 488–501.e22. doi:10.1016/j.cell.2018.08.062

Lalonde M, Chartrand P. 2020. TERRA, a multifaceted regulator of telomerase activity at telomeres. *J Mol Biol* **432:** 4232–4243. doi:10.1016/j.jmb.2020.02.004

Lawson ARJ, Abascal F, Coorens THH, Hooks Y, O'Neill L, Latimer C, Raine K, Sanders MA, Warren AY, Mahbubani KTA, et al. 2020. Extensive heterogeneity in somatic mutation and selection in the human bladder. *Science* **370:** 75–82. doi:10.1126/science.aba8347

Lee-Six H, Olafsson S, Ellis P, Osborne RJ, Sanders MA, Moore L, Georgakopoulos N, Torrente F, Noorani A, Goddard M, et al. 2019. The landscape of somatic mutation in normal colorectal epithelial cells. *Nature* **574:** 532–537. doi:10.1038/s41586-019-1672-7

Letsolo BT, Rowson J, Baird DM. 2010. Fusion of short telomeres in human cells is characterized by extensive deletion and microhomology, and can result in complex rearrangements. *Nucleic Acids Res* **38:** 1841–1852. doi:10.1093/nar/gkp1183

Lex K, Maia Gil M, Lopes-Bastos B, Figueira M, Marzullo M, Giannetti K, Carvalho T, Ferreira MG. 2020. Telomere shortening produces an inflammatory environment that increases tumor incidence in zebrafish. *Proc Natl Acad Sci* **117:** 15066–15074. doi:10.1073/pnas.1920049117

Li B, Oestreich S, de Lange T. 2000. Identification of human Rap1: implications for telomere evolution. *Cell* **101:** 471–483. doi:10.1016/S0092-8674(00)80858-2

Li Y, Schwab C, Ryan S, Papaemmanuil E, Robinson HM, Jacobs P, Moorman AV, Dyer S, Borrow J, Griffiths M, et al. 2014. Constitutional and somatic rearrangement of chromosome 21 in acute lymphoblastic leukaemia. *Nature* **508**: 98–102. doi:10.1038/nature13115

Li F, Deng Z, Zhang L, Wu C, Jin Y, Hwang I, Vladimirova O, Xu L, Yang L, Lu B, et al. 2019. ATRX loss induces telomere dysfunction and necessitates induction of alternative lengthening of telomeres during human cell immortalization. *EMBO J* **38**: e96659. doi:10.15252/embj.2017 96659

Lian J, Wu X, He F, Karnak D, Tang W, Meng Y, Xiang D, Ji M, Lawrence TS, Xu L. 2011. A natural BH3 mimetic induces autophagy in apoptosis-resistant prostate cancer via modulating Bcl-2-Beclin1 interaction at endoplasmic reticulum. *Cell Death Differ* **18**: 60–71. doi:10.1038/cdd .2010.74

Liddiard K, Grimstead JW, Cleal K, Evans A, Baird DM. 2021. Tracking telomere fusions through crisis reveals conflict between DNA transcription and the DNA damage response. *NAR Cancer* **3**: zcaa044. doi:10.1093/nar can/zcaa044

Liddiard K, Aston-Evans AN, Cleal K, Hendrickson EA, Baird DM. 2022. POLQ suppresses genome instability and alterations in DNA repeat tract lengths. *NAR Cancer* **4**: zcac020. doi:10.1093/narcan/zcac020

Lin TT, Letsolo BT, Jones RE, Rowson J, Pratt G, Hewamana S, Fegan C, Pepper C, Baird DM. 2010. Telomere dysfunction and fusion during the progression of chronic lymphocytic leukemia: evidence for a telomere crisis. *Blood* **116**: 1899–1907. doi:10.1182/blood-2010-02-272104

Liu H, Zhang H, Wu X, Ma D, Wu J, Wang L, Jiang Y, Fei Y, Zhu C, Tan R, et al. 2018a. Nuclear cGAS suppresses DNA repair and promotes tumorigenesis. *Nature* **563**: 131–136. doi:10.1038/s41586-018-0629-6

Liu S, Kwon M, Mannino M, Yang N, Renda F, Khodjakov A, Pellman D. 2018b. Nuclear envelope assembly defects link mitotic errors to chromothripsis. *Nature* **561**: 551–555. doi:10.1038/s41586-018-0534-z

Llorente B, Smith CE, Symington LS. 2008. Break-induced replication: what is it and what is it for? *Cell Cycle* **7**: 859–864. doi:10.4161/cc.7.7.5613

Lo AWI, Sabatier L, Fouladi B, Pottier G, Ricoul M, Murnane JP. 2002. DNA amplification by breakage/fusion/bridge cycles initiated by spontaneous telomere loss in a human cancer cell line. *Neoplasia* **4**: 531–538. doi:10.1038/sj.neo .7900267

Loayza D, De Lange T. 2003. POT1 as a terminal transducer of TRF1 telomere length control. *Nature* **423**: 1013–1018. doi:10.1038/nature01688

Löbrich M, Jeggo PA. 2007. The impact of a negligent G$_2$/M checkpoint on genomic instability and cancer induction. *Nat Rev Cancer* **7**: 861–869. doi:10.1038/nrc2248

Lototska L, Yue J, Li J, Giraud-Panis M, Songyang Z, Royle NJ, Liti G, Ye J, Gilson E, Mendez-Bermudez A. 2020. Human RAP1 specifically protects telomeres of senescent cells from DNA damage. *EMBO Rep* **21**: e49076. doi:10 .15252/embr.201949076

Lovejoy CA, Li W, Reisenweber S, Thongthip S, Bruno J, de Lange T, De S, Petrini JHJ, Sung PA, Jasin M, et al. 2012. Loss of ATRX, genome instability, and an altered DNA damage response are hallmarks of the alternative length-ening of telomeres pathway. *PLoS Genet* **8**: e1002772. doi:10.1371/journal.pgen.1002772

Luke B, Panza A, Redon S, Iglesias N, Li Z, Lingner J. 2008. The Rat1p 5′ to 3′ exonuclease degrades telomeric repeat-containing RNA and promotes telomere elongation in *Saccharomyces cerevisiae*. *Mol Cell* **32**: 465–477. doi:10 .1016/j.molcel.2008.10.019

Lundblad V, Blackburn EH. 1993. An alternative pathway for yeast telomere maintenance rescues est1⁻ senescence. *Cell* **73**: 347–360. doi:10.1016/0092-8674(93)90234-H

Ly P, Teitz LS, Kim DH, Shoshani O, Skaletsky H, Fachinetti D, Page DC, Cleveland DW. 2017. Selective Y centromere inactivation triggers chromosome shattering in micronuclei and repair by non-homologous end joining. *Nat Cell Biol* **19**: 68–75. doi:10.1038/ncb3450

Ma C, Martin S, Trask B, Hamlin JL. 1993. Sister chromatid fusion initiates amplification of the dihydrofolate reductase gene in Chinese hamster cells. *Genes Dev* **7**: 605–620. doi:10.1101/gad.7.4.605

Machado HE, Mitchell E, Øbro NF, Kübler K, Davies M, Leongamornlert D, Cull A, Maura F, Sanders MA, Cagan ATJ, et al. 2022. Diverse mutational landscapes in human lymphocytes. *Nature* **608**: 724–732. doi:10.1038/s41586-022-05072-7

Maciejowski J, de Lange T. 2017. Telomeres in cancer: tumour suppression and genome instability. *Nat Rev Mol Cell Biol* **18**: 175–186. doi:10.1038/nrm.2016.171

Maciejowski J, Li Y, Bosco N, Campbell PJ, de Lange T. 2015. Chromothripsis and kataegis induced by telomere crisis. *Cell* **163**: 1641–1654. doi:10.1016/j.cell.2015.11.054

Mackenzie KJ, Carroll P, Martin CA, Murina O, Fluteau A, Simpson DJ, Olova N, Sutcliffe H, Rainger JK, Leitch A, et al. 2017. cGAS surveillance of micronuclei links genome instability to innate immunity. *Nature* **548**: 461–465. doi:10.1038/nature23449

Maelfait J, Liverpool L, Bridgeman A, Ragan KB, Upton JW, Rehwinkel J. 2017. Sensing of viral and endogenous RNA by ZBP1/DAI induces necroptosis. *EMBO J* **36**: 2529–2543. doi:10.15252/embj.201796476

Makarov VL, Hirose Y, Langmore JP. 1997. Long G tails at both ends of human chromosomes suggest a C strand degradation mechanism for telomere shortening. *Cell* **88**: 657–666. doi:10.1016/S0092-8674(00)81908-X

Malireddi RKS, Kesavardhana S, Kanneganti TD. 2019. ZBP1 and TAK1: master regulators of NLRP3 inflammasome/pyroptosis, apoptosis, and necroptosis (PAN-optosis). *Front Cell Infect Microbiol* **9**: 406. doi:10.3389/fcimb .2019.00406

Martincorena I, Roshan A, Gerstung M, Ellis P, Van Loo P, McLaren S, Wedge DC, Fullam A, Alexandrov LB, Tubio JM, et al. 2015. Tumor evolution. High burden and pervasive positive selection of somatic mutations in normal human skin. *Science* **348**: 880–886. doi:10.1126/science .aaa6806

Martincorena I, Fowler JC, Wabik A, Lawson ARJ, Abascal F, Hall MWJ, Cagan A, Murai K, Mahbubani K, Stratton MR, et al. 2018. Somatic mutant clones colonize the human esophagus with age. *Science* **362**: 911–917. doi:10 .1126/science.aau3879

Martínez P, Thanasoula M, Muñoz P, Liao C, Tejera A, McNees C, Flores JM, Fernández-Capetillo O, Tarsounas M, Blasco MA. 2009. Increased telomere fragility and fusions resulting from *TRF1* deficiency lead to degenerative pathologies and increased cancer in mice. *Genes Dev* **23:** 2060–2075. doi:10.1101/gad.543509

Martinez P, Thanasoula M, Carlos AR, Gómez-López G, Tejera AM, Schoeftner S, Dominguez O, Pisano DG, Tarsounas M, Blasco MA. 2010. Mammalian Rap1 controls telomere function and gene expression through binding to telomeric and extratelomeric sites. *Nat Cell Biol* **12:** 768–780. doi:10.1038/ncb2081

Maser RS, Wong KK, Sahin E, Xia H, Naylor M, Hedberg HM, Artandi SE, DePinho RA. 2007. DNA-dependent protein kinase catalytic subunit is not required for dysfunctional telomere fusion and checkpoint response in the telomerase-deficient mouse. *Mol Cell Biol* **27:** 2253–2265. doi:10.1128/MCB.01354-06

Mateos-Gomez PA, Gong F, Nair N, Miller KM, Lazzerini-Denchi E, Sfeir A. 2015. Mammalian polymerase θ promotes alternative NHEJ and suppresses recombination. *Nature* **518:** 254–257. doi:10.1038/nature14157

Mateos-Gomez PA, Kent T, Deng SK, McDevitt S, Kashkina E, Hoang TM, Pomerantz RT, Sfeir A. 2017. The helicase domain of Polθ counteracts RPA to promote alt-NHEJ. *Nat Struct Mol Biol* **24:** 1116–1123. doi:10.1038/nsmb.3494

McClintock B. 1939. The behavior in successive nuclear divisions of a chromosome broken at meiosis. *Proc Natl Acad Sci* **25:** 405–416. doi:10.1073/pnas.25.8.405

McClintock B. 1941. The stability of broken ends of chromosomes in *Zea mays*. *Genetics* **26:** 234–282. doi:10.1093/genetics/26.2.234

McElligott R, Wellinger RJ. 1997. The terminal DNA structure of mammalian chromosomes. *EMBO J* **16:** 3705–3714. doi:10.1093/emboj/16.12.3705

McWilliams RR, Wieben ED, Rabe KG, Pedersen KS, Wu Y, Sicotte H, Petersen GM. 2011. Prevalence of CDKN2A mutations in pancreatic cancer patients: implications for genetic counseling. *Eur J Hum Genet* **19:** 472–478. doi:10.1038/ejhg.2010.198

Meeker AK, Hicks JL, Platz EA, March GE, Bennett CJ, Delannoy MJ, De Marzo AM. 2002. Telomere shortening is an early somatic DNA alteration in human prostate tumorigenesis. *Cancer Res* **62:** 6405–6409.

Meeker AK, Hicks JL, Iacobuzio-Donahue CA, Montgomery EA, Westra WH, Chan TY, Ronnett BM, De Marzo AM. 2004. Telomere length abnormalities occur early in the initiation of epithelial carcinogenesis. *Clin Cancer Res* **10:** 3317–3326. doi:10.1158/1078-0432.CCR-0984-03

Meyerson M, Counter CM, Eaton EN, Ellisen LW, Steiner P, Caddle SD, Ziaugra L, Beijersbergen RL, Davidoff MJ, Liu Q, et al. 1997. hEST2, the putative human telomerase catalytic subunit gene, is up-regulated in tumor cells and during immortalization. *Cell* **90:** 785–795. doi:10.1016/S0092-8674(00)80538-3

Meylan E, Curran J, Hofmann K, Moradpour D, Binder M, Bartenschlager R, Tschopp J. 2005. Cardif is an adaptor protein in the RIG-I antiviral pathway and is targeted by hepatitis C virus. *Nature* **437:** 1167–1172. doi:10.1038/nature04193

Meyne J, Ratliff RL, Moyzis RK. 1989. Conservation of the human telomere sequence (TTAGGG)n among vertebrates. *Proc Natl Acad Sci* **86:** 7049–7053. doi:10.1073/pnas.86.18.7049

Miller KN, Victorelli SG, Salmonowicz H, Dasgupta N, Liu T, Passos JF, Adams PD. 2021. Cytoplasmic DNA: sources, sensing, and role in aging and disease. *Cell* **184:** 5506–5526. doi:10.1016/j.cell.2021.09.034

Mitchell E, Spencer Chapman M, Williams N, Dawson KJ, Mende N, Calderbank EF, Jung H, Mitchell T, Coorens THH, Spencer DH, et al. 2022. Clonal dynamics of haematopoiesis across the human lifespan. *Nature* **606:** 343–350. doi:10.1038/s41586-022-04786-y

Mitelman F, Johansson B, Mandahl N, Mertens F. 1997. Clinical significance of cytogenetic findings in solid tumors. *Cancer Genet Cytogenet* **95:** 1–8. doi:10.1016/S0165-4608(96)00252-X

Mohr L, Toufektchan E, von Morgen P, Chu K, Kapoor A, Maciejowski J. 2021. ER-directed TREX1 limits cGAS activation at micronuclei. *Mol Cell* **81:** 724–738.e9. doi:10.1016/j.molcel.2020.12.037

Montalto MC, Phillips JS, Ray FA. 1999. Telomerase activation in human fibroblasts during escape from crisis. *J Cell Physiol* **180:** 46–52. doi:10.1002/(SICI)1097-4652(199907)180:1<46::AID-JCP5>3.0.CO;2-K

Moore L, Leongamornlert D, Coorens THH, Sanders MA, Ellis P, Dentro SC, Dawson KJ, Butler T, Rahbari R, Mitchell TJ, et al. 2020. The mutational landscape of normal human endometrial epithelium. *Nature* **580:** 640–646. doi:10.1038/s41586-020-2214-z

Moravec M, Wischnewski H, Bah A, Hu Y, Liu N, Lafranchi L, King MC, Azzalin CM. 2016. TERRA promotes telomerase-mediated telomere elongation in *Schizosaccharomyces pombe*. *EMBO Rep* **17:** 999–1012. doi:10.15252/embr.201541708

Mustjoki S, Young NS. 2021. Somatic mutations in "Benign" disease. *N Engl J Med* **384:** 2039–2052. doi:10.1056/NEJMra2101920

Nakazawa H, English D, Randell PL, Nakazawa K, Martel N, Armstrong BK, Yamasaki H. 1994. UV and skin cancer: specific p53 gene mutation in normal skin as a biologically relevant exposure measurement. *Proc Natl Acad Sci* **91:** 360–364. doi:10.1073/pnas.91.1.360

Nassour J, Radford R, Correia A, Fusté JM, Schoell B, Jauch A, Shaw RJ, Karlseder J. 2019. Autophagic cell death restricts chromosomal instability during replicative crisis. *Nature* **565:** 659–663. doi:10.1038/s41586-019-0885-0

Nassour J, Schmidt TT, Karlseder J. 2021. Telomeres and cancer: resolving the paradox. *Annu Rev Cancer Biol* **5:** 59–77. doi:10.1146/annurev-cancerbio-050420-023410

Nassour J, Aguiar LG, Correia A, Schmidt TT, Mainz L, Przetocka S, Haggblom C, Tadepalle N, Williams A, Shokhirev MN, et al. 2023. Telomere-to-mitochondria signalling by ZBP1 mediates replicative crisis. *Nature* **614:** 767–773. doi:10.1038/s41586-023-05710-8

Nassour J, Przetocka S, Karlseder J. 2024. Telomeres as hotspots for innate immunity and inflammation. *DNA Repair (Amst)* **133:** 103591. doi:10.1016/j.dnarep.2023.103591

Nergadze SG, Farnung BO, Wischnewski H, Khoriauli L, Vitelli V, Chawla R, Giulotto E, Azzalin CM. 2009. CpG-island promoters drive transcription of human telomeres. *RNA* **15:** 2186–2194. doi:10.1261/rna.1748309

Ngo G, Hyatt S, Grimstead J, Jones R, Hendrickson E, Pepper C, Baird D. 2018. PARP inhibition prevents escape from a telomere-driven crisis and inhibits cell immortalisation.

*Oncotarget* **9:** 37549–37563. doi:10.18632/oncotarget.26 499

Nikitina T, Woodcock CL. 2004. Closed chromatin loops at the ends of chromosomes. *J Cell Biol* **166:** 161–165. doi:10 .1083/jcb.200403118

O'Connor MS, Safari A, Xin H, Liu D, Songyang Z. 2006. A critical role for TPP1 and TIN2 interaction in high-order telomeric complex assembly. *Proc Natl Acad Sci* **103:** 11874–11879. doi:10.1073/pnas.0605303103

O'Hagan RC, Chang S, Maser RS, Mohan R, Artandi SE, Chin L, DePinho RA. 2002. Telomere dysfunction provokes regional amplification and deletion in cancer genomes. *Cancer Cell* **2:** 149–155. doi:10.1016/S1535-6108 (02)00094-6

Okamoto K, Bartocci C, Ouzounov I, Diedrich JK, Yates JR, Denchi EL. 2013. A two-step mechanism for TRF2-mediated chromosome end protection. *Nature* **494:** 502–505. doi:10.1038/nature11873

Oshiumi H. 2020. Recent advances and contradictions in the study of the individual roles of ubiquitin ligases that regulate RIG-I-like receptor-mediated antiviral innate immune responses. *Front Immunol* **11:** 1296. doi:10.3389/ fimmu.2020.01296

Oshiumi H, Miyashita M, Inoue N, Okabe M, Matsumoto M, Seya T. 2010. The ubiquitin ligase Riplet is essential for RIG-I-dependent innate immune responses to RNA virus infection. *Cell Host Microbe* **8:** 496–509. doi:10.1016/j .chom.2010.11.008

Ostroverkhova D, Przytycka TM, Panchenko AR. 2023. Cancer driver mutations: predictions and reality. *Trends Mol Med* **29:** 554–566. doi:10.1016/j.molmed.2023.03.007

O'Sullivan JN, Bronner MP, Brentnall TA, Finley JC, Shen W-T, Emerson S, Emond MJ, Gollahon KA, Moskovitz AH, Crispin DA, et al. 2002. Chromosomal instability in ulcerative colitis is related to telomere shortening. *Nat Genet* **32:** 280–284. doi:10.1038/ng989

Palm W, Hockemeyer D, Kibe T, de Lange T. 2009. Functional dissection of human and mouse POT1 proteins. *Mol Cell Biol* **29:** 471–482. doi:10.1128/MCB.01352-08

Pathare GR, Decout A, Glück S, Cavadini S, Makasheva K, Hovius R, Kempf G, Weiss J, Kozicka Z, Guey B, et al. 2020. Structural mechanism of cGAS inhibition by the nucleosome. *Nature* **587:** 668–672. doi:10.1038/s41586-020-2750-6

Pfeiffer V, Crittin J, Grolimund L, Lingner J. 2013. The THO complex component Thp2 counteracts telomeric R-loops and telomere shortening. *EMBO J* **32:** 2861–2871. doi:10 .1038/emboj.2013.217

Porro A, Feuerhahn S, Reichenbach P, Lingner J. 2010. Molecular dissection of telomeric repeat-containing RNA biogenesis unveils the presence of distinct and multiple regulatory pathways. *Mol Cell Biol* **30:** 4808–4817. doi:10 .1128/MCB.00460-10

Porro A, Feuerhahn S, Delafontaine J, Riethman H, Rougemont J, Lingner J. 2014. Functional characterization of the TERRA transcriptome at damaged telomeres. *Nat Commun* **5:** 5379. doi:10.1038/ncomms6379

Qu X, Yu J, Bhagat G, Furuya N, Hibshoosh H, Troxel A, Rosen J, Eskelinen E, Mizushima N, Ohsumi Y, et al. 2003. Promotion of tumorigenesis by heterozygous disruption of the beclin 1 autophagy gene. *J Clin Invest* **112:** 1809–1820. doi:10.1172/JCI20039

Rai R, Zheng H, He H, Luo Y, Multani A, Carpenter PB, Chang S. 2010. The function of classical and alternative non-homologous end-joining pathways in the fusion of dysfunctional telomeres. *EMBO J* **29:** 2598–2610. doi:10 .1038/emboj.2010.142

Raices M, Verdun RE, Compton SA, Haggblom CI, Griffith JD, Dillin A, Karlseder J. 2008. *C. elegans* telomeres contain G-strand and C-strand overhangs that are bound by distinct proteins. *Cell* **132:** 745–757. doi:10.1016/j.cell .2007.12.039

Rausch T, Jones DTW, Zapatka M, Stütz AM, Zichner T, Weischenfeldt J, Jäger N, Remke M, Shih D, Northcott PA, et al. 2012. Genome sequencing of pediatric medulloblastoma links catastrophic DNA rearrangements with TP53 mutations. *Cell* **148:** 59–71. doi:10.1016/j.cell.2011 .12.013

Ray FA, Kraemer PM. 1993. Iterative chromosome mutation and selection as a mechanism of complete transformation of human diploid fibroblasts by SV40 T antigen. *Carcinogenesis* **14:** 1511–1516. doi:10.1093/carcin/14.8.1511

Rebsamen M, Heinz LX, Meylan E, Michallet M-C, Schroder K, Hofmann K, Vazquez J, Benedict CA, Tschopp J. 2009. DAI/ZBP1 recruits RIP1 and RIP3 through RIP homotypic interaction motifs to activate NF-kappaB. *EMBO Rep* **10:** 916–922. doi:10.1038/embor.2009.109

Reeves MQ, Balmain A. 2024. Mutations, bottlenecks, and clonal sweeps: how environmental carcinogens and genomic changes shape clonal evolution during tumor progression. *Cold Spring Harb Perspect Med* **14:** a041388. doi:10.1101/cshperspect.a041388

Revy P, Kannengiesser C, Bertuch AA. 2023. Genetics of human telomere biology disorders. *Nat Rev Genet* **24:** 86–108. doi:10.1038/s41576-022-00527-z

Ribes-Zamora A, Indiviglio SM, Mihalek I, Williams CL, Bertuch AA. 2013. TRF2 interaction with Ku heterotetramerization interface gives insight into c-NHEJ prevention at human telomeres. *Cell Rep* **5:** 194–206. doi:10.1016/j .celrep.2013.08.040

Riboni R, Casati A, Nardo T, Zaccaro E, Ferretti L, Nuzzo F, Mondello C. 1997. Telomeric fusions in cultured human fibroblasts as a source of genomic instability. *Cancer Genet Cytogenet* **95:** 130–136. doi:10.1016/S0165-4608(96) 00248-8

Roger L, Jones RE, Heppel NH, Williams GT, Sampson JR, Baird DM. 2013. Extensive telomere erosion in the initiation of colorectal adenomas and its association with chromosomal instability. *J Natl Cancer Inst* **105:** 1202–1211. doi:10.1093/jnci/djt191

Romero-Zamora D, Rogers S, Low RRJ, Page SG, Lane BJE, Kosaka S, Robinson AB, French L, Lamm N, Ishikawa F, et al. 2025. A CPC-shelterin-BTR axis regulates mitotic telomere deprotection. *Nat Commun* **16:** 2277. doi:10 .1038/s41467-025-57456-8

Rongvaux A, Jackson R, Harman CCD, Li T, West AP, de Zoete MR, Wu Y, Yordy B, Lakhani SA, Kuan C-Y, et al. 2014. Apoptotic caspases prevent the induction of type I interferons by mitochondrial DNA. *Cell* **159:** 1563–1577. doi:10.1016/j.cell.2014.11.037

Russell RC, Guan K. 2022. The multifaceted role of autophagy in cancer. *EMBO J* **41:** e110031. doi:10.15252/embj .2021110031

Cite this article as *Cold Spring Harb Perspect Biol* doi: 10.1101/cshperspect.a041688

Sack GH. 1981. Human cell transformation by simian virus 40—a review. *In Vitro* **17:** 1–19. doi:10.1007/BF02618025

Sarek G, Kotsantis P, Ruis P, Van Ly D, Margalef P, Borel V, Zheng X-F, Flynn HR, Snijders AP, Chowdhury D, et al. 2019. CDK phosphorylation of TRF2 controls t-loop dynamics during the cell cycle. *Nature* **575:** 523–527. doi:10.1038/s41586-019-1744-8

Sarthy J, Bae NS, Scrafford J, Baumann P. 2009. Human RAP1 inhibits non-homologous end joining at telomeres. *EMBO J* **28:** 3390–3399. doi:10.1038/emboj.2009.275

Savy T, Flanders L, Karpanasamy T, Sun M, Gerlinger M. 2025. Cancer evolution: from Darwin to the extended evolutionary synthesis. *Trends Cancer* **11:** 204–215. doi:10.1016/j.trecan.2025.01.001

Schoeftner S, Blasco MA. 2008. Developmentally regulated transcription of mammalian telomeres by DNA-dependent RNA polymerase II. *Nat Cell Biol* **10:** 228–236. doi:10.1038/ncb1685

Seth RB, Sun L, Ea CK, Chen ZJ. 2005. Identification and characterization of MAVS, a mitochondrial antiviral signaling protein that activates NF-kappaB and IRF 3. *Cell* **122:** 669–682. doi:10.1016/j.cell.2005.08.012

Sfeir A, de Lange T. 2012. Removal of shelterin reveals the telomere end-protection problem. *Science* **336:** 593–597. doi:10.1126/science.1218498

Sfeir A, Kosiyatrakul ST, Hockemeyer D, MacRae SL, Karlseder J, Schildkraut CL, de Lange T. 2009. Mammalian telomeres resemble fragile sites and require TRF1 for efficient replication. *Cell* **138:** 90–103. doi:10.1016/j.cell.2009.06.021

Sfeir A, Kabir S, van Overbeek M, Celli GB, de Lange T. 2010. Loss of Rap1 induces telomere recombination in the absence of NHEJ or a DNA damage signal. *Science* **327:** 1657–1661. doi:10.1126/science.1185100

Shah A. 2024. Rethinking cancer initiation: the role of large-scale mutational events. *Genes Chromosomes Cancer* **63:** e23213. doi:10.1002/gcc.23213

Shay JW, Wright WE. 1989. Quantitation of the frequency of immortalization of normal human diploid fibroblasts by SV40 large T-antigen. *Exp Cell Res* **184:** 109–118. doi:10.1016/0014-4827(89)90369-8

Shay JW, Pereira-Smith OM, Wright WE. 1991a. A role for both RB and p53 in the regulation of human cellular senescence. *Exp Cell Res* **196:** 33–39. doi:10.1016/0014-4827(91)90453-2

Shay JW, Wright WE, Werbin H. 1991b. Defining the molecular mechanisms of human cell immortalization. *Biochim Biophys Acta* **1072:** 1–7. doi:10.1016/0304-419x(91)90003-4

Shay JW, Van Der Haegen BA, Ying Y, Wright WE. 1993. The frequency of immortalization of human fibroblasts and mammary epithelial cells transfected with SV40 large T-antigen. *Exp Cell Res* **209:** 45–52. doi:10.1006/excr.1993.1283

Shi R, Weng J, Zhao L, Li X, Gao T, Kong J. 2012. Excessive autophagy contributes to neuron death in cerebral ischemia. *CNS Neurosci Ther* **18:** 250–260. doi:10.1111/j.1755-5949.2012.00295.x

Shih IM, Zhou W, Goodman SN, Lengauer C, Kinzler KW, Vogelstein B. 2001. Evidence that genetic instability occurs at an early stage of colorectal tumorigenesis. *Cancer Res* **61:** 818–822.

Silva B, Pentz R, Figueira AM, Arora R, Lee YW, Hodson C, Wischnewski H, Deans AJ, Azzalin CM. 2019. FANCM limits ALT activity by restricting telomeric replication stress induced by deregulated BLM and R-loops. *Nat Commun* **10:** 2253. doi:10.1038/s41467-019-10179-z

Smith KA, Gorman PA, Stark MB, Groves RP, Stark GR. 1990. Distinctive chromosomal structures are formed very early in the amplification of CAD genes in Syrian hamster cells. *Cell* **63:** 1219–1227. doi:10.1016/0092-8674(90)90417-D

Smith K, Adank M, Kauff N, Lafaro K, Boyd J, Lee JB, Hudis C, Offit K, Robson A. 2007. BRCA mutations in women with ductal carcinoma in situ. *Clin Cancer Res* **13:** 4306–4310. doi:10.1158/1078-0432.CCR-07-0146

Smogorzewska A, de Lange T. 2002. Different telomere damage signaling pathways in human and mouse cells. *EMBO J* **21:** 4338–4348. doi:10.1093/emboj/cdf433

Smogorzewska A, van Steensel B, Bianchi A, Oelmann S, Schaefer MR, Schnapp G, de Lange T. 2000. Control of human telomere length by TRF1 and TRF2. *Mol Cell Biol* **20:** 1659–1668. doi:10.1128/MCB.20.5.1659-1668.2000

Smogorzewska A, Karlseder J, Holtgreve-Grez H, Jauch A, de Lange T. 2002. DNA ligase IV-dependent NHEJ of deprotected mammalian telomeres in $G_1$ and $G_2$. *Curr Biol* **12:** 1635–1644. doi:10.1016/S0960-9822(02)01179-X

Sottoriva A, Kang H, Ma Z, Graham TA, Salomon MP, Zhao J, Marjoram P, Siegmund K, Press MF, Shibata D, et al. 2015. A Big Bang model of human colorectal tumor growth. *Nat Genet* **47:** 209–216. doi:10.1038/ng.3214

Sridharan H, Ragan KB, Guo H, Gilley RP, Landsteiner VJ, Kaiser WJ, Upton JW. 2017. Murine cytomegalovirus IE3-dependent transcription is required for DAI/ZBP1-mediated necroptosis. *EMBO Rep* **18:** 1429–1441. doi:10.15252/embr.201743947

Stephens PJ, Greenman CD, Fu B, Yang F, Bignell GR, Mudie LJ, Pleasance ED, Lau KW, Beare D, Stebbings LA, et al. 2011. Massive genomic rearrangement acquired in a single catastrophic event during cancer development. *Cell* **144:** 27–40. doi:10.1016/j.cell.2010.11.055

Stewart SA, Ben-Porath I, Carey VJ, O'Connor BF, Hahn WC, Weinberg RA. 2003. Erosion of the telomeric single-strand overhang at replicative senescence. *Nat Genet* **33:** 492–496. doi:10.1038/ng1127

Suda K, Nakaoka H, Yoshihara K, Ishiguro T, Tamura R, Mori Y, Yamawaki K, Adachi S, Takahashi T, Kase H, et al. 2018. Clonal expansion and diversification of cancer-associated mutations in endometriosis and normal endometrium. *Cell Rep* **24:** 1777–1789. doi:10.1016/j.celrep.2018.07.037

Sun R, Hu Z, Sottoriva A, Graham TA, Harpak A, Ma Z, Fischer JM, Shibata D, Curtis C. 2017. Between-region genetic divergence reflects the mode and tempo of tumor evolution. *Nat Genet* **49:** 1015–1024. doi:10.1038/ng.3891

Taffoni C, Steer A, Marines J, Chamma H, Vila IK, Laguette N. 2021. Nucleic acid immunity and DNA damage response: new friends and old foes. *Front Immunol* **12:** 660560. doi:10.3389/fimmu.2021.660560

Takai H, Smogorzewska A, de Lange T. 2003. DNA damage foci at dysfunctional telomeres. *Curr Biol* **13:** 1549–1556. doi:10.1016/S0960-9822(03)00542-6

Takai KK, Kibe T, Donigian JR, Frescas D, de Lange T. 2011. Telomere protection by TPP1/POT1 requires tethering to TIN2. *Mol Cell* **44:** 647–659. doi:10.1016/j.molcel.2011.08.043

Takaoka A, Wang Z, Choi MK, Yanai H, Negishi H, Ban T, Lu Y, Miyagishi M, Kodama T, Honda K, et al. 2007. DAI (DLM-1/ZBP1) is a cytosolic DNA sensor and an activator of innate immune response. *Nature* **448:** 501–505. doi:10.1038/nature06013

Tanaka H, Bergstrom DA, Yao MC, Tapscott SJ. 2005. Widespread and nonrandom distribution of DNA palindromes in cancer cells provides a structural platform for subsequent gene amplification. *Nat Genet* **37:** 320–327. doi:10.1038/ng1515

Tang B, Li Y, Qi G, Yuan S, Wang Z, Yu S, Li B, He S. 2015. Clinicopathological significance of CDKN2A promoter hypermethylation frequency with pancreatic cancer. *Sci Rep* **5:** 13563. doi:10.1038/srep13563

Tang J, Fewings E, Chang D, Zeng H, Liu S, Jorapur A, Belote RL, McNeal AS, Tan TM, Yeh I, et al. 2020. The genomic landscapes of individual melanocytes from human skin. *Nature* **586:** 600–605. doi:10.1038/s41586-020-2785-8

Tankimanova M, Capper R, Letsolo BT, Rowson J, Jones RE, Britt-Compton B, Taylor AMR, Baird DM. 2012. Mre11 modulates the fidelity of fusion between short telomeres in human cells. *Nucleic Acids Res* **40:** 2518–2526. doi:10.1093/nar/gkr1117

Terradas M, Martín M, Tusell L, Genescà A. 2010. Genetic activities in micronuclei: is the DNA entrapped in micronuclei lost for the cell? *Mutat Res* **705:** 60–67. doi:10.1016/j.mrrev.2010.03.004

Thapa RJ, Ingram JP, Ragan KB, Nogusa S, Boyd DF, Benitez AA, Sridharan H, Kosoff R, Shubina M, Landsteiner VJ, et al. 2016. DAI senses influenza A virus genomic RNA and activates RIPK3-dependent cell death. *Cell Host Microbe* **20:** 674–681. doi:10.1016/j.chom.2016.09.014

Tomalka JA, Suthar MS, Diamond MS, Sekaly RP. 2022. Innate antiviral immunity: how prior exposures can guide future responses. *Trends Immunol* **43:** 696–705. doi:10.1016/j.it.2022.07.001

Tomáška Ľ, Cesare AJ, AlTurki TM, Griffith JD. 2020. Twenty years of t-loops: a case study for the importance of collaboration in molecular biology. *DNA Repair (Amst)* **94:** 102901. doi:10.1016/j.dnarep.2020.102901

Tsao JL, Yatabe Y, Salovaara R, Järvinen HJ, Mecklin JP, Aaltonen LA, Tavaré S, Shibata D. 2000. Genetic reconstruction of individual colorectal tumor histories. *Proc Natl Acad Sci* **97:** 1236–1241. doi:10.1073/pnas.97.3.1236

Tsuyama N, Miura M, Kitahira M, Ishibashi S, Ide T. 1991. SV40 T-antigen is required for maintenance of immortal growth in SV40-transformed human fibroblasts. *Cell Struct Funct* **16:** 55–62. doi:10.1247/csf.16.55

Umbreit NT, Zhang CZ, Lynch LD, Blaine LJ, Cheng AM, Tourdot R, Sun L, Almubarak HF, Judge K, Mitchell TJ, et al. 2020. Mechanisms generating cancer genome complexity from a single cell division error. *Science* **368:** eaba0712. doi:10.1126/science.aba0712

van Heek NT, Meeker AK, Kern SE, Yeo CJ, Lillemoe KD, Cameron JL, Offerhaus GJA, Hicks JL, Wilentz RE, Goggins MG, et al. 2002. Telomere shortening is nearly universal in pancreatic intraepithelial neoplasia. *Am J Pathol* **161:** 1541–1547. doi:10.1016/S0002-9440(10)64432-X

Van Ly D, Low RRJ, Frölich S, Bartolec TK, Kafer GR, Pickett HA, Gaus K, Cesare AJ. 2018. Telomere loop dynamics in chromosome end protection. *Mol Cell* **71:** 510–525.e6. doi:10.1016/j.molcel.2018.06.025

van Steensel B, de Lange T. 1997. Control of telomere length by the human telomeric protein TRF1. *Nature* **385:** 740–743. doi:10.1038/385740a0

van Steensel B, Smogorzewska A, de Lange T. 1998. TRF2 protects human telomeres from end-to-end fusions. *Cell* **92:** 401–413. doi:10.1016/S0092-8674(00)80932-0

Veldman T, Etheridge KT, Counter CM. 2004. Loss of hPot1 function leads to telomere instability and a cut-like phenotype. *Curr Biol* **14:** 2264–2270. doi:10.1016/j.cub.2004.12.031

Vendramin R, Litchfield K, Swanton C. 2021. Cancer evolution: Darwin and beyond. *EMBO J* **40:** e108389. doi:10.15252/embj.2021108389

Voss V, Senft C, Lang V, Ronellenfitsch MW, Steinbach JP, Seifert V, Kögel D. 2010. The pan-Bcl-2 inhibitor (−)-gossypol triggers autophagic cell death in malignant glioma. *Mol Cancer Res MCR* **8:** 1002–1016. doi:10.1158/1541-7786.MCR-09-0562

Vukovic B, Beheshti B, Park P, Lim G, Bayani J, Zielenska M, Squire JA. 2007. Correlating breakage-fusion-bridge events with the overall chromosomal instability and in vitro karyotype evolution in prostate cancer. *Cytogenet Genome Res* **116:** 1–11. doi:10.1159/000097411

Wang Z, Lieberman PM. 2016. The crosstalk of telomere dysfunction and inflammation through cell-free TERRA containing exosomes. *RNA Biol* **13:** 690–695. doi:10.1080/15476286.2016.1203503

Wang Z, Deng Z, Dahmane N, Tsai K, Wang P, Williams DR, Kossenkov AV, Showe LC, Zhang R, Huang Q, et al. 2015. Telomeric repeat-containing RNA (TERRA) constitutes a nucleoprotein component of extracellular inflammatory exosomes. *Proc Natl Acad Sci* **112:** E6293–E6300. doi:10.1073/pnas.1505962112

Wang R, Li H, Wu J, Cai ZY, Li B, Ni H, Qiu X, Chen H, Liu W, Yang ZH, et al. 2020. Gut stem cell necroptosis by genome instability triggers bowel inflammation. *Nature* **580:** 386–390. doi:10.1038/s41586-020-2127-x

Wang Y, Robinson PS, Coorens THH, Moore L, Lee-Six H, Noorani A, Sanders MA, Jung H, Katainen R, Heuschkel R, et al. 2023. APOBEC mutagenesis is a common process in normal human small intestine. *Nat Genet* **55:** 246–254. doi:10.1038/s41588-022-01296-5

Wei W, Sedivy JM. 1999. Differentiation between senescence (M1) and crisis (M2) in human fibroblast cultures. *Exp Cell Res* **253:** 519–522. doi:10.1006/excr.1999.4665

West AP, Khoury-Hanold W, Staron M, Tal MC, Pineda CM, Lang SM, Bestwick M, Duguay BA, Raimundo N, MacDuff DA, et al. 2015. Mitochondrial DNA stress primes the antiviral innate immune response. *Nature* **520:** 553–557. doi:10.1038/nature14156

Wijewardhane N, Dressler L, Ciccarelli FD. 2021. Normal somatic mutations in cancer transformation. *Cancer Cell* **39:** 125–129. doi:10.1016/j.ccell.2020.11.002

Williams MJ, Werner B, Heide T, Curtis C, Barnes CP, Sottoriva A, Graham TA. 2018. Quantification of subclonal selection in cancer from bulk sequencing data. *Nat Genet* **50:** 895–903. doi:10.1038/s41588-018-0128-6

Wright WE, Shay JW. 1992. The two-stage mechanism controlling cellular senescence and immortalization. *Exp Gerontol* **27:** 383–389. doi:10.1016/0531-5565(92)90069-C

Wright WE, Shay JW. 2001. Cellular senescence as a tumor-protection mechanism: the essential role of counting. *Curr Opin Genet Dev* **11:** 98–103. doi:10.1016/S0959-437X(00)00163-5

Wright WE, Pereira-Smith OM, Shay JW. 1989. Reversible cellular senescence: implications for immortalization of normal human diploid fibroblasts. *Mol Cell Biol* **9:** 3088–3092. doi:10.1128/mcb.9.7.3088-3092.1989

Wu L, Multani AS, He H, Cosme-Blanco W, Deng Y, Deng JM, Bachilo O, Pathak S, Tahara H, Bailey SM, et al. 2006. Pot1 deficiency initiates DNA damage checkpoint activation and aberrant homologous recombination at telomeres. *Cell* **126:** 49–62. doi:10.1016/j.cell.2006.05.037

Xu LG, Wang YY, Han KJ, Li LY, Zhai Z, Shu HB. 2005. VISA is an adapter protein required for virus-triggered IFN-beta signaling. *Mol Cell* **19:** 727–740. doi:10.1016/j.molcel.2005.08.014

Yang S, Wang X, Contino G, Liesa M, Sahin E, Ying H, Bause A, Li Y, Stommel JM, Dell'Antonio G, et al. 2011. Pancreatic cancers require autophagy for tumor growth. *Genes Dev* **25:** 717–729. doi:10.1101/gad.2016111

Ye JZS, Hockemeyer D, Krutchinsky AN, Loayza D, Hooper SM, Chait BT, de Lange T. 2004. POT1-interacting protein PIP1: a telomere length regulator that recruits POT1 to the TIN2/TRF1 complex. *Genes Dev* **18:** 1649–1654. doi:10.1101/gad.1215404

Yokoyama A, Kakiuchi N, Yoshizato T, Nannya Y, Suzuki H, Takeuchi Y, Shiozawa Y, Sato Y, Aoki K, Kim SK, et al. 2019. Age-related remodelling of oesophageal epithelia by mutated cancer drivers. *Nature* **565:** 312–317. doi:10.1038/s41586-018-0811-x

Zhang H, Zhang Y, Mathew R, Beaudoin B, Taylor RW, Mizushima N, White E, Jin S. 2007. Autophagy defect increases chromosome instability under metabolic stress. *Cancer Res* **67:** 4916–4916.

Zhang CZ, Spektor A, Cornils H, Francis JM, Jackson EK, Liu S, Meyerson M, Pellman D. 2015. Chromothripsis from DNA damage in micronuclei. *Nature* **522:** 179–184. doi:10.1038/nature14493

Zhang T, Yin C, Boyd DF, Quarato G, Ingram JP, Shubina M, Ragan KB, Ishizuka T, Crawford JC, Tummers B, et al. 2020. Influenza virus Z-RNAs induce ZBP1-mediated necroptosis. *Cell* **180:** 1115–1129.e13. doi:10.1016/j.cell.2020.02.050

# The Role of Microhomology-Mediated End Joining (MMEJ) at Dysfunctional Telomeres

**David Billing[1,2] and Agnel Sfeir[2]**

[1]Department of Radiation Oncology, Memorial Sloan Kettering Cancer Center, New York, New York 10065, USA

[2]Molecular Biology Program, Sloan Kettering Institute, Memorial Sloan Kettering Cancer Center, New York, New York 10065, USA

*Correspondence:* sfeira@mskcc.org

DNA double-strand break (DSB) repair pathways are crucial for maintaining genome stability and cell viability. However, these pathways can mistakenly recognize chromosome ends as DNA breaks, leading to adverse outcomes such as telomere fusions and malignant transformation. The shelterin complex protects telomeres from activation of DNA repair pathways by inhibiting nonhomologous end joining (NHEJ), homologous recombination (HR), and microhomology-mediated end joining (MMEJ). The focus of this paper is on MMEJ, an error-prone DSB repair pathway characterized by short insertions and deletions flanked by sequence homology. MMEJ is critical in mediating telomere fusions in cells lacking the shelterin complex and at critically short telomeres. Furthermore, studies suggest that MMEJ is the preferred pathway for repairing intratelomeric DSBs and facilitates escape from telomere crisis. Targeting MMEJ to prevent telomere fusions in hematologic malignancies is of potential therapeutic value.

DNA double-strand breaks (DSBs) are highly toxic DNA lesions that involve complete severance of the DNA double helix, compromising the continuity of chromosomes. DSBs can arise from exogenous sources of DNA damage, such as chemotherapeutic agents and ionizing radiation, and are generated endogenously during normal cellular function through replication fork collapse, transcription–replication conflicts, and telomere dysfunction (Jackson and Bartek 2009). Programmed DSBs are crucial for physiological processes, including V(D)J recombination, class switch recombination, and meiosis. Eukaryotic cells have evolved specialized repair pathways to fix DSBs with minimal changes to the coding sequence, including homologous recombination (HR), nonhomologous end joining (NHEJ), and microhomology-mediated end joining (MMEJ). These pathways differ in genetic makeup, the fidelity of the final repair product, and the cell-cycle phases during which they are preferentially deployed. Failure of these pathways or aberrant repair of DSBs can lead to a wide variety of aberrations—deletions, insertions, translocations, and duplications—capable of inducing cell death or promoting malignant transformation.

DSB repair pathways must also navigate the unique challenges imposed by the linear ends of eukaryotic chromosomes. Engaging DSB repair pathways inadvertently at chromosome ends can lead to adverse outcomes, including chromosome fusions, which can cause genomic instability and ultimately result in malignant transformation, cell-cycle arrest, or cell death. Therefore, cells must distinguish the natural ends of linear chromosomes from DSBs and prevent activation of DNA repair. The solution to this "telomere end-protection problem" is inherent to the protective shelterin complex that binds telomeres in mammalian cells and prevents activation of the DNA damage response (DDR) (de Lange 2009, 2018) (briefly summarized below). While telomere deprotection under physiologic conditions can have dire cellular consequences and ultimately manifest in human pathology, the directed manipulation of telomere ends has been used to decipher DNA repair pathways. Specifically, deleting subunits of the Shelterin complex helped identify core components of the DNA end-joining machinery. In this paper, we focus on mutagenic DNA repair by MMEJ and explore how investigating telomeres provides critical mechanistic insight into this previously poorly understood pathway. We also discuss the implication of MMEJ activity on genome stability and telomere dysfunction (see Fig. 1).

## CANONICAL DSB REPAIR BY NHEJ AND HR

Cells rely on multiple DSB repair pathways to preserve genome stability. Each pathway is characterized by its unique set of factors, the timing of the repair process during the cell cycle, and the degree of accuracy upon restoring the DNA. NHEJ is an error-prone DSB repair pathway that operates during all cell cycle stages but predominates during $G_1$ and in nondividing cells (see Fig. 1; Lieber 2008; Karanam et al. 2012; Stinson and Loparo 2021). NHEJ functions by joining broken DNA ends together with fast kinetics. It is triggered when DSBs are recognized by the Ku protein complex, composed of a heterodimer of Ku70 and Ku80 (Mimori and Hardin 1986; de Vries et al. 1989; Paillard and Strauss 1991). The

Ku complex recruits and activates the DNA-dependent protein kinase catalytic subunit (DNA-PKcs), which together forms the DNA-dependent protein kinase (DNA-PK) complex (West et al. 1998). DNA-PK phosphorylates numerous downstream factors that ultimately promote the ligation of the DSB ends through the DNA ligase IV-XRCC4-XLF complex (Chiruvella et al. 2013).

When direct ligation of DNA ends is impeded by incompatible DNA overhangs (i.e., non-blunt DNA ends) and chemical modifications, additional processing steps are required to complete DSB repair by NHEJ. This can be accomplished by activating the endonuclease Artemis by DNA-PKcs, allowing for the elimination of single-stranded DNA (ssDNA) overhangs, thus leading to deletions in the final repair product (Goodarzi et al. 2006; Chang et al. 2015; Pannunzio et al. 2018). Additionally, the Pol X family polymerases DNA Pol μ and DNA Pol λ engage with the Ku complex to introduce nucleotides in a template-independent manner, resulting in short insertions at the repair junctions (Bertocci et al. 2006; Bebenek et al. 2014; Moon et al. 2014; Pannunzio et al. 2018). A critical component during NHEJ is 53BP1, which promotes end joining by blocking DNA end resection (Bunting et al. 2010; Escribano-Díaz et al. 2013; Zimmermann et al. 2013), a crucial step for HR and MMEJ. NHEJ is inherently error-prone, susceptible to introducing small deletions, and can lead to translocations (Gillert et al. 1999; Richardson and Jasin 2000; Elliott et al. 2005).

Unlike NHEJ, HR is a largely error-free repair pathway, as it uses undamaged homologous chromosomes, typically the sister chromatid, as a template to copy sequences when repairing the break (see Fig. 1). HR is confined to the S and $G_2$ phases of the cell cycle when sister chromatids are available (Ranjha et al. 2018) and is initiated through the formation of 3′ ssDNA overhangs by the resection of 5′ DNA ends. This critical DNA end-resection step is initiated by the activity of the MRN complex (MRE11, RAD50, NBS1) and CtIP, with subsequent extensive resection by Exo1 and DNA2 (Sartori et al. 2007; Mimitou and Symington 2008; Zhu et al. 2008; Nimonkar et al. 2011). The breast cancer sus-

Cite this article as *Cold Spring Harb Perspect Biol* doi: 10.1101/cshperspect.a041687

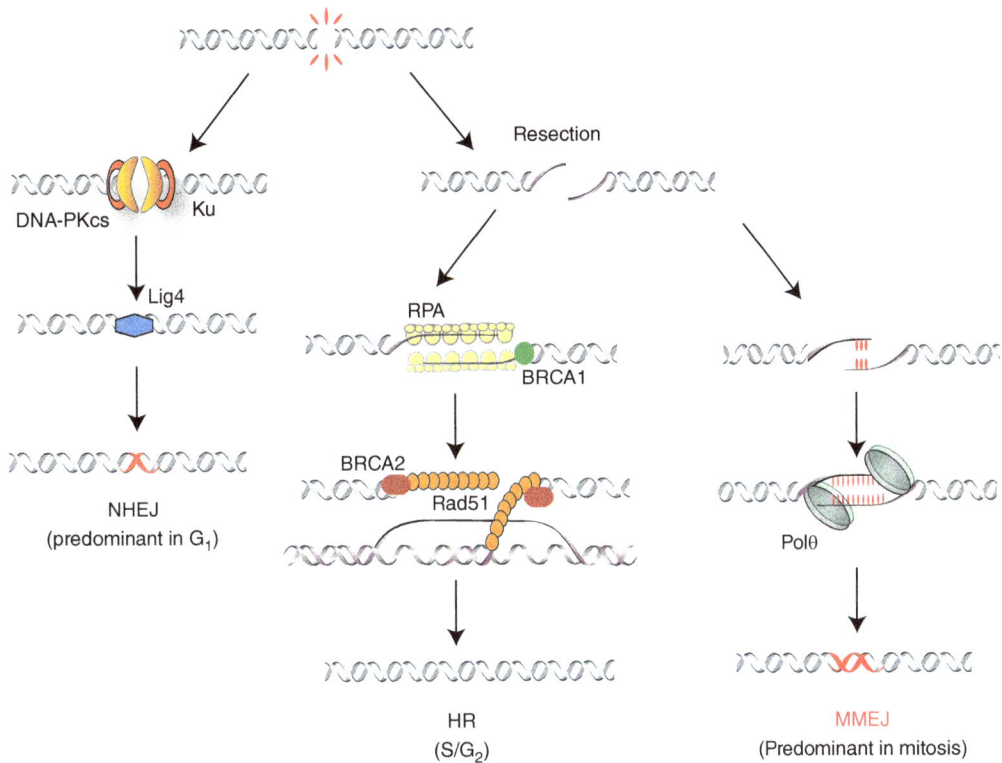

**Figure 1.** DNA double-strand break (DSB) repair pathways. DNA DSB repair pathways vary by cell-cycle preference, genetic makeup, and repair fidelity. (*Left*) Nonhomologous end joining (NHEJ) is the predominant repair pathway in the $G_1$ and $G_0$ phases. The Ku heterodimer binds to DSB ends and activates the DNA-dependent protein kinase catalytic subunit to form the DNA-dependent protein kinase complex, which promotes the direct ligation of DNA ends through the action of Ligase 4, resulting in deletions or translocations. Microhomology-mediated end joining (MMEJ, *right*) and homologous recombination (HR, *center*) share the initial step of 5′ to 3′ end resection, generating 3′ overhangs. In HR, BRCA1 promotes end resection, and replication protein A (RPA) binds the single-stranded DNA overhangs. BRCA2 facilitates the exchange of RPA for the Rad51 recombinase, which invades the donor molecule and performs a homology search, creating a D-loop. The undamaged donor is used as a template for error-free repair in the S and $G_2$ phases of the cell cycle. MMEJ functions primarily in mitosis and anneals short stretches of sequence homology exposed by end resection to anneal DSB ends. Polθ supports the annealing of microhomologies and uses the annealed ends as primers to repair the break, generating short insertions and deletions flanked by sequence homology.

ceptibility gene (*BRCA1*) forms a complex with MRN and CtIP, further promoting end resection (Yu et al. 1998; Zhong et al. 1999; Chen et al. 2008; Yun and Hiom 2009). The resection process, regulated by cyclin-dependent kinase (CDK) phosphorylation of CtIP, is thus temporally controlled to coincide with the S and $G_2$ cell-cycle phases (Huertas et al. 2008; Huertas and Jackson 2009).

Resected DNA is then coated with the ssDNA-binding protein replication protein A

(RPA), which is then replaced by RAD51 recombinase to form a nucleoprotein filament aided by the breast cancer susceptibility gene BRCA2 (Moynahan et al. 2001; Lisby et al. 2004). The RAD51-coated filament invades the undamaged template, typically on the sister chromatid, forming a displacement loop (D-loop) structure and enabling error-free repair by using a homologous sequence as a template. Depending on how the intermediate structure, known as a Holliday junction, is resolved, repair by HR can result in the exchange of genetic material between sister chromatids (crossover) or noncrossover repair through the displacement of the ssDNA (Jasin and Rothstein 2013). Notably, mutations in HR components such as BRCA1/2 can increase reliance on error-prone repair mechanisms. Such shifts in repair pathway preference can contribute to genomic instability and potentially promote tumorigenesis by accumulating DNA damage and disrupting cellular regulatory pathways for growth and differentiation (Roy et al. 2011).

## MICROHOMOLOGY-MEDIATED END JOINING, AN INHERENTLY MUTAGENIC REPAIR PATHWAY

Microhomology-mediated end joining (MMEJ), also known as alternative end joining (Alt-EJ) and θ-mediated end joining (TMEJ), is an error-prone DSB pathway that becomes especially critical in cells that cannot perform HR and NHEJ (see Fig. 1). As its name implies, MMEJ uses stretches of microhomology flanking the DSB sites. The annealing of microhomology guides the repair of DSBs, often resulting in the hallmark insertion or deletion mutations (indels). In most organisms, MMEJ is primarily catalyzed by DNA polymerase θ (POLθ) encoded by the gene POLQ (Chan et al. 2010; Ceccaldi et al. 2015; Mateos-Gomez et al. 2015), which is responsible for the characteristic insertions. The MMEJ pathway was first identified as a residual DSB repair activity in yeast cells lacking NHEJ (Boulton and Jackson 1996) and was initially perceived as an "alternative" or "backup" repair pathway operating when NHEJ and HR were disrupted. However, emerging studies have suggested that MMEJ functions in settings where NHEJ and HR are functional. Furthermore, MMEJ is the repair pathway of choice during mitosis when both HR and NHEJ are suppressed (Brambati et al. 2023; Gelot et al. 2023).

Mechanistically, MMEJ and HR share the initial step of DNA end resection to generate 3′ ssDNA overhangs (see Fig. 2). The MRN complex and CtIP orchestrate resection in mammalian cells, followed by coating of the ssDNA with RPA (Truong et al. 2013). POLθ, which is comprised of a helicase domain and a polymerase domain connected by an unstructured linker domain, uses the ATPase activity to displace RPA and promote the annealing of flanking microhomologies (Newman et al. 2015; Mateos-Gomez et al. 2017; Zerio et al. 2024). Using annealed DSB ends as primers, the polymerase domain of Polθ synthesizes short stretches of nascent DNA leading to insertions (Zahn et al. 2015). Polθ lacks a proofreading activity typical of other A-family DNA polymerases, increasing the likelihood of introducing errors when repairing damaged DNA (Seki et al. 2004). Synthesis of longer stretches of DNA requires hand-off to a more processive DNA polymerase such as Polδ (Stroik et al. 2023). The repair reaction is completed by removing any resulting 3′ flaps, which is proposed to be catalyzed by the APE2 endonuclease (Fleury et al. 2023) and the Polδ exonuclease domain (Stroik et al. 2023), followed by DNA ligation primarily using the DNA ligase 3–XRCC1 complex (Wang et al. 2005; Simsek and Jasin 2011).

The defining feature of MMEJ is a mutational signature characterized by short insertions and deletions flanked by microhomology. Microhomology length can range from 1 or 2 nt in C. elegans and human cells (Koole et al. 2014; Kent et al. 2015), to 6–20 nt in budding yeast (Lee and Lee 2007). Importantly, it has been estimated that in ~90% of DSBs, microhomologies of at least three base pairs are likely found within 15 bp of the break ends. This suggests that most DSBs could theoretically be repaired using MMEJ (Carvajal-Garcia et al. 2020; Ramsden et al. 2022). Often, the homologous sequences are not perfectly aligned at the break ends, creating 3′ flaps. Trimming of the flaps by APE2 endonuclease and the Polδ exonuclease domain (Fleury et al. 2023; Stroik et al. 2023) culminates in the characteristic short deletions associated with MMEJ.

Templated insertions are less common than small deletions but are a distinguishing feature of MMEJ activity (Schimmel et al. 2019). MMEJ drives insertions through at least two independent mechanisms by using sequence homology surrounding the break site. First, using minimal base pairing of microhomology across the DSB,

Polθ initiates DNA synthesis by aligning microhomologies at the DSB. However, the enzyme might disengage prematurely, leaving the ssDNA gap partially filled. The elongated DNA strand could then realign using sequence simi-larity found elsewhere on the same strand, leading Polθ to start a new round of elongation and ssDNA gap filling, resulting in a templated insertion (van Schendel et al. 2016; Khodaverdian et al. 2017). The second pathway involves a "snap-back" reaction, wherein one DSB end temporarily binds to adjacent microhomology in *cis* (Kent et al. 2016). This sequence then serves as a template for Polθ-driven elongation, effectively duplicating a segment of the *cis* strand. After dissociation of annealing in *cis*, the extended self-copied DNA end can be used for microhomology search in *trans*, culminating in a templated insertion (Kent et al. 2016).

## MMEJ IS THE PRIMARY REPAIR PATHWAY IN MITOSIS

Despite its mutagenicity, MMEJ is critical to preventing the catastrophic consequences of DSBs when the canonical DSB repair pathways are unavailable. Specifically, MMEJ is essential when repair by NHEJ and HR is inactivated. Accordingly, loss of MMEJ compromises the viability of cells lacking BRCA1, BRCA2, and other HR genes, and POLQ is synthetic lethal with LIG4 and other NHEJ factors (Ceccaldi et al. 2015; Mateos-Gomez et al. 2015; Feng

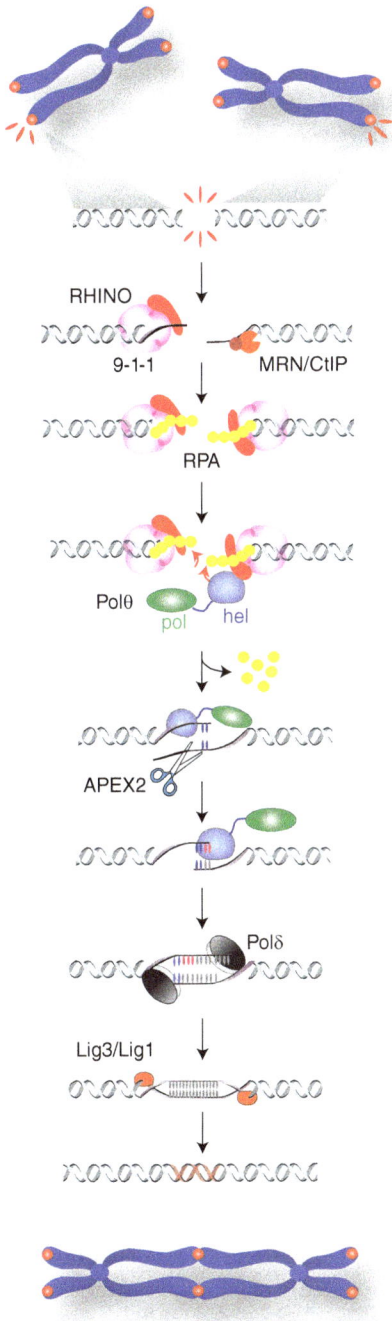

**Figure 2.** Mechanism of microhomology-mediated end joining (MMEJ). Repair of double-strand breaks in mitosis or fusion of critically shortened telomeres is catalyzed by MMEJ. The MRN complex and CtIP catalyze 5′ to 3′ end resection, exposing 3′ single-stranded DNA overhangs in the S and $G_2$ phases of the cell cycle. The 9-1-1 complex binds to DNA ends and is carried into mitosis, where RHINO binds and facilitates the recruitment of Polθ. Once at the DNA break, the helicase domain of Polθ removes replication protein A (RPA) and mediates the annealing of microhomology across the break ends. Any 3′ flaps generated by annealing of sequence not located directly at the break ends are trimmed by the nuclease apurinic/apyrimidic endodeoxyribonuclease 2 (APEX2). The polymerase domain of Polθ synthesizes short stretches of nascent DNA followed by a handoff to the more processive Polδ for the synthesis of longer sequences. The repair process is completed by ligating the DNA ends via Lig3/Lig1.

et al. 2019). This synthetic lethal relationship is currently being exploited in the clinic, as small molecule inhibitors of Polθ are being tested clinically for treating cancers with mutations in HR factors (Zatreanu et al. 2021; Zhou et al. 2021; Bubenik et al. 2022).

Beyond being a backup for NHEJ and HR, MMEJ is the primary pathway in mitosis. HR and NHEJ are suppressed during mitosis through phosphorylation-mediated inactivation of 53BP1 and BRCA2 by CDK1 and PLK1 (Esashi et al. 2005; Lee et al. 2014; Orthwein et al. 2014). Underreplicated DNA or unrepaired DNA damage enters mitosis, where it is repaired by MMEJ (Deng et al. 2019; Heijink et al. 2022; Brambati et al. 2023; Gelot et al. 2023). MMEJ is targeted to mitosis through two different mechanisms. First, Polθ foci accumulation on mitotic chromosomes depends on the mitotic kinase PLK1, facilitating its interaction with TOPBP1 (Gelot et al. 2023). A separate mechanism for targeting MMEJ activity to mitosis was identified through CRISPR-based dropout screens and is mediated by the Rad9–Rad1–Hus1 (9-1-1) complex and their interacting partner RHINO (Hussmann et al. 2021; Brambati et al. 2023). RHINO, which is expressed only in mitosis, is phosphorylated by PLK1, thereby facilitating its interaction with Polθ. Critically, inhibition of Polθ during mitosis, but not during S phase, leads to synthetic lethality in HR-deficient cells (Brambati et al. 2023; Gelot et al. 2023), suggesting that the activity of Polθ in mitosis is central for resolving the unrepaired damage.

The finding that MMEJ acts primarily in mitosis has shifted the perception from a backup pathway to a pathway of last resort—repairing unresolved breaks in mitosis to avoid cell division with a damaged genome, even at the cost of increased mutation.

## THE TELOMERE END-PROTECTION PROBLEM

Work by Hermann Muller and Barbara McClintock revealed that chromosomes tend to fuse at sites of DNA breaks. Their findings suggested that chromosome ends require specialized mechanisms to prevent activation of DNA damage signaling and repair machinery and avoid the detrimental effects of DNA breaks. This is known as the telomere end-protection problem. It was later appreciated that this protection is provided by a specialized, ubiquitously present complex known as shelterin (see Fig. 3). Human shelterin is a complex of six proteins: TRF1, TRF2, POT1, RAP1, TIN2, and TPP1 (de Lange 2005). TRF1 and TRF2 are homodimers that bind to double-stranded TTAGGG-containing telomeric DNA (Bianchi et al. 1999). The telomere 3′ overhang is recognized and bound by POT1 (Baumann and Cech 2001; Lei et al. 2004) and anchored to the rest of the shelterin complex by TPP1 (previously called TINT1 [Houghtaling et al. 2004], PTOP [Liu et al. 2004], and PIP1 [Ye et al. 2004b]). TIN2 is the linchpin that anchors TRF1, TRF2, and TPP1 (Liu et al. 2004; Ye et al. 2004a). Completing the complex, Rap1 interacts with and is recruited to telomeres via TRF2 (Li et al. 2000).

TRF2 blocks NHEJ at chromosome ends as expression of a dominant negative allele (van Steensel et al. 1998), Cre-mediated deletion of TRF2 in MEFS (Celli and de Lange 2005), and CRISPR-Cas9 deletion (Kim et al. 2017), unleash NHEJ activity and lead to chromosome end-to-end fusions. Tethering RAP1 to telomeres in cells depleted of TRF2 rescues telomere fusion, implicating the conserved shelterin component in blocking end joining. However, deletion of RAP1 did not lead to end-to-end fusion (Martinez et al. 2010; Sfeir et al. 2010; Chen et al. 2011). It has been shown that TRF2 blocks NHEJ by facilitating the invasion of the 3′ ssDNA overhang back into the preceding double-stranded telomeric DNA, forming a T-loop (Griffith et al. 1999; Doksani et al. 2013; Sarek et al. 2019). The T-loop conceals the telomere terminus from the DSB repair machinery, particularly NHEJ, and inhibits DNA damage signaling mediated by the master regulator kinases ataxia telangiectasia mutated (ATM) and ataxia telangiectasia and Rad3 related (ATR). Activation of the ATM kinase is prevented by TRF2 through T-loop formation mediated by its dimerization domain. TRF2 also blocks NHEJ downstream from ATM through direct inhibition of the E3 ubiquitin ligase RNF168 via the inhibitor of the DNA damage response (iDDR) region (Karlseder et al. 2004; Denchi and de Lange 2007; Okamoto et al. 2013). Although a T-loop could potentially form a structure resembling a Holliday junction, which in turn might be improperly repaired by HR processes, this risk is mitigated by TRF2, preventing inappropriate intertelomeric recombination and ensuring the integrity of the telomeres (Wang et al. 2004). TRF1 and POT1 prevent ATR kinase activity (Denchi and de Lange 2007; Sfeir et al. 2009). Together, the shelterin complex prevents activation of the DNA damage signaling master regulator kinases ATM and ATR, preventing inappropriate utilization of DSB repair pathways at telomeres.

## SHELTERIN DEPLETION UNLEASHES MMEJ ACTIVITY

The shelterin complex also blocks MMEJ, albeit in a redundant fashion. Deletion of the multiple shelterin subunits results in the derepression of MMEJ at telomeres (Rai et al. 2010; Sfeir and de Lange 2012). However, the deprotection of chromosomes ends upon shelterin deletions does not fully unleash MMEJ. This is likely due to the action of the Ku heterodimer, as deletion of Ku results in MMEJ activity at telomeres (Sfeir and de Lange 2012). Ku inhibits MMEJ activation by competing with PARP1 for DNA-end binding and preventing activation of end resection (Wang et al. 2006; Mimitou and Symington 2010; Cheng et al. 2011).

Studies using mouse embryonic fibroblasts (MEFs) with conditional alleles for various shelterin subunits have enhanced our understanding of how shelterin regulates DDR and repair mechanisms at the ends of chromosomes. The absence of TRF1 and TRF2 leads to the destabilization of the shelterin complex, culminating in the complete loss of all shelterin subunits from telomeres. The loss of shelterin triggers DNA damage signaling through ATM and ATR, leading to chromosomes fusing end-to-end primarily using NHEJ (Sfeir and de Lange 2012). However, disabling NHEJ components did not prevent telomere fusions. Specifically, upon deletion of the NHEJ factor Ku80 in shelterin-free cells, 65% of telomeres were engaged in end-to-

end fusions of telomeres, implicating alternative DNA repair pathways. Inhibition of MMEJ factors PARP1 and Lig3 led to a significant reduction in the end-to-end telomere fusions, suggesting that MMEJ promotes telomere fusions at deprotected telomeres when NHEJ is compromised (Sfeir and de Lange 2012).

## MMEJ PROMOTES CHROMOSOME FUSIONS IN CELLS WITH CRITICALLY SHORT TELOMERES

Deprotected telomeres are potent substrates for DSB repair pathways and result from successive cellular division, where telomeric DNA is lost due to the incomplete replication of linear DNA ends as well as nucleolytic end processing of the leading strand, a problem known as the "telomere end-replication problem" (Watson 1972; Olovnikov 1973; Takai et al. 2024). When a subset of telomeres becomes too short to recruit the shelterin complex, DNA damage signaling is activated, leading to cellular senescence. Inhibition of cell cycle checkpoints through p53 mutation leads to the bypass of senescence and further cellular division, resulting in telomere attrition, dysfunction, chromosome fusions, anaphase bridges, breakage fusion breakage cycles, and overall genome instability during a stage known as telomere crisis (Kaul et al. 2012; Hayashi et al. 2015; Lazzerini-Denchi and Sfeir 2016). Additionally, telomere crisis activates innate immune signaling, which cells must evade to achieve neoplastic transformation (Nassour et al. 2019, 2023). The ability to escape telomere crisis either by reactivating telomerase or activating alternative lengthening of telomeres (ALT) is a crucial step in malignant transformation (Hendrickson and Baird 2015).

Whereas NHEJ was predominantly responsible for telomere fusions in TRF2-deficient cells (van Steensel et al. 1998; Celli and de Lange 2005; Dimitrova et al. 2008; Rai et al. 2010), early studies suggested that chromosome end-to-end fusions at critically shortened telomeres were independent of NHEJ. A study performed in mice carrying an inactivating mutation in telomerase leading to progressive telomere shortening demonstrated that telomere fusions could

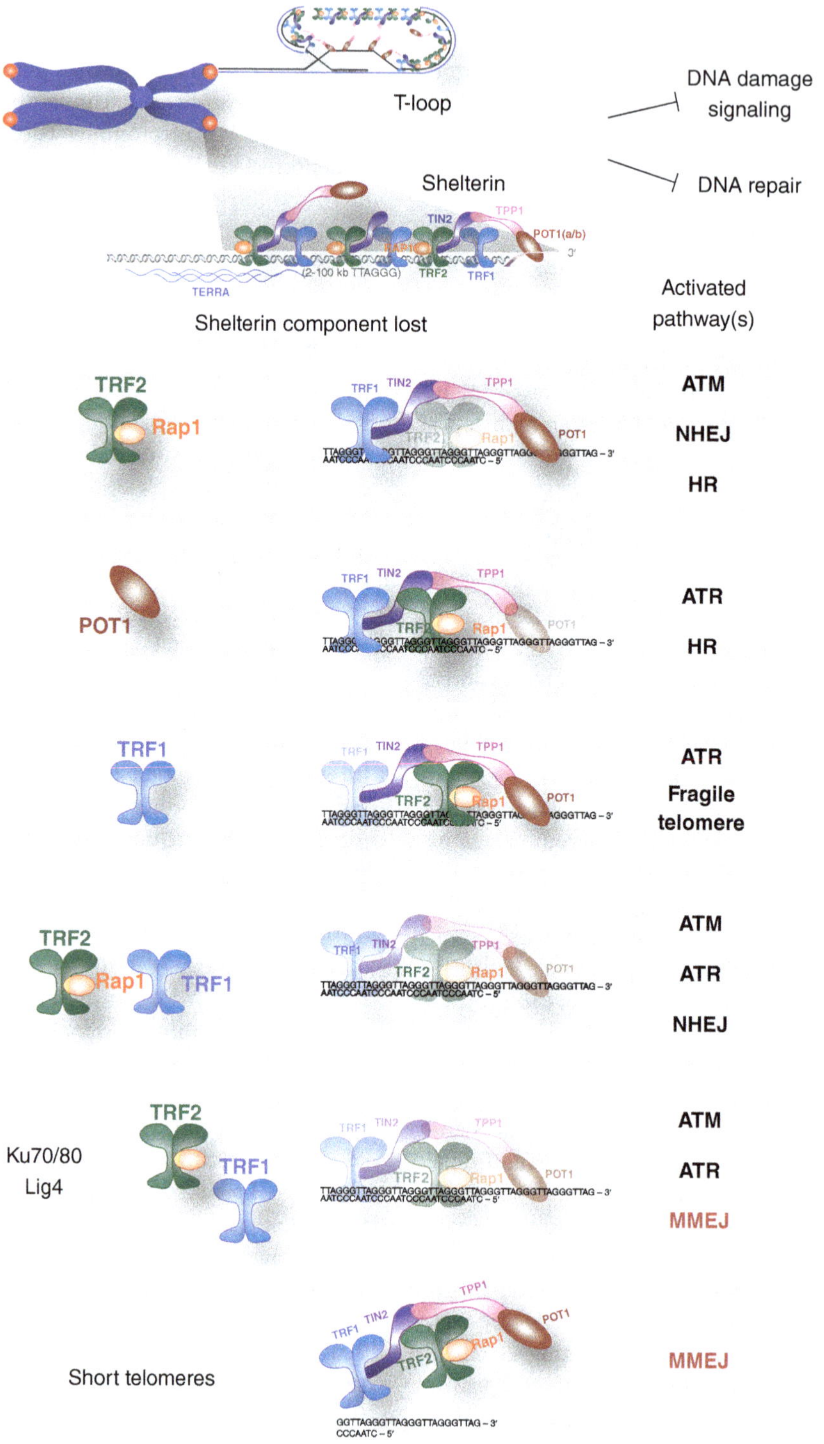

**Figure 3.** (*See following page for legend.*)

occur even in the absence of the key NHEJ factors DNA-PKcs and LIG4 once critical shortening of telomeres occurred (Maser et al. 2007). Subsequently, the mechanistic differences between telomere fusions at telomeres deprotected by shelterin deletions versus critical shortening were defined by a series of experiments using mouse intestinal epithelial cells as a model (Rai et al. 2010). Mouse intestinal epithelial cells divide rapidly, and inactivation of telomerase in these cells causes progressive shortening of telomeres until crisis (Rai et al. 2010). In contrast to previous observations in TRF2-depleted MEFs (Celli and de Lange 2005), deletion of NHEJ factors in gastrointestinal (GI) cells of a telomerase null ($mTerc^{-/-}$) mouse has no significant effect on the number of telomere end-to-end fusions, demonstrating that NHEJ is dispensable in this context.

Further evidence for a role for MMEJ in the fusion of critically shortened telomeres came from studies in human fetal fibroblasts that bypass telomere shortening-induced senescence by inactivation of p53. Specifically, a single-molecule PCR method that measures allele-specific telomere length, termed single-telomere length analysis (STELA), enables the precise measurement of telomere lengths (Capper et al. 2007). It was later amended to detect and follow the progression of telomere fusions over time. In the human fetal fibroblast model, STELA revealed telomere shortening with successive cell divisions until they reached a threshold where they could no longer be detected due to increased fusion events (Capper et al. 2007). By analyzing

the sequence context at the fusion junctions, particularly between the chromosomes XpYp and 17p, comparisons could be made with established mutational hallmarks of various DNA repair processes. A significant finding was that most fusions entailed deletions extending into adjacent regions of the telomeres (Capper et al. 2007). Notably, in 89% of the cases examined, a sequence homology averaging 3.1 nt was present at the fusion points, which aligns with the typical characteristics of MMEJ. This observation was subsequently confirmed across several chromosomal ends, revealing a prevalent MMEJ pattern at eroded telomeres (Letsolo et al. 2010).

Conclusive evidence linking MMEJ to the fusion of shortened telomeres was based on genetic experiments targeting critical components of the DNA repair pathways. HCT116 colon cancer cells expressing a dominant negative *hTERT* allele (hTERT-DN) were targeted to induce MMEJ deficiency (*LIG3* knockout) and NHEJ deficiency (*LIG4* knockout) (Jones et al. 2014). Control cells underwent a telomere crisis, evidenced by telomere fusions and genomic instability, and eventually escaped the crisis by reactivating hTERT. Cells lacking either *LIG3* or *LIG4* did not show telomere and chromosomal fusions during the crisis; however, only $LIG4^{-/-}$ cells overcame the crisis. These results implicate MMEJ in circumventing the telomere crisis. Analyses of the fusion events revealed that whereas interchromosomal fusions were more common in the presence of NHEJ, intrachromosomal fusions were more

Figure 3. Shelterin complex interactions with DNA repair factors. (Top) The shelterin complex protects DNA ends from DNA repair factors and prevents activation of DNA damage signaling. The human shelterin complex comprises TRF1, TRF2, RAP1, TIN2, TPP1, and POT1. The shelterin complex facilitates the formation of a T-loop. In this specialized structure, the 3′ single-stranded DNA (ssDNA) overhang at the end of the telomere invades the preceding double-stranded telomeric DNA to conceal the DNA ends from DNA repair machinery. (Bottom) Loss of shelterin components results in the activation of DNA repair pathways. Loss of TRF2 (and its partner RAP1) results in ataxia telangiectasia mutated (ATM) activation. Additionally, TRF2 prevents HR factors from recognizing the T-loop structure and nonhomologous end joining (NHEJ) activation by inhibiting RNF168. POT1 and TRF1 prevent activation of ATR. Deletion of TRF2 and TRF1 together leads to activation of the ATM and ATR kinases, leading to telomere fusions through the NHEJ pathway. Further deletion of NHEJ pathway components in addition to TRF1/2 leads to telomere fusions through the microhomology-mediated end joining (MMEJ) pathway. MMEJ mediates the fusion of critically shortened telomeres.

frequent when MMEJ was operative. These findings suggest that while NHEJ and MMEJ trigger chromosome fusions, MMEJ provides a distinct selective advantage critical for escaping crisis. Although the exact nature of the selective advantage that MMEJ provides is unclear, it is speculated that sister chromatid fusions may be less toxic than interchromosomal fusions and more likely to lead to the amplification of genes that promote crisis escape (Hendrickson and Baird 2015).

## MMEJ FACILITATES TELOMERE FUSIONS IN HUMAN CANCER MODELS

To explore whether the patterns of chromosomal fusions linked to MMEJ in cells with critically short telomeres observed in vitro apply in vivo, clinical samples from patients with chronic lymphocytic leukemia (CLL) were examined (Lin et al. 2010). CLL is an ideal system to study chromosomal fusions, as the disease is known to harbor a variety of fusion events that can drive malignancy. Additionally, CLL follows a variable clinical course; in some cases, remaining indolent without the need for treatment for many years, and in others, it progresses rapidly and compromises survival, with the presence or absence of different fusion events predicting patient outcomes. Similar to what had been observed in the human fibroblasts (Capper et al. 2007), B cells from CLL patients displayed progressive telomere erosion extending into the subtelomeric region with flanking microhomology detected by STELA and single-molecule fusion PCR, highlighting a role for MMEJ in promoting chromosomal fusions. Significantly, the degree of telomere dysfunction was found to be of prognostic value, even in patients with early CLL, as patients with short telomeres display worse overall survival (Lin et al. 2014). Similarly, telomere dysfunction with chromosomal fusions harboring an MMEJ signature has been observed in colon polyps and colorectal cancer (Roger et al. 2013). Together, these data suggest that telomere dysfunction and subsequent MMEJ-mediated chromosomal fusions can be critical in the malignant transformation of human cancers.

## DEPROTECTED TELOMERES AS A SYSTEM TO UNRAVEL MMEJ FACTORS

Engineering telomere deprotection has proven to be a powerful strategy for uncovering novel MMEJ factors. The challenge of studying MMEJ lies in its typically low activity at DSBs, which makes MMEJ events challenging to detect. As a result, we lack of robust and quantitative readout for MMEJ. Identifying MMEJ signatures at repair junctions often necessitates complex computational analysis of next-generation sequencing data. It is further complicated by the infrequency of MMEJ events, which NHEJ repair can overshadow. Therefore, creating conditions that either favor MMEJ or amplify its activity is crucial for enhancing its detection. Engineered telomere deprotection through depletion of shelterin and NHEJ suppression via Ku80 deletion overcame these challenges and provided a discernable and quantifiable indicator of MMEJ through telomere fusions (Sfeir and de Lange 2012).

Many key MMEJ factors have been discovered or experimentally validated using this telomere fusion assay in TRF1/2 and Ku80-deficient cells. Particularly, Polθ was identified as the critical factor of MMEJ, as its depletion reduced telomere fusions (Mateos-Gomez et al. 2015). Polθ depletion did not impact the frequency of NHEJ-mediated fusions at artificially deprotected telomeres. These results underscore the importance of finding experimental contexts where MMEJ is the preferred repair pathway. This insight into Polθ function in MMEJ at deprotected telomeres has also been extended to nontelomeric loci through other methods, including CRISPR/Cas9-induced breaks combined with mutational signature analysis, fluorescent reporter systems, and translocation fusion point sequence analysis (Simsek et al. 2011; Ceccaldi et al. 2015; Mateos-Gomez et al. 2015; Wyatt et al. 2016; Schimmel et al. 2023; Wimberger et al. 2023). Together, these data show the role of POLQ as the critical driver of MMEJ-mediated telomere fusions and, more broadly, the capability of using artificially deprotected telomeres to identify and study MMEJ factors.

In addition to Polθ, deprotected telomeres were a robust system to identify and validate other MMEJ factors, thereby significantly enhancing our mechanistic understanding of this mutagenic repair. The nuclease apurinic/apyrimidic endo-deoxyribonuclease 2 (APEX2, encoding the protein APE2), which functions in MMEJ by trimming 3′ flaps created during repair (see Fig. 2) was validated as an MMEJ factor using the TRF2/Ku80-deficient telomere fusion assay. Specifically, APE2 was identified as a putative MMEJ factor in a CRISPR/Cas9 dropout screen in HR-deficient cell lines (BRCA1 and PALB2 mutant) (Mengwasser et al. 2019; Álvarez-Quilón et al. 2020; Fleury et al. 2023). APE2 was directly linked to MMEJ by depleting the gene in cells lacking TRF2 and Ku80, and demonstrating that MMEJ-dependent telomere fusions were abrogated compared to cells with wild-type APE2. Additionally, knockout of Polθ in APE2-deficient cells had no additive effect on telomere fusions in TRF2/Ku80-deficient cells, indicating that Polθ and APE2 were epistatic.

More recently, RHNO1 and the 9-1-1 complex, responsible for targeting MMEJ activity to mitosis (see Fig. 2), were identified as putative MMEJ factors using CRISPR/Cas9-based genetic screens (Hussmann et al. 2021; Brambati et al. 2023). Depletion of RHNO1 or 9-1-1 complex factors potently decreased telomere fusions in TRF1/2 Ku80-deficient cells, validating them as bona fide MMEJ factors (Brambati et al. 2023). In conclusion, exploiting the telomere end-protection problem through the depletion of shelterin has been instrumental in uncovering the mechanistic basis of MMEJ factors and has led to the identification of critical MMEJ components, including POLθ, APE2, RHNO1, and 9-1-1.

## MMEJ IS THE PREFERRED DSB REPAIR PATHWAY FOR INTRATELOMERIC DSBs

The shelterin complex at telomeres suppresses pathways that detect and repair DNA breaks to protect telomere ends. Nonetheless, if a DSB were to occur within a telomeric region, it raises the question of whether essential repair processes would be hindered. Furthermore, should repair occur at telomeric sites, might the mechanism differ from that at nontelomeric locations?

Evidence suggests that DSB repair at telomeric loci might indeed be inhibited, as indicated by the persistence of the marker of DNA damage, γH2AX, at telomeres after ionizing radiation, long after resolution at other genomic sites (Fumagalli et al. 2012). Repair of induced DSBs within subtelomeric regions also appears less efficient than in the non-subtelomeric areas (Miller et al. 2011).

A critical tool to study DSB repair, specifically at telomeres, is the TRF1-FOK1 fusion protein (Tang et al. 2013). By fusing the non-sequence-specific restriction endonuclease FokI to the shelterin complex component TRF1, DSBs can be specifically induced at telomeric regions as evidenced by colocalization of DNA damage markers γH2AX, BRCA1, and 53BP1 at telomere repeats. DNA damage at telomeres causes the accumulation of telomere dysfunction-induced foci (TIF) containing DNA repair factors indicating telomere-specific damage (Takai et al. 2003). The presence and persistence of TIF can be used as a marker of DNA damage at telomeric lesions.

When used to induce telomeric damage in ALT⁺ cells that lacked telomerase, TRF-FOK1 triggered interchromosomal recombination and telomere clustering, suggesting a preference for an HR-based mechanism for DSB repair in this subset of cells with altered telomeric function (Cho et al. 2014). However, the same treatment did not induce telomere clustering and recombination in telomerase-positive cells, suggesting different DSB resolution mechanisms. Furthermore, the accumulation of TIF upon induced telomeric DSBs in telomerase-positive cells was dependent of ATM signaling (Doksani and de Lange 2016). These results indicated that TRF2 binding is insufficient to inhibit ATM activation at a DSB within the telomere. Intriguingly, depletion of LIG4 does not impact the levels of DNA damage signaling, suggesting that NHEJ is dispensable for the repair of DSBs at telomeres. Instead, depletion or inhibition of the MMEJ factors PARP1 and LIG3 significantly delays the resolution of TIF after break induction (Doksani and de Lange 2016). Taken together, these results suggest that the repair of telomere-internal DSBs is driven by MMEJ and not prevented by the presence of the shelterin complex. Microhomology in the context of telomeric repeats likely favors MMEJ in a region of the genome where small insertions or deletions can be readily tolerated. In support of this hypothesis, loss of Polθ results in DNA repeat tract length instability (Liddiard et al. 2022). On the other hand, NHEJ may be more likely to form interchromosomal fusions, leading to more catastrophic outcomes for the cell (Maciejowski et al. 2015; Doksani 2019).

## CONCLUDING REMARKS

DNA repair pathways are critical for maintaining the stability of our genetic material. However, these processes are suppressed at chromosome ends through the action of the shelterin complex. Deprotection of telomeres, through either genetic manipulation of shelterin or as a result of telomere attrition, leads to activation of DNA damage signaling and telomere fusion by NHEJ and MMEJ. MMEJ appears to be the primary driver of fusion events at critically shortened telomeres, both in animal models and in the context of human cancer cells, thus implicating this pathway in fostering genome instability in the early stages of cancer. Engineered deprotection of telomeres combined with NHEJ suppression to facilitate MMEJ-mediated telomere fusion advanced our mechanistic understanding of MMEJ. The recent development of inhibitors targeting Polθ (Sfeir 2024; Sfeir et al. 2024) raises the intriguing possibility of suppressing MMEJ-mediated chromosomal rearrangements in patients with indolent or premalignant lesions, thereby improving patient outcomes.

## ACKNOWLEDGMENTS

A.S. is a cofounder, consultant, and shareholder in Repare Therapeutics.

## REFERENCES

Álvarez-Quilón A, Wojtaszek JL, Mathieu MC, Patel T, Appel CD, Hustedt N, Rossi SE, Wallace BD, Setiaputra D, Adam S, et al. 2020. Endogenous DNA 3′ blocks are vulnerabilities for BRCA1 and BRCA2 deficiency and are reversed by the APE2 nuclease. *Mol Cell* 78: 1152–1165. e8. doi:10.1016/j.molcel.2020.05.021

Baumann P, Cech TR. 2001. Pot1, the putative telomere end-binding protein in fission yeast and humans. *Science* 292: 1171–1175. doi:10.1126/science.1060036

Bebenek K, Pedersen LC, Kunkel TA. 2014. Structure-function studies of DNA polymerase λ. *Biochemistry* 53: 2781–2792. doi:10.1021/bi4017236

Bertocci B, De Smet A, Weill JC, Reynaud C-A. 2006. Nonoverlapping functions of DNA polymerases mu, lambda, and terminal deoxynucleotidyltransferase during immunoglobulin V(D)J recombination in vivo. *Immunity* 25: 31–41. doi:10.1016/j.immuni.2006.04.013

Bianchi A, Stansel RM, Fairall L, Griffith JD, Rhodes D, de Lange T. 1999. TRF1 binds a bipartite telomeric site with extreme spatial flexibility. *EMBO J* 18: 5735–5744. doi:10.1093/emboj/18.20.5735

Boulton SJ, Jackson SP. 1996. Saccharomyces cerevisiae Ku70 potentiates illegitimate DNA double-strand break repair and serves as a barrier to error-prone DNA repair pathways. *EMBO J* 15: 5093–5103. doi:10.1002/j.1460-2075.1996.tb00890.x

Brambati A, Sacco O, Porcella S, Heyza J, Kareh M, Schmidt JC, Sfeir A. 2023. RHINO directs MMEJ to repair DNA breaks in mitosis. *Science* 381: 653–660. doi:10.1126/science.adh3694

Bubenik M, Mader P, Mochirian P, Vallée F, Clark J, Truchon JF, Perryman AL, Pau V, Kurinov I, Zahn KE, et al. 2022. Identification of RP-6685, an orally bioavailable compound that inhibits the DNA polymerase activity of Polθ. *J Med Chem* 65: 13198–13215. doi:10.1021/acs.jmedchem.2c00998

Bunting SF, Callén E, Wong N, Chen HT, Polato F, Gunn A, Bothmer A, Feldhahn N, Fernandez-Capetillo O, Cao L, et al. 2010. 53BP1 inhibits homologous recombination in Brca1-deficient cells by blocking resection of DNA breaks. *Cell* 141: 243–254. doi:10.1016/j.cell.2010.03.012

Capper R, Britt-Compton B, Tankimanova M, Rowson J, Letsolo B, Man S, Haughton M, Baird DM. 2007. The nature of telomere fusion and a definition of the critical telomere length in human cells. *Genes Dev* 21: 2495–2508. doi:10.1101/gad.439107

Carvajal-Garcia J, Cho JE, Carvajal-Garcia P, Feng W, Wood RD, Sekelsky J, Gupta GP, Roberts SA, Ramsden DA. 2020. Mechanistic basis for microhomology identification and genome scarring by polymerase θ. *Proc Natl Acad Sci* 117: 8476–8485. doi:10.1073/pnas.1921791117

Ceccaldi R, Liu JC, Amunugama R, Hajdu I, Primack B, Petalcorin MIR, O'Connor KW, Konstantinopoulos PA, Elledge SJ, Boulton SJ, et al. 2015. Homologous-recombination-deficient tumours are dependent on Polθ-mediated repair. *Nature* 518: 258–262. doi:10.1038/nature14184

Celli GB, de Lange T. 2005. DNA processing is not required for ATM-mediated telomere damage response after TRF2 deletion. *Nat Cell Biol* 7: 712–718. doi:10.1038/ncb1275

Chan SH, Yu AM, McVey M. 2010. Dual roles for DNA Polymerase θ in alternative end-joining repair of double-strand breaks in *Drosophila*. *PLOS Genet* 6: e1001005. doi:10.1371/journal.pgen.1001005

Chang HHY, Watanabe G, Lieber MR. 2015. Unifying the DNA end-processing roles of the Artemis nuclease: Ku-dependent Artemis resection at blunt DNA ends. *J Biol Chem* **290:** 24036–24050. doi:10.1074/jbc.M115.680900

Chen L, Nievera CJ, Lee AYL, Wu X. 2008. Cell cycle-dependent complex formation of BRCA1.CtIP.MRN is important for DNA double-strand break repair. *J Biol Chem* **283:** 7713–7720. doi:10.1074/jbc.M710245200

Chen Y, Rai R, Zhou ZR, Kanoh J, Ribeyre C, Yang Y, Zheng H, Damay P, Wang F, Tsujii H, et al. 2011. A conserved motif within RAP1 has diversified roles in telomere protection and regulation in different organisms. *Nat Struct Mol Biol* **18:** 213–221. doi:10.1038/nsmb.1974

Cheng Q, Barboule N, Frit P, Gomez D, Bombarde O, Couderc B, Ren GS, Salles B, Calsou P. 2011. Ku counteracts mobilization of PARP1 and MRN in chromatin damaged with DNA double-strand breaks. *Nucleic Acids Res* **39:** 9605–9619. doi:10.1093/nar/gkr656

Chiruvella KK, Liang Z, Wilson TE. 2013. Repair of double-strand breaks by end joining. *Cold Spring Harb Perspect Biol* **5:** a012757. doi:10.1101/cshperspect.a012757

Cho NW, Dilley RL, Lampson MA, Greenberg RA. 2014. Interchromosomal homology searches drive directional ALT telomere movement and synapsis. *Cell* **159:** 108–121. doi:10.1016/j.cell.2014.08.030

de Lange T. 2005. Shelterin: the protein complex that shapes and safeguards human telomeres. *Genes Dev* **19:** 2100–2110. doi:10.1101/gad.1346005

de Lange T. 2009. How telomeres solve the end-protection problem. *Science* **326:** 948–952. doi:10.1126/science.1170633

de Lange T. 2018. Shelterin-mediated telomere protection. *Annu Rev Genet* **52:** 223–247. doi:10.1146/annurev-genet-032918-021921

Denchi EL, de Lange T. 2007. Protection of telomeres through independent control of ATM and ATR by TRF2 and POT1. *Nature* **448:** 1068–1071. doi:10.1038/nature06065

Deng L, Wu RA, Sonneville R, Kochenova OV, Labib K, Pellman D, Walter JC. 2019. Mitotic CDK promotes replisome disassembly, fork breakage, and complex DNA rearrangements. *Mol Cell* **73:** 915–929.e6. doi:10.1016/j.molcel.2018.12.021

de Vries E, van Driel W, Bergsma WG, Arnberg AC, van der Vliet PC. 1989. Hela nuclear protein recognizing DNA termini and translocating on DNA forming a regular DNA–multimeric protein complex. *J Mol Biol* **208:** 65–78. doi:10.1016/0022-2836(89)90088-0

Dimitrova N, Chen YCM, Spector DL, de Lange T. 2008. 53BP1 promotes non-homologous end joining of telomeres by increasing chromatin mobility. *Nature* **456:** 524–528. doi:10.1038/nature07433

Doksani Y. 2019. The response to DNA damage at telomeric repeats and its consequences for telomere function. *Genes (Basel)* **10:** 318. doi:10.3390/genes10040318

Doksani Y, de Lange T. 2016. Telomere-internal double-strand breaks are repaired by homologous recombination and PARP1/Lig3-dependent end-joining. *Cell Rep* **17:** 1646–1656. doi:10.1016/j.celrep.2016.10.008

Doksani Y, Wu JY, de Lange T, Zhuang X. 2013. Super-resolution fluorescence imaging of telomeres reveals TRF2-dependent T-loop formation. *Cell* **155:** 345–356. doi:10.1016/j.cell.2013.09.048

Elliott B, Richardson C, Jasin M. 2005. Chromosomal translocation mechanisms at intronic alu elements in mammalian cells. *Mol Cell* **17:** 885–894. doi:10.1016/j.molcel.2005.02.028

Esashi F, Christ N, Gannon J, Liu Y, Hunt T, Jasin M, West SC. 2005. CDK-dependent phosphorylation of BRCA2 as a regulatory mechanism for recombinational repair. *Nature* **434:** 598–604. doi:10.1038/nature03404

Escribano-Díaz C, Orthwein A, Fradet-Turcotte A, Xing M, Young JTF, Tkáč J, Cook MA, Rosebrock AP, Munro M, Canny MD, et al. 2013. A cell cycle-dependent regulatory circuit composed of 53BP1-RIF1 and BRCA1-CtIP controls DNA repair pathway choice. *Mol Cell* **49:** 872–883. doi:10.1016/j.molcel.2013.01.001

Feng W, Simpson DA, Carvajal-Garcia J, Price BA, Kumar RJ, Mose LE, Wood RD, Rashid N, Purvis JE, Parker JS, et al. 2019. Genetic determinants of cellular addiction to DNA polymerase θ. *Nat Commun* **10:** 4286. doi:10.1038/s41467-019-12234-1

Fleury H, MacEachern MK, Stiefel CM, Anand R, Sempeck C, Nebenfuehr B, Maurer-Alcalá K, Ball K, Proctor B, Belan O, et al. 2023. The APE2 nuclease is essential for DNA double-strand break repair by microhomology-mediated end joining. *Mol Cell* **83:** 1429–1445.e8. doi:10.1016/j.molcel.2023.03.017

Fumagalli M, Rossiello F, Clerici M, Barozzi S, Cittaro D, Kaplunov JM, Bucci G, Dobreva M, Matti V, Beausejour CM, et al. 2012. Telomeric DNA damage is irreparable and causes persistent DNA-damage-response activation. *Nat Cell Biol* **14:** 355–365. doi:10.1038/ncb2466

Gelot C, Kovacs MT, Miron S, Mylne E, Haan A, Boeffard-Dosierre L, Ghouil R, Popova T, Dingli F, Loew D, et al. 2023. Polθ is phosphorylated by PLK1 to repair double-strand breaks in mitosis. *Nature* **621:** 415–422. doi:10.1038/s41586-023-06506-6

Gillert E, Leis T, Repp R, Reichel M, Hösch A, Breitenlohner I, Angermüller S, Borkhardt A, Harbott J, Lampert F, et al. 1999. A DNA damage repair mechanism is involved in the origin of chromosomal translocations t(4;11) in primary leukemic cells. *Oncogene* **18:** 4663–4671. doi:10.1038/sj.onc.1202842

Goodarzi AA, Yu Y, Riballo E, Douglas P, Walker SA, Ye R, Härer C, Marchetti C, Morrice N, Jeggo PA, et al. 2006. DNA-PK autophosphorylation facilitates Artemis endonuclease activity. *EMBO J* **25:** 3880–3889. doi:10.1038/sj.emboj.7601255

Griffith JD, Comeau L, Rosenfield S, Stansel RM, Bianchi A, Moss H, de Lange T. 1999. Mammalian telomeres end in a large duplex loop. *Cell* **97:** 503–514. doi:10.1016/S0092-8674(00)80760-6

Hayashi MT, Cesare AJ, Rivera T, Karlseder J. 2015. Cell death during crisis is mediated by mitotic telomere deprotection. *Nature* **522:** 492–496. doi:10.1038/nature14513

Heijink AM, Stok C, Porubsky D, Manolika EM, de Kanter JK, Kok YP, Everts M, de Boer HR, Audrey A, Bakker FJ, et al. 2022. Sister chromatid exchanges induced by perturbed replication can form independently of BRCA1, BRCA2 and RAD51. *Nat Commun* **13:** 6722. doi:10.1038/s41467-022-34519-8

Hendrickson EA, Baird DM. 2015. Alternative end joining, clonal evolution, and escape from a telomere-driven crisis. *Mol Cell Oncol* **2:** e975623. doi:10.4161/23723556.2014.975623

Houghtaling BR, Cuttonaro L, Chang W, Smith S. 2004. A dynamic molecular link between the telomere length regulator TRF1 and the chromosome end protector TRF2. *Curr Biol* **14:** 1621–1631. doi:10.1016/j.cub.2004.08.052

Huertas P, Jackson SP. 2009. Human CtIP mediates cell cycle control of DNA end resection and double strand break repair. *J Biol Chem* **284:** 9558–9565. doi:10.1074/jbc.M808906200

Huertas P, Cortés-Ledesma F, Sartori AA, Aguilera A, Jackson SP. 2008. CDK targets Sae2 to control DNA-end resection and homologous recombination. *Nature* **455:** 689–692. doi:10.1038/nature07215

Hussmann JA, Ling J, Ravisankar P, Yan J, Cirincione A, Xu A, Simpson D, Yang D, Bothmer A, Cotta-Ramusino C, et al. 2021. Mapping the genetic landscape of DNA double-strand break repair. *Cell* **184:** 5653–5669.e25. doi:10.1016/j.cell.2021.10.002

Jackson SP, Bartek J. 2009. The DNA-damage response in human biology and disease. *Nature* **461:** 1071–1078. doi:10.1038/nature08467

Jasin M, Rothstein R. 2013. Repair of strand breaks by homologous recombination. *Cold Spring Harb Perspect Biol* **5:** a012740. doi:10.1101/cshperspect.a012740

Jones RE, Oh S, Grimstead JW, Zimbric J, Roger L, Heppel NH, Ashelford KE, Liddiard K, Hendrickson EA, Baird DM. 2014. Escape from telomere-driven crisis is DNA ligase III dependent. *Cell Rep* **8:** 1063–1076. doi:10.1016/j.celrep.2014.07.007

Karanam K, Kafri R, Loewer A, Lahav G. 2012. Quantitative live cell imaging reveals a gradual shift between DNA repair mechanisms and a maximal use of HR in mid S phase. *Mol Cell* **47:** 320–329. doi:10.1016/j.molcel.2012.05.052

Karlseder J, Hoke K, Mirzoeva OK, Bakkenist C, Kastan MB, Petrini JHJ, de Lange T. 2004. The telomeric protein TRF2 binds the ATM kinase and can inhibit the ATM-dependent DNA damage response. *PLoS Biol* **2:** E240. doi:10.1371/journal.pbio.0020240

Kaul Z, Cesare AJ, Huschtscha LI, Neumann AA, Reddel RR. 2012. Five dysfunctional telomeres predict onset of senescence in human cells. *EMBO Rep* **13:** 52–59. doi:10.1038/embor.2011.227

Kent T, Chandramouly G, McDevitt SM, Ozdemir AY, Pomerantz RT. 2015. Mechanism of microhomology-mediated end-joining promoted by human DNA polymerase θ. *Nat Struct Mol Biol* **22:** 230–237. doi:10.1038/nsmb.2961

Kent T, Mateos-Gomez PA, Sfeir A, Pomerantz RT. 2016. Polymerase θ is a robust terminal transferase that oscillates between three different mechanisms during end-joining. *eLife* **5:** e13740. doi:10.7554/eLife.13740

Khodaverdian VY, Hanscom T, Yu AM, Yu TL, Mak V, Brown AJ, Roberts SA, McVey M. 2017. Secondary structure forming sequences drive SD-MMEJ repair of DNA double-strand breaks. *Nucleic Acids Res* **45:** 12848–12861. doi:10.1093/nar/gkx1056

Kim H, Li F, He Q, Deng T, Xu J, Jin F, Coarfa C, Putluri N, Liu D, Songyang Z. 2017. Systematic analysis of human telomeric dysfunction using inducible telosome/shelterin CRISPR/Cas9 knockout cells. *Cell Discov* **3:** 17034. doi:10.1038/celldisc.2017.34

Koole W, van Schendel R, Karambelas AE, van Heteren JT, Okihara KL, Tijsterman M. 2014. A polymerase θ-dependent repair pathway suppresses extensive genomic instability at endogenous G4 DNA sites. *Nat Commun* **5:** 3216. doi:10.1038/ncomms4216

Lazzerini-Denchi E, Sfeir A. 2016. Stop pulling my strings—what telomeres taught us about the DNA damage response. *Nat Rev Mol Cell Biol* **17:** 364–378. doi:10.1038/nrm.2016.43

Lee K, Lee SE. 2007. *Saccharomyces cerevisiae* Sae2- and Tel1-dependent single-strand DNA formation at DNA break promotes microhomology-mediated end joining. *Genetics* **176:** 2003–2014. doi:10.1534/genetics.107.076539

Lee DH, Acharya SS, Kwon M, Drane P, Guan Y, Adelmant G, Kalev P, Shah J, Pellman D, Marto JA, et al. 2014. Dephosphorylation enables the recruitment of 53BP1 to double-strand DNA breaks. *Mol Cell* **54:** 512–525. doi:10.1016/j.molcel.2014.03.020

Lei M, Podell ER, Cech TR. 2004. Structure of human POT1 bound to telomeric single-stranded DNA provides a model for chromosome end-protection. *Nat Struct Mol Biol* **11:** 1223–1229. doi:10.1038/nsmb867

Letsolo BT, Rowson J, Baird DM. 2010. Fusion of short telomeres in human cells is characterized by extensive deletion and microhomology, and can result in complex rearrangements. *Nucleic Acids Res* **38:** 1841–1852. doi:10.1093/nar/gkp1183

Li B, Oestreich S, de Lange T. 2000. Identification of human Rap1: implications for telomere evolution. *Cell* **101:** 471–483. doi:10.1016/S0092-8674(00)80858-2

Liddiard K, Aston-Evans AN, Cleal K, Hendrickson EA, Baird DM. 2022. POLQ suppresses genome instability and alterations in DNA repeat tract lengths. *NAR Cancer* **4:** zcac020. doi:10.1093/narcan/zcac020

Lieber MR. 2008. The mechanism of human nonhomologous DNA end joining. *J Biol Chem* **283:** 1–5. doi:10.1074/jbc.R700039200

Lin TT, Letsolo BT, Jones RE, Rowson J, Pratt G, Hewamana S, Fegan C, Pepper C, Baird DM. 2010. Telomere dysfunction and fusion during the progression of chronic lymphocytic leukemia: evidence for a telomere crisis. *Blood* **116:** 1899–1907. doi:10.1182/blood-2010-02-272104

Lin TT, Norris K, Heppel NH, Pratt G, Allan JM, Allsup DJ, Bailey J, Cawkwell L, Hills R, Grimstead JW, et al. 2014. Telomere dysfunction accurately predicts clinical outcome in chronic lymphocytic leukaemia, even in patients with early stage disease. *Br J Haematol* **167:** 214–223. doi:10.1111/bjh.13023

Lisby M, Barlow JH, Burgess RC, Rothstein R. 2004. Choreography of the DNA damage response: spatiotemporal relationships among checkpoint and repair proteins. *Cell* **118:** 699–713. doi:10.1016/j.cell.2004.08.015

Liu D, Safari A, O'Connor MS, Chan DW, Laegeler A, Qin J, Songyang Z. 2004. PTOP interacts with POT1 and regulates its localization to telomeres. *Nat Cell Biol* **6:** 673–680. doi:10.1038/ncb1142

Maciejowski J, Li Y, Bosco N, Campbell PJ, de Lange T. 2015. Chromothripsis and kataegis induced by telomere crisis. *Cell* **163:** 1641–1654. doi:10.1016/j.cell.2015.11.054

Martinez P, Thanasoula M, Carlos AR, Gómez-López G, Tejera AM, Schoeftner S, Dominguez O, Pisano DG, Tarsounas M, Blasco MA. 2010. Mammalian Rap1 controls telomere function and gene expression through binding to telomeric and extratelomeric sites. *Nat Cell Biol* **12:** 768–780. doi:10.1038/ncb2081

Maser RS, Wong KK, Sahin E, Xia H, Naylor M, Hedberg HM, Artandi SE, DePinho RA. 2007. DNA-dependent protein kinase catalytic subunit is not required for dysfunctional telomere fusion and checkpoint response in the telomerase-deficient mouse. *Mol Cell Biol* **27:** 2253–2265. doi:10.1128/MCB.01354-06

Mateos-Gomez PA, Gong F, Nair N, Miller KM, Lazzerini-Denchi E, Sfeir A. 2015. Mammalian polymerase θ promotes alternative NHEJ and suppresses recombination. *Nature* **518:** 254–257. doi:10.1038/nature14157

Mateos-Gomez PA, Kent T, Deng SK, McDevitt S, Kashkina E, Hoang TM, Pomerantz RT, Sfeir A. 2017. The helicase domain of Polθ counteracts RPA to promote alt-NHEJ. *Nat Struct Mol Biol* **24:** 1116–1123. doi:10.1038/nsmb.3494

Mengwasser KE, Adeyemi RO, Leng Y, Choi MY, Clairmont C, D'Andrea AD, Elledge SJ. 2019. Genetic screens reveal FEN1 and APEX2 as BRCA2 synthetic lethal targets. *Mol Cell* **73:** 885–899.e6. doi:10.1016/j.molcel.2018.12.008

Miller D, Reynolds GE, Mejia R, Stark JM, Murnane JP. 2011. Subtelomeric regions in mammalian cells are deficient in DNA double-strand break repair. *DNA Repair (Amst)* **10:** 536–544. doi:10.1016/j.dnarep.2011.03.001

Mimitou EP, Symington LS. 2008. Sae2, Exo1 and Sgs1 collaborate in DNA double-strand break processing. *Nature* **455:** 770–774. doi:10.1038/nature07312

Mimitou EP, Symington LS. 2010. Ku prevents Exo1 and Sgs1-dependent resection of DNA ends in the absence of a functional MRX complex or Sae2. *EMBO J* **29:** 3358–3369. doi:10.1038/emboj.2010.193

Mimori T, Hardin JA. 1986. Mechanism of interaction between Ku protein and DNA. *J Biol Chem* **261:** 10375–10379. doi:10.1016/S0021-9258(18)67534-9

Moon AF, Pryor JM, Ramsden DA, Kunkel TA, Bebenek K, Pedersen LC. 2014. Sustained active site rigidity during synthesis by human DNA polymerase µ. *Nat Struct Mol Biol* **21:** 253–260. doi:10.1038/nsmb.2766

Moynahan ME, Pierce AJ, Jasin M. 2001. BRCA2 is required for homology-directed repair of chromosomal breaks. *Mol Cell* **7:** 263–272. doi:10.1016/S1097-2765(01)00174-5

Nassour J, Radford R, Correia A, Fusté JM, Schoell B, Jauch A, Shaw RJ, Karlseder J. 2019. Autophagic cell death restricts chromosomal instability during replicative crisis. *Nature* **565:** 659–663. doi:10.1038/s41586-019-0885-0

Nassour J, Aguiar LG, Correia A, Schmidt TT, Mainz L, Przetocka S, Haggblom C, Tadepalle N, Williams A, Shokhirev MN, et al. 2023. Telomere-to-mitochondria signalling by ZBP1 mediates replicative crisis. *Nature* **614:** 767–773. doi:10.1038/s41586-023-05710-8

Newman JA, Cooper CDO, Aitkenhead H, Gileadi O. 2015. Structure of the helicase domain of DNA polymerase θ reveals a possible role in the microhomology-mediated end-joining pathway. *Struct Lond Engl 1993* **23:** 2319–2330. doi:10.1016/j.str.2015.10.014

Nimonkar AV, Genschel J, Kinoshita E, Polaczek P, Campbell JL, Wyman C, Modrich P, Kowalczykowski SC. 2011. BLM–DNA2–RPA–MRN and EXO1–BLM–RPA–MRN constitute two DNA end resection machineries for human DNA break repair. *Genes Dev* **25:** 350–362. doi:10.1101/gad.2003811

Okamoto K, Bartocci C, Ouzounov I, Diedrich JK, Yates JR III, Denchi EL. 2013. A two-step mechanism for TRF2-mediated chromosome-end protection. *Nature* **494:** 502–505. doi:10.1038/nature11873

Olovnikov AM. 1973. A theory of marginotomy: the incomplete copying of template margin in enzymic synthesis of polynucleotides and biological significance of the phenomenon. *J Theor Biol* **41:** 181–190. doi:10.1016/0022-5193(73)90198-7

Orthwein A, Fradet-Turcotte A, Noordermeer SM, Canny MD, Brun CM, Strecker J, Escribano-Diaz C, Durocher D. 2014. Mitosis inhibits DNA double-strand break repair to guard against telomere fusions. *Science* **344:** 189–193. doi:10.1126/science.1248024

Paillard S, Strauss F. 1991. Analysis of the mechanism of interaction of simian Ku protein with DNA. *Nucleic Acids Res* **19:** 5619–5624. doi:10.1093/nar/19.20.5619

Pannunzio NR, Watanabe G, Lieber MR. 2018. Nonhomologous DNA end-joining for repair of DNA double-strand breaks. *J Biol Chem* **293:** 10512–10523. doi:10.1074/jbc.TM117.000374

Rai R, Zheng H, He H, Luo Y, Multani A, Carpenter PB, Chang S. 2010. The function of classical and alternative non-homologous end-joining pathways in the fusion of dysfunctional telomeres. *EMBO J* **29:** 2598–2610. doi:10.1038/emboj.2010.142

Ramsden DA, Carvajal-Garcia J, Gupta GP. 2022. Mechanism, cellular functions and cancer roles of polymerase-θ-mediated DNA end joining. *Nat Rev Mol Cell Biol* **23:** 125–140. doi:10.1038/s41580-021-00405-2

Ranjha L, Howard SM, Cejka P. 2018. Main steps in DNA double-strand break repair: an introduction to homologous recombination and related processes. *Chromosoma* **127:** 187–214. doi:10.1007/s00412-017-0658-1

Richardson C, Jasin M. 2000. Frequent chromosomal translocations induced by DNA double-strand breaks. *Nature* **405:** 697–700. doi:10.1038/35015097

Roger L, Jones RE, Heppel NH, Williams GT, Sampson JR, Baird DM. 2013. Extensive telomere erosion in the initiation of colorectal adenomas and its association with chromosomal instability. *JNCI J Natl Cancer Inst* **105:** 1202–1211. doi:10.1093/jnci/djt191

Roy R, Chun J, Powell SN. 2011. BRCA1 and BRCA2: different roles in a common pathway of genome protection. *Nat Rev Cancer* **12:** 68–78. doi:10.1038/nrc3181

Sarek G, Kotsantis P, Ruis P, Van Ly D, Margalef P, Borel V, Zheng XF, Flynn HR, Snijders AP, Chowdhury D, et al. 2019. CDK phosphorylation of TRF2 controls T-loop dynamics during the cell cycle. *Nature* **575:** 523–527. doi:10.1038/s41586-019-1744-8

Sartori AA, Lukas C, Coates J, Mistrik M, Fu S, Bartek J, Baer R, Lukas J, Jackson SP. 2007. Human CtIP promotes DNA end resection. *Nature* **450:** 509–514. doi:10.1038/nature06337

Schimmel J, van Schendel R, den Dunnen JT, Tijsterman M. 2019. Templated insertions: a smoking gun for polymerase θ-mediated end joining. *Trends Genet TIG* **35:** 632–644. doi:10.1016/j.tig.2019.06.001

Schimmel J, Muñoz-Subirana N, Kool H, Van Schendel R, Van Der Vlies S, Kamp JA, De Vrij FMS, Kushner SA, Smith GCM, Boulton SJ, et al. 2023. Modulating mutational outcomes and improving precise gene editing at CRISPR-Cas9-induced breaks by chemical inhibition of end-joining pathways. *Cell Rep* **42:** 112019. doi:10.1016/j.celrep.2023.112019

Seki M, Masutani C, Yang LW, Schuffert A, Iwai S, Bahar I, Wood RD. 2004. High-efficiency bypass of DNA damage by human DNA polymerase Q. *EMBO J* **23:** 4484–4494. doi:10.1038/sj.emboj.7600424

Sfeir A. 2024. Obscure DNA sequences unveil a new cancer target. *Nat Struct Mol Biol* **31:** 1311–1312. doi:10.1038/s41594-024-01347-x

Sfeir A, de Lange T. 2012. Removal of shelterin reveals the telomere end-protection problem. *Science* **336:** 593–597. doi:10.1126/science.1218498

Sfeir A, Kosiyatrakul ST, Hockemeyer D, MacRae SL, Karlseder J, Schildkraut CL, de Lange T. 2009. Mammalian telomeres resemble fragile sites and require TRF1 for efficient replication. *Cell* **138:** 90–103. doi:10.1016/j.cell.2009.06.021

Sfeir A, Kabir S, van Overbeek M, Celli GB, de Lange T. 2010. Loss of Rap1 induces telomere recombination in the absence of NHEJ or a DNA damage signal. *Science* **327:** 1657–1661. doi:10.1126/science.1185100

Sfeir A, Tijsterman M, McVey M. 2024. Microhomology-mediated end joining chronicles: tracing the evolutionary footprints of genome protection. *Annu Rev Cell Dev Biol* doi:10.1146/annurev-cellbio-111822-014426

Simsek D, Jasin M. 2011. DNA ligase III. *Cell Cycle* **10:** 3636–3644. doi:10.4161/cc.10.21.18094

Simsek D, Brunet E, Wong SYW, Katyal S, Gao Y, McKinnon PJ, Lou J, Zhang L, Li J, Rebar EJ, et al. 2011. DNA ligase III promotes alternative nonhomologous end-joining during chromosomal translocation formation. *PLoS Genet* **7:** e1002080. doi:10.1371/journal.pgen.1002080

Stinson BM, Loparo JJ. 2021. Repair of DNA double-strand breaks by the nonhomologous end joining pathway. *Annu Rev Biochem* **90:** 137–164. doi:10.1146/annurev-biochem-080320-110356

Stroik S, Carvajal-Garcia J, Gupta D, Edwards A, Luthman A, Wyatt DW, Dannenberg RL, Feng W, Kunkel TA, Gupta GP, et al. 2023. Stepwise requirements for polymerases δ and θ in θ-mediated end joining. *Nature* **623:** 836–841. doi:10.1038/s41586-023-06729-7

Takai H, Smogorzewska A, de Lange T. 2003. DNA damage foci at dysfunctional telomeres. *Curr Biol CB* **13:** 1549–1556. doi:10.1016/S0960-9822(03)00542-6

Takai H, Aria V, Borges P, Yeeles JTP, de Lange T. 2024. CST–polymerase α-primase solves a second telomere end-replication problem. *Nature* **627:** 664–670. doi:10.1038/s41586-024-07137-1

Tang J, Cho NW, Cui G, Manion EM, Shanbhag NM, Botuyan MV, Mer G, Greenberg RA. 2013. Acetylation limits 53BP1 association with damaged chromatin to promote homologous recombination. *Nat Struct Mol Biol* **20:** 317–325. doi:10.1038/nsmb.2499

Truong LN, Li Y, Shi LZ, Hwang PYH, He J, Wang H, Razavian N, Berns MW, Wu X. 2013. Microhomology-mediated end joining and homologous recombination share the initial end resection step to repair DNA double-strand breaks in mammalian cells. *Proc Natl Acad Sci* **110:** 7720–7725. doi:10.1073/pnas.1213431110

van Schendel R, van Heteren J, Welten R, Tijsterman M. 2016. Genomic scars generated by polymerase θ reveal the versatile mechanism of alternative end-joining. *PLoS Genet* **12:** e1006368. doi:10.1371/journal.pgen.1006368

van Steensel B, Smogorzewska A, de Lange T. 1998. TRF2 protects human telomeres from end-to-end fusions. *Cell* **92:** 401–413. doi:10.1016/S0092-8674(00)80932-0

Wang RC, Smogorzewska A, de Lange T. 2004. Homologous recombination generates T-loop-sized deletions at human telomeres. *Cell* **119:** 355–368. doi:10.1016/j.cell.2004.10.011

Wang H, Rosidi B, Perrault R, Wang M, Zhang L, Windhofer F, Iliakis G. 2005. DNA ligase III as a candidate component of backup pathways of nonhomologous end joining. *Cancer Res* **65:** 4020–4030. doi:10.1158/0008-5472.CAN-04-3055

Wang M, Wu W, Wu W, Rosidi B, Zhang L, Wang H, Iliakis G. 2006. PARP-1 and Ku compete for repair of DNA double strand breaks by distinct NHEJ pathways. *Nucleic Acids Res* **34:** 6170–6182. doi:10.1093/nar/gkl840

Watson JD. 1972. Origin of concatemeric T7DNA. *Nature New Biol* **239:** 197–201. doi:10.1038/newbio239197a0

West RB, Yaneva M, Lieber MR. 1998. Productive and nonproductive complexes of Ku and DNA-dependent protein kinase at DNA termini. *Mol Cell Biol* **18:** 5908–5920. doi:10.1128/MCB.18.10.5908

Wimberger S, Akrap N, Firth M, Brengdahl J, Engberg S, Schwinn MK, Slater MR, Lundin A, Hsieh PP, Li S, et al. 2023. Simultaneous inhibition of DNA-PK and Polθ improves integration efficiency and precision of genome editing. *Nat Commun* **14:** 4761. doi:10.1038/s41467-023-40344-4

Wyatt DW, Feng W, Conlin MP, Yousefzadeh MJ, Roberts SA, Mieczkowski P, Wood RD, Gupta GP, Ramsden DA. 2016. Essential roles for polymerase θ-mediated end joining in the repair of chromosome breaks. *Mol Cell* **63:** 662–673. doi:10.1016/j.molcel.2016.06.020

Ye JZS, Donigian JR, van Overbeek M, Loayza D, Luo Y, Krutchinsky AN, Chait BT, de Lange T. 2004a. TIN2 binds TRF1 and TRF2 simultaneously and stabilizes the TRF2 complex on telomeres. *J Biol Chem* **279:** 47264–47271. doi:10.1074/jbc.M409047200

Ye JZS, Hockemeyer D, Krutchinsky AN, Loayza D, Hooper SM, Chait BT, de Lange T. 2004b. POT1-interacting protein PIP1: a telomere length regulator that recruits POT1 to the TIN2/TRF1 complex. *Genes Dev* **18:** 1649–1654. doi:10.1101/gad.1215404

Yu X, Wu LC, Bowcock AM, Aronheim A, Baer R. 1998. The C-terminal (BRCT) domains of BRCA1 interact in vivo with CtIP, a protein implicated in the CtBP pathway of transcriptional repression. *J Biol Chem* **273:** 25388–25392. doi:10.1074/jbc.273.39.25388

Yun MH, Hiom K. 2009. CtIP-BRCA1 modulates the choice of DNA double-strand break repair pathway throughout the cell cycle. *Nature* **459:** 460–463. doi:10.1038/nature 07955

Zahn KE, Averill AM, Aller P, Wood RD, Doublié S. 2015. Human DNA polymerase θ grasps the primer terminus to mediate DNA repair. *Nat Struct Mol Biol* **22:** 304–311. doi:10.1038/nsmb.2993

Zatreanu D, Robinson HMR, Alkhatib O, Boursier M, Finch H, Geo L, Grande D, Grinkevich V, Heald RA, Langdon S, et al. 2021. Polθ inhibitors elicit BRCA-gene synthetic lethality and target PARP inhibitor resistance. *Nat Commun* **12:** 3636. doi:10.1038/s41467-021-23463-8

Zerio CJ, Bai Y, Sosa-Alvarado BA, Guzi T, Lander GC. 2024. Human polymerase θ helicase positions DNA microhomologies for double-strand break repair. bioRxiv doi:10.1101/2024.04.26.591388

Zhong Q, Chen CF, Li S, Chen Y, Wang CC, Xiao J, Chen PL, Sharp ZD, Lee WH. 1999. Association of BRCA1 with the hRad50-hMre11-p95 complex and the DNA damage response. *Science* **285:** 747–750. doi:10.1126/science.285 .5428.747

Zhou J, Gelot C, Pantelidou C, Li A, Yücel H, Davis RE, Färkkilä A, Kochupurakkal B, Syed A, Shapiro GI, et al. 2021. A first-in-class polymerase θ inhibitor selectively targets homologous-recombination-deficient tumors. *Nat Cancer* **2:** 598–610. doi:10.1038/s43018-021-00203-x

Zhu Z, Chung WH, Shim EY, Lee SE, Ira G. 2008. Sgs1 helicase and two nucleases Dna2 and Exo1 resect DNA double-strand break ends. *Cell* **134:** 981–994. doi:10.1016/j.cell.2008.08.037

Zimmermann M, Lottersberger F, Buonomo SB, Sfeir A, de Lange T. 2013. 53BP1 regulates DSB repair using Rif1 to control 5′ end resection. *Science* **339:** 700–704. doi:10.1126/science.1231573

# Chromosome Ends in Motion: Telomeres as Hazards and Hubs in Meiosis

Rahul Thadani, Noah Johnson,[1] and Julia Promisel Cooper

Department of Biochemistry and Molecular Genetics, University of Colorado Anschutz Medical Campus, Aurora, Colorado 80045, USA

*Correspondence:* julia.p.cooper@cuanschutz.edu

Beyond their well-known roles in chromosome end protection, telomeres play critical roles in ensuring the fidelity of meiosis, the specialized cell division underlying sexual reproduction. Central to this process is the conserved telomere bouquet, a polarized nuclear arrangement in which telomeres cluster beneath the centrosome. The telomere bouquet orchestrates movements of meiotic chromosomes that facilitate pairing and recombination between homologous chromosomes, the defining events of meiosis. Here, we review both this canonical function and newly discovered meiotic telomere functions. We focus on three species—fission yeast, budding yeast, and mouse—that highlight both general principles and novel insights likely to be broadly applicable across eukaryotes. We propose that these diverse telomere functions provided early eukaryotes with a powerful adaptive advantage, contributing to the evolutionary success of linear chromosomes.

Cellular immortality and sexual reproduction are deeply intertwined. Meiosis generates the genetic diversity that fuels evolution while maintaining ploidy across generations. Haploid gametes arise from diploid progenitor cells that are sustained by an immortal germline. Telomeres play several unique and essential roles in ensuring the fidelity of this process, extending their repertoire beyond end protection to include dynamic reorganization of the genome during meiotic prophase, regulation of cell cycle transitions, and coordination of chromosome structure with nuclear architecture.

Indeed, the persistence of linear chromosomes across eukaryotic evolution, despite the challenges imposed by the end replication prob-lem and the end protection problem, may reflect selective pressures imposed by meiosis. Unlike mitosis, meiosis requires that homologous chromosomes align and recombine in the crowded environment of the nucleus. By clustering into the telomere bouquet, telomeres anchor chromosomes to the nuclear envelope (NE), connect them to the cytoskeleton, and generate the nuclear movements that guide the homology search. Viewed from this perspective, the evolution of linear chromosomes created not only the problem of end protection but also a solution to the challenges of sexual reproduction. Telomeres became not only guardians of chromosome ends but also guides for large-scale movements of whole chromosomes. Their roles in

---

[1]Present address: Biology Graduate Program, University of Oregon, Eugene, OR 97403.

Cite this article as *Cold Spring Harb Perspect Biol* doi: 10.1101/cshperspect.a041705

pairing, signaling, and orchestrating nuclear reorganization may have provided early eukaryotes a powerful evolutionary advantage. The conservation of bouquet formation from yeasts to plants and mammals underscores the evolutionary success of these meiotic functions.

## OVERVIEW OF THE MEIOTIC PROGRAM

Meiosis reduces ploidy through a single round of DNA replication followed by two successive divisions: a reductional meiosis I (MI), in which homologous chromosomes segregate while sister chromatids remain attached, and an equational, mitosis-like meiosis II (MII), in which sister chromatids separate. An extended prophase dominates the MI timeline (Bennett 1977). During this period, homologous chromosome pairs align, and homologous recombination (HR) occurs (Fig. 1). These processes depend on a striking level of nuclear reorganization that, in many organisms, is driven by a dynamic reconfiguration of telomeres and their attachment to the NE.

Following this dynamic prophase, many meiocytes enter periods of relative dormancy. In mammals, for example, oocytes arrest for months (in mouse) or decades (in humans) at diplotene and later undergo a second programmed arrest at metaphase II until fertilization. Such extended arrests pose unique challenges for telomere maintenance, as any damage sustained over this period has the potential to generate an embryo with a curtailed starting telomere length. Indeed, telomere length resetting in germline stem cells appears to be critical for germline immortality. In mice, unusually high telomerase activity is detected in

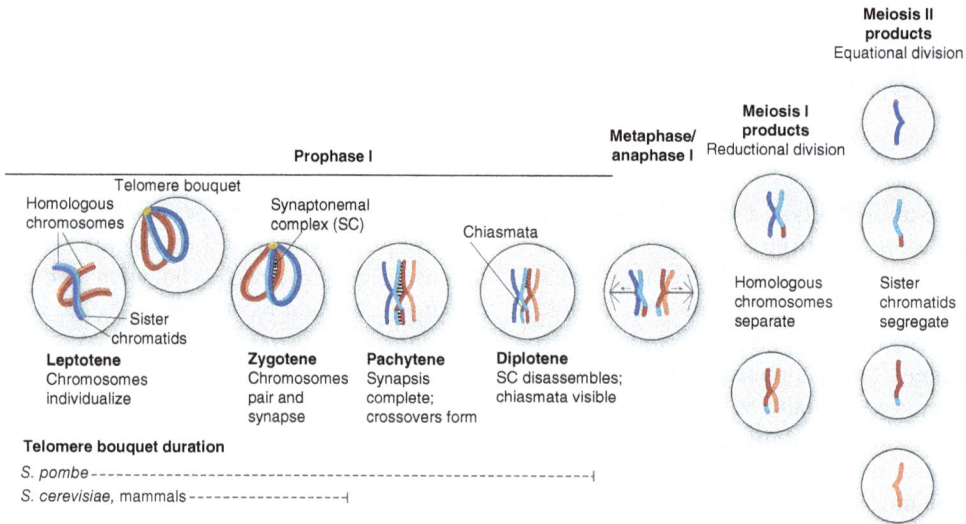

Figure 1. Stages of meiosis. Meiosis comprises a single round of DNA replication followed by two successive cell divisions. An extended prophase I following DNA replication is classically divided into several stages based on cytology and chromosome organization. Leptotene: Chromosomes individualize, condense, and organize into axes, accompanied by initiation of DNA double-strand breaks (DSBs) and telomere bouquet formation. Zygotene: Chromosomes pair and align along their lengths (synapse); the synaptonemal complex (SC), a filamentous protein structure, forms between homologs and stabilizes pairing. Pachytene: Synapsis is complete; DSBs are repaired, and crossovers form between homologs. Diplotene: SC disassembles; homologs begin to separate but are held together by visible chiasmata at sites of crossover formation. Prophase I is followed by a reductional first meiotic division when homologs separate. A second equational division segregates sister chromatids, giving rise to four haploid gametes.

Cite this article as *Cold Spring Harb Perspect Biol* doi: 10.1101/cshperspect.a041705

undifferentiated spermatogonia, which contain the functional germline stem cells, and may counteract age-associated telomere attrition (Pech et al. 2015). Intriguingly, during early embryonic divisions, maternal telomeres are reset to paternal length, using a recombination-based mechanism that depends on sperm- and oocyte-specific epigenetic states established during gametogenesis (Jeon et al. 2025). In budding yeast, recent work suggests a bouquet-dependent resetting of short telomeres to wild-type lengths (Sidarava et al. 2025). In humans, telomere length has emerged as one of the strongest predictors of fertility (Robinson et al. 2024). Thus, meiotic telomere functions encompass both architectural and genome maintenance roles.

## THE TELOMERE BOUQUET

The telomere bouquet, a polarized arrangement of chromosomes in which telomeres gather at the nuclear periphery beneath the centrosome, was first described as a feature of salamander spermatogenesis by Gustav Eisen (1900) and later characterized in remarkable detail during flatworm oogenesis (Gelei 1921; Scherthan 2001). The bouquet is conserved across diverse eukaryotic species ranging from budding yeast, fission yeast, and zebrafish to mouse, humans, and plants such as maize and wheat (Gelei 1921; Dresser and Giroux 1988; Chikashige et al. 1994; Scherthan et al. 1996; Bass et al. 1997; Cooper et al. 1998; Hayashi et al. 1998; Trelles-Sticken et al. 1999; Cowan et al. 2001; Saito et al. 2014; Shibuya et al. 2015; Blokhina et al. 2019). Despite variation in timing and persistence, the bouquet's core canonical function is conserved: promoting homologous pairing and recombination. In species with more tenuous homolog contacts (e.g., fission yeast, which lacks a synaptonemal complex [SC]), the bouquet persists longer, facilitating repeated encounters between homologs. In zebrafish, homolog synapsis initiates at chromosome ends, with bouquet positioning aided by a specialized structure known as the zygotene cilium, which appears to be conserved in the mouse (Mytlis et al. 2022). Across systems, the bouquet also enables rapid chromosome movements and even whole-nucleus rota-

tions, facilitating chromosome individualization, pairing, synapsis, and disentangling. Recently reviewed studies in *Caenorhabditis elegans* and zebrafish have been particularly fruitful in elucidating the principles underlying the topologically complex problem of meiotic chromosome disentanglement (Olaya et al. 2024).

Mechanistically, meiotic chromosome movements are powered by the cytoskeleton, actin in budding yeast and microtubules in most other systems, and are transmitted to chromosomes through the conserved linker of nucleoskeleton and cytoskeleton (LINC) complex. The LINC complex bridges the NE via SUN-domain proteins in the inner membrane and KASH-domain proteins in the outer membrane, connecting chromosomes to cytoskeletal motors (Fig. 2).

In some taxa, other chromosomal landmarks play roles analogous to the telomere bouquet. For instance, pairing centers, located near but not at chromosome ends, mediate homolog alignment in *C. elegans* (MacQueen et al. 2005). In *Drosophila melanogaster* (Takeo et al. 2011), centromeres cluster in a bouquet-like fashion. Indeed, the bouquet has been proposed to have arisen early in meiotic evolution, with subsequent functional specialization (Zickler and Kleckner 2016).

With this foundation, we turn to three model systems that have provided complementary insights into telomere function in meiosis: budding yeast, fission yeast, and mouse. Each of these models has lent unique perspectives that are nonetheless relevant to all eukaryotes and in some cases illuminated cell biology principles that apply to mitotically dividing cells as well.

## FISSION YEAST

Studies of telomere function in the fission yeast *Schizosaccharomyces pombe* have illuminated not only the canonical role of the bouquet in promoting chromosome movement and homolog alignment, but also functions of the bouquet that extend far beyond promoting HR.

The *S. pombe* bouquet persists throughout meiotic prophase, connecting all chromosomes to the cytoskeleton through the LINC complex, and pulling the chromosomes into dramatic os-

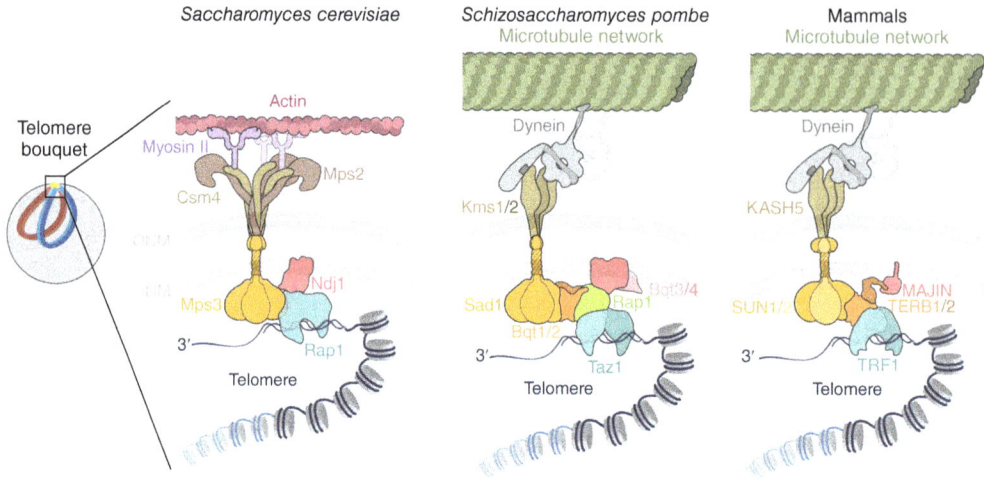

**Figure 2.** The telomere bouquet and linker of nucleoskeleton and cytoskeleton (LINC) complex. Molecular organization of the telomere bouquet and conserved LINC complex in budding yeast, fission yeast, and mammals, showing how telomere-binding proteins are tethered to the nuclear envelope (NE) and cytoskeletal forces are transduced to chromosomes via transmembrane SUN/KASH protein connections.

cillatory movements that traverse the meiocyte and are known as "horsetail" movements. These oscillations facilitate homolog alignment. Furthermore, the bouquet's consequences extend to nuclear architecture, cell cycle regulation, and centromere maintenance.

## Nuclear Reorganization—The Telomere–Centromere Switch

The *S. pombe* bouquet inverts the interphase chromosome architecture (Chikashige et al. 1994, 1997). In mitotically proliferating cells, interphase centromeres interact with the LINC complex, which is concentrated beneath the cytosol-facing centrosome, while telomeres form clusters at distal sites on the NE. Upon meiotic entry, this pattern is reversed: Telomeres gather at the LINC-enriched site beneath the centrosome, while centromeres disengage and disperse.

This telomere–centromere switch occurs via a stepwise process. Telomeric DNA is bound by the double-strand telomere-binding protein Taz1 (orthologous to mammalian telomere repeat-binding factor TRF1/TRF2), which in turn binds Rap1 (Cooper et al. 1997, 1998; Chikashige and Hiraoka 2001; Kanoh and Ishikawa

2001). Early in meiotic prophase, the LINC inner NE component Sad1 interacts with telomeres via the meiosis-specific Rap1-interactors Bqt1 and Bqt2, at the distal NE sites where interphase telomere clustering occurs (Chikashige et al. 2006). This triggers formation of the so-called "telocentrosome" in which the γ-tubulin complex and motor proteins (dynein and kinesin) localize to the cytoplasmic face of these telomeric attachment sites (Yoshida et al. 2013). These telocentrosomes then ferry the telomeres to the centrosome-adjacent region, where the LINC complex is concentrated.

Bouquet formation is not entirely concerted; rather, individual telomeres relocate asynchronously, arriving at the LINC region at different times (Klutstein et al. 2015). This asynchronous arrival results in transient colocalization of some telomeres and centromeres, a rare nuclear configuration, before full bouquet formation and complete centromere disengagement.

## Cell Cycle Control Via Centromere and Telomere Localization

Elegant insights into cell cycle control have emerged from studies in which the telomere–centromere switch was ectopically induced in mitotically cycling cells, leveraging mutations that cause partial centromere detachment from the LINC region (complete detachment is lethal; see below) (Jiménez-Martín et al. 2025). At those centromeres that detach, the outer kinetochore is disassembled. Moreover, detachment elicits widespread chromatin remodeling events that are normally restricted to meiosis. These phenotypes can be suppressed by artificially tethering centromeres back to the LINC complex, demonstrating that centromere positioning is central to maintaining a "mitotic" state.

Mechanistically, centromeric release from the LINC complex triggers liberation of the fission yeast cyclin-dependent kinase 1 (CDK1) Cdc2 from centromeres, allowing its accumulation at telomeres. Thus, nuclear microdomains formed at the centrosome–LINC region may be crucial for regulating cell cycle transitions: In mitotic cells, centromeres concentrate CDK1, while in meiotic cells, telomeres assume this role. This spatial handoff may provide a mechanism by which telomere relocalization regulates meiotic entry.

## Aligning Homologous Chromosomes for Recombination

The horsetail movements generated by bouquet–LINC connections and cytoplasmic dynein are essential for both the initiation and refinement of homolog pairing (Yamamoto et al. 1999; Ding et al. 2004; Saito et al. 2005; Tanaka et al. 2005; Fennell et al. 2015). Loss of dynein abrogates these oscillations, delaying pairing and increasing ectopic recombination. These findings support a dual role: Motion facilitates the initiation of homolog recognition, while preventing prolonged, nonspecific associations that promote entanglements. Quantitative analyses suggest that oscillations create cycles of chromosome "breathing" that encourage correct alignment while limiting excess recombination intermediates (Chacón et al. 2016).

Complementing this mechanical means of promoting homolog pairing, work from Hiraoka and colleagues has revealed the meiosis-specific accumulation of long noncoding RNAs (lncRNAs) at specific chromosomal loci, three

of which have been defined so far. These lncRNAs interact with a set of RNA-binding proteins known as *sme2* RNA-associated proteins (Smps) to create focal sites of robust pairing between homologs (Ding et al. 2012, 2019). Disruption of either the lncRNA coding regions or Smps leads to the loss of robust pairing at these loci. Smps harbor intrinsically disordered domains and undergo liquid–liquid phase separation in vitro in the presence of cognate lncRNAs (Ding et al. 2024). In vivo, robust pairing interactions are hypersensitive to 1,6-hexanediol, supporting a condensate model. Thus, bouquet-driven motion brings chromosomes into proximity, while lncRNA–protein condensates act as molecular "glue" that stabilizes specific homolog contacts.

## Beyond Alignment: Licensing NE Breakdown and Spindle Formation

In addition to promoting homolog pairing, the bouquet plays two essential and surprising roles: (1) licensing NE breakdown, a prerequisite for spindle assembly, and (2) maintaining centromeric integrity.

While mammalian cells undergo complete NE breakdown at the onset of mitosis and meiosis, the *S. pombe* NE is partially degraded at the region just beneath the centrosome, allowing the formation of intranuclear spindles. The mechanisms triggering initiation of NE breakdown, whether complete in mammalian cells or partial in *S. pombe*, have remained largely undefined. Strikingly, *S. pombe* cells lacking the telomere bouquet frequently exhibit failed meiotic divisions marked by monopolar or unstable spindles that confer defective chromosome segregation (Tomita and Cooper 2007). Mechanistic studies revealed that these defects stem from failed NE breakdown and that positioning of the bouquet in the LINC region is essential to confer remodeling of the adjacent NE (Fernández-Álvarez et al. 2016, 2017). In wild-type cells, NE breakdown occurs at this LINC-concentrated site shortly after bouquet dissolution, suggesting that the bouquet leaves a mark that licenses subsequent local NE remodeling.

Notably, spindle failure in bouquet-defective cells is incompletely penetrant: 50% of such cells still form functional bipolar spindles (Tomita and Cooper 2007). This stems from the partial reassociation of centromeres with the LINC region in cells lacking the bouquet; centromeres can substitute for telomeres in performing this meiotic function (Fennell et al. 2015). This discovery led to the observation that in mitotically dividing cells, where centromeres naturally cluster beneath the centrosome, centromeres fulfill this role in promoting local NE breakdown (Fernández-Álvarez et al. 2016). Thus, control of NE breakdown can be conferred by either of the two most prominent chromosomal landmarks, telomeres or centromeres, potentially coupling information about chromosome state to spindle formation. These observations raise the possibility that untimely NE breakdown may be a thus far-unappreciated effect of telomere or centromere dysfunction.

The common feature that enables (meiotic) telomeres and (mitotic) centromeres to trigger local NE remodeling remains unresolved. Although both loci share heterochromatic marks, these appear to be dispensable. A key possibility alluded to above is that both centromeres and meiotic telomeres provide a platform for local CDK1 accumulation, which may in turn trigger local NE breakdown (Moiseeva et al. 2017; Jiménez-Martín et al. 2025). In mammalian cells, NE disassembly requires CDK1-mediated phosphorylation of lamins and nuclear pore complex subunits (Gerace and Blobel 1980; Laurell et al. 2011), yet the triggers that recruit and activate this CDK1 activity remain obscure. Determination of the relevant telomeric and centromeric features that would allow events like CDK1 capture and/or NE breakdown capacity remains a compelling frontier.

## Telomeric Control of Meiotic Centromere Reassembly

Examination of bouquet-defective meioses in which spindle formation was rescued by centromere–LINC contacts revealed a second unanticipated bouquet function: reassembly of kinetochores, which have a small but significant tendency to be dismantled upon meiotic entry (Klutstein et al. 2015). The centromere-specific histone H3 variant CenpA is evicted from 3%–5% of all centromeres in bouquet-defective cells; this is accompanied by the loss of the entire kinetochore and flanking pericentric heterochromatin. In fission yeast, this rate of kinetochore loss causes failure in 12%–15% of meioses; in humans, this rate would be lethal to all meioses.

Two foundational meiosis-specific proteins drive this dismantlement: Spo11, the endonuclease that initiates HR by creating DNA double-strand breaks (DSBs), and the meiosis-specific Rec8 cohesin complex subunit (Hou et al. 2021). Spo11's effect depends on the chromatin remodeler Hrp3 and requires only DNA binding—DSB formation is dispensable. Rec8's centromeric activity depends on the RSC chromatin remodeler. How Spo11 and Rec8 dismantle centromeres, whether this reflects a facet of their normal meiotic functions, and whether such activity contributes to cancers that misexpress them, remain intriguing questions.

Emerging evidence suggests that telomeric heterochromatin contributes in *trans* to the reassembly of kinetochores following Spo11/Rec8-mediated dismantlement. This may occur indirectly, by promoting formation of pericentric heterochromatin that in turn stimulates nearby CenpA assembly, or directly at the kinetochore itself. Both models extend the principle that heterochromatin is required for de novo centromere establishment (Folco et al. 2008). Thus, meiotic kinetochore reassembly by telomeres provides a powerful tool for elucidating the so-far mysterious mechanisms by which heterochromatin promotes centromere assembly.

## Summary: The Bouquet as a Multifunctional Hub

In *S. pombe*, the bouquet emerges as a multifunctional hub. By tethering chromosomes, it promotes motion and pairing. By clustering beneath the centrosome, it licenses NE breakdown and spindle assembly. By concentrating heterochromatic telomeric factors, it promotes kinetochore reassembly. Thus, a structure likely preserved for its role in homolog pairing has been co-opted for additional regulatory functions that coordinate chromosome dynamics with cell cycle transitions.

## BUDDING YEAST

The genetic tractability and conservation of meiotic principles have made the budding yeast *Saccharomyces cerevisiae* a mainstay for meiosis research (Ramesh et al. 2005). Budding yeast meiosis generates four haploid spores that remain connected, allowing genetic outcomes to be tracked at single-meiosis resolution. Unlike *S. pombe*, *S. cerevisiae* chromosomes are fully synapsed by a canonical SC, and phenomena such as crossover (CO) interference, wherein the formation of one CO reduces the likelihood of another occurring nearby, are also present (Zickler and Kleckner 1998; Zickler 1999). Telomere clustering into a bouquet is conserved, although in contrast to fission yeast and mammalian systems, the budding yeast bouquet connects, via the LINC complex, to myosin motors and the actin cytoskeleton rather than to cytoplasmic microtubules. Studies in this organism have shaped our understanding of both bouquet function and the mechanics of telomere-led chromosome motion.

### LINC Complex Assembly and Specialization

Like other eukaryotes, budding yeast use the conserved LINC complex to connect telomeres with cytoskeletal forces (Bone and Starr 2016). Budding yeast have one canonical SUN domain protein, Mps3, but lack a traditional KASH domain protein (Jaspersen et al. 2006; Fan et al. 2020). Csm4, a meiosis-specific outer NE protein, was identified as a member of the LINC complex (Conrad et al. 2008; Wanat et al. 2008). However, only four amino acids of the carboxy-terminal domain of Csm4 cross into the NE lumen, suggesting the involvement of an additional factor, which was shown to be a Csm4 paralog, Mps2, initially implicated in mitotic centrosome duplication and NE insertion (Muñoz-Centeno et al. 1999; Fan et al. 2020). Mps3 interacts with Mps2 and Csm4 to form a noncanonical "telomere-associated LINC com-

plex" (t-LINC), as well as with the meiosis-specific telomere protein Ndj1 to form the bouquet (Conrad et al. 1997, 2007; Fan et al. 2020). This highlights a recurrent principle of meiotic LINC complexes: Many comprise mitotic proteins in tandem with meiosis-specific proteins (Lee et al. 2020a,b; Fan et al. 2022).

While the *S. pombe* bouquet appears as one discrete focus beneath the centrosome, budding yeast LINC complexes are dispersed throughout the NE, and the telomere bouquet, which is visualized as patches harboring multiple gathered Mps3 foci, exists only transiently (Conrad et al. 2007; Bommi et al. 2019; Prasada Rao et al. 2021). Moreover, not all telomere–LINC associations interface directly with force-generation machinery; rather, a lead telomere connects with the actin cytoskeleton and provides the platform for a cascade of loose associations with other chromosomes (Nozaki et al. 2021). Thus, homolog pairing depends on the coordinated action of multiple dynamic telomere–NE contacts rather than a single enduring cluster.

### Regulation of Bouquet Formation and Dissolution

As *S. cerevisiae* bouquet associations are transient and not all telomeres are simultaneously engaged with motors, their assembly and dissolution must be carefully choreographed. Regulation involves cycles of LINC complex assembly/disassembly, posttranslational modifications, and control by cohesin and chromatin state.

The cohesin complex is a conserved ring structure that shapes genome architecture and establishes sister–chromatid cohesion by topologically encircling sister chromatids (Uhlmann 2025). In meiosis, specialized cohesin complexes containing the meiosis-specific Rec8 kleisin subunit canonically maintain cohesion between sister chromatids until their separation in MII. In addition, alternative cohesin complexes participate in numerous aspects of meiotic chromosome organization, including SC formation and HR (Lee and Hirano 2011; Ishiguro 2019).

Budding yeast Rec8 has been implicated in LINC complex stability as well as in bouquet formation and resolution (Trelles-Sticken et al.

2005). In the absence of the Wapl cohesin release factor Wpl1, aberrant Mps3 foci persist into MI, altering telomere-led motion and recombination outcomes (Challa et al. 2016). Loss of Wpl1 also increases Rec8 accumulation at telomere clusters, suggesting that cohesin holocomplexes, not just Rec8, contribute to telomere clustering. Thus, cohesin regulates the dynamics of LINC–telomere associations.

Phosphoproteomic data indicate that the luminal region of Mps3, which controls its NE localization, can be phosphorylated by both CDK and Dbf4-dependent kinase (DDK), altering interactions between Mps3 and negatively charged phospholipids in reconstituted liposomes. In vivo, phosphodeficient alleles of Mps3 confer delayed dissolution of the bouquet. While CDK and DDK are unlikely to access the nuclear lumen directly, phosphorylation of interacting partners such as Rec8 may regulate Mps3 localization indirectly (Prasada Rao et al. 2021). These findings highlight the multilayered regulation required to coordinate telomere clustering with meiotic progression.

## Rapid Prophase Movements

The telomere-led movements driven by interaction between the budding yeast bouquet and the actin cytoskeleton are known as rapid meiotic prophase movements (RPMs). These movements are thought to safeguard meiotic fidelity by three complementary mechanisms: stimulating encounters between homologs, resolving entanglements between chromosome pairs, and minimizing nonhomologous associations (Lee et al. 2012). In the absence of these motions, HR is reduced and ectopic interactions increase, compromising chromosome segregation.

Live-cell imaging and biophysical approaches have revealed distinct categories of telomere motion: (1) pauses, or long-lasting periods (5–15 sec) of no net motion; (2) rapid bursts of motion where the monitored telomere is directly coupled to the actin machinery; and (3) longer, slower trajectories driven indirectly by other telomeres. Most telomeres spend the majority of prophase in a paused state; of the telomeres undergoing movement, ~17% are directly actin-coupled. These quantitative definitions provide a framework to dissect how the frequency and type of motion affect pairing and synapsis. For instance, excessive rapid motion could increase entanglement levels or hamper chromosome synapsis, whereas intermittent bursts combined with indirect motion may represent a balance that allows optimal chromosome alignment (Nozaki et al. 2021).

Mutations in Ndj1 or in core t-LINC components (Csm4 and Mps3) abolish RPMs and bouquet formation (Conrad et al. 2008; Wanat et al. 2008; Lee et al. 2012, 2020a), underscoring their centrality. Together, these findings suggest that, as in S. pombe, telomere-led oscillations in budding yeast serve a dual purpose: initiating homolog contacts while also editing them by destabilizing inappropriate interactions.

## Chromatin Factors Modulating Telomere Dynamics

Beyond LINC components, global chromatin modulators have been implicated in budding yeast meiotic telomere dynamics. For instance, the histone variant H2A.Z, deposited by the SWI2/SNF2-related 1 (SWR1) complex, modifies the chromatin landscape by replacing H2A at specific loci (González-Arranz et al. 2018, 2020). The absence of H2A.Z leads to checkpoint misregulation and defects in meiotic progression, potentially by misregulating kinases that influence t-LINC assembly (González-Arranz et al. 2018; Prasada Rao et al. 2021). Intriguingly, H2A.Z localizes not only to telomeres but also to the centrosome itself, in a SWR1-independent manner. Thus, H2A.Z appears to play a noncanonical, nucleosome-independent role in promoting RPMs. While the mechanism remains unclear, H2A.Z may interact directly with Mps3, fortifying the connection between the force production machinery and the ends of chromosomes (González-Arranz et al. 2020).

## Summary: Lessons from Budding Yeast

Budding yeast studies highlight several general principles of meiotic telomere biology. First,

LINC complexes can be reconfigured with meiosis-specific components (Ndj1 and Csm4) and existing mitotic proteins (Mps2 and Mps3) to achieve telomere tethering. Second, telomere-led movements, although actin-driven rather than microtubule-driven, serve the conserved purpose of balancing homolog alignment with the prevention of aberrant interactions. Third, chromatin factors such as cohesin and histone variants fine-tune the dynamics of bouquet assembly and dissolution. Thus, budding yeast reinforces the concept that telomeres are not passive chromosome ends but dynamic organizers of meiotic nuclear architecture, using solutions that are distinct yet conceptually parallel to those in *S. pombe*.

## MOUSE

Studies of mammalian meiosis have uncovered both conserved strategies and surprising innovations in how telomeres mediate chromosome dynamics. In particular, mouse spermatocytes revealed a specialized mechanism of telomere tethering, "cap exchange," that replaces canonical shelterin with a meiosis-specific complex, telomere repeat binding bouquet formation protein 1/2 (TERB1/2)–membrane-anchored junction protein (MAJIN). Mouse studies also emphasize the structural and protective roles of cohesin at telomeres, echoing themes observed at centromeres.

### Telomere Cap Exchange

Rather than relying on shelterin for NE attachment, mouse telomeres undergo a meiotic switch in which TRF1 is displaced by TERB1/2–MAJIN. TERB1 contains both a TRF-like Myb domain and a TRF1-binding domain, allowing it to form a 1:1 stoichiometric complex with TRF1 in early prophase (Shibuya et al. 2014). TERB2 and MAJIN are meiosis-specific TERB1 interactors. While MAJIN lacks obvious sequence similarity to other bouquet proteins, it resembles fission yeast Bqt4 in harboring a single transmembrane domain at its carboxyl terminus and a basic putative DNA-binding helix–turn–helix domain.

The cap exchange model proposes that TERB1/2–MAJIN complexes are preassembled and sequestered at the NE by MAJIN. During leptotene and zygotene, telomeres are recruited via TRF1–TERB1 interactions, initiating bouquet formation. As prophase progresses, rising CDK activity phosphorylates TERB1 at the TRF1-binding interface, weakening this interaction and displacing shelterin. Telomeres are then stably tethered by TERB1/2–MAJIN, likely achieved through direct DNA binding by MAJIN, since the TERB1 Myb domain lacks DNA-binding activity. Thus, a transient TRF1–TERB1 bridge recruits telomeres, but the interface must later be destabilized to permit full shelterin displacement and stable TERB1/2–MAJIN anchoring (Shibuya et al. 2014, 2015; Pendlebury et al. 2017).

TERB1 connects the bouquet with the LINC complex by binding SUN1 in the inner NE; in turn, SUN1 binds KASH5, which crosses the outer NE to interact with dynein and microtubules. These connections transduce cytoskeletal forces into nuclear movements along microtubule "tracks" that are continuous with the cytoskeletal network (Ding et al. 2007; Morimoto et al. 2012; Lee et al. 2015). Disruption of the TERB1–TRF1 interface compromises telomere attachment to the NE, while strengthening this interface through a structure-guided point mutation on TRF1 also interferes with cap exchange, underlining the requirement for a dynamic shelterin capture/displacement cycle for functional bouquet formation (Pendlebury et al. 2017). Disruption of TERB1, TERB2, or MAJIN leads to spermatocyte arrest in leptotene or zygotene, incomplete synapsis, and accumulation of unrepaired DNA breaks. Structural comparisons of the mouse and human TERB1–TERB2 and TERB2–MAJIN interfaces show that the interactions are preserved (Dunce et al. 2018; Wang et al. 2019), while phylogeny suggests conservation across metazoans (da Cruz et al. 2020). Thus, the mouse TERB1/2–MAJIN pathway represents a conserved strategy that couples NE anchoring (via MAJIN) to cytoskeletal force transduction (via SUN1/KASH5), a chromosome movement–generating module reminiscent of the Bqt1-4/Sad1/Kms2 system in fission yeast.

More than simply providing a backdrop for force transduction to attached telomeres, NE composition has a profound impact on meiotic chromosome structure and function. Male mouse germ cells maintain a highly fluid cell membrane through action of the conserved membrane fluidity sensor adiponectin receptor 2 (AdipoR2). In the absence of regulated lipid synthesis, the cytoskeletal forces that normally power telomere-led chromosome movement instead produce NE invaginations, displacing telomeres to the interior of a rigid nucleus. These defects lead to a failure of homolog synapsis, HR, and ultimately meiotic progression (Zhang et al. 2024).

## Telomere–Cohesin Interactions and Structural Integrity

Mouse studies also emphasize the structural reinforcement of telomeres by meiosis-specific cohesins, which contribute to telomere integrity under mechanical stress. In mitotically proliferating cells, the cohesin ring core complex comprises structural maintenance of chromosomes (SMC)1α and SMC3 proteins, capped by the kleisin subunit RAD21; accessory subunits such as SA1/SA2 modulate the core complex. During meiosis, alternative components such as SMC1β, the kleisins REC8 and RAD21L, and the accessory subunit SA3 replace their mitotic counterparts (Ishiguro 2019).

SMC1β is particularly critical for meiotic telomere function. Mouse meiocytes deficient in SMC1β show reduced association of telomeres with the NE as well as damaged and foreshortened telomeres (Adelfalk et al. 2009). Intriguingly, while many cohesin functions can be achieved interchangeably by SMC1α or SMC1β, the meiotic telomere protection function of cohesin appears to specifically require SMC1β, which differs from SMC1α only in possessing a basic carboxyl terminus that has single-stranded DNA (ssDNA)-binding activity. This observation suggests a role for SMC1β in protecting telomeres from excessive recombination or promoting protective telomere structures such as t-loops. Consistent with the idea of SMC1β favoring a protective telomeric

configuration, $Smc1β^{-/-}$ spermatocytes show increased transcription of noncoding telomere repeat containing RNA (TERRA), and RNA–DNA hybrids (R-loops) at chromosome ends (Biswas et al. 2023).

Recruitment of cohesin to telomeres occurs via TERB1, whose Myb domain interacts with the meiosis-specific SA3 subunit. This coupling between TERB1 and cohesin likely buttresses telomeres against forces generated by prophase chromosome movements, reminiscent of cohesin's role at centromeres in resisting spindle pulling forces and creating the tension necessary for sister centromere biorientation (Shibuya et al. 2014). Indeed, deletion of the TERB1 Myb domain provides a means of interrogating cohesin function at telomeres without the pleiotropic meiotic defects that stem from TERB1 deficiency. $Terb1^{Δmyb/Δmyb}$ spermatocytes retain normal NE attachment, HR, and fertility but display telomere shortening as well as split or bridged shelterin signals, implicating meiotic cohesin in the protection of telomeres from physical erosion (Zhang et al. 2022).

## Summary: Mammalian Innovations

Mouse meiosis has revealed innovations layered on conserved telomere functions. With cap exchange, mammals remodel shelterin into a meiosis-specific tethering system, coupling telomeres to LINC-mediated force generation. Telomere–cohesin cooperation underscores their structural role in protecting chromosome ends in the milieu of prophase dynamics, paralleling cohesin's role at centromeres and recalling observations made in budding yeast. Hence, while the bouquet is broadly conserved, each lineage has evolved a unique molecular solution to stabilize telomere function in meiosis.

## CHROMOSOMAL LANDMARKS AS HR HOT OR COLD SPOTS

Programmed DSB formation is a hallmark of meiosis. DSBs are not randomly distributed along chromosomes but instead occur with elevated frequency at hot spots, determined by a combination of chromatin structure, histone modifications, and sequence-specific protein binding. These DSBs are repaired by HR, with homologs as a preferred repair template. A subset of DSBs is repaired as CO events, which give rise to chiasmata that hold homologs together, while the majority are repaired as non-crossover (NCO) events, arising from nonreciprocal repair of chromosome arms. The formation of DSBs, their HR-mediated repair, and the ensuing distribution of COs are intricately regulated and influenced by chromosomal landmarks (Arter and Keeney 2024).

Centromeres have long been recognized as suppressors of COs, inhibiting DSB formation or favoring NCO repair (Beadle 1932; Pazhayam et al. 2021). This suppression is beneficial as centromeric HR could interfere with spindle attachment, leading to chromosome missegregation and aneuploidy. In budding yeast, DSBs are vanishingly rare within 1–3 kb of centromeres and below the genomic average in a larger 5–10 kb pericentric region (Pan et al. 2011). The CCAN$^{Ctf19}$ kinetochore subcomplex underlies this DSB suppression, while cohesin inhibits COs over a broader ~20 kb pericentric region (Vincenten et al. 2015). In fission yeast, pericentric DSBs are suppressed by the heterochromatin protein HP1$^{Swi6}$, which recruits cohesin complexes that contain Rec8 but lack another meiotic cohesin subunit, SA3$^{Rec11}$, which is confined to chromosome arms. HP1$^{Swi6}$ further directly inhibits SA3$^{Rec11}$, an activator of DSB formation (Nambiar and Smith 2018). In mouse (Li et al. 2019) and humans (Palsson et al. 2025), NCOs are favored over COs near centromeres through thus far poorly understood mechanisms, with a stronger centromere suppression effect observed in males.

Like at centromeres, COs at telomeres have the potential to disrupt vital protein interactions, such as those that enable telomere–NE tethering and cohesin recruitment and might be expected to be suppressed. This is indeed the case in budding yeast, where DSBs are suppressed (3.5-fold lower than the genomic average) in a ~20 kb region adjacent to chromosome ends; NCOs also appear to be favored over COs for repair in this region. However, both DSB and CO frequency are elevated in the ~100 kb sub-

telomeric region centromere proximal to the "cold" subtelomeric domain (Chen et al. 2008; Pan et al. 2011; Subramanian et al. 2019). In contrast, mammalian subtelomeres appear to be hot spots of recombination, driven in large part by early replication of these regions (Pratto et al. 2021). In humans, NCO resolution is favored over the terminal ~1 Mb of the chromosome in males and over the ~5 Mb telomere-adjacent region in females (Palsson et al. 2025). In mouse, COs do not seem to be suppressed near chromosome ends; instead, both CO and NCO events are frequent (Li et al. 2019). Conceivably, as long as COs are avoided at telomere repeat sequences, elevated recombination rates at subtelomeres could be beneficial, for instance, by facilitating homolog pairing; indeed, several early cytological observations (Gelei 1921) identified sites of synapsis initiation close to telomeres. As a consequence, in part, of elevated meiotic recombination rates, subtelomeres have emerged as some of the most rapidly evolving regions of eukaryotic genomes, often harboring genes that co-opt their plasticity to enable rapid adaptive evolution (Mefford and Trask 2002).

## CONCLUSIONS

From yeast to mammals, telomeres emerge as more than passive chromosome ends: They are dynamic organizers of meiotic architecture. The conservation of bouquet formation across eukaryotes highlights its central role in pairing homologs, while each lineage reveals additional, often surprising, elaborations.

In *S. pombe*, the bouquet persists throughout prophase, coupling telomeres to horsetail oscillations and co-opting their spatial concentration to license NE breakdown and promote kinetochore reassembly. In *S. cerevisiae*, noncanonical LINC complexes tether telomeres to actin motors, generating rapid prophase movements whose regulation depends on cohesins and chromatin state. In mammals, telomere cap exchange replaces shelterin with TERB1/2–MAJIN to achieve NE anchoring, while a specialized cohesin module reinforces telomere structure against mechanical stress.

Together, these perspectives underscore two unifying principles. First, telomere attachment to the NE provides a versatile platform for transmitting cytoskeletal forces, reshaping the genome in ways that facilitate homolog recognition and recombination. Second, once assembled, the bouquet serves as a hub where telomere factors are concentrated and co-opted for broader regulatory roles, from licensing NE remodeling to stabilizing chromosome structures.

The persistence of linear chromosomes across eukaryotic evolution, despite their inherent challenges, may in part reflect the opportunities they afford in meiosis: telomeres as guardians, anchors, and guides. Understanding how these roles are integrated across species not only illuminates meiotic fidelity but also offers insight into broader principles of nuclear organization and genome stability.

## ACKNOWLEDGMENTS

We thank Maria Diaz de la Loza for her help with illustrations.

## REFERENCES

Adelfalk C, Janschek J, Revenkova E, Blei C, Liebe B, Göb E, Alsheimer M, Benavente R, de Boer E, Novak I, et al. 2009. Cohesin SMC1β protects telomeres in meiocytes. *J Cell Biol* **187**: 185–199. doi:10.1083/jcb.200808016

Arter M, Keeney S. 2024. Divergence and conservation of the meiotic recombination machinery. *Nat Rev Genet* **25**: 309–325. doi:10.1038/s41576-023-00669-8

Bass HW, Marshall WF, Sedat JW, Agard DA, Cande WZ. 1997. Telomeres cluster de novo before the initiation of synapsis: a three-dimensional spatial analysis of telomere positions before and during meiotic prophase. *J Cell Biol* **137**: 5–18. doi:10.1083/jcb.137.1.5

Beadle GW. 1932. A possible influence of the spindle fibre on crossing-over in *Drosophila*. *Proc Natl Acad Sci* **18**: 160–165. doi:10.1073/pnas.18.2.160

Bennett MD. 1977. The time and duration of meiosis. *Philos Trans R Soc Lond B Biol Sci* **277**: 201–226. doi:10.1098/rstb.1977.0012

Biswas U, Mallik TD, Pschirer J, Lesche M, Sameith K, Jessberger R. 2023. Cohesin SMC1β promotes closed chromatin and controls TERRA expression at spermatocyte telomeres. *Life Sci Alliance* **6**: e202201798. doi:10.26508/lsa.202201798

Blokhina YP, Nguyen AD, Draper BW, Burgess SM. 2019. The telomere bouquet is a hub where meiotic double-strand breaks, synapsis, and stable homolog juxtaposition

are coordinated in the zebrafish, *Danio rerio*. *PLoS Genet* **15**: e1007730. doi:10.1371/journal.pgen.1007730

Bommi JR, Rao H, Challa K, Higashide M, Shinmyozu K, Nakayama JI, Shinohara M, Shinohara A. 2019. Meiosis-specific cohesin component, Rec8, promotes the localization of Mps3 SUN domain protein on the nuclear envelope. *Genes Cells* **24**: 94–106. doi:10.1111/gtc.12653

Bone CR, Starr DA. 2016. Nuclear migration events throughout development. *J Cell Sci* **129**: 1951–1961. doi:10.1242/jcs.179788

Chacón MR, Delivani P, Tolić IM. 2016. Meiotic nuclear oscillations are necessary to avoid excessive chromosome associations. *Cell Rep* **17**: 1632–1645. doi:10.1016/j.celrep.2016.10.014

Challa K, Lee MS, Shinohara M, Kim KP, Shinohara A. 2016. Rad61/Wpl1 (Wapl), a cohesin regulator, controls chromosome compaction during meiosis. *Nucleic Acids Res* **44**: 3190–3203. doi:10.1093/nar/gkw034

Chen SY, Tsubouchi T, Rockmill B, Sandler JS, Richards DR, Vader G, Hochwagen A, Roeder GS, Fung JC. 2008. Global analysis of the meiotic crossover landscape. *Dev Cell* **15**: 401–415. doi:10.1016/j.devcel.2008.07.006

Chikashige Y, Hiraoka Y. 2001. Telomere binding of the Rap1 protein is required for meiosis in fission yeast. *Curr Biol* **11**: 1618–1623. doi:10.1016/S0960-9822(01)00457-2

Chikashige Y, Ding DQ, Funabiki H, Haraguchi T, Mashiko S, Yanagida M, Hiraoka Y. 1994. Telomere-led premeiotic chromosome movement in fission yeast. *Science* **264**: 270–273. doi:10.1126/science.8146661

Chikashige Y, Ding DQ, Imai Y, Yamamoto M, Haraguchi T, Hiraoka Y. 1997. Meiotic nuclear reorganization: switching the position of centromeres and telomeres in the fission yeast *Schizosaccharomyces pombe*. *EMBO J* **16**: 193–202. doi:10.1093/emboj/16.1.193

Chikashige Y, Tsutsumi C, Yamane M, Okamasa K, Haraguchi T, Hiraoka Y. 2006. Meiotic proteins Bqt1 and Bqt2 tether telomeres to form the bouquet arrangement of chromosomes. *Cell* **125**: 59–69. doi:10.1016/j.cell.2006.01.048

Conrad MN, Dominguez AM, Dresser ME. 1997. Ndj1p, a meiotic telomere protein required for normal chromosome synapsis and segregation in yeast. *Science* **276**: 1252–1255. doi:10.1126/science.276.5316.1252

Conrad MN, Lee CY, Wilkerson JL, Dresser ME. 2007. MPS3 mediates meiotic bouquet formation in *Saccharomyces cerevisiae*. *Proc Natl Acad Sci* **104**: 8863–8868. doi:10.1073/pnas.0606165104

Conrad MN, Lee CY, Chao G, Shinohara M, Kosaka H, Shinohara A, Conchello JA, Dresser ME. 2008. Rapid telomere movement in meiotic prophase is promoted by NDJ1, MPS3, and CSM4 and is modulated by recombination. *Cell* **133**: 1175–1187. doi:10.1016/j.cell.2008.04.047

Cooper JP, Nimmo ER, Allshire RC, Cech TR. 1997. Regulation of telomere length and function by a Myb-domain protein in fission yeast. *Nature* **385**: 744–747. doi:10.1038/385744a0

Cooper JP, Watanabe Y, Nurse P. 1998. Fission yeast Taz1 protein is required for meiotic telomere clustering and recombination. *Nature* **392**: 828–831. doi:10.1038/33947

Cowan CR, Carlton PM, Cande WZ. 2001. The polar arrangement of telomeres in interphase and meiosis. Rabl organization and the bouquet. *Plant Physiol* **125:** 532–538. doi:10.1104/pp.125.2.532

da Cruz I, Brochier-Armanet C, Benavente R. 2020. The TERB1-TERB2-MAJIN complex of mouse meiotic telomeres dates back to the common ancestor of metazoans. *BMC Evol Biol* **20:** 55. doi:10.1186/s12862-020-01612-9

Ding DQ, Yamamoto A, Haraguchi T, Hiraoka Y. 2004. Dynamics of homologous chromosome pairing during meiotic prophase in fission yeast. *Dev Cell* **6:** 329–341. doi:10.1016/S1534-5807(04)00059-0

Ding X, Xu R, Yu J, Xu T, Zhuang Y, Han M. 2007. SUN1 is required for telomere attachment to nuclear envelope and gametogenesis in mice. *Dev Cell* **12:** 863–872. doi:10.1016/j.devcel.2007.03.018

Ding DQ, Okamasa K, Yamane M, Tsutsumi C, Haraguchi T, Yamamoto M, Hiraoka Y. 2012. Meiosis-specific noncoding RNA mediates robust pairing of homologous chromosomes in meiosis. *Science* **336:** 732–736. doi:10.1126/science.1219518

Ding DQ, Okamasa K, Katou Y, Oya E, Nakayama JI, Chikashige Y, Shirahige K, Haraguchi T, Hiraoka Y. 2019. Chromosome-associated RNA–protein complexes promote pairing of homologous chromosomes during meiosis in *Schizosaccharomyces pombe*. *Nat Commun* **10:** 5598. doi:10.1038/s41467-019-13609-0

Ding DQ, Okamasa K, Yoshimura Y, Matsuda A, Yamamoto TG, Hiraoka Y, Nakayama JI. 2024. Proteins and noncoding RNAs that promote homologous chromosome recognition and pairing in fission yeast meiosis undergo condensate formation in vitro. *FASEB J* **38:** e70163. doi:10.1096/fj.202302563RR

Dresser ME, Giroux CN. 1988. Meiotic chromosome behavior in spread preparations of yeast. *J Cell Biol* **106:** 567–573. doi:10.1083/jcb.106.3.567

Dunce JM, Milburn AE, Gurusaran M, da Cruz I, Sen LT, Benavente R, Davies OR. 2018. Structural basis of meiotic telomere attachment to the nuclear envelope by MAJIN-TERB2-TERB1. *Nat Commun* **9:** 5355. doi:10.1038/s41467-018-07794-7

Eisen G. 1900. The spermatogenesis of batrachoseps. polymorphous spermatogonia, auxocytes, and spermatocytes. *J Morphol* **17:** 1–117. doi:10.1002/jmor.1050170102

Fan J, Jin H, Koch BA, Yu HG. 2020. Mps2 links Csm4 and Mps3 to form a telomere-associated LINC complex in budding yeast. *Life Sci Alliance* **3:** e202000824. doi:10.26508/lsa.202000824

Fan J, Sun Z, Wang Y. 2022. The assembly of a noncanonical LINC complex in *Saccharomyces cerevisiae*. *Curr Genet* **68:** 91–96. doi:10.1007/s00294-021-01220-0

Fennell A, Fernández-Álvarez A, Tomita K, Cooper JP. 2015. Telomeres and centromeres have interchangeable roles in promoting meiotic spindle formation. *J Cell Biol* **208:** 415–428. doi:10.1083/jcb.201409058

Fernández-Álvarez A, Cooper JP. 2017. The functionally elusive Rabl chromosome configuration directly regulates nuclear membrane remodeling at mitotic onset. *Cell Cycle* **16:** 1392–1396. doi:10.1080/15384101.2017.1338986

Fernández-Álvarez A, Bez C, O'Toole Eileen T, Morphew M, Cooper Julia P. 2016. Mitotic nuclear envelope breakdown and spindle nucleation are controlled by interphase contacts between centromeres and the nuclear envelope. *Dev Cell* **39:** 544–559. doi:10.1016/j.devcel.2016.10.021

Folco HD, Pidoux AL, Urano T, Allshire RC. 2008. Heterochromatin and RNAi are required to establish CENP-A chromatin at centromeres. *Science* **319:** 94–97. doi:10.1126/science.1150944

Gelei J. 1921. Weitere studien über die oogenese des *Dendrocoelum lacteum*. II: Die längskonjugation der chromosomen. *Arch für Zellforsch* **16:** 88–169.

Gerace L, Blobel G. 1980. The nuclear envelope lamina is reversibly depolymerized during mitosis. *Cell* **19:** 277–287. doi:10.1016/0092-8674(80)90409-2

González-Arranz S, Cavero S, Morillo-Huesca M, Andújar E, Prrez-Alegre M, Prado F, San-Segundo P. 2018. Functional impact of the H2A.Z histone variant during meiosis in *Saccharomyces cerevisiae*. *Genetics* **209:** 997–1015. doi:10.1534/genetics.118.301110

González-Arranz S, Gardner JM, Yu Z, Patel NJ, Heldrich J, Santos B, Carballo JA, Jaspersen SL, Hochwagen A, San-Segundo PA. 2020. SWR1-Independent association of H2A.Z to the LINC complex promotes meiotic chromosome motion. *Front Cell Dev Biol* **8:** 594092. doi:10.3389/fcell.2020.594092

Hayashi A, Ogawa H, Kohno K, Gasser SM, Hiraoka Y. 1998. Meiotic behaviours of chromosomes and microtubules in budding yeast: relocalization of centromeres and telomeres during meiotic prophase. *Genes Cells* **3:** 587–601. doi:10.1046/j.1365-2443.1998.00215.x

Hou H, Kyriacou E, Thadani R, Klutstein M, Chapman JH, Cooper JP. 2021. Centromeres are dismantled by foundational meiotic proteins Spo11 and Rec8. *Nature* **591:** 671–676. doi:10.1038/s41586-021-03279-8

Ishiguro KI. 2019. The cohesin complex in mammalian meiosis. *Genes Cells* **24:** 6–30. doi:10.1111/gtc.12652

Jaspersen SL, Martin AE, Glazko G, Giddings TH Jr, Morgan G, Mushegian A, Winey M. 2006. The Sad1-UNC-84 homology domain in Mps3 interacts with Mps2 to connect the spindle pole body with the nuclear envelope. *J Cell Biol* **174:** 665–675. doi:10.1083/jcb.200601062

Jeon HJ, Levine MT, Lampson MA. 2025. A parent-of-origin effect on embryonic telomere elongation determines telomere length inheritance. *Curr Biol* **35:** 5081–5089.e3. doi:10.1016/j.cub.2025.08.052

Jiménez-Martín A, Pineda-Santaella A, Martín-García R, Esteban-Villafañe R, Matarrese A, Pinto-Cruz J, Camacho-Cabañas S, León-Periñán D, Terrizzano A, Daga RR, et al. 2025. Centromere positioning orchestrates telomere bouquet formation and the initiation of meiotic differentiation. *Nat Commun* **16:** 837. doi:10.1038/s41467-025-56049-9

Kanoh J, Ishikawa F. 2001. Sprap1 and spRif1, recruited to telomeres by Taz1, are essential for telomere function in fission yeast. *Curr Biol* **11:** 1624–1630. doi:10.1016/S0960-9822(01)00503-6

Klutstein M, Fennell A, Fernández-Álvarez A, Cooper JP. 2015. The telomere bouquet regulates meiotic centromere assembly. *Nat Cell Biol* **17:** 458–469. doi:10.1038/ncb3132

Laurell E, Beck K, Krupina K, Theerthagiri G, Bodenmiller B, Horvath P, Aebersold R, Antonin W, Kutay U. 2011. Phosphorylation of Nup98 by multiple kinases is crucial for NPC disassembly during mitotic entry. *Cell* **144:** 539–550. doi:10.1016/j.cell.2011.01.012

Lee J, Hirano T. 2011. RAD21L, a novel cohesin subunit implicated in linking homologous chromosomes in mammalian meiosis. *J Cell Biol* **192:** 263–276. doi:10.1083/jcb.201008005

Lee CY, Conrad MN, Dresser ME. 2012. Meiotic chromosome pairing is promoted by telomere-led chromosome movements independent of bouquet formation. *PLoS Genet* **8:** e1002730. doi:10.1371/journal.pgen.1002730

Lee CY, Horn Henning F, Stewart Colin L, Burke B, Bolcun-Filas E, Schimenti John C, Dresser Michael E, Pezza Roberto J. 2015. Mechanism and regulation of rapid telomere prophase movements in mouse meiotic chromosomes. *Cell Rep* **11:** 551–563. doi:10.1016/j.celrep.2015.03.045

Lee CY, Bisig CG, Conrad MM, Ditamo Y, de Almeida LP, Dresser ME, Pezza RJ. 2020a. Extranuclear structural components that mediate dynamic chromosome movements in yeast meiosis. *Curr Biol* **30:** 1207–1216.e4. doi:10.1016/j.cub.2020.01.054

Lee CY, Bisig CG, Conrad MN, Ditamo Y, Previato de Almeida L, Dresser ME, Pezza RJ. 2020b. Telomere-led meiotic chromosome movements: recent update in structure and function. *Nucleus* **11:** 111–116. doi:10.1080/19491034.2020.1769456

Li R, Bitoun E, Altemose N, Davies RW, Davies B, Myers SR. 2019. A high-resolution map of non-crossover events reveals impacts of genetic diversity on mammalian meiotic recombination. *Nat Commun* **10:** 3900. doi:10.1038/s41467-019-11675-y

MacQueen AJ, Phillips CM, Bhalla N, Weiser P, Villeneuve AM, Dernburg AF. 2005. Chromosome sites play dual roles to establish homologous synapsis during meiosis in *C. elegans*. *Cell* **123:** 1037–1050. doi:10.1016/j.cell.2005.09.034

Mefford HC, Trask BJ. 2002. The complex structure and dynamic evolution of human subtelomeres. *Nat Rev Genet* **3:** 91–102. doi:10.1038/nrg727

Moiseeva V, Amelina H, Collopy LC, Armstrong CA, Pearson SR, Tomita K. 2017. The telomere bouquet facilitates meiotic prophase progression and exit in fission yeast. *Cell Discov* **3:** 17041. doi:10.1038/celldisc.2017.41

Morimoto A, Shibuya H, Zhu X, Kim J, Ishiguro K, Han M, Watanabe Y. 2012. A conserved KASH domain protein associates with telomeres, SUN1, and dynactin during mammalian meiosis. *J Cell Biol* **198:** 165–172. doi:10.1083/jcb.201204085

Muñoz-Centeno MC, McBratney S, Monterrosa A, Byers B, Mann C, Winey M. 1999. *Saccharomyces cerevisiae MPS2* encodes a membrane protein localized at the spindle pole body and the nuclear envelope. *Mol Biol Cell* **10:** 2393–2406. doi:10.1091/mbc.10.7.2393

Mytlis A, Kumar V, Qiu T, Deis R, Hart N, Levy K, Masek M, Shawahny A, Ahmad A, Eitan H, et al. 2022. Control of meiotic chromosomal bouquet and germ cell morphogenesis by the zygotene cilium. *Science* **376:** eabh3104. doi:10.1126/science.abh3104

Nambiar M, Smith GR. 2018. Pericentromere-specific cohesin complex prevents meiotic pericentric DNA double-strand breaks and lethal crossovers. *Mol Cell* **71:** 540–553. doi:10.1016/j.molcel.2018.06.035

Nozaki T, Chang F, Weiner B, Kleckner N. 2021. High temporal resolution 3D live-cell imaging of budding yeast meiosis defines discontinuous actin/telomere-mediated chromosome motion, correlated nuclear envelope deformation and actin filament dynamics. *Front Cell Dev Biol* **9:** 687132. doi:10.3389/fcell.2021.687132

Olaya I, Burgess SM, Rog O. 2024. Formation and resolution of meiotic chromosome entanglements and interlocks. *J Cell Sci* **137:** jcs262004. doi:10.1242/jcs.262004

Palsson G, Hardarson MT, Jonsson H, Steinthorsdottir V, Stefansson OA, Eggertsson HP, Gudjonsson SA, Olason PI, Gylfason A, Masson G, et al. 2025. Complete human recombination maps. *Nature* **639:** 700–707. doi:10.1038/s41586-024-08450-5

Pan J, Sasaki M, Kniewel R, Murakami H, Blitzblau Hannah G, Tischfield Sam E, Zhu X, Neale Matthew J, Jasin M, Socci Nicholas D, et al. 2011. A hierarchical combination of factors shapes the genome-wide topography of yeast meiotic recombination initiation. *Cell* **144:** 719–731. doi:10.1016/j.cell.2011.02.009

Pazhayam NM, Turcotte CA, Sekelsky J. 2021. Meiotic crossover patterning. *Front Cell Dev Biol* **9:** 681123. doi:10.3389/fcell.2021.681123

Pech MF, Garbuzov A, Hasegawa K, Sukhwani M, Zhang RJ, Benayoun BA, Brockman SA, Lin S, Brunet A, Orwig KE, et al. 2015. High telomerase is a hallmark of undifferentiated spermatogonia and is required for maintenance of male germline stem cells. *Genes Dev* **29:** 2420–2434. doi:10.1101/gad.271783.115

Pendlebury DF, Fujiwara Y, Tesmer VM, Smith EM, Shibuya H, Watanabe Y, Nandakumar J. 2017. Dissecting the telomere–inner nuclear membrane interface formed in meiosis. *Nat Struct Mol Biol* **24:** 1064–1072. doi:10.1038/nsmb.3493

Prasada Rao HB, Sato T, Challa K, Fujita Y, Shinohara M, Shinohara A. 2021. Phosphorylation of luminal region of the SUN-domain protein Mps3 promotes nuclear envelope localization during meiosis. *eLife* **10:** e63119. doi:10.7554/eLife.63119

Pratto F, Brick K, Cheng G, Lam KWG, Cloutier JM, Dahiya D, Wellard SR, Jordan PW, Camerini-Otero RD. 2021. Meiotic recombination mirrors patterns of germline replication in mice and humans. *Cell* **184:** 4251–4267. doi:10.1016/j.cell.2021.06.025

Ramesh MA, Malik SB, Logsdon JM Jr. 2005. A phylogenomic inventory of meiotic genes: evidence for sex in *Giardia* and an early eukaryotic origin of meiosis. *Curr Biol* **15:** 185–191. doi:10.1016/j.cub.2005.01.003

Robinson LG, Kalmbach K, Sumerfield O, Nomani W, Wang F, Liu L, Keefe DL. 2024. Telomere dynamics and reproduction. *Fertil Steril* **121:** 4–11. doi:10.1016/j.fertnstert.2023.11.012

Saito TT, Tougan T, Okuzaki D, Kasama T, Nojima H. 2005. Mcp6, a meiosis-specific coiled-coil protein of *Schizosaccharomyces pombe*, localizes to the spindle pole body and is required for horsetail movement and recombination. *J Cell Sci* **118:** 447–459. doi:10.1242/jcs.01629

Saito K, Sakai C, Kawasaki T, Sakai N. 2014. Telomere distribution pattern and synapsis initiation during spermatogenesis in zebrafish. *Dev Dyn* **243:** 1448–1456. doi:10.1002/dvdy.24166

Scherthan H. 2001. A bouquet makes ends meet. *Nat Rev Mol Cell Biol* **2:** 621–627. doi:10.1038/35085086

Scherthan H, Weich S, Schwegler H, Heyting C, Härle M, Cremer T. 1996. Centromere and telomere movements during early meiotic prophase of mouse and man are associated with the onset of chromosome pairing. *J Cell Biol* **134:** 1109–1125. doi:10.1083/jcb.134.5.1109

Shibuya H, Ishiguro KI, Watanabe Y. 2014. The TRF1-binding protein TERB1 promotes chromosome movement and telomere rigidity in meiosis. *Nat Cell Biol* **16:** 145–156. doi:10.1038/ncb2896

Shibuya H, Hernández-Hernández A, Morimoto A, Negishi L, Höög C, Watanabe Y. 2015. MAJIN links telomeric DNA to the nuclear membrane by exchanging telomere cap. *Cell* **163:** 1252–1266. doi:10.1016/j.cell.2015.10.030

Sidarava V, Mearns S, Lydall D. 2025. Long telomere inheritance through budding yeast sexual cycles. *Genetics* **231:** iyaf129. doi:10.1093/genetics/iyaf129

Subramanian VV, Zhu X, Markowitz TE, Vale-Silva LA, San-Segundo PA, Hollingsworth NM, Keeney S, Hochwagen A. 2019. Persistent DNA-break potential near telomeres increases initiation of meiotic recombination on short chromosomes. *Nat Commun* **10:** 970. doi:10.1038/s41467-019-08875-x

Takeo S, Lake Cathleen M, Morais-de-Sá E, Sunkel Cláudio E, Hawley RS. 2011. Synaptonemal complex-dependent centromeric clustering and the initiation of synapsis in *Drosophila* oocytes. *Curr Biol* **21:** 1845–1851. doi:10.1016/j.cub.2011.09.044

Tanaka K, Kohda T, Yamashita A, Nonaka N, Yamamoto M. 2005. Hrs1p/Mcp6p on the meiotic SPB organizes astral microtubule arrays for oscillatory nuclear movement. *Curr Biol* **15:** 1479–1486. doi:10.1016/j.cub.2005.07.058

Tomita K, Cooper JP. 2007. The telomere bouquet controls the meiotic spindle. *Cell* **130:** 113–126. doi:10.1016/j.cell.2007.05.024

Trelles-Sticken E, Loidl J, Scherthan H. 1999. Bouquet formation in budding yeast: initiation of recombination is not required for meiotic telomere clustering. *J Cell Sci* **112:** 651–658. doi:10.1242/jcs.112.5.651

Trelles-Sticken E, Adelfalk C, Loidl J, Scherthan H. 2005. Meiotic telomere clustering requires actin for its formation and cohesin for its resolution. *J Cell Biol* **170:** 213–223. doi:10.1083/jcb.200501042

Uhlmann F. 2025. A unified model for cohesin function in sister chromatid cohesion and chromatin loop formation. *Mol Cell* **85:** 1058–1071. doi:10.1016/j.molcel.2025.02.005

Vincenten N, Kuhl LM, Lam I, Oke A, Kerr ARW, Hochwagen A, Fung J, Keeney S, Vader G, Marston AL. 2015. The kinetochore prevents centromere-proximal crossover recombination during meiosis. *eLife* **4:** e10850. doi:10.7554/eLife.10850

Wanat JJ, Kim KP, Koszul R, Zanders S, Weiner B, Kleckner N, Alani E. 2008. Csm4, in collaboration with Ndj1, mediates telomere-led chromosome dynamics and recombination during yeast meiosis. *PLoS Genet* **4:** e1000188. doi:10.1371/journal.pgen.1000188

Wang Y, Chen Y, Chen J, Wang L, Nie L, Long J, Chang H, Wu J, Huang C, Lei M. 2019. The meiotic TERB1-TERB2–MAJIN complex tethers telomeres to the nuclear envelope. *Nat Commun* **10:** 564. doi:10.1038/s41467-019-08437-1

Yamamoto A, West RR, McIntosh JR, Hiraoka Y. 1999. A cytoplasmic dynein heavy chain is required for oscillatory nuclear movement of meiotic prophase and efficient meiotic recombination in fission yeast. *J Cell Biol* **145:** 1233–1250. doi:10.1083/jcb.145.6.1233

Yoshida M, Katsuyama S, Tateho K, Nakamura H, Miyoshi J, Ohba T, Matsuhara H, Miki F, Okazaki K, Haraguchi T, et al. 2013. Microtubule-organizing center formation at telomeres induces meiotic telomere clustering. *J Cell Biol* **200:** 385–395. doi:10.1083/jcb.201207168

Zhang K, Tarczykowska A, Gupta DK, Pendlebury DF, Zuckerman C, Nandakumar J, Shibuya H. 2022. The TERB1 MYB domain suppresses telomere erosion in meiotic prophase I. *Cell Rep* **38:** 110289. doi:10.1016/j.celrep.2021.110289

Zhang J, Ruiz M, Bergh PO, Henricsson M, Stojanović N, Devkota R, Henn M, Bohlooly-Y M, Hernández-Hernández A, Alsheimer M, et al. 2024. Regulation of meiotic telomere dynamics through membrane fluidity promoted by AdipoR2-ELOVL2. *Nat Commun* **15:** 2315. doi:10.1038/s41467-024-46718-6

Zickler D. 1999. The synaptonemal complex: a structure necessary for pairing, recombination or organization of the meiotic chromosome? *J Soc Biol* **193:** 17–22. doi:10.1051/jbio/1999193010017

Zickler D, Kleckner N. 1998. The leptotene–zygotene transition of meiosis. *Annu Rev Genet* **32:** 619–697. doi:10.1146/annurev.genet.32.1.619

Zickler D, Kleckner N. 2016. A few of our favorite things: pairing, the bouquet, crossover interference and evolution of meiosis. *Semin Cell Dev Biol* **54:** 135–148. doi:10.1016/j.semcdb.2016.02.024

# Structural Biology of Telomerase and Associated Factors

Zala Sekne,[1] Patryk Ludzia,[1] Sebastian Balch,[1] and Thi Hoang Duong Nguyen

MRC Laboratory of Molecular Biology, Trumpington, Cambridge CB2 0QH, United Kingdom

*Correspondence:* knguyen@mrc-lmb.cam.ac.uk

Telomerase ribonucleoprotein (RNP) plays a crucial role in maintaining telomere length by processively adding telomeric repeats to the 3′ ends of chromosomes. Telomerase activation is linked to cancer, while mutations that compromise telomerase function result in diseases such as dyskeratosis congenita. The synthesis of telomeric repeats necessitates two core telomerase components: telomerase reverse transcriptase (TERT) and telomerase RNA (TER). However, cellular telomerase holoenzymes encompass a diverse range of protein factors, both constitutively and transiently interacting. These factors are integral to telomerase assembly or regulation at telomeres. This review emphasizes recent advancements in structural studies of telomerase holoenzymes and their associated factors from *Tetrahymena thermophila*, *Saccharomyces cerevisiae*, *Schizosaccharomyces pombe*, and humans. These studies have significantly deepened our molecular understanding not only of the mechanism underlying telomeric repeat synthesis but also of the biological roles of telomerase-associated proteins.

Telomeres are nucleoprotein structures found at the ends of eukaryotic chromosomes. Telomeres generally consist of repetitive G-rich DNA sequences and associated proteins (Cech 2004; Wellinger and Zakian 2012; Lim and Cech 2021). Vital for genomic stability, telomeres possess a unique capability to shield the chromosome ends from nucleolytic degradation and interchromosomal fusion events (Arnoult and Karlseder 2015; de Lange 2018). Despite this protective function, linear DNA ends encounter a challenge during DNA replication, as they cannot be fully replicated by DNA-dependent DNA polymerases (Waga and Stillman 1998). This phenomenon, known as the "end-replication" problem, results in the gradual attrition of telomeres with each successive round of cell division (Levy et al. 1992; Soudet et al. 2014). Compromised telomere function precipitates genomic instability, cellular senescence, and, ultimately, apoptosis (Verdun and Karlseder 2007).

To offset the loss of telomeres during replication, a specialized reverse transcriptase (RT), called telomerase, adds telomeric DNA repeats to the 3′ termini of chromosomes (Greider and Blackburn 1985, 1987; Blackburn and Collins 2010). Unicellular organisms, such as *Tetrahymena thermophila* and *Saccharomyces cerevisiae*, constitutively express telomerase to enable

---

[1]These authors contributed equally to this work, and each of them has the right to list their name first in a curriculum vitae.

Cite this article as *Cold Spring Harb Perspect Biol* doi: 10.1101/cshperspect.a041697

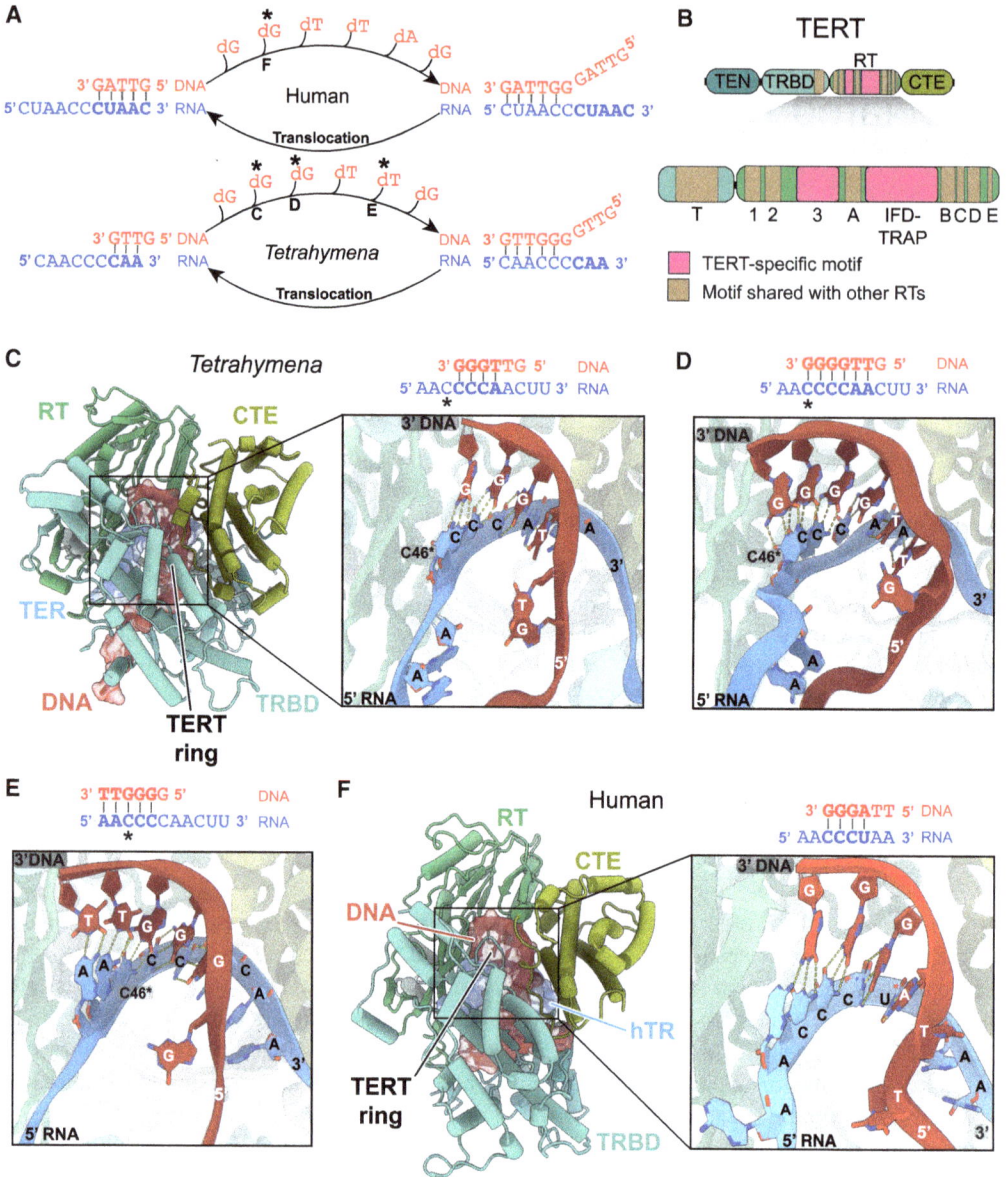

**Figure 1.** The catalytic core of telomerase and the repeat addition processivity (RAP) catalytic cycle. (*A*) Schematics of the human (*top*) and *Tetrahymena* (*bottom*) telomerase catalytic cycles. The alignment region of the template RNA is shown in bold; the remainder of the template RNA represents the template region. The asterisks show the states of the catalytic cycles, where structures have been captured, and the letters represent the corresponding panels of this figure showing the structure. Base-pairing between the RNA template and the substrate DNA is hypothetical. (*B*) General domain schematic of telomerase reverse transcriptase (TERT) with conserved motifs highlighted. Motifs shared with other RTs are colored in brown and motifs specific to TERT are colored in magenta. (TEN) Telomerase essential amino-terminal domain, (TRBD) telomerase RNA-binding domain, (RT) reverse transcriptase, (IFD-TRAP) insertion in fingers subdomain, (CTE) carboxy-terminal extension. (*C*) *Tetrahymena* TERT with domains colored as indicated (PDB 7LMA) (He et al. 2021). A close-up view of the active site shows base-pairing interactions between the RNA template and the DNA substrate. (*Continued*)

unlimited proliferation (Lundblad and Szostak 1989; Min and Collins 2009; Upton et al. 2014). In higher eukaryotes, including humans, telomerase activity is primarily confined to highly proliferative cells such as germ cells and stem cells while being absent in most somatic cells (Harley et al. 1994; Kim et al. 1994). This tight regulation prevents abnormal telomere elongation, which may lead to cellular immortalization and oncogenesis. Indeed, telomerase is up-regulated in nearly 90% of human cancers, facilitating cellular proliferation (Kim et al. 1994; Shay and Bacchetti 1997). Additionally, telomerase deficiency has been identified as a hallmark of patients with a range of premature aging disorders such as dyskeratosis congenita, Hoyeraal–Hreidarsson syndrome, aplastic anemia, and pulmonary fibrosis (Heiss et al. 1998; Mitchell et al. 1999b; Armanios and Blackburn 2012; Holohan et al. 2014; Sarek et al. 2015). Understanding the structure, function, and regulation of telomerase is, therefore, critical for elucidating the molecular mechanisms underlying these diseases and developing therapeutic interventions targeting telomerase (Shay and Wright 2006; Shay 2016).

At the core of telomerase activity lie two essential components: telomerase reverse transcriptase (TERT), a catalytic protein subunit, and telomerase RNA (TER or human telomerase RNA [hTR] in humans) (Fig. 1B; Greider and Blackburn 1989; Lingner et al. 1997b; Weinrich et al. 1997). TERT acts as the catalytic subunit containing active site motifs, and TER provides the template for repeat addition (Fig. 1A; Lingner et al. 1997b; Collins 2009). TERT motifs are shared with conventional viral RTs, including a catalytic triad of aspartic acid residues that coordinate magnesium ions during nucleotide addition (Steitz 1999). While TERT and TER are sufficient for reconstituting telomerase activity in vitro (Weinrich et al. 1997; Gandhi and Collins 1998; Zappulla et al. 2005), cellular telomerase holoenzymes comprise numerous other proteins that play crucial roles in telomerase biogenesis, localization, and regulation (Collins 2006; Egan and Collins 2012; Schmidt and Cech 2015; Roake and Artandi 2020). The significant divergence of TER across eukaryotes results in different requirements for biogenesis and the involvement of a multitude of species-specific TER-binding proteins (Collins 2009; Egan and Collins 2012; Podlevsky and Chen 2016). Like TER, telomeric proteins exhibit significant divergence throughout the phylogenetic tree of life (Podlevsky et al. 2008). They impose further regulation on telomerase, including recruitment, activation, and termination of telomerase activity at telomeres (Wu et al. 2017b). Additionally, telomerase possesses a unique feature compared to viral RTs known as repeat addition processivity (RAP), which enables telomerase to add multiple telomeric repeats in a single DNA-binding event (Fig. 1A;

---

Figure 1. (*Continued*) Note that nucleotide C46 of telomerase RNA (TER) is flipped into the active site and ready to base pair with an incoming dNTP. Only nucleotides (nt) 44–51 of TER are shown for simplicity. (*D*) Close-up view of the *Tetrahymena* telomerase active site where a dG is added to the DNA substrate compared to *C* (PDB 6D6V) (He et al. 2021). The added dG base pairs with C46 and previous base pairs are maintained. (*E*) Close-up view of the *Tetrahymena* telomerase active site. This structure represents a state of telomerase, where 5 nucleotides of the telomeric repeat have been incorporated into the DNA substrate (PDB 7LMB) (He et al. 2021). A comparison with panel *D* shows how A44 and A45 of TER have been flipped into the active site and formed base pairs with the DNA. The base-pairing between the DNA substrate and residues C49, A50, and A51 of TER is lost as the RNA has been pushed through the active site. In the schematics of DNA–RNA base-pairing shown in panels *C–E*, nucleotide C46 is indicated with an asterisk. (*F*) Human TERT with domains colored as indicated in the domain schematic (PDB 7QXA) (Sekne et al. 2022). A close-up view of the active site shows base-pairing interactions between the RNA template and the DNA substrate. Only nucleotides 48–55 of human telomerase RNA (hTR) are shown for simplicity. Note that in the view shown in *C* and *F*, the TEN domain and IFD-TRAP are hidden behind the TERT ring. For panels *C–F*, RNA is shown in blue (black labels), DNA is shown in red (white labels), with base-pairing interactions depicted as yellow dotted lines.

Collins 2011; Wu et al. 2017b). While the molecular mechanism of RAP is yet to be fully understood, elements from both TERT and TER, as well as telomerase regulatory factors, have been shown to influence RAP (Wu et al. 2017b).

Decades of genetic and biochemical studies have identified numerous telomerase holoenzyme subunits, particularly in the three commonly studied phylogenetic groups: ciliates, vertebrates, and fungi (Nguyen et al. 2019). Until recently, X-ray crystallography and nuclear magnetic resonance (NMR) have been instrumental in yielding structures of TERT and TER domains and individual telomerase-associated proteins (Wang et al. 2019). Technological advancements in single-particle cryo-electron microscopy (cryo-EM) have enabled high-resolution structure determination of *Tetrahymena* and human telomerase holoenzymes (Ghanim et al. 2021; He et al. 2021; Nguyen 2021; He and Feigon 2022). These structures not only provide molecular insights into the roles of individual subunits within the holoenzymes but also unveiled new telomerase subunits. Despite the evolutionary divergence between ciliate and vertebrate telomerases, striking parallels can be drawn between the two systems, suggesting convergence in molecular functions of TER motifs and associated proteins. Moreover, structures of telomerase with regulatory factors have been resolved, providing a molecular understanding of telomerase regulation (He et al. 2022b; Liu et al. 2022; Sekne et al. 2022). To date, no structures of yeast telomerase holoenzyme have been reported, but structures of some yeast telomerase-associated factors are available (Lue 2013; Chen et al. 2018a; Basu et al. 2021; Wang et al. 2023a).

This review aims to provide an overview of structural studies conducted on telomerase and its associated factors. We highlight how the insights gained from structural analyses have contributed to our mechanistic understanding of telomerase function and regulation while also raising new questions. A wide variety of model organisms have been used in the field, and their importance cannot be understated. However, this review will primarily discuss findings from studies conducted on *Tetrahymena*, humans,

*S. cerevisiae*, and *Schizosaccharomyces pombe*, with occasional reference to other relevant model organisms. We show the diverse structural landscapes of telomerase and its regulatory mechanisms across different species, shedding light on both conserved features and species-specific adaptations.

## THE CORE OF TELOMERASE HOLOENZYMES: TERT AND TER

TERT exhibits strong domain conservation across eukaryotes, comprising a "telomerase essential amino-terminal" (TEN) domain, a telomerase RNA-binding domain (TRBD), an RT domain, and a carboxy-terminal extension (CTE) (Fig. 1B). TERT associates with TER to form the canonical polymerase right-handed ring structure. In addition to the shared motifs with other RTs, the RT domain also contains TERT-specific motifs, including the "insertion in fingers subdomain" (IFD) and motif 3 (Fig. 1B), both of which contribute to RAP (Lue et al. 2003; Xie et al. 2010). Biochemical studies have delineated distinct functions of the TEN, TRBD, and CTE domains. The TEN domain is essential for RAP and telomerase recruitment to telomeres (Zaug et al. 2010; Robart and Collins 2011; Sexton et al. 2012; Schmidt et al. 2014). The TRBD facilitates TER binding, which is essential for telomerase assembly, while the CTE forms the polymerase thumb domain and participates in DNA substrate binding (Hossain et al. 2002; Moriarty et al. 2004; Bley et al. 2011; Huang et al. 2014; Wu and Collins 2014).

The general architecture of TERT domains was revealed first by the crystal structure of *Tribolium castaneum* TERT and later by the cryo-EM structures of human and *Tetrahymena* telomerase (Gillis et al. 2008; Jiang et al. 2018; Ghanim et al. 2021). The TRBD, RT, and CTE of TERT adopt a unique ring-like architecture, termed the TERT ring (Fig. 1C,F). The formation of the TERT ring is facilitated by extensive interactions between the TRBD and CTE, enclosing the DNA-template duplex (Mitchell et al. 2010; Jiang et al. 2018; Ghanim et al. 2021). Notably, *Tribolium* TERT lacks the TEN domain, which is flexibly tethered to the TRBD-RT-CTE domains

Cite this article as *Cold Spring Harb Perspect Biol* doi: 10.1101/cshperspect.a041697

in *Tetrahymena* and human telomerases, forming a "lid" on top of the TERT ring (Figs. 2C and 3C; Jiang et al. 2018; Ghanim et al. 2021; Lue and Autexier 2023). The IFD was first found to form a multistranded β-sheet motif in *Tetrahymena* telomerase structure and named TRAP due to its function of trapping TER (Fig. 3C; Jiang et al. 2018). This fold was also later found in human TERT (Ghanim et al. 2021). Hence, we refer to the IFD as the IFD-TRAP. IFD-TRAP stabilizes the 5′ terminal DNA nucleotides in the telome-

rase active site of both human and *Tetrahymena* telomerase (Figs. 2C and 3C; He et al. 2021; Sekne et al. 2022).

Distinct from other RTs and DNA polymerases, TERT relies on two conserved structural motifs in TER for telomeric repeat synthesis: the pseudoknot/template (PK/t) domain and stem-terminus element (STE) (Figs. 2C and 3C; Tesmer et al. 1999; Mitchell and Collins 2000; Chen and Greider 2004; Theimer and Feigon 2006). The PK/t domain harbors the DNA

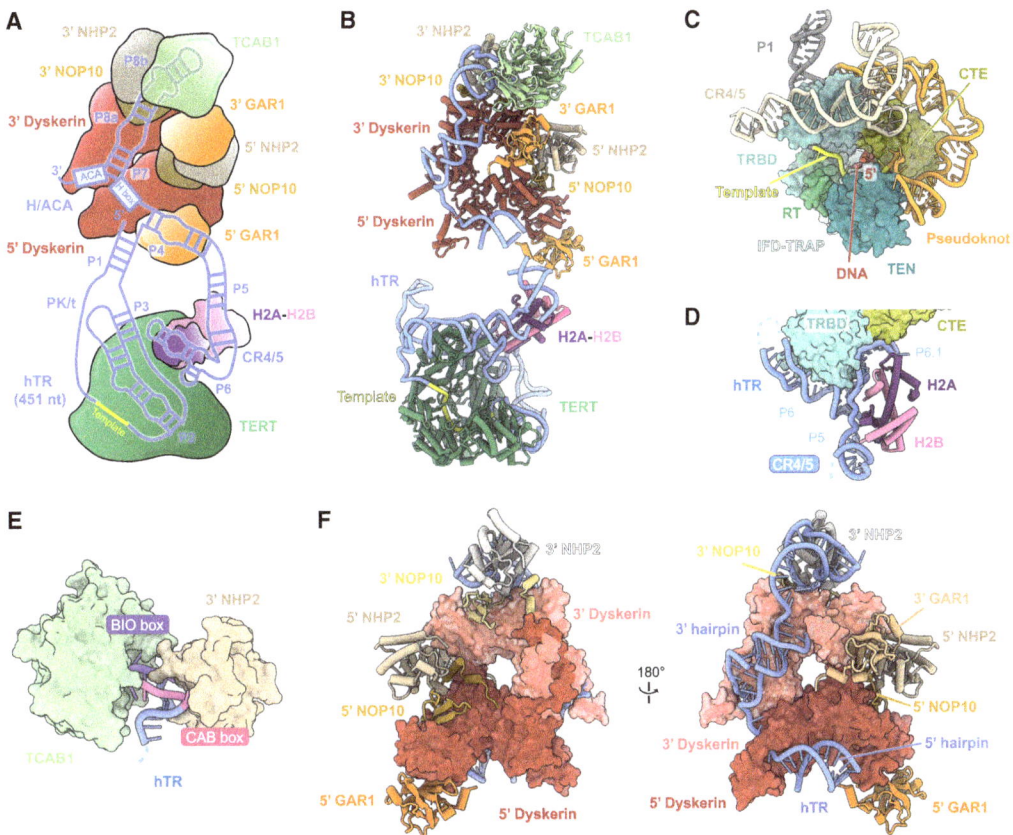

**Figure 2.** Human telomerase ribonucleoprotein (RNP). (*A*) Schematic of the human telomerase holoenzyme, corresponding to the structure shown in *B*. (*B*) Structure of the human telomerase holoenzyme (PDB 7QXA and 8OUE) (Sekne et al. 2022; Ghanim et al. 2024). (*C*) Structure of the human telomerase catalytic core depicting the organization of telomerase reverse transcriptase (TERT) and the PK/t and CR4/5 domains of hTR (PDB 7BG9) (Ghanim et al. 2021). (*D*) Interactions that the CR4/5 domain of hTR makes with the TRBD and CTE of TERT and the histone H2A-H2B dimer (PDB 7BG9) (Ghanim et al. 2021). (*E*) Interactions formed between the BIO and CAB boxes of hTR with TCAB1 and the 3′ NHP2 (PDB 8OUE) (Ghanim et al. 2024). (*F*) Structure of the human telomerase H/ACA RNP (PDB 8OUE) (Ghanim et al. 2024). TCAB1 is not shown for simplicity. The assembly of the 5′ H/ACA heterotetramer and the 3′ H/ACA heterotetramer on the 5′ and 3′ hairpin of hTR, respectively, is shown.

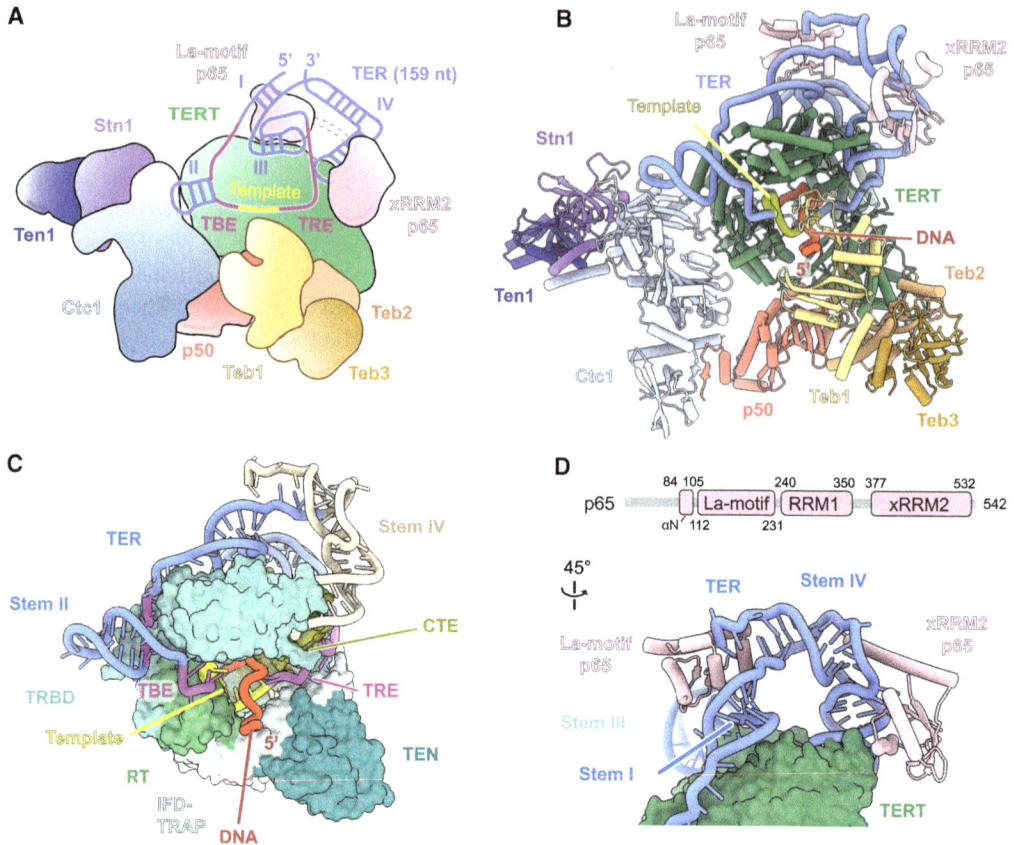

**Figure 3.** *Tetrahymena* telomerase ribonucleoprotein (RNP). (*A*) Schematic of the *Tetrahymena* telomerase holoenzyme with p50, and TEB and Ctc1-Stn1-Ten1 complexes. (*B*) Structure of the *Tetrahymena* telomerase holoenzyme (PDB 7UY5) (He et al. 2022b). (*C*) Structure of the *Tetrahymena* telomerase catalytic core depicting the organization of telomerase reverse transcriptase (TERT) and telomerase RNA (TER) domains (PDB 7UY5) (He et al. 2022b). For simplicity, other proteins are not shown. (*D*) Interactions between TERT, stem-loops I, III, and IV of TER, and xRRM2 and La-motif of p65 (PDB 7UY5) (He et al. 2022b). Domain architecture of p65 is shown as schematic. (αN) Amino-terminal α-helix, (RRM1) RNA recognition motif 1, (xRRM2) RNA recognition motif 2.

synthesis template and a highly conserved pseudoknot fold, wrapping around TRBD and CTE to position the RNA template in the telomerase active site (Figs. 2C and 3C; Jiang et al. 2018; Ghanim et al. 2021). The STE, found as either a simple stem-loop in ciliates or a more complex three-way junction in yeasts and vertebrates (conserved regions 4 and 5 [CR4/5]), interacts with TRBD and CTE (Figs. 2C and 3C; Chen et al. 2000; Lai et al. 2003; Mason et al. 2003; Zappulla and Cech 2004). Stem-loop IV of *Tetrahymena* STE and the equivalent P6.1 stem-

loop in human and yeast STE insert into the interface between TRBD and CTE of TERT, closing the TERT ring (Figs. 2C and 3C). These intricate TERT-TER interactions elucidate the structural and biochemical necessity of PK/t and STE for telomerase catalytic activity.

## A Catalytic Cycle of Telomeric Repeat Addition

The unique ability of telomerase to achieve RAP necessitates an intricate catalytic cycle. The cycle

Cite this article as *Cold Spring Harb Perspect Biol* doi: 10.1101/cshperspect.a041697

is composed of the addition of a single telomeric repeat followed by realignment of the substrate DNA to the RNA template for another repeat addition (Fig. 1A; Wu et al. 2017b). The mechanism of this catalytic cycle has been extensively dissected in numerous biochemical and biophysical studies (Hardy et al. 2001; Förstemann and Lingner 2005; Berman et al. 2011; Qi et al. 2012; Brown et al. 2014; Wu and Collins 2014; Akiyama et al. 2015; Yang and Lee 2015; Wu et al. 2017a). Recently, cryo-EM structures of DNA-bound human telomerase at the elongation stage of the catalytic cycle and *Tetrahymena* telomerase at different stages of the catalytic cycle have allowed comparison of TERT and RNA–DNA duplex conformations (Figs. 1C–F, 2B, and 3B; Ghanim et al. 2021; He et al. 2021; Wan et al. 2021). These structures reveal the details of how telomerase handles its DNA substrate.

A common feature of all these structures is the formation of a relatively short DNA–RNA template duplex of no more than 6 base pairs (bp) in the active site of TERT (Fig. 1C–F; Ghanim et al. 2021; He et al. 2021; Liu et al. 2022). The conservation of a short DNA–RNA template duplex is likely crucial for RAP. An overly stable duplex presents a high energy barrier to overcome at the end of each round of telomeric repeat synthesis. The length of the DNA–RNA duplex is limited by residue L980 of human TERT, which acts as a zipper head and restricts the formation of a longer duplex (Wan et al. 2021). At different stages of the telomerase catalytic cycle in *Tetrahymena*, there are no significant conformational changes to any part of the TERT ring, and a similar conformation of the RNA is maintained in the active site (Fig 1C–E; He et al. 2021). As one 5′ RNA base moves into the active site for nucleotide addition to the 3′ end of the DNA substrate, a 3′ RNA base is subsequently flipped out of the active site cavity at the other end of the RNA template. This allows the RNA conformation to be maintained in the active site.

*Tetrahymena* telomerase uses regions flanking the RNA template to define the template region and align its DNA substrate (Autexier and Greider 1995; Lai et al. 2002; Miller and Collins 2002; Richards et al. 2006). The regions at the 5′ and 3′ ends of the RNA template are named the template boundary element (TBE) and the template recognition element (TRE), respectively (Fig. 3A,C). These regions of RNA show some differences in conformation at the different steps of the catalytic cycle, consistent with a passive movement of RNA sliding through the telomerase active site (He et al. 2021). The 3′ TRE passes through a positively charged surface of TERT, lined by residues from the IFD-TRAP and CTE domains (Fig. 3C). Since mutagenesis of these residues decreases RAP, this surface could act as a valve-like mechanism to prevent back tracking of the RNA into the active site.

In contrast to *Tetrahymena* telomerase, human telomerase relies on base-pairing between the DNA substrate and RNA template for template definition rather than boundary elements (Qi et al. 2012; Wu and Collins 2014; Wu et al. 2017b). Another notable disparity between human and *Tetrahymena* telomerase lies in the role of the TEN domain of TERT in DNA binding. Recent structures of human telomerase have revealed a DNA "anchor site" on the TEN domain of TERT, previously hypothesized to be crucial for telomerase processivity (Lue 2005; Jacobs et al. 2006; Wyatt et al. 2007; Sealey et al. 2010; Jurczyluk et al. 2011; Robart and Collins 2011; Akiyama et al. 2015; Sekne et al. 2022). However, in *Tetrahymena* telomerase, no interactions occur between the DNA and the TEN domain of TERT (Fig. 3C; Jiang et al. 2018). Therefore, *Tetrahymena* likely employs alternative mechanisms to retain the DNA substrate on telomerase for RAP.

While multiple structures of *Tetrahymena* telomerase have provided intriguing mechanistic details on the telomerase catalytic cycle (He et al. 2021), large conformational changes are likely required for translocation of the DNA substrate and realignment to the RNA template. Further studies, possibly at a single-molecule level, are needed to further probe the dynamics and precise molecular details of translocation. Given the differences between human and *Tetrahymena* telomerase, it would be interesting to see how structures of human telomerase compare at different stages of the catalytic cycle. Additionally, telomerases from *S. cerevisiae* and

*S. pombe* exhibit significantly lower processivity than their human and *Tetrahymena* counterparts (Cohn and Blackburn 1995; Lue and Peng 1997). Capturing the structures of the yeast telomerase complexes with DNA substrates at various stages of their catalytic cycle would provide invaluable insight into molecular features that limit their processivity.

## HUMAN TELOMERASE AND ASSOCIATED FACTORS

### Overall Structure of Human Telomerase

The overall architecture and composition of human telomerase holoenzyme were first revealed by the cryo-EM structure of human telomerase determined at 8 Å resolution and then refined in more details by subsequent high-resolution cryo-EM structures of the complex (Nguyen et al. 2018; Ghanim et al. 2021; Wan et al. 2021). Unlike the more compact shape observed in *Tetrahymena* telomerase, human telomerase exhibits an elongated dumbbell configuration. In this arrangement, TERT and a histone H2A-H2B dimer are segregated to one end, referred to as the catalytic core, while two sets of H and ACA box (H/ACA) proteins and a telomerase Cajal body localization factor (TCAB1) occupy the opposite end (Fig. 2A,B; Ghanim et al. 2021; Wan et al. 2021). hTR acts as a flexible tether, connecting the two ends (Fig. 2A,B). The catalytic core is responsible for DNA synthesis, as described in detail above. Within the catalytic core, TERT closely interacts with the conserved PK/t and CR4/5 domains of hTR (Fig. 2C). On the other side, the H/ACA proteins and TCAB1 bind to the 3′ H/ACA domain of hTR, forming a complete H/ACA ribonucleoprotein (RNP) complex. Further discussion will delve into the associated factors of human telomerase, including holoenzyme components, as well as recruitment and termination factors, providing additional insights into the intricate workings of the complex.

### Histones H2A-H2B

A histone H2A-H2B dimer was first identified as part of human telomerase in the 3.8 Å structure of the telomerase catalytic core (Ghanim et al. 2021), a finding corroborated by subsequent structures of human telomerase (Wan et al. 2021; Liu et al. 2022; Sekne et al. 2022). The H2A-H2B dimer binds the P5 and P6.1 stem-loops within the CR4/5 domain of hTR, collectively forming a Y-shaped three-way junction with the P6 stem-loop (Fig. 2D; Ghanim et al. 2021). Notably, the importance of the P6.1 stem-loop for telomerase activity has been underscored in previous studies (Chen et al. 2002; Robart and Collins 2010), yet it adopts heterogeneous conformations in the absence of TERT (Palka et al. 2020). Upon assembling into telomerase, the P6.1 stem is stabilized by both TERT and the histone H2A-H2B dimer (Ghanim et al. 2021).

During the assembly of *Tetrahymena* telomerase, p65 binds and aids the folding of stem-loop IV (see below for more details) (Fig. 3D), which is functionally equivalent to the P6.1 stem of hTR (Berman et al. 2010; Jiang et al. 2018). It is possible that the histone H2A-H2B dimer serves a similar role as p65 during the human telomerase assembly (Ghanim et al. 2021). Additionally, the histone H2A-H2B dimer may facilitate the coupling of DNA replication with telomere extension by anchoring telomerase to the newly synthesized DNA via interactions with the partially formed histone H3-H4 tetramer (Ghanim et al. 2021). Given that only a subset of cryo-EM telomerase particles are associated with the histone dimer, their interactions may be influenced by the cell cycle and/or subcellular localization of telomerase. Therefore, numerous questions persist regarding the precise role of the histone proteins in telomere biology.

### The H/ACA Proteins and TCAB1

The 3′ end of hTR shares characteristics with the H/ACA RNA family, featuring two tandem RNA hairpin structures (5′ and 3′ hairpin), a conserved H box between them, and an ACA box at the 3′ end of the second hairpin (Fig. 2A,B,F; Mitchell et al. 1999a; Egan and Collins 2010). Like other H/ACA RNA family members, such as the small nucleolar RNAs (snoRNAs) and small Cajal body RNAs (scaRNAs), each RNA hairpin associates with an H/ACA heterotetramer of dyskerin, GAR1, NHP2, and NOP10 (Meier 2005; Ye 2007; Hamma and Ferre-D'Amare 2010). We refer to the H/ACA proteins as either 5′ or 3′ depending on the RNA hairpin of hTR that they bind to. Recent cryo-EM structures of telomerase H/ACA RNP, particularly at an unprecedented 2.7 Å resolution, elucidate molecular interactions formed between the H/ACA RNP subunits (Ghanim et al. 2021, 2024; Liu et al. 2022).

Within the telomerase H/ACA RNP, the two heterotetramers asymmetrically bind to the hTR (Nguyen et al. 2018; Ghanim et al. 2021, 2024). The 3′ heterotetramer engages with the 3′ hairpin of hTR through dyskerin, NHP2, NOP10, and TCAB1 (Fig. 2B,E,F), resembling structures seen in the archaeal single-hairpin RNPs (Li and Ye 2006; Ghanim et al. 2021, 2024). In contrast, the 5′ heterotetramer uses only dyskerin to bind to the atypical 5′ hairpin, which extends from the H/ACA RNP into the catalytic core (Fig. 2B, F). This suggests that the 5′ heterotetramer is less stably bound to hTR than the 3′ heterotetramer. Cryo-EM structures, however, reveal that the dyskerin, NHP2, and NOP10 subunits of the 5′ heterotetramer form multiple interactions with the 3′ dyskerin and 3′ GAR1 to enhance its association within telomerase (Fig. 2F; Ghanim et al. 2021, 2024; Wan et al. 2021). The extensive interheterotetramer interactions likely stabilize binding of the 5′ heterotetramer to hTR. This is also supported by the observation that disrupting the 3′ hairpin adversely impacts telomerase activity and hTR accumulation, whereas the disruption of the 5′ hairpin does not (Egan and Collins 2010).

Furthermore, mutations in the H/ACA proteins lead to dyskeratosis congenita, a disease associated with impaired telomere maintenance. Many mutations in patients with dyskeratosis congenita and aplastic anemia cluster at the interfaces between the two H/ACA heterotetramers (Heiss et al. 1998; Mitchell et al. 1999b; Angrisani et al. 2014; Sarek et al. 2015; Armanios 2022). Dyskerin, which is the key orchestrator of the interheterotetramer interactions, serves as a prominent mutation hotspot (Nguyen et al. 2018; Ghanim et al. 2021, 2024). The structural insight into the underlying molecular pathology

of these disease mutations has great potential for future structure-guided drug design. Therapeutics could be developed to either stabilize or disrupt telomerase to treat premature aging diseases or cancer, respectively.

TCAB1 is essential for hTR accumulation in the Cajal bodies (Tycowski et al. 2009; Venteicher et al. 2009). In the human telomerase structure, TCAB1 binds to the terminal stem-loop of hTR harboring the conserved CAB and BIO boxes (Fig. 2A,E; Ghanim et al. 2021; Wan et al. 2021). In the absence of TCAB1, TERT and hTR localize to different nuclear compartments (Klump et al. 2023), and hTR is misfolded (Chen et al. 2018b). However, telomere maintenance remains unaffected by the absence of TCAB1 or coilin, an essential structural component of Cajal bodies (Vogan et al. 2016). Moreover, the loss of TCAB1 does not impact cellular hTR levels (Venteicher et al. 2009). Depletion of coilin results in defects in telomerase recruitment, which can be rectified by overexpression of telomerase (Stern et al. 2012). In contrast, in the absence of TCAB1, overexpression of telomerase fails to rescue the recruitment defect. Therefore, further studies are needed to fully understand the role of TCAB1 in telomere maintenance.

Members of the H/ACA RNP family use the dyskerin subunit for pseudouridylation of ribosomal and spliceosomal RNA, a posttranscriptional modification required for cell survival (Ganot et al. 1997; Carlile et al. 2014; Li et al. 2015; Garus and Autexier 2021). Prior to the first cryo-EM structure of telomerase H/ACA RNP, no double-hairpin eukaryotic H/ACA RNP structures were available (Nguyen et al. 2018). Besides providing insights into telomerase assembly, the recent 2.7 Å cryo-EM structure offers a basis for understanding eukaryotic pseudouridylation by the H/ACA RNPs (Ghanim et al. 2024). While the insights into pseudouridylation are outside the scope of this review, hTR was observed to act as a pseudosubstrate. This could be an autoinhibitory mechanism to prevent telomerase from performing undesired pseudouridylation of cellular RNAs (Ghanim et al. 2024). It can also explain why telomerase has never been shown to have pseudouridylation activity.

## Recruitment and Processivity Factors: TPP1 and POT1

Human telomerase is highly regulated by telomere-associated proteins, notably the multiprotein complex shelterin (Hockemeyer and Collins 2015; Schmidt and Cech 2015). Shelterin consists of double-stranded DNA-binding proteins, TRF1 and TRF2, the latter also binding to RAP1. TRF1 and TRF2 are bridged via TIN2 to a single-stranded (ss)-binding protein POT1 and telomerase recruitment factor TPP1 (Fig. 4B; de Lange 2018). Shelterin plays a dual role in regulating telomerase activity at telomeres. Shelterin promotes the formation of telomeric loops (T-loops), in which the telomeric ssDNA overhang invades the double-stranded telomeric DNA. T-loops limit telomerase access to the 3′ telomeric end (Griffith et al. 1999). On the other hand, telomerase is recruited and stimulated through transient interaction with the TPP1-POT1 heterodimer (Fig. 4B,C; Lei et al. 2005; Wang et al. 2007; Nandakumar et al. 2012; Sexton et al. 2012, 2014; Zhong et al. 2012; Pike et al. 2019). While depletion of TPP1 or TIN2 reduces telomerase recruitment to telomeres, depletion of POT1 does not affect telomerase localization to telomeres (Abreu et al. 2010). Despite this evidence of POT1 being dispensable for TPP1-mediated telomerase recruitment (Abreu et al. 2010), other biochemical studies suggest that POT1 binds the 3′ end of telomeric DNA and tethers TPP1 nearby (Baumann and Cech 2001; Ye et al. 2004). This allows the amino-terminal OB-fold of TPP1 to bind telomerase and position the telomeric DNA substrate in the telomerase active site (Nandakumar et al. 2012; Sexton et al. 2012; Zhong et al. 2012).

Only recently were structures of human telomerase with TPP1 and POT1 determined by cryo-EM (Liu et al. 2022; Sekne et al. 2022). These structures revealed the full extent of the interactions between the OB-fold domain of TPP1 and the TEN domain and the IFD-TRAP of TERT (Fig. 4B,C). In agreement with previous biochemical and genetic data, the interactions are guided by the TPP1 glutamate (E) and leucine (L)-rich (TEL) patch and the amino terminus of OB-fold (NOB) of TPP1 (Armbrus-

ter et al. 2001; Sealey et al. 2010; Nandakumar et al. 2012; Sexton et al. 2012; Zhong et al. 2012; Grill et al. 2018).

POT1 was initially thought to be indirectly associated with telomerase via TPP1 and the telomeric DNA. However, in the human telomerase-TPP1-POT1 structure, POT1 directly binds and stabilizes the otherwise flexible TEN domain of TERT (Fig. 4C; Sekne et al. 2022). The structure reveals direct binding between the TEN domain and the 5′ end of the DNA substrate, allowing the identification of the DNA anchor site on TERT (Fig. 4C). Notably, the DNA anchor site involves residues not only in the TEN domain as initially thought, but also in the IFD-TRAP region (Sekne et al. 2022). Numerous known and disease-related mutations are located at this interface, leading to telomerase with a lower RAP (Armbruster et al. 2001; Robart and Collins 2010; Zaug et al. 2010; Sexton et al. 2012). Residues in POT1 at or near the interface with telomerase are also mutated in tumors (Martínez-Jiménez et al. 2020). However, their effects on telomere maintenance have not been fully characterized. Together with previous kinetic studies (Latrick and Cech 2010), these structural insights suggest that TPP1-POT1 enhances telomerase processivity by stabilizing the DNA substrate on telomerase.

Paradoxically, POT1 has also been shown to inhibit telomerase by competing for the DNA substrate (Kelleher et al. 2005; Xu et al. 2019). Deletion or depletion of POT1 and mutations that reduce or abolish the ability of POT1 to bind telomeric ssDNA result in abnormally elongated telomeres (Loayza and De Lange 2003; Ye et al. 2004; Hockemeyer et al. 2006; Zhong et al. 2012; Glousker et al. 2020). The reported human telomerase-TPP1-POT1 structure likely represents a downstream state in which the 3′ end of DNA is already poised for extension by telomerase. Furthermore, TIN2 has also been suggested to stimulate telomerase processivity together with TPP1 and POT1 (Pike et al. 2019). Despite being included in the cryo-EM sample, TIN2 was unresolved in the cryo-EM structure (Sekne et al. 2022). Therefore, it remains to be explored how the 3′ end of the telomeric DNA is being negotiated

Cite this article as *Cold Spring Harb Perspect Biol* doi: 10.1101/cshperspect.a041697

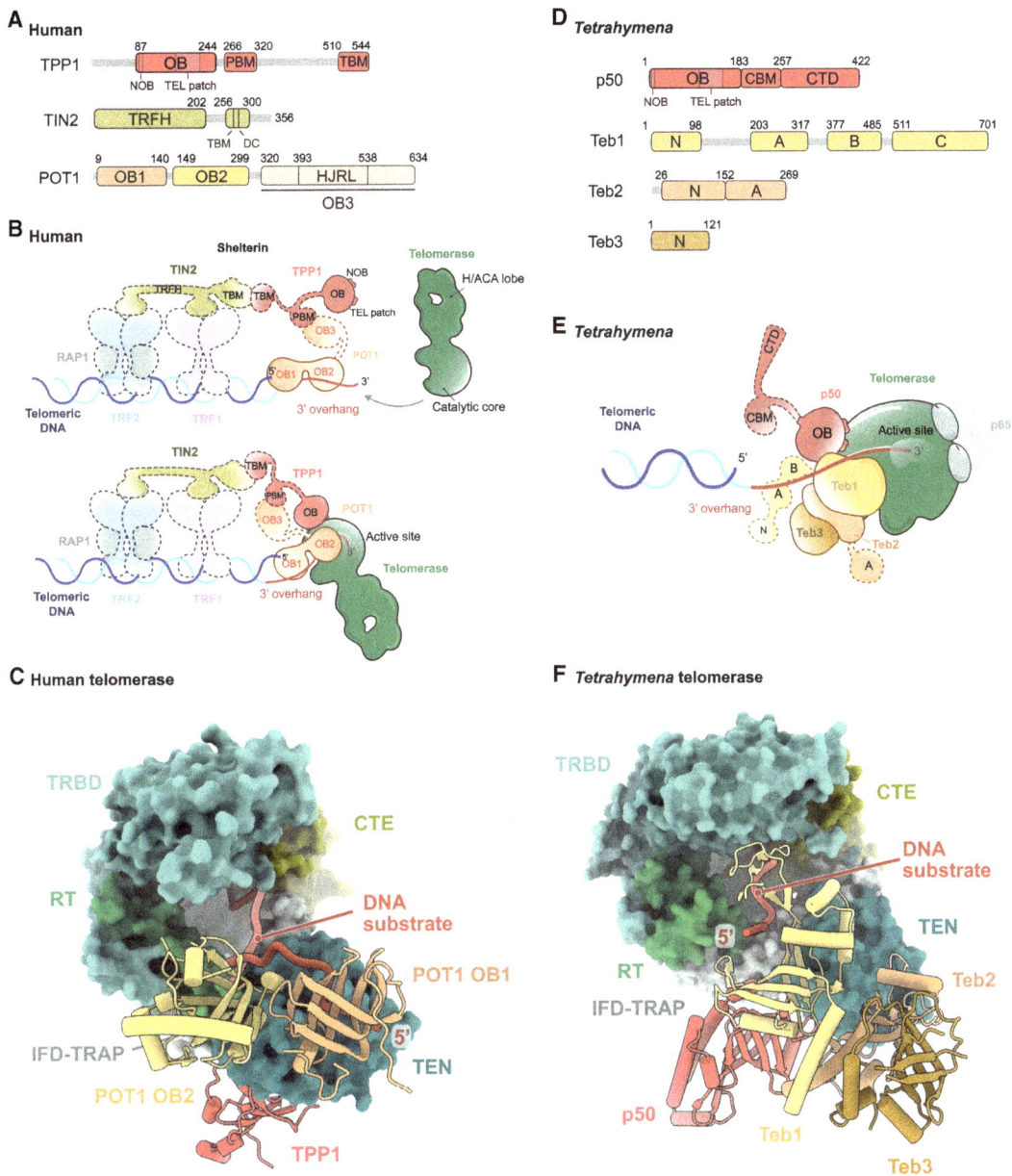

**Figure 4.** Telomerase recruitment in humans and *Tetrahymena*. (*A*) Domain architecture of human shelterin components TPP1, TIN2, and POT1. (OB) Oligosaccharide/oligonucleotide-binding fold, (PBM) POT1-binding motif, (TBM) TIN2-binding motif on TPP1 or TPP1-binding motif on TIN2, (TRFH) TRF homology, (HJRL) Holliday-junction resolvase-like. (*B*) Schematic of human shelterin and recruitment of telomerase by shelterin components TPP1 and POT1. Domains not observed in the structure shown in panel *C* are outlined with a dashed line. (*C*) Structure of substrate-bound human telomerase catalytic core with TPP1 and POT1 (PDB 7QXB) (Sekne et al. 2022). For simplicity, only telomerase reverse transcriptase (TERT) domains of the telomerase catalytic core are shown, and hTR is not shown. (*D*) Domain architecture of the *Tetrahymena* p50, Teb1, Teb2, and Teb3 proteins of the TEB complex. (OB) Oligosaccharide/oligonucleotide-binding fold, (CBM) CST-binding motif, (CTD) carboxy-terminal domain. (*E*) Schematic of *Tetrahymena* telomerase recruitment to telomeres by the p50 and TEB complex. Domains not observed in the structure shown in panel *F* are outlined with a dashed line. (*F*) Structure of substrate-bound *Tetrahymena* telomerase catalytic core with p50 and TEB complex (PDB 7LMA) (He et al. 2021). For simplicity, only TERT domains, p50, and TEB complex of the telomerase catalytic core are shown, and TER is not shown.

between POT1 and telomerase and what the role of TIN2 in telomerase regulation is.

## Termination Factors: CTC1-STN1-TEN1 (CST) Complex

The CTC1-STN1-TEN1 (CST) complex is a conserved ssDNA-binding complex that associates with telomeres from yeast to humans (Price et al. 2010). Previous biochemical and recent structural studies indicate that TPP1 and POT1 recruit the telomerase termination factor CST in humans (Chen et al. 2012; Takai et al. 2016; Cai et al. 2023; Wang et al. 2023a). CST inhibits telomerase and recruits Polα-primase for the extension of the C-strand at telomeres (Cai and de Lange 2023). Various cryo-EM structures have confirmed that CST binds telomeric ssDNA via the OB-F and OB-G domains of CTC1, which form the carboxyl terminus of CTC1 together with the OB-D and OB-E domains (Fig. 5A; Lim et al. 2020; Cai et al. 2022; He et al. 2022a). The carboxyl terminus of CTC1 binds the amino terminus of STN1 (STN1-n), which bridges CTC1 with the TEN1 (Fig. 5A). The carboxy-terminal domain of STN1 (STN1-c) forms alternative conformations known as the "up-head" or "down-arm" (Fig. 5A). The "up-head" conformation is associated with the monomeric CST, while the shift to the "down-arm" conformation supports the CST decameric assembly, which is promoted by ssDNA binding (Lim et al. 2020). Thus, the CST decamer was proposed to function as a nucleosome-like unit for G-rich ssDNA (Lim et al. 2020).

The structure of human CST in complex with TPP1-POT1 showed that the OB2 domain of POT1 interacts with STN1-c and CTC1 at the DNA-binding site of CST (Cai et al. 2023). In this conformation, POT1 obstructs the DNA-binding site on the CTC1, while the DNA-binding interface of POT1 faces the solvent. Hence, the mechanism of DNA handover from POT1 to the CST complex for switching between G- to C-strand synthesis remains elusive. Furthermore, the binding sites of telomerase and CST on TPP1-POT1 are not mutually exclusive, suggesting that CST does not inhibit telomerase by competing for TPP1-POT1 binding. This is in agreement with the *Tetrahymena* telomerase structure in which CST and p50-TEB complex (homologous to TPP1-POT1) coexist (see below for more details) (Fig. 5C; He et al. 2022b). Mutants of CST with reduced DNA affinity inhibit telomerase more poorly, suggesting that CST-mediated inhibition of telomerase relies on the DNA binding (Zaug et al. 2021). Additionally, Zaug et al. (2021) demonstrated that CST cannot inhibit an ongoing human telomerase extension but can prevent telomerase initiation by binding to the ssDNA. CST inhibits telomerase likely by a more complex mechanism than just by ssDNA sequestration. Therefore, the precise mechanisms governing telomerase termination at telomeres remain unclear.

## *TETRAHYMENA* TELOMERASE AND ASSOCIATED FACTORS

### Overall Structure of *Tetrahymena* Telomerase and p65

Telomerase was first discovered in *Tetrahymena* and has since served as a pivotal model system for exploring the structure and function of telomerase (Greider and Blackburn 1985). Over the years, a series of cryo-EM structures of *Tetrahymena* telomerase have unveiled its overall architecture and associated components, corroborating much of the previous biochemical research (Jiang et al. 2015, 2018; He et al. 2021). In essence, *Tetrahymena* telomerase holoenzyme consists of TER, the TERT subunit, and eight accessory proteins (Fig. 3A,B). The "core" of the RNP is made up of TER, TERT, and the La-related protein p65. The complete holoenzyme structure elucidates two major binding sites of p65 with TER as shown in Figure 3D (Jiang et al. 2018). Previous crystal structures highlighted the interaction of the p65 carboxyl terminus with stem-loop IV of TER, where p65 induces a large conformational change (105° bend) in TER to facilitate telomerase assembly (Singh et al. 2012). More recent cryo-EM analysis also identifies p65 binding at the 5' end and poly(U) 3' end of TER, underscoring the pivotal role of p65 in reshaping TER for telomerase assembly (He et al. 2021). This structural evi-

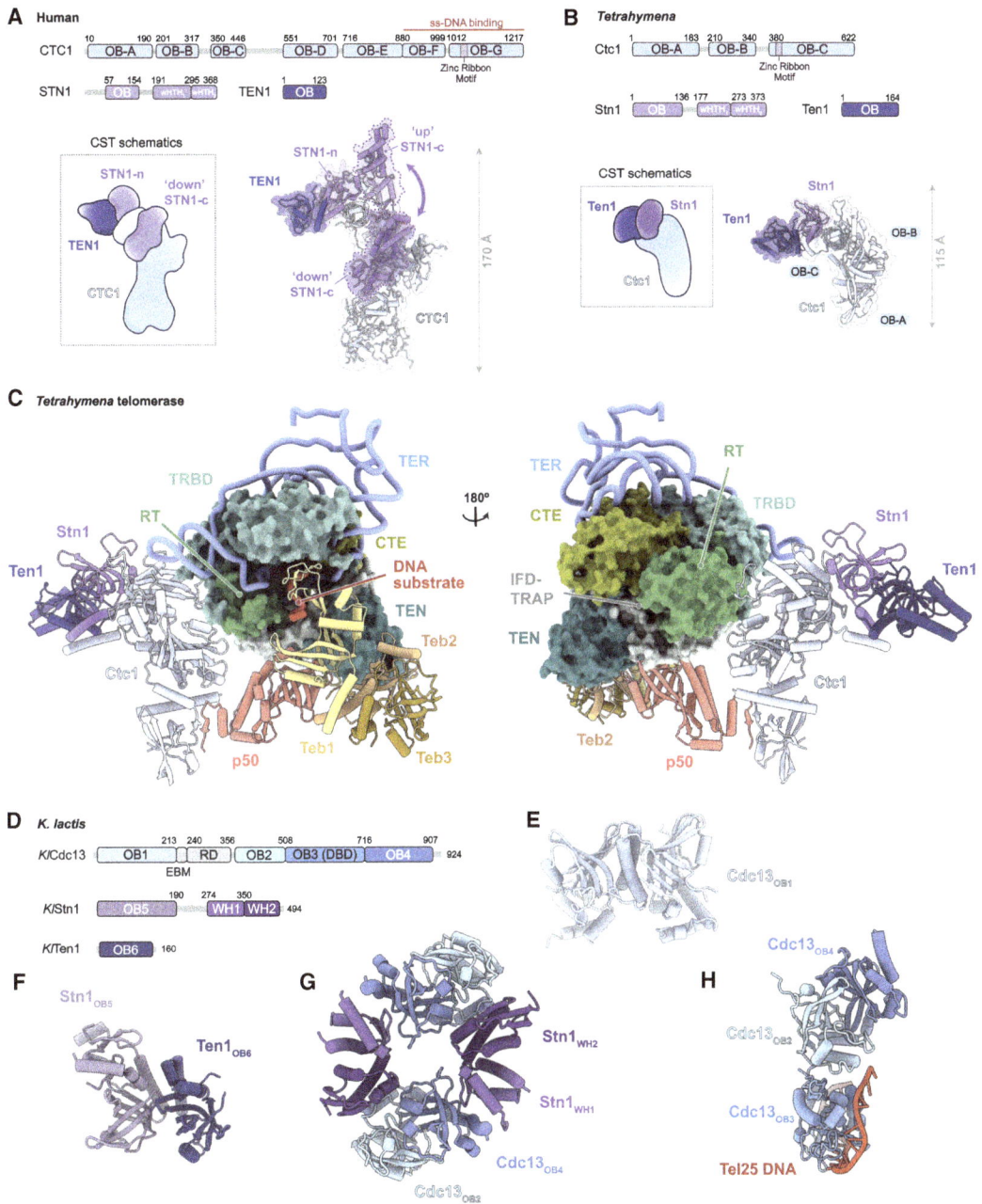

**Figure 5.** Telomerase termination factors. (*A*) Domain architecture, schematic, and model of the human CTC1, STN1, and TEN1 complex (PDB 6W6W and 8SOK) (Lim et al. 2020; Cai et al. 2023). (OB) Oligosaccharide/oligonucleotide-binding fold, (wHTH) winged helix-turn-helix motif. Model of the CTC1-STN1-TEN1 (CST) monomer shows the STN1-c domain in the "up" and "down" conformation. (*B*) Domain architecture, schematic, and structure of the *Tetrahymena* Ctc1, Stn1, and Ten1 complex (PDB 7UY5) (He et al. 2022b). (*C*) Structure of *Tetrahymena* telomerase holoenzyme including p50, TEB, and CST complexes (PDB 7UY5) (He et al. 2022b). For clarity, p65 is not shown in the figure. (*D*) Domain architecture of the *Kluyveromyces lactis* CST subunits. (EBM) Est1-binding motif, (RD) recruitment domain, (DBD) DNA-binding domain. (*E–H*) Crystal structures of OB1 domain of *S. cerevisiae* Cdc13 (PDB 3OIP), the OB domain of *K. lactis* Stn1 in complex with Ten1 (PDB 6LBU), OB2 and OB4 domains of *K. lactis* Cdc13 in complex with the winged helix (WH) domain of Stn1 (PDB 6LBT), OB2-OB4 domains of *K. lactis* Cdc13 in complex with 25 nt ssDNA (Tel25: 5′-ACGGATTTGATTAGG TATGTGGTGT-3′) (PDB 6LBR) (Ge et al. 2020).

dence supports previous in vivo research where depletion of p65 caused decreased TER accumulation (Witkin and Collins 2004). Although p65 could not be entirely resolved in the cryo-EM map, a combination of focused classification of the cryo-EM maps, NMR, and Rosetta modeling enabled the determination of the full-length p65 structure (Wang et al. 2023b). This structure reveals additional interactions between p65 and TER, and provides insight into how p65 promotes telomerase assembly by chaperoning TER.

The remaining subunits in the *Tetrahymena* telomerase holoenzyme have roles in telomerase regulation. They include the human TPP1 homolog p50, the heterotrimeric TEB complex, composed of Teb1, Teb2, and Teb3 proteins, orthologous to the human POT1 protein. Another heterotrimeric complex, containing the proteins p75, p45, and p19 (also known as Ctc1, Stn1, and Ten1), also forms part of the holoenzyme. These proteins are equivalent to the human CST complex. Unlike human telomerase, where they interact more transiently, these regulatory factors of telomerase form part of the *Tetrahymena* holoenzyme (Fig. 3A,B).

## Recruitment and Processivity Factors: p50 and the TEB Complex

p50 contacts TERT at the TEN domain and the IFD-TRAP motif in a very similar manner to TPP1 binding to human telomerase (see above) (Fig. 4C,F; Jiang et al. 2018; He et al. 2021; Liu et al. 2022; Sekne et al. 2022). TPP1 recruits human telomerase through its TEL patch and NOB (Nandakumar et al. 2012; Sexton et al. 2012; Zhong et al. 2012; Grill et al. 2018). Residues equivalent to the TEL patch and NOB of TPP1 correlate to the binding interface of p50 to TERT in *Tetrahymena* (He et al. 2021). The Teb1 subunit of the TEB complex is made up of four OB-fold domains and is analogous to the largest subunit of replication protein A (RPA) (Fig. 4D,E; Upton et al. 2017). Unlike RPA, Teb1 has specificity for telomeric DNA (Zeng et al. 2011). Only the carboxy-terminal OB-fold of Teb1 (Teb1-C) is observed in the cryo-EM density of *Tetrahymena* telomerase (Fig. 4D–F;

Jiang et al. 2018). Teb1-C interacts with both p50 and the TEN domain of TERT. It also binds the 5′ end of the DNA substrate as it exits the TERT ring (Jiang et al. 2018).

In vitro, telomerase activity assays show that the addition of p50 to the core RNP (TERT and TER only) enhanced telomerase activity and RAP, which was further increased upon the addition of p50 with Teb1 (Hong et al. 2013). The p50-TEB complex is comparable to the TPP1-POT1 complex, which enhances the RAP of human telomerase (Wang et al. 2007; Pike et al. 2019; Sekne et al. 2022). Therefore, it has been proposed that p50 enhances RAP indirectly through the recruitment of the ssDNA-binding protein Teb1 and stabilization of the TEN domain and IFD-TRAP of telomerase. The Teb1-C domain allows the DNA substrate to be retained and threaded through the TEB complex, thereby further enhancing RAP. This supports previous biochemical findings that show that deletion of Teb1-C significantly decreases RAP (Min and Collins 2010). Mutations of Teb1 residues at the ssDNA-binding interface also impair RAP (Zeng et al. 2011).

The other subunits of the TEB complex, Teb2 and Teb3, do not influence RAP and are not specific to telomerase since they also function in RPA (Upton et al. 2017). Interestingly, the full trimeric TEB complex shows greater enhancement of RAP compared to Teb1 alone (Upton et al. 2017). In the *Tetrahymena* telomerase structure, Teb2 directly interacts with the TEN domain of TERT (Jiang et al. 2018). Therefore, it is possible that Teb2 and Teb3 play a role in stabilizing Teb1 to facilitate its interaction with the DNA substrate.

## Termination Factors: Ctc1-Stn1-Ten1

Within the *Tetrahymena* telomerase, the CST complex is highly flexible and thus poorly resolved in the earlier holoenzyme structures (Jiang et al. 2018; He et al. 2021). Extensive data analysis has recently elucidated the structure of *Tetrahymena* telomerase with a well-resolved CST complex (He et al. 2022a). The *Tetrahymena* CST directly interacts with the RT domain and TRBD of the TERT ring with its

Ctc1 subunit (Fig. 5B,C). Despite only being composed of three OB-folds, compared to seven in the human homolog, *Tetrahymena* Ctc1 is the largest subunit of CST and forms the only interaction with the rest of the telomerase holoenzyme.

The OB-A domain of Ctc1 interacts with the Ctc1-binding motif (CBM) of p50 (Figs. 3A and 5C). However, previous findings have suggested that the unresolved carboxyl terminus of p50 in the cryo-EM structure is also crucial for its interaction with Ctc1 (Hong et al. 2013). This interaction between the flexible carboxyl terminus of p50 and Ctc1 has been confirmed by NMR studies (He et al. 2022b). Given the flexible nature of the p50 carboxyl terminus, and therefore its interaction with CST, it has been proposed that the structure reported by He et al. (2022b) may represent one of several possible conformations adopted by the CST complex. The p50-Ctc1 interaction likely acts as a hinge point. Moreover, the human CST complex has also been reported to adopt different conformations (Cai and de Lange 2023). High-resolution structures of *Tetrahymena* telomerase highlight the central role of p50 in the binding of the accessory factors of the telomerase holoenzyme and can therefore be considered a "central hub" for binding, as proposed previously by Hong et al. (2013).

## TELOMERASE HOLOENZYMES IN *S. Cerevisiae* AND *S. Pombe*

### Telomerase Holoenzyme Composition

The composition of telomerase in yeast exhibits significant divergence from that of other eukaryotic species (Fig. 6A,C). For instance, yeast TER, known as TLC1 in *S. cerevisiae* or TER1 in *S. pombe*, is more than twice the size (~1200 nt) of hTR (Singer and Gottschling 1994; Leonardi et al. 2008). Despite this discrepancy in size and sequence, TER in yeast contains four functionally conserved components: a template, a pseudoknot domain, an STE, and a stabilizing 3' end (Zappulla 2020).

The core of budding yeast telomerase comprises the catalytic subunit Est2 and TLC1 that harbors the template sequence for reverse transcription (Egan and Collins 2012). While Est2 and TLC1 are sufficient to reconstitute telomerase activity in vitro (Zappulla et al. 2005), additional subunits are indispensable for telomerase functions in vivo. These subunits include essential factors Est1, Est3, and Pop1/Pop6/Pop7 (components shared with P/MRP RNase), as well as regulatory factors, including the Ku70/80 (or Ku for simplicity) heterodimer and the Sm7 heteroheptamer (SmB1, SmD1-3, SmE, SmF, and SmG) (Fig. 6A; Lundblad and Szostak 1989; Lendvay et al. 1996; Lingner et al. 1997a; Zappulla and Cech 2004; Lemieux et al. 2016).

Est1 orchestrates the recruitment of telomerase to telomeres by interacting with the single-stranded telomeric DNA-binding protein Cdc13, while Est3 exhibits cell-cycle regulation and shares structural similarities with human TPP1 (Lendvay et al. 1996; Evans and Lundblad 1999; Hughes et al. 2000; Lee et al. 2010). The Ku heterodimer regulates TLC1 retention in the nucleus and facilitates telomerase recruitment during the $G_1$ phase of the cell cycle (Evans and Lundblad 1999; Fisher et al. 2004; Hass and Zappulla 2015). Sm proteins bind and stabilize the 3' end of TLC1, forming a heptameric ring (Seto et al. 1999). Comprising Pop1, Pop6, and Pop7, the Pop proteins associate with the P3-like domain of TLC1 and also stabilize the association of Est1 and Est2 with TLC1 (Lemieux et al. 2016).

In fission yeast, the telomerase complex comprises the catalytic subunit Trt1, TER1, and additional protein factors, including Est1, the Lsm2-8 complex, and Pof8 (also known as Lar7) (Fig. 6C; Nakamura et al. 1997; Beernink et al. 2003; Leonardi et al. 2008; Webb and Zakian 2008; Tang et al. 2012). Est1 interacts with TER1 using a 14-3-3-like domain and forms a complex with Ccq1, which facilitates telomerase recruitment to telomeres (Fig. 6D; Webb and Zakian 2012). Like Sm proteins in budding yeast, Lsm2-8 complex is important for stabilization of the TER1 3' end (Tang et al. 2012); *S. pombe* Pof8 is a member of the conserved LARP7 family and shares significant sequence and structural similarities with *Tetrahymena* p65 and human LARP7 (Collopy et al. 2018;

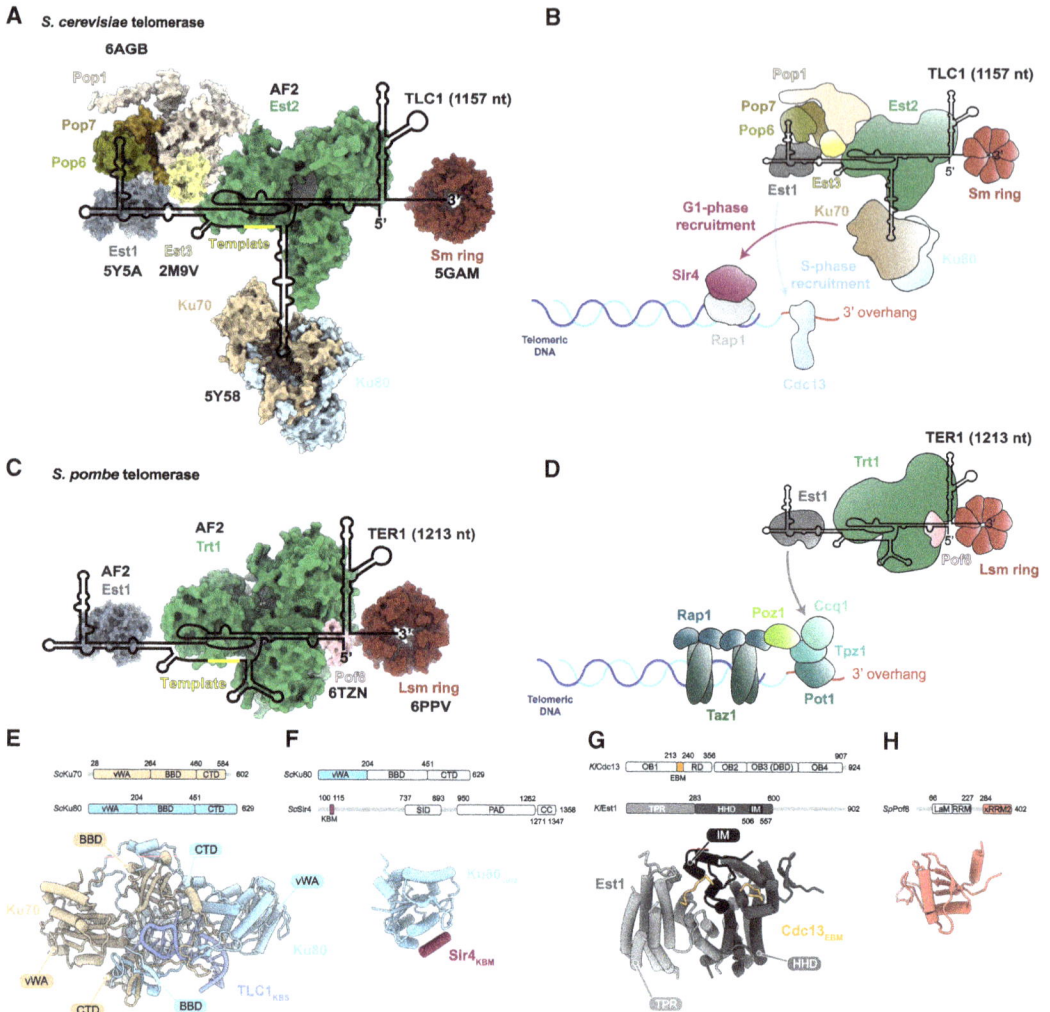

**Figure 6.** Telomerase holoenzymes in *Saccharomyces cerevisiae* and *Schizosaccharomyces pombe* and recruitment factors. (*A*) Structural model of *S. cerevisiae* telomerase holoenzyme prepared using available crystal, cryo-EM, nuclear magnetic resonance (NMR) structures, and AlphaFold2 models (Jumper et al. 2021). Telomerase RNA (TLC1) is drawn as a black line to reflect its secondary structure. Deposition numbers for available structures are annotated in the figure as PDB codes (Rao et al. 2014; Nguyen et al. 2016; Chen et al. 2018a; Lan et al. 2018), while models generated with the AlphaFold2 are indicated in the figure as AF2. Structures shown are not scaled and their position on TLC1 is proposed based on published work. (*B*) Schematic representation of telomerase recruitment in *S. cerevisiae*. Budding yeast recruits and activates telomerase at telomeres using two distinct pathways: one mediated by the interaction between Est1 and Cdc13, which is dominant in S phase, another through Ku80-Sir4 interaction that occurs in $G_1$ phase. Illustrations are not drawn to scale. (*C*) Structural model of *S. pombe* telomerase holoenzyme composed of available crystal structures and AlphaFold2 predicted models. Telomerase RNA (TER1) is drawn as a black line to reflect its secondary structure. Deposition numbers for available structures are annotated in the figure as PDB codes (Montemayor et al. 2020; Basu et al. 2021), while models generated with the AlphaFold2 are indicated in the figure as AF2. Structures shown are not scaled and their position on TER1 is proposed based on published work. (*D*) Schematic representation of telomerase recruitment in *S. pombe*. Fission yeast telomerase recruitment is mediated by shelterin component Tpz1 and its binding partner Ccq1. The latter, when phosphorylated, interacts with telomerase subunit Est1. (*E–H*) Domain architectures and crystal structures of the *S. cerevisiae* Ku70/80 in complex with TLC1$_{KBS}$ (PDB 5Y58), the vWA domain of the *S. cerevisiae* Ku80 in complex with the Ku-binding motif (KBM) of Sir4 (PDB 5Y59), the *Kluyveromyces lactis* Est1 in complex with the Est1-binding motif (EBM) of the *K. lactis* Cdc13 (PDB 5Y5A) and *S. pombe* Pof8 carboxy-terminal RRM domain (PDB 6TZN), respectively (Chen et al. 2018a; Basu et al. 2021). (*Continued*)

Mennie et al. 2018; Páez-Moscoso et al. 2018). The loss of Pof8 leads to telomerase assembly defects and critically short telomeres (Collopy et al. 2018; Mennie et al. 2018; Páez-Moscoso et al. 2018). The carboxy-terminal RNA recognition motif (RRM) domain of Pof8 has been structurally characterized (Fig. 6H; Hu et al. 2020; Basu et al. 2021).

Despite the lack of yeast telomerase structures, certain components such as Ku70/80, Sm and Lsm heptamers, and Pop1/Pop6/Pop7 have been structurally characterized as part of other complexes (Fig. 6A,C; Nguyen et al. 2016; Lan et al. 2018; Montemayor et al. 2020). Like *Tetrahymena* and human telomerase, studying the structures of yeast telomerase holoenzymes could unveil previously unidentified subunits.

## Recruitment Factors in Yeast

Recruitment of telomerase to telomeres in budding yeast is regulated by the cell cycle through two distinct pathways (Fig. 6B). One pathway involves the interaction between the Ku70/80 heterodimer and Sir4, a subunit of the silent information regulator (SIR) complex during the $G_1$ phase (Roy et al. 2004; Hass and Zappulla 2015). The SIR complex, in turn, interacts with the double-stranded telomeric DNA-binding protein Rap1 (Roy et al. 2004; Hass and Zappulla 2015). In the late S phase, another pathway is activated, wherein interactions occur between Est1 and the 3′ single-stranded telomeric DNA-binding protein Cdc13 (Evans and Lundblad 1999; Chan et al. 2008; Hass and Zappulla 2015). The structural basis of these interactions was elucidated through the crystal structures of the Est1 and the Ku70/80 heterodimer bound with their respective partners (Fig. 6F,G; Chen et al. 2018a).

The crystal structure of Ku70/80 in complex with a fragment of TLC1 containing the Ku-binding site ($TLC1_{KBS}$) reveals that $TLC1_{KBS}$ adopts a short, bulged stem-loop structure, specifically recognized by Ku70/80 (Fig. 6E; Chen et al. 2018a). This interaction is crucial for the nuclear retention of TLC1 in yeast cells. To provide further structural insights into this recruitment mechanism, the crystal structure of Ku80 with a Sir4 peptide-containing Ku-binding motif (KBM) was also determined (Fig. 6F). In this structure, the Sir4 peptide binds a hydrophobic groove within Ku80. Mutations in Ku80 and Sir4 that disrupt these interactions in cells result in shorter telomeres compared to the wild-type. Additionally, diminished telomerase recruitment is observed, underscoring the role of Ku80-Sir4 interactions in Ku-mediated telomerase recruitment to telomeres (Chen et al. 2018a). Furthermore, the binding sites of TLC1 and Sir4 on Ku70/80 are not mutually exclusive. A structure model of the $TLC1_{KBS}$-$Ku70/80$-$Sir4_{KBM}$ complex was thus proposed to provide the structural basis of telomerase recruitment to telomeres to double-stranded telomeric DNA by the Sir4.

To investigate the mechanism of telomerase recruitment to the 3′ telomeric overhang by Cdc13, Chen et al. (2018a) determined the structure of the amino-terminal fragment of Est1 in complex with the Est1-binding motif (EBM) of Cdc13 from the dairy yeast *Kluyveromyces lactis* (Fig. 6G). The structure of the complex revealed that Cdc13 folds into two separate patches, each engaging with distinct pockets of Est1. Only one of these two interaction interfaces proved to be indispensable for Est1-Cdc13 interactions. Disrupting this interface led to a decreased association of telomerase with telomeres during the late S phase. Moreover, interfering with both the Ku70/80-Sir4 and Est1-Cdc13 interactions has an additive effect on telomere shortening in cells, suggesting the coordinated interplay between the two pathways in telomere maintenance (Chen et al. 2018a).

**Figure 6.** (*Continued*) (vWA) von Willebrand factor type A domain, (BBD) a β-barrel domain, (CTD) an extended carboxy-terminal α-helical domain, (SID) Sir2-interacting domain, (PAD) partitioning and anchoring domain, (CC) coiled coil, (OB) oligosaccharide/oligonucleotide-binding fold, (EBM) Est1-binding motif, (RD) recruitment domain, (DBD) DNA-binding domain, (TPR) tetratricopeptide repeat, (HHD) helical-hairpin domain, (IM) insertion motif, (LaM) La motif, (RRM) RNA recognition motif. Domains not present in the structures are shown in light gray.

Recruiting telomerase to telomeres in fission yeast involves shelterin, reminiscent of the human system (Fig. 6D; Hu et al. 2016). Fission yeast shelterin comprises six proteins: Taz1, Rap1, Poz1, Tpz1, Pot1, and Ccq1 (Chen 2019). Taz1 binds double-stranded telomeric DNA, while Pot1 binds single-stranded 3′ overhang (Cooper et al. 1997; Baumann and Cech 2001). These proteins are connected by Rap1, Poz1, and Tpz1 (Kanoh and Ishikawa 2001; Harland et al. 2014). Tpz1 and Poz1 are the respective homologs of the human TPP1 and POT1. Telomerase recruitment begins in late S phase when Ccq1, bound by Tpz1 at telomeres, undergoes phosphorylation mediated by Tel1 and Rad3 kinases, orthologs of human ATM and ATR kinases, respectively (Hu et al. 2016). This phosphorylation event facilitates interaction between Ccq1 and the telomerase component Est1 (Fig. 6D).

Interestingly, Ccq1-Est1 and Ccq1-Tpz1 interactions are mutually exclusive (Armstrong et al. 2014), and the Est1-TER1 interaction can disrupt the Ccq1-Est1 interaction (Webb and Zakian 2012; Armstrong et al. 2014). Additionally, an association of Est1 with telomeres requires Trt1 (*S. pombe* TERT) and TER1 and occurs downstream from Ccq1-Est1 interactions (Webb and Zakian 2012). These findings suggest an additional pathway for telomerase recruitment in fission yeast. The TEL-patch region within the OB-fold domain of Tpz1 is crucial for telomere homeostasis and Trt1 association with telomeres (Hu et al. 2016). Telomerase recruitment was proposed to involve the cooperation between two telomere-telomerase interfaces: the cell-cycle-regulated Ccq1-Est1 interaction and the Tpz1-Trt1 interaction (Hu et al. 2016). However, the structural basis of telomerase recruitment by Ccq1 and Tpz1 still remains elusive.

In budding yeast, the CST complex comprises Cdc13, Stn1, and Ten1, and it specifically binds the 3′ G-rich overhang of telomeres. While Stn1 and Ten1 exhibit high structural conservation, budding yeast Cdc13 and human CTC1 share no sequence identity (Fig. 5A,D). In contrast to human CST, which inhibits telomerase activity at telomeres, budding yeast Cdc13 serves as a recruitment platform for telomerase (see above for more details) (Soudet et al. 2014; Chen et al. 2018a). Furthermore, CST regulates telomerase activity when telomeres reach a certain length, and its Stn1-Ten1 module is essential for telomere capping (Puglisi et al. 2008). Similar to humans and *Tetrahymena*, the yeast CST complex recruits DNA Polα-primase for C-strand synthesis (Grossi et al. 2004; Lue et al. 2014). Crystal structures of the *K. lactis* Cdc13 OB1 domain, Cdc13-ssDNA, Cdc13-Stn1, and Stn1-Ten1 complexes have been determined, revealing interactions within the yeast CST complex (Fig. 5E–H; Ge et al. 2020).

The structure of the three carboxy-terminal OB-fold domains of Cdc13 (OB2-OB4) bound to a telomeric ssDNA has unveiled extensive interactions between the OB3 domain and the ssDNA and intramolecular hydrophobic interactions between the OB2 and OB4 domains, identified to form homodimers in earlier studies (Fig. 5H; Yu et al. 2012; Mason et al. 2013; Ge et al. 2020). Previous work identified mutations in the OB2 domain of Cdc13 that result in the deregulation of telomere length and growth defects (Mason et al. 2013). Although these mutations were proposed to disrupt OB2 dimerization, in the recent structure of Cdc13 bound to ssDNA, they were found to interfere with the interactions between OB2 and OB4 domains (Ge et al. 2020). Additionally, both OB2 and OB4 domains are required for interactions with Stn1, as both OB-fold domains facilitate interaction with the winged helix (WH) domains of Stn1 (Fig. 5G). The OB-fold domains of Stn1 and Ten1 pack against each other as revealed by the Stn1$_{OB}$-Ten1 structure (Fig. 5F). The Stn1-Ten1 interactions are crucial for telomere capping, in agreement with a previous observation that Stn1 and Ten1 can bypass Cdc13 for telomere capping in budding yeast. Coupled with functional data, these structures suggest that the yeast CST complex forms a dimer (Ge et al. 2020). Future structural characterization of the full yeast CST complex would enhance our understanding of the distinct roles of CST in telomere protection and telomerase regulation.

   Cite this article as *Cold Spring Harb Perspect Biol* doi: 10.1101/cshperspect.a041697

## CONCLUDING REMARKS

Nearly four decades since the discovery of telomerase (Greider and Blackburn 1985), the field has been enriched with the first atomic structures of both *Tetrahymena* and human telomerase. These structures provide unprecedented insights into telomerase mechanism and illuminate the roles of numerous telomerase-associated factors. However, many aspects relating telomerase assembly, regulation, and evolution remain to be explored as well as interactions with other cellular machineries. Additionally, structural snapshots will greatly benefit from dynamic studies and visualization in a more physiological context. The rapid progress in machine learning–based structure prediction, fluorescence-based imaging, and electron tomography, as highlighted by recent advancements (Jumper et al. 2021; Nogales and Mahamid 2024), poses great promise for the next significant breakthrough in structural studies of telomere maintenance beyond telomerase.

## ACKNOWLEDGMENTS

We thank Hongmiao Hu, Inga Hochheiser, and Sigurdur Thorkelsson for critical comments on the manuscript; and George Ghanim for help with figures. T.H.D.N. is supported by a UKRI-Medical Research Council grant (MC_UP_1201/19), a Wellcome Trust Career Development Grant (226015/Z/22/Z), and an EMBO Young Investigator Program Award.

## REFERENCES

Abreu E, Aritonovska E, Reichenbach P, Cristofari G, Culp B, Terns RM, Lingner J, Terns MP. 2010. TIN2-tethered TPP1 recruits human telomerase to telomeres in vivo. *Mol Cell Biol* **30**: 2971–2982. doi:10.1128/MCB.00240-10

Akiyama BM, Parks JW, Stone MD. 2015. The telomerase essential N-terminal domain promotes DNA synthesis by stabilizing short RNA-DNA hybrids. *Nucleic Acids Res* **43**: 5537–5549. doi:10.1093/nar/gkv406

Angrisani A, Vicidomini R, Turano M, Furia M. 2014. Human dyskerin: beyond telomeres. *Biol Chem* **395**: 593–610. doi:10.1515/hsz-2013-0287

Armanios M. 2022. The role of telomeres in human disease. *Annu Rev Genomics Hum Genet* **23**: 363–381. doi:10.1146/annurev-genom-010422-091101

Armanios M, Blackburn EH. 2012. The telomere syndromes. *Nat Rev Genet* **13**: 693–704. doi:10.1038/nrg3246

Armbruster BN, Banik SSR, Guo C, Smith AC, Counter CM. 2001. N-terminal domains of the human telomerase catalytic subunit required for enzyme activity in vivo. *Mol Cell Biol* **21**: 7775–7786. doi:10.1128/MCB.21.22.7775-7786.2001

Armstrong CA, Pearson SR, Amelina H, Moiseeva V, Tomita K. 2014. Telomerase activation after recruitment in fission yeast. *Curr Biol* **24**: 2006–2011. doi:10.1016/j.cub.2014.07.035

Arnoult N, Karlseder J. 2015. Complex interactions between the DNA-damage response and mammalian telomeres. *Nat Struct Mol Biol* **22**: 859–866. doi:10.1038/nsmb.3092

Autexier C, Greider CW. 1995. Boundary elements of the *Tetrahymena* telomerase RNA template and alignment domains. *Genes Dev* **9**: 2227–2239. doi:10.1101/gad.9.18.2227

Basu R, Eichhorn CD, Cheng R, Peterson RD, Feigon J. 2021. Structure of *S. pombe* telomerase protein Pof8 C-terminal domain is an xRRM conserved among LARP7 proteins. *RNA Biol* **18**: 1181–1192. doi:10.1080/15476286.2020.1836891

Baumann P, Cech TR. 2001. Pot1, the putative telomere end-binding protein in fission yeast and humans. *Science* **292**: 1171–1175. doi:10.1126/science.1060036

Beernink HT, Miller K, Deshpande A, Bucher P, Cooper JP. 2003. Telomere maintenance in fission yeast requires an Est1 ortholog. *Curr Biol* **13**: 575–580. doi:10.1016/S0960-9822(03)00169-6

Berman AJ, Gooding AR, Cech TR. 2010. *Tetrahymena* telomerase protein p65 induces conformational changes throughout telomerase RNA (TER) and rescues telomerase reverse transcriptase and TER assembly mutants. *Mol Cell Biol* **30**: 4965–4976. doi:10.1128/MCB.00827-10

Berman AJ, Akiyama BM, Stone MD, Cech TR. 2011. The RNA accordion model for template positioning by telomerase RNA during telomeric DNA synthesis. *Nat Struct Mol Biol* **18**: 1371–1375. doi:10.1038/nsmb.2174

Blackburn EH, Collins K. 2010. Telomerase: an RNP enzyme synthesizes DNA. In *RNA worlds* (ed. Gesteland RF, Atkins JF, Cech TR), pp. 205–213. Cold Spring Harbor Laboratory Press, Cold Spring Harbor, NY.

Bley CJ, Qi X, Rand DP, Borges CR, Nelson RW, Chen JJ. 2011. RNA-protein binding interface in the telomerase ribonucleoprotein. *Proc Natl Acad Sci* **108**: 20333–20338. doi:10.1073/pnas.1100270108

Brown AF, Podlevsky JD, Qi X, Chen Y, Xie M, Chen JJ. 2014. A self-regulating template in human telomerase. *Proc Natl Acad Sci* **111**: 11311–11316. doi:10.1073/pnas.1402531111

Cai SW, de Lange T. 2023. CST-Polα/primase: the second telomere maintenance machine. *Genes Dev* **37**: 555–569. doi:10.1101/gad.350479.123

Cai SW, Zinder JC, Svetlov V, Bush MW, Nudler E, Walz T, de Lange T. 2022. Cryo-EM structure of the human CST-Polα/primase complex in a recruitment state. *Nat Struct Mol Biol* **29**: 813–819. doi:10.1038/s41594-022-00766-y

Cai SW, Takai H, Walz T, de Lange T. 2023. POT1 recruits and regulates CST–Polα/Primase at human telomeres. bioRxiv doi:10.1101/2023.05.08.539880

Carlile TM, Rojas-Duran MF, Zinshteyn B, Shin H, Bartoli KM, Gilbert WV. 2014. Pseudouridine profiling reveals regulated mRNA pseudouridylation in yeast and human cells. *Nature* **515**: 143–146. doi:10.1038/nature13802

Cech TR. 2004. Beginning to understand the end of the chromosome. *Cell* **116**: 273–279. doi:10.1016/S0092-8674(04)00038-8

Chan A, Boulé JB, Zakian VA. 2008. Two pathways recruit telomerase to *Saccharomyces cerevisiae* telomeres. *PLoS Genet* **4**: e1000236. doi:10.1371/journal.pgen.1000236

Chen Y. 2019. The structural biology of the shelterin complex. *Biol Chem* **400**: 457–466. doi:10.1515/hsz-2018-0368

Chen JL, Greider CW. 2004. An emerging consensus for telomerase RNA structure. *Proc Natl Acad Sci* **101**: 14683–14684. doi:10.1073/pnas.0406204101

Chen JL, Blasco MA, Greider CW. 2000. Secondary structure of vertebrate telomerase RNA. *Cell* **100**: 503–514. doi:10.1016/S0092-8674(00)80687-X

Chen JL, Opperman KK, Greider CW. 2002. A critical stem-loop structure in the CR4-CR5 domain of mammalian telomerase RNA. *Nucleic Acids Res* **30**: 592–597. doi:10.1093/nar/30.2.592

Chen LY, Redon S, Lingner J. 2012. The human CST complex is a terminator of telomerase activity. *Nature* **488**: 540–544. doi:10.1038/nature11269

Chen H, Xue J, Churikov D, Hass EP, Shi S, Lemon LD, Luciano P, Bertuch AA, Zappulla DC, Géli V, et al. 2018a. Structural insights into yeast telomerase recruitment to telomeres. *Cell* **172**: 331–343.e13. doi:10.1016/j.cell.2017.12.008

Chen L, Roake CM, Freund A, Batista PJ, Tian S, Yin YA, Gajera CR, Lin S, Lee B, Pech MF, et al. 2018b. An activity switch in human telomerase based on RNA conformation and shaped by TCAB1. *Cell* **174**: 218–230.e13. doi:10.1016/j.cell.2018.04.039

Cohn M, Blackburn EH. 1995. Telomerase in yeast. *Science* **269**: 396–400. doi:10.1126/science.7618104

Collins K. 2006. The biogenesis and regulation of telomerase holoenzymes. *Nat Rev Mol Cell Biol* **7**: 484–494. doi:10.1038/nrm1961

Collins K. 2009. Forms and functions of telomerase RNA. In *Non-protein coding RNAs* (ed. Walter NG, Woodson SA, Batey RT), pp. 285–301. Springer, Berlin.

Collins K. 2011. Single-stranded DNA repeat synthesis by telomerase. *Curr Opin Chem Biol* **15**: 643–648. doi:10.1016/j.cbpa.2011.07.011

Collopy LC, Ware TL, Goncalves T, í Kongsstovu S, Yang Q, Amelina H, Pinder C, Alenazi A, Moiseeva V, Pearson SR, et al. 2018. LARP7 family proteins have conserved function in telomerase assembly. *Nat Commun* **9**: 557. doi:10.1038/s41467-017-02296-4

Cooper JP, Nimmo ER, Allshire RC, Cech TR. 1997. Regulation of telomere length and function by a Myb-domain protein in fission yeast. *Nature* **385**: 744–747. doi:10.1038/385744a0

de Lange T. 2018. Shelterin-mediated telomere protection. *Annu Rev Genet* **52**: 223–247. doi:10.1146/annurev-genet-032918-021921

Egan ED, Collins K. 2010. Specificity and stoichiometry of subunit interactions in the human telomerase holoen-zyme assembled in vivo. *Mol Cell Biol* **30**: 2775–2786. doi:10.1128/MCB.00151-10

Egan ED, Collins K. 2012. Biogenesis of telomerase ribonucleoproteins. *RNA* **18**: 1747–1759. doi:10.1261/rna.034629.112

Evans SK, Lundblad V. 1999. Est1 and Cdc13 as comediators of telomerase access. *Science* **286**: 117–120. doi:10.1126/science.286.5437.117

Fisher TS, Taggart AK, Zakian VA. 2004. Cell cycle-dependent regulation of yeast telomerase by Ku. *Nat Struct Mol Biol* **11**: 1198–1205. doi:10.1038/nsmb854

Förstemann K, Lingner J. 2005. Telomerase limits the extent of base pairing between template RNA and telomeric DNA. *EMBO Rep* **6**: 361–366. doi:10.1038/sj.embor.7400374

Gandhi L, Collins K. 1998. Interaction of recombinant *Tetrahymena* telomerase proteins p80 and p95 with telomerase RNA and telomeric DNA substrates. *Genes Dev* **12**: 721–733. doi:10.1101/gad.12.5.721

Ganot P, Bortolin ML, Kiss T. 1997. Site-specific pseudouridine formation in preribosomal RNA is guided by small nucleolar RNAs. *Cell* **89**: 799–809. doi:10.1016/S0092-8674(00)80263-9

Garus A, Autexier C. 2021. Dyskerin: an essential pseudouridine synthase with multifaceted roles in ribosome biogenesis, splicing, and telomere maintenance. *RNA* **27**: 1441–1458. doi:10.1261/rna.078953.121

Ge Y, Wu Z, Chen H, Zhong Q, Shi S, Li G, Wu J, Lei M. 2020. Structural insights into telomere protection and homeostasis regulation by yeast CST complex. *Nat Struct Mol Biol* **27**: 752–762. doi:10.1038/s41594-020-0459-8

Ghanim GE, Fountain AJ, van Roon AM, Rangan R, Das R, Collins K, Nguyen THH. 2021. Structure of human telomerase holoenzyme with bound telomeric DNA. *Nature* **593**: 449–453. doi:10.1038/s41586-021-03415-4

Ghanim GE, Sekne Z, Balch S, van Roon AMM, Nguyen THD. 2024. 2.7 Å cryo-EM structure of human telomerase H/ACA ribonucleoprotein. *Nat Commun* **15**: 746. doi:10.1038/s41467-024-45002-x

Gillis AJ, Schuller AP, Skordalakes E. 2008. Structure of the *Tribolium castaneum* telomerase catalytic subunit TERT. *Nature* **455**: 633–637. doi:10.1038/nature07283

Glousker G, Briod AS, Quadroni M, Lingner J. 2020. Human shelterin protein POT1 prevents severe telomere instability induced by homology-directed DNA repair. *EMBO J* **39**: e104500. doi:10.15252/embj.2020104500

Greider CW, Blackburn EH. 1985. Identification of a specific telomere terminal transferase activity in *Tetrahymena* extracts. *Cell* **43**: 405–413. doi:10.1016/0092-8674(85)90170-9

Greider CW, Blackburn EH. 1987. The telomere terminal transferase of *Tetrahymena* is a ribonucleoprotein enzyme with two kinds of primer specificity. *Cell* **51**: 887–898. doi:10.1016/0092-8674(87)90576-9

Greider CW, Blackburn EH. 1989. A telomeric sequence in the RNA of *Tetrahymena* telomerase required for telomere repeat synthesis. *Nature* **337**: 331–337. doi:10.1038/337331a0

Griffith JD, Comeau L, Rosenfield S, Stansel RM, Bianchi A, Moss H, de Lange T. 1999. Mammalian telomeres end in a large duplex loop. *Cell* **97**: 503–514. doi:10.1016/S0092-8674(00)80760-6

Grill S, Tesmer VM, Nandakumar J. 2018. The N terminus of the OB domain of telomere protein TPP1 is critical for telomerase action. *Cell Rep* **22**: 1132–1140. doi:10.1016/j.celrep.2018.01.012

Grossi S, Puglisi A, Dmitriev PV, Lopes M, Shore D. 2004. Pol12, the B subunit of DNA polymerase α, functions in both telomere capping and length regulation. *Genes Dev* **18**: 992–1006. doi:10.1101/gad.300004

Hamma T, Ferre-D'Amare AR. 2010. The box H/ACA ribonucleoprotein complex: interplay of RNA and protein structures in post-transcriptional RNA modification. *J Biol Chem* **285**: 805–809. doi:10.1074/jbc.R109.076893

Hardy CD, Schultz CS, Collins K. 2001. Requirements for the dGTP-dependent repeat addition processivity of recombinant *Tetrahymena* telomerase. *J Biol Chem* **276**: 4863–4871. doi:10.1074/jbc.M005158200

Harland JL, Chang YT, Moser BA, Nakamura TM. 2014. Tpz1-Ccq1 and Tpz1-Poz1 interactions within fission yeast shelterin modulate Ccq1 Thr93 phosphorylation and telomerase recruitment. *PLoS Genet* **10**: e1004708. doi:10.1371/journal.pgen.1004708

Harley CB, Kim NW, Prowse KR, Weinrich SL, Hirsch KS, West MD, Bacchetti S, Hirte HW, Counter CM, Greider CW, et al. 1994. Telomerase, cell immortality, and cancer. *Cold Spring Harb Symp Quant Biol* **59**: 307–315. doi:10.1101/SQB.1994.059.01.035

Hass EP, Zappulla DC. 2015. The Ku subunit of telomerase binds Sir4 to recruit telomerase to lengthen telomeres in *S. cerevisiae*. *eLife* **4**: e07750. doi:10.7554/eLife.07750

He Y, Feigon J. 2022. Telomerase structural biology comes of age. *Curr Opin Struct Biol* **76**: 102446. doi:10.1016/j.sbi.2022.102446

He Y, Wang Y, Liu B, Helmling C, Sušac L, Cheng R, Zhou ZH, Feigon J. 2021. Structures of telomerase at several steps of telomere repeat synthesis. *Nature* **593**: 454–459. doi:10.1038/s41586-021-03529-9

He Q, Lin X, Chavez BL, Agrawal S, Lusk BL, Lim CJ. 2022a. Structures of the human CST-Polα-primase complex bound to telomere templates. *Nature* **608**: 826–832. doi:10.1038/s41586-022-05040-1

He Y, Song H, Chan H, Liu B, Wang Y, Sušac L, Zhou ZH, Feigon J. 2022b. Structure of *Tetrahymena* telomerase-bound CST with polymerase α-primase. *Nature* **608**: 813–818. doi:10.1038/s41586-022-04931-7

Heiss NS, Knight SW, Vulliamy TJ, Klauck SM, Wiemann S, Mason PJ, Poustka A, Dokal I. 1998. X-linked dyskeratosis congenita is caused by mutations in a highly conserved gene with putative nucleolar functions. *Nat Genet* **19**: 32–38. doi:10.1038/ng0598-32

Hockemeyer D, Collins K. 2015. Control of telomerase action at human telomeres. *Nat Struct Mol Biol* **22**: 848–852. doi:10.1038/nsmb.3083

Hockemeyer D, Daniels JP, Takai H, de Lange T. 2006. Recent expansion of the telomeric complex in rodents: two distinct POT1 proteins protect mouse telomeres. *Cell* **126**: 63–77. doi:10.1016/j.cell.2006.04.044

Holohan B, Wright WE, Shay JW. 2014. Telomeropathies: an emerging spectrum disorder. *J Cell Biol* **205**: 289–299. doi:10.1083/jcb.201401012

Hong K, Upton H, Miracco EJ, Jiang J, Zhou ZH, Feigon J, Collins K. 2013. *Tetrahymena* telomerase holoenzyme assembly, activation, and inhibition by domains of the p50 central hub. *Mol Cell Biol* **33**: 3962–3971. doi:10.1128/MCB.00792-13

Hossain S, Singh S, Lue NF. 2002. Functional analysis of the C-terminal extension of telomerase reverse transcriptase. *J Biol Chem* **277**: 36174–36180. doi:10.1074/jbc.M201976200

Hu X, Liu J, Jun HI, Kim JK, Qiao F. 2016. Multi-step coordination of telomerase recruitment in fission yeast through two coupled telomere-telomerase interfaces. *eLife* **5**: e15470. doi:10.7554/eLife.15470

Hu X, Kim JK, Yu C, Jun HI, Liu J, Sankaran B, Huang L, Qiao F. 2020. Quality-control mechanism for telomerase RNA folding in the cell. *Cell Rep* **33**: 108568. doi:10.1016/j.celrep.2020.108568

Huang J, Brown AF, Wu J, Xue J, Bley CJ, Rand DP, Wu L, Zhang R, Chen JJ, Lei M. 2014. Structural basis for protein-RNA recognition in telomerase. *Nat Struct Mol Biol* **21**: 507–512. doi:10.1038/nsmb.2819

Hughes TR, Evans SK, Weilbaecher RG, Lundblad V. 2000. The Est3 protein is a subunit of yeast telomerase. *Curr Biol* **10**: 809–812. doi:10.1016/S0960-9822(00)00562-5

Jacobs SA, Podell ER, Cech TR. 2006. Crystal structure of the essential N-terminal domain of telomerase reverse transcriptase. *Nat Struct Mol Biol* **13**: 218–225. doi:10.1038/nsmb1054

Jiang J, Chan H, Cash DD, Miracco EJ, Ogorzalek Loo RR, Upton HE, Cascio D, O'Brien Johnson R, Collins K, Loo JA, et al. 2015. Structure of *Tetrahymena* telomerase reveals previously unknown subunits, functions, and interactions. *Science* **350**: aab4070. doi:10.1126/science.aab4070

Jiang J, Wang Y, Sušac L, Chan H, Basu R, Zhou ZH, Feigon J. 2018. Structure of telomerase with telomeric DNA. *Cell* **173**: 1179–1190.e13. doi:10.1016/j.cell.2018.04.038

Jumper J, Evans R, Pritzel A, Green T, Figurnov M, Ronneberger O, Tunyasuvunakool K, Bates R, Žídek A, Potapenko A, et al. 2021. Highly accurate protein structure prediction with AlphaFold. *Nature* **596**: 583–589. doi:10.1038/s41586-021-03819-2

Jurczyluk J, Nouwens AS, Holien JK, Adams TE, Lovrecz GO, Parker MW, Cohen SB, Bryan TM. 2011. Direct involvement of the TEN domain at the active site of human telomerase. *Nucleic Acids Res* **39**: 1774–1788. doi:10.1093/nar/gkq1083

Kanoh J, Ishikawa F. 2001. Sprap1 and spRif1, recruited to telomeres by Taz1, are essential for telomere function in fission yeast. *Curr Biol* **11**: 1624–1630. doi:10.1016/S0960-9822(01)00503-6

Kelleher C, Kurth I, Lingner J. 2005. Human protection of telomeres 1 (POT1) is a negative regulator of telomerase activity in vitro. *Mol Cell Biol* **25**: 808–818. doi:10.1128/MCB.25.2.808-818.2005

Kim NW, Piatyszek MA, Prowse KR, Harley CB, West MD, Ho PLC, Coviello GM, Wright WE, Weinrich SL, Shay JW. 1994. Specific association of human telomerase activity with immortal cells and cancer. *Science* **266**: 2011–2015. doi:10.1126/science.7605428

Klump BM, Perez GI, Patrick EM, Adams-Boone K, Cohen SB, Han L, Yu K, Schmidt JC. 2023. TCAB1 prevents nucleolar accumulation of the telomerase RNA to facilitate telomerase assembly. *Cell Rep* **42:** 112577. doi:10 .1016/j.celrep.2023.112577

Lai CK, Miller MC, Collins K. 2002. Template boundary definition in *Tetrahymena* telomerase. *Genes Dev* **16:** 415–420. doi:10.1101/gad.962602

Lai CK, Miller MC, Collins K. 2003. Roles for RNA in telomerase nucleotide and repeat addition processivity. *Mol Cell* **11:** 1673–1683. doi:10.1016/S1097-2765(03)00232-6

Lan P, Tan M, Zhang Y, Niu S, Chen J, Shi S, Qiu S, Wang X, Peng X, Cai G, et al. 2018. Structural insight into precursor tRNA processing by yeast ribonuclease P. *Science* **362:** eaat6678. doi:10.1126/science.aat6678

Latrick CM, Cech TR. 2010. POT1-TPP1 enhances telomerase processivity by slowing primer dissociation and aiding translocation. *EMBO J* **29:** 924–933. doi:10.1038/emboj.2009.409

Lee J, Mandell EK, Rao T, Wuttke DS, Lundblad V. 2010. Investigating the role of the Est3 protein in yeast telomere replication. *Nucleic Acids Res* **38:** 2279–2290. doi:10 .1093/nar/gkp1173

Lei M, Zaug AJ, Podell ER, Cech TR. 2005. Switching human telomerase on and off with hPOT1 protein in vitro. *J Biol Chem* **280:** 20449–20456. doi:10.1074/jbc.M502212200

Lemieux B, Laterreur N, Perederina A, Noël JF, Dubois ML, Krasilnikov AS, Wellinger RJ. 2016. Active yeast telomerase shares subunits with ribonucleoproteins RNase P and RNase MRP. *Cell* **165:** 1171–1181. doi:10.1016/j.cell.2016 .04.018

Lendvay TS, Morris DK, Sah J, Balasubramamian B, Lundblad V. 1996. Senescence mutants of *Saccharomyces cerevisiae* with a defect in telomere replication identify three additional *EST* genes. *Genetics* **144:** 1399–1412. doi:10 .1093/genetics/144.4.1399

Leonardi J, Box JA, Bunch JT, Baumann P. 2008. TER1, the RNA subunit of fission yeast telomerase. *Nat Struct Mol Biol* **15:** 26–33. doi:10.1038/nsmb1343

Levy MZ, Allsopp RC, Futcher AB, Greider CW, Harley CB. 1992. Telomere end-replication problem and cell aging. *J Mol Biol* **225:** 951–960. doi:10.1016/0022-2836(92) 90096-3

Li L, Ye K. 2006. Crystal structure of an H/ACA box ribonucleoprotein particle. *Nature* **443:** 302–307. doi:10 .1038/nature05151

Li X, Zhu P, Ma S, Song J, Bai J, Sun F, Yi C. 2015. Chemical pulldown reveals dynamic pseudouridylation of the mammalian transcriptome. *Nat Chem Biol* **11:** 592–597. doi:10.1038/nchembio.1836

Lim CJ, Cech TR. 2021. Shaping human telomeres: from shelterin and CST complexes to telomeric chromatin organization. *Nat Rev Mol Cell Biol* **22:** 283–298. doi:10 .1038/s41580-021-00328-y

Lim CJ, Barbour AT, Zaug AJ, Goodrich KJ, McKay AE, Wuttke DS, Cech TR. 2020. The structure of human CST reveals a decameric assembly bound to telomeric DNA. *Science* **368:** 1081–1085. doi:10.1126/science.aaz 9649

Lingner J, Cech TR, Hughes TR, Lundblad V. 1997a. Three ever shorter telomere (*EST*) genes are dispensable for in vitro yeast telomerase activity. *Proc Natl Acad Sci* **94:** 11190–11195. doi:10.1073/pnas.94.21.11190

Lingner J, Hughes TR, Shevchenko A, Mann M, Lundblad V, Cech TR. 1997b. Reverse transcriptase motifs in the catalytic subunit of telomerase. *Science* **276:** 561–567. doi:10 .1126/science.276.5312.561

Liu B, He Y, Wang Y, Song H, Zhou ZH, Feigon J. 2022. Structure of active human telomerase with telomere shelterin protein TPP1. *Nature* **604:** 578–583. doi:10.1038/ s41586-022-04582-8

Loayza D, De Lange T. 2003. POT1 as a terminal transducer of TRF1 telomere length control. *Nature* **423:** 1013–1018. doi:10.1038/nature01688

Lue NF. 2005. A physical and functional constituent of telomerase anchor site. *J Biol Chem* **280:** 26586–26591. doi:10.1074/jbc.M503028200

Lue NF. 2013. *Yeast telomerases: structure, mechanism and regulation*. Landes Bioscience, Austin, TX.

Lue NF, Autexier C. 2023. Orchestrating nucleic acid-protein interactions at chromosome ends: telomerase mechanisms come into focus. *Nat Struct Mol Biol* **30:** 878–890. doi:10.1038/s41594-023-01022-7

Lue NF, Peng Y. 1997. Identification and characterization of a telomerase activity from *Schizosaccaromyces pombe*. *Nucleic Acids Res* **25:** 4331–4337. doi:10.1093/nar/25.21 .4331

Lue NF, Lin YC, Mian IS. 2003. A conserved telomerase motif within the catalytic domain of telomerase reverse transcriptase is specifically required for repeat addition processivity. *Mol Cell Biol* **23:** 8440–8449. doi:10.1128/ MCB.23.23.8440-8449.2003

Lue NF, Chan J, Wright WE, Hurwitz J. 2014. The CDC13-STN1-TEN1 complex stimulates Polα activity by promoting RNA priming and primase-to-polymerase switch. *Nat Commun* **5:** 5762. doi:10.1038/ncomms6762

Lundblad V, Szostak JW. 1989. A mutant with a defect in telomere elongation leads to senescence in yeast. *Cell* **57:** 633–643. doi:10.1016/0092-8674(89)90132-3

Martínez-Jiménez F, Muiños F, Sentís I, Deu-Pons J, Reyes-Salazar I, Arnedo-Pac C, Mularoni L, Pich O, Bonet J, Kranas H, et al. 2020. A compendium of mutational cancer driver genes. *Nat Rev Cancer* **20:** 555–572. doi:10 .1038/s41568-020-0290-x

Mason DX, Goneska E, Greider CW. 2003. Stem-loop IV of *tetrahymena* telomerase RNA stimulates processivity in trans. *Mol Cell Biol* **23:** 5606–5613. doi:10.1128/MCB.23 .16.5606-5613.2003

Mason M, Wanat JJ, Harper S, Schultz DC, Speicher DW, Johnson FB, Skordalakes E. 2013. Cdc13 OB2 dimerization required for productive Stn1 binding and efficient telomere maintenance. *Structure* **21:** 109–120. doi:10 .1016/j.str.2012.10.012

Meier UT. 2005. The many facets of H/ACA ribonucleoproteins. *Chromosoma* **114:** 1–14. doi:10.1007/s00412-005-0333-9

Mennie AK, Moser BA, Nakamura TM. 2018. LARP7-like protein Pof8 regulates telomerase assembly and poly(A) +TERRA expression in fission yeast. *Nat Commun* **9:** 586. doi:10.1038/s41467-018-02874-0

Miller MC, Collins K. 2002. Telomerase recognizes its template by using an adjacent RNA motif. *Proc Natl Acad Sci* **99:** 6585–6590. doi:10.1073/pnas.102024699

Min B, Collins K. 2009. An RPA-related sequence-specific DNA-binding subunit of telomerase holoenzyme is required for elongation processivity and telomere maintenance. *Mol Cell* **36:** 609–619. doi:10.1016/j.molcel.2009.09.041

Min B, Collins K. 2010. Multiple mechanisms for elongation processivity within the reconstituted *Tetrahymena* telomerase holoenzyme. *J Biol Chem* **285:** 16434–16443. doi:10.1074/jbc.M110.119172

Mitchell JR, Collins K. 2000. Human telomerase activation requires two independent interactions between telomerase RNA and telomerase reverse transcriptase. *Mol Cell* **6:** 361–371. doi:10.1016/S1097-2765(00)00036-8

Mitchell JR, Cheng J, Collins K. 1999a. A box H/ACA small nucleolar RNA-like domain at the human telomerase RNA 3′ end. *Mol Cell Biol* **19:** 567–576. doi:10.1128/MCB.19.1.567

Mitchell JR, Wood E, Collins K. 1999b. A telomerase component is defective in the human disease dyskeratosis congenita. *Nature* **402:** 551–555. doi:10.1038/990141

Mitchell M, Gillis A, Futahashi M, Fujiwara H, Skordalakes E. 2010. Structural basis for telomerase catalytic subunit TERT binding to RNA template and telomeric DNA. *Nat Struct Mol Biol* **17:** 513–518. doi:10.1038/nsmb.1777

Montemayor EJ, Virta JM, Hayes SM, Nomura Y, Brow DA, Butcher SE. 2020. Molecular basis for the distinct cellular functions of the Lsm1-7 and Lsm2-8 complexes. *RNA* **26:** 1400–1413. doi:10.1261/rna.075879.120

Moriarty TJ, Marie-Egyptienne DT, Autexier C. 2004. Functional organization of repeat addition processivity and DNA synthesis determinants in the human telomerase multimer. *Mol Cell Biol* **24:** 3720–3733. doi:10.1128/MCB.24.9.3720-3733.2004

Nakamura TM, Morin GB, Chapman KB, Weinrich SL, Andrews WH, Lingner J, Harley CB, Cech TR. 1997. Telomerase catalytic subunit homologs from fission yeast and human. *Science* **277:** 955–959. doi:10.1126/science.277.5328.955

Nandakumar J, Bell CF, Weidenfeld I, Zaug AJ, Leinwand LA, Cech TR. 2012. The TEL patch of telomere protein TPP1 mediates telomerase recruitment and processivity. *Nature* **492:** 285–289. doi:10.1038/nature11648

Nguyen THD. 2021. Structural biology of human telomerase: progress and prospects. *Biochem Soc Trans* **49:** 1927–1939. doi:10.1042/BST20200042

Nguyen THD, Galej WP, Bai XC, Oubridge C, Newman AJ, Scheres SHW, Nagai K. 2016. Cryo-EM structure of the yeast U4/U6.U5 tri-snRNP at 3.7 Å resolution. *Nature* **530:** 298–302. doi:10.1038/nature16940

Nguyen THD, Tam J, Wu RA, Greber BJ, Toso D, Nogales E, Collins K. 2018. Cryo-EM structure of substrate-bound human telomerase holoenzyme. *Nature* **557:** 190–195. doi:10.1038/s41586-018-0062-x

Nguyen THD, Collins K, Nogales E. 2019. Telomerase structures and regulation: shedding light on the chromosome end. *Curr Opin Struct Biol* **55:** 185–193. doi:10.1016/j.sbi.2019.04.009

Nogales E, Mahamid J. 2024. Bridging structural and cell biology with cryo-electron microscopy. *Nature* **628:** 47–56. doi:10.1038/s41586-024-07198-2

Páez-Moscoso DJ, Pan L, Sigauke RF, Schroeder MR, Tang W, Baumann P. 2018. Pof8 is a La-related protein and a constitutive component of telomerase in fission yeast. *Nat Commun* **9:** 587. doi:10.1038/s41467-017-02284-8

Palka C, Forino N, Hentschel J, Das R, Stone MD. 2020. Folding heterogeneity in the essential human telomerase RNA three-way junction. *RNA* **26:** 1787–1800. doi:10.1261/rna.077255.120

Pike AM, Strong MA, Ouyang JPT, Greider CW. 2019. TIN2 functions with TPP1/POT1 to stimulate telomerase processivity. *Mol Cell Biol* **39:** e00593. doi:10.1128/MCB.00593-18

Podlevsky JD, Chen JJL. 2016. Evolutionary perspectives of telomerase RNA structure and function. *RNA Biol* **13:** 720–732. doi:10.1080/15476286.2016.1205768

Podlevsky JD, Bley CJ, Omana RV, Qi X, Chen JJL. 2008. The telomerase database. *Nucleic Acids Res* **36:** D339–D343. doi:10.1093/nar/gkm700

Price CM, Boltz KA, Chaiken MF, Stewart JA, Beilstein MA, Shippen DE. 2010. Evolution of CST function in telomere maintenance. *Cell Cycle* **9:** 3157–3165. doi:10.4161/cc.9.16.12547

Puglisi A, Bianchi A, Lemmens L, Damay P, Shore D. 2008. Distinct roles for yeast Stn1 in telomere capping and telomerase inhibition. *EMBO J* **27:** 2328–2339. doi:10.1038/emboj.2008.158

Qi X, Xie M, Brown AF, Bley CJ, Podlevsky JD, Chen JJL. 2012. RNA/DNA hybrid binding affinity determines telomerase template-translocation efficiency. *EMBO J* **31:** 150–161. doi:10.1038/emboj.2011.363

Rao T, Lubin JW, Armstrong GS, Tucey TM, Lundblad V, Wuttke DS. 2014. Structure of Est3 reveals a bimodal surface with differential roles in telomere replication. *Proc Natl Acad Sci* **111:** 214–218. doi:10.1073/pnas.1316453111

Richards RJ, Wu H, Trantirek L, O'Connor CM, Collins K, Feigon J. 2006. Structural study of elements of *Tetrahymena* telomerase RNA stem-loop IV domain important for function. *RNA* **12:** 1475–1485. doi:10.1261/rna.112306

Roake CM, Artandi SE. 2020. Regulation of human telomerase in homeostasis and disease. *Nat Rev Mol Cell Biol* **21:** 384–397. doi:10.1038/s41580-020-0234-z

Robart AR, Collins K. 2010. Investigation of human telomerase holoenzyme assembly, activity, and processivity using disease-linked subunit variants. *J Biol Chem* **285:** 4375–4386. doi:10.1074/jbc.M109.088575

Robart AR, Collins K. 2011. Human telomerase domain interactions capture DNA for TEN domain-dependent processive elongation. *Mol Cell* **42:** 308–318. doi:10.1016/j.molcel.2011.03.012

Roy R, Meier B, McAinsh AD, Feldmann HM, Jackson SP. 2004. Separation-of-function mutants of yeast Ku80 reveal a Yku80p-Sir4p interaction involved in telomeric silencing. *J Biol Chem* **279:** 86–94. doi:10.1074/jbc.M306841200

Sarek G, Marzec P, Margalef P, Boulton SJ. 2015. Molecular basis of telomere dysfunction in human genetic diseases. *Nat Struct Mol Biol* **22:** 867–874. doi:10.1038/nsmb.3093

Schmidt JC, Cech TR. 2015. Human telomerase: biogenesis, trafficking, recruitment, and activation. *Genes Dev* **29:** 1095–1105. doi:10.1101/gad.263863.115

Schmidt JC, Dalby AB, Cech TR. 2014. Identification of human TERT elements necessary for telomerase recruitment to telomeres. *eLife* **3:** e03563. doi:10.7554/eLife .03563

Sealey DC, Zheng L, Taboski MA, Cruickshank J, Ikura M, Harrington LA. 2010. The N-terminus of hTERT contains a DNA-binding domain and is required for telomerase activity and cellular immortalization. *Nucleic Acids Res* **38:** 2019–2035. doi:10.1093/nar/gkp1160

Sekne Z, Ghanim GE, van Roon A-M, Nguyen THD. 2022. Structural basis of human telomerase recruitment by TPP1-POT1. *Science* **375:** 1173–1176. doi:10.1126/sci ence.abn6840

Seto AG, Zaug AJ, Sobel SG, Wolin SL, Cech TR. 1999. *Saccharomyces cerevisiae* telomerase is an Sm small nuclear ribonucleoprotein particle. *Nature* **401:** 177–180. doi:10.1038/43694

Sexton AN, Youmans DT, Collins K. 2012. Specificity requirements for human telomere protein interaction with telomerase holoenzyme. *J Biol Chem* **287:** 34455–34464. doi:10.1074/jbc.M112.394767

Sexton AN, Regalado SG, Lai CS, Cost GJ, O'Neil CM, Urnov FD, Gregory PD, Jaenisch R, Collins K, Hockemeyer D. 2014. Genetic and molecular identification of three human TPP1 functions in telomerase action: recruitment, activation, and homeostasis set point regulation. *Genes Dev* **28:** 1885–1899. doi:10.1101/gad.246819.114

Shay JW. 2016. Role of telomeres and telomerase in aging and cancer. *Cancer Discov* **6:** 584–593. doi:10.1158/2159-8290.CD-16-0062

Shay JW, Bacchetti S. 1997. A survey of telomerase activity in human cancer. *Eur J Cancer* **33:** 787–791. doi:10.1016/ S0959-8049(97)00062-2

Shay JW, Wright WE. 2006. Telomerase therapeutics for cancer: challenges and new directions. *Nat Rev Drug Discov* **5:** 577–584. doi:10.1038/nrd2081

Singer MS, Gottschling DE. 1994. *TLC1:* template RNA component of *Saccharomyces cerevisiae* telomerase. *Science* **266:** 404–409. doi:10.1126/science.7545955

Singh M, Wang Z, Koo BK, Patel A, Cascio D, Collins K, Feigon J. 2012. Structural basis for telomerase RNA recognition and RNP assembly by the holoenzyme La family protein p65. *Mol Cell* **47:** 16–26. doi:10.1016/j.molcel .2012.05.018

Soudet J, Jolivet P, Teixeira MT. 2014. Elucidation of the DNA end-replication problem in *Saccharomyces cerevisiae. Mol Cell* **53:** 954–964. doi:10.1016/j.molcel.2014.02 .030

Steitz TA. 1999. DNA polymerases: structural diversity and common mechanisms. *J Biol Chem* **274:** 17395–17398. doi:10.1074/jbc.274.25.17395

Stern JL, Zyner KG, Pickett HA, Cohen SB, Bryan TM. 2012. Telomerase recruitment requires both TCAB1 and Cajal bodies independently. *Mol Cell Biol* **32:** 2384–2395. doi:10.1128/MCB.00379-12

Takai H, Jenkinson E, Kabir S, Babul-Hirji R, Najm-Tehrani N, Chitayat DA, Crow YJ, de Lange T. 2016. A POT1 mutation implicates defective telomere end fill-in and telomere truncations in Coats plus. *Genes Dev* **30:** 812–826. doi:10.1101/gad.276873.115

Tang W, Kannan R, Blanchette M, Baumann P. 2012. Telomerase RNA biogenesis involves sequential binding by Sm and Lsm complexes. *Nature* **484:** 260–264. doi:10 .1038/nature10924

Tesmer VM, Ford LP, Holt SE, Frank BC, Yi X, Aisner DL, Ouellette M, Shay JW, Wright WE. 1999. Two inactive fragments of the integral RNA cooperate to assemble active telomerase with the human protein catalytic subunit (hTERT) in vitro. *Mol Cell Biol* **19:** 6207–6216. doi:10 .1128/MCB.19.9.6207

Theimer CA, Feigon J. 2006. Structure and function of telomerase RNA. *Curr Opin Struct Biol* **16:** 307–318. doi:10 .1016/j.sbi.2006.05.005

Tycowski KT, Shu MD, Kukoyi A, Steitz JA. 2009. A conserved WD40 protein binds the Cajal body localization signal of scaRNP particles. *Mol Cell* **34:** 47–57. doi:10 .1016/j.molcel.2009.02.020

Upton HE, Hong K, Collins K. 2014. Direct single-stranded DNA binding by Teb1 mediates the recruitment of *Tetrahymena thermophila* telomerase to telomeres. *Mol Cell Biol* **34:** 4200–4212. doi:10.1128/MCB.01030-14

Upton HE, Chan H, Feigon J, Collins K. 2017. Shared subunits of *tetrahymena* telomerase holoenzyme and replication protein a have different functions in different cellular complexes. *J Biol Chem* **292:** 217–228. doi:10.1074/ jbc.M116.763664

Venteicher AS, Abreu EB, Meng Z, McCann KE, Terns RM, Veenstra TD, Terns MP, Artandi SE. 2009. A human telomerase holoenzyme protein required for Cajal body localization and telomere synthesis. *Science* **323:** 644–648. doi:10.1126/science.1165357

Verdun RE, Karlseder J. 2007. Replication and protection of telomeres. *Nature* **447:** 924–931. doi:10.1038/nature 05976

Vogan JM, Zhang X, Youmans DT, Regalado SG, Johnson JZ, Hockemeyer D, Collins K. 2016. Minimized human telomerase maintains telomeres and resolves endogenous roles of H/ACA proteins, TCAB1, and Cajal bodies. *eLife* **5:** e18221. doi:10.7554/eLife.18221

Waga S, Stillman B. 1998. The DNA replication fork in eukaryotic cells. *Annu Rev Biochem* **67:** 721–751. doi:10 .1146/annurev.biochem.67.1.721

Wan F, Ding Y, Zhang Y, Wu Z, Li S, Yang L, Yan X, Lan P, Li G, Wu J, et al. 2021. Zipper head mechanism of telomere synthesis by human telomerase. *Cell Res* **31:** 1275–1290. doi:10.1038/s41422-021-00586-7

Wang F, Podell ER, Zaug AJ, Yang Y, Baciu P, Cech TR, Lei M. 2007. The POT1-TPP1 telomere complex is a telomerase processivity factor. *Nature* **445:** 506–510. doi:10 .1038/nature05454

Wang Y, Sušac L, Feigon J. 2019. Structural biology of telomerase. *Cold Spring Harb Perspect Biol* **11:** a032383. doi:10.1101/cshperspect.a032383

Wang H, Ma T, Zhang X, Chen W, Lan Y, Kuang G, Hsu SJ, He Z, Chen Y, Stewart J, et al. 2023a. CTC1 OB-B interaction with TPP1 terminates telomerase and prevents telomere overextension. *Nucleic Acids Res* **51:** 4914–4928. doi:10.1093/nar/gkad237

Wang Y, He Y, Wang Y, Yang Y, Singh M, Eichhorn CD, Cheng X, Jiang YX, Zhou ZH, Feigon J. 2023b. Structure of LARP7 protein p65–telomerase RNA complex in telomerase revealed by Cryo-EM and NMR. *J Mol Biol* **435**: 168044. doi:10.1016/j.jmb.2023.168044

Webb CJ, Zakian VA. 2008. Identification and characterization of the *Schizosaccharomyces pombe* TER1 telomerase RNA. *Nat Struct Mol Biol* **15**: 34–42. doi:10.1038/nsmb1354

Webb CJ, Zakian VA. 2012. *Schizosaccharomyces pombe* Ccq1 and TER1 bind the 14-3-3-like domain of Est1, which promotes and stabilizes telomerase-telomere association. *Genes Dev* **26**: 82–91. doi:10.1101/gad.181826.111

Weinrich SL, Pruzan R, Ma L, Ouellette M, Tesmer VM, Holt SE, Bodnar AG, Lichsteiner S, Kim NW, Trager JB, et al. 1997. Reconstitution of human telomerase with the template RNA component hTR and the catalytic protein subunit hTRT. *Nat Genet* **17**: 498–502. doi:10.1038/ng1297-498

Wellinger RJ, Zakian VA. 2012. Everything you ever wanted to know about *Saccharomyces cerevisiae* telomeres: beginning to end. *Genetics* **191**: 1073–1105. doi:10.1534/genetics.111.137851

Witkin KL, Collins K. 2004. Holoenzyme proteins required for the physiological assembly and activity of telomerase. *Genes Dev* **18**: 1107–1118. doi:10.1101/gad.1201704

Wu RA, Collins K. 2014. Human telomerase specialization for repeat synthesis by unique handling of primer-template duplex. *EMBO J* **33**: 921–935. doi:10.1002/embj.201387205

Wu RA, Tam J, Collins K. 2017a. DNA-binding determinants and cellular thresholds for human telomerase repeat addition processivity. *EMBO J* **36**: 1908–1927. doi:10.15252/embj.201796887

Wu RA, Upton HE, Vogan JM, Collins K. 2017b. Telomerase mechanism of telomere synthesis. *Annu Rev Biochem* **86**: 439–460. doi:10.1146/annurev-biochem-061516-045019

Wyatt HD, Lobb DA, Beattie TL. 2007. Characterization of physical and functional anchor site interactions in human telomerase. *Mol Cell Biol* **27**: 3226–3240. doi:10.1128/MCB.02368-06

Xie M, Podlevsky JD, Qi X, Bley CJ, Chen JJ. 2010. A novel motif in telomerase reverse transcriptase regulates telomere repeat addition rate and processivity. *Nucleic Acids Res* **38**: 1982–1996. doi:10.1093/nar/gkp1198

Xu M, Kiselar J, Whited TL, Hernandez-Sanchez W, Taylor DJ. 2019. POT1-TPP1 differentially regulates telomerase via POT1 His266 and as a function of single-stranded telomere DNA length. *Proc Natl Acad Sci* **116**: 23527–23533. doi:10.1073/pnas.1905381116

Yang W, Lee YS. 2015. A DNA-hairpin model for repeat-addition processivity in telomere synthesis. *Nat Struct Mol Biol* **22**: 844–847. doi:10.1038/nsmb.3098

Ye K. 2007. H/ACA guide RNAs, proteins and complexes. *Curr Opin Struct Biol* **17**: 287–292. doi:10.1016/j.sbi.2007.05.012

Ye JZ, Hockemeyer D, Krutchinsky AN, Loayza D, Hooper SM, Chait BT, de Lange T. 2004. POT1-interacting protein PIP1: a telomere length regulator that recruits POT1 to the TIN2/TRF1 complex. *Genes Dev* **18**: 1649–1654. doi:10.1101/gad.1215404

Yu EY, Sun J, Lei M, Lue NF. 2012. Analyses of *Candida* Cdc13 orthologues revealed a novel OB fold dimer arrangement, dimerization-assisted DNA binding, and substantial structural differences between Cdc13 and RPA70. *Mol Cell Biol* **32**: 186–198. doi:10.1128/MCB.05875-11

Zappulla DC. 2020. Yeast telomerase RNA flexibly scaffolds protein subunits: results and repercussions. *Molecules* **25**: 2750. doi:10.3390/molecules25122750

Zappulla DC, Cech TR. 2004. Yeast telomerase RNA: a flexible scaffold for protein subunits. *Proc Natl Acad Sci* **101**: 10024–10029. doi:10.1073/pnas.0403641101

Zappulla DC, Goodrich K, Cech TR. 2005. A miniature yeast telomerase RNA functions in vivo and reconstitutes activity in vitro. *Nat Struct Mol Biol* **12**: 1072–1077. doi:10.1038/nsmb1019

Zaug AJ, Podell ER, Nandakumar J, Cech TR. 2010. Functional interaction between telomere protein TPP1 and telomerase. *Genes Dev* **24**: 613–622. doi:10.1101/gad.1881810

Zaug AJ, Lim CJ, Olson CL, Carilli MT, Goodrich KJ, Wuttke DS, Cech TR. 2021. CST does not evict elongating telomerase but prevents initiation by ssDNA binding. *Nucleic Acids Res* **49**: 11653–11665. doi:10.1093/nar/gkab942

Zeng Z, Min B, Huang J, Hong K, Yang Y, Collins K, Lei M. 2011. Structural basis for *Tetrahymena* telomerase processivity factor Teb1 binding to single-stranded telomeric-repeat DNA. *Proc Natl Acad Sci* **108**: 20357–20361. doi:10.1073/pnas.1113624108

Zhong FL, Batista LF, Freund A, Pech MF, Venteicher AS, Artandi SE. 2012. TPP1 OB-fold domain controls telomere maintenance by recruiting telomerase to chromosome ends. *Cell* **150**: 481–494. doi:10.1016/j.cell.2012.07.012

# Biogenesis and Regulation of Telomerase during Development and Cancer

Lu Chen[1] and Luis Francisco Zirnberger Batista[2,3]

[1]Cancer Signaling and Epigenetics Program and Cancer Epigenetics Institute, Institute for Cancer Research, Fox Chase Cancer Center, Philadelphia, Pennsylvania 19111, USA

[2]Department of Medicine, Washington University in St. Louis, St. Louis, Missouri 63110, USA

[3]Center for Genome Integrity, Siteman Cancer Center, Washington University in St. Louis, St. Louis, Missouri 63110, USA

Correspondence: lu.chen@fccc.edu; lbatista@wustl.edu

Telomerase is a large ribonucleoprotein complex responsible for the addition of telomeric DNA repeats to chromosomal ends. Telomerase is composed of core and accessory components that work in coordination to ensure telomere length is maintained during development and in specific cell types. Telomerase activity is tightly regulated and is strongly increased in most tumor cells. On the other hand, loss-of-function mutations either in accessory factors or in core components of the complex impact telomere maintenance and cause a large spectrum of severe phenotypes, typically described as telomere biology disorders. A central element for efficient telomerase function is the proper biogenesis and assembly of the holoenzyme. Here, we discuss our current understanding of these processes and how they modulate telomerase efficiency. We consider how these processes are influenced by the specific subcellular localization of different telomerase components during different stages of the assembly of the holoenzyme. We describe the tremendous progress made in this area over the last decade and how recently discovered aspects of telomerase biogenesis can be exploited clinically, to actively benefit patients suffering from telomere biology disorders.

Telomeres are specialized nucleoprotein complexes that are essential for genomic stability. By preventing chromosomal ends from being recognized as DNA double-stranded breaks, telomeres prevent the erroneous activation of DNA damage response (DDR) mechanisms that would culminate in the accumulation of chromosomal fusions and genetic instability (Lazzerini-Denchi and Sfeir 2016; de Lange 2018). Telomeres are composed of double-stranded TTAGGG repeats that vary in length depending on species and cell type (Demanelis et al. 2020), followed by a 3′ end single-stranded region known as the G overhang (Blackburn 2005). Telomere length is influenced by each cell's replicative history, as conventional DNA polymerases are unable to fully replicate linear DNA molecules (Harley et al. 1990), and also by telomere rapid deletion mechanisms, including trimming by different nucleases (Pickett et al. 2011; Rivera et al. 2017). This leads to a gradual shortening of telomeres over time that is commonly referred to as the "end-replication" problem (Levy et al. 1992). Progressive telomere shortening can be counterbalanced by telomerase, a ribonucleoprotein (RNP) complex that is able to elongate telomeres, thereby preventing gradual telomere shortening over time (Greider and Blackburn 1985; Lee et al. 1998; Blackburn and Collins 2011). Indeed, while telomerase allows for continued cellular expansion, cellular proliferation in the absence of telomerase culminates in the activation of DDR responses that trigger cellular senescence and death (d'Adda di Fagagna et al. 2003; Herbig et al. 2004). In humans, however, telomerase activity is mostly restricted to germ cells as well as somatic stem and progenitor cell populations and not found in differentiated cell types (Wright et al. 1996). However, ~85% of human cancers show reactivation of telomerase expression, which allows continuous proliferation of tumor cells (Bodnar et al. 1998). On the other hand, loss-of-function mutations that impact different aspects of telomerase cause accelerated telomere shortening and instability and are found in a series of genetic diseases collectively referred to as telomere biology disorders (Revy et al. 2023).

These findings highlight the essential role of telomerase for cellular and organismal viability. A fundamental aspect for efficient telomerase function is the correct assembly of the telomerase RNP complex and its efficient recruitment to telomeres. Here, we discuss the molecular mechanisms that regulate the biogenesis of telomerase components, the assembly of the RNP complex, as well as the subcellular localization of where these events take place. We consider how these different processes are intertwined to allow efficient telomerase activity and the continuous maintenance of telomere length over time.

## THE BIOGENESIS OF HUMAN TELOMERASE

The last few years have seen tremendous progress in our understanding of telomerase biogenesis, led primarily by advances in cryogenic electron microscopy (cryo-EM) (Nguyen et al. 2018; Ghanim et al. 2021; Wan et al. 2021; Liu et al. 2022), live cell imaging at single-molecule sensitivity (Schmidt et al. 2016, 2018; Laprade

et al. 2020), and nascent RNA end-sequencing (Roake et al. 2019). Additionally, the discovery of novel mutations in genes not previously associated with telomere biology disorders contributed to our knowledge of the molecular series of events that are necessary for efficient telomerase function (Dhanraj et al. 2015; Stuart et al. 2015; Tummala et al. 2015; Gable et al. 2019). We will incorporate these recent findings to provide an up-to-date view of telomerase formation and assembly. While in vitro telomerase activity only requires the telomerase reverse transcriptase (TERT) and the telomerase RNA (TR) components, in vivo the catalytically active telomerase RNP is far larger and is composed not only of TERT and TR, but also by the telomerase Cajal body protein 1 (TCAB1), and two copies of each of the proteins that comprise the dyskerin complex: dyskerin (DKC1), NOP10, NHP2, and GAR1 (Fig. 1). Moreover, recent data obtained from cryo-EM identified an H2A–H2B dimer directly bound to an essential motif of TR, which indicates these could be part of the telomerase complex, and modulate TR function (Ghanim et al. 2021; Liu et al. 2022). For ease of reading, we will describe the role of these different components in telomerase biogenesis separately and detail the sequential steps of events that culminate in the assembly of a functional telomerase complex that is recruited to DNA and able to efficiently elongate telomeres.

## The Biogenesis, Structure, and Localization of hTR and Its Associated Components

The sequence of events necessary for telomerase assembly and function revolves around its RNA component—TR. Among different species, TRs serve as the platform for the RNP complex assembly and as a template for the reverse transcriptase function of TERT. The TR component of telomerase also plays a central role in telomerase accumulation and localization in vertebrates, through its different domains and association with different components of the telomerase complex. The structure and size of TR components vary significantly between different species, ranging from ~150 nt in ciliates

Telomerase biogenesis

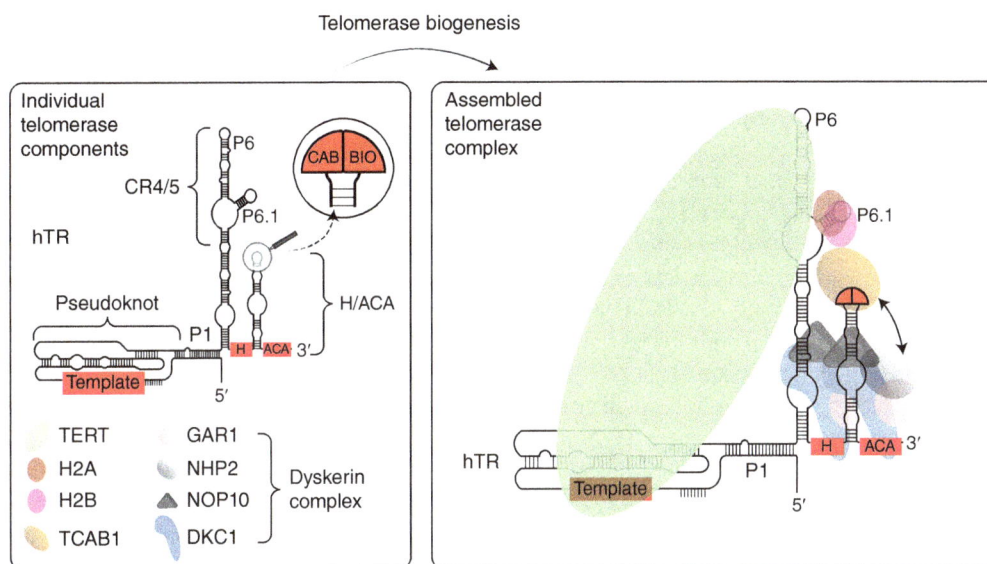

**Figure 1.** The assembled telomerase complex and its individual components. While telomerase activity in vitro can be achieved only with telomerase reverse transcriptase (TERT) and human telomerase RNA (hTR), in vivo the assembled telomerase ribonucleoprotein (RNP) is a large structure, comprised of 12 individual components, all of which regulate telomerase function. hTR plays a central role in the assembly and overall organization of the RNP, directly binding to TERT through its CR4/5 domain and template-pseudoknot regions. Additionally, hTR directly binds to telomerase Cajal body protein 1 (TCAB1) through its Cajal body box (CAB) box and to the dyskerin complex (DKC1, NOP10, NHP2, GAR1) through its H/ACA domain. Recent cryogenic electron microscopy (cryo-EM) data show that histones H2A and H2B bind to telomerase through the P6 and P6.1 loops within the CR4/5 domain of hTR. Correct assembly of this complex requires cellular trafficking through different nuclear compartments and is necessary for efficient engagement with telomeres. Recent advances in cryo-EM have helped pinpoint the interactions between different members of this RNP (such as a direct interaction between TCAB1 and NHP2). Mutations in several of these components are found in patients with telomere biology disorders and are discussed in the text.

to ~450 nt in vertebrates, and more than 1300 nt in yeast (Theimer and Feigon 2006; Podlevsky et al. 2008). Here, we provide a detailed analysis of the biogenesis and function of the human telomerase RNA (hTR) component.

## hTR Structure

In humans, mature hTR molecules are 451 nt long (Fig. 1). hTR is composed of two separate lobes: the H/ACA lobe that contains the H/ACA domain and is bound by TCAB1 and two sets of the dyskerin tetramer complex, and the catalytic lobe, which includes the pseudoknot-template domains and is directly bound to TERT. These two separate lobes are connected by the conserved regions 4 and 5 (CR4/5) of hTR, where

the P6 and P6.1 loops are located (Fig. 1). The CR4/5 domain of hTR directly interacts with TERT, independently from the template domain, through its P6–P6.1 hairpin region (Zhang et al. 2011). Recently, it was described that histone H2A–H2B dimers are also found bound to the CR4/5 domain of hTR (Ghanim et al. 2021; Liu et al. 2022), suggesting a role for these histones in the folding, and therefore function, of hTR. The H/ACA domain sits on hTR's 3′ end and is configured in a "hairpin–hinge–hairpin–tail" arrangement. The "hinge" is formed by the H box consensus sequence (5′-AGAGGA-3′), which is then followed by a 5′-ACA-3′ sequence located 3 nt upstream of hTR's 3′ end (Mitchell et al. 1999). This H/ACA box domain is also shared with small nucleolar (sno) and small Cajal body (sca) RNA molecules, which act as guide RNAs in the site-specific pseudouridylation of ribosomal RNAs and small nuclear RNAs, respectively (Borchardt et al. 2020). However, to date, no pseudouridylation targets of hTR have been reported, suggesting that the H/ACA domain of hTR could function solely as a stability factor for hTR (discussed below). The H/ACA lobe also contains a stem-loop structure that holds a 4-nt-long Cajal body box (CAB) motif that binds to TCAB1, and a biogenesis-promoting box (BIO box) motif that is involved in hTR stability and accumulation (Egan and Collins 2012; Ketele et al. 2016). On the opposite side of the molecule, at hTR's 5′ end, sits its catalytic lobe, which represents the largest functional domain of the molecule. This region is divided into three segments, a large pseudoknot loop that directly binds to TERT, a short template region that is complementary to telomeric DNA, and the P1 stem region, which serves as a template boundary element (Fig. 1; Zhang et al. 2011). Mutations in the different regions of the catalytic domain are the most common mutations in hTR found in telomere biology syndrome patients (Revy et al. 2023).

## hTR Biogenesis and Assembly of the Telomerase Ribonucleoprotein

While the majority of snoRNAs and scaRNAs are contained within introns of mRNAs and transcribed along with their host genes, hTR is unique in that it is transcribed individually from a dedicated promoter (Feng et al. 1995). However, similarly to snoRNAs and scaRNAs, the dyskerin complex associates with hTR cotranscriptionally and is necessary for its stability (Darzacq et al. 2006). Each of the "hairpins" in the H/ACA domain of hTR initially associates with a heterotetramer composed of dyskerin, NOP10, NHP2, and NAF1. While dyskerin and NOP10 bind directly to hTR, NHP2, and NAF1 bind to dyskerin itself (Egan and Collins 2012; Qin and Autexier 2021). The binding of dyskerin to NOP10, NHP2, and NAF1 happens before its association with hTR. As NAF1 has been shown to bind to nascent H/ACA sno-

RNAs (Fatica et al. 2002), this can help explain the cotranscriptional binding of the dyskerin complex to hTR. Binding to the dyskerin complex is facilitated by the BIO box region in hTR and is essential for hTR stability. Indeed, pathogenic mutations in dyskerin, NOP10, NHP2, and NAF1 reduce hTR levels and have been identified in telomere biology syndrome patients, further illustrating the vital role of this complex for telomerase function (Revy et al. 2023).

At a later stage of the maturation process, and after binding of hTR to the dyskerin complex, NAF1 is substituted by GAR1 (Leulliot et al. 2007), an event that takes place in Cajal bodies (CBs) (Darzacq et al. 2006), nuclear compartments that are scaffolded by coilin and play a central role in the biogenesis of snRNAs and snoRNAs (Neugebauer 2017). The recruitment of hTR to CBs is performed by TCAB1, which binds to the CAB domain of hTR and is a core component of the telomerase complex (Venteicher et al. 2009). A recent high-resolution cryo-EM structure of the H/ACA domain of hTR revealed that TCAB1 directly interacts with the 3′ end of NHP2 (Ghanim et al. 2024). This structure of the H/ACA domain of hTR meticulously mapped the interactions between hTR and the two dyskerin heterotetramers and showed that these interact extensively with one another via the two DKC1 subunits (Ghanim et al. 2024). Additional proteins are required for proper hTR biogenesis but are associated with the telomerase complex only transiently. These include SHQ1, which binds to the RNA-binding region of dyskerin (Walbott et al. 2011) before its binding to NAF1 and hTR (Grozdanov et al. 2009). This binding, which happens in the cytoplasm, prevents the premature association of dyskerin with hTR and prevents nonspecific binding of RNAs to dyskerin. In this multistep process of hTR biogenesis, SHQ1 is then removed (upon nuclear import) from dyskerin by the AAA$^+$ ATPases pontin and reptin (Machado-Pinilla et al. 2012), which are themselves essential for hTR accumulation and telomerase activity (Venteicher et al. 2008). This sequential step of events necessary for telomerase RNP biogenesis is additionally controlled by proteins that regulate hTR localization and cellular trafficking and will be discussed in more detail in the following section.

## hTR Localization and Trafficking

Multiple lines of evidence establish correct subcellular localization as a central regulatory mechanism for telomerase biogenesis and activity in human cells (Fig. 2, middle column). Telomere elongation by telomerase requires this complex to eventually associate with DNA. However, before that, correct telomerase assembly involves dynamic nuclear trafficking in which many steps revolve around hTR maturation and are dependent on the different proteins that bind to hTR at different stages of its biogenesis. While at any given time, most hTR molecules are found freely diffusing around the cellular nucleus (Schmidt et al. 2016, 2018; Laprade et al. 2020), telomerase associates with CBs in human cells (similarly to other scaRNPs). This association is dependent on the binding of TCAB1 to the CAB box region of hTR (Venteicher et al. 2009; Laprade et al. 2020). Replacement of NAF1 with GAR1 in the telomerase complex most likely happens immediately before hTR association to TCAB1 and consequent translocation to CBs (Darzacq et al. 2006), and TCAB1 has been shown to directly interact with GAR1 (Ghanim et al. 2021), highlighting a highly coordinated assembly/trafficking process. TCAB1 folding and action is in turn dependent on the TRiC chaperonin (Freund et al. 2014), and loss of either TriC or TCAB1 leads to telomerase mislocalization and reduced catalytic activity, without reducing hTR levels (Venteicher et al. 2009; Freund et al. 2014; Chen et al. 2018). Indeed, while in wild-type (WT) cells hTR is transiently associated with the nucleolus, deletion of TCAB1 causes a significant accumulation of hTR in nucleoli (Fig. 2, left column; Venteicher et al. 2009; Klump et al. 2023). Accordingly, it has been recently demonstrated that TCAB1 binding to hTR prevents its trafficking into nucleoli (Klump et al. 2023), which indicates that hTR molecules found in nucleoli might represent immature precursors (Nguyen et al. 2015).

Recent evidence also shows that the recruitment of hTR to telomeres does not depend on its localization to CBs (Laprade et al. 2020), which is substantiated by the fact that the ablation of CBs (through depletion of coilin) does not impair the trafficking of hTR to telomeres or telomerase function (Stern et al. 2012; Chen et al. 2015). However, TCAB1 has also been shown to control telomerase catalytic activity by promoting the proper folding of hTR's CR4/5 region, which increases its binding to TERT (Chen et al. 2018). Interestingly, hTR association with TERT reduces its residence time in CBs, which is proposed to facilitate its sequential recruitment to telomeres (Laprade et al. 2020). While the molecular events leading to telomerase recruitment to telomeres will be discussed in a separate section below, these results show the dynamic process of hTR trafficking in the nucleus, and how proper hTR localization determines telomerase efficiency in human cells, which is corroborated by mutations in TCAB1 leading to mislocalized telomerase and causing severe phenotypes in patients suffering from telomere biology disorders (Zhong et al. 2011).

## Posttranscriptional Modifications Are Essential Regulators of hTR Localization, Levels, and Function

Over the last few years, different posttranscriptional modifications to both the 5′ and 3′ end of hTR have emerged as major regulators of hTR biogenesis and telomerase activity (Fig. 3). Importantly, these modifications are performed by pathways that can be modulated to restore telomerase activity and telomere homeostasis in cells harboring mutations that cause reduced levels of hTR, creating new possibilities for the clinical management of patients suffering with different telomere biology syndromes. These posttranscriptional modifications are discussed below.

### The methylation status of the 5′ cap regulates hTR levels and localization. RNA polymerase II (RNA Pol II) transcripts acquire a monomethylguanosine (MMG) modification on their 5′ end, referred to as 7-methylguanylate cap (m$^7$G). This 5′ end capping process happens cotran-

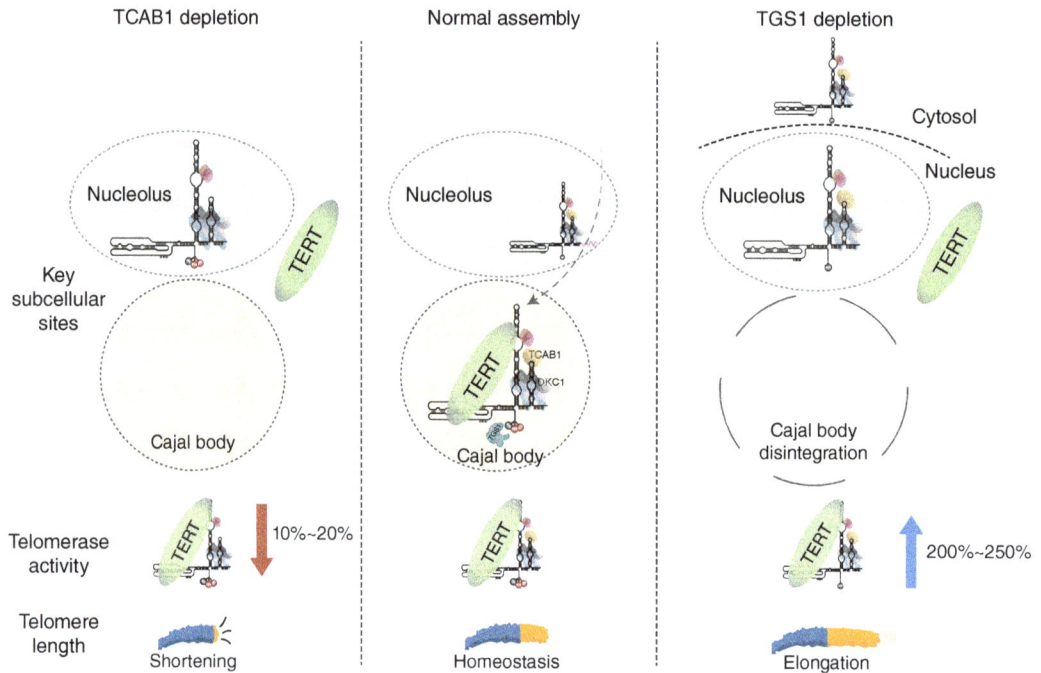

**Figure 2.** Telomerase biogenesis is intimately linked to human telomerase RNA (hTR) trafficking. In unperturbed human cells (*center* column), telomerase components rapidly diffuse across nuclear space but show prolonged residency within Cajal bodies (CBs). Telomerase reverse transcriptase (TERT) is largely excluded from nucleoli while hTR can transit through nucleoli. 3′ extended hTR has been preferentially detected in the nucleolus (Nguyen et al. 2015). The precise mechanism and location by which TERT–hTR assembles remain speculative. TCAB1 (hTR chaperon, *left* panel) and trymethylguanosine synthase 1 (TGS1) (RNA cap hypermethylase, *right* panel) are concentrated in CBs. When telomerase Cajal body protein 1 (TCAB1) is depleted (*left* column) cellular hTR mislocalizes to nucleoli from CBs, partitioning into a TERT-excluded compartment. This is consistent with the markedly reduced telomerase activity (10%–20% when compared to wild-type [WT] cells) and telomere shortening in TCAB1-deficient cells. However, nucleolar localization per se does not impede telomerase biogenesis, as there seems to exist an alternative pathway for nucleolar hTR to assemble with TERT productively. In TGS1-deficient cells (*right* column), hTR shows cytosolic retention, nucleolar accumulation, and a higher steady-state level, boosting telomerase biogenesis (150%–200% when compared to WT), and robust telomere elongation. How compromised CBs (in TCAB1 and TGS1-deficient cells) lead to opposite telomerase functional outcomes remains to be determined.

scriptionally and regulates different processes on these RNA molecules, including nuclear export and degradation (Ramanathan et al. 2016). While in snRNAs formation of an MMG cap recruits the cap-binding complex (CBC), the adaptor protein PHAX, and the nuclear export protein CRM1 for cytoplasmic export, MMG-capped snoRNAs are directed by PHAX directly into CBs (Izaurralde et al. 1995; Ohno et al. 2000). Once in CBs, MMG caps of snoRNAs are hypermethylated to N2, 2, 7 trimethylguanosine (TMG) caps by trimethylguanosine syn-

thase 1 (TGS1). TMG-capped snoRNAs are then recruited to nucleoli independently from CRM1 (Pradet-Balade et al. 2011). While the mechanisms that specifically control hTR export and localization are not fully described, it has recently been shown that depletion of TGS1 (and therefore prevention of TMG cap formation) increases hTR association to both the CBC complex and Sm chaperone proteins, which causes an increase of mature hTR levels in the nucleus and cytoplasm of cells (Fig. 2, right column; Chen et al. 2020). MMG-capped hTR re-

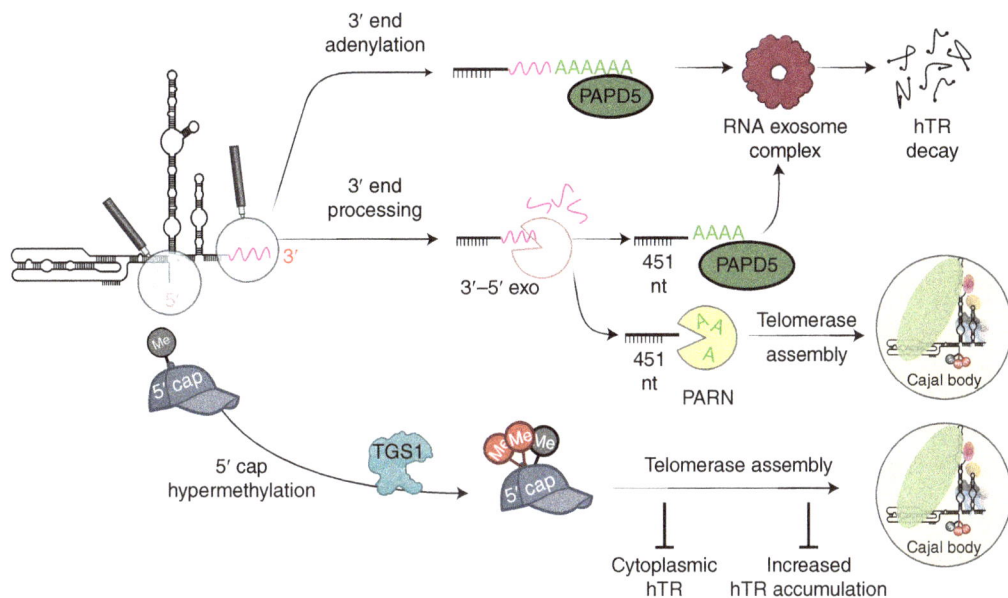

**Figure 3.** Posttranscriptional modifications of human telomerase RNA (hTR) regulate its decay and modulate the assembly and function of telomerase. hTR is initially transcribed as longer, 3′ end extended immature precursors (up to 1500 nt) that are polyadenylated by PAPD5, leading to their exonucleolytic decay by the exosome. Alternatively, these 3′ end extended immature precursors can be trimmed down to mature, 451 nt hTR molecules by 3′ exoribonucleases. PAPD5 can also adenylate fully trimmed hTR molecules to promote their 3′–5′ degradation, which is increased in patient cells with mutations in DKC1 and poly(A)-specific ribonuclease (PARN). The binding of hTR to the dyskerin complex (through its H/ACA domain) increases hTR stability. Binding to telomerase Cajal body protein 1 (TCAB1) facilitates telomerase trafficking to Cajal bodies (CBs), which aids in the formation of functional telomerase complexes. Nascent hTR molecules are transcribed with a 5′ monomethyl-guanosine (MMG) cap that is then hypermethylated by trymethylguanosine synthase 1 (TGS1) to form an N2, 2, 7 trimethylguanosine (TMG) cap. In the absence of TGS1, hTR retains its 5′ end MMG cap, and a fraction is exported into the cytoplasm and total hTR levels increase after TGS1 silencing, possibly because mislocalization protects hTR from degradation. How, and if, the 3′ and 5′ end processing of hTR molecules are coordinated remains to be determined.

mains functional and able to bind to TERT, and TGS1-depleted cells show increased telomerase activity and telomere elongation when compared to their WT counterparts (Chen et al. 2020). Although the molecular mechanisms causing hTR accumulation in TGS1-depleted cells remain obscure, these results indicate that TGS1 silencing could be a potential avenue for telomere elongation in cells that harbor mutations that impair telomerase activity due to reduced hTR levels (Batista et al. 2022). Indeed, the chemical inhibition of TGS1 with sinefungin has recently been shown to increase hTR levels and telomerase activity in human cells (Buemi

et al. 2022; Galati et al. 2022), including cells from patients with telomere biology disorders (Galati et al. 2022).

*Correct 3′ end processing is a major determinant of hTR accumulation and telomerase activity.* While mature hTR comprises a 451-nt-long molecule, at any given time hTR molecules exist as a pool of different transcripts that vary in length, due to 3′ end extensions (Goldfarb and Cech 2013; Roake et al. 2019). These 3′ end extended molecules are encoded by the hTR locus, and it has been suggested that they are formed by readthrough of RNA Pol II (Qin and Autexier 2021). While the regulation of hTR transcription termination in human cells remains mostly unknown (as unlike other snoRNAs, hTR is not contained in introns of host genes), it has recently been shown that the integrator complex regulates termination of hTR transcription (Rubtsova et al. 2019), as is the case for other RNA classes transcribed by RNA Pol II (Mendoza-Figueroa et al. 2020). While the majority of hTR is composed of mature, 451 nt molecules, nascent RNA end-sequencing has recently revealed that the majority of hTR is initially transcribed as longer precursors, which are then processed into 451 nt molecules (Roake et al. 2019).

Following transcription, hTR precursors are rapidly polyadenylated at their 3′ end by the noncanonical poly(A) polymerase PAPD5 (Fig. 3; Moon et al. 2015; Tseng et al. 2015; Shukla et al. 2016). PAPD5 is part of the TRAMP complex, which is composed also of the helicase MTR4 and ZCCHC7, a zinc-finger protein that binds to RNA (Schmid and Jensen 2019). The TRAMP complex facilitates the surveillance and degradation of abnormal RNAs by the RNA exosome complex, which represents the main 3′–5′ ribonuclease in human cells and acts on most types of nuclear and cytoplasmic RNAs (Chlebowski et al. 2013). Indeed, the addition of 3′ end poly(A) tails by PAPD5 quickly leads to exosome-mediated degradation of hTR (Moon et al. 2015; Tseng et al. 2015; Shukla et al. 2016). The recruitment of the RNA exosome complex is performed by the nuclear exosome targeting (NEXT) complex, which is composed of a dimer of MTR4, ZCCHC8, and RBM7 (Gerlach et al. 2022). Depletion of either ZCCHC8 or RBM7 causes increased levels of 3′ end adenylated hTR precursors (Tseng et al. 2015), and mutations in ZCCHC8 have recently been found in pulmonary fibrosis patients with short telomeres (Gable et al. 2019). Cells harboring ZCCHC8 mutations showed an accumulation of 3′ end extended hTR, at the expense of mature hTR molecules, leading to reduced telomerase activity (Gable et al. 2019). These results indicate that exosome targeting is a major regulator of hTR maturation and function. Interestingly, the 3′ and 5′ end posttranscriptional modifications to

hTR may act in a coordinated fashion, as the CBC complex, which binds to MMG-capped hTR molecules, also recruits the exosome RNA decay machinery through NEXT (Tseng et al. 2015). More research is necessary to clarify how these processes are coordinated.

Working in opposition to PAPD5, poly(A)-specific ribonuclease (PARN) deadenylates hTR precursors, preventing their degradation by the exosome and leading to the accumulation of mature, functional hTR molecules (Moon et al. 2015; Shukla et al. 2016; Roake et al. 2019). Moreover, it has been suggested that PARN could, in addition to deadenylating hTR precursors, also directly trim hTR 3′ end transcripts to generate mature, 451 nt hTR molecules (Tseng et al. 2018). The role of PARN in hTR biogenesis was initially found through the discovery of clinically relevant PARN mutations that caused significant telomere shortening in patients. These patients show low levels of mature hTR, resulting in reduced telomerase activity and telomere shortening (Dhanraj et al. 2015; Stuart et al. 2015; Tummala et al. 2015). The direct correlation between exosome-mediated decay of 3′ poly(A) tailed hTR precursors and telomerase activity opens the possibility that the inhibition of 3′ end tailing by PAPD5 could be explored as a clinical alternative for patients harboring mutations that cause reduced accumulation of mature, functional hTR. Indeed, the genetic inhibition of PAPD5 in cells harboring mutations in DKC1 or PARN reduces the abundance of 3′ extended hTR molecules, and increases the number of deadenylated, functional hTR molecules (Boyraz et al. 2016; Shukla et al. 2016). Moreover, the genetic inhibition of PAPD5 restores hematopoietic development and output in cells with mutant DKC1 (Fok et al. 2019), rescuing, therefore, a major phenotype found in telomere biology disorders. Adding support to this hypothesis, the pharmacological inhibition of PAPD5 was recently shown to increase mature levels of hTR, increase telomerase activity, promote telomere lengthening, and improve hematopoietic development in both DKC1 and PARN mutant models (Nagpal et al. 2020; Shukla et al. 2020).

Finally, while the opposing roles of PARN and PAPD5 during hTR biogenesis are well established, it is interesting that in cells depleted of both (PARN and PADP5), the maturation kinetics of hTR is similar to WT cells while 3′ end adenylation of precursors is significantly reduced (Roake et al. 2019). This indicates the possibility of an additional 3′–5′ exonuclease able to deadenylate hTR and promote its maturation. A possible candidate is TOE1, an exonuclease that localizes to CBs and deadenylates different noncoding RNA classes (Son et al. 2018). Indeed, TOE1-deficient cells show an increase in oligoadenylated hTR precursors and impaired telomerase activity (Machado-Pinilla et al. 2012).

In conclusion, the biogenesis of hTR is a highly coordinated process, involving multiple proteins and protein complexes, with successive steps occurring sequentially at different subnuclear localizations. While central to telomerase biogenesis and activity, and despite tremendous progress in recent years, many open questions remain on the regulators of the hTR maturation process and how these can be exploited for targeted treatment of telomere biology disorders.

## Regulation of hTERT Expression and Levels

While hTR is ubiquitously expressed in human cells, human TERT (hTERT) expression is restricted to specific cell types and developmental stages, often serving as a limiting factor for telomerase assembly and activity in human tissues (Bodnar et al. 1998). While high levels of telomerase are found in the germline, during embryogenesis (in the inner cell mass of the blastocyst), and in various progenitor/stem cell compartments, telomerase is quickly silenced as development progresses and cells differentiate into functional lineages (Wright et al. 1996). On the contrary, cellular reprogramming experiments demonstrated that telomerase is reactivated in induced pluripotent stem cells generated from human fibroblasts, consolidating the developmental stage as a major regulator of hTERT expression (Takahashi et al. 2007; Agarwal et al. 2010; Batista et al. 2011). While often seen as a cancer prevention mechanism, as cells with no telomerase show progressive telomere shortening that results in senescence, the molecular mechanisms regulating hTERT suppression during tissue development and formation remain poorly understood. Similarly, mechanisms regulating hTERT activation during tumorigenesis are not yet completely described, despite significant progress over the last few years (Yuan et al. 2019; Lorbeer and Hockemeyer 2020). In this section, we summarize different events that contribute to the regulation of hTERT expression, and therefore to the biogenesis of telomerase in human settings. A more comprehensive view of this subject can be found in Martin and Hockmeyer (2025).

### Genetic and Epigenetic Factors that Regulate hTERT Transcription

The hTERT gene is ~40 kb long, consisting of 16 exons (15 introns) localized in the short arm of chromosome 5 (5p.15:33) (Fig. 4; Yuan et al. 2019). It has a single promoter that is rich in binding sites for various transcription factors including Myc, Klf4, and Sp1. While these have been shown to regulate hTERT transcription, their expression does not completely rescue hTERT silencing in terminally differentiated human somatic cells (Greenberg et al. 1999; Wu et al. 1999; Oh et al. 2001; Wong et al. 2010). However, hTERT is reactivated in most human cancers, and different molecular mechanisms account for this, including integration of exogenous tumor viral regulators (including Epstein–Barr virus [EBV], and cytomegalovirus [CMV]) (Bellon and Nicot 2008), and hTERT copy number variations (Gay-Bellile et al. 2017). In addition, hypermethylation of the hTERT promoter region has recently been associated with hTERT up-regulation in human cancer. This hypermethylation is found specifically in a region immediately upstream of the TERT core promoter that contains 52 different CpG sites and is now described as the TERT hypermethylated oncological region (THOR) (Lee et al. 2019). This epigenetic regulation of hTERT expression, where hypomethylated THOR is associated with hTERT repression and hyperme-

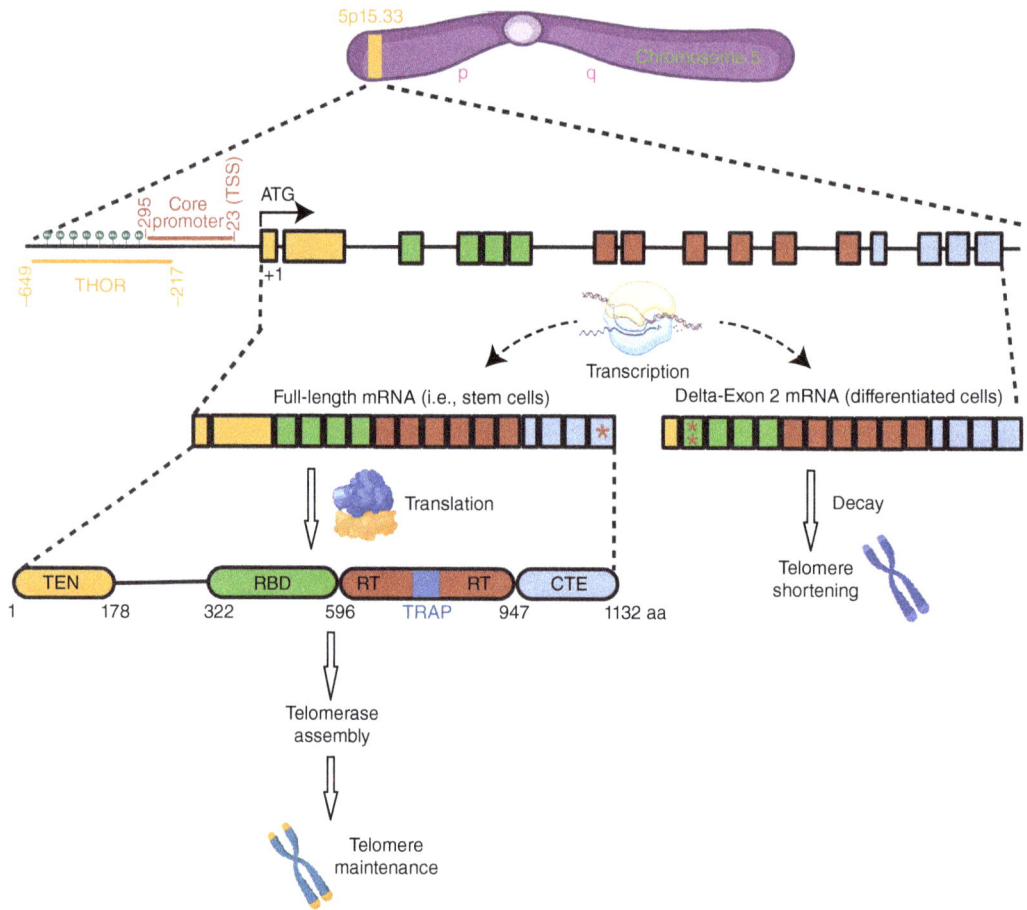

**Figure 4.** Regulation of telomerase reverse transcriptase (TERT) transcription is a major determinant of telomerase assembly. Human TERT (hTERT) consists of 16 exons localized in the short arm of chromosome 5 (5p.15:33), and its transcription is silenced in most somatic cells. Different factors contribute to hTERT expression during development in germ and somatic stem cells, as well as its reactivation in most human cancers from different origins. The most common mechanism for hTERT reactivation in cancer is the acquisition of somatic TERT promoter mutations (TPMs) in specific positions of the TERT proximal promoter ($-57$; $-124$ and $-146$, in relation to ATG). Additionally, hypermethylation of the hTERT promoter in the TERT hypermethylated oncological region (THOR) also causes hTERT up-regulation in cancer. During development, hTERT levels are controlled by alternative splicing, where terminally differentiated cells show loss of Exon 2, causing a frameshift mutation that creates two premature stop codons in Exon 3, which are associated with increased decay. In stem and pluripotent cells, hTERT molecules are fully transcribed, and protected from exacerbated decay, TERT is translated into an 1132 amino acid protein, and incorporated into telomerase. (TSS) Transcription start site, (TEN) telomerase essential amino-terminal domain, (RBD) RNA-binding domain, (RT domain) reverse transcriptase domain, (CTE) carboxy-terminal extension.

thylated THOR is found in cancer samples that show hTERT reactivation, has been found in a variety of cancer types (Lee et al. 2019) and potentially accounts for a significant mechanism leading to telomerase formation and activity during tumorigenesis (Lee et al. 2021). Once

transcribed, hTERT transiently binds to the chaperone heat shock protein 90 (Hsp90), which increases its stability and folding, facilitating its binding to hTR and modulating telomerase activity (Holt et al. 1999; Forsythe et al. 2001).

Cite this article as *Cold Spring Harb Perspect Biol* doi: 10.1101/cshperspect.a041692

## Somatic Mutations in the TERT Promoter Robustly Increase hTERT Levels and Telomerase Activity

While the mechanisms described above are commonly observed in tumor samples, the most frequently observed mechanism of hTERT reactivation in human cancer is the acquisition of somatic mutations in specific positions of the TERT proximal promoter (−57; −124 and −146, in relation to the ATG transcriptional start site). These mutations (referred to as TERT promoter mutations [TPMs]), were initially found in melanoma (Horn et al. 2013; Huang et al. 2013), but have now been found in multiple tumor types that are usually derived from cells with reduced self-renewal, including gliomas and hepatocellular carcinomas (HCCs) (Killela et al. 2013). In fact, a comprehensive analysis of 31 cancer types from The Cancer Genome Atlas (TCGA) established that TPMs represent the most common noncoding mutation in cancer and one of the most frequent mutations found in cancer overall, with a staggering number of 27% of all analyzed samples harboring a TPM (Barthel et al. 2017). Functionally, TPMs create de novo binding sites for ETS (E26 transformation specific) transcription factors, most predominantly the GA-binding protein α and β (GABPα and GABPβ) (Bell et al. 2015). This binding happens as a tetramer (with two GABPα/β being formed), and the inhibition of either GABPα or GABPβ reduces hTERT expression in cancer cells harboring mutations in the TERT promoter (Bell et al. 2015; Mancini et al. 2018). Mutations in the TERT promoter are heterozygous, and the allele harboring the mutation is the only transcriptionally active allele, showing histone marks typically associated with active chromatin (while the allele harboring the WT promoter sequence retains marks of epigenetic silencing) (Huang et al. 2015; Stern et al. 2015). Mutations in the TERT promoter occur early during tumorigenesis (Nault and Zucman-Rossi 2016), and expression analysis shows that up-regulation of hTERT levels in cells harboring these mutations is gradual and associated with the accumulation of critically short telomeres (Chiba et al. 2017).

Combined, these results indicate that TPMs occur early during carcinogenesis where they initially increase cellular life span by lengthening of critically short telomeres and then promote continued cellular proliferation (Lorbeer and Hockemeyer 2020).

## Alternative Splicing Regulates hTERT Levels during Development

hTERT levels are not only regulated transcriptionally (as discussed above) but also posttranscriptionally. Transcriptome analysis of hTERT has revealed more than 20 alternatively spliced isoforms (Sæbøe-Larssen et al. 2006; Bollmann 2013) that are mostly predicted to encode catalytically inactive proteins. Interestingly, different hTERT isoforms have been associated with different cancer types (both telomerase positive and negative) and developmental stages, suggesting that they play important roles in hTERT regulation and telomerase activity. Indeed, several lines of evidence suggest that alternative splicing regulates telomerase activity during human development (Ulaner et al. 1998; Brenner et al. 1999) and cancer (Fajkus et al. 2003; Petrenko et al. 2010). The hTERT protein contains four domains (the telomerase amino-terminal [TEN] domain; the TR-binding domain [TRBD]; the reverse transcriptase domain [RT domain], and the carboxy-terminal extension [CTE]) (Fig. 4). Most alternatively spliced hTERT mRNAs differ in exons that encode the RT domain and are, therefore, predicted to generate catalytically inactive proteins. However, some of these splice variants have been shown to be translated into protein and to negatively regulate telomerase activity in human cells (Zhu et al. 2014), or even to regulate cellular processes outside telomere elongation (Bollmann 2013). Recently, targeted sequencing techniques that allow analysis of low abundance transcripts identified hTERT reads spanning the junction of exon 1 to exon 3 (indicating loss of exon 2–hTERT-ΔEx2) in human terminally differentiated cells, but not in their undifferentiated, isogenic counterparts (Penev et al. 2021). hTERT-ΔEx2 was enriched in cell types that lack telomerase activity, and reduced in pluripotent cells with high telomerase levels, including embryonic and induced pluripotent stem cells. The absence of exon 2 causes a frameshift that creates two tandem premature stop codons in exon 3, which have been predicted to trigger nonsense-mediated RNA decay (Withers et al. 2012; Hug et al. 2016) and cause rapid decay of hTERT transcripts. The forced inclusion of this exon during stem cell differentiation prevents telomerase silencing during tissue development (Penev et al. 2021). This posttranscriptional regulation of hTERT, therefore, directly modulates telomerase activity during development and synergizes with transcriptional mechanisms to promote telomerase biogenesis and function in human stem cells (Barranco 2021; Penev et al. 2021).

## Telomerase Recruitment to Telomeres

An essential step for telomere elongation is the recruitment of a functional telomerase complex to telomeric DNA. However, the low abundance of telomerase and telomeres in human cells (estimated at ~240 telomerase molecules and 184 telomeres in the late S phase of the cell cycle) (Xi and Cech 2014) indicates that this process does not happen by simple diffusion. Rather, extensive research shows that telomerase is actively recruited to telomeres, where its action is regulated to promote homeostasis of telomere length. As we discuss in previous sections, the assembly of the telomerase RNP is dependent on the expression of its multiple components, happens at multiple nuclear sites, and with regulated timing. In this section, we discuss how telomerase is recruited to telomeres and how telomere length is regulated in human cells.

### Localization to Cajal Bodies Increases Telomerase Recruitment to Telomeres

Multiple lines of evidence indicate that TCAB1 facilitates the recruitment of telomerase to telomeric DNA (Tycowski et al. 2009; Venteicher et al. 2009). Accordingly, mutations in TCAB1 are found in patients suffering with telomere biology syndromes (Zhong et al. 2011). TCAB1 binds to the CAB domain of hTR (Fig. 1), and if this interaction is abrogated, telomerase mislocalizes to the nucleolus (Tycowski

et al. 2009; Venteicher et al. 2009). If TCAB1 is absent, hTR does not localize to CBs but is rather tightly associated with the nucleolus (Laprade et al. 2020). However, the importance of telomerase localization to CBs remains unclear, as telomerase in mice cells does not always accumulate in CBs (Tomlinson et al. 2010) and even some human tumor cells (with high levels of telomerase) retain efficient telomere maintenance after disruption of CBs (Chen et al. 2015). These results indicate that telomerase association with CBs could modulate telomerase activity by improving telomerase–telomere interaction and more efficient telomere elongation, particularly in cells with reduced levels of telomerase expression. However, more studies are required to confirm this hypothesis and to precisely pinpoint how the localization of telomerase to CBs promotes its recruitment to telomeres in human cells.

### Shelterin Recruits Telomerase to Telomeres

Telomeric DNA is comprised of TTAGGG repeats extending up to 15 kb length in humans. It can be divided into two separate structures, a double-stranded DNA region that is several kilobases long, followed by a single-stranded 3′ end tail known as the G overhang (Lim and Cech 2021). Telomeres are associated with shelterin, a protein complex composed of six different proteins that bind both the double-stranded and single-stranded telomeric regions (Palm and de Lange 2008). Association with shelterin arranges telomeres into different structures, including end-capped telomeres and telomere loops (Lim and Cech 2021), structures that must be resolved for telomerase to access telomeric overhangs. Interestingly, the recruitment of telomerase to DNA ends is mediated by shelterin, which facilitates the pairing of the 3′ overhang with the hTR template within telomerase (Hockemeyer and Collins 2015).

Shelterin is comprised of Telomeric Repeat binding Factors 1 and 2 (TRF1 and TRF2), Protection of Telomeres 1 (POT1), TRF1-Interacting Nuclear Protein 2 (TIN2), Rap1 (the human ortholog of the yeast Repressor/Activator Protein 1), and TPP1. TRF1 and 2 bind double-stranded telomeric DNA, and POT1 binds to the single-stranded 3′ overhang. Rap1 works as an accessory unit of TRF2. TIN2 directly binds to TRF1 and 2, whereas TPP1 is connected to shelterin via its binding to both TIN2 and POT1 (de Lange 2018). The recruitment of telomerase to telomeres was initially shown to be reduced with the loss of TIN2-anchored TPP1 (Abreu et al. 2010). More recent data show that the recruitment of telomerase to telomeres is mediated by the binding of the TEN domain of TERT to the TEL-patch region of TPP1, which resides on the surface of its oligonucleotide-binding fold domain (Fig. 5; Nandakumar et al. 2012; Sexton et al. 2012; Zhong et al. 2012). Mutations in the TEL-patch domain of TPP1 lead to telomere shortening in telomerase-positive cells, and have been found in patients with telomere biology disorders, highlighting the importance of this interaction for telomerase recruitment (Guo et al. 2014; Kocak et al. 2014; Bertrand et al. 2024). Additionally, mutations in the amino-terminal OB-fold domain of TPP1 have also been identified in telomere biology disorder patients (Tummala et al. 2018), and this

region of the protein has been shown to be essential for telomerase processivity and recruitment to telomeres (Grill et al. 2018). The critical role of TPP1 for telomerase action at telomeres has been confirmed by high-resolution live cell imaging experiments that also show that before forming a stable association with telomeres, telomerase probes telomeres thousands of times during the S phase of the cell cycle. Both of these transient interactions and stable interactions require TPP1-TERT binding (Schmidt et al. 2016, 2018). Finally, these functional studies have been confirmed by recent structural studies that defined the interaction between TPP1 and telomerase that facilitate recruitment and processivity (Liu et al. 2022; Sekne et al. 2022).

The role of shelterin in regulating telomerase and telomere length has been further solidified by recent genome-wide sequencing efforts in different types of human cancers. These have identified heterozygous mutations in POT1 as a recurrent event in tumorigenesis (Wu et al. 2020). These cancer-associated mutations in POT1 do not cause deprotection of telomeres

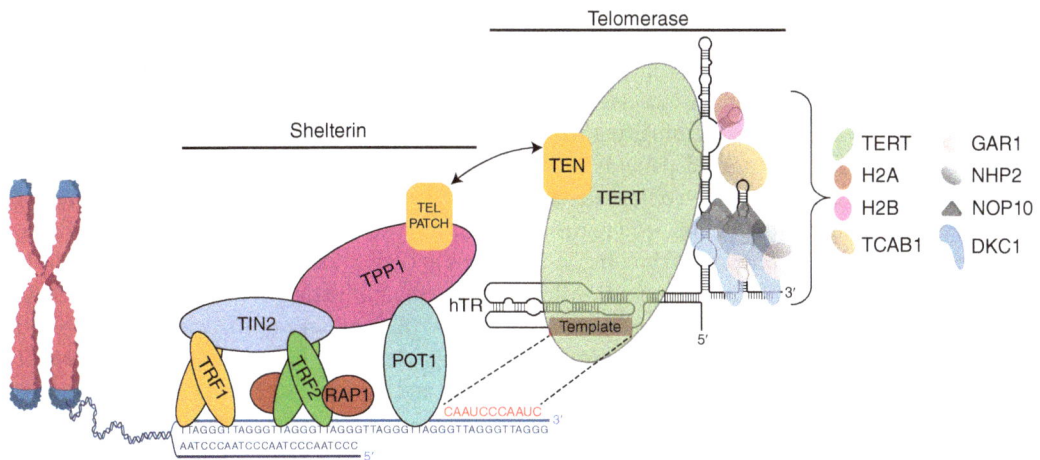

Figure 5. Shelterin recruits telomerase to promote telomere elongation. The shelterin complex directly binds to telomeric DNA. While Telomeric Repeat binding Factors 1 and 2 (TRF1 and TRF2) bind to the double-stranded region of telomeres, Protection of Telomeres 1 (POT1) binds to the single-stranded 3′ overhang DNA. Repressor/Activator Protein 1 (RAP1) binds directly to TRF2, while TRF1-Interacting Nuclear Protein 2 (TIN2) binds to both TRF1 and TRF2. TPP1 binds to TIN2 and to POT1. The TEL-PATCH domain of TPP1 directly binds to telomerase reverse transcriptase (TERT) through its amino-terminal (TEN) domain, and recruits telomerase to telomeres. The template region of hTR binds to the telomeric 3′ end strand.

and do not activate DDRs but rather seem to be selected for and persist over time (Kim et al. 2021). In fact, heterozygous mutations in POT1 have recently been shown to cause elongated telomeres in patients with clonal hematopoiesis and a wide range of benign and malignant neoplasms (DeBoy et al. 2023). The increased risk of cancer in these patients seems to be directly related to their increased capacity to sustain telomere length over time, leading to extended cellular longevity. These results show that while exacerbated telomere shortening is associated with increased genetic instability and DNA damage in patients suffering with telomere biology disorders such as dyskeratosis congenita (Revy et al. 2023), enhanced telomere maintenance, leading to exacerbated telomere elongation, is also detrimental for correct tissue maintenance over time.

Finally, as the protection of single-stranded telomere overhangs by POT1–TPP1 requires its binding to the central shelterin component TIN2 (Takai et al. 2011), it is not surprising that mutations in TIN2 are also found in patients with telomere biology disorders (Revy et al. 2023). These mutations do not reduce telomerase activity and do not impact TIN2 localization but disrupt TPP1-dependent recruitment of telomerase to telomeres, leading to exacerbated telomere shortening in patients (Yang et al. 2011). These results indicate that TIN2 might act as a telomerase stimulating factor, and that mutations found in patients, instead of causing direct deprotection of telomeres through defective shelterin function, lead to disease due to reduced telomerase action in patients (Frank et al. 2015). In agreement with this possibility, we know that TIN2 cooperates with TPP1–POT1 to act as a telomerase-stimulatory factor (Pike et al. 2019). Of note, mice harboring a clinically relevant TIN2 mutation showed telomere defects that were, at least partially, independent of telomerase, as they were also observed in telomerase-negative cells (Frescas and de Lange 2014). These results indicate more research is necessary to establish precisely how mutations in TIN2 cause severe phenotypes in patients, and which molecular strategies can be used to mitigate these defects.

## CONCLUSIONS AND PERSPECTIVES

Research regarding telomerase biogenesis and assembly has benefited significantly from recent developments in single-molecule imaging and improvements in cryo-EM resolution. Individual telomerase molecules can now be tracked in human living cells, with their precise subcellular localization pinpointed during different stages of RNP maturation and trafficking to telomeres. Adding to this, the enormous amount of data generated by high-throughput sequencing allowed for the identification of mutations in genes not previously implicated in telomerase biogenesis, such as PARN and ZCCHC8. These results have also helped increase our knowledge on the several steps necessary for telomerase biogenesis, in particular regarding its RNA component, hTR.

Telomerase facilitates continuous cellular proliferation while, in unperturbed settings, mitigating tumorigenesis. Not surprisingly, its biogenesis and regulation are highly coordinated, involving different subcellular compartments and multiple protein complexes that ensure telomerase molecules are assembled before the decay of its RNA component, and able to eventually attach to, and elongate, telomeres. Similarly, the tight control of expression of its RT component adds an additional layer of security to prevent widespread telomere elongation in human tissues. These events happen in a multistep process that includes the assembly of the H/ACA lobe of telomerase, which is then followed by posttranscriptional modifications in both 5′ and 3′ end of hTR, and finally by its association with the RT component of telomerase. A summary of the many factors influencing telomerase biogenesis, assembly, recruitment, and function at telomeres can be found in Figure 6.

The relevance of the tight control of telomerase biogenesis becomes clear when mutations compromise this highly coordinated process. On one hand, increased levels of TERT, caused by mutations in its promoter, gene amplifications, or epigenetic mechanisms are found in most human cancers. On the other hand, mutations that compromise telomerase

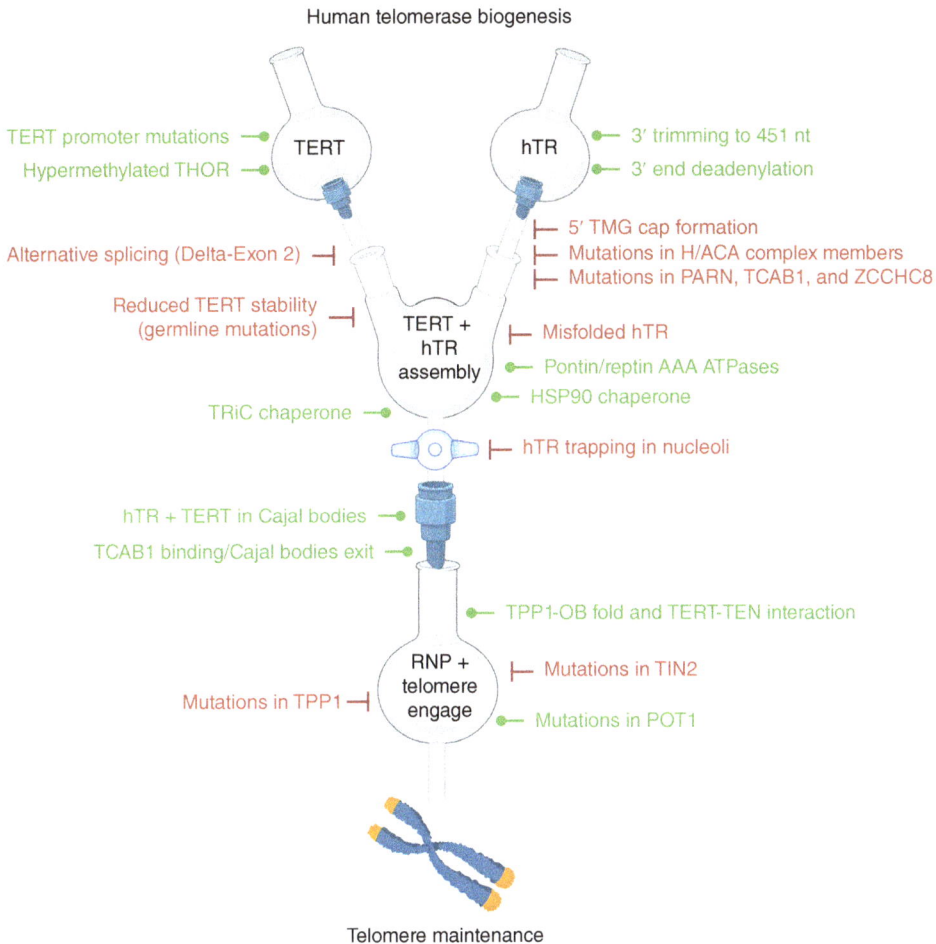

**Figure 6.** Stepwise formation of the telomerase complex. Telomerase biogenesis is a complex multistep process that directly modulates telomerase function and telomere maintenance. Many different factors, described in this article, modulate the formation of an active telomerase ribonucleoprotein, able to extend telomeres and therefore mitigate the deleterious consequences of telomere dysfunctional for organismal life. These factors act during different stages of telomerase biogenesis, either stimulating telomerase formation (depicted in green), or inhibiting the formation of active telomerase complexes (depicted in red).

biogenesis, in particular mutations that compromise the stability and localization of hTR, are the most common genetic alterations found in patients with telomere biology disorders. However, the fact that telomerase biogenesis and assembly is a complex pathway opens an array of possibilities that can be exploited for clinical intervention, such as the inhibition of PAPD5 and TGS1 in patients harboring mutations that compromise hTR stability (Batista et al. 2022).

Finally, while our knowledge of telomerase biogenesis has significantly increased over the last decade, several aspects of this process remain obscure. For instance, a possible functional link between the 5′ and 3′ end posttranscriptional modifications of hTR could shed light on how these pathways act to promote its stability in different cell types. Moreover, several lines of evidence indicate that an additional pathway for the 3′ end processing of hTR must exist and could also have clinical implications. Moreover, several aspects regarding the timing and location of different steps during biogenesis and to what extent specific subcellular location sites determine telomere homeostasis remain unknown. These will likely be achieved through in-depth structural analysis of the intermediates formed during telomerase biogenesis and trafficking. Similarly, additional research is necessary to determine more precisely how telomerase interacts with and is regulated by other telomere-binding proteins, including shelterin and the CST complex.

Collectively, we hope we have demonstrated that while tremendous amounts of data have been generated since the initial discovery of telomerase, much remains to be determined on how this fascinating RNP is formed, assembled, and directed to telomeres, where it plays an essential role for cellular and organismal viability.

## ACKNOWLEDGMENTS

The authors thank all authors cited in this work and apologize to all colleagues whose work could not be cited due to space restrictions. We thank Miguel Godinho Ferreira, Jayakrishnan Nandakumar, and Abby Green, as well as members of our laboratories for critical reading of this manuscript. L.C. is supported by the NIH (R35GM150538). L.F.Z.B. is supported by the NIH (R01CA258386, R01HL174789, and R01HL172961), the DOD (BM200111 and BM230053), the American Cancer Society, and the Siteman Cancer Center.

## REFERENCES

*Reference is also in this subject collection.

Abreu E, Aritonovska E, Reichenbach P, Cristofari G, Culp B, Terns RM, Lingner J, Terns MP. 2010. TIN2-tethered TPP1 recruits human telomerase to telomeres in vivo. *Mol Cell Biol* 30: 2971–2982. doi:10.1128/MCB.00240-10

Agarwal S, Loh YH, McLoughlin EM, Huang J, Park IH, Miller JD, Huo H, Okuka M, Dos Reis RM, Loewer S, et al. 2010. Telomere elongation in induced pluripotent stem cells from dyskeratosis congenita patients. *Nature* 464: 292–296. doi:10.1038/nature08792

Barranco C. 2021. Alternative splicing regulates telomerase repression. *Nat Rev Genet* 22: 414.

Barthel FP, Wei W, Tang M, Martinez-Ledesma E, Hu X, Amin SB, Akdemir KC, Seth S, Song X, Wang Q, et al.

2017. Systematic analysis of telomere length and somatic alterations in 31 cancer types. *Nat Genet* 49: 349–357. doi:10.1038/ng.3781

Batista LF, Pech MF, Zhong FL, Nguyen HN, Xie KT, Zaug AJ, Crary SM, Choi J, Sebastiano V, Cherry A, et al. 2011. Telomere shortening and loss of self-renewal in dyskeratosis congenita induced pluripotent stem cells. *Nature* 474: 399–402. doi:10.1038/nature10084

Batista LFZ, Dokal I, Parker R. 2022. Telomere biology disorders: time for moving towards the clinic? *Trends Mol Med* 28: 882–891. doi:10.1016/j.molmed.2022.08.001

Bell RJ, Rube HT, Kreig A, Mancini A, Fouse SD, Nagarajan RP, Choi S, Hong C, He D, Pekmezci M, et al. 2015. Cancer. The transcription factor GABP selectively binds and activates the mutant TERT promoter in cancer. *Science* 348: 1036–1039. doi:10.1126/science.aab0015

Bellon M, Nicot C. 2008. Regulation of telomerase and telomeres: human tumor viruses take control. *J Natl Cancer Inst* 100: 98–108. doi:10.1093/jnci/djm269

Bertrand A, Ba I, Kermasson L, Pirabakaran V, Chable N, Lainey E, Ménard C, Kallel F, Picard C, Hadiji S, et al. 2024. Characterization of novel mutations in the TEL-patch domain of the telomeric factor TPP1 associated with telomere biology disorders. *Hum Mol Genet* 33: 612–623. doi:10.1093/hmg/ddad210

Blackburn EH. 2005. Telomeres and telomerase: their mechanisms of action and the effects of altering their functions. *FEBS Lett* 579: 859–862. doi:10.1016/j.febslet.2004.11.036

Blackburn EH, Collins K. 2011. Telomerase: an RNP enzyme synthesizes DNA. *Cold Spring Harb Perspect Biol* 3: a003558. doi:10.1101/cshperspect.a003558

Bodnar AG, Ouellette M, Frolkis M, Holt SE, Chiu CP, Morin GB, Harley CB, Shay JW, Lichtsteiner S, Wright WE. 1998. Extension of life-span by introduction of telomerase into normal human cells. *Science* 279: 349–352. doi:10.1126/science.279.5349.349

Bollmann FM. 2013. Physiological and pathological significance of human telomerase reverse transcriptase splice variants. *Biochimie* 95: 1965–1970. doi:10.1016/j.biochi.2013.07.031

Borchardt EK, Martinez NM, Gilbert WV. 2020. Regulation and function of RNA pseudouridylation in human cells. *Annu Rev Genet* 54: 309–336. doi:10.1146/annurev-genet-112618-043830

Boyraz B, Moon DH, Segal M, Muosieyiri MZ, Aykanat A, Tai AK, Cahan P, Agarwal S. 2016. Posttranscriptional manipulation of TERC reverses molecular hallmarks of telomere disease. *J Clin Invest* 126: 3377–3382. doi:10.1172/JCI87547

Brenner CA, Wolny YM, Adler RR, Cohen J. 1999. Alternative splicing of the telomerase catalytic subunit in human oocytes and embryos. *Mol Hum Reprod* 5: 845–850. doi:10.1093/molehr/5.9.845

Buemi V, Schillaci O, Santorsola M, Bonazza D, Broccia PV, Zappone A, Bottin C, Dell'Omo G, Kengne S, Cacchione S, et al. 2022. TGS1 mediates 2,2,7-trimethyl guanosine capping of the human telomerase RNA to direct telomerase dependent telomere maintenance. *Nat Commun* 13: 2302. doi:10.1038/s41467-022-29907-z

Chen Y, Deng Z, Jiang S, Hu Q, Liu H, Songyang Z, Ma W, Chen S, Zhao Y. 2015. Human cells lacking coilin and

Cajal bodies are proficient in telomerase assembly, trafficking and telomere maintenance. *Nucleic Acids Res* **43**: 385–395. doi:10.1093/nar/gku1277

Chen L, Roake CM, Freund A, Batista PJ, Tian S, Yin YA, Gajera CR, Lin S, Lee B, Pech MF, et al. 2018. An activity switch in human telomerase based on RNA conformation and shaped by TCAB1. *Cell* **174**: 218–230.e13. doi:10.1016/j.cell.2018.04.039

Chen L, Roake CM, Galati A, Bavasso F, Micheli E, Saggio I, Schoeftner S, Cacchione S, Gatti M, Artandi SE, et al. 2020. Loss of human TGS1 hypermethylase promotes increased telomerase RNA and telomere elongation. *Cell Rep* **30**: 1358–1372.e5. doi:10.1016/j.celrep.2020.01.004

Chiba K, Lorbeer FK, Shain AH, McSwiggen DT, Schruf E, Oh A, Ryu J, Darzacq X, Bastian BC, Hockemeyer D. 2017. Mutations in the promoter of the telomerase gene *TERT* contribute to tumorigenesis by a two-step mechanism. *Science* **357**: 1416–1420. doi:10.1126/science.aao0535

Chlebowski A, Lubas M, Jensen TH, Dziembowski A. 2013. RNA decay machines: the exosome. *Biochim Biophys Acta* **1829**: 552–560. doi:10.1016/j.bbagrm.2013.01.006

D'Adda di Fagagna F, Reaper PM, Clay-Farrace L, Fiegler H, Carr P, Von Zglinicki T, Saretzki G, Carter NP, Jackson SP. 2003. A DNA damage checkpoint response in telomere-initiated senescence. *Nature* **426**: 194–198. doi:10.1038/nature02118

Darzacq X, Kittur N, Roy S, Shav-Tal Y, Singer RH, Meier UT. 2006. Stepwise RNP assembly at the site of H/ACA RNA transcription in human cells. *J Cell Biol* **173**: 207–218. doi:10.1083/jcb.200601105

DeBoy EA, Tassia MG, Schratz KE, Yan SM, Cosner ZL, McNally EJ, Gable DL, Xiang Z, Lombard DB, Antonarakis ES, et al. 2023. Familial clonal hematopoiesis in a long telomere syndrome. *N Engl J Med* **388**: 2422–2433. doi:10.1056/NEJMoa2300503

de Lange T. 2018. Shelterin-mediated telomere protection. *Annu Rev Genet* **52**: 223–247. doi:10.1146/annurev-genet-032918-021921

Demanelis K, Jasmine F, Chen LS, Chernoff M, Tong L, Delgado D, Zhang C, Shinkle J, Sabarinathan M, Lin H, et al. 2020. Determinants of telomere length across human tissues. *Science* **369**: eaaz6876. doi:10.1126/science.aaz6876

Dhanraj S, Gunja SM, Deveau AP, Nissbeck M, Boonyawat B, Coombs AJ, Renieri A, Mucciolo M, Marozza A, Buoni S, et al. 2015. Bone marrow failure and developmental delay caused by mutations in poly(A)-specific ribonuclease (*PARN*). *J Med Genet* **52**: 738–748. doi:10.1136/jmedgenet-2015-103292

Egan ED, Collins K. 2012. An enhanced H/ACA RNP assembly mechanism for human telomerase RNA. *Mol Cell Biol* **32**: 2428–2439. doi:10.1128/MCB.00286-12

Fajkus J, Borsky M, Kunicka Z, Kovarikova M, Dvorakova D, Hofmanova J, Kozubik A. 2003. Changes in telomerase activity, expression and splicing in response to differentiation of normal and carcinoma colon cells. *Anticancer Res* **23**: 1605–1612.

Fatica A, Dlakić M, Tollervey D. 2002. Naf1 p is a box H/ACA snoRNP assembly factor. *RNA* **8**: 1502–1514. doi:10.1017/S1355838202022094

Feng J, Funk WD, Wang SS, Weinrich SL, Avilion AA, Chiu CP, Adams RR, Chang E, Allsopp RC, Yu J, et al. 1995. The RNA component of human telomerase. *Science* **269**: 1236–1241. doi:10.1126/science.7544491

Fok WC, Shukla S, Vessoni AT, Brenner KA, Parker R, Sturgeon CM, Batista LFZ. 2019. Posttranscriptional modulation of TERC by PAPD5 inhibition rescues hematopoietic development in dyskeratosis congenita. *Blood* **133**: 1308–1312. doi:10.1182/blood-2018-11-885368

Forsythe HL, Jarvis JL, Turner JW, Elmore LW, Holt SE. 2001. Stable association of hsp90 and p23, but not hsp70, with active human telomerase. *J Biol Chem* **276**: 15571–15574. doi:10.1074/jbc.C100055200

Frank AK, Tran DC, Qu RW, Stohr BA, Segal DJ, Xu L. 2015. The shelterin TIN2 subunit mediates recruitment of telomerase to telomeres. *PLoS Genet* **11**: e1005410. doi:10.1371/journal.pgen.1005410

Frescas D, de Lange T. 2014. A TIN2 dyskeratosis congenita mutation causes telomerase-independent telomere shortening in mice. *Genes Dev* **28**: 153–166. doi:10.1101/gad.233395.113

Freund A, Zhong FL, Venteicher AS, Meng Z, Veenstra TD, Frydman J, Artandi SE. 2014. Proteostatic control of telomerase function through TRiC-mediated folding of TCAB1. *Cell* **159**: 1389–1403. doi:10.1016/j.cell.2014.10.059

Gable DL, Gaysinskaya V, Atik CC, Talbot CC, Kang B, Stanley SE, Pugh EW, Amat-Codina N, Schenk KM, Arcasoy MO, et al. 2019. *ZCCHC8*, the nuclear exosome targeting component, is mutated in familial pulmonary fibrosis and is required for telomerase RNA maturation. *Genes Dev* **33**: 1381–1396. doi:10.1101/gad.326785.119

Galati A, Scatolini L, Micheli E, Bavasso F, Cicconi A, Maccallini P, Chen L, Roake CM, Schoeftner S, Artandi SE, et al. 2022. The *S*-adenosylmethionine analog sinefungin inhibits the trimethylguanosine synthase TGS1 to promote telomerase activity and telomere lengthening. *FEBS Lett* **596**: 42–52. doi:10.1002/1873-3468.14240

Gay-Bellile M, Véronèse L, Combes P, Eymard-Pierre E, Kwiatkowski F, Dauplat MM, Cayre A, Privat M, Abrial C, Bignon YJ, et al. 2017. *TERT* promoter status and gene copy number gains: effect on *TERT* expression and association with prognosis in breast cancer. *Oncotarget* **8**: 77540–77551. doi:10.18632/oncotarget.20560

Gerlach P, Garland W, Lingaraju M, Salerno-Kochan A, Bonneau F, Basquin J, Jensen TH, Conti E. 2022. Structure and regulation of the nuclear exosome targeting complex guides RNA substrates to the exosome. *Mol Cell* **82**: 2505–2518.e7. doi:10.1016/j.molcel.2022.04.011

Ghanim GE, Fountain AJ, van Roon AM, Rangan R, Das R, Collins K, Nguyen THD. 2021. Structure of human telomerase holoenzyme with bound telomeric DNA. *Nature* **593**: 449–453. doi:10.1038/s41586-021-03415-4

Ghanim GE, Sekne Z, Balch S, van Roon AM, Nguyen THD. 2024. 2.7 Å cryo-EM structure of human telomerase H/ACA ribonucleoprotein. *Nat Commun* **15**: 746. doi:10.1038/s41467-024-45002-x

Goldfarb KC, Cech TR. 2013. 3′ terminal diversity of MRP RNA and other human noncoding RNAs revealed by deep sequencing. *BMC Mol Biol* **14**: 23. doi:10.1186/1471-2199-14-23

Greenberg RA, Chin L, Femino A, Lee KH, Gottlieb GJ, Singer RH, Greider CW, DePinho RA. 1999. Short dysfunctional telomeres impair tumorigenesis in the INK4a (Δ2/3) cancer-prone mouse. *Cell* **97:** 515–525. doi:10.1016/S0092-8674(00)80761-8

Greider CW, Blackburn EH. 1985. Identification of a specific telomere terminal transferase activity in *Tetrahymena* extracts. *Cell* **43:** 405–413. doi:10.1016/0092-8674(85)90170-9

Grill S, Tesmer VM, Nandakumar J. 2018. The N terminus of the OB domain of telomere protein TPP1 is critical for telomerase action. *Cell Rep* **22:** 1132–1140. doi:10.1016/j.celrep.2018.01.012

Grozdanov PN, Roy S, Kittur N, Meier UT. 2009. SHQ1 is required prior to NAF1 for assembly of H/ACA small nucleolar and telomerase RNPs. *RNA* **15:** 1188–1197. doi:10.1261/rna.1532109

Guo Y, Kartawinata M, Li J, Pickett HA, Teo J, Kilo T, Barbaro PM, Keating B, Chen Y, Tian L, et al. 2014. Inherited bone marrow failure associated with germline mutation of ACD, the gene encoding telomere protein TPP1. *Blood* **124:** 2767–2774. doi:10.1182/blood-2014-08-596445

Harley CB, Futcher AB, Greider CW. 1990. Telomeres shorten during ageing of human fibroblasts. *Nature* **345:** 458–460. doi:10.1038/345458a0

Herbig U, Jobling WA, Chen BP, Chen DJ, Sedivy JM. 2004. Telomere shortening triggers senescence of human cells through a pathway involving ATM, p53, and p21(CIP1), but not p16(INK4a). *Mol Cell* **14:** 501–513. doi:10.1016/S1097-2765(04)00256-4

Hockemeyer D, Collins K. 2015. Control of telomerase action at human telomeres. *Nat Struct Mol Biol* **22:** 848–852. doi:10.1038/nsmb.3083

Holt SE, Aisner DL, Baur J, Tesmer VM, Dy M, Ouellette M, Trager JB, Morin GB, Toft DO, Shay JW, et al. 1999. Functional requirement of p23 and Hsp90 in telomerase complexes. *Genes Dev* **13:** 817–826. doi:10.1101/gad.13.7.817

Horn S, Figl A, Rachakonda PS, Fischer C, Sucker A, Gast A, Kadel S, Moll I, Nagore E, Hemminki K, et al. 2013. TERT promoter mutations in familial and sporadic melanoma. *Science* **339:** 959–961. doi:10.1126/science.1230062

Huang FW, Hodis E, Xu MJ, Kryukov GV, Chin L, Garraway LA. 2013. Highly recurrent *TERT* promoter mutations in human melanoma. *Science* **339:** 957–959. doi:10.1126/science.1229259

Huang FW, Bielski CM, Rinne ML, Hahn WC, Sellers WR, Stegmeier F, Garraway LA, Kryukov GV. 2015. TERT promoter mutations and monoallelic activation of TERT in cancer. *Oncogenesis* **4:** e176. doi:10.1038/oncsis.2015.39

Hug N, Longman D, Cáceres JF. 2016. Mechanism and regulation of the nonsense-mediated decay pathway. *Nucleic Acids Res* **44:** 1483–1495. doi:10.1093/nar/gkw010

Izaurralde E, Lewis J, Gamberi C, Jarmolowski A, McGuigan C, Mattaj IW. 1995. A cap-binding protein complex mediating U snRNA export. *Nature* **376:** 709–712. doi:10.1038/376709a0

Ketele A, Kiss T, Jády BE. 2016. Human intron-encoded AluACA RNAs and telomerase RNA share a common element promoting RNA accumulation. *RNA Biol* **13:** 1274–1285. doi:10.1080/15476286.2016.1239689

Killela PJ, Reitman ZJ, Jiao Y, Bettegowda C, Agrawal N, Diaz LA, Friedman AH, Friedman H, Gallia GL, Giovanella BC, et al. 2013. *TERT* promoter mutations occur frequently in gliomas and a subset of tumors derived from cells with low rates of self-renewal. *Proc Natl Acad Sci* **110:** 6021–6026. doi:10.1073/pnas.1303607110

Kim WT, Hennick K, Johnson J, Finnerty B, Choo S, Short SB, Drubin C, Forster R, McMaster ML, Hockemeyer D. 2021. Cancer-associated POT1 mutations lead to telomere elongation without induction of a DNA damage response. *EMBO J* **40:** e107346. doi:10.15252/embj.2020107346

Klump BM, Perez GI, Patrick EM, Adams-Boone K, Cohen SB, Han L, Yu K, Schmidt JC. 2023. TCAB1 prevents nucleolar accumulation of the telomerase RNA to facilitate telomerase assembly. *Cell Rep* **42:** 112577. doi:10.1016/j.celrep.2023.112577

Kocak H, Ballew BJ, Bisht K, Eggebeen R, Hicks BD, Suman S, O'Neil A, Giri N; NCI DCEG Cancer Genomics Research Laboratory; NCI DCEG Cancer Sequencing Working Group, et al. 2014. Hoyeraal–Hreidarsson syndrome caused by a germline mutation in the TEL patch of the telomere protein TPP1. *Genes Dev* **28:** 2090–2102. doi:10.1101/gad.248567.114

Laprade H, Querido E, Smith MJ, Guérit D, Crimmins H, Conomos D, Pourret E, Chartrand P, Sfeir A. 2020. Single-molecule imaging of telomerase RNA reveals a recruitment-retention model for telomere elongation. *Mol Cell* **79:** 115–126.e6. doi:10.1016/j.molcel.2020.05.005

Lazzerini-Denchi E, Sfeir A. 2016. Stop pulling my strings—what telomeres taught us about the DNA damage response. *Nat Rev Mol Cell Biol* **17:** 364–378. doi:10.1038/nrm.2016.43

Lee HW, Blasco MA, Gottlieb GJ, Horner JW II, Greider CW, DePinho RA. 1998. Essential role of mouse telomerase in highly proliferative organs. *Nature* **392:** 569–574. doi:10.1038/33345

Lee DD, Leao R, Komosa M, Gallo M, Zhang CH, Lipman T, Remke M, Heidari A, Nunes NM, Apolonio JD, et al. 2019. DNA hypermethylation within TERT promoter upregulates TERT expression in cancer. *J Clin Invest* **129:** 223–229. doi:10.1172/JCI121303

Lee DD, Komosa M, Sudhaman S, Leao R, Zhang CH, Apolonio JD, Hermanns T, Wild PJ, Klocker H, Nassiri F, et al. 2021. Dual role of allele-specific DNA hypermethylation within the TERT promoter in cancer. *J Clin Invest* **131:** e146915. doi:10.1172/JCI146915

Leulliot N, Godin KS, Hoareau-Aveilla C, Quevillon-Cheruel S, Varani G, Henry Y, Van Tilbeurgh H. 2007. The box H/ACA RNP assembly factor Naf1p contains a domain homologous to Gar1p mediating its interaction with Cbf5p. *J Mol Biol* **371:** 1338–1353. doi:10.1016/j.jmb.2007.06.031

Levy MZ, Allsopp RC, Futcher AB, Greider CW, Harley CB. 1992. Telomere end-replication problem and cell aging. *J Mol Biol* **225:** 951–960. doi:10.1016/0022-2836(92)90096-3

Lim CJ, Cech TR. 2021. Shaping human telomeres: from shelterin and CST complexes to telomeric chromatin organization. *Nat Rev Mol Cell Biol* **22:** 283–298. doi:10.1038/s41580-021-00328-y

Liu B, He Y, Wang Y, Song H, Zhou ZH, Feigon J. 2022. Structure of active human telomerase with telomere shelterin protein TPP1. *Nature* **604:** 578–583. doi:10.1038/s41586-022-04582-8

Lorbeer FK, Hockemeyer D. 2020. TERT promoter mutations and telomeres during tumorigenesis. *Curr Opin Genet Dev* **60:** 56–62. doi:10.1016/j.gde.2020.02.001

Machado-Pinilla R, Liger D, Leulliot N, Meier UT. 2012. Mechanism of the AAA⁺ ATPases pontin and reptin in the biogenesis of H/ACA RNPs. *RNA* **18:** 1833–1845. doi:10.1261/rna.034942.112

Mancini A, Xavier-Magalhães A, Woods WS, Nguyen KT, Amen AM, Hayes JL, Fellmann C, Gapinske M, McKinney AM, Hong C, et al. 2018. Disruption of the β1L isoform of GABP reverses glioblastoma replicative immortality in a TERT promoter mutation-dependent manner. *Cancer Cell* **34:** 513–528.e8. doi:10.1016/j.ccell.2018.08.003

* Martin A, Hockemeyer D. 2025. Regulation of human telomerase: from molecular interactions to population genetics. *Cold Spring Harb Perspect Biol* doi:10.1101/cshperspect.a041693

Mendoza-Figueroa MS, Tatomer DC, Wilusz JE. 2020. The integrator complex in transcription and development. *Trends Biochem Sci* **45:** 923–934. doi:10.1016/j.tibs.2020.07.004

Mitchell JR, Cheng J, Collins K. 1999. A box H/ACA small nucleolar RNA-like domain at the human telomerase RNA 3′ end. *Mol Cell Biol* **19:** 567–576. doi:10.1128/MCB.19.1.567

Moon DH, Segal M, Boyraz B, Guinan E, Hofmann I, Cahan P, Tai AK, Agarwal S. 2015. Poly(A)-specific ribonuclease (PARN) mediates 3′-end maturation of the telomerase RNA component. *Nat Genet* **47:** 1482–1488. doi:10.1038/ng.3423

Nagpal N, Wang J, Zeng J, Lo E, Moon DH, Luk K, Braun RO, Burroughs LM, Keel SB, Reilly C, et al. 2020. Small-molecule PAPD5 inhibitors restore telomerase activity in patient stem cells. *Cell Stem Cell* **26:** 896–909.e8. doi:10.1016/j.stem.2020.03.016

Nandakumar J, Bell CF, Weidenfeld I, Zaug AJ, Leinwand LA, Cech TR. 2012. The TEL patch of telomere protein TPP1 mediates telomerase recruitment and processivity. *Nature* **492:** 285–289. doi:10.1038/nature11648

Nault JC, Zucman-Rossi J. 2016. TERT promoter mutations in primary liver tumors. *Clin Res Hepatol Gastroenterol* **40:** 9–14. doi:10.1016/j.clinre.2015.07.006

Neugebauer KM. 2017. Special focus on the Cajal body. *RNA Biol* **14:** 669–670. doi:10.1080/15476286.2017.1316928

Nguyen D, Grenier St-Sauveur V, Bergeron D, Dupuis-Sandoval F, Scott MS, Bachand F. 2015. A polyadenylation-dependent 3′ end maturation pathway is required for the synthesis of the human telomerase RNA. *Cell Rep* **13:** 2244–2257. doi:10.1016/j.celrep.2015.11.003

Nguyen THD, Tam J, Wu RA, Greber BJ, Toso D, Nogales E, Collins K. 2018. Cryo-EM structure of substrate-bound human telomerase holoenzyme. *Nature* **557:** 190–195. doi:10.1038/s41586-018-0062-x

Oh ST, Kyo S, Laimins LA. 2001. Telomerase activation by human papillomavirus type 16 E6 protein: induction of human telomerase reverse transcriptase expression through Myc and GC-rich Sp1 binding sites. *J Virol* **75:** 5559–5566. doi:10.1128/JVI.75.12.5559-5566.2001

Ohno M, Segref A, Bachi A, Wilm M, Mattaj IW. 2000. PHAX, a mediator of U snRNA nuclear export whose activity is regulated by phosphorylation. *Cell* **101:** 187–198. doi:10.1016/S0092-8674(00)80829-6

Palm W, de Lange T. 2008. How shelterin protects mammalian telomeres. *Annu Rev Genet* **42:** 301–334. doi:10.1146/annurev.genet.41.110306.130350

Penev A, Bazley A, Shen M, Boeke JD, Savage SA, Sfeir A. 2021. Alternative splicing is a developmental switch for hTERT expression. *Mol Cell* **81:** 2349–2360.e6. doi:10.1016/j.molcel.2021.03.033

Petrenko AA, Korolenkova LI, Skvortsov DA, Fedorova MD, Skoblov MU, Baranova AV, Zvereva ME, Rubtsova MP, Kisseljov FL. 2010. Cervical intraepithelial neoplasia: telomerase activity and splice pattern of hTERT mRNA. *Biochimie* **92:** 1827–1831. doi:10.1016/j.biochi.2010.07.015

Pickett HA, Henson JD, Au AY, Neumann AA, Reddel RR. 2011. Normal mammalian cells negatively regulate telomere length by telomere trimming. *Hum Mol Genet* **20:** 4684–4692. doi:10.1093/hmg/ddr402

Pike AM, Strong MA, Ouyang JPT, Greider CW. 2019. TIN2 functions with TPP1/POT1 to stimulate telomerase processivity. *Mol Cell Biol* **39:** e00593. doi:10.1128/MCB.00593-18

Podlevsky JD, Bley CJ, Omana RV, Qi X, Chen JJ. 2008. The telomerase database. *Nucleic Acids Res* **36:** D339–D343. doi:10.1093/nar/gkm700

Pradet-Balade B, Girard C, Boulon S, Paul C, Azzag K, Bordonné R, Bertrand E, Verheggen C. 2011. CRM1 controls the composition of nucleoplasmic pre-snoRNA complexes to licence them for nucleolar transport. *EMBO J* **30:** 2205–2218. doi:10.1038/emboj.2011.128

Qin J, Autexier C. 2021. Regulation of human telomerase RNA biogenesis and localization. *RNA Biol* **18:** 305–315. doi:10.1080/15476286.2020.1809196

Ramanathan A, Robb GB, Chan SH. 2016. mRNA capping: biological functions and applications. *Nucleic Acids Res* **44:** 7511–7526. doi:10.1093/nar/gkw551

Revy P, Kannengiesser C, Bertuch AA. 2023. Genetics of human telomere biology disorders. *Nat Rev Genet* **24:** 86–108. doi:10.1038/s41576-022-00527-z

Rivera T, Haggblom C, Cosconati S, Karlseder J. 2017. A balance between elongation and trimming regulates telomere stability in stem cells. *Nat Struct Mol Biol* **24:** 30–39. doi:10.1038/nsmb.3335

Roake CM, Chen L, Chakravarthy AL, Ferrell JE Jr, Raffa GD, Artandi SE. 2019. Disruption of telomerase RNA maturation kinetics precipitates disease. *Mol Cell* **74:** 688–700.e3. doi:10.1016/j.molcel.2019.02.033

Rubtsova MP, Vasilkova DP, Moshareva MA, Malyavko AN, Meerson MB, Zatsepin TS, Naraykina YV, Beletsky AV, Ravin NV, Dontsova OA. 2019. Integrator is a key component of human telomerase RNA biogenesis. *Sci Rep* **9:** 1701. doi:10.1038/s41598-018-38297-6

Sæbøe-Larssen S, Fossberg E, Gaudernack G. 2006. Characterization of novel alternative splicing sites in human telomerase reverse transcriptase (hTERT): analysis of expression and mutual correlation in mRNA isoforms from normal and tumour tissues. *BMC Mol Biol* **7**: 26. doi:10.1186/1471-2199-7-26

Schmid M, Jensen TH. 2019. The nuclear RNA exosome and its cofactors. *Adv Exp Med Biol* **1203**: 113–132. doi:10.1007/978-3-030-31434-7_4

Schmidt JC, Zaug AJ, Cech TR. 2016. Live cell imaging reveals the dynamics of telomerase recruitment to telomeres. *Cell* **166**: 1188–1197.e9. doi:10.1016/j.cell.2016.07.033

Schmidt JC, Zaug AJ, Kufer R, Cech TR. 2018. Dynamics of human telomerase recruitment depend on template-telomere base pairing. *Mol Biol Cell* **29**: 869–880. doi:10.1091/mbc.E17-11-0637

Sekne Z, Ghanim GE, van Roon AM, Nguyen THD. 2022. Structural basis of human telomerase recruitment by TPP1-POT1. *Science* **375**: 1173–1176. doi:10.1126/science.abn6840

Sexton AN, Youmans DT, Collins K. 2012. Specificity requirements for human telomere protein interaction with telomerase holoenzyme. *J Biol Chem* **287**: 34455–34464. doi:10.1074/jbc.M112.394767

Shukla S, Schmidt JC, Goldfarb KC, Cech TR, Parker R. 2016. Inhibition of telomerase RNA decay rescues telomerase deficiency caused by dyskerin or PARN defects. *Nat Struct Mol Biol* **23**: 286–292. doi:10.1038/nsmb.3184

Shukla S, Jeong HC, Sturgeon CM, Parker R, Batista LFZ. 2020. Chemical inhibition of PAPD5/7 rescues telomerase function and hematopoiesis in dyskeratosis congenita. *Blood Adv* **4**: 2717–2722. doi:10.1182/bloodadvances.2020001848

Son A, Park JE, Kim VN. 2018. PARN and TOE1 constitute a 3′ end maturation module for nuclear non-coding RNAs. *Cell Rep* **23**: 888–898. doi:10.1016/j.celrep.2018.03.089

Stern JL, Zyner KG, Pickett HA, Cohen SB, Bryan TM. 2012. Telomerase recruitment requires both TCAB1 and Cajal bodies independently. *Mol Cell Biol* **32**: 2384–2395. doi:10.1128/MCB.00379-12

Stern JL, Theodorescu D, Vogelstein B, Papadopoulos N, Cech TR. 2015. Mutation of the *TERT* promoter, switch to active chromatin, and monoallelic *TERT* expression in multiple cancers. *Genes Dev* **29**: 2219–2224. doi:10.1101/gad.269498.115

Stuart BD, Choi J, Zaidi S, Xing C, Holohan B, Chen R, Choi M, Dharwadkar P, Torres F, Girod CE, et al. 2015. Exome sequencing links mutations in PARN and RTEL1 with familial pulmonary fibrosis and telomere shortening. *Nat Genet* **47**: 512–517. doi:10.1038/ng.3278

Takahashi K, Tanabe K, Ohnuki M, Narita M, Ichisaka T, Tomoda K, Yamanaka S. 2007. Induction of pluripotent stem cells from adult human fibroblasts by defined factors. *Cell* **131**: 861–872. doi:10.1016/j.cell.2007.11.019

Takai KK, Kibe T, Donigian JR, Frescas D, de Lange T. 2011. Telomere protection by TPP1/POT1 requires tethering to TIN2. *Mol Cell* **44**: 647–659. doi:10.1016/j.molcel.2011.08.043

Theimer CA, Feigon J. 2006. Structure and function of telomerase RNA. *Curr Opin Struct Biol* **16**: 307–318. doi:10.1016/j.sbi.2006.05.005

Tomlinson RL, Li J, Culp BR, Terns RM, Terns MP. 2010. A Cajal body-independent pathway for telomerase trafficking in mice. *Exp Cell Res* **316**: 2797–2809. doi:10.1016/j.yexcr.2010.07.001

Tseng CK, Wang HF, Burns AM, Schroeder MR, Gaspari M, Baumann P. 2015. Human telomerase RNA processing and quality control. *Cell Rep* **13**: 2232–2243. doi:10.1016/j.celrep.2015.10.075

Tseng CK, Wang HF, Schroeder MR, Baumann P. 2018. The H/ACA complex disrupts triplex in hTR precursor to permit processing by RRP6 and PARN. *Nat Commun* **9**: 5430. doi:10.1038/s41467-018-07822-6

Tummala H, Walne A, Collopy L, Cardoso S, de la Fuente J, Lawson S, Powell J, Cooper N, Foster A, Mohammed S, et al. 2015. Poly(A)-specific ribonuclease deficiency impacts telomere biology and causes dyskeratosis congenita. *J Clin Invest* **125**: 2151–2160. doi:10.1172/JCI78963

Tummala H, Collopy LC, Walne AJ, Ellison A, Cardoso S, Aksu T, Yarali N, Aslan D, Fikret Akata R, Teo J, et al. 2018. Homozygous OB-fold variants in telomere protein TPP1 are associated with dyskeratosis congenita-like phenotypes. *Blood* **132**: 1349–1353. doi:10.1182/blood-2018-03-837799

Tycowski KT, Shu MD, Kukoyi A, Steitz JA. 2009. A conserved WD40 protein binds the Cajal body localization signal of scaRNP particles. *Mol Cell* **34**: 47–57. doi:10.1016/j.molcel.2009.02.020

Ulaner GA, Hu JF, Vu TH, Giudice LC, Hoffman AR. 1998. Telomerase activity in human development is regulated by human telomerase reverse transcriptase (hTERT) transcription and by alternate splicing of hTERT transcripts. *Cancer Res* **58**: 4168–4172.

Venteicher AS, Meng Z, Mason PJ, Veenstra TD, Artandi SE. 2008. Identification of ATPases pontin and reptin as telomerase components essential for holoenzyme assembly. *Cell* **132**: 945–957. doi:10.1016/j.cell.2008.01.019

Venteicher AS, Abreu EB, Meng Z, McCann KE, Terns RM, Veenstra TD, Terns MP, Artandi SE. 2009. A human telomerase holoenzyme protein required for Cajal body localization and telomere synthesis. *Science* **323**: 644–648. doi:10.1126/science.1165357

Walbott H, Machado-Pinilla R, Liger D, Blaud M, Réty S, Grozdanov PN, Godin K, van Tilbeurgh H, Varani G, Meier UT, et al. 2011. The H/ACA RNP assembly factor SHQ1 functions as an RNA mimic. *Genes Dev* **25**: 2398–2408. doi:10.1101/gad.176834.111

Wan F, Ding Y, Zhang Y, Wu Z, Li S, Yang L, Yan X, Lan P, Li G, Wu J, et al. 2021. Zipper head mechanism of telomere synthesis by human telomerase. *Cell Res* **31**: 1275–1290. doi:10.1038/s41422-021-00586-7

Withers JB, Ashvetiya T, Beemon KL. 2012. Exclusion of exon 2 is a common mRNA splice variant of primate telomerase reverse transcriptases. *PLoS ONE* **7**: e48016. doi:10.1371/journal.pone.0048016

Wong CW, Hou PS, Tseng SF, Chien CL, Wu KJ, Chen HF, Ho HN, Kyo S, Teng SC. 2010. Krüppel-like transcription factor 4 contributes to maintenance of telomerase activity in stem cells. *Stem Cells* **28**: 1510–1517. doi:10.1002/stem.477

Wright WE, Piatyszek MA, Rainey WE, Byrd W, Shay JW. 1996. Telomerase activity in human germline and embryonic tissues and cells. *Dev Genet* **18**: 173–179. doi:10

.1002/(SICI)1520-6408(1996)18:2<173::AID-DVG10>3
.0.CO;2-3

Wu KJ, Grandori C, Amacker M, Simon-Vermot N, Polack A, Lingner J, Dalla-Favera R. 1999. Direct activation of TERT transcription by c-MYC. *Nat Genet* **21:** 220–224. doi:10.1038/6010

Wu Y, Poulos RC, Reddel RR. 2020. Role of POT1 in human cancer. *Cancers (Basel)* **12:** 2739. doi:10.3390/cancers 12102739

Xi L, Cech TR. 2014. Inventory of telomerase components in human cells reveals multiple subpopulations of hTR and hTERT. *Nucleic Acids Res* **42:** 8565–8577. doi:10.1093/nar/gku560

Yang D, He Q, Kim H, Ma W, Songyang Z. 2011. TIN2 protein dyskeratosis congenita missense mutants are defective in association with telomerase. *J Biol Chem* **286:** 23022–23030. doi:10.1074/jbc.M111.225870

Yuan X, Larsson C, Xu D. 2019. Mechanisms underlying the activation of TERT transcription and telomerase activity in human cancer: old actors and new players. *Oncogene* **38:** 6172–6183. doi:10.1038/s41388-019-0872-9

Zhang Q, Kim NK, Feigon J. 2011. Architecture of human telomerase RNA. *Proc Natl Acad Sci* **108:** 20325–20332. doi:10.1073/pnas.1100279108

Zhong F, Savage SA, Shkreli M, Giri N, Jessop L, Myers T, Chen R, Alter BP, Artandi SE. 2011. Disruption of telomerase trafficking by TCAB1 mutation causes dyskeratosis congenita. *Genes Dev* **25:** 11–16. doi:10.1101/gad .2006411

Zhong FL, Batista LF, Freund A, Pech MF, Venteicher AS, Artandi SE. 2012. TPP1 OB-fold domain controls telomere maintenance by recruiting telomerase to chromosome ends. *Cell* **150:** 481–494. doi:10.1016/j.cell.2012.07 .012

Zhu S, Rousseau P, Lauzon C, Gandin V, Topisirovic I, Autexier C. 2014. Inactive C-terminal telomerase reverse transcriptase insertion splicing variants are dominant-negative inhibitors of telomerase. *Biochimie* **101:** 93–103. doi:10.1016/j.biochi.2013.12.023

# Regulation of Human Telomerase: From Molecular Interactions to Population Genetics

## Annika Martin[1] and Dirk Hockemeyer[1,2]

[1]Department of Molecular and Cell Biology; [2]Innovative Genomics Institute, University of California, Berkeley, California 94720, USA

*Correspondence:* hockemeyer@berkeley.edu

Human telomeres play critical roles in protecting chromosome ends and preserving genomic integrity. Telomerase, essential for maintaining telomere length and cellular replicative capacity, is only expressed in a small subset of human cells: stem and progenitor populations. Conversely, most somatic cells' telomeres shorten with each cell division; this shortening provides a potent tumor suppressor mechanism. Thus, telomerase regulation shapes not only cellular life span and differentiation, but also the regenerative capacity and long-term integrity of tissues. Here, we review the current understanding of telomere length control and telomerase regulation in humans, from molecular interactions at chromosome ends to the tissue-specific variation of telomere length dynamics, drawing insight from pluripotent and adult stem cell populations, as well as telomerase dysregulation in cancer and telomere biology disorders.

Human telomeres are repetitive nucleoprotein structures at the ends of linear chromosomes composed of ∼5–15 kb of 5′-TTAGGG-3′ DNA repeats terminating in a single-stranded 3′ overhang structure (Szostak and Blackburn 1982; Moyzis et al. 1988; McElligott and Wellinger 1997). These repeats are bound by a protective complex of proteins aptly named shelterin (de Lange 2018). Telomeres perform three primary functions: protecting chromosomal ends from degradation, suppressing DNA damage response (DDR) signaling, and providing a reservoir of dispensable terminal DNA sequence. Cells require this reservoir of noncoding sequence since each chromosome end shortens by ∼40–200 bp per division in human cells due to the end-replication problem and nucleolytic processing of chromosome ends (Harley et al. 1990; Hastie et al. 1990; Olovnikov 1996; Huffman et al. 2000; Chow et al. 2012; Takai et al. 2024). If this telomere sequence loss is not counteracted by telomerase, proliferative cells will accumulate critically short telomeres, which induce the DDR, triggering replicative senescence or cell death (Hemann et al. 2001; d'Adda di Fagagna et al. 2003; Herbig et al. 2004; de Lange 2009; Cesare and Karlseder 2012). Otherwise, genomic attrition would continue, ultimately compromising the long-term propagation of genetic information to cellular progeny. In humans, the selective repression of telomerase in most somatic cell types has evolved as a strong tumor suppressor mechanism (Counter et al. 1992; Kim et al. 1994; Wright et al. 1996). Restriction of telomerase activity to early development or to specific cell types such as germline cells, somatic stem cells, and other proliferative cell types preserves genomic integrity throughout development, tissue renewal, and reproduction, without compromising chromosome end protection. Thus, telomere biology dictates central aspects of the cellular life span and residence time of human cells (Allsopp et al. 1992; Morrison et al. 1996; Weng et al. 1996; Wright et al. 1996; Bodnar et al. 1998).

Over the last few decades, telomere and telomerase research has identified telomerase regulation and telomere length control as a unifying criteria of stem cells, whereas telomerase silencing is a central feature of cellular differentiation in human development. Consequently, defects in telomere maintenance drive tissue failure syndromes, while inappropriate telomerase activity is linked to the aberrant immortal phenotype of cancer cells. Here, we will review the molecular machinery that regulates telomerase activity from the molecular to organismal scale, beginning with the molecular interactions of telomerase at chromosome ends in telomerase-positive stem cells and cancer cells, progressing to the differences in cell-specific control of active telomerase levels, and concluding with insights from human disease caused by telomere length dysregulation and comparing the tissue-specific impacts at an organismal or population scale.

## THE MOLECULAR LEVEL: REGULATION OF TELOMERASE ACTIVITY AT CHROMOSOME ENDS

In stem cells with active telomerase, telomere lengths are maintained within a relatively narrow window termed the telomere length "set point." Originally identified in yeast (Greider 1996), this set point is particularly evident in human embryonic stem cell (hESC) and induced pluripotent stem cell (iPSC) lines, in which telomeres are maintained at a stable length ranging from ~8 to 15 kb (Thomson et al. 1998; Agarwal et al. 2010). Despite iPSC derivation from differentiated cells with shorter telomeres, the reprogramming process resets telomeres to lengths comparable to those in hESCs (Marion et al. 2009; Batista et al. 2011). Loss of either telomerase reverse transcriptase (TERT) or the RNA component TERC results in gradual telomere shortening and eventual replicative senescence, although rescuing these deficiencies results in elongation back to the wild-type set point (Sexton et al. 2014; Chiba et al. 2015, 2017; Vogan et al. 2016).

Even in immortal cells, telomerase is not abundant. In cancer cell lines, an estimated 50–240 telomerase molecules are sufficient to immortalize a cell, a level in these cells that is substoichiometric to its substrate: the $3'$ end of each chromatid arm (Yi et al. 2001; Cohen et al. 2007; Xi and Cech 2014). Telomerase is recruited to telomeres only in S phase (Hagen et al. 1990; Wright et al. 1999) and preferentially to the shortest telomeres (Ouellette et al. 2000; Hemann et al. 2001), where it acts processively, adding several repeats prior to dissociating from the telomere (Zhao et al. 2011; Patrick et al. 2020). Since human telomeres are generally much longer than the telomere attrition rate in one cell cycle, not every telomere requires elongation during every cell cycle to maintain a homeostatic set point. Therefore, in cells where telomerase is active but limiting, long telomeres continue to shorten while the shortest telomeres are maintained. These observations lead to the model that telomeres toggle between an extendable and nonextendible state: The shortest telomeres are preferentially extended by telomerase but switch to a nonextendible state following elongation (Teixeira et al. 2004).

Together, these data indicate that telomere homeostasis in stem cell populations is not merely a counterbalance to telomere shortening, but a cell-intrinsic process of measuring telomere length, controlling telomerase access to telomere ends, and tuning the extent of telomere elongation to match the cellularly defined set point. While the exact determinant of a telomere's extendible or nonextendible state remains unknown, several key proteins have been identified as critical contributors to telomere length control, and many current models have been proposed for how telomerase selectively elongates the shortest telomeres.

Cite this article as *Cold Spring Harb Perspect Biol* doi: 10.1101/cshperspect.a041693

## The Shelterin Complex: Key Regulators of Telomere Accessibility

Telomere set point regulation is largely mediated through the activity of shelterin, which affects telomere accessibility, telomerase recruitment, and enzymatic processivity (Fig. 1). Shelterin is a six-member protein complex composed of TRF1, TRF2, RAP1, TIN2, TPP1, and POT1. Of these, only RAP1 appears to have minimal effects on telomere length in human cells (Kabir et al. 2014; Lototska et al. 2020). TRF1 and TRF2 anchor the complex to telomeres through sequence-specific binding of telomeric repeats and protect double-stranded telomeric sequence by helping to resolve stalled replication forks and repressing the DDR, respectively (Zhong et al. 1992; Chong et al. 1995; Court et al. 2005; Sfeir et al. 2009; Lin et al. 2014). Loss of these shelterin anchors results in DDR signaling at chromosome ends, ultimately leading to cellular arrest or lethality due to aberrant repair attempts and chromosome end fusions (de Lange 2018). The other DNA-binding member of the complex, POT1, instead binds to the single-stranded overhang portion of the telomere and protects the junction between single-stranded and double-stranded telomeric DNA (Baumann and Cech 2001; Lei et al. 2004; Loayza et al. 2004; Tesmer et al. 2023). There, POT1 represses DDR signaling by excluding the single-stranded DNA-binding complex RPA (Hockemeyer et al. 2005, 2006; Wu et al. 2006; Denchi and de Lange 2007; Gong and de Lange 2010; Flynn et al. 2011). Overexpression and gene knockdown studies show that TRF1, TRF2, and POT1 all negatively regulate telomere length, perhaps due to their role in shielding and

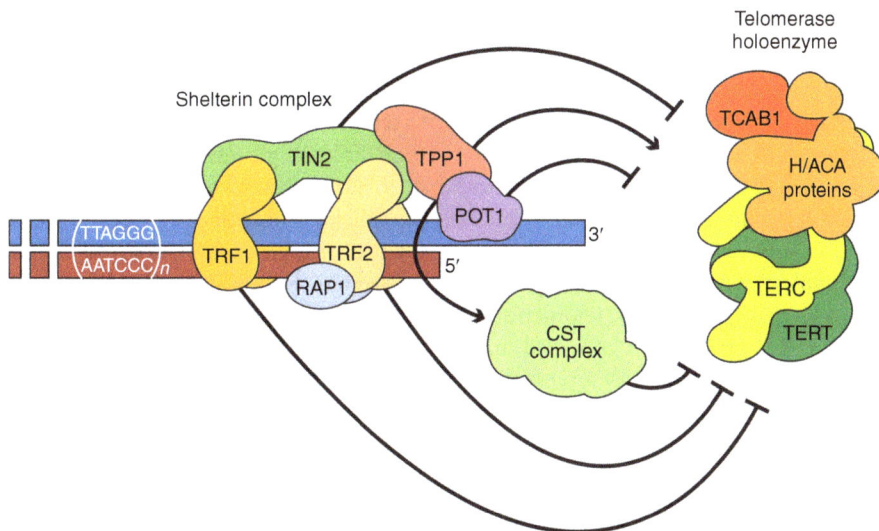

**Figure 1.** Roles of the shelterin complex in telomerase regulation at chromosome ends. The telomerase holoenzyme is composed of a reverse transcriptase (telomerase reverse transcriptase [TERT]), which elongates telomeres using an RNA template (telomerase RNA component [TERC]), mediated by cofactors including TCAB1 and H/ACA proteins. Its access to telomere ends is largely controlled by the six-member shelterin complex. Shelterin is anchored to chromosome ends through the DNA-binding proteins TRF1 and TRF2, which aid in DNA damage response (DDR) protection and sequester chromosome ends, negatively regulating telomere length. These DNA-binding interactions are stabilized by TIN2, which acts as a scaffold, recruiting TPP1 and POT1. POT1 bridges double-stranded and single-stranded DNA, protecting the 3′ telomeric overhang and inhibiting telomerase access. TPP1 directly recruits telomerase through its TEL patch. Following telomere replication and extension, POT1/TPP1 recruit CST, which mediates fill-in synthesis and terminates telomerase activity. This positive and negative regulation of telomerase by shelterin defines the telomere length set point in human stem cells.

sequestering the chromosome end (van Steensel and de Lange 1997; Smogorzewska et al. 2000; Loayza and de Lange 2003; Hockemeyer et al. 2005; Shiekh et al. 2022). Targeting additional shelterin to an individual telomere leads to the shortening of this telomere, establishing that telomere length is counted by a *cis* mechanism (Ancelin et al. 2002).

Conversely, TPP1 directly recruits telomerase to telomere ends by binding to TERT through its TEL patch (Wang et al. 2007; Xin et al. 2007; Abreu et al. 2010; Nandakumar et al. 2012; Zhong et al. 2012; Schmidt et al. 2014; Sexton et al. 2014). When this region is genetically ablated, telomeres shorten at the same rate as TERT knockout cells, suggesting that TPP1 recruitment is the primary mechanism of TERT localization to telomeres (Sexton et al. 2014). Point mutations within the TEL patch show graded effects on telomerase activity directly proportional to TERT-binding affinity (Nandakumar et al. 2012). TPP1 also stimulates telomerase processivity in biochemical assays, an effect increased when in complex with TIN2 and POT1 (Wang et al. 2007; Pike et al. 2019), indicating that there may be an additional conformational effect of TPP1 binding to TERT. This is further supported by mutations distal to the TEL patch interaction domain, which reduce repeat addition processivity in vitro and result in short but stable telomeres (Nandakumar et al. 2012; Grill et al. 2019). Critically, telomeres in these mutant cells appear fully protected and can be rescued through telomerase overexpression, demonstrating that alterations to shelterin can adjust the telomere set point without sacrificing telomere integrity (Boyle et al. 2020).

Since telomerase only elongates the 3′ end of the telomere, faithful replication of the telomere depends on recruitment of fill-in synthesis machinery (Miyake et al. 2009; Surovtseva et al. 2009; Chen et al. 2012; Wu et al. 2012; Kratz and de Lange 2018; Lim and Cech 2021; Cai and de Lange 2023; Takai et al. 2024). This is accomplished through POT1/TPP1 recruitment of the CST complex, an event that may terminate telomerase action at telomeres (Wan et al. 2009; Chen et al. 2012; Wu et al. 2012; Kratz and de Lange 2018). CST recruits the DNA polymerase

α-primase complex to mediate the fill-in synthesis of the C-strand (Casteel et al. 2009; Diotti et al. 2015). Mutants in CST recruitment hyperelongate the telomeric G-strand by telomerase, but fail to maintain telomere homeostasis since the complementary C-strand is not synthesized (Surovtseva et al. 2009; Wang et al. 2012; Feng et al. 2017). The coordination of this overhang handoff from the TPP1/POT1 complex to CST remains a subject of intense investigation, but may be influenced by single-stranded DNA-binding competition and phosphorylation state of POT1 (Chen et al. 2012; Gu et al. 2018; Wang et al. 2023; Cai et al. 2024; Martin et al. 2025).

TIN2, identified by interaction with TRF1, is the structural bridge at the heart of shelterin (Kim et al. 1999). TIN2 is required to stabilize the interaction between TRF1 and TRF2 with telomeric sequence, and is responsible for linking the DNA interaction proteins with the telomerase-interacting TPP1 (Kim et al. 2004; Ye et al. 2004; Kaur et al. 2021; Pan et al. 2021). Loss of TIN2 results in similar telomere deprotection as POT1 knockout, indicating that the primary method of recruitment of TPP1/POT1 to telomere ends is through TIN2 binding rather than direct single-stranded DNA binding (Takai et al. 2011; Frescas and De Lange 2014). Multiple splicing isoforms of TIN2 exist, although the functional roles of these isoforms are still being uncovered (Kaminker et al. 2009; Ishdorj et al. 2017; Pike et al. 2019).

This careful biochemical characterization, genetic perturbation, and ever-improving structural resolution of shelterin establishes that telomere length homeostasis is carefully controlled. However, we still do not fully understand how cells integrate the number of binding sites along the double-stranded portion of the telomere to preferentially allow telomerase extension of the shortest telomeres in the cell.

## Models of Telomere Length Control: The Switch between Extendible and Nonextendible

To address this gap in knowledge, several models have emerged for this integration of shelterin information across the length of the telomere.

Cite this article as *Cold Spring Harb Perspect Biol* doi: 10.1101/cshperspect.a041693

Here, we will divide these models into two categories: direct integration of shelterin-mediated telomerase regulation and indirect temporal control of telomerase activity.

Since shelterin can both activate and repress telomerase activity at chromosome ends, it is possible that varying occupancy of shelterin proteins with known roles in telomerase regulation may drive the switch between extendible and nonextendible states (Fig. 2A). TRF1 and TRF2 have different exchange kinetics and protein abundance at telomeres, whereas TIN2 recruits only substoichiometric TPP1-POT1

(Mattern et al. 2004; Takai et al. 2010), indicating that shelterin assembly on telomeres is likely nonuniform and dynamic. Analyses of purified shelterin complexes support this hypothesis, finding different subunit stoichiometry and dynamic conformational changes between complexes (Lim et al. 2017; Zinder et al. 2022). Quantitative studies also show variation in global shelterin composition as a function of telomere length (Takai et al. 2010; Grolimund et al. 2013). However, the extent to which long and short telomeres within the same cell have different shelterin occupancy and composition is still

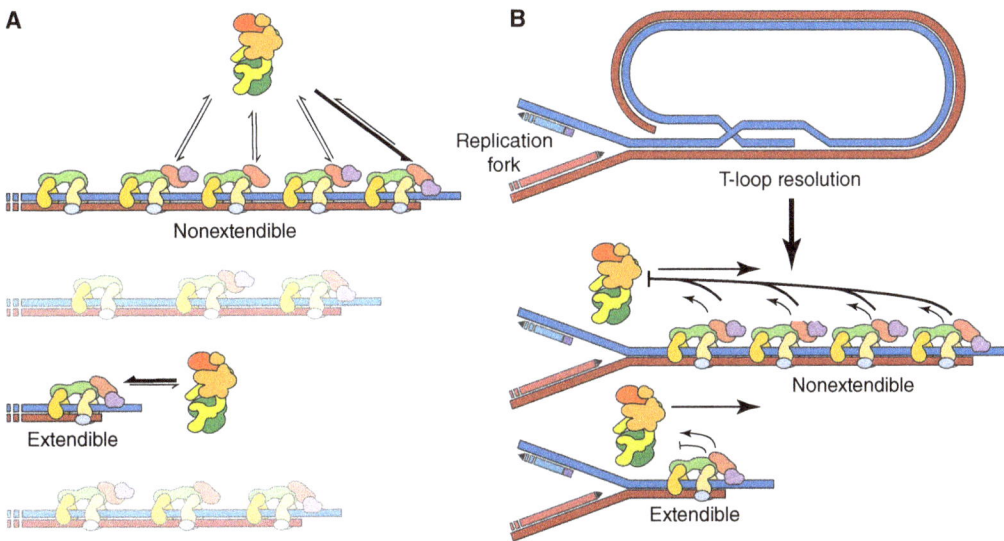

Figure 2. Two categories of extendible/nonextendible models for differential telomerase recruitment. To maintain telomere homeostasis, telomerase preferentially extends the shortest telomeres. Therefore, it is hypothesized that telomeres toggle between extendible and nonextendible states, with short telomeres transitioning from extendible to nonextendible once they are lengthened by telomerase. However, the molecular determinants of what constitutes an extendible state remain unknown. Here, we illustrate two categories of extendible/nonextendible classifications. (A) The extendible state of telomeres may be directly controlled by differential stoichiometry of shelterin components, or by unequal distribution of shelterin along a telomere. Because shelterin proteins both recruit and repress telomerase activity at telomere ends, altering the ratios of these proteins changes the cumulative effect of shelterin on telomerase activity. It may also affect the stability of TPP1-telomerase interactions, since telomerase often briefly samples telomeres, while stable interactions associated with processivity are much rarer. (B) While direct interactions with shelterin are still required for telomerase activity, the preference toward shortest telomeres may be mediated by cell-cycle-dependent factors. One hypothesis suggests that telomerase may travel with replication fork machinery and is more likely to reach the 3′ end of a short telomere due to fewer impediments to DNA replication (such as bound shelterin, G-quadruplexes, nucleosomes, etc.) than in longer telomeres. T-loop resolution may act as another critical regulator of telomerase accessibly. It should be noted that the degree of heterogeneity in shelterin complex stoichimetry is currently unknown and that the models of direct and temporal control of the extendible state are not necessarily mutually exclusive.

not well understood. Similarly, the distribution of shelterin proteins along the length of a single telomere also remains unresolved.

Single-molecule telomerase dynamics indicate two modes of telomerase–telomere interactions mediated by shelterin: short "sampling" interactions where telomerase quickly associates and dissociates from the telomere, and more stable interactions likely associated with telomerase activity (Schmidt et al. 2016). Thus, longer telomeres may simply act as a sponge, promoting many brief interactions and sequestering telomerase from telomere ends. Alternatively, different shelterin compositions enriched on shorter telomeres may promote more stable interactions.

An alternative hypothesis centers around the link between telomerase and telomere replication (Fig. 2B). One model, proposed by Dr. Carol Greider, suggests that telomerase may travel with DNA replication machinery toward telomere ends; there, replication fork challenges such as shelterin binding and G-quadruplex DNA increase the probability of telomerase dissociation from longer telomeres before it can reach its substrate (Greider 2016). Although this has not been directly tested in mammalian cells, telomerase does travel with the "replication band" in ciliate models (Olins et al. 1989; Fang and Cech 1995; Greider 2016). Alternatively, this recruitment may be orchestrated through DDR pathways, as in yeast (Greenwell et al. 1995; Arnerić and Lingner 2007; Chang et al. 2007). ATM and ATR signaling during S phase may increase the colocalization of telomerase with telomeres in a cell-cycle-dependent manner, thereby affecting telomere length regulation, although reports differ regarding the extent of this effect (Pennarun et al. 2010; McKerlie et al. 2012; Lee et al. 2015; Tong et al. 2015).

The T-loop, created when the 3′ overhang invades the preceding telomeric sequence to create a lariat structure mediated by TRF2 (Griffith et al. 1999; Doksani et al. 2013; Timashev and De Lange 2020), may also impede telomere replication and extension. RTEL helicase, responsible for T-loop resolution, is linked to telomere length regulation and resolution of stalled replication forks, where it may prevent cleavage of the looped telomere and aberrant telomerase activity at reversed forks (Uringa et al. 2012; Vannier et al. 2012; Sarek et al. 2015; Margalef et al. 2018; Olivier et al. 2018; Awad et al. 2020). Interestingly, mutations in RTEL result in telomere shortening and severe forms of telomere biology disorders (TBDs) and RTEL variants have recently been implicated in the telomere length increase in *Mus musculus* compared with *Mus spretus* (Kipling and Cooke 1990; Ding et al. 2004; Ballew et al. 2013; Deng et al. 2013; Le Guen et al. 2013; Vannier et al. 2014; Smoom et al. 2023).

Much remains unresolved about shelterin control of telomere length, and new insights into telomerase recruitment through cryo-EM structures continually generate more hypotheses. In these structures, telomerase appears to make previously unreported contact to POT1, gating the active site, as well as a histone H2A-H2B heterodimer within the telomerase holoenzyme (Ghanim et al. 2021; Liu et al. 2022; Sekne et al. 2022). Currently, the function of these interactions remains unknown.

## Beyond Shelterin

While the shelterin complex is the most thoroughly characterized effector of telomerase regulation at chromosome ends, careful control of repeat addition processivity and the influence of telomerase cofactors also play critical roles. Outside these two complexes, other mechanisms of telomere length control may also regulate telomerase activity at telomeres, although their role is not fully understood within the human context.

In humans, telomeric heterochromatin is unique (Déjardin and Kingston 2009; Goldberg et al. 2010; Lewis et al. 2010; Grolimund et al. 2013), and alterations to this chromatin state may impact telomerase access or activity. As demonstrated in other model organisms, epigenetic modification of telomeric chromatin, including methylation of both DNA and histones, also affects telomere length in humans and mice (Hazelrigg et al. 1984; Gottschling et al. 1990; Baur et al. 2001; Gonzalo et al. 2006; Ottaviani et al. 2008; Yehezkel et al. 2008; Robin et al.

2014). Mouse cells deficient for key histone methyltransferases show telomere hyperelongation (García-Cao et al. 2004). Similarly, knockout of all three Rb proteins in mouse embryonic fibroblasts results in loss of H4K20 trimethylation and concurrent telomere elongation independent of telomere deprotection or level of telomerase activity within the cell (García-Cao et al. 2002). Importantly, $TERC^{-/-}$ mice with shortening telomeres show progressive loss of both H3K9 and H4K20 methylation, without effect on telomere binding by TRF1 or TRF2 (Benetti et al. 2007). Together, these data suggest a direct epigenetic connection between telomere length, telomere state, and telomerase activity.

Telomeric repeat-containing RNA (TERRA) provides another link between chromatin state and telomere length maintenance. TERRA are long noncoding RNA molecules expressed in most eukaryotic organisms; they are transcribed from distinct sites within the subtelomeric region of the chromosome and terminate in varying numbers of telomeric repeats (Azzalin et al. 2007; Schoeftner and Blasco 2008; Nergadze et al. 2009). Ranging from ~100 bp to 9000 bp in mammals, they may provide an indirect readout of both telomere length and chromatin state through the number of repeats they contain and the level of TERRA transcription (Azzalin et al. 2007; Arnoult et al. 2012). In vitro, TERRA acts as a potent inhibitor of telomerase activity (Schoeftner and Blasco 2008; Redon et al. 2010) and TERRA overexpression in yeast leads to *cis* telomere shortening through resection of telomeres by exonuclease 1 (Pfeiffer and Lingner 2012). However, in human cells, TERRA does not appear to directly interact with telomerase during a normal cell cycle, perhaps due to sequestration by RNA-binding proteins like hnRNPA1 (de Silanes et al. 2010; Redon et al. 2010). Instead, evidence suggests that TERRA and hnRNPA1 may be involved in coordinating the reassociation of POT1 with single-stranded telomeric DNA after DNA replication (Nandakumar et al. 2010; Flynn et al. 2011; Porro et al. 2014). Thus, TERRA may play a shelterin-dependent or independent role in telomerase regulation.

## THE CELLULAR LEVEL: CONTROL OF ACTIVE TELOMERASE LEVELS

Unlike stem cell populations, most human somatic cells do not have telomerase activity (Günes and Rudolph 2013; Aubert 2014). While the telomere protective effects of the shelterin complex remain essential in somatic cells, setpoint regulation requires telomerase and is therefore relegated entirely to progenitor cell populations (Wright et al. 1996). TERC biogenesis and assembly of the telomerase holoenzyme is a highly complex, carefully regulated process (Collins 2006; Schmidt and Cech 2015; Chen et al. 2018). However, TERC and most telomerase accessory proteins appear to be broadly expressed, whereas TERT expression is restricted to only cells with active telomerase (Liu et al. 2000; Yang et al. 2008; Chiba et al. 2015). Therefore, the critical determinant of telomerase activity at the cellular level is whether TERT is expressed at all.

The human TERT gene is located at the distal portion of the Chromosome 5p arm, ~1.3 Mb from the telomere (Cong et al. 1999; Kim et al. 2016). Its GC-rich, TATA-box-less promoter is composed of ~4 kb CpG island with the TERT translational start codon (ATG) almost directly in the center (Fig. 3; Cong et al. 1999; Takakura et al. 1999). Surprisingly, the distal portion of this promoter appears to be hypermethylated both in reprogrammed iPCSs and cancer cell lines, compared with hypomethylation in somatic cells (Devereux et al. 1999; Takasawa et al. 2018). In fact, hypermethylation of this region is considered a biomarker of poor prognosis in several cancer types, leading to its designation as the TERT hypermethylated oncological region (THOR) (Castelo-Branco et al. 2013; Avin et al. 2019; Lee et al. 2019, 2020). Current hypotheses postulate that the hypermethylated state interferes with constitutive repressor binding, allowing for increased transcription of the locus (Lee et al. 2019). Conversely, the proximal portion of the TERT promoter, spanning ~300 bp surrounding the TSS, is hypomethylated with active chromatin marks like H3K27 histone acetylation in cell types with active TERT expression, including stem and can-

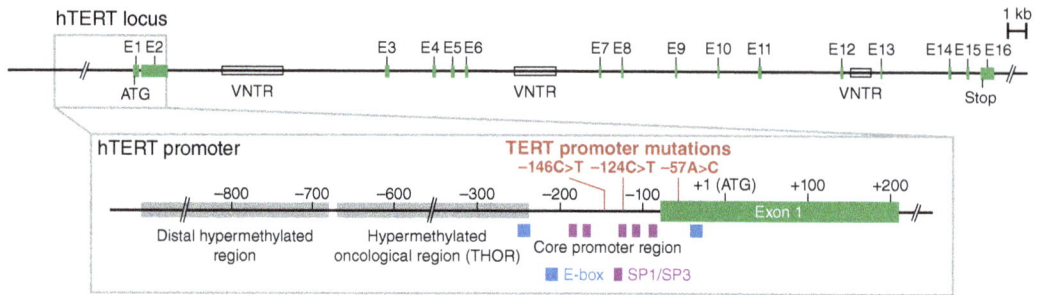

**Figure 3.** Schematic of the telomerase reverse transcriptase (TERT) locus in human cells. The core TERT promoter region spans ~230 bp from the translational start site and contains five SP1/SP3 binding sites and two enhancer box (E-boxes) known to regulate telomerase expression in stem cells. It is also a hotspot for cancer-associated single-nucleotide TERT promoter mutations, which result in de novo ETS factor-binding sites. Immediately adjacent to the core promoter is a ~500 bp, CpG-rich region named the TERT hypermethylated oncological region (THOR), which is preferentially methylated in TERT-positive cancers and stem cells, and unmethylated in normal somatic cells. Distal portions of the TERT promoter are also heavily methylated.

cer cells (Renaud et al. 2007; Zinn et al. 2007; Rowland et al. 2020).

This island of open chromatin directly correlates with TERT expression and disappears upon differentiation (Shin et al. 2003; Cheng et al. 2017a). Therefore, one of the key open questions in TERT regulation is the mechanism by which TERT is repressed. The differentiation of hESCs in vitro results in rapid loss of telomerase activity and an ~1000-fold bulk reduction in TERT expression (Wang et al. 2009; Jia et al. 2011; Chiba et al. 2015). This occurs robustly and repeatedly in vitro regardless of differentiation paradigm and is accompanied by the accumulation of heterochromain marks in the core TERT promoter region (Cheng et al. 2017a). This region contains binding sites for dozens of transcription factors identified to directly interact with the TERT promoter, with varying effects on TERT transcription and often in a cell- or cancer-type-specific manner (Ramlee et al. 2016). Here, we focus on just a few universal core transcription factors required for TERT expression in stem cells and discuss what is known about how TERT is repressed when these cells differentiate.

## c-Myc, Max, and Mxd1

The first class of transcription factors found to directly regulate TERT expression was the MYC/MAX/MAD network, a group of broadly ex-

pressed cofactors that bind to enhancer-box (E-box) promoter sequences and alter transcription in key cellular growth, proliferation, metabolism, and oncogenesis pathways (Wang et al. 1998; Wu et al. 1999). In stem and cancer cells, the c-Myc/Max heterodimer binds two critical E-boxes in the core TERT promoter region to directly activate TERT expression (Wu et al. 1999; Kyo et al. 2000; Goueli and Janknecht 2003). Abolishment of these E-boxes results in reduced promoter activity and insensitivity to c-Myc overexpression, although c-Myc knockdown surprisingly results in promoter activation and spread of active histone marks (Zhao et al. 2014). Therefore, c-Myc may also play a secondary role in maintaining the low TERT expression level in stem cells, perhaps through antagonistic interaction partners such as Miz-1, although this has not been fully explored. During differentiation, E-box occupancy at the TERT promoter shifts from c-Myc/Max to Mxd1/Max, with concurrent histone deacetylation and TERT repression without loss of c-Myc expression (Günes et al. 2000; Xiong and Frasch 2021). Therefore, TERT silencing during differentiation may be partially controlled by competition of Mxd1 and c-Myc for Max cofactors and E-box binding.

## Sp1 and Sp3

Sp1 is another transcription factor with broad expression and effects, but a critical role in

Cite this article as *Cold Spring Harb Perspect Biol* doi: 10.1101/cshperspect.a041693

TERT regulation. Sp1 overexpression leads to up-regulation of TERT, although this effect size varies between cell lines, perhaps dependent on endogenous Sp1 expression levels (Kyo et al. 2000). In human papillomavirus (HPV)-immortalized keratinocytes, five identified Sp1 binding sites completely abolished TERT expression when co-mutated with the two known E-boxes (Oh et al. 2001). Interestingly, c-Myc overexpression in normal primary human keratinocyte cells showed no effect on promoter activity when all five Sp1 sites were mutated, indicating a potential synergetic role between the two factors (Kyo et al. 2000). However, Sp1 and family member Sp3 have also been implicated in TERT repression during differentiation. In human fibroblasts with intact TERT repression, both Sp1 and Sp3 were still shown to bind to the TERT promoter region and associate with histone deacetylases, although it is unclear whether this is causative for repression, or indicative that these factors alone are insufficient to prevent heterochromatinization (Won et al. 2002; Cheng et al. 2015). Posttranslational modification (PTM), including acetylation, of Sp1 is a key determinant of its transcriptional activating or repressing activity (Tan and Khachigian 2009). Therefore, Sp1 and Sp3 may play both activating and repressing roles in TERT regulation, depending on the PTM. It is currently unclear how this effect is changed during differentiation, especially since Sp1 and Sp3 play critical roles in many other cellular pathways.

## Challenges in Decoding TERT Transcriptional Regulation

Beyond these two vignettes, many other putative repressors of TERT have been identified, including CTCF, E2F1, KLF2, MEN1, and others (Crowe et al. 2001; Lin and Elledge 2003; Renaud et al. 2007; Lacerte et al. 2008; Meeran et al. 2010; Zhang et al. 2014; Hara et al. 2015). However, even as more regulators are identified, a consistent model of TERT repression remains evasive. Much of our knowledge of these transcription factors and regulatory networks is derived through the investigation of TERT-positive cancer cell lines. These are the most abundant and genetically tractable TERT-expressing cells but inherently have aberrant telomerase regulation to support immortality. This provides critical insight into oncogenesis but complicates our understanding of normal cellular growth and differentiation, since transcription factors with robust effects in one cancer cell line may show limited or no effects in others. The challenge is threefold: First, TERT expression is very low, even in stem cells and most cancer cells, which makes detection and quantification challenging. Second, cancers may undergo different evolutionary trajectories, resulting in reliance on different factors to maintain TERT expression in their unique genetic environment. Last, even in normal tissues, there may be tissue-specific or non-cell-autonomous mechanisms of TERT transcriptional regulation, including hormonal regulation (Calado et al. 2009). Since most tissues in the human body appear to be TERT negative, it is unlikely that each would accomplish this repression in a unique manner. However, none of the currently identified methods of TERT regulation sufficiently explain this repression.

Adding to this challenge, the transcriptional control of TERT seems to differ between laboratory mice and humans. Although there is clear evidence for tissue-specific regulation of TERT expression and repression upon differentiation, telomerase does appear to be more active in mouse somatic tissues than in humans (Prowse and Greider 1995; Horikawa et al. 2005; Ritz et al. 2005; Wang et al. 2009; Pech et al. 2015; Lin et al. 2018). These differences in TERT silencing cannot be fully explained by the minimal TERT promoter region, as integration of a human core promoter in a mouse TERT context is insufficient to recapitulate TERT silencing upon differentiation and actually increased TERT expression (Cheng et al. 2017a). Therefore, to reverse-engineer the minimal human TERT-repressive elements, Cheng et al. integrated a 160 kb bacterial artificial chromosome containing the TERT locus into mouse cells and demonstrated that this region was sufficient to mimic human TERT repression upon differentiation (Cheng et al. 2017a, 2019, 2024). This indicates that more distal enhancer regions may be required for robust TERT silencing upon differentiation. Endogenous replacement of the mouse TERT 5′ intergenic region as well as introns 2 and 6 with their respective human sequences reduces telomere length in successive generations to near-human levels and more closely mimics human tissue-specific TERT expression (Zhang et al. 2025). Other lines of evidence suggest that these long-range interactions may be connected to the telomere position effect (Kim et al. 2016). However, the critical players in this process are not yet established and it remains possible that several converging and redundant strategies of repression are used in different tissues.

## When TERT Regulation Is Broken: Insights from TERT Promoter Mutations

Despite the heterogeneity of individual cancer cell lines, cross-cancer genetic analysis has established TERT expression as a rate-limiting step to telomerase activation across many cell types. To immortalize, transformed cells must stabilize their telomere reserve. While cells occasionally do this through alternative, recombination-based methods called ALT (Cesare and Reddel 2010; Lu and Pickett 2022), 80%–90% of cancers accomplish this through the reactivation of telomerase (Counter et al. 1992; Kim et al. 1994; Shay and Bacchetti 1997). Surprisingly, mechanisms for this reactivation were only recently understood. The most common noncoding mutations in cancer are the TERT promoter mutations (TPMs) −146C>T, −124C>T, and −57A>C relative to the translational start codon (Horn et al. 2013; Huang et al. 2013; Killela et al. 2013; Fredriksson et al. 2014; Weinhold et al. 2014). These precise nucleotide substitutions are identifiable in up to 20% of cancers, occur mutually exclusively and heterozygously, and are sufficient to enable maintenance of short telomeres and bypass of replicative senescence (Chiba et al. 2015, 2017; Huang et al. 2015; Yuan et al. 2019; Li et al. 2020b; Rheinbay et al. 2020). It is important to note that these mutations are gain-of-function mutations, rather than a loss of normal repressor binding.

TPMs result in de novo ETS factor binding sites, which recruit GABPa/b (Bell et al. 2015). ETS-domain transcription factors are a large family of evolutionarily conserved and broadly expressed transcription factors that drive many critical cellular processes, including proliferation and differentiation (Sharrocks 2001). Of these ETS factors, GABPa/b is unique as it requires heterodimerization of a DNA-binding domain (GABPa) and a transcriptional activating domain (GABPb) (Rosmarin et al. 2004). Knockdown and chromatin capture assays demonstrate that GABPa/b elicits no effect on wild-type TERT alleles, but induces monoallelic relaxation of heterochromatin and transcriptional up-regulation of TERT on alleles with TPMs (Bell et al. 2015; Stern et al. 2015). However, just as in normal TERT regulation, there is likely some degree of tissue or cancer specificity. In a subset of glioblastoma multiforme cells with TPMs, a noncanonical NF-κB subunit, p52, associates specifically with the −124C>T mutation, but not the −146C>T mutation, to drive TERT up-regulation (Li et al. 2015). Similarly, in BRAFV600E cancers with TPMs, the level of TERT activation is due to positive feedback between GABPa/b and phospho-Sp1 derived from constitutive ERK signaling (Wu et al. 2021). Long-range interactions may also play a role in TPM-mediated TERT overexpression (Akıncılar et al. 2016).

Importantly, these TPMs are sufficient to stabilize but not elongate telomeres (Chiba et al. 2015, 2017), perhaps explaining why many telomerase-positive cancers have generally short telomeres (Hayward et al. 2017). This initial stabilization is insufficient to immortalize cells, requiring additional genetic or epigenetic perturbations to fully stabilize chromosome ends (Chiba et al. 2017). Somewhat surprisingly, there is strong evidence for TPM selection very early in oncogenesis, with some premalignant lesions showing similar rates of TPM incidence as their neoplastic counterparts (Wang et al. 1998; Nault et al. 2014; Shain et al. 2015; Cheng et al. 2017b; Hysek et al. 2019; Hosen et al. 2020), indicating that replicative senescence is an early barrier to transformation in many cell types, rather than a fail-safe against already hyperproliferative cells. Analysis of TPM incidences in dysplastic nevi (DN), precursor lesions that can progress to melanoma, delineated two populations: those with TPMs correlated with advanced patient age and shorter telomeres. However, TPMs were not found in nevi with BRAF V600E mutations, the most frequent and strongest oncogenic driver in DN. DNs with BRAF V600E mutations were found in younger patients and had longer telomeres (Lorbeer et al. 2024). These data indicate that oncogenesis can progress through two pathways: First, acquisition of strong oncogenic driver mutations like BRAF V600E leads to oncogene-induced senescence as the first barrier to transformation. Second, if this pathway is not triggered, extended proliferation across organismal age may select for cells that can bypass replicative senescence as a first barrier to transformation (Lorbeer et al. 2024).

The broader TERT promoter/enhancer region is also a hotspot for amplification and other structural rearrangements, indicating that the dosage of TERT expression is also a contributing factor to oncogenic progression. Beyond the three common TPMs, cancer-type-specific TPMs have also been identified at novel loci (Mitchell et al. 2018). Nearly 4% of tumors have an amplification encompassing the TERT locus (Barthel et al. 2017; Yuan et al. 2019), whereas structural rearrangements common in neuroblastoma up-regulate TERT expression by repositioning the TERT locus near strong enhancer elements (Peifer et al. 2015; Valentijn et al. 2015). Uniquely, in a subset of hepatocellular carcinoma and cervical cancers, integration of hepatitis B or HPV enhancers near the TERT promoter appear to drive TERT overexpression (Ferber et al. 2003; Paterlini-Bréchot et al. 2003; Kawai-Kitahata et al. 2016). Even in untransformed cells, co-overexpression of TERT and TERC is sufficient to hyperelongate telomeres (Cristofari and Lingner 2006), immortalizing differentiated cells (Bodnar et al. 1998) and overriding the telomere length set point of telomerase positive cells like hESCs (Chiba et al. 2015). However, cancer cells do not seem to have a true set point like hESCs. Cells from the same tumor type and subclones of cancer cell lines may have substantial variation in telomere length, indicating

that telomere length control in cancer is complex beyond TERT expression levels (Bryan et al. 1998).

## Beyond Transcription

While transcriptional regulation of TERT has long been considered the dominant method of telomerase regulation at the cellular level, several other processes have been implicated in fine-tuning the level of active telomerase in the cell. Telomerase biogenesis and cofactors are discussed in depth elsewhere (Chen and Batista 2025), so we will focus on just a few highlights.

One emerging pathway for telomerase regulation is the alternative splicing of TERT. The TERT gene is composed of 16 exons and 15 introns with ~20 different splicing isoforms detected in vitro (Kilian et al. 1997; Yi et al. 2000; Sæbøe-Larssen et al. 2006; Liu et al. 2017; Subasri et al. 2021). To date, only the full-length isoform of TERT is thought to be catalytically active, as nearly all other isoforms contain disruptions of one of the essential reverse transcriptase domains in exons 4–11 (Kilian et al. 1997; Yi et al. 2000). The three most commonly discussed splicing variations are named α, β, and γ, comprising a loss of exon 6, exons 7/8, and exon 11, respectively, each of which results in the production of a nonfunctional or even dominant-negative protein (Colgin et al. 2000; Yi et al. 2000, 2001; Hisatomi et al. 2003; Sæbøe-Larssen et al. 2006). β Splicing may be controlled by a variable tandem number repeat (VNTR) sequence located in intron 6; when mutated, β splicing is abolished, but can be rescued by restoring proposed interactions between the VNTR locus and distal portions of the pre-mRNA, indicating a role for RNA:RNA pairing in TERT splicing regulation (Wong et al. 2014).

Of note, some estimates indicate that only ~5% of all TERT transcripts are full length (Yi et al. 2001). Thus, alternative splicing has been proposed as a mechanism of telomerase repression during differentiation, acting in concert with transcriptional regulation. In the human fetal kidney, active telomerase levels drop prior to transcriptional silencing due to a shift in splicing isoforms toward the nonfunctional β isoform (Ulaner et al. 2001). Similarly, during the differentiation of iPSCs to neuronal precursor cells, a shift in splicing isoforms precedes transcriptional silencing (Kim et al. 2023). Another variant, exon 2 exclusion (ΔEx2), leads to multiple stop codons in exon 3 and rapid mRNA degradation by nonsense-mediated decay (Penev et al. 2021). ΔEx2 was shown to be much more abundant in differentiated tissues than in stem cells or tissues with known stem cell niches (Penev et al. 2021). However, the dynamics of this splicing shift in human development and cellular differentiation remain unknown. Environmental conditions and cell cycle progression may also affect splicing patterns or temporally regulate mRNA processing (Dumbović et al. 2021; Kim et al. 2023).

Even when a full-length transcript is translated and can be assembled into a functional telomerase holoenzyme, it is heavily restricted in its activity through a variety of PTMs and interaction partners, which affect nuclear localization (Seimiya et al. 2000; Haendeler et al. 2003; Ahmed et al. 2008), stability and degradation (Lee et al. 2010; Jung et al. 2013), enzymatic activity/processivity (Kang et al. 1999; Kharbanda et al. 2000; Chang et al. 2006), and other functions (Liu et al. 2024). Thus, telomerase activity is tightly regulated from transcriptional initiation to termination of catalytic activity.

## BEYOND THE CELL: TISSUE-LEVEL TELOMERASE REGULATION AND INSIGHTS FROM POPULATION GENETICS

Due to the low expression of TERT and the sensitivity required to detect telomerase activity with single-cell resolution, tissue-specific telomerase regulation has been challenging to decode, and we still lack a comprehensive atlas of immortal stem and progenitor populations. The most accessible readout for telomerase regulation is through telomere length measurement. Telomere length shows strong genetic inheritance (Njajou et al. 2007; Broer et al. 2013; As-

ghar et al. 2015), but varies significantly within the population. Despite this inherent heterogeneity, substantial deviation in telomere length above or below an age-matched distribution drives tissue-specific pathology (Savage and Bertuch 2010; Revy et al. 2023). Thus, evidence suggests that telomerase regulation across tissues may vary as a function of proliferative demand, relative stem cell proportion, organismal aging, and other factors.

### Tissue-Specific Regulation of Telomere Length

The most extensive analysis of tissue-specific telomerase regulation has been performed in the hematopoietic system. Hematopoietic stem cells (HSCs), were identified by capacity to repopulate a bone marrow niche and differentiate into various hematopoietic lineages (Morrison and Weissman 1994; Orkin and Zon 2008; Wilson et al. 2008). Early analysis demonstrated that telomerase activity is present in both HSCs and their proliferative progeny, with exceptionally high telomerase activity in germinal center (GC) B cells (Broccoli et al. 1995; Hiyama et al. 1995; Hu et al. 1997; Weng et al. 1997). As leukocytes are easily assayed for telomere length through flow cytometry and fluorescence in situ hybridization (Flow-FISH), mean leukocyte telomere length has become a proxy for organismal telomere length in most human studies (Baerlocher and Lansdorp 2003; Baerlocher et al. 2006). Analysis of these hematopoietic cells highlights key patterns of telomere length dynamics in human tissues (Aubert et al. 2012).

First, mean telomere lengths between individuals vary by ~4 kb, even among age-matched healthy donors (Aubert et al. 2012). Thus, an age-agnostic definition of "too short" and "too long" telomeres still eludes us. While a child with telomeres shorter than the first percentile for telomere length in their age demographic is likely clinically symptomatic, their mean telomere length may be the same as a seemingly healthy 60-year-old.

Second, telomere lengths vary between different cell lineages, with granulocytes and B lymphocytes shortening at a lower rate than memory T cells or mature N/K cells (Rufer et al. 1998; Aubert et al. 2012; Lin et al. 2016). Recent long-read sequencing advances suggest not only cell-specific telomerase regulation, but also chromosome-specific telomere lengths (Karimian et al. 2024; Sanchez et al. 2024; Schmidt et al. 2024), demonstrating that telomere heterogeneity is preserved from populations to individual chromosome ends. This heterogeneity may be attributed to differential telomerase regulation, but further inquiry is needed to mechanistically characterize these differences.

Third, contrary to the stable "set point" of cultured hESCs and iPSCs, the telomere lengths of HSCs decrease over time (Vaziri et al. 1994; Bernitz et al. 2016). Similarly, the telomere lengths of putative HSCs are shorter in adults than in umbilical cord blood, and bone marrow recipients show shorter telomere lengths than their donors (Akiyama et al. 2000; Robertson et al. 2001; Thornley et al. 2002; Aubert et al. 2012). This rate of shortening is also not consistent across organismal lifetime but can be subdivided into three phases: dramatic telomere shortening within the first year of life, moderate shortening between 1 and 18 years, and decreased telomere shortening in adulthood (Frenck et al. 1998; Rufer et al. 1998; Aubert et al. 2012). This indicates the telomeric demand during organismal growth may be substantially different than at homeostasis. Whether age-associated telomere shortening is due to cell-intrinsic restriction of telomerase activity below homeostatic levels or cell-extrinsic factors associated with organismal aging is currently unknown.

Outside the hematopoietic system, tissue-specific telomerase activity and telomere length changes remain understudied. In some tissues, including the aortic heart valve, telomere shortening has been associated with age-dependent disease risk (Kurz et al. 2006; Theodoris et al. 2017; Wang et al. 2022). Recent postmortem analysis of telomere lengths in organ donors reveals that while most tissues show a negative correlation between telomere length and age, inferred telomere shortening rates vary widely between tissues and some tissues show no correlation (Demanelis et al. 2020). Whether this is primarily due to differences in replicative burden, tissue-specific modulation of telomerase activity, or perhaps even a varying set point between adult stem cells of different niches, remains a key question in telomere biology and its relation to human disease.

## Tissue-Level Insight from Defective Telomere Maintenance

While uncovering differential telomere length regulation in healthy tissues remains a challenge, important insight can be derived from telomerase dysregulation in human disease. While aberrant telomerase hyperactivity and improper TERT expression are associated with cancer, mutations resulting in reduced telomerase extension and critically short telomeres lead to TBDs, a class of genetic disorders characterized by failure of highly replicative tissues and encompassing a range of life-threatening conditions, including bone marrow failure, liver and lung disease, cancer, and other complications (Alder and Armanios 2022; Revy et al. 2023). TBDs are caused by mutations in genes responsible for telomere maintenance, leading to shortened or dysfunctional telomeres and reduced cellular replicative capacity. On the other hand, cancer predisposition with long telomeres (CPLT) was recently proposed as a clinical classification for genetic predisposition toward various cancers, including melanoma, thyroid cancer, sarcoma, glioma, and lymphoproliferative neoplasms, which is coupled with telomere length >90th percentile for age (Savage 2025; Savage et al. 2025); CPLT is driven by variants in shelterin complex genes, which result in excessive telomere elongation and increased cellular replicative capacity. Therefore, by comparing both the tissue-specific incidence of TBD phenotypes with the cancer-specific incidence of TPMs, as well as the roles of various telomerase regulatory genes in both TBD and CPLT, patterns of inferred telomerase regulation begin to appear (Fig. 4).

Despite being the most common noncoding mutations in cancer, TPMs are extremely can-

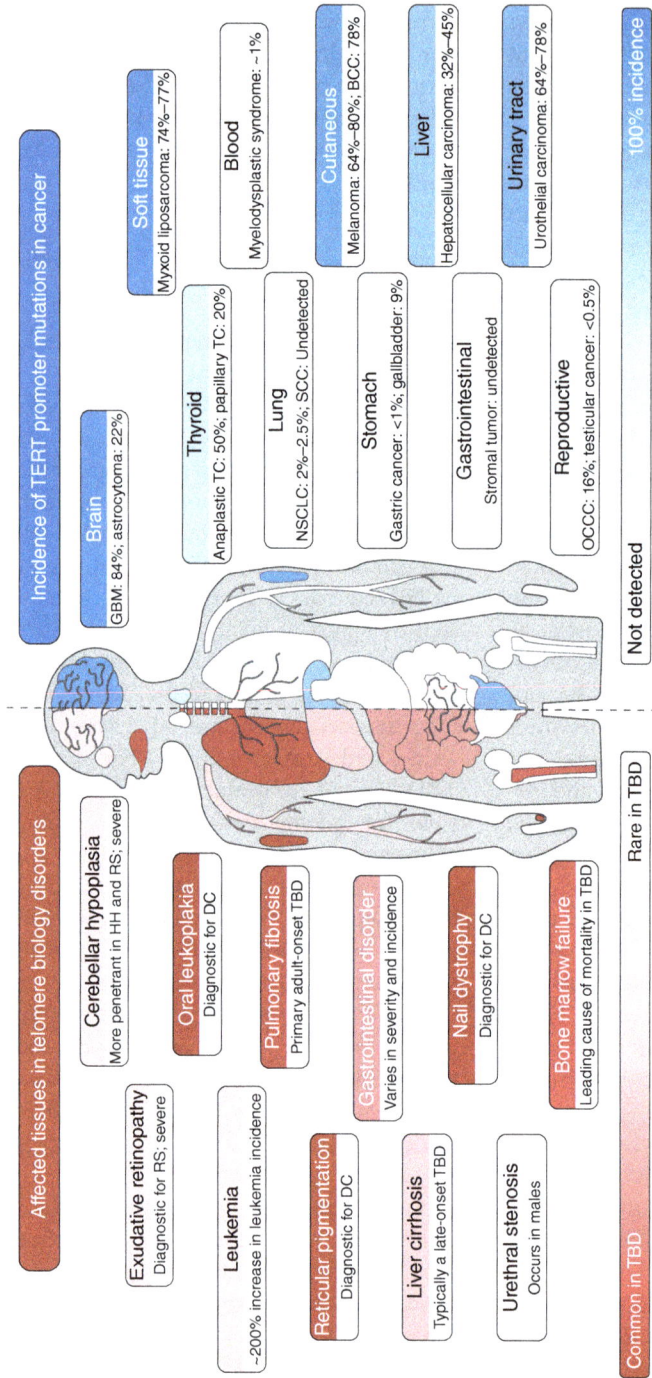

**Figure 4.** Tissue specificity of telomere biology disorder (TBD) symptoms and telomerase reverse transcriptase (TERT) promoter mutation (TPM) incidence in cancer. Even in cases of familial TBD, symptoms show tissue-specific penetrance. The most common symptoms occur in proliferative tissues, likely with a stem cell compartment. These include nail dystrophy, reticular pigmentation, and oral leukoplakia, with bone marrow failure as the leading cause of mortality. Conversely, TPMs occur largely in cancers arising from lowly proliferative tissues without known telomerase activity, including glioblastoma, melanoma, and urothelial carcinoma. Comparison of tissue-specific disease manifestation and mutational burden may shed insight on tissue-specific telomere length control and telomerase regulation. (DC) Dyskeratosis congenita, (HH) Hoyeraal-Hreidarsson syndrome, (RS) Revesz syndrome, (GBM) glioblastoma multiforma, (TC) thyroid carcinoma, (NSCLC) non-small-cell lung cancer, (SCLC) squamous cell carcinoma, (BCC) basal cell carcinoma, (OCCC) ovarian clear cell carcinoma.

cer-specific, with high prevalence in glioblastoma (84%) (Liu et al. 2013; Huang et al. 2015; Olympios et al. 2021), melanoma (64%–80%) (Horn et al. 2013; Huang et al. 2013), urothelial and bladder cancer (~65%) (Killela et al. 2013; Liu et al. 2013), and hepatocellular carcinoma (32%–45%) (Killela et al. 2013; Liu et al. 2013), despite being very rare in intestinal cancers and leukemia (Killela et al. 2013; Vinagre et al. 2013; Huang et al. 2015; Mosrati et al. 2015; Gupta et al. 2021; Nofrini et al. 2021; El Zarif et al. 2024). As discussed, TPMs are likely one of the earliest oncogenic mutations in many cells and are often associated with short telomeres and poorer prognosis. Therefore, TPMs may be indicative of cancer types arising from lowly proliferative cells that lack telomerase activity and would normally undergo replicative senescence (Lorbeer and Hockemeyer 2020).

Similarly, the frequency and severity of TBD symptoms are tissue delineated (Savage and Bertuch 2010). TBD causative mutations have been identified in many telomerase regulators, including the core TERT and TERC components (Vulliamy et al. 2004; Yamaguchi et al. 2005; Armanios et al. 2007), telomerase cofactors (Heiss et al. 1998; Walne et al. 2007; Vulliamy et al. 2008), TERC maturation proteins (Zhong et al. 2011; Tummala et al. 2015; Roake et al. 2019), members of the shelterin complex (Savage et al. 2008; Guo et al. 2014; Kocak et al. 2014; Takai et al. 2016), and other proteins involved in telomere replication and processing (Kermasson et al. 2022; Sharma et al. 2022; Agrawal et al. 2024; Kochman et al. 2024). To date, at least 18 genes have been linked to heritable TBDs (Revy et al. 2023). Severity and age of onset of TBD correlate with the degree of telomere shortening (Alder et al. 2018). TBDs are also diseases of generational anticipation, both within families and tissues. Individuals with TBD-associated germline mutations may never develop clinically relevant TBD, but the reduction of telomerase activity in the germline predisposes each successive generation toward more severe symptoms as the telomere pool undergoes generational attrition (Vulliamy et al. 2004; Armanios et al. 2005). Similarly, while aberrant telomere length maintenance can oc-

cur in a population of cells with no direct clinical manifestation, high proliferative burden on short-telomere progenitor cells may result in disease manifestation in tissues that are largely telomerase negative.

Even in familial TBD with germline mutations, clinical manifestation is tissue-specific. In severe TBD, bone marrow failure is the leading cause of mortality, usually co-occurrent with milder effects on other proliferative tissues, including skin pigmentation defects, nail dystrophy, and oral leukoplakia (Dokal 2011), whereas pulmonary fibrosis is the most common late-manifesting syndrome of telomere dysfunction (Alder and Armanios 2022). Therefore, clinical manifestations of TBDs may indicate tissues with essential telomerase activity or regenerative tissues that harbor a stem cell population.

By comparing the tissue-specificity of TBD symptoms and TPM occurrence in cancer, we can hypothesize telomerase regulation patterns at a systemic level. Hematopoietic cells have known telomerase activity, resulting in strong TBD symptoms and few TPMs. This logic can be extended to other tissues where telomerase regulation is less understood. TPMs are also notably reduced in lung cancer despite clear dependence on telomere maintenance for organismal health (Ma et al. 2014; Alder and Armanios 2022; Yang et al. 2023). This is not to say that no TERT mutations occur within lung cancer, as TERT amplifications are frequent in non-small-cell lung cancer and indicative of a poor prognosis (Zhu et al. 2006; Kang et al. 2008). Therefore, telomerase dosage may be restrictive of cancer progression in the lung, rather than TERT promoter silencing. On the other hand, TPMs are extremely common in glioblastoma (Liu et al. 2013; Huang et al. 2015; Olympios et al. 2021), but substantial neurological effects of TBDs tend to be rare, present in the most severe forms of TBD including Hoyeraal–Hreidarsson (HH) syndrome and Revesz syndrome (RS), and indicative of a severe general loss of replicative potential early in development (Bhala et al. 2019). This perhaps implies that brain tissue is largely telomerase negative and TERT transcription is limiting in oncogenesis.

Cite this article as *Cold Spring Harb Perspect Biol* doi: 10.1101/cshperspect.a041693

In other tissues, the associations are not so clear. While hepatocellular carcinoma has a high incidence of TPMs (Killela et al. 2013; Liu et al. 2013), hepatic morbidities are not uncommon late-onset manifestations of TBD (Kapuria et al. 2019). Further, the liver is a regenerative tissue with a hypothesized stem cell compartment, which presumes telomerase activity (Lin et al. 2018). Thus, whether TPMs activate TERT in a cell-type-specific manner to drive oncogenesis, or whether their effect is dose-dependent remains unclear. While these conclusions are still largely speculative and challenged by tissue heterogeneity and the complexity of human disease, a comprehensive analysis of tissue specificity in TBD symptom presentation, mutational burden, and cancer incidence may hold key information about adult stem cell compartments and the differences in telomerase regulation between tissues.

The clearest examples of telomerase regulation driving tissue-specific disease are observed in genes wherein mutations can result in either TBDs or CLPT (Table 1). The most frequent and often most severe shelterin TBD mutations occur within TIN2, likely due to its role in bridging

**Table 1.** Comparison of cancer-associated and telomere biology disorder (TBD)-associated mutations within the same genes

| Gene product | Cancer-associated mutation(s) | Telomere effects | TBD-associated mutation(s) | Telomere effects |
|---|---|---|---|---|
| TIN2 | Familial heterozygous amino-terminal frame-shift mutations (He et al. 2020; Schmutz et al. 2020) | Elongation | Heterozygous carboxy-terminal mutations within the dyskeratosis congenita (DC) cluster (Savage et al. 2008; Walne et al. 2008) | Shortening |
| | Premature stop codon (Koivuluoma et al. 2023) | Unknown | Exon 6 truncating mutations (Sasa et al. 2012) | Shortening |
| POT1 | Heterozygous mutations disrupting DNA-binding or TPP1-interaction domains (Speedy et al. 2016; Shen et al. 2020) | Elongation | Heterozygous point mutation in CST recruitment domain (Takai et al. 2016) | 3′ Elongation; defective 5′ synthesis |
| | | | Heterozygous point mutation in DNA-binding domain (Kelich et al. 2022) | Shortening |
| TPP1 | Promoter mutations (Chun-on et al. 2022) | Elongation | TEL patch mutations (Guo et al. 2014; Kocak et al. 2014) | Shortening |
| | Nonsense mutations (Aoude et al. 2015) | Predicted shortening | | |
| TERT | Promoter mutations (Horn et al. 2013; Huang et al. 2013; Bell et al. 2015; Chiba et al. 2015, 2017) | Stabilization | Varied and largely monoallelic variants in binding domains, catalytic domain, or regions associated with processivity (Vulliamy et al. 2004; Yamaguchi et al. 2005; Armanios et al. 2007) | Shortening |
| | Amplifications and rearrangements (Barthel et al. 2017; Yuan et al. 2019) | Elongation | | |
| TERC | Copy number amplification (Soder et al. 1997; Sugita et al. 2000) | Predicted elongation | Template sequence point mutations and TERT-binding domain mutations (Armanios et al. 2007; Tsakiri et al. 2007) | Shortening |

Because both short and long telomeres are associated with human disease, careful regulation of telomerase levels and activity is critical for organismal health. This is most clearly illustrated in genes that are mutated both in cancer, a disease often characterized by aberrant telomere hyperactivity, and TBDs, genetic disorders resulting in tissue failure because of critically short telomeres. These genes include telomerase components TERT and TERC, as well as members of the shelterin complex.

the DNA-interacting and telomerase-interacting members of shelterin. The carboxy-terminal portion of TIN2 contains a 30 amino acid region accounting for ~12% of all dyskeratosis congenita (DC) cases, with heterozygous dominant mutations resulting in bone marrow failure due to aberrantly short telomeres (Savage et al. 2008; Frescas and De Lange 2014). Interestingly, this DC cluster does not fall into any known interaction domain and does not appear to affect telomere protection (Yang et al. 2011; Frank et al. 2015; Choo et al. 2022). Therefore, it is unknown whether these mutations result in destabilization of shelterin complex assembly, direct inhibition of telomerase activity, or other interactions at the telomere. However, heterozygous loss-of-function mutations preceding the DC cluster of TIN2 result in telomere elongation and a strong familial cancer predisposition (Schmutz et al. 2020; Shen et al. 2020). In fact, early frame-shift mutations in either the TBD-associated mutant allele or in the wild-type allele may be sufficient to restore telomere length and prevent hematopoietic TBD symptoms (Alder et al. 2015; Choo et al. 2022), highlighting that TIN2-mediated telomerase regulation in HSCs is a careful balance.

The most frequently inherited mutations in CLPT are germline POT1 heterozygous missense mutations, which are associated with hyperelongation of telomeres (Kim et al. 2021) and familial cancer (Robles-Espinoza et al. 2014; Speedy et al. 2016; Chen et al. 2017; Shen et al. 2020). Recent deep scanning mutagenesis of POT1 in human stem cells has demonstrated that some cancer-associated POT1 alleles act as separation-of-function mutants, driving telomere elongation more effectively than frame-shift alleles, while still suppressing DDR signaling at telomere ends below a cell-intrinsic threshold for full ATR pathway activation (Martin et al. 2025). Therefore, this subset of alleles may more accurately be classified as oncogenic activating mutations rather than simple loss-of-function alleles since they retain or gain telomerase stimulatory activity, which is not present in full loss-of-function alleles (Martin et al. 2025). Unlike TPMs, cancer-associated shelterin mutations that result in telomere elongation presume a telomerase-positive background and appear to elongate telomeres in a telomerase-dependent manner. POT1 mutations are prevalent in melanoma (Robles-Espinoza et al. 2014), where TPMs are thought to be one of the earliest transforming mutations, as well as in chronic lymphocytic leukemia (Speedy et al. 2016), a class of cells known to have active telomerase. Similarly, TPP1 promoter mutations synergize with TPMs to overexpress both TPP1 and TERT, leading to elongated telomeres in melanoma (Chun-on et al. 2022).

Human disease is complex, however, and not all TBD and cancer manifestations are easily correlated with telomerase regulation. POT1 missense mutations have also been identified in TBD: not only in cases of familial pulmonary fibrosis, which show classical telomere shortening (Kelich et al. 2022), but also early-onset, severe Coats plus syndrome (Takai et al. 2016). Contrary to canonical TBD, these mutations did not result in critically short leukocyte telomere lengths, but instead showed hyperelongation of the 3′ overhang, the accumulation of telomeric damage-induced foci, and increased frequency of sudden telomere loss (Takai et al. 2016). Since a single critically short telomere is sufficient to induce replicative senescence (Baird et al. 2003; Abdallah et al. 2009), perhaps these events of catastrophic telomere loss, rather than alterations to telomerase regulation, lead to premature senescence and tissue failure. Adding to the challenge of complexity, critically short telomeres are also linked to increased risk of certain cancers, an effect that may arise cell-autonomously through increased genomic instability (Lin et al. 2010; Roger et al. 2013; Maciejowski et al. 2015), or non-cell-autonomously by depleting hematopoiesis and impairing immune surveillance (Schratz et al. 2023).

## Population-Scale Insights into Telomere Length Regulation at an Organismal Level

While telomere lengths vary widely in the population, several trends have emerged in telomere length variation between populations, influenced by both sex and ethnicity. However, because of high heterogeneity in telomere lengths, relatively low sample sizes, and imprecise methods for quantifying telomere length, reports differ as to the relative contributions of these factors, and contributing genetic loci have been challenging to identify (Prescott et al. 2011). Generally, telomere length appears to be longer in women than men, an effect that increases with age and may be due to a hormone response element in the TERT promoter region (Bekaert et al. 2007; Hunt et al. 2008; Calado et al. 2009; Ly et al. 2019; Vyas et al. 2021). Similarly, several independent studies have demonstrated longer telomeres in populations with African ancestry than those with European ancestry (Hunt et al. 2008; Diez Roux et al. 2009). However, telomere length differences may also show dependence on socioeconomic status (Hamad et al. 2016; Needham et al. 2019), perhaps since perceived stress is also correlated with shortening telomere lengths (Epel et al. 2004; Mathur et al. 2016; Lin and Epel 2022). Unfortunately, the conflicting nature of many reports makes it challenging to delineate populational differences in telomere length regulation, and meta-analysis indicates that more precise methods of telomere length measurement, better defined experimental paradigms, and larger sample sizes are necessary for conclusive results (Gardner et al. 2014). The development of long-read sequencing methods for precise telomere length quantification may begin to address these concerns while also enabling the delineation of maternal and paternal inheritance patterns and single chromosome arm resolution of telomere length control (Smoom et al. 2023; Karimian et al. 2024; Sanchez et al. 2024; Schmidt et al. 2024).

Despite these challenges, collaborative efforts to broaden databases of telomere length measurements, and improvements in telomere length estimation from whole-genome sequencing are beginning to enable complex genome-wide association studies (GWAS) with enough power to provide insight into telomere length regulation across populations. One of the largest comparative studies of leukocyte telomere length demonstrated that individuals with a mean telomere length one standard deviation below the mean have a 2.5 year lower life expectancy than individuals one standard deviation above the mean (Codd et al. 2021). GWAS further identified ~36 genes associated with variants correlating with telomere length (Codd et al. 2013; Mangino et al. 2015; Li et al. 2020a; Taub et al. 2022). While several of these genes are known to be involved in telomere regulation, including telomere replication (STN1, a member of CST), telomere end processing (SNMB1), or telomerase RNA processing (PARN), many other candidate genes remain to be tested. One pathway nominated through GWAS and confirmed through CRISPR screening is that of thymidine nucleotide metabolism (Li et al. 2020a; Mannherz and Agarwal 2023). Simply treating cultured cells with an excess of thymidine is sufficient to elongate telomeres in both stem and cancer cells, an effect that is independent from other nucleosides and entirely uncoupled from known mechanisms of telomere length regulation (Mannherz and Agarwal 2023). While population-scale studies to date have left us with more hypotheses than conclusions, they signify a new age of modern genomics in uncovering telomerase regulation in the human system.

Currently, telomere biology is most clearly observed through a very narrow lens. Much is known about the roles of specific molecular players in specific cell systems. Individual members of the shelterin complex are being teased apart at amino acid level genetic resolution and the advent of cryo-electron microscopy allows molecular contacts to be resolved with angstrom precision. However, as technology allows us to advance ever smaller, the broad picture of telomere biology is just barely coming into focus.

## REFERENCES

*Reference is also in this subject collection.

Abdallah P, Luciano P, Runge KW, Lisby M, Géli V, Gilson E, Teixeira MT. 2009. A two-step model for senescence triggered by a single critically short telomere. *Nat Cell Biol* **11**: 988–993. doi:10.1038/ncb1911

Abreu E, Aritonovska E, Reichenbach P, Cristofari G, Culp B, Terns RM, Lingner J, Terns MP. 2010. TIN2-tethered TPP1 recruits human telomerase to telomeres in vivo. *Mol Cell Biol* **30**: 2971–2982. doi:10.1128/MCB.00240-10

Agarwal S, Loh YH, McLoughlin EM, Huang J, Park IH, Miller JD, Huo H, Okuka M, dos Reis RM, Loewer S, et al. 2010. Telomere elongation in induced pluripotent stem cells from dyskeratosis congenita patients. *Nature* **464:** 292–296. doi:10.1038/nature08792

Agrawal S, Lin X, Susvirkar V, O'Connor MS, Chavez BL, Tholkes VR, Abe KM, He Q, Huang X, Lim CJ. 2024. Human replication protein A complex is a telomerase processivity factor essential for telomere maintenance. bioRxiv doi:10.1101/2024.08.16.608355v1

Ahmed S, Passos JF, Birket MJ, Beckmann T, Brings S, Peters H, Birch-Machin MA, von Zglinicki T, Saretzki G. 2008. Telomerase does not counteract telomere shortening but protects mitochondrial function under oxidative stress. *J Cell Sci* **121:** 1046–1053. doi:10.1242/jcs.019372

Akıncılar SC, Khattar E, Boon PLS, Unal B, Fullwood MJ, Tergaonkar V. 2016. Long-range chromatin interactions drive mutant *TERT* promoter activation. *Cancer Discov* **6:** 1276–1291. doi:10.1158/2159-8290.CD-16-0177

Akiyama M, Asai O, Kuraishi Y, Urashima M, Hoshi Y, Sakamaki H, Yabe H, Furukawa T, Yamada O, Mizoguchi H, et al. 2000. Shortening of telomeres in recipients of both autologous and allogeneic hematopoietic stem cell transplantation. *Bone Marrow Transplant* **25:** 441–447. doi:10.1038/sj.bmt.1702144

Alder JK, Armanios M. 2022. Telomere-mediated lung disease. *Physiol Rev* **102:** 1703–1720. doi:10.1152/physrev.00046.2021

Alder JK, Stanley SE, Wagner CL, Hamilton M, Hanumanthu VS, Armanios M. 2015. Exome sequencing identifies mutant *TINF2* in a family with pulmonary fibrosis. *Chest* **147:** 1361–1368. doi:10.1378/chest.14-1947

Alder JK, Hanumanthu VS, Strong MA, DeZern AE, Stanley SE, Takemoto CM, Danilova L, Applegate CD, Bolton SG, Mohr DW, et al. 2018. Diagnostic utility of telomere length testing in a hospital-based setting. *Proc Natl Acad Sci* **115:** E2358–E2365. doi:10.1073/pnas.1720427115

Allsopp RC, Vaziri H, Patterson C, Goldstein S, Younglai EV, Futcher AB, Greider CW, Harley CB. 1992. Telomere length predicts replicative capacity of human fibroblasts. *Proc Natl Acad Sci* **89:** 10114–10118. doi:10.1073/pnas.89.21.10114

Ancelin K, Brunori M, Bauwens S, Koering CE, Brun C, Ricoul M, Pommier JP, Sabatier L, Gilson E. 2002. Targeting assay to study the *cis* functions of human telomeric proteins: evidence for inhibition of telomerase by TRF1 and for activation of telomere degradation by TRF2. *Mol Cell Biol* **22:** 3474–3487. doi:10.1128/MCB.22.10.3474-3487.2002

Aoude LG, Pritchard AL, Robles-Espinoza CD, Wadt K, Harland M, Choi J, Gartside M, Quesada V, Johansson P, Palmer JM, et al. 2015. Nonsense mutations in the shelterin complex genes ACD and TERF2IP in familial melanoma. *JNCI: J Natl Cancer Inst* **107:** dju408.

Armanios M, Chen JL, Chang YPC, Brodsky RA, Hawkins A, Griffin CA, Eshleman JR, Cohen AR, Chakravarti A, Hamosh A, et al. 2005. Haploinsufficiency of telomerase reverse transcriptase leads to anticipation in autosomal dominant dyskeratosis congenita. *Proc Natl Acad Sci* **102:** 15960–15964. doi:10.1073/pnas.0508124102

Armanios MY, Chen JJL, Cogan JD, Alder JK, Ingersoll RG, Markin C, Lawson WE, Xie M, Vulto I, Phillips JA, et al. 2007. Telomerase mutations in families with idiopathic pulmonary fibrosis. *N Engl J Med* **356:** 1317–1326. doi:10.1056/NEJMoa066157

Arnerić M, Lingner J. 2007. Tel1 kinase and subtelomere-bound Tbf1 mediate preferential elongation of short telomeres by telomerase in yeast. *EMBO Rep* **8:** 1080–1085. doi:10.1038/sj.embor.7401082

Arnoult N, Van Beneden A, Decottignies A. 2012. Telomere length regulates TERRA levels through increased trimethylation of telomeric H3K9 and HP1α. *Nat Struct Mol Biol* **19:** 948–956. doi:10.1038/nsmb.2364

Asghar M, Bensch S, Tarka M, Hansson B, Hasselquist D. 2015. Maternal and genetic factors determine early life telomere length. *Proc Biol Sci* **282:** 20142263. doi:10.1098/rspb.2014.2263

Aubert G. 2014. Telomere dynamics and aging. *Prog Mol Biol Transl Sci* **125:** 89–111. doi:10.1016/B978-0-12-397898-1.00004-9

Aubert G, Baerlocher GM, Vulto I, Poon SS, Lansdorp PM. 2012. Collapse of telomere homeostasis in hematopoietic cells caused by heterozygous mutations in telomerase genes. *PLoS Genet* **8:** e1002696. doi:10.1371/journal.pgen.1002696

Avin BA, Wang Y, Gilpatrick T, Workman RE, Lee I, Timp W, Umbricht CB, Zeiger MA. 2019. Characterization of human telomerase reverse transcriptase promoter methylation and transcription factor binding in differentiated thyroid cancer cell lines. *Genes Chromosomes Cancer* **58:** 530–540. doi:10.1002/gcc.22735

Awad A, Glousker G, Lamm N, Tawil S, Hourvitz N, Smoom R, Revy P, Tzfati Y. 2020. Full length RTEL1 is required for the elongation of the single-stranded telomeric overhang by telomerase. *Nucleic Acids Res* **48:** 7239–7251. doi:10.1093/nar/gkaa503

Azzalin CM, Reichenbach P, Khoriauli L, Giulotto E, Lingner J. 2007. Telomeric repeat containing RNA and RNA surveillance factors at mammalian chromosome ends. *Science* **318:** 798–801. doi:10.1126/science.1147182

Baerlocher GM, Lansdorp PM. 2003. Telomere length measurements in leukocyte subsets by automated multicolor flow-FISH. *Cytometry A* **55A:** 1–6. doi:10.1002/cyto.a.10064

Baerlocher GM, Vulto I, de Jong G, Lansdorp PM. 2006. Flow cytometry and FISH to measure the average length of telomeres (flow FISH). *Nat Protoc* **1:** 2365–2376. doi:10.1038/nprot.2006.263

Baird DM, Rowson J, Wynford-Thomas D, Kipling D. 2003. Extensive allelic variation and ultrashort telomeres in senescent human cells. *Nat Genet* **33:** 203–207. doi:10.1038/ng1084

Ballew BJ, Joseph V, De S, Sarek G, Vannier JB, Stracker T, Schrader KA, Small TN, O'Reilly R, Manschreck C, et al. 2013. A recessive founder mutation in regulator of telomere elongation helicase 1, RTEL1, underlies severe immunodeficiency and features of Hoyeraal–Hreidarsson syndrome. *PLoS Genet* **9:** e1003695. doi:10.1371/journal.pgen.1003695

Barthel FP, Wei W, Tang M, Martinez-Ledesma E, Hu X, Amin SB, Akdemir KC, Seth S, Song X, Wang Q, et al. 2017. Systematic analysis of telomere length and somatic alterations in 31 cancer types. *Nat Genet* **49:** 349–357. doi:10.1038/ng.3781

Batista LFZ, Pech MF, Zhong FL, Nguyen HN, Xie KT, Zaug AJ, Crary SM, Choi J, Sebastiano V, Cherry A, et al. 2011. Telomere shortening and loss of self-renewal in dyskeratosis congenita induced pluripotent stem cells. *Nature* 474: 399–402. doi:10.1038/nature10084

Baumann P, Cech TR. 2001. Pot1, the putative telomere end-binding protein in fission yeast and humans. *Science* 292: 1171–1175. doi:10.1126/science.1060036

Baur JA, Zou Y, Shay JW, Wright WE. 2001. Telomere position effect in human cells. *Science* 292: 2075–2077. doi:10.1126/science.1062329

Bekaert S, De Meyer T, Rietzschel ER, De Buyzere ML, De Bacquer D, Langlois M, Segers P, Cooman L, Van Damme P, Cassiman P, et al. 2007. Telomere length and cardiovascular risk factors in a middle-aged population free of overt cardiovascular disease. *Aging Cell* 6: 639–647. doi:10.1111/j.1474-9726.2007.00321.x

Bell RJA, Rube HT, Kreig A, Mancini A, Fouse SD, Nagarajan RP, Choi S, Hong C, He D, Pekmezci M, et al. 2015. The transcription factor GABP selectively binds and activates the mutant TERT promoter in cancer. *Science* 348: 1036–1039. doi:10.1126/science.aab0015

Benetti R, García-Cao M, Blasco MA. 2007. Telomere length regulates the epigenetic status of mammalian telomeres and subtelomeres. *Nat Genet* 39: 243–250. doi:10.1038/ng1952

Bernitz JM, Kim HS, MacArthur B, Sieburg H, Moore K. 2016. Hematopoietic stem cells count and remember self-renewal divisions. *Cell* 167: 1296–1309.e10. doi:10.1016/j.cell.2016.10.022

Bhala S, Best AF, Giri N, Alter BP, Pao M, Gropman A, Baker EH, Savage SA. 2019. CNS manifestations in patients with telomere biology disorders. *Neurol Genet* 5: 370. doi:10.1212/NXG.0000000000000370

Bodnar AG, Ouellette M, Frolkis M, Holt SE, Chiu CP, Morin GB, Harley CB, Shay JW, Lichtsteiner S, Wright WE. 1998. Extension of life-span by introduction of telomerase into normal human cells. *Science* 279: 349–352. doi:10.1126/science.279.5349.349

Boyle JM, Hennick KM, Regalado SG, Vogan JM, Zhang X, Collins K, Hockemeyer D. 2020. Telomere length set point regulation in human pluripotent stem cells critically depends on the shelterin protein TPP1. *Mol Biol Cell* 31: 2583–2596. doi:10.1091/mbc.E19-08-0447

Broccoli D, Young JW, de Lange T. 1995. Telomerase activity in normal and malignant hematopoietic cells. *Proc Natl Acad Sci* 92: 9082–9086. doi:10.1073/pnas.92.20.9082

Broer L, Codd V, Nyholt DR, Deelen J, Mangino M, Willemsen G, Albrecht E, Amin N, Beekman M, De Geus EJC, et al. 2013. Meta-analysis of telomere length in 19 713 subjects reveals high heritability, stronger maternal inheritance and a paternal age effect. *Eur J Hum Genet* 21: 1163–1168. doi:10.1038/ejhg.2012.303

Bryan TM, Englezou A, Dunham MA, Reddel RR. 1998. Telomere length dynamics in telomerase-positive immortal human cell populations. *Exp Cell Res* 239: 370–378. doi:10.1006/excr.1997.3907

Cai SW, de Lange T. 2023. CST–Polα/primase: the second telomere maintenance machine. *Genes Dev* 37: 555–569. doi:10.1101/gad.350479.123

Cai SW, Takai H, Zaug AJ, Dilgen TC, Cech TR, Walz T, de Lange T. 2024. POT1 recruits and regulates CST-Polα/primase at human telomeres. *Cell* 187: 3638–3651.e18. doi:10.1016/j.cell.2024.05.002

Calado RT, Yewdell WT, Wilkerson KL, Regal JA, Kajigaya S, Stratakis CA, Young NS. 2009. Sex hormones, acting on the TERT gene, increase telomerase activity in human primary hematopoietic cells. *Blood* 114: 2236–2243. doi:10.1182/blood-2008-09-178871

Casteel DE, Zhuang S, Zeng Y, Perrino FW, Boss GR, Goulian M, Pilz RB. 2009. A DNA polymerase-α·primase cofactor with homology to replication protein A-32 regulates DNA replication in mammalian cells. *J Biol Chem* 284: 5807–5818. doi:10.1074/jbc.M807593200

Castelo-Branco P, Choufani S, Mack S, Gallagher D, Zhang C, Lipman T, Zhukova N, Walker EJ, Martin D, Merino D, et al. 2013. Methylation of the TERT promoter and risk stratification of childhood brain tumours: an integrative genomic and molecular study. *Lancet Oncol* 14: 534–542. doi:10.1016/S1470-2045(13)70110-4

Cesare AJ, Karlseder J. 2012. A three-state model of telomere control over human proliferative boundaries. *Curr Opin Cell Biol* 24: 731–738. doi:10.1016/j.ceb.2012.08.007

Cesare AJ, Reddel RR. 2010. Alternative lengthening of telomeres: models, mechanisms and implications. *Nat Rev Genet* 11: 319–330. doi:10.1038/nrg2763

Chang JT, Lu YC, Chen YJ, Tseng CP, Chen YL, Fang CW, Cheng AJ. 2006. hTERT phosphorylation by PKC is essential for telomerase holoprotein integrity and enzyme activity in head neck cancer cells. *Br J Cancer* 94: 870–878. doi:10.1038/sj.bjc.6603008

Chang M, Arneric M, Lingner J. 2007. Telomerase repeat addition processivity is increased at critically short telomeres in a Tel1-dependent manner in *Saccharomyces cerevisiae*. *Genes Dev* 21: 2485–2494. doi:10.1101/gad.1588807

* Chen L, Batista LFZ. 2025. Biogenesis and regulation of telomerase during development and cancer. *Cold Spring Harb Perspect Biol* doi:10.1101/cshperspect.a041692

Chen LY, Redon S, Lingner J. 2012. The human CST complex is a terminator of telomerase activity. *Nature* 488: 540–544. doi:10.1038/nature11269

Chen C, Gu P, Wu J, Chen X, Niu S, Sun H, Wu L, Li N, Peng J, Shi S, et al. 2017. Structural insights into POT1-TPP1 interaction and POT1 C-terminal mutations in human cancer. *Nat Commun* 8: 14929. doi:10.1038/ncomms14929

Chen L, Roake CM, Freund A, Batista PJ, Tian S, Yin YA, Gajera CR, Lin S, Lee B, Pech MF, et al. 2018. An activity switch in human telomerase based on RNA conformation and shaped by TCAB1. *Cell* 174: 218–230.e13. doi:10.1016/j.cell.2018.04.039

Cheng D, Zhao Y, Wang S, Jia W, Kang J, Zhu J. 2015. Human telomerase reverse transcriptase (hTERT) transcription requires Sp1/Sp3 binding to the promoter and a permissive chromatin environment. *J Biol Chem* 290: 30193–30203. doi:10.1074/jbc.M115.662221

Cheng D, Wang S, Jia W, Zhao Y, Zhang F, Kang J, Zhu J. 2017a. Regulation of human and mouse telomerase genes by genomic contexts and transcription factors during embryonic stem cell differentiation. *Sci Rep* 7: 1–12. doi:10.1038/s41598-016-0028-x

Cheng L, Montironi R, Lopez-Beltran A. 2017b. TERT promoter mutations occur frequently in urothelial papilloma

Cite this article as *Cold Spring Harb Perspect Biol* doi: 10.1101/cshperspect.a041693

and papillary urothelial neoplasm of low malignant potential. *Eur Urol* **71**: 497–498. doi:10.1016/j.eururo.2016.12.008

Cheng D, Zhao Y, Zhang F, Zhang J, Wang S, Zhu J. 2019. Engineering a humanized telomerase reverse transcriptase gene in mouse embryonic stem cells. *Sci Rep* **9**: 1–11. doi:10.1038/s41598-018-37186-2

Cheng D, Zhang F, Porter KI, Wang S, Zhang H, Davis CJ, Robertson GP, Zhu J. 2024. Humanization of the mouse *Tert* gene reset telomeres to human length. *Res Sq* doi:10.21203/rs.3.rs-3617723/v1

Chiba K, Johnson JZ, Vogan JM, Wagner T, Boyle JM, Hockemeyer D. 2015. Cancer-associated TERT promoter mutations abrogate telomerase silencing. *eLife* **4**: 07918. doi:10.7554/eLife.07918

Chiba K, Lorbeer FK, Shain AH, McSwiggen DT, Schruf E, Oh A, Ryu J, Darzacq X, Bastian BC, Hockemeyer D. 2017. Mutations in the promoter of the telomerase gene *TERT* contribute to tumorigenesis by a two-step mechanism. *Science* **357**: 1416–1420. doi:10.1126/science.aao0535

Chong L, van Steensel B, Broccoli D, Erdjument-Bromage H, Hanish J, Tempst P, de Lange T. 1995. A human telomeric protein. *Science* **270**: 1663–1667. doi:10.1126/science.270.5242.1663

Choo S, Lorbeer FK, Regalado SG, Short SB, Wu SS, Rieser G, Bertuch AA, Hockemeyer D. 2022. Editing *TINF2* as a potential therapeutic approach to restore telomere length in dyskeratosis congenita. *Blood* **140**: 608–618. doi:10.1182/blood.2021013750

Chow TT, Zhao Y, Mak SS, Shay JW, Wright WE. 2012. Early and late steps in telomere overhang processing in normal human cells: the position of the final RNA primer drives telomere shortening. *Genes Dev* **26**: 1167–1178. doi:10.1101/gad.187211.112

Chun-on P, Hinchie AM, Beale HC, Gil Silva AA, Rush E, Sander C, Connelly CJ, Seynnaeve BKN, Kirkwood JM, Vaske OM, et al. 2022. TPP1 promoter mutations cooperate with TERT promoter mutations to lengthen telomeres in melanoma. *Science* **378**: 664–668. doi:10.1126/science.abq0607

Codd V, Nelson CP, Albrecht E, Mangino M, Deelen J, Buxton JL, Hottenga JJ, Fischer K, Esko T, Surakka I, et al. 2013. Identification of seven loci affecting mean telomere length and their association with disease. *Nat Genet* **45**: 422–427. doi:10.1038/ng.2528

Codd V, Wang Q, Allara E, Musicha C, Kaptoge S, Stoma S, Jiang T, Hamby SE, Braund PS, Bountziouka V, et al. 2021. Polygenic basis and biomedical consequences of telomere length variation. *Nat Genet* **53**: 1425–1433. doi:10.1038/s41588-021-00944-6

Cohen SB, Graham ME, Lovrecz GO, Bache N, Robinson PJ, Reddel RR. 2007. Protein composition of catalytically active human telomerase from immortal cells. *Science* **315**: 1850–1853. doi:10.1126/science.1138596

Colgin LM, Wilkinso C, Englezou A, Kilian A, Robinson MO, Reddel RR. 2000. The hTERTα splice variant is a dominant negative inhibitor of telomerase activity. *Neoplasia* **2**: 426–432. doi:10.1038/sj.neo.7900112

Collins K. 2006. The biogenesis and regulation of telomerase holoenzymes. *Nat Rev Mol Cell Biol* **7**: 484–494. doi:10.1038/nrm1961

Cong YS, Wen J, Bacchetti S. 1999. The human telomerase catalytic subunit hTERT: organization of the gene and characterization of the promoter. *Hum Mol Genet* **8**: 137–142. doi:10.1093/hmg/8.1.137

Counter CM, Avilion AA, LeFeuvre CE, Stewart NG, Greider CW, Harley CB, Bacchetti S. 1992. Telomere shortening associated with chromosome instability is arrested in immortal cells which express telomerase activity. *EMBO J* **11**: 1921–1929. doi:10.1002/j.1460-2075.1992.tb05245.x

Court R, Chapman L, Fairall L, Rhodes D. 2005. How the human telomeric proteins TRF1 and TRF2 recognize telomeric DNA: a view from high-resolution crystal structures. *EMBO Rep* **6**: 39–45. doi:10.1038/sj.embor.7400314

Cristofari G, Lingner J. 2006. Telomere length homeostasis requires that telomerase levels are limiting. *EMBO J* **25**: 565–574. doi:10.1038/sj.emboj.7600952

Crowe DL, Nguyen DC, Tsang KJ, Kyo S. 2001. E2F-1 represses transcription of the human telomerase reverse transcriptase gene. *Nucleic Acids Res* **29**: 2789–2794. doi:10.1093/nar/29.13.2789

d'Adda di Fagagna F, Reaper PM, Clay-Farrace L, Fiegler H, Carr P, Von Zglinicki T, Saretzki G, Carter NP, Jackson SP. 2003. A DNA damage checkpoint response in telomere-initiated senescence. *Nature* **426**: 194–198. doi:10.1038/nature02118

Déjardin J, Kingston RE. 2009. Purification of proteins associated with specific genomic Loci. *Cell* **136**: 175–186. doi:10.1016/j.cell.2008.11.045

de Lange T. 2009. How telomeres solve the end-protection problem. *Science* **326**: 948–952. doi:10.1126/science.1170633

de Lange T. 2018. Shelterin-mediated telomere protection. *Annu Rev Genet* **52**: 223–247. doi:10.1146/annurev-genet-032918-021921

Demanelis K, Jasmine F, Chen LS, Chernoff M, Tong L, Delgado D, Zhang C, Shinkle J, Sabarinathan M, Lin H, et al. 2020. Determinants of telomere length across human tissues. *Science* **369**: eaaz6876. doi:10.1126/science.aaz6876

Denchi EL, de Lange T. 2007. Protection of telomeres through independent control of ATM and ATR by TRF2 and POT1. *Nature* **448**: 1068–1071. doi:10.1038/nature06065

Deng Z, Glousker G, Molczan A, Fox AJ, Lamm N, Dheekollu J, Weizman O-E, Schertzer M, Wang Z, Vladimirova O, et al. 2013. Inherited mutations in the helicase RTEL1 cause telomere dysfunction and Hoyeraal–Hreidarsson syndrome. *Proc Natl Acad Sci* **110**: E3408–E3416. doi:10.1073/pnas.1300600110

de Silanes IL, d'Alcontres MS, Blasco MA. 2010. TERRA transcripts are bound by a complex array of RNA-binding proteins. *Nat Commun* **1**: 33. doi:10.1038/ncomms1032

Devereux TR, Horikawa I, Anna CH, Annab LA, Afshari CA, Barrett JC. 1999. DNA methylation analysis of the promoter region of the human telomerase reverse transcriptase (hTERT) gene. *Cancer Res* **59**: 6087–6090.

Diez Roux AV, Ranjit N, Jenny NS, Shea S, Cushman M, Fitzpatrick A, Seeman T. 2009. Race/ethnicity and telomere length in the multi-ethnic study of atherosclerosis. *Aging Cell* **8**: 251–257. doi:10.1111/j.1474-9726.2009.00470.x

Ding H, Schertzer M, Wu X, Gertsenstein M, Selig S, Kammori M, Pourvali R, Poon S, Vulto I, Chavez E, et al. 2004. Regulation of murine telomere length by Rtel: an essential gene encoding a helicase-like protein. *Cell* **117:** 873–886. doi:10.1016/j.cell.2004.05.026

Diotti R, Kalan S, Matveyenko A, Loayza D. 2015. DNA-directed polymerase subunits play a vital role in human telomeric overhang processing. *Mol Cancer Res* **13:** 402–410. doi:10.1158/1541-7786.MCR-14-0381

Dokal I. 2011. Dyskeratosis congenita. *Hematology* **2011:** 480–486. doi:10.1182/asheducation-2011.1.480

Doksani Y, Wu JY, de Lange T, Zhuang X. 2013. Super-resolution fluorescence imaging of telomeres reveals TRF2-dependent T-loop formation. *Cell* **155:** 345–356. doi:10.1016/j.cell.2013.09.048

Dumbović G, Braunschweig U, Langner HK, Smallegan M, Biayna J, Hass EP, Jastrzebska K, Blencowe B, Cech TR, Caruthers MH, et al. 2021. Nuclear compartmentalization of TERT mRNA and TUG1 lncRNA is driven by intron retention. *Nat Commun* **12:** 3308. doi:10.1038/s41467-021-23221-w

El Zarif T, Machaalani M, Nawfal R, Nassar AH, Xie W, Choueiri TK, Pomerantz M. 2024. *TERT* promoter mutations frequency across race, sex, and cancer type. *Oncologist* **29:** 8–14. doi:10.1093/oncolo/oyad208

Epel ES, Blackburn EH, Lin J, Dhabhar FS, Adler NE, Morrow JD, Cawthon RM. 2004. Accelerated telomere shortening in response to life stress. *Proc Natl Acad Sci* **101:** 17312–17315. doi:10.1073/pnas.0407162101

Fang G, Cech TR. 1995. Telomerase RNA localized in the replication band and spherical subnuclear organelles in hypotrichous ciliates. *J Cell Biol* **130:** 243–253. doi:10.1083/jcb.130.2.243

Feng X, Hsu SJ, Kasbek C, Chaiken M, Price CM. 2017. CTC1-mediated C-strand fill-in is an essential step in telomere length maintenance. *Nucleic Acids Res* **45:** 4281–4293. doi:10.1093/nar/gkx125

Ferber MJ, Montoya DP, Yu C, Aderca I, McGee A, Thorland EC, Nagorney DM, Gostout BS, Burgart LJ, Boix L, et al. 2003. Integrations of the hepatitis B virus (HBV) and human papillomavirus (HPV) into the human telomerase reverse transcriptase (hTERT) gene in liver and cervical cancers. *Oncogene* **22:** 3813–3820. doi:10.1038/sj.onc.1206528

Flynn RL, Centore RC, O'Sullivan RJ, Rai R, Tse A, Songyang Z, Chang S, Karlseder J, Zou L. 2011. TERRA and hnRNPA1 orchestrate an RPA-to-POT1 switch on telomeric single-stranded DNA. *Nature* **471:** 532–536. doi:10.1038/nature09772

Frank AK, Tran DC, Qu RW, Stohr BA, Segal DJ, Xu L. 2015. The shelterin TIN2 subunit mediates recruitment of telomerase to telomeres. *PLoS Genet* **11:** e1005410. doi:10.1371/journal.pgen.1005410

Fredriksson NJ, Ny L, Nilsson JA, Larsson E. 2014. Systematic analysis of noncoding somatic mutations and gene expression alterations across 14 tumor types. *Nat Genet* **46:** 1258–1263. doi:10.1038/ng.3141

Frenck RW, Blackburn EH, Shannon KM. 1998. The rate of telomere sequence loss in human leukocytes varies with age. *Proc Natl Acad Sci* **95:** 5607–5610. doi:10.1073/pnas.95.10.5607

Frescas D, De Lange T. 2014. Binding of TPP1 protein to TIN2 protein is required for POT1a,b protein-mediated telomere protection. *J Biol Chem* **289:** 24180–24187. doi:10.1074/jbc.M114.592592

García-Cao M, Gonzalo S, Dean D, Blasco MA. 2002. A role for the Rb family of proteins in controlling telomere length. *Nat Genet* **32:** 415–419. doi:10.1038/ng1011

García-Cao M, O'Sullivan R, Peters AHFM, Jenuwein T, Blasco MA. 2004. Epigenetic regulation of telomere length in mammalian cells by the Suv39h1 and Suv39h2 histone methyltransferases. *Nat Genet* **36:** 94–99. doi:10.1038/ng1278

Gardner M, Bann D, Wiley L, Cooper R, Hardy R, Nitsch D, Martin-Ruiz C, Shiels P, Sayer AA, Barbieri M, et al. 2014. Gender and telomere length: systematic review and meta-analysis. *Exp Gerontol* **51:** 15–27. doi:10.1016/j.exger.2013.12.004

Ghanim GE, Fountain AJ, van Roon A-MM, Rangan R, Das R, Collins K, Nguyen THD. 2021. Structure of human telomerase holoenzyme with bound telomeric DNA. *Nature* **593:** 449–453. doi:10.1038/s41586-021-03415-4

Goldberg AD, Banaszynski LA, Noh KM, Lewis PW, Elsaesser SJ, Stadler S, Dewell S, Law M, Guo X, Li X, et al. 2010. Distinct factors control histone variant H3.3 localization at specific genomic regions. *Cell* **140:** 678–691. doi:10.1016/j.cell.2010.01.003

Gong Y, de Lange T. 2010. A Shld1-controlled POT1a provides support for repression of ATR signaling at telomeres through RPA exclusion. *Mol Cell* **40:** 377–387. doi:10.1016/j.molcel.2010.10.016

Gonzalo S, Jaco I, Fraga MF, Chen T, Li E, Esteller M, Blasco MA. 2006. DNA methyltransferases control telomere length and telomere recombination in mammalian cells. *Nat Cell Biol* **8:** 416–424. doi:10.1038/ncb1386

Gottschling DE, Aparicio OM, Billington BL, Zakian VA. 1990. Position effect at *S. cerevisiae* telomeres: reversible repression of Pol II transcription. *Cell* **63:** 751–762. doi:10.1016/0092-8674(90)90141-Z

Goueli BS, Janknecht R. 2003. Regulation of telomerase reverse transcriptase gene activity by upstream stimulatory factor. *Oncogene* **22:** 8042–8047. doi:10.1038/sj.onc.1206847

Greenwell PW, Kronmal SL, Porter SE, Gassenhuber J, Obermaier B, Petes TD. 1995. *TEL1*, a gene involved in controlling telomere length in *S. cerevisiae*, is homologous to the human ataxia telangiectasia gene. *Cell* **82:** 823–829. doi:10.1016/0092-8674(95)90479-4

Greider CW. 1996. Telomere length regulation. *Annu Rev Biochem* **65:** 337–365. doi:10.1146/annurev.bi.65.070196.002005

Greider CW. 2016. Regulating telomere length from the inside out: the replication fork model. *Genes Dev* **30:** 1483–1491. doi:10.1101/gad.280578.116

Griffith JD, Comeau L, Rosenfield S, Stansel RM, Bianchi A, Moss H, de Lange T. 1999. Mammalian telomeres end in a large duplex loop. *Cell* **97:** 503–514. doi:10.1016/S0092-8674(00)80760-6

Grill S, Bisht K, Tesmer VM, Shami AN, Hammoud SS, Nandakumar J. 2019. Two separation-of-function isoforms of human TPP1 dictate telomerase regulation in somatic and germ cells. *Cell Rep* **27:** 3511–3521.e7. doi:10.1016/j.celrep.2019.05.073

Grolimund L, Aeby E, Hamelin R, Armand F, Chiappe D, Moniatte M, Lingner J. 2013. A quantitative telomeric chromatin isolation protocol identifies different telomeric states. *Nat Commun* **4**: 2848. doi:10.1038/ncomms3848

Gu P, Jia S, Takasugi T, Smith E, Nandakumar J, Hendrickson E, Chang S. 2018. CTC1-STN1 coordinates G- and C-strand synthesis to regulate telomere length. *Aging Cell* **17**: e12783. doi:10.1111/acel.12783

Günes C, Rudolph KL. 2013. The role of telomeres in stem cells and cancer. *Cell* **152**: 390–393. doi:10.1016/j.cell.2013.01.010

Günes C, Lichtsteiner S, Vasserot AP, Englert C. 2000. Expression of the hTERT gene is regulated at the level of transcriptional initiation and repressed by Mad1. *Cancer Res* **60**: 2116–2121.

Guo Y, Kartawinata M, Li J, Pickett HA, Teo J, Kilo T, Barbaro PM, Keating B, Chen Y, Tian L, et al. 2014. Inherited bone marrow failure associated with germline mutation of ACD, the gene encoding telomere protein TPP1. *Blood* **124**: 2767–2774. doi:10.1182/blood-2014-08-596445

Gupta S, Vanderbilt CM, Lin YT, Benhamida JK, Jungbluth AA, Rana S, Momeni-Boroujeni A, Chang JC, Mcfarlane T, Salazar P, et al. 2021. A pan-cancer study of somatic TERT promoter mutations and amplification in 30,773 tumors profiled by clinical genomic sequencing. *J Mol Diagn* **23**: 253–263. doi:10.1016/j.jmoldx.2020.11.003

Haendeler J, Hoffmann J, Brandes RP, Zeiher AM, Dimmeler S. 2003. Hydrogen peroxide triggers nuclear export of telomerase reverse transcriptase via Src kinase family-dependent phosphorylation of tyrosine 707. *Mol Cell Biol* **23**: 4598–4610. doi:10.1128/MCB.23.13.4598-4610.2003

Hagen KGT, Gilbert DM, Willard HF, Cohen SN. 1990. Replication timing of DNA sequences associated with human centromeres and telomeres. *Mol Cell Biol* **10**: 6348–6355. doi:10.1128/mcb.10.12.6348-6355.1990

Hamad R, Tuljapurkar S, Rehkopf DH. 2016. Racial and socioeconomic variation in genetic markers of telomere length: a cross-sectional study of U.S. older adults. *EBioMedicine* **11**: 296–301. doi:10.1016/j.ebiom.2016.08.015

Hara T, Mizuguchi M, Fujii M, Nakamura M. 2015. Krüppel-like factor 2 represses transcription of the telomerase catalytic subunit human telomerase reverse transcriptase (hTERT) in human T cells. *J Biol Chem* **290**: 8758–8763. doi:10.1074/jbc.M114.610386

Harley CB, Futcher AB, Greider CW. 1990. Telomeres shorten during ageing of human fibroblasts. *Nature* **345**: 458–460. doi:10.1038/345458a0

Hastie ND, Dempster M, Dunlop MG, Thompson AM, Green DK, Allshire RC. 1990. Telomere reduction in human colorectal carcinoma and with ageing. *Nature* **346**: 866–868. doi:10.1038/346866a0

Hayward NK, Wilmott JS, Waddell N, Johansson PA, Field MA, Nones K, Patch AM, Kakavand H, Alexandrov LB, Burke H, et al. 2017. Whole-genome landscapes of major melanoma subtypes. *Nature* **545**: 175–180. doi:10.1038/nature22071

Hazelrigg T, Levis R, Rubin GM. 1984. Transformation of *white* locus DNA in *Drosophila*: dosage compensation, *zeste* interaction, and position effects. *Cell* **36**: 469–481. doi:10.1016/0092-8674(84)90240-X

He H, Li W, Comiskey DF, Liyanarachchi S, Nieminen TT, Wang Y, DeLap KE, Brock P, de la Chapelle A. 2020. A truncating germline mutation of TINF2 in individuals with thyroid cancer or melanoma results in longer yelomeres. *Thyroid* **30**: 204–213.

Heiss NS, Knight SW, Vulliamy TJ, Klauck SM, Wiemann S, Mason PJ, Poustka A, Dokal I. 1998. X-linked dyskeratosis congenita is caused by mutations in a highly conserved gene with putative nucleolar functions. *Nat Genet* **19**: 32–38. doi:10.1038/ng0598-32

Hemann MT, Strong MA, Hao LY, Greider CW. 2001. The shortest telomere, not average telomere length, is critical for cell viability and chromosome stability. *Cell* **107**: 67–77. doi:10.1016/S0092-8674(01)00504-9

Herbig U, Jobling WA, Chen BPC, Chen DJ, Sedivy JM. 2004. Telomere shortening triggers senescence of human cells through a pathway involving ATM, p53, and p21 (CIP1), but not p16(INK4a). *Mol Cell* **14**: 501–513. doi:10.1016/S1097-2765(04)00256-4

Hisatomi H, Ohyashiki K, Ohyashiki JH, Nagao K, Kanamaru T, Hirata H, Hibi N, Tsukada Y. 2003. Expression profile of a γ-deletion variant of the human telomerase reverse transcriptase gene. *Neoplasia* **5**: 193–197. doi:10.1016/S1476-5586(03)80051-9

Hiyama K, Hirai Y, Kyoizumi S, Akiyama M, Hiyama E, Piatyszek MA, Shay JW, Ishioka S, Yamakido M. 1995. Activation of telomerase in human lymphocytes and hematopoietic progenitor cells. *J Immunol* **155**: 3711–3715. doi:10.4049/jimmunol.155.8.3711

Hockemeyer D, Sfeir AJ, Shay JW, Wright WE, de Lange T. 2005. POT1 protects telomeres from a transient DNA damage response and determines how human chromosomes end. *EMBO J* **24**: 2667–2678. doi:10.1038/sj.emboj.7600733

Hockemeyer D, Daniels JP, Takai H, de Lange T. 2006. Recent expansion of the telomeric complex in rodents: two distinct POT1 proteins protect mouse telomeres. *Cell* **126**: 63–77. doi:10.1016/j.cell.2006.04.044

Horikawa I, Chiang YJ, Patterson T, Feigenbaum L, Leem SH, Michishita E, Larionov V, Hodes RJ, Barrett JC. 2005. Differential *cis*-regulation of human versus mouse *TERT* gene expression in vivo: identification of a human-specific repressive element. *Proc Natl Acad Sci* **102**: 18437–18442. doi:10.1073/pnas.0508964102

Horn S, Figl A, Rachakonda PS, Fischer C, Sucker A, Gast A, Kadel S, Moll I, Nagore E, Hemminki K, et al. 2013. *TERT* promoter mutations in familial and sporadic melanoma. *Science* **339**: 959–961. doi:10.1126/science.1230062

Hosen MI, Sheikh M, Zvereva M, Scelo G, Forey N, Durand G, Voegele C, Poustchi H, Khoshnia M, Roshandel G, et al. 2020. Urinary TERT promoter mutations are detectable up to 10 years prior to clinical diagnosis of bladder cancer: evidence from the Golestan Cohort Study. *EBioMedicine* **53**: 102643. doi:10.1016/j.ebiom.2020.102643

Hu BT, Lee SC, Marin E, Ryan DH, Insel RA. 1997. Telomerase is up-regulated in human germinal center B cells in vivo and can be re-expressed in memory B cells activated in vitro. *J Immunol* **159**: 1068–1071. doi:10.4049/jimmunol.159.3.1068

Huang FW, Hodis E, Xu MJ, Kryukov GV, Chin L, Garraway LA. 2013. Highly recurrent *TERT* promoter mutations in human melanoma. *Science* **339**: 957–959. doi:10.1126/science.1229259

Huang DS, Wang Z, He XJ, Diplas BH, Yang R, Killela PJ, Meng Q, Ye ZY, Wang W, Jiang XT, et al. 2015. Recurrent TERT promoter mutations identified in a large-scale study of multiple tumour types are associated with increased TERT expression and telomerase activation. *Eur J Cancer* **51:** 969–976. doi:10.1016/j.ejca.2015.03.010

Huffman KE, Levene SD, Tesmer VM, Shay JW, Wright WE. 2000. Telomere shortening is proportional to the size of the G-rich telomeric 3′-overhang. *J Biol Chem* **275:** 19719–19722. doi:10.1074/jbc.M002843200

Hunt SC, Chen W, Gardner JP, Kimura M, Srinivasan SR, Eckfeldt JH, Berenson GS, Aviv A. 2008. Leukocyte telomeres are longer in African Americans than in whites: the National Heart, Lung, and Blood Institute Family Heart Study and the Bogalusa Heart Study. *Aging Cell* **7:** 451–458. doi:10.1111/j.1474-9726.2008.00397.x

Hysek M, Paulsson JO, Jatta K, Shabo I, Stenman A, Höög A, Larsson C, Zedenius J, Juhlin CC. 2019. Clinical routine TERT promoter mutational screening of follicular thyroid tumors of uncertain malignant potential (FT-UMPs): a useful predictor of metastatic disease. *Cancers (Basel)* **11:** 1443. doi:10.3390/cancers11101443

Ishdorj G, Kost SEF, Beiggi S, Zang Y, Gibson SB, Johnston JB. 2017. A novel spliced variant of the TIN2 shelterin is present in chronic lymphocytic leukemia. *Leukemia Res* **59:** 66–74. doi:10.1016/j.leukres.2017.05.017

Jia W, Wang S, Horner JW, Wang N, Wang H, Gunther EJ, DePinho RA, Zhu J. 2011. A BAC transgenic reporter recapitulates in vivo regulation of human telomerase reverse transcriptase in development and tumorigenesis. *FASEB J* **25:** 979–989. doi:10.1096/fj.10-173989

Jung HY, Wang X, Jun S, Park JI. 2013. Dyrk2-associated EDD-DDB1-VprBP E3 ligase inhibits telomerase by TERT degradation. *J Biol Chem* **288:** 7252–7262. doi:10.1074/jbc.M112.416792

Kabir S, Hockemeyer D, de Lange T. 2014. TALEN gene knockouts reveal no requirement for the conserved human shelterin protein Rap1 in telomere protection and length regulation. *Cell Rep* **9:** 1273–1280. doi:10.1016/j.celrep.2014.10.014

Kaminker PG, Kim SH, Desprez PY, Campisi J. 2009. A novel form of the telomere-associated protein TIN2 localizes to the nuclear matrix. *Cell Cycle* **8:** 931–939. doi:10.4161/cc.8.6.7941

Kang SS, Kwon T, Kwon DY, Do SI. 1999. Akt protein kinase enhances human telomerase activity through phosphorylation of telomerase reverse transcriptase subunit. *J Biol Chem* **274:** 13085–13090. doi:10.1074/jbc.274.19.13085

Kang JU, Koo SH, Kwon KC, Park JW, Kim JM. 2008. Gain at chromosomal region 5p15.33, containing TERT, is the most frequent genetic event in early stages of non-small-cell lung cancer. *Cancer Genet Cytogenet* **182:** 1–11. doi:10.1016/j.cancergencyto.2007.12.004

Kapuria D, Ben-Yakov G, Ortolano R, Ho-Cho M, Kalchiem-Dekel O, Takyar V, Lingala S, Gara N, Tana M, Kim YJ, et al. 2019. The spectrum of hepatic involvement in patients with telomere disease. *Hepatology* **69:** 2579–2585. doi:10.1002/hep.30578

Karimian K, Groot A, Huso V, Kahidi R, Tan K-T, Sholes S, Keener R, McDyer JF, Alder JK, Li H, et al. 2024. Human telomere length is chromosome end-specific and conserved across individuals. *Science* **384:** 533–539.

Kaur P, Barnes R, Pan H, Detwiler AC, Liu M, Mahn C, Hall J, Messenger Z, You C, Piehler J, et al. 2021. TIN2 is an architectural protein that facilitates TRF2-mediated *trans*- and *cis*-interactions on telomeric DNA. *Nucleic Acids Res* **49:** 13000–13018. doi:10.1093/nar/gkab1142

Kawai-Kitahata F, Asahina Y, Tanaka S, Kakinuma S, Murakawa M, Nitta S, Watanabe T, Otani S, Taniguchi M, Goto F, et al. 2016. Comprehensive analyses of mutations and hepatitis B virus integration in hepatocellular carcinoma with clinicopathological features. *J Gastroenterol* **51:** 473–486. doi:10.1007/s00535-015-1126-4

Kelich J, Aramburu T, van der Vis JJ, Showe L, Kossenkov A, van der Smagt J, Massink M, Schoemaker A, Hennekam E, Veltkamp M, et al. 2022. Telomere dysfunction implicates POT1 in patients with idiopathic pulmonary fibrosis. *J Exp Med* **219:** e20211681. doi:10.1084/jem.20211681

Kermasson L, Churikov D, Awad A, Smoom R, Lainey E, Touzot F, Audebert-Bellanger S, Haro S, Roger L, Costa E, et al. 2022. Inherited human Apollo deficiency causes severe bone marrow failure and developmental defects. *Blood* **139:** 2427–2440. doi:10.1182/blood.2021010791

Kharbanda S, Kumar V, Dhar S, Pandey P, Chen C, Majumder P, Yuan ZM, Whang Y, Strauss W, Pandita TK, et al. 2000. Regulation of the hTERT telomerase catalytic subunit by the c-Abl tyrosine kinase. *Curr Biol* **10:** 568–575. doi:10.1016/S0960-9822(00)00483-8

Kilian A, Bowtell DD, Abud HE, Hime GR, Venter DJ, Keese PK, Duncan EL, Reddel RR, Jefferson RA. 1997. Isolation of a candidate human telomerase catalytic subunit gene, which reveals complex splicing patterns in different cell types. *Hum Mol Genet* **6:** 2011–2019. doi:10.1093/hmg/6.12.2011

Killela PJ, Reitman ZJ, Jiao Y, Bettegowda C, Agrawal N, Diaz LA, Friedman AH, Friedman H, Gallia GL, Giovanella BC, et al. 2013. *TERT* promoter mutations occur frequently in gliomas and a subset of tumors derived from cells with low rates of self-renewal. *Proc Natl Acad Sci* **110:** 6021–6026. doi:10.1073/pnas.1303607110

Kim NW, Piatyszek MA, Prowse KR, Harley CB, West MD, Ho PLC, Coviello GM, Wright WE, Weinrich SL, Shay JW. 1994. Specific association of human telomerase activity with immortal cells and cancer. *Science* **266:** 2011–2015. doi:10.1126/science.7605428

Kim SH, Kaminker P, Campisi J. 1999. TIN2, a new regulator of telomere length in human cells. *Nat Genet* **23:** 405–412. doi:10.1038/70508

Kim SH, Beausejour C, Davalos AR, Kaminker P, Heo SJ, Campisi J. 2004. TIN2 mediates functions of TRF2 at human telomeres. *J Biol Chem* **279:** 43799–43804. doi:10.1074/jbc.M408650200

Kim W, Ludlow AT, Min J, Robin JD, Stadler G, Mender I, Lai TP, Zhang N, Wright WE, Shay JW. 2016. Regulation of the human telomerase gene TERT by telomere position effect—over long distances (TPE-OLD): implications for aging and cancer. *PLoS Biol* **14:** e2000016. doi:10.1371/journal.pbio.2000016

Kim W, Hennick K, Johnson J, Finnerty B, Choo S, Short SB, Drubin C, Forster R, McMaster ML, Hockemeyer D. 2021. Cancer-associated POT1 mutations lead to telomere elongation without induction of a DNA damage response. *EMBO J* **40:** e107346. doi:10.15252/embj.2020107346

Kim JJ, Sayed ME, Ahn A, Slusher AL, Ying JY, Ludlow AT. 2023. Dynamics of TERT regulation via alternative splicing in stem cells and cancer cells. *PLoS One* **18:** e0289327. doi:10.1371/journal.pone.0289327

Kipling D, Cooke HJ. 1990. Hypervariable ultra-long telomeres in mice. *Nature* **347:** 400–402. doi:10.1038/347400a0

Kocak H, Ballew BJ, Bisht K, Eggebeen R, Hicks BD, Suman S, O'Neil A, Giri N, Maillard I, Alter BP, et al. 2014. Hoyeraal–Hreidarsson syndrome caused by a germline mutation in the TEL patch of the telomere protein TPP1. *Genes Dev* **28:** 2090–2102. doi:10.1101/gad.248567.114

Kochman R, Ba I, Yates M, Pirabakaran V, Gourmelon F, Churikov D, Laffaille M, Kermasson L, Hamelin C, Marois I, et al. 2024. Heterozygous RPA2 variant as a novel genetic cause of telomere biology disorders. *Genes Dev* **38:** 755–771.

Koivuluoma S, Vorimo S, Mattila TM, Tervasmäki A, Kumpula T, Kuismin O, Winqvist R, Moilanen J, Mantere T, Pylkäs K. 2023. Truncating TINF2 p.Tyr312Ter variant and inherited breast cancer susceptibility. *Fam Cancer* **22:** 13–17. doi:10.1007/s10689-022-00295-z

Kratz K, de Lange T. 2018. Protection of telomeres 1 proteins POT1a and POT1b can repress ATR signaling by RPA exclusion, but binding to CST limits ATR repression by POT1b. *J Biol Chem* **293:** 14384–14392. doi:10.1074/jbc.RA118.004598

Kurz DJ, Kloeckener-Gruissem B, Akhmedov A, Eberli FR, Bühler I, Berger W, Bertel O, Lüscher TF. 2006. Degenerative aortic valve stenosis, but not coronary disease, is associated with shorter telomere length in the elderly. *Arterioscler Thromb Vasc Biol* **26:** e114–e117. doi:10.1161/01.ATV.0000222961.24912.6

Kyo S, Takakura M, Taira T, Kanaya T, Itoh H, Yutsudo M, Ariga H, Inoue M. 2000. Sp1 cooperates with c-Myc to activate transcription of the human telomerase reverse transcriptase gene (hTERT). *Nucleic Acids Res* **28:** 669–677. doi:10.1093/nar/28.3.669

Lacerte A, Korah J, Roy M, Yang XJ, Lemay S, Lebrun JJ. 2008. Transforming growth factor-β inhibits telomerase through SMAD3 and E2F transcription factors. *Cell Signal* **20:** 50–59. doi:10.1016/j.cellsig.2007.08.012

Lee JH, Khadka P, Baek SH, Chung IK. 2010. CHIP promotes human telomerase reverse transcriptase degradation and negatively regulates telomerase activity. *J Biol Chem* **285:** 42033–42045. doi:10.1074/jbc.M110.149831

Lee SS, Bohrson C, Pike AM, Wheelan SJ, Greider CW. 2015. ATM kinase is required for telomere elongation in mouse and human cells. *Cell Rep* **13:** 1623–1632. doi:10.1016/j.celrep.2015.10.035

Lee DD, Leão R, Komosa M, Gallo M, Zhang CH, Lipman T, Remke M, Heidari A, Nunes NM, Apolónio JD, et al. 2019. DNA hypermethylation within TERT promoter upregulates TERT expression in cancer. *J Clin Invest* **129:** 223–229. doi:10.1172/JCI121303

Lee DD, Komosa M, Nunes NM, Tabori U. 2020. DNA methylation of the TERT promoter and its impact on human cancer. *Curr Opin Genet Dev* **60:** 17–24. doi:10.1016/j.gde.2020.02.003

Le Guen T, Jullien L, Touzot F, Schertzer M, Gaillard L, Perderiset M, Carpentier W, Nitschke P, Picard C, Couil-lault G, et al. 2013. Human RTEL1 deficiency causes Hoyeraal–Hreidarsson syndrome with short telomeres and genome instability. *Hum Mol Genet* **22:** 3239–3249. doi:10.1093/hmg/ddt178

Lei M, Podell ER, Cech TR. 2004. Structure of human POT1 bound to telomeric single-stranded DNA provides a model for chromosome end-protection. *Nat Struct Mol Biol* **11:** 1223–1229. doi:10.1038/nsmb867

Lewis PW, Elsaesser SJ, Noh K-M, Stadler SC, Allis CD. 2010. Daxx is an H3.3-specific histone chaperone and cooperates with ATRX in replication-independent chromatin assembly at telomeres. *Proc Natl Acad Sci* **107:** 14075–14080. doi:10.1073/pnas.1008850107

Li Y, Zhou QL, Sun W, Chandrasekharan P, Cheng HS, Ying Z, Lakshmanan M, Raju A, Tenen DG, Cheng SY, et al. 2015. Non-canonical NF-κB signalling and ETS1/2 cooperatively drive C250T mutant TERT promoter activation. *Nat Cell Biol* **17:** 1327–1338. doi:10.1038/ncb3240

Li C, Stoma S, Lotta LA, Warner S, Albrecht E, Allione A, Arp PP, Broer L, Buxton JL, Da Silva Couto Alves A, et al. 2020a. Genome-wide association analysis in humans links nucleotide metabolism to leukocyte telomere length. *Am J Hum Genet* **106:** 389–404. doi:10.1016/j.ajhg.2020.02.006

Li X, Qian X, Wang B, Xia Y, Zheng Y, Du L, Xu D, Xing D, DePinho RA, Lu Z. 2020b. Programmable base editing of mutated TERT promoter inhibits brain tumour growth. *Nat Cell Biol* **22:** 282–288. doi:10.1038/s41556-020-0471-6

Lim CJ, Cech TR. 2021. Shaping human telomeres: from shelterin and CST complexes to telomeric chromatin organization. *Nat Rev Mol Cell Biol* **22:** 283–298. doi:10.1038/s41580-021-00328-y

Lim CJ, Zaug AJ, Kim HJ, Cech TR. 2017. Reconstitution of human shelterin complexes reveals unexpected stoichiometry and dual pathways to enhance telomerase processivity. *Nat Commun* **8:** 1075. doi:10.1038/s41467-017-01313-w

Lin SY, Elledge SJ. 2003. Multiple tumor suppressor pathways negatively regulate telomerase. *Cell* **113:** 881–889. doi:10.1016/S0092-8674(03)00430-6

Lin J, Epel E. 2022. Stress and telomere shortening: insights from cellular mechanisms. *Ageing Res Rev* **73:** 101507. doi:10.1016/j.arr.2021.101507

Lin TT, Letsolo BT, Jones RE, Rowson J, Pratt G, Hewamana S, Fegan C, Pepper C, Baird DM. 2010. Telomere dysfunction and fusion during the progression of chronic lymphocytic leukemia: evidence for a telomere crisis. *Blood* **116:** 1899–1907. doi:10.1182/blood-2010-02-272104

Lin J, Countryman P, Buncher N, Kaur P, Longjiang E, Zhang Y, Gibson G, You C, Watkins SC, Piehler J, et al. 2014. TRF1 and TRF2 use different mechanisms to find telomeric DNA but share a novel mechanism to search for protein partners at telomeres. *Nucleic Acids Res* **42:** 2493–2504. doi:10.1093/nar/gkt1132

Lin J, Cheon J, Brown R, Coccia M, Puterman E, Aschbacher K, Sinclair E, Epel E, Blackburn EH. 2016. Systematic and cell type-specific telomere length changes in subsets of lymphocytes. *J Immunol Res* **2016:** e5371050. doi:10.1155/2016/5371050

Lin S, Nascimento EM, Gajera CR, Chen L, Neuhöfer P, Garbuzov A, Wang S, Artandi SE. 2018. Distributed hepatocytes expressing telomerase repopulate the liver in homeostasis and injury. *Nature* **556:** 244–248. doi:10 .1038/s41586-018-0004-7

Liu Y, Snow BE, Hande MP, Yeung D, Erdmann NJ, Wakeham A, Itie A, Siderovski DP, Lansdorp PM, Robinson MO, et al. 2000. The telomerase reverse transcriptase is limiting and necessary for telomerase function in vivo. *Curr Biol* **10:** 1459–1462. doi:10.1016/S0960-9822(00) 00805-8

Liu X, Wu G, Shan Y, Hartmann C, Von Deimling A, Xing M. 2013. Highly prevalent *TERT* promoter mutations in bladder cancer and glioblastoma. *Cell Cycle* **12:** 1637–1638. doi:10.4161/cc.24662

Liu X, Wang Y, Chang G, Wang F, Wang F, Geng X. 2017. Alternative splicing of hTERT pre-mRNA: a potential strategy for the regulation of telomerase activity. *Int J Mol Sci* **18:** 567. doi:10.3390/ijms18030567

Liu B, He Y, Wang Y, Song H, Zhou ZH, Feigon J. 2022. Structure of active human telomerase with telomere shelterin protein TPP1. *Nature* **604:** 578–583. doi:10.1038/ s41586-022-04582-8

Liu M, Zhang Y, Jian Y, Gu L, Zhang D, Zhou H, Wang Y, Xu ZX. 2024. The regulations of telomerase reverse transcriptase (TERT) in cancer. *Cell Death Dis* **15:** 1–12. doi:10 .1038/s41419-023-06384-w

Loayza D, de Lange T. 2003. POT1 as a terminal transducer of TRF1 telomere length control. *Nature* **423:** 1013–1018. doi:10.1038/nature01688

Loayza D, Parsons H, Donigian J, Hoke K, de Lange T. 2004. DNA binding features of human POT1. *J Biol Chem* **279:** 13241–13248. doi:10.1074/jbc.M312309200

Lorbeer FK, Hockemeyer D. 2020. TERT promoter mutations and telomeres during tumorigenesis. *Curr Opin Genet Dev* **60:** 56–62. doi:10.1016/j.gde.2020.02.001

Lorbeer FK, Rieser G, Goel A, Wang M, Oh A, Yeh I, Bastian BC, Hockemeyer D. 2024. Distinct senescence mechanisms restrain progression of dysplastic nevi. *PNAS Nexus* **3:** pgae041. doi:10.1093/pnasnexus/pgae041

Lototska L, Yue J, Li J, Giraud-Panis M, Songyang Z, Royle NJ, Liti G, Ye J, Gilson E, Mendez-Bermudez A. 2020. Human RAP 1 specifically protects telomeres of senescent cells from DNA damage. *EMBO Rep* **21:** e49076. doi:10 .15252/embr.201949076

Lu R, Pickett HA. 2022. Telomeric replication stress: the beginning and the end for alternative lengthening of telomeres cancers. *Open Biol* **12:** 220011. doi:10.1098/rsob .220011

Ly K, Walker C, Berry S, Snell R, Marks E, Thayer Z, Atatoa-Carr P, Morton S. 2019. Telomere length in early childhood is associated with sex and ethnicity. *Sci Rep* **9:** 10359. doi:10.1038/s41598-019-46338-x

Ma X, Gong R, Wang R, Pan Y, Cai D, Pan B, Li Y, Xiang J, Li H, Zhang J, et al. 2014. Recurrent TERT promoter mutations in non-small cell lung cancers. *Lung Cancer* **86:** 369–373. doi:10.1016/j.lungcan.2014.10.009

Maciejowski J, Li Y, Bosco N, Campbell PJ, de Lange T. 2015. Chromothripsis and kataegis induced by telomere crisis. *Cell* **163:** 1641–1654. doi:10.1016/j.cell.2015.11.054

Mangino M, Christiansen L, Stone R, Hunt SC, Horvath K, Eisenberg DTA, Kimura M, Petersen I, Kark JD, Herbig U, et al. 2015. DCAF4, a novel gene associated with leu-

cocyte telomere length. *J Med Genet* **52:** 157–162. doi:10 .1136/jmedgenet-2014-102681

Mannherz W, Agarwal S. 2023. Thymidine nucleotide metabolism controls human telomere length. *Nat Genet* **55:** 568–580. doi:10.1038/s41588-023-01339-5

Margalef P, Kotsantis P, Borel V, Bellelli R, Panier S, Boulton SJ. 2018. Stabilization of reversed replication forks by telomerase drives telomere catastrophe. *Cell* **172:** 439–453.e14. doi:10.1016/j.cell.2017.11.047

Marion RM, Strati K, Li H, Tejera A, Schoeftner S, Ortega S, Serrano M, Blasco MA. 2009. Telomeres acquire embryonic stem cell characteristics in induced pluripotent stem cells. *Cell Stem Cell* **4:** 141–154. doi:10.1016/j.stem.2008 .12.010

Martin A, Schabort J, Bartke-Croughan R, Tran S, Preetham A, Lu R, Ho R, Gao J, Jenkins S, Boyle J, et al. 2025. Active telomere elongation by a subclass of cancer-associated POT1 mutations. *Genes Dev* **39:** 445–462.

Mathur MB, Epel E, Kind S, Desai M, Parks CG, Sandler DP, Khazeni N. 2016. Perceived stress and telomere length: a systematic review, meta-analysis, and methodologic considerations for advancing the field. *Brain Behav Immun* **54:** 158–169. doi:10.1016/j.bbi.2016.02.002

Mattern KA, Swiggers SJJ, Nigg AL, Löwenberg B, Houtsmuller AB, Zijlmans JMJM. 2004. Dynamics of protein binding to telomeres in living cells: implications for telomere structure and function. *Mol Cell Biol* **24:** 5587–5594. doi:10.1128/MCB.24.12.5587-5594.2004

McElligott R, Wellinger RJ. 1997. The terminal DNA structure of mammalian chromosomes. *EMBO J* **16:** 3705–3714. doi:10.1093/emboj/16.12.3705

McKerlie M, Lin S, Zhu XD. 2012. ATM regulates proteasome-dependent subnuclear localization of TRF1, which is important for telomere maintenance. *Nucleic Acids Res* **40:** 3975–3989. doi:10.1093/nar/gks035

Meeran SM, Patel SN, Tollefsbol TO. 2010. Sulforaphane causes epigenetic repression of hTERT expression in human breast cancer cell lines. *PLoS One* **5:** e11457. doi:10 .1371/journal.pone.0011457

Mitchell TJ, Turajlic S, Rowan A, Nicol D, Farmery JHR, O'Brien T, Martincorena I, Tarpey P, Angelopoulos N, Yates LR, et al. 2018. Timing the landmark events in the evolution of clear cell renal cell cancer: TRACERx renal. *Cell* **173:** 611–623.e17. doi:10.1016/j.cell.2018.02.020

Miyake Y, Nakamura M, Nabetani A, Shimamura S, Tamura M, Yonehara S, Saito M, Ishikawa F. 2009. RPA-like mammalian Ctc1-Stn1-Ten1 complex binds to single-stranded DNA and protects telomeres independently of the Pot1 pathway. *Mol Cell* **36:** 193–206. doi:10.1016/j.molcel.2009 .08.009

Morrison SJ, Weissman IL. 1994. The long-term repopulating subset of hematopoietic stem cells is deterministic and isolatable by phenotype. *Immunity* **1:** 661–673. doi:10 .1016/1074-7613(94)90037-X

Morrison SJ, Prowse KR, Ho P, Weissman IL. 1996. Telomerase activity in hematopoietic cells is associated with self-renewal potential. *Immunity* **5:** 207–216. doi:10.1016/ S1074-7613(00)80316-7

Mosrati MA, Willander K, Falk IJ, Hermanson M, Höglund M, Stockelberg D, Wei Y, Lotfi K, Söderkvist P. 2015. Association between TERT promoter polymorphisms and acute myeloid leukemia risk and prognosis. *Oncotarget* **6:** 25109–25120. doi:10.18632/oncotarget.4668

Moyzis RK, Buckingham JM, Cram LS, Dani M, Deaven LL, Jones MD, Meyne J, Ratliff RL, Wu JR. 1988. A highly conserved repetitive DNA sequence, (TTAGGG)*n*, present at the telomeres of human chromosomes. *Proc Natl Acad Scis* **85:** 6622–6626. doi:10.1073/pnas.85.18.6622

Nandakumar J, Podell ER, Cech TR. 2010. How telomeric protein POT1 avoids RNA to achieve specificity for single-stranded DNA. *Proc Natl Acad Sci* **107:** 651–656. doi:10.1073/pnas.0911099107

Nandakumar J, Bell CF, Weidenfeld I, Zaug AJ, Leinwand LA, Cech TR. 2012. The TEL patch of telomere protein TPP1 mediates telomerase recruitment and processivity. *Nature* **492:** 285–289. doi:10.1038/nature11648

Nault JC, Calderaro J, Di Tommaso L, Balabaud C, Zafrani ES, Bioulac-Sage P, Roncalli M, Zucman-Rossi J. 2014. Telomerase reverse transcriptase promoter mutation is an early somatic genetic alteration in the transformation of premalignant nodules in hepatocellular carcinoma on cirrhosis. *Hepatology* **60:** 1983–1992. doi:10.1002/hep.27372

Needham BL, Salerno S, Roberts E, Boss J, Allgood KL, Mukherjee B. 2019. Do black/white differences in telomere length depend on socioeconomic status? *Biodemography Soc Biol* **65:** 287–312. doi:10.1080/19485565.2020.1765734

Nergadze SG, Farnung BO, Wischnewski H, Khoriauli L, Vitelli V, Chawla R, Giulotto E, Azzalin CM. 2009. CpG-island promoters drive transcription of human telomeres. *RNA* **15:** 2186–2194. doi:10.1261/rna.1748309

Njajou OT, Cawthon RM, Damcott CM, Wu SH, Ott S, Garant MJ, Blackburn EH, Mitchell BD, Shuldiner AR, Hsueh WC. 2007. Telomere length is paternally inherited and is associated with parental lifespan. *Proc Natl Acad Sci* **104:** 12135–12139. doi:10.1073/pnas.0702703104

Nofrini V, Matteucci C, Pellanera F, Gorello P, Di Giacomo D, Lema Fernandez AG, Nardelli C, Iannotti T, Brandimarte L, Arniani S, et al. 2021. Activating somatic and germline TERT promoter variants in myeloid malignancies. *Leukemia* **35:** 274–278. doi:10.1038/s41375-020-0837-6

Oh ST, Kyo S, Laimins LA. 2001. Telomerase activation by human papillomavirus type 16 E6 protein: induction of human telomerase reverse transcriptase expression through Myc and GC-rich Sp1 binding sites. *J Virol* **75:** 5559–5566. doi:10.1128/JVI.75.12.5559-5566.2001

Olins DE, Olins AL, Cacheiro LH, Tan EM. 1989. Proliferating cell nuclear antigen/cyclin in the ciliate *Euplotes eurystomus*: localization in the replication band and in micronuclei. *J Cell Biol* **109:** 1399–1410. doi:10.1083/jcb.109.4.1399

Olivier M, Charbonnel C, Amiard S, White CI, Gallego ME. 2018. RAD51 and RTEL1 compensate telomere loss in the absence of telomerase. *Nucleic Acids Res* **46:** 2432–2445. doi:10.1093/nar/gkx1322

Olovnikov AM. 1996. Telomeres, telomerase, and aging: origin of the theory. *Exp Gerontol* **31:** 443–448. doi:10.1016/0531-5565(96)00005-8

Olympios N, Gilard V, Marguet F, Clatot F, Di Fiore F, Fontanilles M. 2021. TERT promoter alterations in glioblas-toma: a systematic review. *Cancers (Basel)* **13:** 1147. doi:10.3390/cancers13051147

Orkin SH, Zon LI. 2008. Hematopoiesis: an evolving paradigm for stem cell biology. *Cell* **132:** 631–644. doi:10.1016/j.cell.2008.01.025

Ottaviani A, Gilson E, Magdinier F. 2008. Telomeric position effect: from the yeast paradigm to human pathologies? *Biochimie* **90:** 93–107. doi:10.1016/j.biochi.2007.07.022

Ouellette MM, Liao M, Herbert BS, Johnson M, Holt SE, Liss HS, Shay JW, Wright WE. 2000. Subsenescent telomere lengths in fibroblasts immortalized by limiting amounts of telomerase. *J Biol Chem* **275:** 10072–10076. doi:10.1074/jbc.275.14.10072

Pan H, Kaur P, Barnes R, Detwiler AC, Sanford SL, Liu M, Xu P, Mahn C, Tang Q, Hao P, et al. 2021. Structure, dynamics, and regulation of TRF1-TIN2-mediated *trans*- and *cis*-interactions on telomeric DNA. *J Biol Chem* **297:** 101080. doi:10.1016/j.jbc.2021.101080

Paterlini-Bréchot P, Saigo K, Murakami Y, Chami M, Gozuacik D, Mugnier C, Lagorce D, Bréchot C. 2003. Hepatitis B virus-related insertional mutagenesis occurs frequently in human liver cancers and recurrently targets human telomerase gene. *Oncogene* **22:** 3911–3916. doi:10.1038/sj.onc.1206492

Patrick EM, Slivka JD, Payne B, Comstock MJ, Schmidt JC. 2020. Observation of processive telomerase catalysis using high-resolution optical tweezers. *Nat Chem Biol* **16:** 801–809. doi:10.1038/s41589-020-0478-0

Pech MF, Garbuzov A, Hasegawa K, Sukhwani M, Zhang RJ, Benayoun BA, Brockman SA, Lin S, Brunet A, Orwig KE, et al. 2015. High telomerase is a hallmark of undifferentiated spermatogonia and is required for maintenance of male germline stem cells. *Genes Dev* **29:** 2420–2434. doi:10.1101/gad.271783.115

Peifer M, Hertwig F, Roels F, Dreidax D, Gartlgruber M, Menon R, Krämer A, Roncaioli JL, Sand F, Heuckmann JM, et al. 2015. Telomerase activation by genomic rearrangements in high-risk neuroblastoma. *Nature* **526:** 700–704. doi:10.1038/nature14980

Penev A, Bazley A, Shen M, Boeke JD, Savage SA, Sfeir A. 2021. Alternative splicing is a developmental switch for hTERT expression. *Mol Cell* **81:** 2349–2360.e6. doi:10.1016/j.molcel.2021.03.033

Pennarun G, Hoffschir F, Revaud D, Granotier C, Gauthier LR, Mailliet P, Biard DS, Boussin FD. 2010. ATR contributes to telomere maintenance in human cells. *Nucleic Acids Res* **38:** 2955–2963. doi:10.1093/nar/gkp1248

Pfeiffer V, Lingner J. 2012. TERRA promotes telomere shortening through exonuclease 1–mediated resection of chromosome ends. *PLoS Genet* **8:** e1002747. doi:10.1371/journal.pgen.1002747

Pike AM, Strong MA, Paul J, Ouyang T, Greider CW. 2019. TIN2 functions with TPP1/POT1 to stimulate telomerase processivity. *Mol Cell Biol* **39:** e00593. doi:10.1128/MCB.00593-18

Porro A, Feuerhahn S, Lingner J. 2014. TERRA-reinforced association of LSD1 with MRE11 promotes processing of uncapped telomeres. *Cell Rep* **6:** 765–776. doi:10.1016/j.celrep.2014.01.022

Prescott J, Kraft P, Chasman DI, Savage SA, Mirabello L, Berndt SI, Weissfeld JL, Han J, Hayes RB, Chanock SJ, et al. 2011. Genome-wide association study of relative

telomere length. *PLoS One* **6:** e19635. doi:10.1371/journal
.pone.0019635

Prowse KR, Greider CW. 1995. Developmental and tissue-
specific regulation of mouse telomerase and telomere
length. *Proc Natl Acad Sci* **92:** 4818–4822. doi:10.1073/
pnas.92.11.4818

Ramlee MK, Wang J, Toh WX, Li S. 2016. Transcription
regulation of the human telomerase reverse transcriptase
(hTERT) gene. *Genes (Basel)* **7:** 50. doi:10.3390/genes
7080050

Redon S, Reichenbach P, Lingner J. 2010. The non-coding
RNA TERRA is a natural ligand and direct inhibitor of
human telomerase. *Nucleic Acids Res* **38:** 5797–5806.
doi:10.1093/nar/gkq296

Renaud S, Loukinov D, Abdullaev Z, Guilleret I, Bosman FT,
Lobanenkov V, Benhattar J. 2007. Dual role of DNA
methylation inside and outside of CTCF-binding regions
in the transcriptional regulation of the telomerase hTERT
gene. *Nucleic Acids Res* **35:** 1245–1256. doi:10.1093/nar/
gkl1125

Revy P, Kannengiesser C, Bertuch AA. 2023. Genetics of
human telomere biology disorders. *Nat Rev Genet* **24:**
86–108. doi:10.1038/s41576-022-00527-z

Rheinbay E, Nielsen MM, Abascal F, Wala JA, Shapira O,
Tiao G, Hornshøj H, Hess JM, Juul RI, Lin Z, et al. 2020.
Analyses of non-coding somatic drivers in 2,658 cancer
whole genomes. *Nature* **578:** 102–111. doi:10.1038/
s41586-020-1965-x

Ritz JM, Kühle O, Riethdorf S, Sipos B, Deppert W, Englert
C, Günes C. 2005. A novel transgenic mouse model re-
veals humanlike regulation of an 8-kbp human *TERT*
gene promoter fragment in normal and tumor tissues.
*Cancer Res* **65:** 1187–1196. doi:10.1158/0008-5472
.CAN-04-3046

Roake CM, Chen L, Chakravarthy AL, Ferrell JE, Raffa GD,
Artandi SE. 2019. Disruption of telomerase RNA maturation kinetics precipitates disease. *Mol Cell* **74:** 688–700.
e3. doi:10.1016/j.molcel.2019.02.033

Robertson JD, Testa NG, Russell NH, Jackson G, Parker AN,
Milligan DW, Stainer C, Chakrabarti S, Dougal M, Cho-
pra R. 2001. Accelerated telomere shortening following
allogeneic transplantation is independent of the cell
source and occurs within the first-year posttransplant.
*Bone Marrow Transplant* **27:** 1283–1286. doi:10.1038/sj
.bmt.1703069

Robin JD, Ludlow AT, Batten K, Magdinier F, Stadler G,
Wagner KR, Shay JW, Wright WE. 2014. Telomere position effect: regulation of gene expression with progressive
telomere shortening over long distances. *Genes Dev* **28:**
2464–2476. doi:10.1101/gad.251041.114

Robles-Espinoza CD, Harland M, Ramsay AJ, Aoude LG,
Quesada V, Ding Z, Pooley KA, Pritchard AL, Tiffen
JC, Petljak M, et al. 2014. POT1 loss-of-function variants
predispose to familial melanoma. *Nature Genet* **46:** 478–
481. doi:10.1038/ng.2947

Roger L, Jones RE, Heppel NH, Williams GT, Sampson JR,
Baird DM. 2013. Extensive telomere erosion in the initiation of colorectal adenomas and its association with
chromosomal instability. *J Natl Cancer Inst* **105:** 1202–
1211. doi:10.1093/jnci/djt191

Rosmarin AG, Resendes KK, Yang Z, McMillan JN, Fleming
SL. 2004. GA-binding protein transcription factor: a review of GABP as an integrator of intracellular signaling
and protein–protein interactions. *Blood Cell Mol Dis* **32:**
143–154. doi:10.1016/j.bcmd.2003.09.005

Rowland TJ, Bonham AJ, Cech TR. 2020. Allele-specific
proximal promoter hypomethylation of the telomerase
reverse transcriptase gene (*TERT*) associates with *TERT*
expression in multiple cancers. *Mol Oncol* **14:** 2358–2374.
doi:10.1002/1878-0261.12786

Rufer N, Dragowska W, Thornbury G, Roosnek E, Lansdorp
PM. 1998. Telomere length dynamics in human lympho-
cyte subpopulations measured by flow cytometry. *Nat
Biotechnol* **16:** 743–747. doi:10.1038/nbt0898-743

Sæbøe-Larssen S, Fossberg E, Gaudernack G. 2006. Characterization of novel alternative splicing sites in human telomerase reverse transcriptase (hTERT): analysis of expression and mutual correlation in mRNA isoforms
from normal and tumour tissues. *BMC Mol Biol* **7:** 26.
doi:10.1186/1471-2199-7-26

Sanchez SE, Gu Y, Wang Y, Golla A, Martin A, Shomali W,
Hockemeyer D, Savage SA, Artandi SE. 2024. Digital telomere measurement by long-read sequencing distinguishes healthy aging from disease. *Nat Commun* **15:** 5148.

Sarek G, Vannier JB, Panier S, Petrini JHJ, Boulton SJ. 2015.
TRF2 recruits RTEL1 to telomeres in S phase to promote
T-loop unwinding. *Mol Cell* **57:** 622–635. doi:10.1016/j
.molcel.2014.12.024

Sasa G, Ribes-Zamora A, Nelson N, Bertuch A. 2012. Three
novel truncating TINF2 mutations causing severe dyskeratosis congenita in early childhood. *Clin Genet* **81:** 470–
478.

* Savage SA. 2025. Telomeres and human disease. *Cold Spring
Harb Perspect Biol* doi:10.1101/cshperspect.a041684

Savage SA, Bertuch AA. 2010. The genetics and clinical
manifestations of telomere biology disorders. *Genet
Med* **12:** 753–764. doi:10.1097/GIM.0b013e3181f415b5

Savage SA, Giri N, Baerlocher GM, Orr N, Lansdorp PM,
Alter BP. 2008. TINF2, a component of the shelterin telomere protection complex, is mutated in dyskeratosis congenita. *Am J Hum Genet* **82:** 501–509. doi:10.1016/j.ajhg
.2007.10.004

Savage SA, Bertuch AA; Team Telomere and the Clinical
Care Consortium for Telomere-Associated Ailments
(CCCTAA). 2025. Different phenotypes with different
endings—telomere biology disorders and cancer predisposition with long telomeres. *Br J Haematol* **206:** 69–73.
doi:10.1111/bjh.19851

Schmidt JC, Cech TR. 2015. Human telomerase: biogenesis,
trafficking, recruitment, and activation. *Genes Dev* **29:**
1095–1105. doi:10.1101/gad.263863.115

Schmidt JC, Dalby AB, Cech TR. 2014. Identification of
human TERT elements necessary for telomerase recruitment to telomeres. *eLife* **3:** e03563. doi:10.7554/eLife
.03563

Schmidt JC, Zaug AJ, Cech TR. 2016. Live cell imaging reveals the dynamics of telomerase recruitment to telomeres. *Cell* **166:** 1188–1197.e9. doi:10.1016/j.cell.2016
.07.033

Schmidt TT, Tyer C, Rughani P, Haggblom C, Jones JR, Dai
X, Frazer KA, Gage FH, Juul S, Hickey S, et al. 2024. High
resolution long-read telomere sequencing reveals dynamic mechanisms in aging and cancer. *Nat Commun* **15:**
5149.

Schmutz I, Mensenkamp AR, Takai KK, Haadsma M, Spruijt L, De Voer RM, Choo SS, Lorbeer FK, Van Grinsven EJ, Hockemeyer D, et al. 2020. Tinf2 is a haploinsufficient tumor suppressor that limits telomere length. *eLife* **9:** 1–20. doi:10.7554/eLife.61235

Schoeftner S, Blasco MA. 2008. Developmentally regulated transcription of mammalian telomeres by DNA-dependent RNA polymerase II. *Nat Cell Biol* **10:** 228–236. doi:10.1038/ncb1685

Schratz KE, Flasch DA, Atik CC, Cosner ZL, Blackford AL, Yang W, Gable DL, Vellanki PJ, Xiang Z, Gaysinskaya V, et al. 2023. T cell immune deficiency rather than chromosome instability predisposes patients with short telomere syndromes to squamous cancers. *Cancer Cell* **41:** 807–817.e6. doi:10.1016/j.ccell.2023.03.005

Seimiya H, Sawada H, Muramatsu Y, Shimizu M, Ohko K, Yamane K, Tsuruo T. 2000. Involvement of 14-3-3 proteins in nuclear localization of telomerase. *EMBO J* **19:** 2652–2661. doi:10.1093/emboj/19.11.2652

Sekne Z, Ghanim GE, van Roon AMM, Nguyen THD. 2022. Structural basis of human telomerase recruitment by TPP1-POT1. *Science* **375:** 1173–1176. doi:10.1126/science.abn6840

Sexton AN, Regalado SG, Lai CS, Cost GJ, O'Neil CM, Urnov FD, Gregory PD, Jaenisch R, Collins K, Hockemeyer D. 2014. Genetic and molecular identification of three human TPP1 functions in telomerase action: recruitment, activation, and homeostasis set point regulation. *Genes Dev* **28:** 1885–1899. doi:10.1101/gad.246819.114

Sfeir A, Kosiyatrakul ST, Hockemeyer D, MacRae SL, Karlseder J, Schildkraut CL, de Lange T. 2009. Mammalian telomeres resemble fragile sites and require TRF1 for efficient replication. *Cell* **138:** 90–103. doi:10.1016/j.cell.2009.06.021

Shain AH, Yeh I, Kovalyshyn I, Sriharan A, Talevich E, Gagnon A, Dummer R, North J, Pincus L, Ruben B, et al. 2015. The genetic evolution of melanoma from precursor lesions. *New Engl J Med* **373:** 1926–1936. doi:10.1056/NEJMoa1502583

Sharma R, Sahoo SS, Honda M, Granger SL, Goodings C, Sanchez L, Künstner A, Busch H, Beier F, Pruett-Miller SM, et al. 2022. Gain-of-function mutations in RPA1 cause a syndrome with short telomeres and somatic genetic rescue. *Blood* **139:** 1039–1051. doi:10.1182/blood.2021011980

Sharrocks AD. 2001. The ETS-domain transcription factor family. *Nat Rev Mol Cell Biol* **2:** 827–837. doi:10.1038/35099076

Shay JW, Bacchetti S. 1997. A survey of telomerase activity in human cancer. *Eur J Cancer* **33:** 787–791. doi:10.1016/S0959-8049(97)00062-2

Shen E, Xiu J, López GY, Bentley R, Jalali A, Heimberger AB, Bainbridge MN, Bondy ML, Walsh KM. 2020. *POT1* mutation spectrum in tumor types commonly diagnosed among *POT1*-associated hereditary cancer syndrome families. *J Med Genet* **57:** 664–670. doi:10.1136/jmedgenet-2019-106657

Shieikh S, Jack A, Saurabh A, Mustafa G, Kodikara SG, Gyawali P, Hoque ME, Pressé S, Yildiz A, Balci H. 2022. Shelterin reduces the accessibility of telomeric overhangs. *Nucleic Acids Res* **50:** 12885–12895. doi:10.1093/nar/gkac1176

Shin KH, Kang MK, Dicterow E, Park NH. 2003. Hypermethylation of the hTERT promoter inhibits the expression of telomerase activity in normal oral fibroblasts and senescent normal oral keratinocytes. *Br J Cancer* **89:** 1473–1478. doi:10.1038/sj.bjc.6601291

Smogorzewska A, van Steensel B, Bianchi A, Oelmann S, Schaefer MR, Schnapp G, de Lange T. 2000. Control of human telomere length by TRF1 and TRF2. *Mol Cell Biol* **20:** 1659–1668. doi:10.1128/MCB.20.5.1659-1668.2000

Smoom R, May CL, Ortiz V, Tigue M, Kolev HM, Rowe M, Reizel Y, Morgan A, Egyes N, Lichtental D, et al. 2023. Telomouse—a mouse model with human-length telomeres generated by a single amino acid change in RTEL1. *Nat Commun* **14:** 6708. doi:10.1038/s41467-023-42534-6

Soder AI, Hoare SF, Muir S, Going JJ, Parkinson EK, Keith WN. 1997. Amplification, increased dosage and in situ expression of the telomerase RNA gene in human cancer. *Oncogene* **14:** 1013–1021.

Speedy HE, Kinnersley B, Chubb D, Broderick P, Law PJ, Litchfield K, Jayne S, Dyer MJS, Dearden C, Follows GA, et al. 2016. Germ line mutations in shelterin complex genes are associated with familial chronic lymphocytic leukemia. *Blood* **128:** 2319–2326. doi:10.1182/blood-2016-01-695692

Stern JL, Theodorescu D, Vogelstein B, Papadopoulos N, Cech TR. 2015. Mutation of the *TERT* promoter, switch to active chromatin, and monoallelic *TERT* expression in multiple cancers. *Genes Dev* **29:** 2219–2224. doi:10.1101/gad.269498.115

Subasri M, Shooshtari P, Watson AJ, Betts DH. 2021. Analysis of TERT Isoforms across TCGA, GTEx and CCLE datasets. *Cancers (Basel)* **13:** 1853. doi:10.3390/cancers13081853

Sugita M, Tanaka N, Davidson S, Sekiya S, Varella-Garcia M, West J, Drabkin HA, Gemmill RM. 2000. Molecular definition of a small amplification domain within 3q26 in tumors of cervix, ovary, and lung. *Cancer Genet Cytogenet* **117:** 9–18.

Surovtseva YV, Churikov D, Boltz KA, Song X, Lamb JC, Warrington R, Leehy K, Heacock M, Price CM, Shippen DE. 2009. Conserved telomere maintenance component 1 interacts with STN1 and maintains chromosome ends in higher eukaryotes. *Mol Cell* **36:** 207–218. doi:10.1016/j.molcel.2009.09.017

Szostak JW, Blackburn EH. 1982. Cloning yeast telomeres on linear plasmid vectors. *Cell* **29:** 245–255. doi:10.1016/0092-8674(82)90109-X

Takai KK, Hooper S, Blackwood S, Gandhi R, de Lange T. 2010. In vivo stoichiometry of shelterin components. *J Biol Chem* **285:** 1457–1467. doi:10.1074/jbc.M109.038026

Takai KK, Kibe T, Donigian JR, Frescas D, de Lange T. 2011. Telomere protection by TPP1/POT1 requires tethering to TIN2. *Mol Cell* **44:** 647–659. doi:10.1016/j.molcel.2011.08.043

Takai H, Jenkinson E, Kabir S, Babul-Hirji R, Najm-Tehrani N, Chitayat DA, Crow YJ, de Lange T. 2016. A POT1 mutation implicates defective telomere end fill-in and telomere truncations in Coats plus. *Genes Dev* **30:** 812–826. doi:10.1101/gad.276873.115

Takai H, Aria V, Borges P, Yeeles JTP, de Lange T. 2024. CST–polymerase α-primase solves a second telomere end-replication problem. *Nature* **627:** 664–670. doi:10.1038/s41586-024-07137-1

Takakura M, Kyo S, Kanaya T, Hirano H, Takeda J, Yutsudo M, Inoue M. 1999. Cloning of human telomerase catalytic subunit (hTERT) gene promoter and identification of proximal core promoter sequences essential for transcriptional activation in immortalized and cancer cells. *Cancer Res* **59:** 551–557.

Takasawa K, Arai Y, Yamazaki-Inoue M, Toyoda M, Akutsu H, Umezawa A, Nishino K. 2018. DNA hypermethylation enhanced telomerase reverse transcriptase expression in human-induced pluripotent stem cells. *Hum Cell* **31:** 78–86. doi:10.1007/s13577-017-0190-x

Tan NY, Khachigian LM. 2009. Sp1 phosphorylation and its regulation of gene transcription. *Mol Cell Biol* **29:** 2483–2488. doi:10.1128/MCB.01828-08

Taub MA, Conomos MP, Keener R, Iyer KR, Weinstock JS, Yanek LR, Lane J, Miller-Fleming TW, Brody JA, Raffield LM, et al. 2022. Genetic determinants of telomere length from 109,122 ancestrally diverse whole-genome sequences in TOPMed. *Cell Genom* **2:** 100084. doi:10.1016/j.xgen.2021.100084

Teixeira MT, Arneric M, Sperisen P, Lingner J. 2004. Telomere length homeostasis is achieved via a switch between telomerase-extendible and -nonextendible states. *Cell* **117:** 323–335. doi:10.1016/S0092-8674(04)00334-4

Tesmer VM, Brenner KA, Nandakumar J. 2023. Human POT1 protects the telomeric ds-ssDNA junction by capping the 5′ end of the chromosome. *Science* **381:** 771–778. doi:10.1126/science.adi2436

Theodoris CV, Mourkioti F, Huang Y, Ranade SS, Liu L, Blau HM, Srivastava D. 2017. Long telomeres protect against age-dependent cardiac disease caused by NOTCH1 haploinsufficiency. *J Clin Invest* **127:** 1683–1688. doi:10.1172/JCI90338

Thomson JA, Itskovitz-Eldor J, Shapiro SS, Waknitz MA, Swiergiel JJ, Marshall VS, Jones JM. 1998. Embryonic stem cell lines derived from human blastocysts. *Science* **282:** 1145–1147. doi:10.1126/science.282.5391.1145

Thornley I, Sutherland R, Wynn R, Nayar R, Sung L, Corpus G, Kiss T, Lipton J, Doyle J, Saunders F, et al. 2002. Early hematopoietic reconstitution after clinical stem cell transplantation: evidence for stochastic stem cell behavior and limited acceleration in telomere loss. *Blood* **99:** 2387–2396. doi:10.1182/blood.V99.7.2387

Timashev LA, De Lange T. 2020. Characterization of T-loop formation by TRF2. *Nucleus* **11:** 164–177. doi:10.1080/19491034.2020.1783782

Tong AS, Stern JL, Sfeir A, Kartawinata M, de Lange T, Zhu XD, Bryan TM. 2015. ATM and ATR signalling regulate the recruitment of human telomerase to telomeres. *Cell Rep* **13:** 1633–1646. doi:10.1016/j.celrep.2015.10.041

Tsakiri KD, Cronkhite JT, Kuan PJ, Xing C, Raghu G, Weissler JC, Rosenblatt RL, Shay JW, Garcia CK. 2007. Adult-onset pulmonary fibrosis caused by mutations in telomerase. *Proc Natl Acad Sci* **104:** 7552–7557.

Tummala H, Walne A, Collopy L, Cardoso S, de la Fuente J, Lawson S, Powell J, Cooper N, Foster A, Mohammed S, et al. 2015. Poly(A)-specific ribonuclease deficiency impacts

telomere biology and causes dyskeratosis congenita. *J Clin Invest* **125:** 2151–2160. doi:10.1172/JCI78963

Ulaner GA, Hu JF, Vu TH, Giudice LC, Hoffman AR. 2001. Tissue-specific alternate splicing of human telomerase reverse transcriptase (hTERT) influences telomere lengths during human development. *Int J Cancer* **91:** 644–649. doi:10.1002/1097-0215(200002)9999:9999<::AID-IJC1103>3.0.CO;2-V

Uringa EJ, Lisaingo K, Pickett HA, Brind'Amour J, Rohde JH, Zelensky A, Essers J, Lansdorp PM. 2012. RTEL1 contributes to DNA replication and repair and telomere maintenance. *Mol Biol Cell* **23:** 2782–2792. doi:10.1091/mbc.e12-03-0179

Valentijn LJ, Koster J, Zwijnenburg DA, Hasselt NE, Van Sluis P, Volckmann R, Van Noesel MM, George RE, Tytgat GAM, Molenaar JJ, et al. 2015. TERT rearrangements are frequent in neuroblastoma and identify aggressive tumors. *Nat Genet* **47:** 1411–1414. doi:10.1038/ng.3438

Vannier JB, Pavicic-Kaltenbrunner V, Petalcorin MIR, Ding H, Boulton SJ. 2012. RTEL1 dismantles T loops and counteracts telomeric G4-DNA to maintain telomere integrity. *Cell* **149:** 795–806. doi:10.1016/j.cell.2012.03.030

Vannier JB, Sarek G, Boulton SJ. 2014. RTEL1: functions of a disease-associated helicase. *Trends Cell Biol* **24:** 416–425. doi:10.1016/j.tcb.2014.01.004

van Steensel B, de Lange T. 1997. Control of telomere length by the human telomeric protein TRF1. *Nature* **385:** 740–743. doi:10.1038/385740a0

Vaziri H, Dragowska W, Allsopp RC, Thomas TE, Harley CB, Lansdorp PM. 1994. Evidence for a mitotic clock in human hematopoietic stem cells: loss of telomeric DNA with age. *Proc Natl Acad Sci* **91:** 9857–9860. doi:10.1073/pnas.91.21.9857

Vinagre J, Almeida A, Pópulo H, Batista R, Lyra J, Pinto V, Coelho R, Celestino R, Prazeres H, Lima L, et al. 2013. Frequency of TERT promoter mutations in human cancers. *Nat Commun* **4:** 2185. doi:10.1038/ncomms3185

Vogan JM, Zhang X, Youmans DT, Regalado SG, Johnson JZ, Hockemeyer D, Collins K. 2016. Minimized human telomerase maintains telomeres and resolves endogenous roles of H/ACA proteins, TCAB1, and Cajal bodies. *eLife* **5:** e18221. doi:10.7554/eLife.18221

Vulliamy T, Marrone A, Szydlo R, Walne A, Mason PJ, Dokal I. 2004. Disease anticipation is associated with progressive telomere shortening in families with dyskeratosis congenita due to mutations in TERC. *Nat Genet* **36:** 447–449. doi:10.1038/ng1346

Vulliamy T, Beswick R, Kirwan M, Marrone A, Digweed M, Walne A, Dokal I. 2008. Mutations in the telomerase component NHP2 cause the premature ageing syndrome dyskeratosis congenita. *Proc Natl Acad Sci* **105:** 8073–8078. doi:10.1073/pnas.0800042105

Vyas CM, Ogata S, Reynolds CF, Mischoulon D, Chang G, Cook NR, Manson JE, Crous-Bou M, De Vivo I, Okereke OI. 2021. Telomere length and its relationships with lifestyle and behavioural factors: variations by sex and race/ethnicity. *Age Ageing* **50:** 838–846. doi:10.1093/ageing/afaa186

Walne AJ, Vulliamy T, Marrone A, Beswick R, Kirwan M, Masunari Y, Al-Qurashi F, Aljurf M, Dokal I. 2007. Genetic heterogeneity in autosomal recessive dyskeratosis congenita with one subtype due to mutations in the telo-

Cite this article as *Cold Spring Harb Perspect Biol* doi: 10.1101/cshperspect.a041693

merase-associated protein NOP10. *Hum Mol Genet* **16:** 1619–1629. doi:10.1093/hmg/ddm111

Walne AJ, Vulliamy T, Beswick R, Kirwan M, Dokal I. 2008. TINF2 mutations result in very short telomeres: analysis of a large cohort of patients with dyskeratosis congenita and related bone marrow failure syndromes. *Blood* **112:** 3594–3600.

Wan M, Qin J, Songyang Z, Liu D. 2009. OB fold-containing protein 1 (OBFC1), a human homolog of yeast Stn1, associates with TPP1 and is implicated in telomere length regulation. *J Biol Chem* **284:** 26725–26731. doi:10.1074/jbc.M109.021105

Wang J, Xie LY, Allan S, Beach D, Hannon GJ. 1998. Myc activates telomerase. *Genes Dev* **12:** 1769–1774. doi:10.1101/gad.12.12.1769

Wang F, Podell ER, Zaug AJ, Yang Y, Baciu P, Cech TR, Lei M. 2007. The POT1–TPP1 telomere complex is a telomerase processivity factor. *Nature* **445:** 506–510. doi:10.1038/nature05454

Wang S, Zhao Y, Hu C, Zhu J. 2009. Differential repression of human and mouse *TERT* genes during cell differentiation. *Nucleic Acids Res* **37:** 2618–2629. doi:10.1093/nar/gkp125

Wang F, Stewart JA, Kasbek C, Zhao Y, Wright WE, Price CM. 2012. Human CST has independent functions during telomere duplex replication and C-strand fill-in. *Cell Rep* **2:** 1096–1103. doi:10.1016/j.celrep.2012.10.007

Wang J, Hao Y, Zhu Z, Liu B, Zhang X, Wei N, Wang T, Lv Y, Xu C, Ma M, et al. 2022. Causality of telomere length associated with calcific aortic valvular stenosis: a Mendelian randomization study. *Front Med* **9:** 1077686. doi:10.3389/fmed.2022.1077686

Wang H, Ma T, Zhang X, Chen W, Lan Y, Kuang G, Hsu SJ, He Z, Chen Y, Stewart J, et al. 2023. CTC1 OB-B interaction with TPP1 terminates telomerase and prevents telomere overextension. *Nucleic Acids Res* **51:** 4914–4928. doi:10.1093/nar/gkad237

Weinhold N, Jacobsen A, Schultz N, Sander C, Lee W. 2014. Genome-wide analysis of noncoding regulatory mutations in cancer. *Nat Genet* **46:** 1160–1165. doi:10.1038/ng.3101

Weng NP, Levine BL, June CH, Hodes RJ. 1996. Regulated expression of telomerase activity in human T lymphocyte development and activation. *J Exp Med* **183:** 2471–2479. doi:10.1084/jem.183.6.2471

Weng N, Granger L, Hodes RJ. 1997. Telomere lengthening and telomerase activation during human B cell differentiation. *Proc Natl Acad Sci* **94:** 10827–10832. doi:10.1073/pnas.94.20.10827

Wilson A, Laurenti E, Oser G, van der Wath RC, Blanco-Bose W, Jaworski M, Offner S, Dunant CF, Eshkind L, Bockamp E, et al. 2008. Hematopoietic stem cells reversibly switch from dormancy to self-renewal during homeostasis and repair. *Cell* **135:** 1118–1129. doi:10.1016/j.cell.2008.10.048

Won J, Yim J, Kim TK. 2002. Sp1 and Sp3 Recruit histone deacetylase to repress transcription of human telomerase reverse transcriptase (hTERT) promoter in normal human somatic cells. *J Biol Chem* **277:** 38230–38238. doi:10.1074/jbc.M206064200

Wong MS, Shay JW, Wright WE. 2014. Regulation of human telomerase splicing by RNA:RNA pairing. *Nat Commun* **5:** 3306. doi:10.1038/ncomms4306

Wright WE, Piatyszek MA, Rainey WE, Byrd W, Shay JW. 1996. Telomerase activity in human germline and embryonic tissues and cells. *Dev Genet* **18:** 173–179. doi:10.1002/(SICI)1520-6408(1996)18:2<173::AID-DVG10>3.0.CO;2-3

Wright WE, Tesmer VM, Liao ML, Shay JW. 1999. Normal human telomeres are not late replicating. *Exp Cell Res* **251:** 492–499. doi:10.1006/excr.1999.4602

Wu KJ, Grandori C, Amacker M, Simon-Vermot N, Polack A, Lingner J, Dalla-Favera R. 1999. Direct activation of TERT transcription by c-MYC. *Nat Genet* **21:** 220–224. doi:10.1038/6010

Wu L, Multani AS, He H, Cosme-Blanco W, Deng Y, Deng JM, Bachilo O, Pathak S, Tahara H, Bailey SM, et al. 2006. Pot1 deficiency initiates DNA damage checkpoint activation and aberrant homologous recombination at telomeres. *Cell* **126:** 49–62. doi:10.1016/j.cell.2006.05.037

Wu P, Takai H, de Lange T. 2012. Telomeric 3′ overhangs derive from resection by Exo1 and apollo and fill-in by POT1b-associated CST. *Cell* **150:** 39–52. doi:10.1016/j.cell.2012.05.026

Wu Y, Shi L, Zhao Y, Chen P, Cui R, Ji M, He N, Wang M, Li G, Hou P. 2021. Synergistic activation of mutant TERT promoter by Sp1 and GABPA in BRAFV600E-driven human cancers. *NPJ Precis Oncol* **5:** 3. doi:10.1038/s41698-020-00140-5

Xi L, Cech TR. 2014. Inventory of telomerase components in human cells reveals multiple subpopulations of hTR and hTERT. *Nucleic Acids Res* **42:** 8565–8577. doi:10.1093/nar/gku560

Xin H, Liu D, Wan M, Safari A, Kim H, Sun W, O'Connor MS, Songyang Z. 2007. TPP1 is a homologue of ciliate TEBP-β and interacts with POT1 to recruit telomerase. *Nature* **445:** 559–562. doi:10.1038/nature05469

Xiong F, Frasch WD. 2021. ωqPCR measures telomere length from single cells in base pair units. *Nucleic Acids Res* **49:** 3719–3734. doi:10.1093/nar/gkab124

Yamaguchi H, Calado RT, Ly H, Kajigaya S, Baerlocher GM, Chanock SJ, Lansdorp PM, Young NS. 2005. Mutations in TERT, the gene for telomerase reverse transcriptase, in aplastic anemia. *N Engl J Med* **352:** 1413–1424. doi:10.1056/NEJMoa042980

Yang C, Przyborski S, Cooke MJ, Zhang X, Stewart R, Anyfantis G, Atkinson SP, Saretzki G, Armstrong L, Lako M. 2008. A key role for telomerase reverse transcriptase unit in modulating human embryonic stem cell proliferation, cell cycle dynamics, and in vitro differentiation. *Stem Cells* **26:** 850–863. doi:10.1634/stemcells.2007-0677

Yang D, He Q, Kim H, Ma W, Songyang Z. 2011. TIN2 protein dyskeratosis congenita missense mutants are defective in association with telomerase. *J Biol Chem* **286:** 23022–23030. doi:10.1074/jbc.M111.225870

Yang L, Wang M, Li N, Yan LD, Zhou W, Yu ZQ, Peng XC, Cai J, Yang YH. 2023. TERT mutations in non-small-cell lung cancer: clinicopathologic features and prognostic implications. *Clin Med Insights Oncol* **17:** 11795549221140781. doi:10.1177/11795549221140781

Ye JZS, Donigian JR, Van Overbeek M, Loayza D, Luo Y, Krutchinsky AN, Chait BT, De Lange T. 2004. TIN2 binds

TRF1 and TRF2 simultaneously and stabilizes the TRF2 complex on telomeres. *J Biol Chem* **279**: 47264–47271. doi:10.1074/jbc.M409047200

Yehezkel S, Segev Y, Viegas-Péquignot E, Skorecki K, Selig S. 2008. Hypomethylation of subtelomeric regions in ICF syndrome is associated with abnormally short telomeres and enhanced transcription from telomeric regions. *Hum Mol Genet* **17**: 2776–2789. doi:10.1093/hmg/ddn177

Yi X, White DM, Aisner DL, Baur JA, Wright WE, Shay JW. 2000. An alternate splicing variant of the human telomerase catalytic subunit inhibits telomerase activity. *Neoplasia* **2**: 433–440. doi:10.1038/sj.neo.7900113

Yi X, Shay JW, Wright WE. 2001. Quantitation of telomerase components and hTERT mRNA splicing patterns in immortal human cells. *Nucleic Acids Res* **29**: 4818–4825. doi:10.1093/nar/29.23.4818

Yuan X, Larsson C, Xu D. 2019. Mechanisms underlying the activation of TERT transcription and telomerase activity in human cancer: old actors and new players. *Oncogene* **38**: 6172–6183. doi:10.1038/s41388-019-0872-9

Zhang Y, Zhang A, Shen C, Zhang B, Rao Z, Wang R, Yang S, Ning S, Mao G, Fang D. 2014. E2F1 acts as a negative feedback regulator of c-Myc-induced hTERT transcription during tumorigenesis. *Oncol Rep* **32**: 1273–1280. doi:10.3892/or.2014.3287

Zhang F, Cheng D, Porter KI, Heck EA, Wang S, Zhang H, Davis CJ, Robertson GP, Zhu J. 2025. Modification of the telomerase gene with human regulatory sequences resets mouse telomeres to human length. *Nat Commun* **16**: 1211.

Zhao Y, Abreu E, Kim J, Stadler G, Eskiocak U, Terns MP, Terns RM, Shay JW, Wright WE. 2011. Processive and distributive extension of human telomeres by telomerase under homeostatic and nonequilibrium condi-

tions. *Mol Cell* **42**: 297–307. doi:10.1016/j.molcel.2011.03.020

Zhao Y, Cheng D, Wang S, Zhu J. 2014. Dual roles of c-Myc in the regulation of hTERT gene. *Nucleic Acids Res* **42**: 10385–10398. doi:10.1093/nar/gku721

Zhong Z, Shiue L, Kaplan S, de Lange T. 1992. A mammalian factor that binds telomeric TTAGGG repeats in vitro. *Mol Cell Biol* **12**: 4834–4843.

Zhong F, Savage SA, Shkreli M, Giri N, Jessop L, Myers T, Chen R, Alter BP, Artandi SE. 2011. Disruption of telomerase trafficking by TCAB1 mutation causes dyskeratosis congenita. *Genes Dev* **25**: 11–16. doi:10.1101/gad.2006411

Zhong FL, Batista LFZ, Freund A, Pech MF, Venteicher AS, Artandi SE. 2012. TPP1 OB-fold domain controls telomere maintenance by recruiting telomerase to chromosome ends. *Cell* **150**: 481–494. doi:10.1016/j.cell.2012.07.012

Zhu CQ, Cutz JC, Liu N, Lau D, Shepherd FA, Squire JA, Tsao M-S. 2006. Amplification of telomerase (hTERT) gene is a poor prognostic marker in non-small-cell lung cancer. *Br J Cancer* **94**: 1452–1459. doi:10.1038/sj.bjc.6603110

Zinder JC, Olinares PDB, Svetlov V, Bush MW, Nudler E, Chait BT, Walz T, de Lange T. 2022. Shelterin is a dimeric complex with extensive structural heterogeneity. *Proc Natl Acad Sci* **119**: e2201662119. doi:10.1073/pnas.2201662119

Zinn RL, Pruitt K, Eguchi S, Baylin SB, Herman JG. 2007. hTERT is expressed in cancer cell lines despite promoter DNA methylation by preservation of unmethylated DNA and active chromatin around the transcription start site. *Cancer Res* **67**: 194–201. doi:10.1158/0008-5472.CAN-06-3396

# Telomerase RNA Shapes the Evolutionary Diversity of Telomerase Ribonucleoproteins (RNPs)

## Julian J.-L. Chen[1] and Raymund J. Wellinger[2]

[1]School of Molecular Sciences, Arizona State University, Tempe, Arizona 85281, USA

[2]Department of Microbiology and Infectious Diseases, Faculty of Medicine and Health Sciences, Université de Sherbrooke, PRAC, Sherbrooke, Québec J1E 4K8, Canada

*Correspondence:* jlchen@asu.edu; raymund.wellinger@usherbrooke.ca

Telomerase emerged in early eukaryotes as a highly specialized reverse transcriptase for maintaining chromosome integrity. The telomerase enzyme contains an integral RNA, providing the template for DNA repeat synthesis. This central telomerase RNA not only provides the template but also contributes to the enzyme's catalytic function and the biogenesis of the ribonucleoprotein. Remarkably, telomerase RNA exhibits significant diversity in sequence, structure, and biogenesis across eukaryotic lineages, a feature that sets it apart from other functional RNAs. In ciliates and plants, telomerase RNA is transcribed by RNA polymerase III, whereas in animals and fungi, it is predominantly transcribed by RNA polymerase II. These differences result in distinct pathways for RNA synthesis, maturation, and trafficking. This work highlights how the diversity in size and structure of telomerase RNAs impacts the complexity and evolution of telomerase ribonucleoproteins, spanning from unicellular eukaryotes to multicellular plants and animals, highlighting telomerase RNA's critical role in telomere biology.

With the discoveries of the telomerase enzyme and the essential functions of its associated RNA, it was derived that telomerase is a telomere-specific reverse transcriptase (Greider and Blackburn 1987). Indeed, in the species studied, the ciliate *Tetrahymena thermophila*, the RNA component of telomerase harbored a templating sequence of ~1.5–2 copies of the species-specific telomeric repeat sequence (Greider and Blackburn 1989). The later discoveries of how telomere maintenance is involved in human aging and cancer etiology spurred an acute interest in understanding the biochemistry, structure, function, and regulation of this enzyme. Today, ~35 years later, we know that certain initially described principles apply to this enzyme almost universally in the eukaryotic kingdom, and we do have the first <4 Å cryo-electron microscopy (EM) structures of human and ciliate telomerase ribonucleoproteins (RNPs) in various states of primer association and DNA synthesis (Ghanim et al. 2021; He et al. 2021; Liu et al. 2022; Sekne et al. 2022). However, research on telomerase structure and function, as well as on its regulation, has not been a straight path. The telomerase RNA

(TR) component (also referred to as TER or telomerase RNA component [TERC] or telomerase component 1 [Tlc1] in budding yeast) was particularly difficult to identify in this regard. It turned out that not much about this RNA is universally or even broadly conserved. Indeed, the astonishing diversity of TRs, while a serious nuisance for discovering such RNAs in new and evolutionarily distant organisms, renders them also an extremely interesting subject of study in terms of evolutionary biology. This diversity is associated with almost any parameter we like to associate with RNAs in general: Gene locus, expression levels, types of RNA polymerase for transcription, size of the transcripts, sequence, and structure of the RNA, as well as $5'$ and $3'$ characteristics are all remarkably variable, suggesting a very high tolerance to changes during evolution. There certainly are a few substructure similarities and commonalities in TRs, the most obvious being the above-mentioned templating sequence flanked by a template boundary-defining element at its $5'$-side. There is also a universal pseudoknot (PK) structure and a structure called conserved region 4/5 (CR4/5) (Fig. 1) (see below). These conserved RNA elements are all needed to orchestrate correct protein associations and reverse transcriptase activity of the enzyme. In contrast to the highly variable TR component, the universal catalytic protein subunit telomerase reverse transcriptase (TERT) (Est2 in budding yeast or Trt1 in fission yeast) has an easily recognizable sequence homology in its reverse transcriptase domain across all species. TERT, together with TR, has been defined as the minimal core of telomerase enzyme that has demonstrable activity in vitro. While the functional TR is expressed from a single gene in most organisms, some plant and insect species harbor multiple TR genes (Závodník et al. 2023). Such gene duplication or multiplication would allow the TR paralogs to tolerate dramatic changes in template sequence or structural domains, which would thus provide the basis for the fast evolution of this RNA and the sequence variation of telomeric DNA repeats in these species.

Here, we will review the current knowledge on TR diversity across many phyla, discuss common features of RNA maturation and RNP assembly, and finish with an up-to-date account of the budding yeast RNP biogenesis from birth to death.

## DISTINCT MECHANISMS FOR TR SYNTHESIS

TR synthesis and maturation vary significantly among eukaryotic groups. These differences arise from the use of distinct transcriptional machineries. In certain eukaryotic lineages such as ciliates, plants, and green algae, TRs are small RNAs (140–210 nt for ciliate TR and 235–347 nt for plant TRs), transcribed by RNA polymerase III, which uses a terminal poly(U) tract for transcription termination and leaves a triphosphate at the $5'$ end (Fig. 1; Greider and Blackburn 1989; McCormick-Graham and Romero 1995; Fajkus et al. 2019, 2021; Song et al. 2019). Conversely, animal TRs (312–559 nt) and fungal TRs (900–2400 nt) are larger RNAs transcribed from a small nuclear RNA (snRNA)–type promoter by RNA polymerase II with a $3'$ end processed by nucleases and the $5'$ end protected by a $5'$-2,2,7-trimethylguanosine (TMG) cap (Chapon et al. 1997; Mitchell et al. 1999; Chen et al. 2000, 2020; Franke et al. 2008; Qi et al. 2013; Podlevsky and Chen 2016; Logeswaran et al. 2021). It is intriguing that distinct transcription enzymes are used for TR synthesis in different branches of the evolutionary tree, which also raises the question of the evolutionary origin of TR. Interestingly, in the basal eukaryote trypanosome, TR is transcribed by RNA Pol II (Gupta et al. 2013), arguing that Pol II might have been the initial transcription machinery for ancient TR, which underwent a subsequent switch to Pol III in the lineages of ciliates and plants. An alternative Pol III–first hypothesis would necessitate multiple independent evolutionary transitions in the trypanosome and animal–fungal lineages, a scenario that would seem rather convoluted. However, the evolutionary transition from RNA Pol II to Pol III transcription may be less restricted than expected, as the two enzymes are not mutually exclusive for some promoters (Rajendra et al. 2024). Interestingly, hymenopteran insects

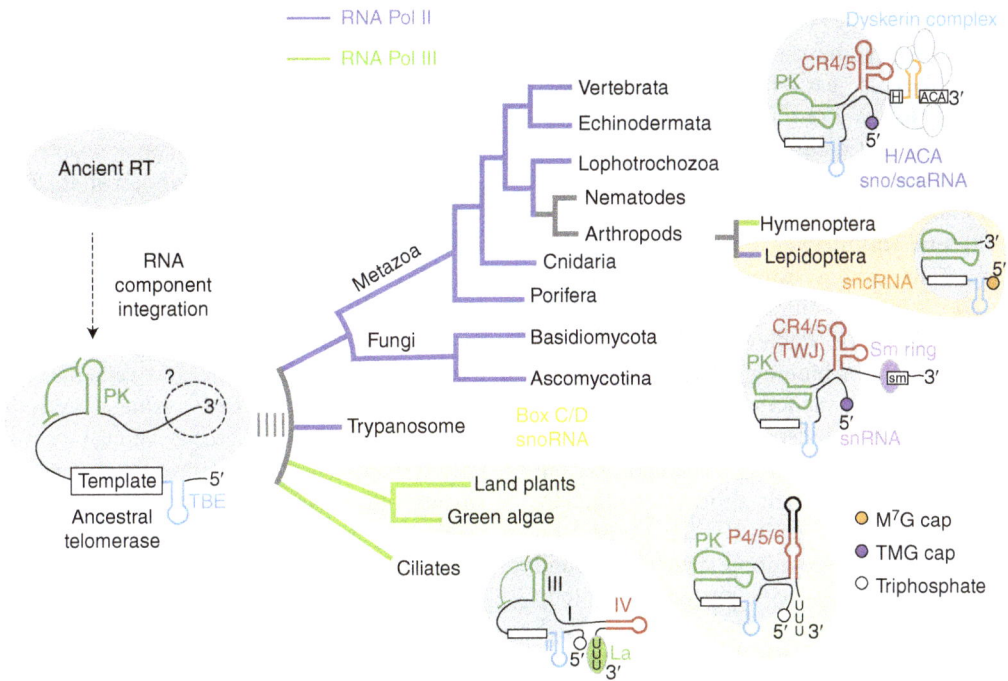

**Figure 1.** Evolutionary diversity of telomerase ribonucleoprotein (RNP). Telomerase RNP presumably emerged from an ancient reverse transcriptase (RT) with a stable RNA molecule internalized to form an RNP complex in early eukaryotes. This internalized telomerase RNA (TR) component would minimally require a template defined by a template boundary element (TBE, blue) and a structural element (i.e., pseudoknot [PK]), for assembly with the catalytic RT component. When the metazoan/fungal lineage branched away from the ciliate/plant lineage, distinct transcription enzymes, RNA polymerase II versus III, were used separately for TR synthesis. Each transcription system gives rise to specific attributes of the transcripts: a trimethylguanosine (TMG) cap (a purple circle) or triphosphate (a white circle) at the 5′ end and a poly(U) tail or a processed end at the 3′ end. The truncated insect TR is a small noncoding RNA (sncRNA) that lacks a cap-hypermethylation mechanism and thus possesses a 5′ 7-methyl-G (M$^7$G) cap (an orange circle) instead. In each evolutionary lineage, TR acquired a specific RNA structure at the 3′ end and the associated protein subunits from existing RNA species for the 3′-end processing and biogenesis pathway. For example, the H/ACA small Cajal body–specific RNA (scaRNA) biogenesis pathway was acquired for animal TRs with the dyskerin complex, small nuclear RNA (snRNA) for fungal TRs with the Sm protein ring complex, box C/D small nucleolar RNA (snoRNA) for trypanosome TR, and Pol III transcribed small RNA for ciliate and plant TRs with the La protein protecting the 3′-poly(U) tail. The processed and stable TRs were then assembled with the telomerase reverse transcriptase (TERT) protein (gray) and other specific telomerase accessory proteins through the universal PK (green) and conserved region 4/5 (CR4/5) (red) (or equivalent elements) domain of TR to produce functional telomerase enzyme.

switched the TR transcription enzyme from Pol II to Pol III (Fajkus et al. 2023), while TRs from the sister order lepidopteran insects continue to use RNA polymerase II for TR synthesis (Fig. 1).

While TR genes in most species are transcribed as individual genes, some TRs are processed from messenger RNA (mRNA) pre-cursors. In the basidiomycete fungus *Ustilago maydis*, TR is not transcribed from an independent gene but processed from the 3′ UTR of a protein-coding mRNA precursor (Logeswaran et al. 2022). This remarkable variability of pathways for the generation of functional TR highlights the extensive evolutionary divergence and adaptability of how the TR can be generated.

Understanding the origins and evolutionary trajectory of TR synthesis will require comprehensive studies in early-branching eukaryotes. However, this is a challenging task due to the highly divergent nature of TR sequences across different species, which renders conclusive gene identification difficult.

Beyond distinctive transcription mechanisms, TRs exhibit extensive size variations due to distinct structural subdomains and 3′-end configurations (Podlevsky et al. 2008). The diversity of RNA-binding proteins associated with various TR subdomains across different eukaryotic groups underscores the inherent flexibility of TR biogenesis. For example, certain long fungal RNAs contain subdomains that are also found on other RNPs (P3-like domain on RNaseP/MRP) or stem-loop structures bound by non-telomerase-specific proteins (yKu loop in yeast) (Peterson et al. 2001; Lemieux et al. 2016). Another major factor contributing to the TR size variations are the structural domains that define the 3′ ends of the TRs (Fig. 1). For example, the smaller TRs found in ciliates and plants require only a short poly(U) tract to define their 3′ ends. In contrast, the larger TR in animals relies on a H/ACA small Cajal body–specific RNA (scaRNA) structural domain, which binds the dyskerin protein complex to accurately define its 3′ end (Jády et al. 2004; Egan and Collins 2010). In the basal eukaryote trypanosome, a box C/D small nucleolar RNA (snoRNA) structural domain is adapted for TR 3′-end maturation (Gupta et al. 2013). In the fungal TRs, an SM site near the 3′ end is bound by the Sm protein ring complex for processing and protection, a mechanism that appears to be shared with the snRNAs (Seto et al. 1999; Tang et al. 2012). As a consequence of the various types of transcription machinery used for TR synthesis, distinct mechanisms will apply to transcription termination and 3′-end processing of precursor transcripts. In ciliates and plants, TR transcription terminates at a poly(U) tract, which is subsequently bound by La protein to protect the RNA from exonuclease degradation (Singh et al. 2012). Conversely, in animals and fungi, RNA polymerase II terminates transcription at sites downstream from the mature end,

and the precursor TR is processed by a 3′–5′ exonuclease to generate the final 3′ end of the TR (Kao et al. 2024). Remarkably, in fission yeasts and some filamentous fungi, an intron splicing-mediated RNA cleavage is adapted to generate the 3′ end of TR (Kannan et al. 2015; Qi et al. 2015).

We surmise that the presence of the variable structural subdomains correlates with a very early, mostly cotranscriptional, binding of the cognate proteins. This would ensure that the transcribed RNA is stable and remains in a form that is primed for later telomerase-specific protein associations. These variations thus may reflect evolutionary adaptations that have occurred within distinct eukaryotic lineages. The origins and evolutionary pathways of these divergent TR biogenesis mechanisms in the major eukaryotic kingdoms—plants, fungi, and animals—remain subjects for further investigation. However, we speculate that in each lineage, RNA synthesis mode and the non-telomerase-specific subdomains in the RNAs were selected for optimal expression level and primary RNP stability. This primary RNP, however, would not be an active telomerase enzyme, as it still lacks the telomerase-specific proteins required for activity. Hence, a speculative, generally applicable hypothesis posits that first, in each lineage, evolutionary pressure was on producing the right amount of a very stable precursor RNP that is primed for telomerase protein association (Fig. 2). This may involve the TR core hijacking genes of abundant RNAs, such as RNA Pol III transcribed small RNA or RNA Pol II transcribed snRNAs and snoRNAs, and results in lineage-specific solutions to ensuring RNA production and stability (Fig. 1). We speculate that such stable precursor RNPs would ensure a basal availability for complete RNP production, if required. Telomerase-specific proteins can still be dynamically regulated such as not to have an overabundance of enzymes. These proteins would associate later in a posttranscriptional step that can happen in the nucleus or cytoplasm, depending on the system (Fig. 2). However, this step of RNP assembly would depend on proper prearrangement of the architecture in the conserved core elements.

Cite this article as *Cold Spring Harb Perspect Biol* doi: 10.1101/cshperspect.a041700

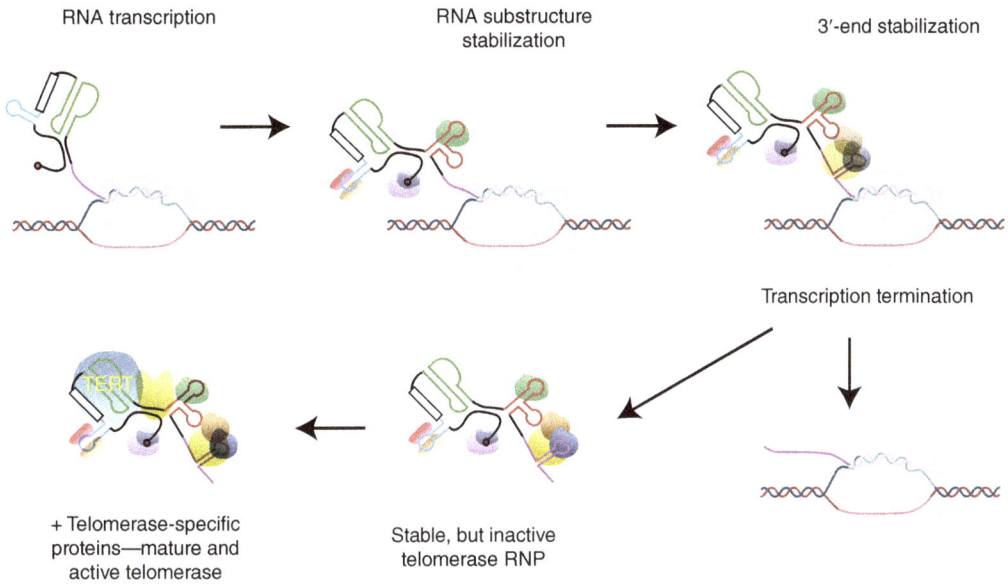

**Figure 2.** A unifying conceptual model for the assembly of the telomerase ribonucleoprotein (RNP). The currently available data suggest that during transcription, telomerase RNAs (TRs) first assemble stable subdomains either directly via a very stable RNA folding or by associating with RNA-binding proteins that stabilize species-specific subdomains (*top middle*). Given that the 3′ end of any RNA is particularly susceptible to degradation, TRs acquired a variety of strong stabilizing features that are dependent on the evolutionary branch (*top right*, see Fig. 1). This event allows transcription termination and release of the primary telomerase RNP that is inactive (*bottom middle*). Only now do the telomerase-specific protein complexes associate with the RNP to create a mature and active telomerase (*bottom left*). Note that this last step of RNP assembly may occur in the cytoplasm or in the nucleus, depending on the organism and the proteins in the cotranscriptionally assembled RNP.

## THE TR STRUCTURAL CORE IS ESSENTIAL FOR TELOMERASE ACTIVITY

In addition to 3′ subdomains for TR biogenesis, the catalytic activity of telomerases across eukaryotes requires two essential TR structural core domains in the 5′ region, namely, the template (T)-PK and the CR4/5, also called three-way junction (TWJ) structure, or stem terminal element (STE) (see Fig. 1). The T/PK domain is responsible for the most basic and essential function of TR, which is to provide the template with a defined boundary for DNA synthesis (Chen and Greider 2003) and a structural element, that is, PK, for assembling with the catalytic TERT protein (Fig. 1; Chen and Greider 2005). The proximal CR4/5 domain plays a vital role in regulating the catalytic activity of the TERT protein (Mitchell and Collins 2000;

Chen et al. 2002). These two domains can independently bind to the TERT protein, facilitating in vitro activity as demonstrated in various organisms, including animals, fungi, plants, ciliates, and flagellates (Podlevsky and Chen 2016). Both the T/PK and CR4/5 domains bind to the TR-binding domain (TRBD) and interact with the carboxy-terminal extension (CTE) domain of TERT (see Chen and Batista 2025; Martin and Hockemeyer 2025). The dependency on these TR domains for telomerase activity, however, varies among groups; vertebrate and fungal telomerases necessitate both domains for full activity, whereas flagellates and echinoderms show partial activity with the T-PK domain alone, suggesting a reduced dependence on the CR4/5 domain (Podlevsky et al. 2016a,b). Similarly, ciliate and plant TRs also require a second TR structural element, stem-loop IV and P4/5/6, respectively, for reconstituting full telomerase activity in vitro (Mason et al. 2003; Song et al. 2019).

The T-PK domain includes structural elements that regulate telomerase function, while the CR4/5 domain displays greater functional and structural diversity. Notably, the vertebrate-type CR4/5 structure is found in basal sponges and filamentous fungi but not in plants, suggesting an ancestral lineage predating the divergence of fungal and metazoan groups. This diversity likely stems from coevolution with the TRBD of TERT. The requirement for two TR domains indicates a universal attribute among telomerases across eukaryotes. The presence of both the canonical CR4/5 and CR4/5-like domains in early branching eukaryotic species suggests an evolutionary adaptation for the TR-dependent catalysis of the TERT protein, which presumably prevents the TERT protein from using a nonspecific RNA template during the early evolution of telomerase RNP (Podlevsky et al. 2016a). For vertebrates and sponges, the conservation of the T-PK, CR4/5, and H/ACA domains supports a common ancestral origin for animal TR, with variations in structure reflecting evolutionary changes across metazoan lineages (Logeswaran et al. 2021).

The animal telomerase RNP has undergone unusual evolutionary renovations in arthropods. Within the arthropod phylum, there is evidence of both retrotransposon-mediated and telomerase-mediated mechanisms for telomere maintenance. Certain clades of insects use retrotransposon mechanisms (Casacuberta 2017), yet many insect taxa exhibit uniform telomeric repeats, suggesting that a telomerase-mediated mechanism may also be at play (Vítková et al. 2005). Recent studies of TRs in hymenopteran (Fajkus et al. 2023) and lepidopteran insects (Chou et al. 2025) reveal intriguing divergence of telomerase in the insect lineages. The insect TR appears to be miniature and degenerate, with the entire CR4/5 and H/ACA domains removed, leaving a minimal TR core required for constituting functional telomerase with the TERT protein (Chou et al. 2025). The insect TERT protein is unique as it is catalytically active in the absence of TR (Gillis et al.

2008; Mitchell et al. 2010; Chou et al. 2025). Moreover, it lacks motifs, such as the TEN domain and TRAP motif, that are conserved in other eukaryotes for telomerase processivity (Wang et al. 2020), suggesting the insect TERT has undergone significant evolutionary changes to accommodate the degenerate insect TR. This miniature insect telomerase RNP exemplifies a simple pathway for the emergence of the first telomerase RNP in early eukaryotes from an ancient RT associated with a simple RNA molecule that harbors a defined template and an RT-binding element (Fig. 1). Further investigation of TRs in additional arthropod species and nematodes is expected to provide deeper insights into the intriguing divergent evolution of telomerase RNPs across eukaryotic kingdoms.

## OVERALL FOLDING OF THE RNAs

Timely and ordered consecutive associations of telomerase protein components strongly depend on the RNA being folded into the right tertiary architecture. In addition, protein associations may require a change in the RNA structure for the next step in RNP assembly. If we are to understand telomerase biogenesis, we will need to know which elements of RNA folding are spontaneously correct, which are modified or stabilized by protein binding, and which could be dead ends. Our speculative model above predicts that evolutionary pressure first is on generating an appropriately prefolded RNA that can be assembled with telomerase-specific proteins later (see Fig. 2). For example, TRs from the ciliates may first fold into a structure quite distinct from the one found in the fully assembled and active RNP. For the *T. thermophila* RNA, the very early binding of the p65 subunit clearly induces a striking bend in stem 4 and affects RNA structure in other places to prime it for TERT association (Prathapam et al. 2005; Stone et al. 2007; Berman et al. 2010; Singh et al. 2012). Moreover, the essential PK structure of the *Tetrahymena* TR may first fold into an alternative structure, which is remodeled by the TERT association (Cash and Feigon 2017). As also predicted by our model, mutations that interfere with this orderly RNA

Cite this article as *Cold Spring Harb Perspect Biol* doi: 10.1101/cshperspect.a041700

refolding have strong negative phenotypes in terms of RNA stability but not necessarily telomerase catalytic activity. In other words, even if the prefolded TR RNA is not in an optimal structure, the final binding of TERT may induce the desired mature form, but this step may occur much less frequently, and the RNA often degrades before this can happen. Thus, even the short ciliate TR of ∼160 nt needs proper protein-aided restructuring during maturation and RNP assembly. In the mature functional RNP, the RNA structure seems quite homogeneous with only minor restructuring, even during the catalytic cycle of DNA synthesis (He et al. 2021).

For human RNA, a similar scenario is conceptually postulated. The recently determined cryo-EM structures of the assembled and active enzyme overall show a two-domain architecture for the RNP. The RNA 5′-terminal domain comprises the known critical T/PK and CR4/5-TWJ elements for TERT binding and telomerase enzymatic activity. The 3′-terminal domain contains a complete sno/scaRNA module associated with two copies of the dyskerin/NOP10/NHP2/Gar1 heterotetrameric complex plus the TCAB1 protein (Nguyen et al. 2018; Ghanim et al. 2021; Liu et al. 2022; Sekne et al. 2022). Despite the fact that the 5′ domain of the RNA is synthesized first, it is currently thought that the proteins for the 3′-sno/sca domain of the RNA are assembled first (Schmidt and Cech 2015). This cotranscriptional part of the assembly pathway is postulated to follow the mechanisms proposed for other sno/scaRNP assemblies and involves chaperoning proteins as well as protein exchanges on intermediary complexes (Massenet et al. 2017). After this assembly, the telomerase-specific TERT protein associates with its binding location. In terms of the RNA structure, there is increasing evidence that at least some parts of the nascent human TR (hTR) RNA can fold into alternative structures, which are then modified in the final maturation steps. For example, there is evidence that the TCAB protein association is required to keep parts of the 3′-sno/scaRNA structure and the CR4/5 in a conformation that spurs high telomerase activity (Chen et al. 2018). Similarly, sno/scaRNP assembly interrupts tertiary RNA interactions

of the 3′-extended pre-hTR RNAs and promotes 3′ resection in a way that creates a mature form (Tseng et al. 2018). A recent overview using analyses of single-molecule populations inside cells also came to the conclusion that at least the CR4/5 domain and the T/PK area in human cells exist in alternative conformations in specific cell types (Forino et al. 2025). At present, it is not clear whether these alternative foldings are dynamic enough to be bound by TERT or other cognate proteins and changed into the one found in the mature RNP or whether they are dead-end molecules bound for degradation.

The yeast TRs are much larger, and while phylogenetically based secondary structures have been developed (Dandjinou et al. 2004; Zappulla et al. 2005; Brown et al. 2007; Gunisova et al. 2009), direct comprehensive structural information is not yet available. For the *Saccharomyces* species, the secondary structure predictions did concur with previous domain determinations for the binding of Est1 and Est2 (TERT) (Livengood et al. 2002; Seto et al. 2002), as well as yKu association (Peterson et al. 2001). Limited direct biochemical and some genetic verification of the predictions overall confirmed the structures (Chappell and Lundblad 2004; Dandjinou et al. 2004; Lin et al. 2004; Laterreur et al. 2018). However, some uncertainty on how the T/PK region folds and how the CR4/5 (TWJ) area is arranged remains (Chen and Greider 2004). This latter region, although present and essential for telomerase activity and TERT association in many other systems, is not essential for *Saccharomyces cerevisiae* telomerase (Livengood et al. 2002). Given the scarcity of direct structural data, however, it is difficult to discuss folding dynamics for the large 1158 nt yeast TR. As we discuss in some detail below, yeast telomerase RNP maturation also occurs along the lines of our generalized model (Fig. 2). Chaperoning, structuring, and transport proteins associate first with the nascent RNA to keep it in a proper RNA architecture for the telomerase-specific proteins to associate later in the cytoplasm. It is worth noting that in the absence of yKu, the TR is mislocalized to the cytoplasm, suggesting that maturation steps occurring there or nuclear reimport are difficult to complete. These consequences will contribute to the severe telomere shortening phenotype observed (Gallardo et al. 2008). However, it remains to be determined whether this potential difficulty late in RNP assembly/transport is due to a structural problem on the RNA or some other issue associated with the absence of yKu.

Thus, the common theme for telomerase RNP generation, assembly, and maturation emerging from all systems appears to be that lineage-specific telomerase accessory proteins are the first to associate with the TRs, during or right after transcription (Fig. 2). These proteins, while very different in the various systems studied, have all evolved to structure and stabilize RNA domains on other cognate RNAs that are very old in evolutionary terms. Hence, one could say that chaperoning and structuring "professionals" associate with TRs first to set the proper RNA architecture for the telomerase-specific proteins and RNP functions. Whether alternative folding in the RNA is tolerated, because it is dynamic enough to still turn into the proper structure or whether there are such structures that are dead ends to be eliminated, remains to be determined. The same must be said for exploiting RNA-folding switches for regulatory purposes. In all cases that are known in enough detail today, the final RNA architecture in the active RNP appears to be quite homogenous. Thus, while there certainly are a multitude of pathways and chaperoning proteins for the RNA structuring problem, there appears to be just one end point, the species-specific end structure.

## A CASE STUDY OF TELOMERASE RNP ASSEMBLY: THE Tlc1 RNA IN BUDDING YEASTS

As laid out in the paragraphs above, TR synthesis, maturation, and RNP assembly vary widely, depending on the evolutionary branches of organisms studied. Our current knowledge of these processes occurring in mammalian cells is summarized in other sections of the book (see Chen and Batista 2025; Martin and Hockemeyer 2025). Here we will summarize what we presently know about these issues in budding yeast (*S. cerevisiae*), with cross references to fission yeast (*Schizosaccharomyces pombe*) as appropriate.

### Transcription

In budding yeast, the TR is an 1158-nt-long noncoding RNA (lncRNA) generated by RNA polymerase II from a cell cycle–regulated promoter (Fig. 3, step 1; Chapon et al. 1997; Dionne et al. 2013). RNA polymerase II–associated factors (PAF1c) also contribute to the final steady-state level of the RNA, but it remains unclear whether their effect is direct or indirect (Mozdy et al. 2008). The cell cycle–regulated transcription is mediated by specific transcription enhancer motifs in the promoter of Tlc1 that upregulate a whole set of S phase–specific genes (Koch and Nasmyth 1994; Spellman et al. 1998; Dionne et al. 2013). Transcriptional activity from the *TLC1* promoter is very low, however, and cell cycle–dependent changes in Tlc1 RNA levels are undetectable in global RNA preparations (Mozdy and Cech 2006; Dionne et al. 2013). Low overall transcription is most likely physiologically important, given that both over- and underexpression of the TR cause strong negative phenotypes (Singer and Gottschling 1994; Mozdy et al. 2008; Dionne et al. 2013). The RNA itself is present at roughly 15–20 molecules per cell, but the actual number varies enormously from 0 to more than 30 (Mozdy and Cech 2006; Bajon et al. 2015). Nevertheless, given that the actual guestimates for the number of telomerase-specific proteins are significantly higher than for the RNA (Tuzon et al. 2011) and the fact that *TCL1* gene expression is haplo-insufficient (Mozdy and Cech 2006; Rowland et al. 2019), the TR is the limiting component for the telomerase RNP in budding yeast.

Cell cycle–dependent regulation of TR transcription has not been described in human or *S. pombe* cells. At least for humans, hTR abundance also varies greatly between different cells and their physiological status (normal or transformed cancer cells), but it is not necessarily the limiting factor for telomerase assembly (Xi and

Cech 2014; Rowland et al. 2019). It is interesting to note that in human cells, where it is thought that the TERT protein is the limiting factor, transcription of the mRNA of TERT was extremely heterogeneous as for the Tlc1 RNA in yeast (Rowland et al. 2019).

## The 5′ and 3′ Ends of the RNA after Transcription

Transcripts generated by RNA polymerase II will be modified at their 5′ end to carry a 7-methyl-G ($M^7G$) cap, which is bound by the cap-binding complex (CBC), composed of two proteins called Sto1/Cbp20. Recent results showed that the CBC is indeed also bound to the yeast TR (Neumann et al. 2023), which is thought to be the beginning of setting up an export-competent telomerase RNP (Fig. 3, step 2).

The 3′-end formation of the Tlc1 RNA is surprisingly complex and may reflect the mixture of class-identifying factors found on the RNP. The most important feature at the 3′ end is a canonical sequence for binding of the $Sm_7$ ring complex, a characteristic typical for snRNAs (Seto et al. 1999). This site is just 7–8 nt upstream of the eventual 3′ end at nt 1158 (Bosoy et al. 2003; Dandjinou et al. 2004; Larose et al. 2007) (note that nucleotide numbering starts from +1 at the 5′ end as determined in Dandjinou et al. 2004). Also similar to snRNAs, the initial Tlc1 RNA transcript is longer than the mature form and, therefore, is subject to processing events (Chapon et al. 1997; Jamonnak et al. 2011; Noël et al. 2012). While there are extended polyadenylated forms of Tlc1, the most important extended Tlc1 transcripts are only ~40–50 nt longer than the mature form and are oligoadenylated (Jamonnak et al. 2011). Transcriptional termination for these RNAs involves the Nrd1/Nab3 pathway, and consensus sequences for the binding of these factors have been described in the short extension between nt 1158 and 1210 (Noël et al. 2012; Neumann et al. 2023). Indeed, the Tlc1 oligoadenylation on sites 1195 and 1210 very precisely matches the location of the Nrd1 consensus (Jamonnak et al. 2011; Noël et al. 2012). Further-

more, Nab2 binding on the extension of these RNAs is essential for further export factor assembly (Fig. 3, step 2; Neumann et al. 2023).

Altogether, the 3′-end formation of the yeast TR appears to result from intermingled pathways, namely, both the canonical mRNA pathway and the Nrd1/Nab3-snRNA pathway being active. It is important to note that the extended forms are not processed in the nucleus before export, presumably because that would remove the binding site for the Nab2 protein and impede the formation of the export-competent RNP.

In *S. pombe*, the TR is also an RNA Pol II transcript, possibly bound by the CBC. However, its 3′ formation uses another mechanism entirely that is based on a half-splicing reaction (Box et al. 2008). In essence, only one *trans*-esterification occurs on the RNA, and the 3′-end lariat RNA is lost. Variable mechanisms for how the second splicing step is prevented were reported for several yeast and filamentous fungal species, indicating that there may be multiple origins for this mechanism (Kannan et al. 2015; Qi et al. 2015). Also, counter the expectation that the RNA is an $Sm_7$-bound snRNA, the mature RNA harbors an $LSm_7$ ring near the 3′ end (Tang et al. 2012).

## Telomerase Trafficking

Yeast TRs can be found in the cytoplasm to where they are transported via multiple pathways in parallel (Teixeira et al. 2002; Gallardo and Chartrand 2008; Gallardo et al. 2011). First, the export from the nucleus depends on the major Xpo1(Crm1) export pathway that other noncoding RNAs (ncRNAs), such as snRNAs, also use. In addition, the Mex67/Mtr2 pathway, usually associated with the export of mRNAs, contributes prominently to telomerase export (Vasianovich et al. 2020; Neumann et al. 2023). While there are RNAs that use one or the other pathway exclusively, there are increasing examples of RNPs, such as the telomerase RNP, that use a combination of these export pathways (Nostramo and Hopper 2020; Bartle et al. 2022).

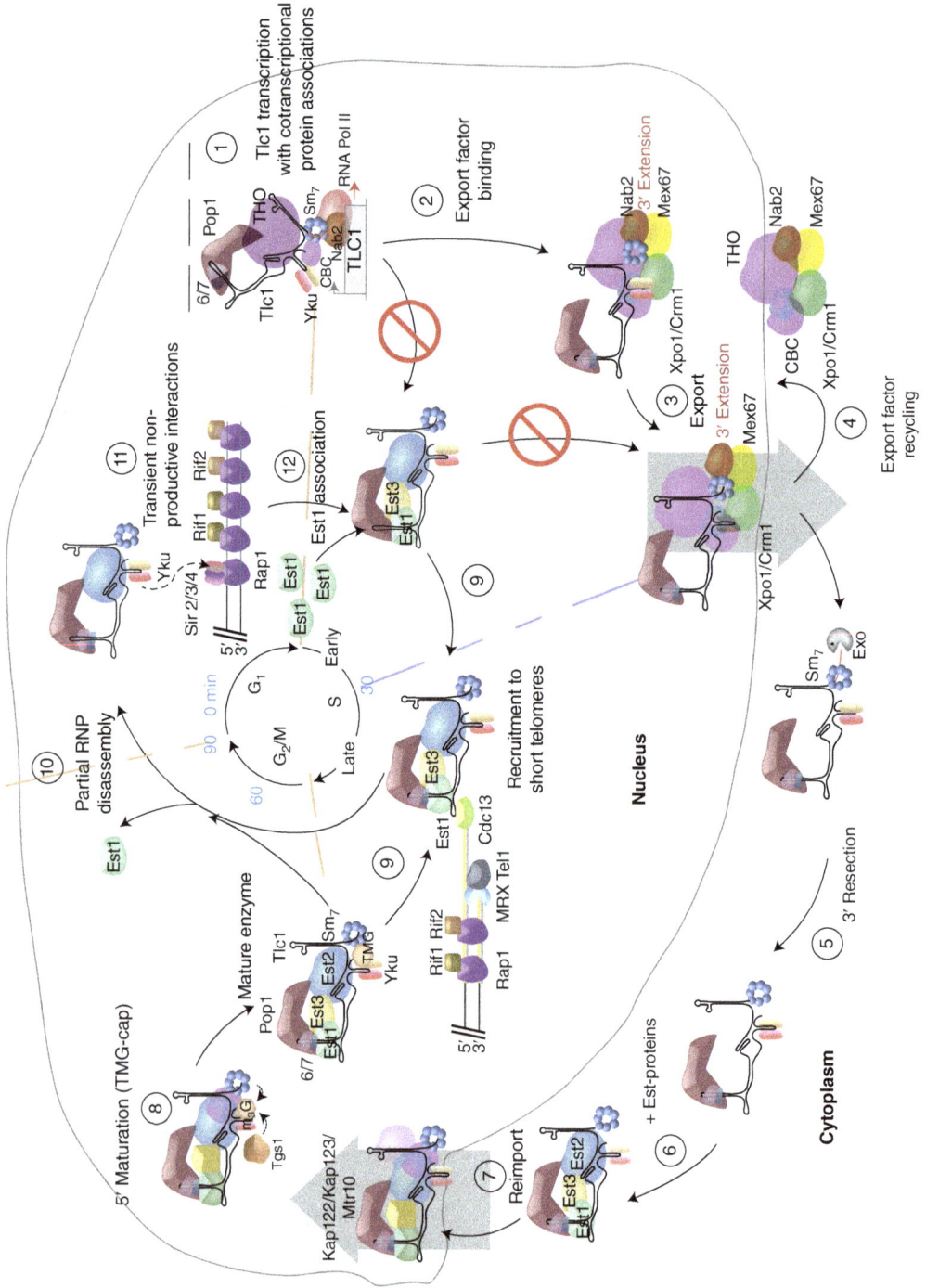

**Figure 3.** (*See following page for legend.*)

The CBC at the 5′ end is thought to interact directly with Xpo1(Crm1), as it does for snRNAs (Becker et al. 2019), and it also interacts with proteins of the THO complex (Eyboulet et al. 2020). At the 3′ end, a cotranscriptional association of the Nab2 protein to the extended RNA serves as an adaptor for the Mex67/Mtr2 proteins (Neumann et al. 2023), as well as for the THO complex (Batisse et al. 2009; Eyboulet et al. 2020). Therefore, the 5′ end, as well as the short extension on the 3′ end, serves as a base and may be bridged by the THO complex to assemble an export-competent RNP. The predicted secondary structure of the RNA puts the 5′ and 3′ ends in very close proximity, which could be helpful for this mechanism. However, the cotranscriptionally assembled complex does not contain proteins that are essential for telomerase recruitment or enzymatic activity (Est1 or Est2), and the RNP is enzymatically inactive (Neumann et al. 2023). Remarkably, the yKu proteins, the Pop6/Pop7, and Pop1 proteins all associate with the Tlc1 RNA during transcription, which is thought to help keep the RNA in conformation, allowing a mature RNP setup. Finally, recent RNA chromatin immunoprecipitation (ChIP) experiments also showed that the $Sm_7$-ring proteins bind cotranscriptionally to many RNAs with the cognate recognition sequence near the 3′ end (Neumann et al. 2023). On Tlc1, this $Sm_7$ ring is associated with the extended precursor RNA, but its presence is required for RNA stabilization only in the cytoplasm, not in the nucleus (Vasianovich et al. 2020).

Once this complex is assembled, it is exported to the cytoplasm (Fig. 3, step 3). Similar to what is known for other RNPs, it is expected that many of the export factors, including the Nab2 complex, will be disassembled on the cytoplasmic side and recycled (Fig. 3, step 4). This in turn will expose the short 3′ extension, which will be resected, generating the mature 3′ end (Fig. 3, step 5). The actual exonuclease for this step has not been identified, but coimmunoprecipitation experiments on fractionated cell extracts strongly suggest that RNA 3′ processing precedes binding of the Est1, Est2, and, most likely, Est3 proteins (Fig. 3, step 6; Neumann et al. 2023). This major protein swap and 3′-end processing prime the RNP for reentry into the nucleus, which presumably is mediated by karyopherin-β proteins Kap122/Kap123 and the importin Mtr10 (Fig. 3, step 7; Ferrezuelo et al. 2002; Gallardo and Chartrand 2008). There are no TMG-capped RNAs detected in the cytoplasm (Neumann et al. 2023), suggesting that this 5′ modification occurs after reentry into the nucleus (Fig. 3, step 8), specifically in the nucleolus, given that the yeast TMG synthase (Tgs1) is a nucleolar enzyme (Mouaikel et al. 2002). This final modification of the TR is thought to use the same mechanism as the TMG capping of snRNAs and snoRNAs in yeast, which involves a molecular interaction between $Sm_7$ ring proteins and the Tgs1 enzyme (Mouaikel et al. 2002). We speculate that the initial short 3′ extension with the Nab2 protein on it may prevent TMG capping before nuclear export by occluding the $Sm_7$–Tgs1 interaction.

After TMG capping in the nucleolus, the telomerase RNP is fully assembled and returns to the nucleoplasm (Fig. 3, mature enzyme) for acting on telomeres (Fig. 3, step 9). As we have established above, the TR is transcribed during the $G_1$–S transition. Using an inducible tagging system to specifically track newly transcribed Tlc1 RNA in time course experiments, we observed that freshly transcribed RNA took

---

Figure 3. A case study: maturation and assembly steps of the budding yeast telomerase ribonucleoproteins (RNP). The drawing depicts the stages and currently known intermediates during the telomerase RNP maturation in *Saccharomyces cerevisiae*. Gray background: nucleus; white background: cytoplasm. Numbers in circles depict individual events that have been experimentally documented. Note that the RNP during the first intranuclear phase (steps 1–3) is inactive and must pass through the cytoplasm to acquire telomerase-specific subunits. In the cytoplasm, resection at the 3′ end is an irreversible event and must be completed before Est2 (telomerase reverse transcriptase [TERT]) association (step 5 always before step 6). For this reason, we propose that the mature RNP after step 8 cannot assemble into an export-competent version anymore and remains in the nucleus for the rest of its life span. For details, see the text.

~60 min from the time of transcription to reentry into the nucleus (Vasianovich et al. 2020). It is, therefore, possible that newly assembled enzymes reenter the nucleus only after the S phase after transcription. Furthermore, there is evidence that telomerase partly disassembles in $G_2/M$ phases. Specifically, as the level of the Est1 protein is reduced at the entry of $G_1$, it may be released from the RNP (Fig. 3, step 10; Osterhage et al. 2006). However, *EST1* transcription is cell cycle regulated and occurs in late G/S. Therefore, newly synthesized Est1 can be assembled in the next S phase with the rest of the telomerase RNP (Fig. 3, step 12) to be recruited to telomeres again (Fig. 3, step 9). In the interval without Est1, telomerase appears stable and can make transient, nonproductive associations with telomeres that are mediated by Yku80–Sir4 interactions (Fig. 3, step 11; Peterson et al. 2001; Roy et al. 2004). Importantly, the existing data do show that the telomerase RNP only makes one passage through the cytoplasm (Neumann et al. 2023). This single passage is ensured by the unidirectional movement of the RNP that is brought about by reversible (change of protein components) but also irreversible changes (nucleolytic processing of the 3' end; TMG capping at the 5' end) during telomerase trafficking. The end result of this dynamic process is a nucleoplasmic telomerase that cannot reassemble to an export-competent RNP and, therefore, is stuck there for the remainder of its life span (Gallardo et al. 2011; Neumann et al. 2023).

This comprehensive time- and space-resolved model for the life of the yeast telomerase RNP can also explain disparate results from the literature. For example, there is some evidence to suggest that telomerase disassembly in $G_2/M$ includes a loss of the Est1 and Est2 proteins (Tucey and Lundblad 2014). However, the Est2-lacking RNPs could simply be new telomerases that have not yet received Est2 (see Fig. 3 before step 6). However, some key questions remain to be answered. At present, it is not clear whether the passage of the RNA through the cytoplasm is an obligatory step for RNP assembly. It remains possible that the RNP could be matured entirely in the nucleus and that the cytoplasmic passage is a nonessential step caused by the RNA being dealt with as any other RNA polymerase II product. Next, we doubt that all proteins of the export-RNP have been identified. The dynamics of the Est3 protein associations is still unclear. This protein is thought not to bind Tlc1 RNA directly but to require prior Est1 and Est2 association. One possibility is that Est3 parallels Est1 dynamics on and off the RNP, but we have very little evidence for that. How is the reimport of the assembled RNP from the cytoplasm to the nucleus regulated? While it seems that the Kap122/Kap123 as well as the karyopherin Mtr10 are involved, the mechanistic details remain to be discovered. Finally, posttranslational modifications on proteins certainly affect the biochemical properties and behavior of the telomerase RNP (Osterhage et al. 2006; Lin et al. 2015; Eyboulet et al. 2020), but there is the potential for much more in this regard.

The fission yeast TR may complete its processing and protein assembly completely in the nucleus, but presently a passage through the cytoplasm cannot be excluded (Páez-Moscoso et al. 2022; Porat et al. 2022). The available evidence suggests that first, a number of structural RNA-binding proteins that are not specific for the telomerase RNP (Pof8/Bmc1/Thc1; $Sm_7$ ring) associate with the TR during or right after transcription. During maturation, the $Sm_7$ ring proteins spur the formation of a TMG cap at the 5' end, but they are eventually exchanged for an $Lsm_7$ ring at the 3' end (Tang et al. 2012). The Pof8/Bmc1/Thc1 proteins will aid in this switch at the 3' end and, at the same time, are required for a stable association of the *S. pombe* catalytic component Trt1 (Páez-Moscoso et al. 2022; Porat et al. 2022). Again, we consider it very much possible that the full complement of telomerase components is not yet known, and hitherto unknown proteins may affect the biogenesis pathway.

## CONCLUDING REMARKS

In summary, the exploration of TR diversity and its role in the evolution of telomerase RNP has greatly advanced our understanding of this critical enzyme in telomere biology over the past few decades. The impressive variety in the structure and biogenesis of TR components contributes to the unique composition and regulation of telomerase RNPs across different eukaryotic species. While this diversity presents certain research challenges, it also reveals some conserved features, particularly in the core association with the crucial catalytic subunit, TERT. The interplay between conservation and diversity among telomerase components offers valuable insights into the evolutionary biology of RNP enzymes and how telomerase functions. We feel it is remarkable that the RNA component of an RNP present in virtually all eukaryotic species and having an evolutionary conserved function is so diverse. Further research into this diversity as well as the conserved features of the RNA may yield new insights into the pathways of RNA evolution and function. Thus, the unique features of species-specific RNPs may also open therapeutic avenues in humans that were not previously suspected.

## ACKNOWLEDGMENTS

We thank many colleagues in the field for constructive discussions. Work in our laboratories was supported by a foundation grant from the Canadian Institutes for Health Research to R.J.W. (FDN154315) and funds from the Canadian Research Chair in Telomere Biology to R.J.W. R.J.W. is also a member of the Center for Research on Aging. This work was also supported by a National Science Foundation (NSF) grant (MCB2046798) and a National Institutes of Health (NIH) grant (GM149864) to J.J.-L.C.

## REFERENCES

*Reference is also in this subject collection.

Bajon E, Laterreur N, Wellinger RJ. 2015. A single templating RNA in yeast telomerase. *Cell Rep* **12:** 441–448. doi:10 .1016/j.celrep.2015.06.045

Bartle L, Vasianovich Y, Wellinger RJ. 2022. Maturation and shuttling of the yeast telomerase RNP: assembling something new using recycled parts. *Curr Genet* **68:** 3–14. doi:10.1007/s00294-021-01210-2

Batisse J, Batisse C, Budd A, Böttcher B, Hurt E. 2009. Purification of nuclear poly(A)-binding protein Nab2 re-

veals association with the yeast transcriptome and a messenger ribonucleoprotein core structure. *J Biol Chem* **284:** 34911–34917. doi:10.1074/jbc.M109.062034

Becker D, Hirsch AG, Bender L, Lingner T, Salinas G, Krebber H. 2019. Nuclear pre-snRNA export is an essential quality assurance mechanism for functional spliceosomes. *Cell Rep* **27:** 3199–3214.e3. doi:10.1016/j.celrep .2019.05.031

Berman AJ, Gooding AR, Cech TR. 2010. *Tetrahymena* telomerase protein p65 induces conformational changes throughout telomerase RNA (TER) and rescues telomerase reverse transcriptase and TER assembly mutants. *Mol Cell Biol* **30:** 4965–4976. doi:10.1128/MCB.00827-10

Bosoy D, Peng Y, Mian IS, Lue NF. 2003. Conserved N-terminal motifs of telomerase reverse transcriptase required for ribonucleoprotein assembly in vivo. *J Biol Chem* **278:** 3882–3890. doi:10.1074/jbc.M210645200

Box JA, Bunch JT, Tang W, Baumann P. 2008. Spliceosomal cleavage generates the 3′ end of telomerase RNA. *Nature* **456:** 910–914. doi:10.1038/nature07584

Brown Y, Abraham M, Pearl S, Kabaha MM, Elboher E, Tzfati Y. 2007. A critical three-way junction is conserved in budding yeast and vertebrate telomerase RNAs. *Nucleic Acids Res* **35:** 6280–6289. doi:10.1093/nar/gkm713

Casacuberta E. 2017. *Drosophila*: retrotransposons making up telomeres. *Viruses* **9:** 192. doi:10.3390/v9070192

Cash DD, Feigon J. 2017. Structure and folding of the *Tetrahymena* telomerase RNA pseudoknot. *Nucleic Acids Res* **45:** 482–495. doi:10.1093/nar/gkw1153

Chapon C, Cech TR, Zaug AJ. 1997. Polyadenylation of telomerase RNA in budding yeast. *RNA* **3:** 1337–1351.

Chappell AS, Lundblad V. 2004. Structural elements required for association of the *Saccharomyces cerevisiae* telomerase RNA with the Est2 reverse transcriptase. *Mol Cell Biol* **24:** 7720–7736. doi:10.1128/MCB.24.17.7720-7736 .2004

*Chen L, Batista LFZ. 2025. Biogenesis and regulation of telomerase during development and cancer. *Cold Spring Harb Perspect Biol* doi:10.1101/cshperspect.a041692

Chen JL, Greider CW. 2003. Template boundary definition in mammalian telomerase. *Genes Dev* **17:** 2747–2752. doi:10.1101/gad.1140303

Chen JL, Greider CW. 2004. An emerging consensus for telomerase RNA structure. *Proc Natl Acad Sci* **101:** 14683–14684. doi:10.1073/pnas.0406204101

Chen JL, Greider CW. 2005. Functional analysis of the pseudoknot structure in human telomerase RNA. *Proc Natl Acad Sci* **102:** 8080–8085. doi:10.1073/pnas.0502259102

Chen JL, Blasco MA, Greider CW. 2000. Secondary structure of vertebrate telomerase RNA. *Cell* **100:** 503–514. doi:10 .1016/S0092-8674(00)80687-X

Chen JL, Opperman KK, Greider CW. 2002. A critical stem-loop structure in the CR4-CR5 domain of mammalian telomerase RNA. *Nucleic Acids Res* **30:** 592–597. doi:10 .1093/nar/30.2.592

Chen L, Roake CM, Freund A, Batista PJ, Tian S, Yin YA, Gajera CR, Lin S, Lee B, Pech MF, et al. 2018. An activity switch in human telomerase based on RNA conformation and shaped by TCAB1. *Cell* **174:** 218–230.e13. doi:10 .1016/j.cell.2018.04.039

Chen L, Roake CM, Galati A, Bavasso F, Micheli E, Saggio I, Schoeftner S, Cacchione S, Gatti M, Artandi SE, et al. 2020. Loss of human TGS1 hypermethylase promotes increased telomerase RNA and telomere elongation. *Cell Rep* **30:** 1358–1372.e5. doi:10.1016/j.celrep.2020.01.004

Chou YS, Logeswaran D, Chow CN, Dunn PL, Podlevsky JD, Liu T, Akhter K, Chen JJL. 2025. A degenerate telomerase RNA directs telomeric DNA synthesis in lepidopteran insects. *Proc Natl Acad Sci* **122:** e2424443122. doi:10.1073/pnas.2424443122

Dandjinou AT, Lévesque N, Larose S, Lucier JF, Abou Elela S, Wellinger RJ. 2004. A phylogenetically based secondary structure for the yeast telomerase RNA. *Curr Biol* **14:** 1148–1158. doi:10.1016/j.cub.2004.05.054

Dionne I, Larose S, Dandjinou AT, Abou Elela S, Wellinger RJ. 2013. Cell cycle-dependent transcription factors control the expression of yeast telomerase RNA. *RNA* **19:** 992–1002. doi:10.1261/rna.037663.112

Egan ED, Collins K. 2010. Specificity and stoichiometry of subunit interactions in the human telomerase holoenzyme assembled in vivo. *Mol Cell Biol* **30:** 2775–2786. doi:10.1128/MCB.00151-10

Eyboulet F, Jeronimo C, Côté J, Robert F. 2020. The deubiquitylase Ubp15 couples transcription to mRNA export. *eLife* **9:** e61264. doi:10.7554/eLife.61264

Fajkus P, Peška V, Závodník M, Fojtová M, Fulnečková J, Dobias S, Kilar A, Dvořáčková M, Zachová D, Nečasová I, et al. 2019. Telomerase RNAs in land plants. *Nucleic Acids Res* **47:** 9842–9856. doi:10.1093/nar/gkz695

Fajkus P, Kilar A, Nelson ADL, Holá M, Peška V, Goffová I, Fojtová M, Zachová D, Fulnečková J, Fajkus J. 2021. Evolution of plant telomerase RNAs: farther to the past, deeper to the roots. *Nucleic Acids Res* **49:** 7680–7694. doi:10.1093/nar/gkab545

Fajkus P, Adámik M, Nelson ADL, Kilar AM, Franek M, Bubeník M, Frydrychová RC, Votavová A, Sýkorová E, Fajkus J, et al. 2023. Telomerase RNA in Hymenoptera (Insecta) switched to plant/ciliate-like biogenesis. *Nucleic Acids Res* **51:** 420–433. doi:10.1093/nar/gkac1202

Ferrezuelo F, Steiner B, Aldea M, Futcher B. 2002. Biogenesis of yeast telomerase depends on the importin mtr10. *Mol Cell Biol* **22:** 6046–6055. doi:10.1128/MCB.22.17.6046-6055.2002

Forino NM, Woo JZ, Zaug AJ, Jimenez AG, Edelson E, Cech TR, Rouskin S, Stone MD. 2025. Telomerase RNA structural heterogeneity in living human cells detected by DMS-MaPseq. *Nat Commun* **16:** 925. doi:10.1038/s41467-025-56149-6

Franke J, Gehlen J, Ehrenhofer-Murray AE. 2008. Hypermethylation of yeast telomerase RNA by the snRNA and snoRNA methyltransferase Tgs1. *J Cell Sci* **121:** 3553–3560. doi:10.1242/jcs.033308

Gallardo F, Chartrand P. 2008. Telomerase biogenesis: the long road before getting to the end. *RNA Biol* **5:** 212–215. doi:10.4161/rna.7115

Gallardo F, Olivier C, Dandjinou AT, Wellinger RJ, Chartrand P. 2008. TLC1 RNA nucleo-cytoplasmic trafficking links telomerase biogenesis to its recruitment to telomeres. *EMBO J* **27:** 748–757. doi:10.1038/emboj.2008.21

Gallardo F, Laterreur N, Cusanelli E, Ouenzar F, Querido E, Wellinger RJ, Chartrand P. 2011. Live cell imaging of telomerase RNA dynamics reveals cell cycle-dependent clustering of telomerase at elongating telomeres. *Mol Cell* **44:** 819–827. doi:10.1016/j.molcel.2011.09.020

Ghanim GE, Fountain AJ, van Roon AM, Rangan R, Das R, Collins K, Nguyen THD. 2021. Structure of human telomerase holoenzyme with bound telomeric DNA. *Nature* **593:** 449–453. doi:10.1038/s41586-021-03415-4

Gillis AJ, Schuller AP, Skordalakes E. 2008. Structure of the *Tribolium castaneum* telomerase catalytic subunit TERT. *Nature* **455:** 633–637. doi:10.1038/nature07283

Greider CW, Blackburn EH. 1987. The telomere terminal transferase of *Tetrahymena* is a ribonucleoprotein enzyme with two kinds of primer specificity. *Cell* **51:** 887–898. doi:10.1016/0092-8674(87)90576-9

Greider CW, Blackburn EH. 1989. A telomeric sequence in the RNA of Tetrahymena telomerase required for telomere repeat synthesis. *Nature* **337:** 331–337. doi:10.1038/337331a0

Gunisova S, Elboher E, Nosek J, Gorkovoy V, Brown Y, Lucier JF, Laterreur N, Wellinger RJ, Tzfati Y, Tomaska L. 2009. Identification and comparative analysis of telomerase RNAs from *Candida* species reveal conservation of functional elements. *RNA* **15:** 546–559. doi:10.1261/rna.1194009

Gupta SK, Kolet L, Doniger T, Biswas VK, Unger R, Tzfati Y, Michaeli S. 2013. The *Trypanosoma brucei* telomerase RNA (TER) homologue binds core proteins of the C/D snoRNA family. *FEBS Lett* **587:** 1399–1404. doi:10.1016/j.febslet.2013.03.017

He Y, Wang Y, Liu B, Helmling C, Sušac L, Cheng R, Zhou ZH, Feigon J. 2021. Structures of telomerase at several steps of telomere repeat synthesis. *Nature* **593:** 454–459. doi:10.1038/s41586-021-03529-9

Jády BE, Bertrand E, Kiss T. 2004. Human telomerase RNA and box H/ACA scaRNAs share a common Cajal body-specific localization signal. *J Cell Biol* **164:** 647–652. doi:10.1083/jcb.200310138

Jamonnak N, Creamer TJ, Darby MM, Schaughency P, Wheelan SJ, Corden JL. 2011. Yeast Nrd1, Nab3, and Sen1 transcriptome-wide binding maps suggest multiple roles in post-transcriptional RNA processing. *RNA* **17:** 2011–2025. doi:10.1261/rna.2840711

Kannan R, Helston RM, Dannebaum RO, Baumann P. 2015. Diverse mechanisms for spliceosome-mediated 3′ end processing of telomerase RNA. *Nat Commun* **6:** 6104. doi:10.1038/ncomms7104

Kao TL, Huang YC, Chen YH, Baumann P, Tseng CK. 2024. LARP3, LARP7, and MePCE are involved in the early stage of human telomerase RNA biogenesis. *Nat Commun* **15:** 5955. doi:10.1038/s41467-024-50422-w

Koch C, Nasmyth K. 1994. Cell cycle regulated transcription in yeast. *Curr Opin Cell Biol* **6:** 451–459. doi:10.1016/0955-0674(94)90039-6

Larose S, Laterreur N, Ghazal G, Gagnon J, Wellinger RJ, Elela SA. 2007. RNase III–dependent regulation of yeast telomerase. *J Biol Chem* **282:** 4373–4381. doi:10.1074/jbc.M607145200

Laterreur N, Lemieux B, Neumann H, Berger-Dancause JC, Lafontaine D, Wellinger RJ. 2018. The yeast telomerase module for telomere recruitment requires a specific RNA architecture. *RNA* **24:** 1067–1079. doi:10.1261/rna.066696.118

Cite this article as *Cold Spring Harb Perspect Biol* doi: 10.1101/cshperspect.a041700

Lemieux B, Laterreur N, Perederina A, Noël JF, Dubois ML, Krasilnikov AS, Wellinger RJ. 2016. Active yeast telomerase shares subunits with ribonucleoproteins RNase P and RNase MRP. *Cell* **165**: 1171–1181. doi:10.1016/j.cell.2016.04.018

Lin J, Ly H, Hussain A, Abraham M, Pearl S, Tzfati Y, Parslow TG, Blackburn EH. 2004. A universal telomerase RNA core structure includes structured motifs required for binding the telomerase reverse transcriptase protein. *Proc Natl Acad Sci* **101**: 14713–14718. doi:10.1073/pnas.0405879101

Lin KW, McDonald KR, Guise AJ, Chan A, Cristea IM, Zakian VA. 2015. Proteomics of yeast telomerase identified Cdc48-Npl4-Ufd1 and Ufd4 as regulators of Est1 and telomere length. *Nat Commun* **6**: 8290. doi:10.1038/ncomms9290

Liu B, He Y, Wang Y, Song H, Zhou ZH, Feigon J. 2022. Structure of active human telomerase with telomere shelterin protein TPP1. *Nature* **604**: 578–583. doi:10.1038/s41586-022-04582-8

Livengood AJ, Zaug AJ, Cech TR. 2002. Essential regions of *Saccharomyces cerevisiae* telomerase RNA: separate elements for Est1p and Est2p interaction. *Mol Cell Biol* **22**: 2366–2374. doi:10.1128/MCB.22.7.2366-2374.2002

Logeswaran D, Li Y, Podlevsky JD, Chen JJ. 2021. Monophyletic origin and divergent evolution of animal telomerase RNA. *Mol Biol Evol* **38**: 215–228. doi:10.1093/molbev/msaa203

Logeswaran D, Li Y, Akhter K, Podlevsky JD, Olson TL, Forsberg K, Chen JJ. 2022. Biogenesis of telomerase RNA from a protein-coding mRNA precursor. *Proc Natl Acad Sci* **119**: e2204636119. doi:10.1073/pnas.2204636119

* Martin A, Hockemeyer D. 2025. Regulation of human telomerase: from molecular interactions to population genetics. *Cold Spring Harb Perspect Biol* doi: 10.1101/cshperspect.a041693

Mason DX, Goneska E, Greider CW. 2003. Stem-loop IV of *Tetrahymena* telomerase RNA stimulates processivity in trans. *Mol Cell Biol* **23**: 5606–5613. doi:10.1128/MCB.23.16.5606-5613.2003

Massenet S, Bertrand E, Verheggen C. 2017. Assembly and trafficking of box C/D and H/ACA snoRNPs. *RNA Biol* **14**: 680–692. doi:10.1080/15476286.2016.1243646

McCormick-Graham M, Romero DP. 1995. Ciliate telomerase RNA structural features. *Nucleic Acids Res* **23**: 1091–1097. doi:10.1093/nar/23.7.1091

Mitchell JR, Collins K. 2000. Human telomerase activation requires two independent interactions between telomerase RNA and telomerase reverse transcriptase. *Mol Cell* **6**: 361–371. doi:10.1016/S1097-2765(00)00036-8

Mitchell JR, Cheng J, Collins K. 1999. A box H/ACA small nucleolar RNA-like domain at the human telomerase RNA 3′ end. *Mol Cell Biol* **19**: 567–576. doi:10.1128/MCB.19.1.567

Mitchell M, Gillis A, Futahashi M, Fujiwara H, Skordalakes E. 2010. Structural basis for telomerase catalytic subunit TERT binding to RNA template and telomeric DNA. *Nat Struct Mol Biol* **17**: 513–518. doi:10.1038/nsmb.1777

Mouaikel J, Verheggen C, Bertrand E, Tazi J, Bordonné R. 2002. Hypermethylation of the cap structure of both yeast snRNAs and snoRNAs requires a conserved methyltransferase that is localized to the nucleolus. *Mol Cell* **9**: 891–901. doi:10.1016/S1097-2765(02)00484-7

Mozdy AD, Cech TR. 2006. Low abundance of telomerase in yeast: implications for telomerase haploinsufficiency. *RNA* **12**: 1721–1737. doi:10.1261/rna.134706

Mozdy AD, Podell ER, Cech TR. 2008. Multiple yeast genes, including Paf1 complex genes, affect telomere length via telomerase RNA abundance. *Mol Cell Biol* **28**: 4152–4161. doi:10.1128/MCB.00512-08

Neumann H, Bartle L, Bonnell E, Wellinger RJ. 2023. Ratcheted transport and sequential assembly of the yeast telomerase RNP. *Cell Rep* **42**: 113565. doi:10.1016/j.celrep.2023.113565

Nguyen THD, Tam J, Wu RA, Greber BJ, Toso D, Nogales E, Collins K. 2018. Cryo-EM structure of substrate-bound human telomerase holoenzyme. *Nature* **557**: 190–195. doi:10.1038/s41586-018-0062-x

Noël JF, Larose S, Abou Elela S, Wellinger RJ. 2012. Budding yeast telomerase RNA transcription termination is dictated by the Nrd1/Nab3 non-coding RNA termination pathway. *Nucleic Acids Res* **40**: 5625–5636. doi:10.1093/nar/gks200

Nostramo RT, Hopper AK. 2020. A novel assay provides insight into tRNAPhe retrograde nuclear import and re-export in *S. cerevisiae*. *Nucleic Acids Res* **48**: 11577–11588. doi:10.1093/nar/gkaa879

Osterhage JL, Talley JM, Friedman KL. 2006. Proteasome-dependent degradation of Est1p regulates the cell cycle-restricted assembly of telomerase in *Saccharomyces cerevisiae*. *Nat Struct Mol Biol* **13**: 720–728. doi:10.1038/nsmb1125

Páez-Moscoso DJ, Ho DV, Pan L, Hildebrand K, Jensen KL, Levy MJ, Florens L, Baumann P. 2022. A putative cap binding protein and the methyl phosphate capping enzyme Bin3/MePCE function in telomerase biogenesis. *Nat Commun* **13**: 1067. doi:10.1038/s41467-022-28545-9

Peterson SE, Stellwagen AE, Diede SJ, Singer MS, Haimberger ZW, Johnson CO, Tzoneva M, Gottschling DE. 2001. The function of a stem-loop in telomerase RNA is linked to the DNA repair protein Ku. *Nat Genet* **27**: 64–67. doi:10.1038/83778

Podlevsky JD, Chen JJL. 2016. Evolutionary perspectives of telomerase RNA structure and function. *RNA Biol* **13**: 720–732. doi:10.1080/15476286.2016.1205768

Podlevsky JD, Bley CJ, Omana RV, Qi X, Chen JJL. 2008. The telomerase database. *Nucleic Acids Res* **36**: D339–D343. doi:10.1093/nar/gkm700

Podlevsky JD, Li Y, Chen JJL. 2016a. The functional requirement of two structural domains within telomerase RNA emerged early in eukaryotes. *Nucleic Acids Res* **44**: 9891–9901. doi:10.1093/nar/gkw605

Podlevsky JD, Li Y, Chen JJL. 2016b. Structure and function of echinoderm telomerase RNA. *RNA* **22**: 204–215. doi:10.1261/rna.053280.115

Porat J, El Baidouri M, Grigull J, Deragon JM, Bayfield MA. 2022. The methyl phosphate capping enzyme Bmc1/Bin3 is a stable component of the fission yeast telomerase holoenzyme. *Nat Commun* **13**: 1277. doi:10.1038/s41467-022-28985-3

Prathapam R, Witkin KL, O'Connor CM, Collins K. 2005. A telomerase holoenzyme protein enhances telomerase

RNA assembly with telomerase reverse transcriptase. *Nat Struct Mol Biol* **12:** 252–257. doi:10.1038/nsmb900

Qi X, Li Y, Honda S, Hoffmann S, Marz M, Mosig A, Podlevsky JD, Stadler PF, Selker EU, Chen JJL. 2013. The common ancestral core of vertebrate and fungal telomerase RNAs. *Nucleic Acids Res* **41:** 450–462. doi:10.1093/nar/gks980

Qi X, Rand DP, Podlevsky JD, Li Y, Mosig A, Stadler PF, Chen JJL. 2015. Prevalent and distinct spliceosomal 3′-end processing mechanisms for fungal telomerase RNA. *Nat Commun* **6:** 6105. doi:10.1038/ncomms7105

Rajendra K, Cheng R, Zhou S, Lizarazo S, Smith DJ, Van Bortle K. 2024. Evidence of RNA polymerase III recruitment and transcription at protein-coding gene promoters. *Mol Cell* **84:** 4111–4124.e5. doi:10.1016/j.molcel.2024.09.019

Rowland TJ, Dumbović G, Hass EP, Rinn JL, Cech TR. 2019. Single-cell imaging reveals unexpected heterogeneity of telomerase reverse transcriptase expression across human cancer cell lines. *Proc Natl Acad Sci* **116:** 18488–18497. doi:10.1073/pnas.1908275116

Roy R, Meier B, McAinsh AD, Feldmann HM, Jackson SP. 2004. Separation-of-function mutants of yeast Ku80 reveal a Yku80p-Sir4p interaction involved in telomeric silencing. *J Biol Chem* **279:** 86–94. doi:10.1074/jbc.M306841200

Schmidt JC, Cech TR. 2015. Human telomerase: biogenesis, trafficking, recruitment, and activation. *Genes Dev* **29:** 1095–1105. doi:10.1101/gad.263863.115

Sekne Z, Ghanim GE, van Roon AM, Nguyen THD. 2022. Structural basis of human telomerase recruitment by TPP1-POT1. *Science* **375:** 1173–1176. doi:10.1126/science.abn6840

Seto AG, Zaug AJ, Sobel SG, Wolin SL, Cech TR. 1999. *Saccharomyces cerevisiae* telomerase is an Sm small nuclear ribonucleoprotein particle. *Nature* **401:** 177–180. doi:10.1038/43694

Seto AG, Livengood AJ, Tzfati Y, Blackburn EH, Cech TR. 2002. A bulged stem tethers Est1p to telomerase RNA in budding yeast. *Genes Dev* **16:** 2800–2812. doi:10.1101/gad.1029302

Singer MS, Gottschling DE. 1994. *TLC1*: template RNA component of *Saccharomyces cerevisiae* telomerase. *Science* **266:** 404–409. doi:10.1126/science.7545955

Singh M, Wang Z, Koo BK, Patel A, Cascio D, Collins K, Feigon J. 2012. Structural basis for telomerase RNA recognition and RNP assembly by the holoenzyme La family protein p65. *Mol Cell* **47:** 16–26. doi:10.1016/j.molcel.2012.05.018

Song J, Logeswaran D, Castillo-González C, Li Y, Bose S, Aklilu BB, Ma Z, Polkhovskiy A, Chen JJ, Shippen DE. 2019. The conserved structure of plant telomerase RNA provides the missing link for an evolutionary pathway from ciliates to humans. *Proc Natl Acad Sci* **116:** 24542–24550. doi:10.1073/pnas.1915312116

Spellman PT, Sherlock G, Zhang MQ, Iyer VR, Anders K, Eisen MB, Brown PO, Botstein D, Futcher B. 1998. Comprehensive identification of cell cycle-regulated genes of the yeast *Saccharomyces cerevisiae* by microarray hybridization. *Mol Biol Cell* **9:** 3273–3297. doi:10.1091/mbc.9.12.3273

Stone MD, Mihalusova M, O'Connor CM, Prathapam R, Collins K, Zhuang X. 2007. Stepwise protein-mediated RNA folding directs assembly of telomerase ribonucleoprotein. *Nature* **446:** 458–461. doi:10.1038/nature05600

Tang W, Kannan R, Blanchette M, Baumann P. 2012. Telomerase RNA biogenesis involves sequential binding by Sm and Lsm complexes. *Nature* **484:** 260–264. doi:10.1038/nature10924

Teixeira MT, Förstemann K, Gasser SM, Lingner J. 2002. Intracellular trafficking of yeast telomerase components. *EMBO Rep* **3:** 652–659. doi:10.1093/embo-reports/kvf133

Tseng CK, Wang HF, Schroeder MR, Baumann P. 2018. The H/ACA complex disrupts triplex in hTR precursor to permit processing by RRP6 and PARN. *Nat Commun* **9:** 5430. doi:10.1038/s41467-018-07822-6

Tucey TM, Lundblad V. 2014. Regulated assembly and disassembly of the yeast telomerase quaternary complex. *Genes Dev* **28:** 2077–2089. doi:10.1101/gad.246256.114

Tuzon CT, Wu Y, Chan A, Zakian VA. 2011. The *Saccharomyces cerevisiae* telomerase subunit Est3 binds telomeres in a cell cycle– and Est1-dependent manner and interacts directly with Est1 in vitro. *PLoS Genet* **7:** e1002060. doi:10.1371/journal.pgen.1002060

Vasianovich Y, Bajon E, Wellinger RJ. 2020. Telomerase biogenesis requires a novel Mex67 function and a cytoplasmic association with the Sm7 complex. *eLife* **9:** e60000. doi:10.7554/eLife.60000

Vítková M, Král J, Traut W, Zrzavý J, Marec F. 2005. The evolutionary origin of insect telomeric repeats, (TTAGG)$_n$. *Chromosome Res* **13:** 145–156. doi:10.1007/s10577-005-7721-0

Wang Y, Gallagher-Jones M, Sušac L, Song H, Feigon J. 2020. A structurally conserved human and *Tetrahymena* telomerase catalytic core. *Proc Natl Acad Sci* **117:** 31078–31087. doi:10.1073/pnas.2011684117

Xi L, Cech TR. 2014. Inventory of telomerase components in human cells reveals multiple subpopulations of hTR and hTERT. *Nucleic Acids Res* **42:** 8565–8577. doi:10.1093/nar/gku560

Zappulla DC, Goodrich K, Cech TR. 2005. A miniature yeast telomerase RNA functions in vivo and reconstitutes activity in vitro. *Nat Struct Mol Biol* **12:** 1072–1077. doi:10.1038/nsmb1019

Závodník M, Fajkus P, Franek M, Kopecký D, Garcia S, Dodsworth S, Orejuela A, Kilar A, Ptáček J, Mátl M, et al. 2023. Telomerase RNA gene paralogs in plants—the usual pathway to unusual telomeres. *New Phytol* **239:** 2353–2366. doi:10.1111/nph.19110

# Telomeres and Human Disease

Sharon A. Savage

Clinical Genetics Branch, Division of Cancer Epidemiology and Genetics, National Cancer Institute, National Institutes of Health, Bethesda, Maryland 20892-6772, USA

*Correspondence:* savagesh@mail.nih.gov

Telomeres, the long nucleotide repeats, and protein complex at chromosome ends, are central to genomic integrity. Telomere length (TL) varies widely between populations due to germline genetics, environmental exposures, and other factors. Very short telomeres caused by pathogenic germline variants in telomere maintenance genes cause the telomere biology disorders, a spectrum of life-threatening conditions including bone marrow failure, liver and lung disease, cancer, and other complications. Cancer predisposition with long telomeres is caused by rare pathogenic germline variants in components of the shelterin telomere protection protein complex and associated primarily with elevated risk of melanoma, thyroid cancer, sarcoma, and lymphoproliferative malignancies. In the middle, studies of the general population at risk of common illnesses, such as cardiovascular disease and cancer, have found statistically significant differences in TL but uncertain clinical applicability. This work reviews connections between telomere biology and human disease focusing on similarities and differences across the phenotypic spectrum.

Connections between telomere structure, function, and length are of great interest in the context of human disease because of their intrinsic roles in genomic integrity, cell division, and cell death. Telomeres, long $(TTAGGG)_n$ repeats and a protein complex at chromosome ends, shorten with each cell division due to the inability of DNA polymerase to fully replicate DNA ends during cell division. Consequently, telomeres have been studied as a biomarker of aging. The first connection between human genetic disease and telomere biology came with the 1998 discovery of germline mutations in dyskerin, a key component of the telomerase enzyme complex, in the X-linked recessive form of dyskeratosis congenita (DC) (Heiss et al. 1998; Mitchell et al. 1999). Since then, advances in genetics, telomere length (TL) measurement, and studies of rare and common diseases have led to a much broader understanding of the complex role telomeres play in human aging and disease etiology. This work reviews the connections between rare and common germline genetic variants and the resultant clinical manifestations with the goal of encouraging additional mechanistic research defining the connections between aberrant telomere biology and human disease.

## TELOMERES AND THE HUMAN LIFE SPAN

TL has been widely studied in the context of normal aging and in association with a bevy of rare and common diseases. TL varies widely in human populations and consistently declines with age with a more rapid decline in childhood than middle age. The population variability in lymphocyte TL ranges from ~8200 to 12,600 bp in newborns and 4600 to 8900 bp in 40-year-old healthy individuals as measured by flow cytometry with fluorescent in situ hybridization (flow FISH) (Baerlocher and Lansdorp 2003; Alter et al. 2012). TL studies using Southern blot estimated the annual rate of TL change at −27 bp with a wide range of −2900 to +740 change in the number of base pairs per year (Ye et al. 2023). This variability in human TL is attributed to numerous intrinsic and extrinsic factors including inheritance of rare and common germline variants, epigenetic regulation, environmental exposures, and other factors still to be discovered (Aviv and Shay 2018; Savage 2018; Chakravarti et al. 2021; Codd et al. 2021; Niewisch et al. 2022; Rossiello et al. 2022; Revy et al. 2023).

Telomere shortening imposes limits on cellular replicative capacity and thus there is an intrinsic limit to life span (Dong et al. 2016; Lenart and Vaupel 2017; Rozing et al. 2017). Work by Steenstrup et al. (2017) assessed leukocyte TL by Southern blot in more than 12,000 individuals, conducted computer simulations for age-related TL decline and mortality rates for each population, and compared these data with that of individuals with DC and their family members. They found that TL <5 kbp was the threshold below which the probability of survival substantially declines and termed this the "telomeric brink." The authors assessed average TL by Southern blot but were not able to address differences in TL by specific chromosome arm, which is important because the shortest telomere triggers cellular senescence or apoptosis, not the average TL (Hemann et al. 2001). Notably, studies of single TL using other methods (described below) found some individuals with telomeres ≪5 kbp (Lai et al. 2022, 2023; Raj et al. 2023; Karimian et al. 2024). A telomeric brink as a concept is mechanistically plausible based on the current understanding of telomere biology and cellular replicative capacity but requires more study.

## DETERMINING TELOMERE LENGTH

Measuring TL by a variety of methods has been widely used to assess the multifaceted role(s) of telomere biology in human disease. It is important to understand the strengths and limitations of different telomere measurement methods before determining their relevance (or lack thereof) to disease.

### Measuring Telomeres

The many methods of determining TL have been reviewed in detail elsewhere (Aubert et al. 2012; Lee et al. 2017; Lai et al. 2018; Ferrer et al. 2023b). In brief, the laboratory gold standard terminal restriction fragment (TRF) assay uses specific restriction enzymes and Southern blotting to measure TL but is relatively limited by the need for micrograms of high-quality DNA. TRF is difficult to use in a high-throughput manner and has not been widely adopted by the clinical or epidemiology communities, with a few exceptions (Toupance et al. 2017; Arbeev et al. 2020). The advent of qPCR as a high-throughput method for measuring TL by comparing the telomere copy number with that of a single copy gene made TL measurement feasible for large, population-scale studies. qPCR TL measurement remains a valuable tool provided the appropriate quality control measures (e.g., consistent DNA extraction methods, DNA storage, and batch effects) are considered (Gadalla et al. 2016; Dagnall et al. 2017; Lindrose et al. 2021).

Flow FISH TL measurement in lymphocytes was the first diagnostic assay established for telomere biology disorders (TBDs) (Alter et al. 2007). It allows for TL measurement in total granulocytes, total lymphocytes, and four lymphocyte subsets (naive T cells, memory T cells, B cells, and natural killer [NK] cells) providing opportunities to also understand TL in hematopoiesis (Baerlocher and Lansdorp 2003; Baerlocher et al. 2007; Ouyang et al. 2007). Telomeres <1st percentile for age are highly sensitive and specific for diagnosing DC and related TBDs (discussed further below) (Alter et al. 2007, 2012; Gadalla et al. 2016; Alder et al. 2018).

Cite this article as *Cold Spring Harb Perspect Biol* doi: 10.1101/cshperspect.a041684

There has been long-standing interest in measuring the length of single telomeres and methods have recently improved to make this feasible. STELA (single TL analysis), which measures TL at chromosomes 17p and XpYp was one of the first methods developed (Baird et al. 2003). HT-STELA, a high-throughput version of STELA, has been shown to be useful in diagnosing TBDs (Norris et al. 2021). Telomere shortest length assay (TeSLA) can determine the distribution of TL across all chromosomes, including that of ultrashort telomeres, but is a labor-intensive PCR and gel-based method requiring relatively large quantities of high-quality DNA (Lai et al. 2017, 2023; Raj et al. 2023).

Significant advances in next-generation sequencing technologies have resulted in several exciting studies using long-read whole genome sequencing and specialized bioinformatic algorithms to determine the TL of specific chromosome arms (Karimian et al. 2024; Sanchez et al. 2024; Schmidt et al. 2024). These approaches have the potential to improve the diagnosis of TBDs, to improve understanding of longitudinal changes in TL, and inform future studies aimed at understanding telomeres and genomic stability.

## Genetically Inferring Telomere Length

Genome-wide association studies (GWAS) seeking to understand the contribution of common germline variants to telomere biology in populations have discovered numerous single-nucleotide polymorphisms (SNPs) across the genome associated with TL (Prescott et al. 2011; Codd et al. 2013, 2021; Mangino et al. 2015; Telomeres Mendelian Randomization Collaboration et al. 2017; Delgado et al. 2018; Allaire et al. 2023). The coding and noncoding regions associated with TL included known telomere biology genes (e.g., *RTEL1* and *TERT*) and numerous other new regions. Although most GWAS used qPCR to measure TL, the breadth and depth of consistency in findings at the population level outweigh its lack of sensitivity in measuring very short telomeres. The largest TL GWAS to date used qPCR TL and genome-wide genotype data from 464,716 UK Biobank participants to confirm 30 known loci and discover 108 others (Codd et al. 2021). This study also estimated the heritability of leukocyte relative TL at 8.1% in that population.

TL GWAS discoveries led many to use Mendelian randomization methods to develop a polygenic score (PGS) in lieu of actually measuring telomeres. Combinations of SNPs can be used to compute a polygenic score (PGS) of baseline TL between different groups (Machiela et al. 2017; Telomeres Mendelian Randomization Collaboration et al. 2017; Codd et al. 2021; Chen et al. 2023). In this way, PGSs have the potential to advance understanding of the role of baseline TL in human disease while avoiding the challenges and pitfalls of TL measurement and the effects of aging and environmental exposures on telomeres (Polygenic Risk Score Task Force of the International Common Disease 2021; Wand et al. 2023). For example, Brown et al. (2022) measured TL, performed a GWAS, and did a targeted somatic sequencing in a study of myelofibrosis, a myeloproliferative neoplasm associated with somatic JAK2$^{V617F}$ mutations and high risk of progression to acute myeloid leukemia (AML). They found that although qPCR-measured TL was shorter in cases with myelofibrosis than in controls, the cases had longer PGS TL, suggesting that their cells had an increased replicative capacity from birth that could have allowed for continued clonal expansion after the occurrence of certain somatic mutations.

## TELOMERES AND COMMON DISEASE

### The Advent of Telomere Molecular Epidemiology

The 2002 publication of qPCR as a high-throughput method to measure TL spawned a plethora of studies seeking to understand TL associations with many common diseases and human phenotypes (Cawthon 2002; Martin-Ruiz et al. 2015; Lindrose et al. 2021). Many early epidemiology studies of TL, regardless of phenotype, were limited by statistical power due to small sample sizes and ascertainment biases, especially in cancer, because cases were obtained at or after diagnosis when disease- and/or treatment-related factors could contribute to changes in TL (Wentzensen et al. 2011). The Telomere Research Network (TRN) was created in 2019 to facilitate "the collaboration between basic telomere biologists, population and exposure researchers, and other scientists across disciplines to advance interdisciplinary research on telomeres as sentinels of environment exposure, psychosocial stress, and disease susceptibility." The TRN has reviewed the strengths and limitations of TL reporting across population studies and developed recommendations for sample collection and storage and qPCR TL assays (available at trn.tulane.edu) (Lindrose et al. 2021).

For illustrative purposes, a few examples of recent studies and reviews of TL and cancer, cardiovascular disease, psychosocial well-being,

and other phenotypes are described below (Fig. 1). These examples are not meant to be inclusive of the entire field but provide some background and the foundation for further reading.

## Cancer

The first studies seeking to determine whether germline TL, typically derived from blood or buccal cell DNA, was associated with risk of certain cancers were published more than 20 years ago (Wu et al. 2003). Initially, short telomeres appeared to be associated with an increased risk of head and neck squamous cell carcinoma (HNSCC), bladder cancer, and possibly breast and lung cancers (Wu et al. 2003; Shen et al. 2011; Prescott et al. 2012). Subsequent studies failed to replicate those findings and as germline TL in other cancers was studied, key limitations of many association studies were

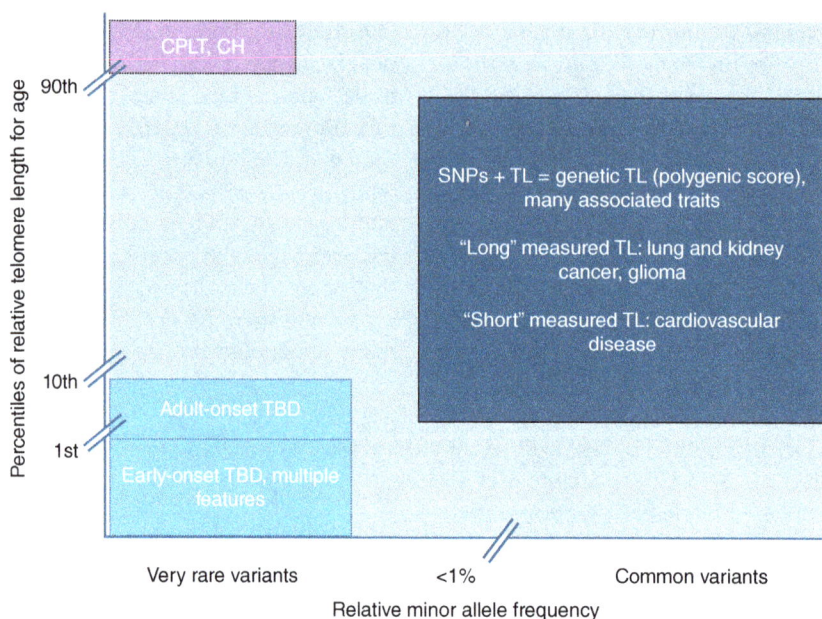

Figure 1. Schematic illustrating relative telomere lengths (TLs) and germline variant minor allele frequency in relation to human phenotypes. Percentiles of TL are estimates based on lymphocyte flow cytometry with fluorescence in situ hybridization measures for the telomere biology disorders (TBDs). Flow FISH TL has not been systematically studied as a possible diagnostic test in patients with cancer predisposition with long telomeres (CPLTs) or clonal hematopoiesis (CH). Variant allele frequencies are meant to show the rarity of variants associated with TBDs and CPLT with the common diseases associated with common single-nucleotide polymorphisms.

noted including ascertainment biases (i.e., obtaining DNA samples after cancer diagnosis and/or therapy that could affect TL), limited statistical power, and technical issues related to qPCR and other TL measurement methods (Wentzensen et al. 2011; Savage 2024).

The data are now relatively consistent. Long telomeres are associated with an increased risk of lung cancer, melanoma, sarcoma, prostate cancer, hematological cancers, and non-Hodgkin lymphoma (Zhang et al. 2015; Rode et al. 2016; Barthel et al. 2017; Ballinger et al. 2023; Chen et al. 2023; Liu et al. 2023). Notably, the contribution of rare germline mutations in shelterin complex genes likely contributes to some of the findings in these association studies—see the section Cancer Predisposition with Long Telomeres. Breast cancer risk and TL do not appear to be related. Bladder, esophageal, and gastric cancers appear to be associated with shorter telomeres. Additional information can be found in Telomeres Mendelian Randomization Collaboration et al. (2017), the meta-analysis of 73 published TL and cancer association studies (Giaccherini et al. 2021), in the study of 472,174 UK Biobank participants (Codd et al. 2021), and systematic review and meta-analysis of Mendelian randomization studies (Chen et al. 2023).

## Cardiovascular Disease

Case-control and population-based studies of cardiovascular disease have been some of the more consistent studies to show a relationship between short buccal or blood cell telomeres and disease risk (Fitzpatrick et al. 2007; Said et al. 2017; Révész et al. 2018; Arbeev et al. 2020; Codd et al. 2021). Early studies hypothesized that telomeres could be a sentinel of the accumulated burden of age-related oxidative and inflammatory stresses, which, in turn, could be connected with atherosclerosis, a common age-related disease (Yin and Pickering 2023). Small studies finding short leukocyte telomeres in people with coronary artery disease, atherosclerosis, and related diseases were replicated in numerous large studies. The 2014 meta-analysis of 15 cohort and 12 case-control studies found

that short telomeres were associated with increased risk of stroke, myocardial infarction, and type 2 diabetes (D'Mello et al. 2015).

Using leukocyte TL data from 472,432 individuals in the UK Biobank, Schneider et al. (2022) assessed relationships between TL and overall mortality and several other noncancer contributors to mortality. They confirmed the work of many others, finding increased cardiovascular mortality in people with shorter telomeres. However, like most of the association studies described in this section, the hazard ratio was small, 1.09 (95% confidence interval [CI] 1.06–1.12). Subsequent studies over the last decade have similarly confirmed these findings as reviewed (Yin and Pickering 2023).

## Other Phenotypes

A complete review of the studies seeking to determine whether germline TL is associated with a multitude of human conditions is well beyond the scope of this paper. A few examples of meta-analyses and systematic reviews are noted here.

A meta-analysis of 17 studies of social support and TL did not find statistically significant associations (Montoya and Uchino 2023). Another systematic review and meta-analysis of 543 associations from 138 studies found a weak association between shorter telomeres and greater life adversity but noted there was evidence of publication bias and that most studies were underpowered (Pepper et al. 2018). A study of the relationship(s) between leukocyte TL and health behaviors and modifiable traits in 422,797 UK Biobank participants found an array of statistically significant associations but that healthy behaviors explained <0.2% of TL variability (Bountziouka et al. 2022). The systematic review and meta-analysis of 62 studies with 310 outcomes and 396 Mendelian randomization associations found TL associated with an array of health outcomes, including short telomeres and inflammatory diseases (Chen et al. 2023).

## TELOMERE BIOLOGY DISORDERS (TBDs)

The term "telomere biology disorder" or "TBD" encompasses the spectrum of germline genetic disorders caused by rare germline pathogenic variants (i.e., mutations) in genes essential in telomere maintenance and stability resulting in short telomeres for age and a spectrum of phenotypes (Fig. 1; Tables 1 and 2; Niewisch and Savage 2019; Tummala et al. 2022; Revy et al. 2023; Savage et al. 2024). DC, the first described TBD, is classically diagnosed by the presence of the mucocutaneous triad of oral leukoplakia, nail dysplasia, and abnormal skin pigmentation or the presence of two of three of the triad and bone marrow failure (BMF) (Fig. 2; Heiss et al. 1998; Mitchell et al. 1999; Vulliamy et al. 2006). Patients with DC are also at high risk of pulmonary fibrosis, AML, myelodysplastic syndrome (MDS), HNSCC, cryptogenic liver disease, stenosis of the urethra, esophagus, or lacrimal ducts, avascular necrosis of the hips or shoulders, and many other complications (Alter et al. 2018; Niewisch and Savage 2019; Tummala et al. 2022; Revy et al. 2023; Schratz et al. 2023; Vittal et al. 2023). Subsequent discoveries of germline mutations in telomerase (*TERT*) and its RNA component, *TERC*, as autosomal dominant (AD) causes of classic DC and in patients and families with apparently isolated BMF or pulmonary fibrosis suggested a wider phenotypic spectrum and uncovered variable phenotypic penetrance and expressivity (Tables 1 and 2; Yamaguchi et al. 2003, 2005; Marrone et al. 2004; Vulliamy et al. 2004; Tsakiri et al. 2007; Diaz de Leon et al. 2010; Niewisch and Savage 2019; Niewisch et al. 2022; Tummala et al. 2022; Revy et al. 2023).

Lymphocyte telomeres <1st percentile for age measured by flow FISH is highly sensitive and specific for differentiating patients with DC from their unaffected relatives and those with other inherited BMF syndromes and is the primary diagnostic test for TBDs (Alter et al. 2008, 2012, 2018; Tometten et al. 2023). HT-STELA is a relatively new diagnostic test for TBDs and has not yet been widely implemented but holds promise to make diagnosis quicker and easier (Norris et al. 2021). Defining a TL cutoff based on age led to the discovery of germline *TINF2* mutations as a relatively common cause of TBDs before the age of whole exome sequencing (Savage et al. 2008). Specifically, variants in exon 6 of

TIN2 (encoded by *TINF2*) were identified by using very short TL as the phenotype in a large family instead of their variable clinical presentations (Savage et al. 2008). Subsequent studies using related approaches and next-generation sequencing over the last decade have greatly expanded the understanding of the genetic causes of TBDs to include at least 18 different genes (Table 1; Walne et al. 2007, 2013; Vulliamy et al. 2008; Zhong et al. 2011; Anderson et al. 2012; Polvi et al. 2012; Ballew et al. 2013a,b; Le Guen et al. 2013; Guo et al. 2014; Kocak et al. 2014a; Moon et al. 2015; Tummala et al. 2015; Simon et al. 2016; Stanley et al. 2016; Takai et al. 2016; Gable et al. 2019; Toufektchan et al. 2020; Kermasson et al. 2022; Sharma et al. 2022; Revy et al. 2023).

### Genotype–Phenotype Associations

The clinical manifestations of TBDs can be very complex based on the affected gene and mode of inheritance of the disease. TBD features can develop at any age and do not always develop in a consistent sequence (Walne et al. 2016; Alter et al. 2018; Ward et al. 2018; Schratz et al. 2020; Himes et al. 2021; Mangaonkar et al. 2021; Niewisch et al. 2022; Tometten et al. 2023). Classic DC, Hoyeraal–Hreidarsson syndrome (HH), and Coats plus typically present early in childhood, whereas middle aged or older adults may present with just one feature of a TBD (e.g., isolated pulmonary fibrosis). The most common adult-onset TBDs are caused by heterozygous pathogenic variants in *TERT*, *TERC*, *RTEL1*, or *PARN*. The clinical manifestations in these patients usually occur in adulthood but the spectrum can vary. Although they may present with only one TBD feature, such patients remain at risk of other complications, especially BMF if lung transplant is required for pulmonary fibrosis (Silhan et al. 2014; Newton et al. 2017).

Patients with HH are often diagnosed as infants or toddlers with cerebellar hypoplasia resulting in ataxia and developmental delay, difficult-to-diagnose immunodeficiency, and intrauterine growth restriction in addition to features of DC (Hreidarsson et al. 1988; Knight

**Table 1.** Genes with rare germline variants implicated in telomere biology disorders

| Functional group | Gene (protein name and abbreviation) | Function | Consequences of mutation(s) | Disease subtype(s) | Mode of inheritance | Telomere length |
|---|---|---|---|---|---|---|
| Telomerase enzyme complex | DKC1 (dyskerin, DKC1) | Telomerase assembly, TERC stability | Reduced TERC stability and telomerase activity | DC, HH, PF Female carriers may have subtle findings | XLR | Very short |
| | NAF1 (nuclear assembly factor 1 ribonucleoprotein [NAF1]) | Telomerase assembly, TERC stability | Reduced TERC stability and telomerase activity | PF, LD, MDS | AD | Short |
| | NHP2 (NOLA2 nucleolar protein family A, member 2 [NHP2]) | Telomerase assembly, TERC stability | Reduced TERC stability and telomerase activity | DC | AR | Very short |
| | NOP 10 (NOLA nuclear protein family A, member 3 [NOP10]) | Telomerase assembly, TERC stability | Reduced TERC stability and telomerase activity | DC | AR | Very short |
| | TERC (encodes an RNA: hTR, human telomerase RNA component [TERC]) | Telomere elongation | Reduction of telomerase activity | DC, BMF, PF, LD, MDS, AML, HH | AD | Short to very short |
| | TERT (telomerase reverse transcriptase [TERT]) | Telomere elongation/ telomerase recruitment | Reduction telomerase recruitment, processivity, and/ or activity | DC, BMF, PF, LD, MDS, AML | AD | Short to very short |
| | | | | HH | AR | Very short |
| Shelterin | ACD (telomere protection protein 1 [TPP1]) | Telomerase recruitment, activity, and processivity | Impaired telomerase recruitment | BMF | AD | Short |
| | | | | HH | AR | Very short |
| | POT1 (protection of telomeres 1 [POT1]) | Interaction with CST complex, negative telomerase regulation, telomere protection from DDR | Defective telomerase regulation, dysfunctional telomere replication | CP | AR | Short |
| | TINF2 (TERF1 [TRF1]-interacting nuclear factor 2 [TIN2]) | Telomerase regulation, sister telomere cohesion, telomere protection from DDR | Multifactorial disruption of telomere maintenance | DC, HH, RS, PF | AD, de novo | Very short |

*Continued*

**Table 1.** *Continued*

| Functional group | Gene (protein name and abbreviation) | Function | Consequences of mutation(s) | Disease subtype(s) | Mode of inheritance | Telomere length |
|---|---|---|---|---|---|---|
| CST | *CTC1* (conserved telomere maintenance component 1 [CTC1]) | CST-complex/C-strand fill in, telomere replication | Impaired telomere replication, fragile telomeres | DC, CP | AR | Short to very short |
| | *STN1* (STN1 subunit of CST-complex subunit [STN1]) | CST-complex/C-strand fill in, telomere replication | Impaired telomere replication | CP | AR | Short to very short |
| Functionally connected with telomere biology through different interactions | *DCLRE1B* (DNA cross-link repair 1B, Apollo) | Repair of interstrand DNA cross-links interacts with TRF2 | Impaired DNA repair, telomere instability | DC, HH | AR | Normal |
| | *PARN* (poly(A)-specific ribonuclease [PARN]) | Associated with telomerase complex/TERC RNA maturation and stabilization | Reduced TERC stability and telomerase activity | PF | AD | Short |
| | | | | DC, HH | AR | Very short |
| | *RPA1* (replication protein A1 [RPA1]) | DNA repair | Impaired telomere maintenance | PF, BMF | AD | Short |
| | *RTEL1* (regulator of telomere elongation helicase 1 [RTEL1]) | Telomeric DNA replication/repair, T-loop stability and unwinding, prevention of telomere loss during cell division | Impaired telomere replication/stability, impaired telomere stability | PF, BMF, DC | AD | Short |
| | | | | DC, HH | AR | Very short |
| | *WRAP53* (telomere Cajal body associated protein 1 [TCAB1]) | Associated with telomerase complex/telomerase trafficking through Cajal bodies and recruitment | Impaired telomerase trafficking though Cajal body and recruitment to telomeres | DC, HH | AR | Very short |
| | *ZCCHC8* (zinc finger CCHC-type containing 8 [ZCCHC8]) | hTR maturation and stability | Reduced hTR maturation and stability | BMF, PF | AD | Short |
| Suggested connections but less direct | *MDM4* (MDM4 regulator of p53 [MDM4]) | Inhibition of p53 activity | Hyperactivation of p53 | BMF, MDS, HNSCC | AD | Short |

*Continued*

Cite this article as *Cold Spring Harb Perspect Biol* doi: 10.1101/cshperspect.a041684

**Table 1.** *Continued*

| Functional group | Gene (protein name and abbreviation) | Function | Consequences of mutation(s) | Disease subtype(s) | Mode of inheritance | Telomere length |
|---|---|---|---|---|---|---|
| | *NPM1* (nucleophosmin/ nucleoplasmin family member 1 [NPM1]) | Ribosomal protein assembly and transport | Impaired ribosomal RNA maturation, altering hTR stability | BMF | AD | Short |
| | TYMS-ENOSF1/ TYMS | De novo nucleotide synthesis, thymidine nucleotide metabolism | Impaired telomerase regulation | DC | Digenic | Short |

Telomere length listed here is based on age-adjusted comparisons from publications listed in the text and is not meant to be quantitative but to give the reader the context by which telomere length is affected by the germline variant(s). (Table updated and adapted from Niewisch and Savage 2019.)

(AD) Autosomal dominant, (AML) acute myeloid leukemia, (AR) autosomal recessive, (BMF) bone marrow failure, (CP) Coats plus syndrome, (CPLT) cancer predisposition with long telomeres, (DC) dyskeratosis congenita, (DDR) DNA damage response, (HH) Hoyeraal–Hreidarsson syndrome, (LF) liver disease, (MDS) myelodysplastic syndrome, (PF) pulmonary fibrosis, (RS) Revesz syndrome.

et al. 1999; Le Guen et al. 2013; Kocak et al. 2014b; Rolles et al. 2023). Bilateral exudative retinopathy, intracranial calcifications, and intrauterine growth restriction are features of Revesz syndrome and Coats plus (Tolmie et al. 1988; Revesz et al. 1992; Crow et al. 2004). Additionally, patients with Coats plus typically have bone healing abnormalities and gastrointestinal bleeding due to vascular telangiectasias. Notably, as patients with DC, HH, and Revesz syndrome are living longer thanks to improvements in supportive care, additional clinical manifestations are now developing to include gastrointestinal telangiectasias and pulmonary arteriovenous malformations like those seen in Coats plus (Gorgy et al. 2015; Samuel et al. 2015; Khincha et al. 2017; Himes et al. 2021). Importantly, adolescents and young adults with these severe TBDs are also at risk of HNSCC and other cancers that are not typical for their age group (Alter et al. 2018; Schratz et al. 2020; Niewisch et al. 2022).

Most reports of TBD-related clinical manifestations have been summaries of different registries, cohorts, or reviews of the literature (Vieri et al. 2021; Tummala et al. 2022; Revy et al. 2023; Schratz et al. 2023; Tometten et al. 2023). Stat-

istically robust genotype–phenotype association studies are especially challenging in the TBDs due to the large numbers of genes involved with few recurrent variants in unrelated individuals, and variable clinical manifestations. We addressed this challenge by grouping the 231 patients in our cohort by mode of inheritance, autosomal recessive (AR), X-linked recessive (XLR, *DKC1*), and AD (Niewisch et al. 2022). *TINF2* was considered separately because it frequently occurs as a monoallelic de novo germline variant with early onset clinical manifestations, although a few *TINF2* families are reported (Savage et al. 2008; Walne et al. 2008).

We found that patients with AR/XLR or *TINF2* disease were diagnosed at younger ages than those with AD disease (median 11.3 [range 0–45.9] vs. 8.1 [range 0–71.6] vs. 36.1 [range 0.7–69.4] yr, respectively) (Niewisch et al. 2022). Patients with AR/XLR or *TINF2* disease were more likely to have telomeres <1st percentile for age compared with AD patients whose telomeres were more likely in the 1st to 10th percentile (odds ratio [OR] 13.5, 95% CI 4.52–58.54, $P < 0.01$). As clinically expected, the median overall survival was better in those with AD disease (64.9 yr [95% CI 59.8–67.6]) than AR/

**Table 2.** Subtypes of telomere biology disorders (TBDs) and their primary manifestations

| Disorder | Diagnostic clinical features | Telomere length |
|---|---|---|
| Dyskeratosis congenita (DC) | At least two of the three mucocutaneous triad features (nail dysplasia, abnormal skin pigmentation, and/or oral leukoplakia) OR one of the triad plus bone marrow failure (BMF) | <1st percentile |
| Coats plus syndrome (CP) | Bilateral exudative retinopathy, intracranial calcifications, gastrointestinal (GI) bleeding, abnormal bone healing | ≤1st percentile |
| Hoyeraal–Hreidarsson syndrome (HH) | Features consistent with DC plus cerebellar hypoplasia: may also have immunodeficiency and/or IUGR | <<1st percentile |
| Revesz syndrome (RS) | Features consistent with DC plus bilateral exudative retinopathy: may also have intracranial calcifications and/or IUGR | <<1st percentile |
| Isolated telomere biology disorder (TBD): usually (but not always) presents in adulthood with one or two features; there may be a family history of related manifestations | | |
| | BMF with negative Fanconi anemia testing | <10th percentile |
| | Cryptogenic liver disease not related to infection, alcohol, or drug use | <10th percentile |
| | Head/neck squamous cell carcinoma occurring in individuals ≤40 yr of age in the absence of smoking or alcohol use and with negative Fanconi anemia testing | <10th percentile |
| | Interstitial lung disease, idiopathic pulmonary fibrosis | <10th percentile |

Telomere length in age-adjusted percentiles is based on lymphocyte telomere length measured by flow cytometry with in situ hybridization and compiled from the literature as described in the text.

(GI) Gastrointestinal, (IUGR) intrauterine growth restriction.

XLR (37.9 yr [95% CI 13.5–47.3]) or *TINF2* (27.8 yr [95% CI 19.1–39.1]). At the last follow-up, 42% of the patients had died due to lung disease (pulmonary fibrosis or pulmonary arteriovenous malformations, 27.8%), cancer (15.5%), or BMF (15.5%).

## Cancer in Telomere Biology Disorders

DC and related TBDs are recognized cancer predisposition disorders with early estimates suggesting patients with DC had 10-fold increased risk of cancer compared with the general population (Alter et al. 2009, 2010). Larger, longitudinal studies conducted by NCI investigators found the overall risk is approximately fourfold higher than the general population and that this risk is primarily attributed to tongue squamous

cell carcinoma (SCC) (216-fold higher) and MDS/AML (74-fold higher) (Alter et al. 2016). An independent study of primarily adult patients with TBDs by the group at Hopkins also reported elevated risks of oral cancer (16.2-fold) (Schratz et al. 2023). Each cohort had different ascertainment with the NCI cohort, likely skewed toward ascertainment of participants interested in participating in cancer studies and individuals with earlier age at onset, whereas the Hopkins study included more individuals with heterozygous TBDs and older ages than the NCI study. The Hopkins study postulated that common cancers, such as colorectal or lung cancer, were not associated with TBDs. However, competing risks of early mortality and limited statistical power in all studies of TBD-related cancer make it nearly impossible

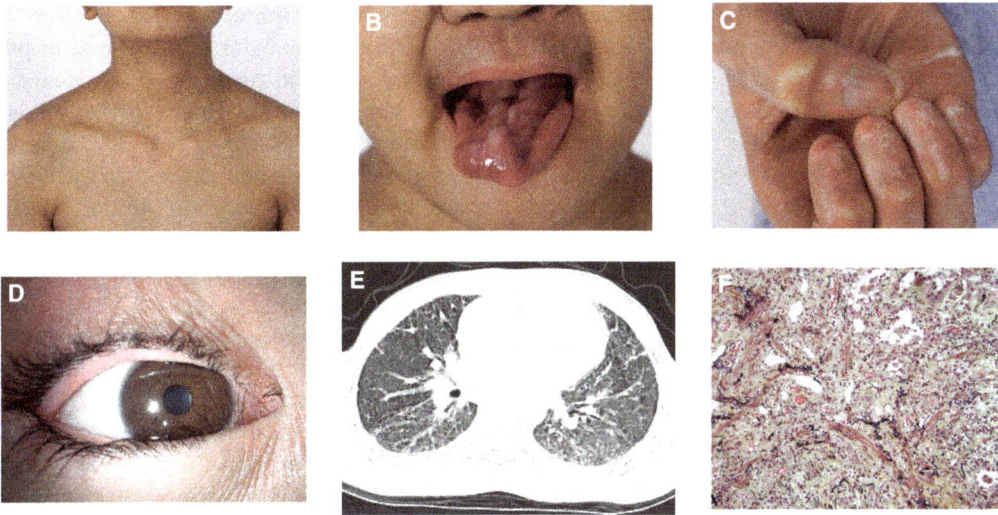

**Figure 2.** Features of classic dyskeratosis congenita, the prototypic telomere biology disorder. A Hispanic male (NCI-204-1) presented with bone marrow failure as a young child and underwent successful hematopoietic cell transplantation at 4 years of age. Over the next several years, he developed the mucocutaneous triad and was found to have a *TINF2* pathogenic variant on skin fibroblast DNA sequencing. He developed pulmonary fibrosis at 13 years of age and underwent successful lung transplantation. Photos of the mucocutaneous triad were obtained at 12 years of age and show (*A*) hyper- and hypopigmented skin; (*B*) oral leukoplakia with geographic tongue; and (*C*) severe nail dysplasia and palmar hyperkeratosis. Additional complications for this patient include (*D*) abnormal eyelash growth resulting in corneal abrasion and scarring; (*E*) bilateral, diffuse areas of ground glass opacities and fibrosis of the lung parenchyma from chest computerized tomography obtained at 11 years 7 months; and (*F*) diffuse fibrosis of explanted lung obtained at 13 years of age. (*E* and *F* were previously published in Giri et al. 2011 and reprinted under the terms of the Creative Commons Attribution License.)

to be confident that patients with TBDs have lower incidence of most solid cancers as suggested by the authors (Schratz et al. 2023).

The mechanism(s) of cancer predisposition in patients with TBDs are not well understood. The primary hypothesis has been that very short telomeres result in chromosomal instability and promote carcinogenesis through the acquisition of somatic mutations and chromosomal rearrangements (Savage et al. 2024). One study connected the SCC with reduced CD4$^+$ or total lymphocyte counts in 13 patients with TBDs with similar findings in a TERC knockout mouse model (Schratz et al. 2023). Additional profiling of solid malignancies in patients with TBDs is required to fully understand the underlying pathobiology.

There is a growing body of literature on acquired somatic mutations in the hematopoietic compartment across the spectrum of TBDs, some of which are adaptive and others maladaptive (Revy et al. 2019; Lasho and Patnaik 2024). Adaptive somatic mutations (i.e., somatic reversion) are those resulting in an improvement of the phenotype, including in hematopoiesis. Such somatic mutations have been reported in patients with TBDs due to germline variants in *DKC1*, *RPA1*, *TERC*, *TERT*, and *TINF2* (Maryoung et al. 2017; Gutierrez-Rodrigues et al. 2019; Revy et al. 2019; Sharma et al. 2022). Conversely, maladaptive somatic mutations have been reported in the context of MDS and clonal hematopoiesis (CH) in numerous populations, including those with TBDs. CH has been reported to be approximately threefold higher in TBDs than healthy individuals, although those studies are somewhat skewed toward older TBD patients with some degree of BMF (Schratz et al. 2020). Patients with TBDs may have a predilection to acquire somatic hematopoietic mutations in *U2AF1*, a key RNA splicing gene associated with MDS in other populations (Schratz et al. 2020, 2021; Ferrer et al. 2023a; Gutierrez-Rodrigues et al. 2024). Somatic mutations have also been reported in the *TERT* promoter, *POT1*, *PPM1D*, *SMC1A*, and *TP53*, as well as *TET2*, *DNMT3A*, and *ASXL1*, the typical MDS-associated epigenetic regulator genes (Schratz et al. 2021; Ferrer et al. 2022, 2023a; Gutierrez-Rodrigues et al. 2024) Notably, somatic mutations in *POT1*, *PPM1D*, and the *TERT* promoter were not associated with cancer development (Gutierrez-Rodrigues et al. 2024). Large longitudinal studies are required to fully understand the development and evolution of CH and its relationship to MDS/AML and outcomes in the TBDs.

## Clinical Management

Patients with TBDs are at high risk of multiple progressive and life-threatening complications. Multidisciplinary care with a medical home is required to optimize patient outcomes and well-being. Specific clinical care recommendations are available in the TBDs: Diagnosis and Management Guidelines published in 2022 by Team Telomere, the DC/TBD patient support and advocacy group. (These Guidelines include the latest expert recommendations and can be freely obtained here: teamtelomere.org/diagnosis-management-guidelines.) Recent reviews detailing approaches to management are also available (Vieri et al. 2021; Savage 2022; Tummala et al. 2022; Niewisch et al. 2023; Revy et al. 2023).

In brief, regular surveillance for BMF, pulmonary fibrosis, liver disease, osteopenia, HNSCC, and other cancers is recommended. Some patients will require more intensive management than others based on the genetic etiology and their specific disease manifestations. Hematopoietic cell transplantation (HCT) using a nonmyeloablative regimen is relatively well-tolerated for patients with limited comorbidities. Some patients may choose oral androgens, such as danazol, to improve blood counts in lieu or to delay HCT (Townsley et al. 2016; Khincha et al. 2018; Vieri et al. 2020; Kirschner

et al. 2021; Thompson et al. 2022; Clé et al. 2023). Anti-inflammatories often used in pulmonary fibrosis do not appear to benefit patients with TBDs and may reduce survival (Newton et al. 2019; Zhang et al. 2023). Antifibrotics, such as pirfenidone, may marginally improve lung function but more data are needed before recommending their routine use (Justet et al. 2021). Individuals with TBDs in need of a liver transplant appear to have outcomes similar to individuals without TBDs and should not be excluded from consideration of liver transplant (Wang et al. 2024).

Only three clinical trials are listed as specifically recruiting participants with TBDs as of October 2024 (clinicaltrials.gov)—two on danazol and TL (NCT03312400 and NCT04638517), and one gene therapy trial (NCT04211714). Another study is focused on understanding the psychosocial challenges of living with TBDs (NCT04959188). The HCT trial (ClinicalTrials.gov identifier: NCT01659606) is complete and closed to new accrual. The clinical research community strongly recommends that patients with TBDs be included in clinical trials of pulmonary disease, liver disease, BMF, and cancer. We are hopeful that the ongoing work of the Clinical Care Consortium of Telomere-Associated Ailments (CCCTAA) an international consortium of clinicians and scientists who agreed to share de-identified data to learn more about TBD manifestations will improve outcomes as we understand more about the full spectrum of TBDs and the consequences of very short telomeres (Higgs et al. 2019).

## CANCER PREDISPOSITION WITH LONG TELOMERES

In 2014, the telomere research field was likely surprised by the discovery of long blood and buccal cell telomeres caused by rare deleterious germline variants in *POT1* in families with melanoma (Robles-Espinoza et al. 2014; Shi et al. 2014; Trigueros-Motos 2014). Telomere dysfunction and long telomeres had been previously reported in chronic lymphocytic leukemia (CLL) (Lin et al. 2014) and subsequently germline *POT1* mutations were identified in 2016 as a putative cause of familial CLL (Speedy et al. 2016). To date, rare deleterious germline variants have been reported in five of the six shelterin proteins as associated with long telomeres and a spectrum of cancers (Tables 1 and 3).

## POT1 Tumor Predisposition

The term "POT1 tumor predisposition (POT1-TPD)" appears to have first been used in the 2020 version of the GeneReviews with the same title (Henry et al. 1993) and appropriately encompasses the spectrum of malignancies associated with rare deleterious variants in *POT1* resulting in long telomeres. *POT1*-associated cancers are summarized in Table 3 and have been reviewed in detail elsewhere (Herrera-Mullar et al. 2023; Zade and Khattar 2023). Germline *POT1* variants in families associated with elevated melanoma risk have been well-established, even after accounting for the most common melanoma risk factor, ultraviolet light exposure (Robles-Espinoza et al. 2014; Shi et al. 2014; Trigueros-Motos 2014; Nathan et al. 2021; Simonin-Wilmer et al. 2023). Patients with heterozygous *POT1* germline variants were more likely to have the spitzoid melanoma morphology than those with the more common familial melanoma etiology, *CDKN2A*, or *CDK4* variants (Sargen et al. 2020; Goldstein et al. 2023).

The hematopoietic phenotype associated with germline *POT1* variants has expanded since the first study reported an ∼3.6-fold higher incidence of CLL in POT1 heterozygotes than the noncarriers (Speedy et al. 2016). *POT1* var-iants have been identified in families with Hodgkin lymphoma (McMaster et al. 2018) and in individuals with CH (DeBoy et al. 2023). Notably, functional studies of POT1 germline variants identified in Hodgkin lymphoma families found that these mutations do not trigger a DNA damage response but do lead to telomere elongation (Kim et al. 2021).

Whole exome sequencing of a Li–Fraumeni-like family with cardiac angiosarcoma identified a heterozygous germline *POT1* variant (p.R117C) associated with long telomeres and increased telomere fragility. Larger studies of individuals with sarcoma identified several other individuals with POT1 variants, further solidifying the association of POT1 with cancer etiology (Ballinger et al. 2023).

## Other Shelterin Proteins

In addition to POT1, four other components of the six protein shelterin telomere protection complex have germline variants associated with the development of similar cancers. Like the POT1 story, rare deleterious variants in *ACD* and *TERF2IP* were first reported in familial melanoma and associated with long telomeres (Aoude et al. 2015; Malińska et al. 2020; Goldstein et al. 2023). Interestingly, a rare truncating *TINF2* variant resulting in longer telomeres in a family with papillary thyroid cancer and melanoma has also been reported (He et al. 2020). Subsequent studies have confirmed that certain *TINF2* variants do predispose to long telomeres and melanoma, papillary thyroid cancer, and sarcoma (DeBoy et al. 2023; Jensen et al.

**Table 3.** Cancer predisposition with long telomeres (CPLT) associated genes and cancer types

| Gene name | Protein name | Melanoma | Sarcoma | Thyroid cancer | Chronic lymphocytic leukemia | Hodgkin lymphoma | Clonal hematopoiesis |
|---|---|---|---|---|---|---|---|
| *ACD* | TPP1 | ▓ | | | | | |
| *POT1* | POT1 | ▓ | ▓ | ▓ | ▓ | ▓ | ▓ |
| *TERF1* | TRF1 | ▓ | ▓ | | | | |
| *TERF2IP* | RAP1 | ▓ | | | | | |
| *TINF2* | TIN2 | ▓ | ▓ | ▓ | | | |

Gray boxes indicate the occurrence of at least one case of cancer with a rare germline coding variant in the gene. OMIM refers to POT1 tumor predisposition as tumor predisposition syndrome 3 (TPDS3, MIM #615848).

2023). Ballinger et al. (2023) identified rare deleterious variants in *TERF1*, *TINF2*, and *TERF2IP*, in addition to *POT1*, in individuals with sarcoma and some of these patients also had a personal or family history of melanoma, thyroid, or other cancers. TL was longer in the carriers than in other patients with sarcoma in that cohort.

## WHAT'S IN A NAME

The designation of cancer CPLT is proposed to define the set of cancer-prone germline genetic disorders caused by rare deleterious variants in telomere biology genes, predominately those of the shelterin telomere protection complex resulting in longer than average telomeres (Savage et al. 2024). The distinction of this as an entity separate from TBDs is important because those with CPLT do not have (and do not appear to develop) the TBD features described above but do have an elevated risk of certain cancers and delayed hair graying (a clinically insignificant phenotype). The underlying biology is also distinct. The longer telomeres associated with CPLT variants maintain telomere end-capping function and enhance cellular replicative capacity. In contrast, the telomeres present in the TBDs result in a profound reduction in cellular life span induced by shortened and/or dysfunctional telomeres, which drives a range of organ pathologies and cancers distinct from those observed in CPLT. Moreover, while most TBD-associated germline variants result in very short telomeres, some variants cause telomere dysfunction in the context of TLs within a normal range (Kermasson et al. 2022). Thus, the inclusion of biology in the term TBD.

## FUTURE DIRECTIONS

Basic science, clinical, and epidemiology studies show that maintaining TL is a delicate balancing act affected by cellular intrinsic and extrinsic factors. Disruption of this balance in relatively small ways through common germline variants or certain environmental exposures can be associated with an increased risk of common diseases. Larger disruptions to the telomere balance

are caused by rare pathogenic germline variants resulting in very short telomeres for age and the TBDs, or long telomeres and CPLT.

Numerous unanswered questions remain across the spectrum of telomere-related phenotypes, for example:

1. Can TL testing aid in diagnosing CPLT?

2. What is a reasonable approach to cancer screening for people with CPLT?

3. Adults with TBDs often have flow FISH telomeres in the 1st to 10th percentile for age. Is there a better diagnostic test?

4. Are there systemic therapies to ameliorate TBD clinical manifestations?

5. Are polygenic TL scores based on common single-nucleotide variants applicable to understanding disease risk(s) in the clinic?

6. What are the optimal approaches to clinically curate germline genetic variants in telomere maintenance genes?

Multidisciplinary, international collaborative studies between clinicians, basic scientists, epidemiologists, and those affected by TBDs or CPLT are required to address these and many other questions as we strive to thoroughly understand telomere biology and its essential roles in health and well-being.

## COMPETING INTEREST STATEMENT

I declare no relevant conflicts of interest.

## ACKNOWLEDGMENTS

I am grateful to all patients and their families without whom scientific advances in telomere biology would not be possible. This work was supported by the intramural research program of the Division of Cancer Epidemiology and Genetics, the National Cancer Institute, and the National Institutes of Health.

## REFERENCES

Alder JK, Hanumanthu VS, Strong MA, DeZern AE, Stanley SE, Takemoto CM, Danilova L, Applegate CD, Bolton SG, Mohr DW, et al. 2018. Diagnostic utility of telomere

 Cite this article as *Cold Spring Harb Perspect Biol* doi: 10.1101/cshperspect.a041684

length testing in a hospital-based setting. *Proc Natl Acad Sci* 115: E2358–E2365. doi:10.1073/pnas.172042 7115

Allaire P, He J, Mayer J, Moat L, Gerstenberger P, Wilhorn R, Strutz S, Kim DSL, Zeng C, Cox N, et al. 2023. Genetic and clinical determinants of telomere length. *HGG Adv* 4: 100201. doi:10.1016/j.xhgg.2023.100201

Alter BP, Baerlocher GM, Savage SA, Chanock SJ, Weksler BB, Willner JP, Peters JA, Giri N, Lansdorp PM. 2007. Very short telomere length by flow fluorescence in situ hybridization identifies patients with dyskeratosis congenita. *Blood* 110: 1439–1447. doi:10.1182/blood-2007-02-075598

Alter BP, Baerlocher G, Giri N, Lansdorp PM, Savage SA. 2008. Very short telomeres are characteristic of dyskeratosis congenita and not other inherited bone marrow failure syndromes. *Blood* 112: 1044. doi:10.1182/blood .V112.11.1044.1044

Alter BP, Giri N, Savage SA, Rosenberg PS. 2009. Cancer in dyskeratosis congenita. *Blood* 113: 6549–6557. doi:10 .1182/blood-2008-12-192880

Alter BP, Giri N, Savage SA, Peters JA, Loud JT, Leathwood L, Carr AG, Greene MH, Rosenberg PS. 2010. Malignancies and survival patterns in the National Cancer Institute inherited bone marrow failure syndromes cohort study. *Br J Haematol* 150: 179–188. doi:10.1111/j.1365-2141 .2010.08212.x

Alter BP, Rosenberg PS, Giri N, Baerlocher GM, Lansdorp PM, Savage SA. 2012. Telomere length is associated with disease severity and declines with age in dyskeratosis congenita. *Haematologica* 97: 353–359. doi:10.3324/haema tol.2011.055269

Alter BP, Giri N, Savage SA, Rosenberg PS. 2016. Cancer in the National Cancer Institute inherited bone marrow failure syndrome cohort after 15 years of follow-up. *Blood* 128: 334. doi:10.1182/blood.V128.22.334.334

Alter BP, Giri N, Savage SA, Rosenberg PS. 2018. Cancer in the National Cancer Institute inherited bone marrow failure syndrome cohort after fifteen years of follow-up. *Haematologica* 103: 30–39. doi:10.3324/haematol.2017.178 111

Anderson BH, Kasher PR, Mayer J, Szynkiewicz M, Jenkinson EM, Bhaskar SS, Urquhart JE, Daly SB, Dickerson JE, O'Sullivan J, et al. 2012. Mutations in CTC1, encoding conserved telomere maintenance component 1, cause Coats plus. *Nat Genet* 44: 338. doi:10.1038/ng.1084

Aoude LG, Pritchard AL, Robles-Espinoza CD, Wadt K, Harland M, Choi J, Gartside M, Quesada V, Johansson P, Palmer JM, et al. 2015. Nonsense mutations in the shelterin complex genes ACD and TERF2IP in familial melanoma. *J Natl Cancer Inst* 107: dju408. doi:10.1093/ jnci/dju408

Arbeev KG, Verhulst S, Steenstrup T, Kark JD, Bagley O, Kooperberg C, Reiner AP, Hwang SJ, Levy D, Fitzpatrick AL, et al. 2020. Association of leukocyte telomere length with mortality among adult participants in 3 longitudinal studies. *JAMA Netw Open* 3: e200023. doi:10.1001/jama networkopen.2020.0023

Aubert G, Hills M, Lansdorp PM. 2012. Telomere length measurement—caveats and a critical assessment of the available technologies and tools. *Mutat Res* 730: 59–67. doi:10.1016/j.mrfmmm.2011.04.003

Aviv A, Shay JW. 2018. Reflections on telomere dynamics and ageing-related diseases in humans. *Philos Trans R Soc Lond B Biol Sci* 373: 20160436. doi:10.1098/rstb.2016 .0436

Baerlocher GM, Lansdorp PM. 2003. Telomere length measurements in leukocyte subsets by automated multicolor flow-FISH. *Cytometry A* 55: 1–6. doi:10.1002/cyto.a .10064

Baerlocher GM, Rice K, Vulto I, Lansdorp PM. 2007. Longitudinal data on telomere length in leukocytes from newborn baboons support a marked drop in stem cell turnover around 1 year of age. *Aging Cell* 6: 121–123. doi:10 .1111/j.1474-9726.2006.00254.x

Baird DM, Rowson J, Wynford-Thomas D, Kipling D. 2003. Extensive allelic variation and ultrashort telomeres in senescent human cells. *Nat Genet* 33: 203–207. doi:10.1038/ ng1084

Ballew BJ, Joseph V, De S, Sarek G, Vannier JB, Stracker T, Schrader KA, Small TN, O'Reilly R, Manschreck C, et al. 2013a. A recessive founder mutation in regulator of telomere elongation helicase 1, RTEL1, underlies severe immunodeficiency and features of Hoyeraal–Hreidarsson syndrome. *PLoS Genet* 9: e1003695. doi:10.1371/journal .pgen.1003695

Ballew BJ, Yeager M, Jacobs K, Giri N, Boland J, Burdett L, Alter BP, Savage SA. 2013b. Germline mutations of regulator of telomere elongation helicase 1, RTEL1, in Dyskeratosis congenita. *Hum Genet* 132: 473–480. doi:10 .1007/s00439-013-1265-8

Ballinger ML, Pattnaik S, Mundra PA, Zaheed M, Rath E, Priestley P, Baber J, Ray-Coquard I, Isambert N, Causeret S, et al. 2023. Heritable defects in telomere and mitotic function selectively predispose to sarcomas. *Science* 379: 253–260. doi:10.1126/science.abj4784

Barthel FP, Wei W, Tang M, Martinez-Ledesma E, Hu X, Amin SB, Akdemir KC, Seth S, Song X, Wang Q, et al. 2017. Systematic analysis of telomere length and somatic alterations in 31 cancer types. *Nat Genet* 49: 349–357. doi:10.1038/ng.3781

Bountziouka V, Musicha C, Allara E, Kaptoge S, Wang Q, Angelantonio ED, Butterworth AS, Thompson JR, Danesh JN, Wood AM, et al. 2022. Modifiable traits, healthy behaviours, and leukocyte telomere length: a population-based study in UK Biobank. *Lancet Healthy Longev* 3: e321–e331. doi:10.1016/S2666-7568(22)00072-1

Brown DW, Zhou W, Wang Y, Jones K, Luo W, Dagnall C, Teshome K, Klein A, Chang T, Lin SH, et al. 2022. Germline-somatic JAK2 interactions are associated with clonal expansion in myelofibrosis. *Nat Commun* 13: 5284. doi:10.1038/s41467-022-32986-7

Cawthon RM. 2002. Telomere measurement by quantitative PCR. *Nucleic Acids Res* 30: e47. doi:10.1093/nar/30.10 .e47

Chakravarti D, LaBella KA, DePinho RA. 2021. Telomeres: history, health, and hallmarks of aging. *Cell* 184: 306–322. doi:10.1016/j.cell.2020.12.028

Chen B, Yan Y, Wang H, Xu J. 2023. Association between genetically determined telomere length and health-related outcomes: a systematic review and meta-analysis of Mendelian randomization studies. *Aging Cell* 22: e13874. doi:10.1111/acel.13874

Clé DV, Catto LFB, Gutierrez-Rodrigues F, Donaires FS, Pinto AL, Santana BA, Darrigo LG, Valera ET, Koenigkam-Santos M, Baddini-Martinez J, et al. 2023. Effects of nandrolone decanoate on telomere length and clinical outcome in patients with telomeropathies: a prospective trial. *Haematologica* **108:** 1300–1312. doi:10.3324/haematol.2022.281808

Codd V, Nelson CP, Albrecht E, Mangino M, Deelen J, Buxton JL, Hottenga JJ, Fischer K, Esko T, Surakka I, et al. 2013. Identification of seven loci affecting mean telomere length and their association with disease. *Nat Genet* **45:** 422–427. doi:10.1038/ng.2528

Codd V, Wang Q, Allara E, Musicha C, Kaptoge S, Stoma S, Jiang T, Hamby SE, Braund PS, Bountziouka V, et al. 2021. Polygenic basis and biomedical consequences of telomere length variation. *Nat Genet* **53:** 1425–1433. doi:10.1038/s41588-021-00944-6

Crow YJ, McMenamin J, Haenggeli CA, Hadley DM, Tirupathi S, Treacy EP, Zuberi SM, Browne BH, Tolmie JL, Stephenson JB. 2004. Coats' plus: a progressive familial syndrome of bilateral Coats' disease, characteristic cerebral calcification, leukoencephalopathy, slow pre- and postnatal linear growth and defects of bone marrow and integument. *Neuropediatrics* **35:** 10–19. doi:10.1055/s-2003-43552

Dagnall CL, Hicks B, Teshome K, Hutchinson AA, Gadalla SM, Khincha PP, Yeager M, Savage SA. 2017. Effect of pre-analytic variables on the reproducibility of qPCR relative telomere length measurement. *PLoS ONE* **12:** e0184098. doi:10.1371/journal.pone.0184098

DeBoy EA, Tassia MG, Schratz KE, Yan SM, Cosner ZL, McNally EJ, Gable DL, Xiang Z, Lombard DB, Antonarakis ES, et al. 2023. Familial clonal hematopoiesis in a long telomere syndrome. *N Engl J Med* **388:** 2422–2433. doi:10.1056/NEJMoa2300503

Delgado DA, Zhang C, Chen LS, Gao J, Roy S, Shinkle J, Sabarinathan M, Argos M, Tong L, Ahmed A, et al. 2018. Genome-wide association study of telomere length among South Asians identifies a second RTEL1 association signal. *J Med Genet* **55:** 64–71. doi:10.1136/jmedgenet-2017-104922

Diaz de Leon A, Cronkhite JT, Katzenstein AL, Godwin JD, Raghu G, Glazer CS, Rosenblatt RL, Girod CE, Garrity ER, Xing C, et al. 2010. Telomere lengths, pulmonary fibrosis and telomerase (TERT) mutations. *PLoS ONE* **5:** e10680. doi:10.1371/journal.pone.0010680

D'Mello MJ, Ross SA, Briel M, Anand SS, Gerstein H, Paré G. 2015. Association between shortened leukocyte telomere length and cardiometabolic outcomes: systematic review and meta-analysis. *Circ Cardiovasc Genet* **8:** 82–90. doi:10.1161/CIRCGENETICS.113.000485

Dong X, Milholland B, Vijg J. 2016. Evidence for a limit to human lifespan. *Nature* **538:** 257–259. doi:10.1038/nature19793

Ferrer A, Mangaonkar AA, Patnaik MM. 2022. Clonal hematopoiesis and myeloid neoplasms in the context of telomere biology disorders. *Curr Hematol Malig Rep* **17:** 61–68. doi:10.1007/s11899-022-00662-8

Ferrer A, Lasho T, Fernandez JA, Steinauer NP, Simon RA, Finke CM, Carmona EM, Wylam ME, Ongie LJ, Burnap BN, et al. 2023a. Patients with telomere biology disorders show context specific somatic mosaic states with high frequency of *U2AF1* variants. *Am J Hematol* **98:** E357–E359. doi:10.1002/ajh.27086

Ferrer A, Stephens ZD, Kocher JA. 2023b. Experimental and computational approaches to measure telomere length: recent advances and future directions. *Curr Hematol Malig Rep* **18:** 284–291. doi:10.1007/s11899-023-00717-4

Fitzpatrick AL, Kronmal RA, Gardner JP, Psaty BM, Jenny NS, Tracy RP, Walston J, Kimura M, Aviv A. 2007. Leukocyte telomere length and cardiovascular disease in the cardiovascular health study. *Am J Epidemiol* **165:** 14–21. doi:10.1093/aje/kwj346

Gable DL, Gaysinskaya V, Atik CC, Talbot CC, Kang B, Stanley SE, Pugh EW, Amat-Codina N, Schenk KM, Arcasoy MO, et al. 2019. *ZCCHC8*, the nuclear exosome targeting component, is mutated in familial pulmonary fibrosis and is required for telomerase RNA maturation. *Genes Dev* **33:** 1381–1396. doi:10.1101/gad.326785.119

Gadalla SM, Khincha PP, Katki HA, Giri N, Wong JY, Spellman S, Yanovski JA, Han JC, De Vivo I, Alter BP, et al. 2016. The limitations of qPCR telomere length measurement in diagnosing dyskeratosis congenita. *Mol Genet Genom Med* **4:** 475–479. doi:10.1002/mgg3.220

Giaccherini M, Gentiluomo M, Fornili M, Lucenteforte E, Baglietto L, Campa D. 2021. Association between telomere length and mitochondrial copy number and cancer risk in humans: a meta-analysis on more than 300,000 individuals. *Crit Rev Oncol Hematol* **167:** 103510. doi:10.1016/j.critrevonc.2021.103510

Giri N, Lee R, Faro A, Huddleston CB, White FV, Alter BP, Savage SA. 2011. Lung transplantation for pulmonary fibrosis in dyskeratosis congenita: case report and systematic literature review. *BMC Blood Disord* **11:** 3. doi:10.1186/1471-2326-11-3

Goldstein AM, Qin R, Chu EY, Elder DE, Massi D, Adams DJ, Harms PW, Robles-Espinoza CD, Newton-Bishop JA, Bishop DT, et al. 2023. Association of germline variants in telomere maintenance genes (POT1, TERF2IP, ACD, and TERT) with spitzoid morphology in familial melanoma: a multi-center case series. *JAAD Int* **11:** 43–51. doi:10.1016/j.jdin.2023.01.013

Gorgy AI, Jonassaint NL, Stanley SE, Koteish A, DeZern AE, Walter JE, Sopha SC, Hamilton JP, Hoover-Fong J, Chen AR, et al. 2015. Hepatopulmonary syndrome is a frequent cause of dyspnea in the short telomere disorders. *Chest* **148:** 1019–1026. doi:10.1378/chest.15-0825

Guo Y, Kartawinata M, Li J, Pickett HA, Teo J, Kilo T, Barbaro PM, Keating B, Chen Y, Tian L, et al. 2014. Inherited bone marrow failure associated with germline mutation of ACD, the gene encoding telomere protein TPP1. *Blood* **124:** 2767–2774. doi:10.1182/blood-2014-08-596445

Gutierrez-Rodrigues F, Donaires FS, Pinto A, Vicente A, Dillon LW, Clé DV, Santana BA, Piroznia M, Ibanez M, Townsley DM, et al. 2019. Pathogenic TERT promoter variants in telomere diseases. *Genet Med* **21:** 1594–1602. doi:10.1038/s41436-018-0385-x

Gutierrez-Rodrigues F, Groarke EM, Thongon N, Rodriguez-Sevilla JJ, Catto LFB, Niewisch MR, Shalhoub RN, McReynolds LJ, Clé DV, Patel BA, et al. 2024. Clonal landscape and clinical outcomes of telomere biology disorders: somatic rescuing and cancer mutations. *Blood* **144:** 2402–2416. doi:10.1182/blood.2024025023

Cite this article as *Cold Spring Harb Perspect Biol* doi: 10.1101/cshperspect.a041684

He H, Li W, Comiskey DF, Liyanarachchi S, Nieminen TT, Wang Y, DeLap KE, Brock P, de la Chapelle A. 2020. A truncating germline mutation of *TINF2* in individuals with thyroid cancer or melanoma results in longer telomeres. *Thyroid* 30: 204–213. doi:10.1089/thy.2019.0156

Heiss NS, Knight SW, Vulliamy TJ, Klauck SM, Wiemann S, Mason PJ, Poustka A, Dokal I. 1998. X-linked dyskeratosis congenita is caused by mutations in a highly conserved gene with putative nucleolar functions. *Nat Genet* 19: 32–38. doi:10.1038/ng0598-32

Hemann MT, Strong MA, Hao LY, Greider CW. 2001. The shortest telomere, not average telomere length, is critical for cell viability and chromosome stability. *Cell* 107: 67–77. doi:10.1016/S0092-8674(01)00504-9

Henry ML, Osborne J, Else T. 1993. *POT1* tumor predisposition. In *Genereviews* (ed. Adam MP, Feldman J, Mirzaa GM, et al.). University of Washington, Seattle.

Herrera-Mullar J, Fulk K, Brannan T, Yussuf A, Polfus L, Richardson ME, Horton C. 2023. Characterization of POT1 tumor predisposition syndrome: tumor prevalence in a clinically diverse hereditary cancer cohort. *Genet Med* 25: 100937. doi:10.1016/j.gim.2023.100937

Higgs C, Crow YJ, Adams DM, Chang E, Hayes D, Herbig U, Huang JN, Himes R, Jajoo K, Johnson FB, et al. 2019. Understanding the evolving phenotype of vascular complications in telomere biology disorders. *Angiogenesis* 22: 95–102. doi:10.1007/s10456-018-9640-7

Himes RW, Chiou EH, Queliza K, Shouval DS, Somech R, Agarwal S, Jajoo K, Ziegler DS, Kratz CP, Huang J, et al. 2021. Gastrointestinal hemorrhage: a manifestation of the telomere biology disorders. *J Pediatr* 230: 55–61.e4. doi:10.1016/j.jpeds.2020.09.038

Hreidarsson S, Kristjansson K, Johannesson G, Johannsson JH. 1988. A syndrome of progressive pancytopenia with microcephaly, cerebellar hypoplasia and growth failure. *Acta Paediatr Scand* 77: 773–775. doi:10.1111/j.1651-2227.1988.tb10751.x

Jensen MR, Jelsig AM, Gerdes AM, Holmich LR, Kainu KH, Lorentzen HF, Hansen MH, Bak M, Johansson PA, Hayward NK, et al. 2023. TINF2 is a major susceptibility gene in Danish patients with multiple primary melanoma. *HGG Adv* 4: 100225. doi:10.1016/j.xhgg.2023.100225

Justet A, Klay D, Porcher R, Cottin V, Ahmad K, Molina Molina M, Nunes H, Reynaud-Gaubert M, Naccache JM, Manali E, et al. 2021. Safety and efficacy of pirfenidone and nintedanib in patients with idiopathic pulmonary fibrosis and carrying a telomere-related gene mutation. *Eur Respir J* 57: 2003198. doi:10.1183/13993003.03198-2020

Karimian K, Groot A, Huso V, Kahidi R, Tan KT, Sholes S, Keener R, McDyer JF, Alder JK, Li H, et al. 2024. Human telomere length is chromosome end-specific and conserved across individuals. *Science* 384: 533–539. doi:10.1126/science.ado0431

Kermasson L, Churikov D, Awad A, Smoom R, Lainey E, Touzot F, Audebert-Bellanger S, Haro S, Roger L, Costa E, et al. 2022. Inherited human Apollo deficiency causes severe bone marrow failure and developmental defects. *Blood* 139: 2427–2440. doi:10.1182/blood.2021010791

Khincha PP, Bertuch AA, Agarwal S, Townsley DM, Young NS, Keel S, Shimamura A, Boulad F, Simoneau T, Justino H, et al. 2017. Pulmonary arteriovenous malformations: an uncharacterised phenotype of dyskeratosis congenita and related telomere biology disorders. *Eur Respir J* 49: 1601640. doi:10.1183/13993003.01640-2016

Khincha PP, Bertuch AA, Gadalla SM, Giri N, Alter BP, Savage SA. 2018. Similar telomere attrition rates in androgen-treated and untreated patients with dyskeratosis congenita. *Blood Adv* 2: 1243–1249. doi:10.1182/bloodadvances.2018016964

Kim WT, Hennick K, Johnson J, Finnerty B, Choo S, Short SB, Drubin C, Forster R, McMaster ML, Hockemeyer D. 2021. Cancer-associated POT1 mutations lead to telomere elongation without induction of a DNA damage response. *EMBO J* 40: e107346. doi:10.15252/embj.2020107346

Kirschner M, Vieri M, Kricheldorf K, Ferreira MSV, Wlodarski MW, Schwarz M, Balabanov S, Rolles B, Isfort S, Koschmieder S, et al. 2021. Androgen derivatives improve blood counts and elongate telomere length in adult cryptic dyskeratosis congenita. *Br J Haematol* 193: 669–673. doi:10.1111/bjh.16997

Knight SW, Heiss NS, Vulliamy TJ, Aalfs CM, McMahon C, Richmond P, Jones A, Hennekam RC, Poustka A, Mason PJ, et al. 1999. Unexplained aplastic anaemia, immunodeficiency, and cerebellar hypoplasia (Hoyeraal–Hreidarsson syndrome) due to mutations in the dyskeratosis congenita gene, *DKC1*. *Br J Haematol* 107: 335–339. doi:10.1046/j.1365-2141.1999.01690.x

Kocak H, Ballew BJ, Bisht K, Eggebeen R, Hicks BD, Suman S, O'Neil A, Giri N; NCI DCEG Cancer Genomics Research Laboratory; NCI DCEG Cancer Sequencing Working Group, et al. 2014a. Hoyeraal–Heidarsson syndrome caused by a germline mutation in the TEL patch of the telomere protein TPP1. *Genes Dev* 28: 2090–2102. doi:10.1101/gad.248567.114

Kocak H, Ballew BJ, Bisht K, Eggebeen R, Hicks BD, Suman S, O'Neil A, Giri N, Maillard I, Alter BP, et al. 2014b. Hoyeraal–Hreidarsson syndrome caused by a germline mutation in the TEL patch of the telomere protein TPP1. *Genes Dev* 28: 2090–2102. doi:10.1101/gad.248567.114

Lai TP, Zhang N, Noh J, Mender I, Tedone E, Huang E, Wright WE, Danuser G, Shay JW. 2017. A method for measuring the distribution of the shortest telomeres in cells and tissues. *Nat Commun* 8: 1356. doi:10.1038/s41467-017-01291-z

Lai TP, Wright WE, Shay JW. 2018. Comparison of telomere length measurement methods. *Philos Trans R Soc Lond B Biol Sci* 373: 20160451. doi:10.1098/rstb.2016.0451

Lai TP, Verhulst S, Dagnall CL, Hutchinson A, Spellman SR, Howard A, Katki HA, Levine JE, Saber W, Aviv A, et al. 2022. Decoupling blood telomere length from age in recipients of allogeneic hematopoietic cell transplant in the BMT-CTN 1202. *Front Immunol* 13: 966301. doi:10.3389/fimmu.2022.966301

Lai TP, Verhulst S, Savage SA, Gadalla SM, Benetos A, Toupance S, Factor-Litvak P, Susser E, Aviv A. 2023. Buildup from birth onward of short telomeres in human hematopoietic cells. *Aging Cell* 22: e13844. doi:10.1111/acel.13844

Lasho T, Patnaik MM. 2024. Adaptive and maladaptive clonal hematopoiesis in telomere biology disorders. *Curr*

*Hematol Malig Rep* **19:** 35–44. doi:10.1007/s11899-023-00719-2

Lee M, Napier CE, Yang SF, Arthur JW, Reddel RR, Pickett HA. 2017. Comparative analysis of whole genome sequencing-based telomere length measurement techniques. *Methods* **114:** 4–15. doi:10.1016/j.ymeth.2016.08.008

Le Guen T, Jullien L, Touzot F, Schertzer M, Gaillard L, Perderiset M, Carpentier W, Nitschke P, Picard C, Couillault G, et al. 2013. Human RTEL1 deficiency causes Hoyeraal–Hreidarsson syndrome with short telomeres and genome instability. *Hum Mol Genet* **22:** 3239–3249. doi:10.1093/hmg/ddt178

Lenart A, Vaupel JW. 2017. Questionable evidence for a limit to human lifespan. *Nature* **546:** E13–E14. doi:10.1038/nature22790

Lin TT, Norris K, Heppel NH, Pratt G, Allan JM, Allsup DJ, Bailey J, Cawkwell L, Hills R, Grimstead JW, et al. 2014. Telomere dysfunction accurately predicts clinical outcome in chronic lymphocytic leukaemia, even in patients with early stage disease. *Br J Haematol* **167:** 214–223. doi:10.1111/bjh.13023

Lindrose AR, McLester-Davis LWY, Tristano RI, Kataria L, Gadalla SM, Eisenberg DTA, Verhulst S, Drury S. 2021. Method comparison studies of telomere length measurement using qPCR approaches: a critical appraisal of the literature. *PLoS ONE* **16:** e0245582. doi:10.1371/journal.pone.0245582

Liu M, Lan Y, Zhang H, Zhang X, Wu M, Yang L, Zhou J, Tong M, Leng L, Zheng H, et al. 2023. Telomere length is associated with increased risk of cutaneous melanoma: a Mendelian randomization study. *Melanoma Res* **33:** 475–481. doi:10.1097/CMR.0000000000000917

Machiela MJ, Hofmann JN, Carreras-Torres R, Brown KM, Johansson M, Wang Z, Foll M, Li P, Rothman N, Savage SA, et al. 2017. Genetic variants related to longer telomere length are associated with increased risk of renal cell carcinoma. *Eur Urol* **72:** 747–754. doi:10.1016/j.eururo.2017.07.015

Malińska K, Deptuła J, Rogoza-Janiszewska E, Górski B, Scott R, Rudnicka H, Kashyap A, Domagala P, Hybiak J, Masojć B, et al. 2020. Constitutional variants in POT1, TERF2IP, and ACD genes in patients with melanoma in the Polish population. *Eur J Cancer Prev* **29:** 511–519. doi:10.1097/CEJ.0000000000000633

Mangaonkar AA, Ferrer A, Vairo FPE, Hammel CW, Prochnow C, Gangat N, Hogan WJ, Litzow MR, Peters SG, Scott JP, et al. 2021. Clinical and molecular correlates from a predominantly adult cohort of patients with short telomere lengths. *Blood Cancer J* **11:** 170. doi:10.1038/s41408-021-00564-7

Mangino M, Christiansen L, Stone R, Hunt SC, Horvath K, Eisenberg DT, Kimura M, Petersen I, Kark JD, Herbig U, et al. 2015. DCAF4, a novel gene associated with leucocyte telomere length. *J Med Genet* **52:** 157–162. doi:10.1136/jmedgenet-2014-102681

Marrone A, Stevens D, Vulliamy T, Dokal I, Mason PJ. 2004. Heterozygous telomerase RNA mutations found in dyskeratosis congenita and aplastic anemia reduce telomerase activity via haploinsufficiency. *Blood* **104:** 3936–3942. doi:10.1182/blood-2004-05-1829

Martin-Ruiz CM, Baird D, Roger L, Boukamp P, Krunic D, Cawthon R, Dokter MM, van der Harst P, Bekaert S, de Meyer T, et al. 2015. Reproducibility of telomere length assessment: an international collaborative study. *Int J Epidemiol* **44:** 1673–1683. doi:10.1093/ije/dyu191

Maryoung L, Yue Y, Young A, Newton CA, Barba C, van Oers NS, Wang RC, Garcia CK. 2017. Somatic mutations in telomerase promoter counterbalance germline loss-of-function mutations. *J Clin Invest* **127:** 982–986. doi:10.1172/JCI91161

McMaster ML, Sun C, Landi MT, Savage SA, Rotunno M, Yang XR, Jones K, Vogt A, Hutchinson A, Zhu B, et al. 2018. Germline mutations in *protection of telomeres 1* in two families with Hodgkin lymphoma. *Br J Haematol* **181:** 372–377. doi:10.1111/bjh.15203

Mitchell JR, Wood E, Collins K. 1999. A telomerase component is defective in the human disease dyskeratosis congenita. *Nature* **402:** 551–555. doi:10.1038/990141

Montoya M, Uchino BN. 2023. Social support and telomere length: a meta-analysis. *J Behav Med* **46:** 556–565. doi:10.1007/s10865-022-00389-0

Moon DH, Segal M, Boyraz B, Guinan E, Hofmann I, Cahan P, Tai AK, Agarwal S. 2015. Poly(A)-specific ribonuclease (PARN) mediates 3′-end maturation of the telomerase RNA component. *Nat Genet* **47:** 1482–1488. doi:10.1038/ng.3423

Nathan V, Palmer JM, Johansson PA, Hamilton HR, Warrier SK, Glasson W, McGrath LA, Kahl VFS, Vasireddy RS, Pickett HA, et al. 2021. Loss-of-function variants in POT1 predispose to uveal melanoma. *J Med Genet* **58:** 234–236. doi:10.1136/jmedgenet-2020-107098

Newton CA, Kozlitina J, Lines JR, Kaza V, Torres F, Garcia CK. 2017. Telomere length in patients with pulmonary fibrosis associated with chronic lung allograft dysfunction and post-lung transplantation survival. *J Heart Lung Transplant* **36:** 845–853. doi:10.1016/j.healun.2017.02.005

Newton CA, Zhang D, Oldham JM, Kozlitina J, Ma SF, Martinez FJ, Raghu G, Noth I, Garcia CK. 2019. Telomere length and use of immunosuppressive medications in idiopathic pulmonary fibrosis. *Am J Respir Crit Care Med* **200:** 336–347. doi:10.1164/rccm.201809-1646OC

Niewisch MR, Savage SA. 2019. An update on the biology and management of dyskeratosis congenita and related telomere biology disorders. *Expert Rev Hematol* **12:** 1037–1052. doi:10.1080/17474086.2019.1662720

Niewisch MR, Giri N, McReynolds LJ, Alsaggaf R, Bhala S, Alter BP, Savage SA. 2022. Disease progression and clinical outcomes in telomere biology disorders. *Blood* **139:** 1807–1819. doi:10.1182/blood.2021013523

Niewisch MR, Beier F, Savage SA. 2023. Clinical manifestations of telomere biology disorders in adults. *Hematology Am Soc Hematol Educ Program* **2023:** 563–572. doi:10.1182/hematology.2023000490

Norris K, Walne AJ, Ponsford MJ, Cleal K, Grimstead JW, Ellison A, Alnajar J, Dokal I, Vulliamy T, Baird DM. 2021. High-throughput STELA provides a rapid test for the diagnosis of telomere biology disorders. *Hum Genet* **140:** 945–955. doi:10.1007/s00439-021-02257-4

Ouyang Q, Baerlocher G, Vulto I, Lansdorp PM. 2007. Telomere length in human natural killer cell subsets. *Ann NY Acad Sci* **1106:** 240–252. doi:10.1196/annals.1392.001

Cite this article as *Cold Spring Harb Perspect Biol* doi: 10.1101/cshperspect.a041684

Pepper GV, Bateson M, Nettle D. 2018. Telomeres as integrative markers of exposure to stress and adversity: a systematic review and meta-analysis. *R Soc Open Sci* **5:** 180744. doi:10.1098/rsos.180744

Polvi A, Linnankivi T, Kivelä T, Herva R, Keating JP, Mäkitie O, Pareyson D, Vainionpää L, Lahtinen J, Hovatta I, et al. 2012. Mutations in CTC1, encoding the CTS telomere maintenance complex component 1, cause cerebroretinal microangiopathy with calcifications and cysts. *Am J Hum Genet* **90:** 540–549. doi:10.1016/j.ajhg.2012.02.002

Polygenic Risk Score Task Force of the International Common Disease Alliance. 2021. Responsible use of polygenic risk scores in the clinic: potential benefits, risks and gaps. *Nat Med* **27:** 1876–1884. doi:10.1038/s41591-021-01549-6

Prescott J, Kraft P, Chasman D, Savage S, Mirabello L, Berndt S, Weissfeld J, Han J, Hayes R, Chanock S, et al. 2011. Genome-wide association study of relative telomere length. *PLoS ONE* **6:** e19635. doi:10.1371/journal.pone.0019635

Prescott J, Wentzensen IM, Savage SA, De Vivo I. 2012. Epidemiologic evidence for a role of telomere dysfunction in cancer etiology. *Mutat Res* **730:** 75–84. doi:10.1016/j.mrfmmm.2011.06.009

Raj HA, Lai TP, Niewisch MR, Giri N, Wang Y, Spellman SR, Aviv A, Gadalla SM, Savage SA. 2023. The distribution and accumulation of the shortest telomeres in telomere biology disorders. *Br J Haematol* **203:** 820–828. doi:10.1111/bjh.18945

Revesz T, Fletcher S, al-Gazali LI, DeBuse P. 1992. Bilateral retinopathy, aplastic anaemia, and central nervous system abnormalities: a new syndrome? *J Med Genet* **29:** 673–675. doi:10.1136/jmg.29.9.673

Révész D, Verhoeven JE, Picard M, Lin J, Sidney S, Epel ES, Penninx B, Puterman E. 2018. Associations between cellular aging markers and metabolic syndrome: findings from the CARDIA study. *J Clin Endocrinol Metab* **103:** 148–157. doi:10.1210/jc.2017-01625

Revy P, Kannengiesser C, Fischer A. 2019. Somatic genetic rescue in Mendelian haematopoietic diseases. *Nat Rev Genet* **20:** 582–598. doi:10.1038/s41576-019-0139-x

Revy P, Kannengiesser C, Bertuch AA. 2023. Genetics of human telomere biology disorders. *Nat Rev Genet* **24:** 86–108. doi:10.1038/s41576-022-00527-z

Robles-Espinoza CD, Harland M, Ramsay AJ, Aoude LG, Quesada V, Ding Z, Pooley KA, Pritchard AL, Tiffen JC, Petljak M, et al. 2014. POT1 loss-of-function variants predispose to familial melanoma. *Nat Genet* **46:** 478–481. doi:10.1038/ng.2947

Rode L, Nordestgaard BG, Bojesen SE. 2016. Long telomeres and cancer risk among 95,568 individuals from the general population. *Int J Epidemiol* **45:** 1634–1643. doi:10.1093/ije/dyw179

Rolles B, Caballero-Oteyza A, Proietti M, Goldacker S, Warnatz K, Camacho-Ordonez N, Prader S, Schmid JP, Vieri M, Isfort S, et al. 2023. Telomere biology disorders may manifest as common variable immunodeficiency (CVID). *Clin Immunol* **257:** 109837. doi:10.1016/j.clim.2023.109837

Rossiello F, Jurk D, Passos JF, d'Adda di Fagagna F. 2022. Telomere dysfunction in ageing and age-related diseases.

*Nat Cell Biol* **24:** 135–147. doi:10.1038/s41556-022-00842-x

Rozing MP, Kirkwood TBL, Westendorp RGJ. 2017. Is there evidence for a limit to human lifespan? *Nature* **546:** E11–E12. doi:10.1038/nature22788

Said MA, Eppinga RN, Hagemeijer Y, Verweij N, van der Harst P. 2017. Telomere length and risk of cardiovascular disease and cancer. *J Am Coll Cardiol* **70:** 506–507. doi:10.1016/j.jacc.2017.05.044

Samuel BP, Duffner UA, Abdel-Mageed AS, Vettukattil JJ. 2015. Pulmonary arteriovenous malformations in dyskeratosis congenita. *Pediatr Dermatol* **32:** e165–e166. doi:10.1111/pde.12589

Sanchez SE, Gu Y, Wang Y, Golla A, Martin A, Shomali W, Hockemeyer D, Savage SA, Artandi SE. 2024. Digital telomere measurement by long-read sequencing distinguishes healthy aging from disease. *Nat Commun* **15:** 5148. doi:10.1038/s41467-024-49007-4

Sargen MR, Calista D, Elder DE, Massi D, Chu EY, Potrony M, Pfeiffer RM, Carrera C, Aguilera P, Alos L, et al. 2020. Histologic features of melanoma associated with germline mutations of CDKN2A, CDK4, and POT1 in melanoma-prone families from the United States, Italy, and Spain. *J Am Acad Dermatol* **83:** 860–869. doi:10.1016/j.jaad.2020.03.100

Savage SA. 2018. Beginning at the ends: telomeres and human disease. *F1000Res* **7:** 524. doi:10.12688/f1000research.14068.1

Savage SA. 2022. Dyskeratosis congenita and telomere biology disorders. *Hematology Am Soc Hematol Educ Program* **2022:** 637–648. doi:10.1182/hematology.2022000394

Savage SA. 2024. Telomere length and cancer risk: finding Goldilocks. *Biogerontology* **25:** 265–278. doi:10.1007/s10522-023-10080-9

Savage SA, Giri N, Baerlocher GM, Orr N, Lansdorp PM, Alter BP. 2008. TINF2, a component of the shelterin telomere protection complex, is mutated in dyskeratosis congenita. *Am J Hum Genet* **82:** 501–509. doi:10.1016/j.ajhg.2007.10.004

Savage SA, Bertuch AA, Team T; Team Telomere and the Clinical Care Consortium for Telomere-Associated Ailments (CCCTA). 2024. Different phenotypes with different endings—telomere biology disorders and cancer predisposition with long telomeres. *Br J Haematol* doi:10.1111/bjh.19851

Schmidt TT, Tyer C, Rughani P, Haggblom C, Jones JR, Dai X, Frazer KA, Gage FH, Juul S, Hickey S, et al. 2024. High resolution long-read telomere sequencing reveals dynamic mechanisms in aging and cancer. *Nat Commun* **15:** 5149. doi:10.1038/s41467-024-48917-7

Schneider CV, Schneider KM, Teumer A, Rudolph KL, Hartmann D, Rader DJ, Strnad P. 2022. Association of telomere length with risk of disease and mortality. *JAMA Intern Med* **182:** 291–300. doi:10.1001/jamainternmed.2021.7804

Schratz KE, Haley L, Danoff SK, Blackford AL, DeZern AE, Gocke CD, Duffield AS, Armanios M. 2020. Cancer spectrum and outcomes in the Mendelian short telomere syndromes. *Blood* **135:** 1946–1956. doi:10.1182/blood.2019003264

Schratz KE, Gaysinskaya V, Cosner ZL, DeBoy EA, Xiang Z, Kasch-Semenza L, Florea L, Shah PD, Armanios M. 2021. Somatic reversion impacts myelodysplastic syndromes and acute myeloid leukemia evolution in the short telomere disorders. *J Clin Invest* **131**: e147598. doi:10.1172/JCI147598

Schratz KE, Flasch DA, Atik CC, Cosner ZL, Blackford AL, Yang W, Gable DL, Vellanki PJ, Xiang Z, Gaysinskaya V, et al. 2023. T cell immune deficiency rather than chromosome instability predisposes patients with short telomere syndromes to squamous cancers. *Cancer Cell* **41**: 807–817.e6. doi:10.1016/j.ccell.2023.03.005

Sharma R, Sahoo SS, Honda M, Granger SL, Goodings C, Sanchez L, Künstner A, Busch H, Beier F, Pruett-Miller SM, et al. 2022. Gain-of-function mutations in RPA1 cause a syndrome with short telomeres and somatic genetic rescue. *Blood* **139**: 1039–1051. doi:10.1182/blood.2021011980

Shen M, Cawthon R, Rothman N, Weinstein SJ, Virtamo J, Hosgood HD III, Hu W, Lim U, Albanes D, Lan Q. 2011. A prospective study of telomere length measured by monochrome multiplex quantitative PCR and risk of lung cancer. *Lung Cancer* **73**: 133–137. doi:10.1016/j.lungcan.2010.11.009

Shi J, Yang XR, Ballew B, Rotunno M, Calista D, Fargnoli MC, Ghiorzo P, Bressac-de Paillerets B, Nagore E, Avril MF, et al. 2014. Rare missense variants in POT1 predispose to familial cutaneous malignant melanoma. *Nat Genet* **46**: 482–486. doi:10.1038/ng.2941

Silhan LL, Shah PD, Chambers DC, Snyder LD, Riise GC, Wagner CL, Hellström-Lindberg E, Orens JB, Mewton JF, Danoff SK, et al. 2014. Lung transplantation in telomerase mutation carriers with pulmonary fibrosis. *Eur Respir J* **44**: 178–187. doi:10.1183/09031936.00060014

Simon AJ, Lev A, Zhang Y, Weiss B, Rylova A, Eyal E, Kol N, Barel O, Cesarkas K, Soudack M, et al. 2016. Mutations in STN1 cause Coats plus syndrome and are associated with genomic and telomere defects. *J Exp Med* **213**: 1429–1440. doi:10.1084/jem.20151618

Simonin-Wilmer I, Ossio R, Leddin EM, Harland M, Pooley KA, Martil de la Garza MG, Obolenski S, Hewinson J, Wong CC, Iyer V, et al. 2023. Population-based analysis of POT1 variants in a cutaneous melanoma case-control cohort. *J Med Genet* **60**: 692–696. doi:10.1136/jmg-2022-108776

Speedy HE, Kinnersley B, Chubb D, Broderick P, Law PJ, Litchfield K, Jayne S, Dyer MJS, Dearden C, Follows GA, et al. 2016. Germ line mutations in shelterin complex genes are associated with familial chronic lymphocytic leukemia. *Blood* **128**: 2319–2326. doi:10.1182/blood-2016-01-695692

Stanley SE, Gable DL, Wagner CL, Carlile TM, Hanumanthu VS, Podlevsky JD, Khalil SE, DeZern AE, Rojas-Duran MF, Applegate CD, et al. 2016. Loss-of-function mutations in the RNA biogenesis factor NAF1 predispose to pulmonary fibrosis-emphysema. *Sci Transl Med* **8**: 351ra107. doi:10.1126/scitranslmed.aaf7837

Steenstrup T, Kark JD, Verhulst S, Thinggaard M, Hjelmborg JVB, Dalgård C, Kyvik KO, Christiansen L, Mangino M, Spector TD, et al. 2017. Telomeres and the natural lifespan limit in humans. *Aging* **9**: 1130–1142. doi:10.18632/aging.101216

Takai H, Jenkinson E, Kabir S, Babul-Hirji R, Najm-Tehrani N, Chitayat DA, Crow YJ, de Lange T. 2016. A POT1 mutation implicates defective telomere end fill-in and telomere truncations in Coats plus. *Genes Dev* **30**: 812–826. doi:10.1101/gad.276873.115

Telomeres Mendelian Randomization Collaboration; Haycock PC, Burgess S, Nounu A, Zheng J, Okoli GN, Bowden J, Wade KH, Timpson NJ, Evans DM, et al. 2017. Association between telomere length and risk of cancer and non-neoplastic diseases: a Mendelian randomization study. *JAMA Oncol* **3**: 636–651. doi:10.1001/jamaoncol.2017.2316

Thompson MB, Muldoon D, de Andrade KC, Giri N, Alter BP, Savage SA, Shamburek RD, Khincha PP. 2022. Lipoprotein particle alterations due to androgen therapy in individuals with dyskeratosis congenita. *EBioMedicine* **75**: 103760. doi:10.1016/j.ebiom.2021.103760

Tolmie JL, Browne BH, McGettrick PM, Stephenson JBP. 1988. A familial syndrome with coats' reaction retinal angiomas, hair and nail defects and intracranial calcification. *Eye* **2**: 297–303. doi:10.1038/eye.1988.56

Tometten M, Kirschner M, Meyer R, Begemann M, Halfmeyer I, Vieri M, Kricheldorf K, Maurer A, Platzbecker U, Radsak M, et al. 2023. Identification of adult patients with classical dyskeratosis congenita or cryptic telomere biology disorder by telomere length screening using age-modified criteria. *Hemasphere* **7**: e874. doi:10.1097/HS9.0000000000000874

Toufektchan E, Lejour V, Durand R, Giri N, Draskovic I, Bardot B, Laplante F, Jaber S, Alter BP, Londono-Vallejo JA, et al. 2020. Germline mutation of MDM4, a major p53 regulator, in a familial syndrome of defective telomere maintenance. *Sci Adv* **6**: eaay3511. doi:10.1126/sciadv.aay3511

Toupance S, Labat C, Temmar M, Rossignol P, Kimura M, Aviv A, Benetos A. 2017. Short telomeres, but not telomere attrition rates, are associated with carotid atherosclerosis. *Hypertension* **70**: 420–425. doi:10.1161/HYPERTENSIONAHA.117.09354

Townsley DM, Dumitriu B, Young NS. 2016. Danazol treatment for telomere diseases. *N Engl J Med* **375**: 1095–1096. doi:10.1056/NEJMc1607752

Trigueros-Motos L. 2014. Mutations in POT1 predispose to familial cutaneous malignant melanoma. *Clin Genet* **86**: 217–218. doi:10.1111/cge.12416

Tsakiri KD, Cronkhite JT, Kuan PJ, Xing C, Raghu G, Weissler JC, Rosenblatt RL, Shay JW, Garcia CK. 2007. Adult-onset pulmonary fibrosis caused by mutations in telomerase. *Proc Natl Acad Sci* **104**: 7552–7557. doi:10.1073/pnas.0701009104

Tummala H, Walne A, Collopy L, Cardoso S, de la Fuente J, Lawson S, Powell J, Cooper N, Foster A, Mohammed S, et al. 2015. Poly(A)-specific ribonuclease deficiency impacts telomere biology and causes dyskeratosis congenita. *J Clin Invest* **125**: 2151–2160. doi:10.1172/JCI78963

Tummala H, Walne A, Dokal I. 2022. The biology and management of dyskeratosis congenita and related disorders of telomeres. *Expert Rev Hematol* **15**: 685–696. doi:10.1080/17474086.2022.2108784

Vieri M, Kirschner M, Tometten M, Abels A, Rolles B, Isfort S, Panse J, Brümmendorf TH, Beier F. 2020. Comparable effects of the androgen derivatives danazol, oxymetholone and nandrolone on telomerase activity in human primary hematopoietic cells from patients with dyskeratosis congenita. *Int J Mol Sci* **21:** 7196. doi:10.3390/ijms21197196

Vieri M, Brümmendorf TH, Beier F. 2021. Treatment of telomeropathies. *Best Pract Res Clin Haematol* **34:** 101282. doi:10.1016/j.beha.2021.101282

Vittal A, Niewisch MR, Bhala S, Kudaravalli P, Rahman F, Hercun J, Kleiner DE, Savage SA, Koh C, Heller T, et al. 2023. Progression of liver disease and portal hypertension in dyskeratosis congenita and related telomere biology disorders. *Hepatology* **78:** 1777–1787. doi:10.1097/HEP.0000000000000461

Vulliamy T, Marrone A, Szydlo R, Walne A, Mason PJ, Dokal I. 2004. Disease anticipation is associated with progressive telomere shortening in families with dyskeratosis congenita due to mutations in TERC. *Nat Genet* **36:** 447–449. doi:10.1038/ng1346

Vulliamy TJ, Marrone A, Knight SW, Walne A, Mason PJ, Dokal I. 2006. Mutations in dyskeratosis congenita: their impact on telomere length and the diversity of clinical presentation. *Blood* **107:** 2680–2685. doi:10.1182/blood-2005-07-2622

Vulliamy T, Beswick R, Kirwan M, Marrone A, Digweed M, Walne A, Dokal I. 2008. Mutations in the telomerase component NHP2 cause the premature ageing syndrome dyskeratosis congenita. *Proc Natl Acad Sci* **105:** 8073–8078. doi:10.1073/pnas.0800042105

Walne AJ, Vulliamy T, Marrone A, Beswick R, Kirwan M, Masunari Y, Al-Qurashi FH, Aljurf M, Dokal I. 2007. Genetic heterogeneity in autosomal recessive dyskeratosis congenita with one subtype due to mutations in the telomerase-associated protein NOP10. *Hum Mol Genet* **16:** 1619–1629. doi:10.1093/hmg/ddm111

Walne AJ, Vulliamy T, Beswick R, Kirwan M, Dokal I. 2008. TINF2 mutations result in very short telomeres: analysis of a large cohort of patients with dyskeratosis congenita and related bone marrow failure syndromes. *Blood* **112:** 3594–3600. doi:10.1182/blood-2008-05-153445

Walne AJ, Vulliamy T, Kirwan M, Plagnol V, Dokal I. 2013. Constitutional mutations in RTEL1 cause severe dyskeratosis congenita. *Am J Hum Genet* **92:** 448–453. doi:10.1016/j.ajhg.2013.02.001

Walne AJ, Collopy L, Cardoso S, Ellison A, Plagnol V, Albayrak C, Albayrak D, Kilic SS, Patiroglu T, Akar H, et al. 2016. Marked overlap of four genetic syndromes with dyskeratosis congenita confounds clinical diagnosis. *Haematologica* **101:** 1180–1189. doi:10.3324/haematol.2016.147769

Wand H, Kalia SS, Helm BM, Suckiel SA, Brockman D, Vriesen N, Goudar RK, Austin J, Yanes T. 2023. Clinical genetic counseling and translation considerations for polygenic scores in personalized risk assessments: a practice resource from the National Society of Genetic Counselors. *J Genet Couns* **32:** 558–575. doi:10.1002/jgc4.1668

Wang YM, Kaj-Carbaidwala B, Lane A, Agarwal S, Beier F, Bertuch A, Borovsky KA, Brennan SK, Calado RT, Catto LFB, et al. 2024. Liver disease and transplantation in telomere biology disorders: an international multicenter cohort. *Hepatol Commun* **8:** e0462. doi:10.1097/HC9.0000000000000462

Ward SC, Savage SA, Giri N, Alter BP, Rosenberg PS, Pichard DC, Cowen EW. 2018. Beyond the triad: inheritance, mucocutaneous phenotype, and mortality in a cohort of patients with dyskeratosis congenita. *J Am Acad Dermatol* **78:** 804–806. doi:10.1016/j.jaad.2017.10.017

Wentzensen IM, Mirabello L, Pfeiffer RM, Savage SA. 2011. The association of telomere length and cancer: a meta-analysis. *Cancer Epidemiol Biomarkers Prev* **20:** 1238–1250. doi:10.1158/1055-9965.EPI-11-0005

Wu X, Amos CI, Zhu Y, Zhao H, Grossman BH, Shay JW, Luo S, Hong WK, Spitz MR. 2003. Telomere dysfunction: a potential cancer predisposition factor. *J Natl Cancer Inst* **95:** 1211–1218. doi:10.1093/jnci/djg011

Yamaguchi H, Baerlocher GM, Lansdorp PM, Chanock SJ, Nunez O, Sloand E, Young NS. 2003. Mutations of the human telomerase RNA gene (TERC) in aplastic anemia and myelodysplastic syndrome. *Blood* **102:** 916–918. doi:10.1182/blood-2003-01-0335

Yamaguchi H, Calado RT, Ly H, Kajigaya S, Baerlocher GM, Chanock SJ, Lansdorp PM, Young NS. 2005. Mutations in TERT, the gene for telomerase reverse transcriptase, in aplastic anemia. *N Engl J Med* **352:** 1413–1424. doi:10.1056/NEJMoa042980

Ye Q, Apsley AT, Etzel L, Hastings WJ, Kozlosky JT, Walker C, Wolf SE, Shalev I. 2023. Telomere length and chronological age across the human lifespan: a systematic review and meta-analysis of 414 study samples including 743,019 individuals. *Ageing Res Rev* **90:** 102031. doi:10.1016/j.arr.2023.102031

Yin H, Pickering JG. 2023. Telomere length: implications for atherogenesis. *Curr Atheroscler Rep* **25:** 95–103. doi:10.1007/s11883-023-01082-6

Zade NH, Khattar E. 2023. POT1 mutations cause differential effects on telomere length leading to opposing disease phenotypes. *J Cell Physiol* **238:** 1237–1255. doi:10.1002/jcp.31034

Zhang C, Doherty JA, Burgess S, Hung RJ, Lindström S, Kraft P, Gong J, Amos CI, Sellers TA, Monteiro AN, et al. 2015. Genetic determinants of telomere length and risk of common cancers: a Mendelian randomization study. *Hum Mol Genet* **24:** 5356–5366. doi:10.1093/hmg/ddv252

Zhang D, Adegunsoye A, Oldham JM, Kozlitina J, Garcia N, Poonawalla M, Strykowski R, Linderholm AL, Ley B, Ma SF, et al. 2023. Telomere length and immunosuppression in non-idiopathic pulmonary fibrosis interstitial lung disease. *Eur Respir J* **62:** 2300441. doi:10.1183/13993003.00441-2023

Zhong F, Savage SA, Shkreli M, Giri N, Jessop L, Myers T, Chen R, Alter BP, Artandi SE. 2011. Disruption of telomerase trafficking by TCAB1 mutation causes dyskeratosis congenita. *Genes Dev* **25:** 11–16. doi:10.1101/gad.2006411

# Telomerase in Cancer Therapeutics

Silvia Siteni,[1,2] Anthony Grichuk,[1,2] and Jerry W. Shay[1]

[1]University of Texas Southwestern Medical Center, Department of Cell Biology, Dallas, Texas 75390, USA

*Correspondence:* jerry.shay@utsouthwestern.edu

While silent in normal differentiated human tissues, telomerase is reactivated in most human cancers. Thus, telomerase is an almost universal oncology target. This update describes preclinical and clinical advancements using a variety of approaches to target telomerase. These include direct telomerase inhibitors, G-quadruplex DNA-interacting ligands, telomerase-based vaccine platforms, telomerase promoter-driven attenuated viruses, and telomerase-mediated telomere targeting approaches. While imetelstat has been recently approved by the Food and Drug Administration (FDA), several other approaches are in late-stage clinical development. The pros and cons of the major approaches will be reviewed.

Telomere shortening is a well-established hallmark of aging that, at the cellular level, leads to replicative senescence by progressive telomere shortening when at least one telomere on one chromosome is recognized as DNA damage that is not repaired (Hemann et al. 2001; Zou et al. 2004). Senescence can be overcome by the activation of telomerase or alternative lengthening of telomeres (ALT) providing unlimited proliferation in somatic cells, a well-established hallmark of advanced cancer (Bodnar et al. 1998). In long-lived species, such as humans, repression of telomerase during fetal development occurs in most somatic cells to prevent unlimited proliferation (Wright et al. 1996). As a result, telomere shortening-induced growth arrest provides an effective initial barrier to prevent cancer progression. Since the discovery of telomerase, there have been multiple attempts to target cancer in preclinical animal models and in human clinical trials using telomere and telomerase inhibitors (Fig. 1). The rate-limiting step in telomerase activation is the telomerase reverse transcriptase (TERT) gene. Derepression of the TERT gene occurs through several mechanisms, such as TERT promoter mutations, leading to deregulated transcriptional signaling and posttranscriptional/translational regulation. In addition, telomerase activation can occur via TERT genomic amplification, viral insertions driving TERT regulation, and TERT chromosomal rearrangements. The main approaches to targeting telomerase and telomeres include G-quadruplex (G4) DNA-interacting ligands (TMPyP4 and BRACO-19), direct telomerase inhibitors (BIBR-1532), telomerase vaccines (GRNVAC1, GV1001, VX-001, INVAC-1, UV1), and TERT promoter driving the expression of a suicide gene such as a proapoptosis gene or attenuated replication-competent virus (OBP-301 also known as Telomelysin). Additional anticancer approaches include MST-312, a synthetic compound related to EGCG (tea catechin) that leads to progressive telo-

---

[2]These authors contributed equally to this work.

Cite this article as *Cold Spring Harb Perspect Biol* doi: 10.1101/cshperspect.a041703

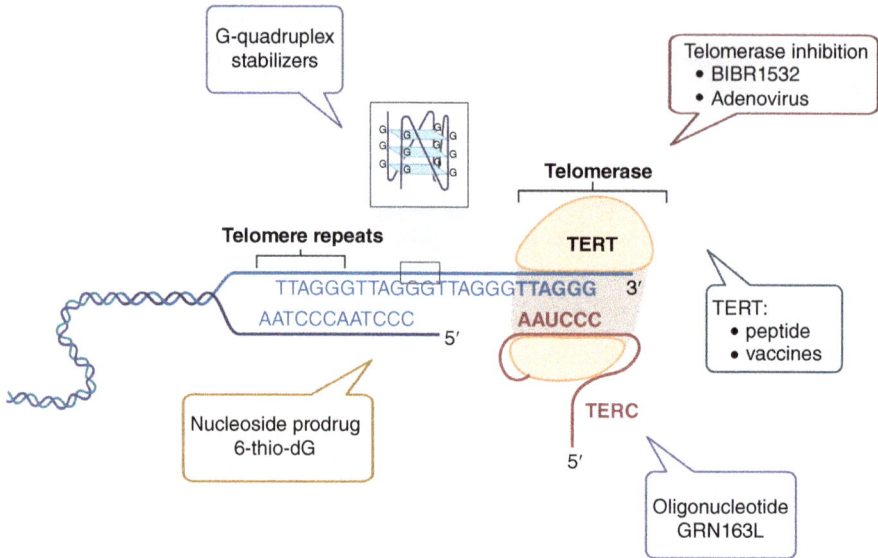

**Figure 1.** Telomeres and telomerase as therapeutic targets in cancer treatment. BIBR1532 inhibits telomerase activity specifically targeting the human telomerase reverse transcriptase (hTERT) component. The adenovirus Telomelysin selectively replicates in tumor cells expressing telomerase reverse transcriptase (TERT), causing cell rupture and inducing an immunogenic response. GRNVAC1 is a dendritic base vaccine using the hTERT mRNA, which is introduced into immature dendritic cells and subsequently induced to mature by cytokine treatment, leading to an adaptive immune response. GV1001 is a 16-amino acid peptide comprising a sequence from the human enzyme TERT used to activate the immune system to recognize cancer cells expressing hTERT. The UV1 vaccine is composed of three long synthetic peptides, representing 60 amino acids of the reverse transcriptase subunit of hTERT that can induce a specific T-cell response against hTERT. G-quadruplex stabilizers are small molecules targeting the quadruplexes at telomeres due to the high frequency of guanine, creating a spatial obstacle to telomerase activity and telomere length maintenance. The oligonucleotide GRN163L (imetelstat) directed against the 11-base hTERC, prevents the interaction of hTERT with the 3′ telomere overhang. 6-thio-dG is a prodrug preferentially incorporated into the G-rich strand of the telomere by telomerase, resulting in a rapid telomere uncapping. (Figure created with BioRender.com.)

mere shortening (Seimiya et al. 2002), oligonucleotides that target the functional RNA component (TERC) of the telomerase holoenzyme (GRN163L/imetelstat), and a small molecule telomerase-mediated telomere uncapping drug (6-thio-2′deoxyguanosine/6-thio-dG). Most of these have been the subject of earlier reviews (Keith et al. 2007; Jafri et al. 2016; Sugarman et al. 2019; Seimiya 2020; Gao and Pickett 2022) and will only be briefly updated. However, we will discuss molecular mechanisms, preclinical and clinical performances, and cancer cell versus normal cell discrimination. In this review article, we will provide insights into the progress and failures of these approaches, provide an update on ongoing preclinical and clinical studies,

and highlight the challenges and future directions in the field.

## BIBR1532: SELECTIVE INHIBITOR OF HUMAN TELOMERASE

BIBR1532 (2-[(E)-3-naphtalen-2-yl-but-2-eno-ylamino]-benzoic acid) is a potent and specific nonnucleosidic inhibitor of human telomerase reverse transcriptase (hTERT) (Damm et al. 2001; Pascolo et al. 2002). BIBR1532 inhibits telomerase activity by specifically binding to the conserved hydrophobic pocket (FVYL) motif in the active site of hTERT (Damm et al. 2001). BIBR1532 inhibits native and recombinant human telomerase with an $IC_{50}$ of

Cite this article as *Cold Spring Harb Perspect Biol* doi: 10.1101/cshperspect.a041703

~100 nm in cell-free assays but only in the mid-micromolar range using cell-based experiments. Thus, BIBR1532 is a novel class of telomerase inhibitor that interferes with telomerase processivity with mechanistic similarities to non-nucleosidic inhibitors of HIV1 reverse transcriptase (Bryan et al. 2015). While low concentrations of BIBR1532 are not toxic to telomerase-expressing lung cancer cells, there is a dose-dependent cytotoxicity at high concentrations (Ding et al. 2019). In xenografts, at 100 mg/kg (oral gavage), there is a partial tumor reduction effect over 60 days of treatment. In addition, even at lower concentrations of BIBR1532, there is an increase in the therapeutic efficacy when combined with radiation exposure in vitro and in vivo by enhancing mitotic catastrophe leading to apoptosis and/or senescence in leukemia, lung cancer, and breast cancer (El-Daly et al. 2005; Bashash et al. 2013; Ding et al. 2019; Doğan et al. 2019; Nasrollahzadeh et al. 2020). Overall, these studies demonstrate that BIBR1532 inhibits telomerase function and could be a promising and safe therapeutic strategy in combination with other therapeutic agents. Even though BIBR1532 was reported in 2001, as far as can be determined, no human clinical trials were initiated.

## G-QUADRUPLEX DNA-INTERACTING LIGANDS

G4s are guanine-rich four-stranded DNA secondary structures that form at sites such as telomeres and at gene promoters, for example, the MYC gene. There is substantial evidence for the formation of G4s in living cells (Di Antonio et al. 2020; Summers et al. 2021; Esnault et al. 2023) and the involvement of G4 structures in multiple cellular processes (including as transcriptional repressors) that make these attractive targets for drug design (Tian et al. 2018). Therefore, compounds recognizing and interacting with G4s are of interest as potential anticancer therapies. While G4 small molecules can inhibit telomerase activity, this may be an indirect effect in most instances. For example, the telomerase (hTERT) promoter has G4 sequences and an E-box DNA motif (c-Myc binding site), so any ligand that affects Myc binding will also affect telomerase activity. G4 ligands are more likely to behave as inhibitors of telomeres by induction of replication stress leading to genomic instability since there are multiple gene promoters (~40%) with G4 sequences, including c-MYC, c-KIT, and K-RAS (Huppert and Balasubramanian 2007; Monsen et al. 2023). Thus, most G4 ligands behave similarly to chemotherapy approaches and will affect dividing telomerase silent normal and telomerase-expressing cancer cells. Even though several studies report G4 ligand effects on telomerase, telomeres, telomere-binding proteins, and T-loop stability (Riou et al. 2002; Gomez et al. 2010; Zhou et al. 2016), the lack of cancer-specific selectivity results in sufficient side effects that most clinical trials have been halted. For example, BRACO-19, RHPS4, quarfloxin, and APTO-253 are no longer in clinical development due to lack of efficacy or toxicity issues (Lim et al. 2010; Iachettini et al. 2013; Ohanian et al. 2021). However, CX-5461 has achieved recent safety, tolerability, and pharmacokinetics results, and remains in clinical evaluation (Jin et al. 2023). In summary, recent approaches on synthetic G4 DNA-interacting ligands may allow more selective targeting of G4s as an approach toward the development of highly effective anticancer drugs, especially in combination with other cancer therapeutics.

## TELOMERASE-BASED VACCINE PLATFORMS

Tumor cell–associated antigens (neoantigens) should preferentially trigger innate and adaptive anticancer immune responses but have a restricted expression pattern in normal cells to minimize tolerance and potential autoimmune side effects. Thus, it is reasonable to assess if telomerase is an almost universal antitumor antigen since few normal cells express telomerase. A challenge when considering developing a telomerase vaccine is the occurrence of central tolerance instead of adaptive immune responses. Other challenges include off-target effects (e.g., normal telomerase-expressing cells) and the development of autoimmunity. However, if one can identify tumor-specific mutations that give rise to tumor neoantigens (altered peptides) as targets for vaccination, then it is worth exploring. Even though telomerase is expressed in most tumors, due to tumor cell heterogeneity and the existence of the ALT pathway, it may be difficult to produce an almost universal tumor vaccine. GRNVAC1 is a dendritic base vaccine that was produced by electrophoresis of hTERT mRNA along with a lysosomal targeting sequence (LAMP-1) into immature dendritic cells (DCs), which were then made into mature DCs using cytokines. AML patients with clinical responses in first-line therapy or in early relapse were vaccinated with $10^7$ autologous cells intradermally and repeated weekly. A phase II trial was completed to evaluate the safety and tolerability of the vaccination regimen. A total of 21 patients with AML received GRNVAC1. Nineteen patients were in clinical remission, and after vaccination, 13 (81%) remained in remission when historically, in this patient population, only 45% would have remained in remission (Khoury et al. 2010). Overall, the vaccine showed promise and was well tolerated with no major treatment-related toxicities. While these clinical studies were completed in 2010, no follow-up studies have been reported.

GV1001 (also known as Riavaxtm or Tertomotide), a hTERT catalytic subunit-derived 16-mer peptide, was developed as a novel anticancer vaccine against various cancers including pancreatic cancer. Multiple clinical trials were conducted with GV1001 in patients with pancreatic cancer and treated with both chemotherapy and GV1001. While many patients had immune responses, none were associated with clinical improvement (reviewed by Negrini et al. 2020). More recently, a therapeutic telomerase vaccine (UV1 consisting of three synthetic peptides covering 54 amino acids in the TERT active site) was combined with immunotherapy (pembrolizumab) in melanoma patients with poor responses compared to checkpoint inhibition alone. This phase I clinical trial conferred substantial outcome improvements with encouraging safety results and a phase II trial is in progress (Ellingsen et al. 2023).

There are other therapies in clinical development and in total there have been more than 33 telomerase vaccine clinical trials since 2000, as recently reviewed (Negrini et al. 2020; Ellingsen et al. 2021), but none have received FDA approval. While these hTERT-vaccine clinical trials are promising, new strategies are needed to improve clinical efficacy going forward. In summary, what has emerged is that combining telomerase peptide vaccination protocols with other standard of care cancer therapies (such as immunotherapy, radiation therapy, and other chemotherapy approaches) may be the direction to pursue. Thus, there remains promise that these clinical trials may show additive or even synergetic effects (Ellingsen et al. 2023). Knowledge about antitumor immunity is rapidly advancing, so this remains a very promising approach.

## TELOMERASE PROMOTER-DRIVEN ATTENUATED ADENOVIRUS

OBP-301 is a gene-modified oncolytic adenovirus (E1A and E1B genes) that selectively replicates in telomerase-expressing cells (Fig. 2; Kawashima et al. 2004). In addition, viral replication may induce immunogenic responses increasing infiltration of effector CD8$^+$ T cells into the tumor microenvironment (TME). In preclinical studies, OBP-301 was effective in several types of cancer cells, and there were no safety concerns in toxicological studies. A phase I clinical trial for solid tumors indicated that adverse events were mild to moderate and transient, and there was a sign of efficacy showing tumor shrinkage in eight out of 12 evaluable patients. In Japan, an investigator-initiated clinical trial for unresectable, chemotherapy-resistant, locally advanced esophageal cancer in combination with radiotherapy was conducted. The trial resulted in eight out of 13 patients showing complete responses, with CD8$^+$ T-cell infiltration observed in the tumor area (Tanabe et al. 2019). OBP-301 then underwent a phase I study for esophageal cancer (multiple intratumoral injections) in combination with ionizing radiation therapy, that resulted in clinical benefits to patients not eligible for standard treatments (Shirakawa et al. 2021). This resulted in orphan drug designation and led to a phase II multicenter study for esophageal cancer

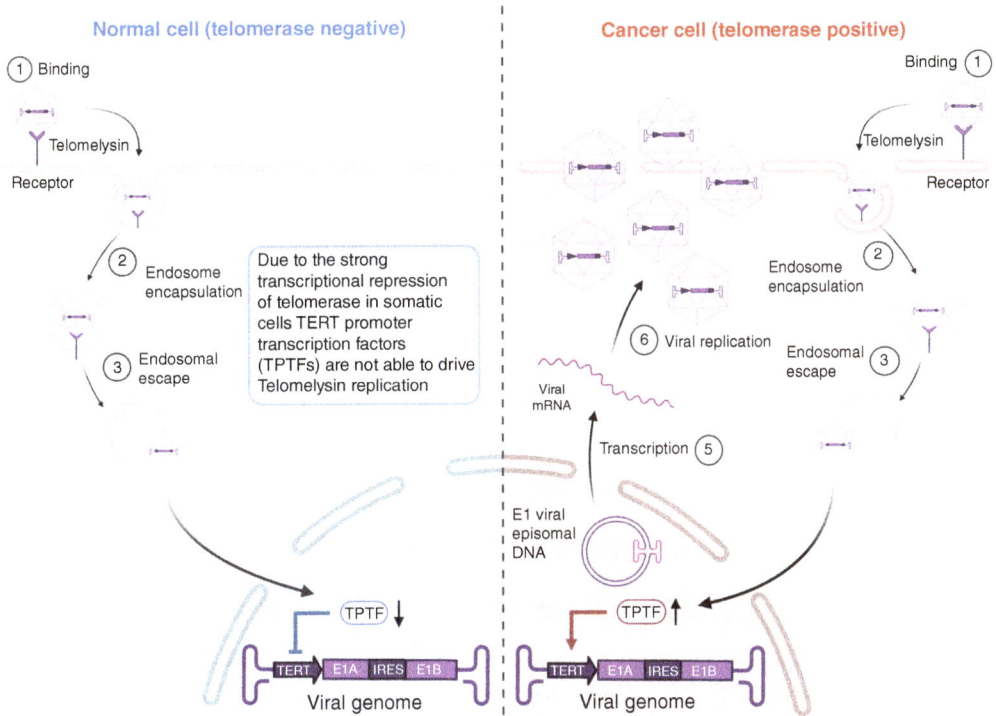

**Figure 2.** OBP-301, known as Telomelysin, is a modified adenovirus that replicates specifically in cancer cells expressing telomerase. The viral genome is engineered to contain a portion of the human telomerase reverse transcriptase (hTERT) promoter and E1A and E1B adenoviral genes linked with the internal ribosome entry site (IRES). The exceptionally low presence of telomerase reverse transcriptase (TERT) expression in normal cells, does not allow adenovirus replication (*left* panel), whereas in telomerase-expressing cancer cells, the OBP-1 hTERT promoter is activated, allowing virus replication and consequent lysis of tumor cells, and spreading to adjacent cells (*right* panel). (Figure created with BioRender.com.)

in combination with immunotherapy (Shah et al. 2023). The clinical trial with OBP-301 viral particle intratumoral injection was safe. Furthermore, in combination with pembrolizumab. OBP-301 had some durable responses in patients thus demonstrating activity in immunotherapy refractory disease (Shah et al. 2023). In summary, oncolytic adenoviruses are a promising area for future research and clinical trials, but effective results using systemic therapy have not been reported, which may limit the general utility of this approach.

## IMETELSTAT (GRN163L)

GRN163 is an oligonucleotide directed against the 11-base hTERC functional RNA template re-

gion of telomerase, but as originally evaluated had limitations related to the low ability to permeate the membranes of cells. This obstacle was resolved by modifying GRN163 to incorporate a lipidated 13-mer thio-phosphoramidate (GRN163L, now called imetelstat) that binds hTERC RNA sequences with improved selectivity, stability, enhanced biodistribution, potency, and improved pharmacological properties. GRN163L has a sequence complementary to the hTERC template sequence and binds with high affinity, preventing the interaction of hTERT with the 3′ telomere overhang, thus avoiding the annealing and polymerization activity of telomerase (Tomita and Collopy 2018). The lipid palmitate moiety is built into the 13-mer oligonucleotide synthesis eliminating the neces-

**A**

TAGGGTTAGACAA-3'

R=-(CH2)₁₃CH3 (palmitoyl)

GRN163L (imetelstat)

**B**

6-thio-2'-deoxyguanosine (THIO)

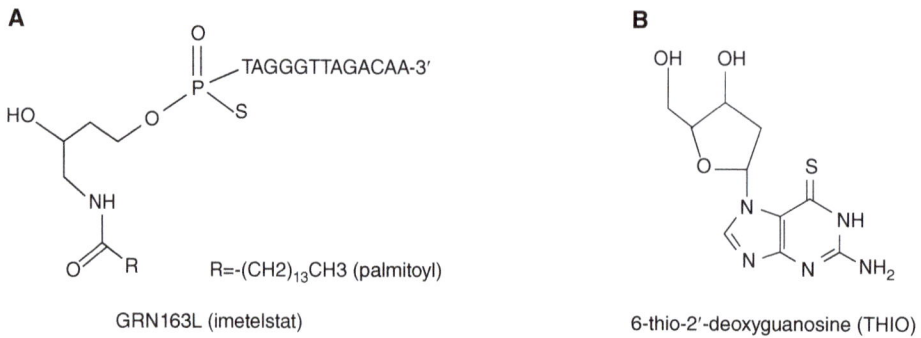

**Figure 3.** The chemical structures of imetelstat and 6-thio-2'deoxyguanosine (THIO). (*A*) Imetelstat is an oligonucleotide complementary to the 11-base human telomerase reverse transcriptase (hTERC) functional RNA template region of telomerase, conjugated to a lipidated 13-mer thio-phosphoramidate. (*B*) THIO is a modified guanine incorporated at telomeres by telomerase, due to telomerase lack of proofreading ability. (Created with ChemDraw.)

sity to use a lipid carrier for GRN163L to enter cells (Figs. 3A and 4; Herbert et al. 2005).

GRN163L's high-affinity binding has been demonstrated in a large panel of different human cell lines, showing IC50 values between 0.15 and 1.35 μM (Herbert et al. 2005). A single treatment with GRN163L (1 μM) inhibits telomerase activity for ~72 h (Dikmen et al. 2005; Hochreiter et al. 2006; Marian et al. 2010a).

GRN163L not only inhibits telomerase, leading to progressive telomere shortening compared to the control, but it also eventually increases DNA damage responses (DDR) due to double-strand DNA breaks (DSBs) induced by telomere shortening. Concomitant treatment with an ATM inhibitor, GRN163L, and etoposide (inhibitor of topoisomerase II), increases the cytotoxic effects compared to etoposide as a monotherapy in

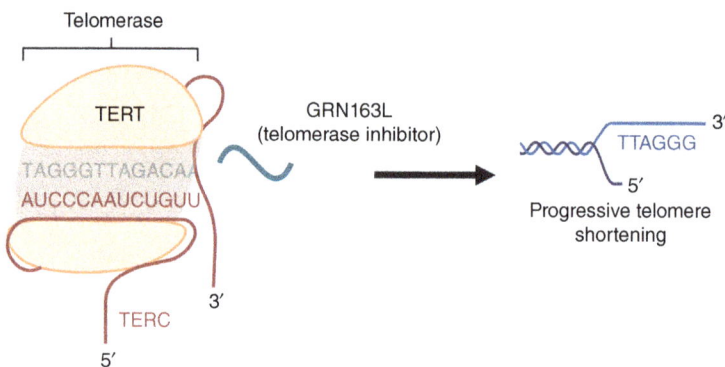

**Figure 4.** GRN163L (imetelstat). GRN163L (green) is an oligonucleotide complementary to human telomerase reverse transcriptase (hTERC) sequence (red), stabilized by a lipidated 13-mer thio-phosphoramidate, conferring a better biodistribution of the molecule. GRN163L binding to the hTERC template sequence constitutes an obstacle to the interaction of hTERT and interactions with the 3' telomere overhang. This leads to progressive shortening of telomeres. Thus, imetelstat increases DNA damage at the shortest telomeres, inducing morphological changes and reducing the tumor burden by apoptosis. Imetelstat (RYTELO) was recently approved for the treatment of adult patients with lower-risk myelodysplastic syndrome (MDS) with transfusion-dependent anemia. (Figure created with BioRender.com.)

breast and colorectal cancer (Tamakawa et al. 2010), as well as in high-risk neuroblastoma (Fischer-Mertens et al. 2022). Similar results were obtained via a combination treatment of GRN163L and Irinotecan (a topoisomerase I inhibitor), showing a synergistic increase in cytotoxicity (Tamakawa et al. 2010). In esophageal cancer cells, administration of GRN163L 3 days before 2 Gy ionizing irradiation, increased the number of γH2AX and 53BP1 foci (Wu et al. 2012) and enhanced tumor radiosensitivity (Wu et al. 2017). In summary, a large variety of different cancer and cancer stem cells in vitro showed similar biological effects when treated with GRN163L/imetelstat, mainly a decrease in cell viability after a lag period depending on initial telomere length. This includes morphological and genomic changes, telomeric DNA damage, and in some studies tumor shrinkage in vivo (Haddley 2012; Barszczyk et al. 2014; Hu et al. 2014).

One concern about a direct telomerase inhibitor is the lag period between the initiation of treatment and when the telomeres become short enough to induce cell death. There have been several long term in vitro and in vivo reports with imetelstat. For example, the number of myeloma cells and lung cancer cells decreases in vitro after 3–5 weeks of continuous treatment. The telomere average at the end of this period was 2.4-fold reduced in myeloma cells, with a 40% increase of telomere-signal free ends. Other investigators (Hochreiter et al. 2006) found that breast cancer cells did not show changes in population doublings compared to other cancer cell lines, whereas telomerase activity was inhibited, and telomere length progressively declined. Long-term treatment with GRN163L in breast cancer cell lines inhibited cell growth after 6–12 weeks, depending on the initial telomere length. In these experiments, cell cultures were treated every 3–4 days resulting in progressive telomere shortening. In a panel of prostate cancer cell lines and glioblastoma stem cells, long-term treatment with imetelstat inhibited telomerase activity in a dose-dependent manner, and also led to progressive telomere shortening (Marian et al. 2010a,b). Finally, a study on the effects of imetelstat in a panel of 64 non-small-cell lung

cancers (NSCLCs) with a telomere length ranging between 1.5 and 10 kb and treated with imetelstat (3 μM), observed no correlation between colony formation and telomere length at the initiation of treatment. However, a correlation between telomere length quartiles and sensitivity to the drug was observed. Furthermore, imetelstat efficacy of treatment depends on initial telomere length and treatment duration in vivo (Frink et al. 2016). However, upon cessation of treatment, residual cells rapidly regrew their telomeres back to their original length (Frink et al. 2016). Recently, a novel mechanism of action of imetelstat was discovered. In a preclinical study on acute myeloid leukemia (AML), imetelstat-induced ferroptosis, decreasing AML cell numbers and delaying tumor recurrence when coupled with chemotherapeutic agents that induce oxidative stress (Bruedigam et al. 2024).

Imetelstat alone showed significant telomerase inhibition and tumor shrinkage in an orthotopic model of glioblastoma, whereas its combination with ionizing radiation significantly increased the overall survival of mice (Ferrandon et al. 2015; reviewed by Berardinelli et al. 2017). Since hsp90 is required in telomerase assembly, 17AAG (an hsp90 inhibitor) resulted in the inhibition of telomerase activity in myeloma cells. In another study, myeloma cells were injected subcutaneously in the interscapular space, allowing tumor formation. When GRN163L (45 mg/kg) was administered intraperitoneally daily for 3 weeks, tumor size was significantly reduced (Shammas et al. 2008). In another experimental approach, tumor cells were injected into mice and subsequently treated (5 and 15 mg/kg GRN163L) three times per week for 3 weeks. Tumors treated with 5 mg/kg resulted in smaller size compared to the controls, whereas the 15 mg/kg treated mice showed no tumor formation. When the treatment was suspended, residual tumors grew back (Dikmen et al. 2005; Gryaznov et al. 2007; Goldblatt et al. 2009). In subcutaneous tumors, GRN163L decreased tumor size significantly, on average 10-fold smaller size compared to control (Marian et al. 2010a). Overall, these results support the concept that a significant lag time will be required to affect tumor cell growth following treatment with GRN163L, and that the amount of telomerase inhibition and starting length of tumor telomeres correlate with tumor control.

Irrespective of the preclinical results, GRN163L (imetelstat) has been evaluated in clinical trials, with only partial success in myeloid malignancies (Waksal et al. 2023). In contrast, myelofibrosis (MF) is characterized by bone marrow fibrosis, cytopenia, functional and structural alteration of megakaryocytes, including an enlarged spleen. Imetelstat may selectively induce apoptosis in myelofibrotic cells, including progenitor stem cells, without affecting normal cells (Wang et al. 2018). It is unclear whether this is due to direct telomerase inhibition or an off-target effect of imetelstat. In a phase II clinical trial (IMbarc NCT02426086), the administration of imetelstat reduced the spleen volume, and the total symptoms score in some of the patient cohort (Mascarenhas et al. 2021). Changes in telomere length were not observed. However, the initial telomere length, particularly those patients with shorter telomeres, was associated with a higher response to the treatment. The limitations to the study were associated with an increase of cytopenia. To reduce these side effects but keep the benefits of the treatment, imetelstat is currently being evaluated in JAKi R/R (relapse/refractory) patients. This trial, named IMpactMF, is currently in a phase III (NCT04576156) trial.

Myelodysplastic syndrome (MDS) is referred to as a disorder where the hematopoietic stem cells are inefficient, resulting in cytopenia (Cazzola and Malcovati 2005). In a phase I clinical trial, nine patients with MDS were treated with imetelstat for 4 weeks, resulting in 38% of patients reaching transfusion independence (TI), and one patient showed a significant reduction in spleen size (Tefferi et al. 2015). Subsequently, in a phase II trial, MDS patients that were transfusion-dependent were treated with imetelstat, of which 38 out of 59 patients did not harbor a del5q aberration. These patients are at higher risk of developing AML compared to those harboring the del5q aberration. TI was reached after 8 weeks and lasted 86 weeks in the group without the del5q. Importantly, in the non-del5q group, a correlation was observed be-

tween telomerase activity decrease and the ability to become TI (Santini 2021). A phase III trial (Imerge NCT02598661) showed a TI of 8 weeks in 40% of MDS patients treated with imetelstat compared to 15% of those treated with placebo (Platzbecker et al. 2024). Recently, the FDA oncologic drugs advisory committee indicated clinical benefit in the phase III iMerge trial, and FDA approval occurred June 6, 2024 for this indication. Clinical trials for essential thrombocythemia and multiple myeloma were not as successful as the previous ones, and did not advance (Hussain et al. 2023; Waksal et al. 2023). However, a phase II clinical trial (IMpress NCT05583552) is ongoing for AML. Finally, imetelstat failed in a clinical trial to improve NSCLC treatment (Chiappori et al. 2015). In summary, while imetelstat (RYTELO) was approved for the treatment of adult patients with lower-risk MDS with transfusion-dependent anemia, the verdict is still out about other potential indications going forward.

## 6-thio-dG (6-thio-2′-deoxyguanosine ALSO TERMED THIO)

6-thio-dG, a small molecule nucleoside prodrug (Fig. 3B), was first investigated in the mid-1960s and was also under investigation as part of a chemotherapy screen conducted in the 1960–1970s (LePage et al. 1964; Omura et al. 1977; Douglass et al. 1978). 6-thio-dG was originally evaluated in clinical trials with some positive results but did not progress for several reasons (treatment time, extremely high doses, and lacking any knowledge of mechanisms of action). This resulted in 6-thio-dG being set aside for years (Omura et al. 1977; Higgins et al. 1985). Then, in 1985, Carol Greider, a graduate student in Elizabeth Blackburn's laboratory, identified telomerase in Tetrahymena and termed it telomere terminal transferase, currently referred to as telomerase (Greider and Blackburn 1985). Subsequent landmark papers regarding telomerase, telomere biology (Kim et al. 1994; Meyerson et al. 1997; Shay and Bacchetti 1997), and telomere capping/shelterin complex subunits such as TRF1 and TRF2 (Zhong et al. 1992) provided valuable insights that gave rise to new potential use for 6-thio-dG. The minimal components of the telomerase ribonucleoprotein holoenzyme are the reverse transcriptase TERT (the catalytic component of telomerase), and TERC (the RNA component that acts as the template for telomere extension). This new knowledge that telomeres were G-rich, and that telomerase synthesized addition of new telomere repeats, suggested the prodrug 6-thio-dG may be preferentially incorporated into the G-rich strand of the telomere by telomerase (Fig. 5). Also, since telomerase does not have proofreading capabilities, 6-thio-dG, a modified guanine, was not removed from incor-

poration into the telomere. The incorporation of 6-thio-dG into telomeres leads to rapid telomere uncapping (due to loss of t-loop formation), and to the telomere being recognized as DNA damage. This leads to chromosome fusions, aberrant mitosis, production of micronuclei, and cell death. As a result, 6-thio-dG was reexamined with a new target in mind, telomerase. In vitro and in vivo experiments reported 6-thio-dG as a viable and effective anticancer-therapeutic (Mender et al. 2015) with increased specificity in telomerase-positive cancers. 6-thio-dG's success at targeting cancer cells more specifically depended on telomerase activity, showing little

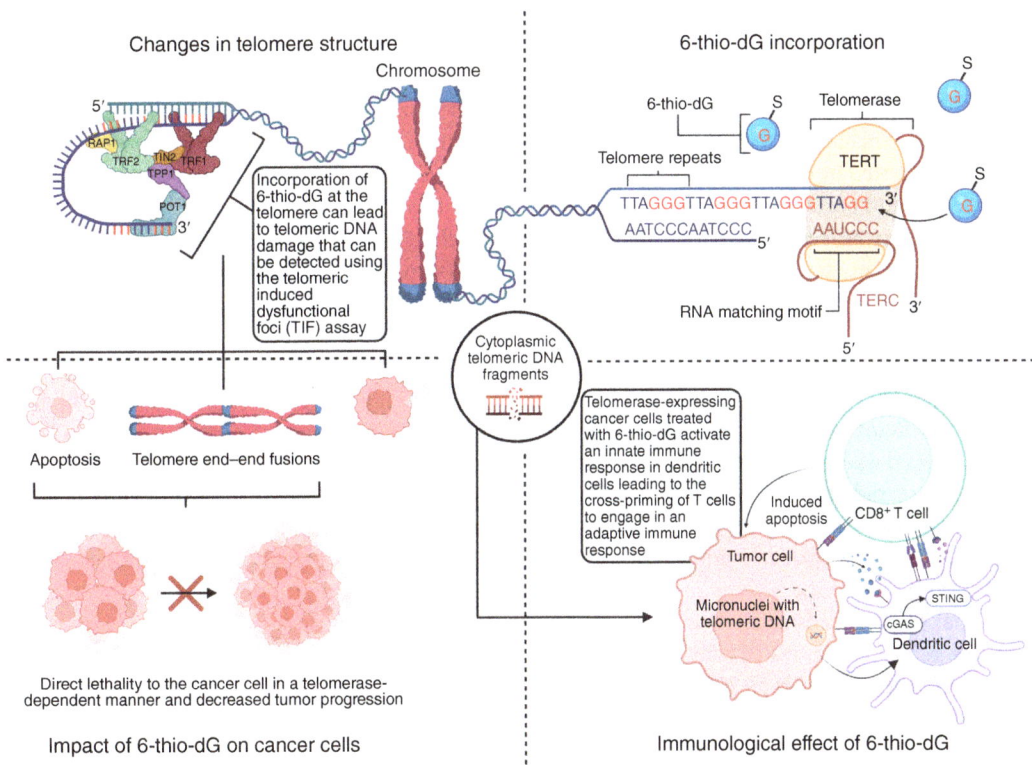

**Figure 5.** THIO (6-thio-2′-deoxyguanosine). THIO is a nucleoside analog prodrug, which competes with guanosine incorporation at telomeres by telomerase (*upper-right* panel), inducing a destabilization of telomeres. Therefore, after THIO incorporation into telomeres cancer cells recognize telomeres as DNA damage, activating the DNA damage response (DDR) (*upper-left* panel). The damage at telomere generates "sticky ends" with the consequent formation of dicentric chromosomes, increasing genomic instability and leading to apoptosis (*bottom-left* panel). However, apoptosis induced by THIO generates DNA and telomeric fragments via micronuclei formation, triggering the cGAS-STING pathway. Altered telomere/protein fragments and other potential neoantigens are taken up by dendritic cells that are then able to cross-prime CD8+ T cells inducing an adaptive immunogenic response (*bottom-right* panel). (Figure created with BioRender.com.)

to no effect on the viability of primary telomerase-negative cells. Additionally, 6-thioguanine, a metabolite of 6-thio-dG, does not have the same telomerase-dependent antitumor effect. This may be due to 6-thioguanine having other intracellular biochemical and metabolic roles contributing to the difference in toxicity and adverse events compared to 6-thio-dG. Furthermore, 6-thio-dG induces telomeric DNA damage as demonstrated by the presence of telomere dysfunction-induced foci (TIF) analysis (Yu et al. 2021; Piñeiro-Hermida et al. 2023; Eglenen-Polat et al. 2024). Since telomeres are only ~1/6000th of the human genome, any nonspecific damage at telomeres would be rare.

Finally, the rapid effect of 6-thio-dG in cancer cells results from telomere uncapping in telomerase-positive cells and the subsequent change in telomere DNA structure because of 6-thio-dG incorporation. As a result, telomerase-positive cancer cells, regardless of telomere length, are affected by 6-thio-dG in a rapid treatment period compared to direct telomerase inhibitors such as imetelstat or BIBR1532. Thus, this strategy was not to target telomerase directly but to introduce a modified guanine into cells so that telomerase would preferentially incorporate it into telomeric DNA. An altered nucleotide incorporated into telomeres may not bind to shelterin proteins efficiently, disrupting t-loop formation, and was predicted to lead to telomere dysfunction, rapid cell death, and potentially the release of neoantigens.

Since 85%–90% of all cancers are telomerase positive, and telomerase activity is not detected in normal somatic cells except for proliferating germline cells and subsets of cells in the blood, skin, and intestine (Jafri et al. 2016), 6-thio-dG's effect is relatively cancer-specific. As a result, 6-thio-dG was further studied in a telomerase-dependent manner to understand the molecular, biochemical, immunological, and antitumor effects. This resulted in the discovery of biomarkers that showed sensitivity to 6-thio-dG, such as DNA damage at telomeres and the absolute number of a cell surface nucleobase transporter, SLC43A3, that correlated with sensitivity of 6-thio-dG in lung cancer (Mender et al. 2020a). While the potential of 6-thio-dG has become more apparent in recent years, the main factor that makes 6-thio-dG such a compelling telomerase-dependent antitumor therapeutic is its immunological effects on the TME.

The immunological effects of 6-thio-dG may be partly due to the release of modified telomeric damaged fragments. These fragments exist initially as micronuclei that lose nuclear envelops and are free DNA/protein fragments that are released from the cell and can subsequently trigger an innate immune response in antigen-presenting cells (e.g., DCs). The immune response exhibited by cancer cells affected by 6-thio-dG-induced cytoplasmic modified telomeric DNA is a result of interferon type one (IFN-1) responses in antigen-presenting cells such as DCs in a STING-dependent manner to mount an innate immune response that leads to cross-priming of naive T cells to become effector CD8+ T cells (Mender et al. 2020b). Moreover, because of 6-thio-dG's effect on modulating an innate immune response, the question arises of why there is a lack of central tolerance. One possibility is that the modified telomere fragments are not recognized as self (normal damaged telomeres) and therefore mount both an innate and adaptive immune response. This has led to the observation that some tumors, which are normally resistant to immune checkpoint blockade, treated with 6-thio-dG and anti-PD-L1, are able to be re-sensitized to this form of therapy, and even further improved the antitumor effects of 6-thio-dG (Mender et al. 2018, 2020b, 2023). These findings not only give rise to other potential combination therapeutic strategies worth investigating with 6-thio-dG but are also the foundation for the newest round of clinical trials performed with 6-thio-dG (THIO) for lung cancer.

It is well established that acquired resistance to immune therapy such as nivolumab (a monoclonal antibody for PD-1) is a common hallmark of current first-line lung cancer (NSCLC) treatments, leaving patients suffering from this type of cancer with few treatment strategies to combat disease progression (Topalian et al. 2019). Even when immune therapies are combined with chemotherapy and demonstrated to be more effective than either immune therapy or chemotherapy as monotherapies for NSCLC patients, overall survival is still poor. Thus, additional therapeutic improvements are needed to minimize systemic toxicities, treatment-induced adverse effects, and improve long-term disease control.

Due to these challenges, a phase II clinical trial (MAIA Biotechnology, THIO/6-thio-dG) is currently under evaluation as a viable therapeutic in sequence with cemiplimab (Libtayo), an FDA-approved PD-1 monoclonal antibody as a therapeutic strategy for treating NSCLC patients. The phase II trial was designed to treat patients who had previously failed one or two prior lines of immunotherapy with or without chemotherapy and had documented disease progression at study entry (Ciuleanu et al. 2022). Expectations associated with this study were centered around the preclinical results (Mender et al. 2020b) that leading in with 3 days of THIO (with low systemic toxicity) would initiate an innate and potentially adaptive immune response. Stopping THIO treatment after 3 days would allow for the development of an adaptive immune response without adversely affecting proliferating T cells. Due to THIO's effect on inducing telomere DNA damage, cancer cells rapidly lose their proliferative capacity and induce a cGAS-STING-dependent innate and adaptive immune response. Sequentially combined with traditional immune therapy (cemiplimab), these effects are the basis for the phase II clinical trial. While the clinical trial is ongoing, 77 patients have already been enrolled in the study. In 2024, ~57 patients had one post-baseline scan with 88% showing disease control rates (DCRs) (defined using the following parameters: complete response [CR], partial response [PR], or stable disease [SD] per RECIST 1.1 at cycle 3). Based on phase II clinical trial's preliminary results, the DCR of patients who failed one line of therapy was 82%; additionally, patients that had two previous lines of therapy had a DCR of 88% and even two patients with three previous lines of therapy had a DCR of 100% for an overall total DCR of 88% across all patients, a drastic improvement in comparison to current therapeutic options (Maia Biotechnology, maiabiotech.com).

While the trial is ongoing, the safety profile of this sequential therapeutic strategy is well tolerated, with the adverse events being grade 1–2 and no dose-limiting toxicities. Finally, a well-characterized marker of THIO efficacy, γH2AX, is increased in a telomerase-mediated telomere-specific manner using TRF1 as a marker for telomeric DNA in patients post-THIO treatment and before cemiplimab treatment. These results are interpreted to suggest that THIO treatment is effective within the first 4 days of administration and provides valuable insight as to how the timing of sequential therapy might play a significant role in enhancing the therapeutic effect of THIO. For example, subsequent transient activation of telomerase in immune cells for outgrowth to combat the tumor is susceptible to THIO incorporation, so only providing THIO 3 days every 3 weeks may prevent the killing of the vast majority of telomerase expressing and expanding cytotoxic T cells. Also, commonly used standard-of-care therapeutics, such as radiation and chemotherapy, rely on targeting actively proliferating cells. As a result, carefully designed sequential therapeutic options surrounding THIO and secondary therapeutics to enhance treatment outcomes are currently under investigation.

## CONCLUSIONS

Mechanisms of telomere maintenance and our understanding of the role of telomerase and the ALT pathway (a telomerase-independent telomere maintenance mechanism) have significantly improved in recent years. It is becoming better understood how cancer cells regulate different molecular events involved in telomere maintenance to maintain proliferative capacity, thus leading to novel approaches to more selectively target these pathways in cancer. Recent insights into the regulation and control of telomerase activity at telomeres (Chen and Zirnberger Batista 2024; Feigon 2024; Martin and Hockemeyer 2024; Sekne et al. 2024) and the ALT pathway (O'Sullivan and Greenberg 2024; Pickett 2024; Reddel 2024) are reviewed by others. In this review, we have covered most of the promising cancer inhibitors that have been developed against telomerase, including the highly attractive hTERT vaccines and imetelstat (recently FDA approved). Finally, 6-thio-dG has received orphan drug designation by the FDA for liver, brain, and small-cell lung cancers, so optimism toward future approvals is increasing.

## ACKNOWLEDGMENTS

The laboratory is supported by grants from the National Cancer Institute (NCI) (U19CA 264385, P50CA070907, and P30CA142543). University of Texas Southwestern (UTSW) has licensed THIO to Maia Biotechnology and J.W.S. and S.S. declare a potential conflict of interest. Images prepared with ChemDraw and BioRender.com.

## REFERENCES

*Reference is also in this subject collection.*

Barszczyk M, Buczkowicz P, Castelo-Branco P, Mack SC, Ramaswamy V, Mangerel J, Agnihotri S, Remke M, Golbourn B, Pajovic S, et al. 2014. Telomerase inhibition abolishes the tumorigenicity of pediatric ependymoma tumor-initiating cells. *Acta Neuropathol* **128**: 863–877. doi:10.1007/s00401-014-1327-6

Bashash D, Ghaffari SH, Mirzaee R, Alimoghaddam K, Ghavamzadeh A. 2013. Telomerase inhibition by non-nucleosidic compound BIBR1532 causes rapid cell death in pre-B acute lymphoblastic leukemia cells. *Leuk Lymphoma* **54**: 561–568. doi:10.3109/10428194.2012.704034

Berardinelli F, Coluzzi E, Sgura A, Antoccia A. 2017. Targeting telomerase and telomeres to enhance ionizing radiation effects in in vitro and in vivo cancer models. *Mutat Res Rev Mutat Res* **773**: 204–219. doi:10.1016/j.mrrev.2017.02.004

Bodnar AG, Ouellette M, Frolkis M, Holt SE, Chiu CP, Morin GB, Harley CB, Shay JW, Lichtsteiner S, Wright WE. 1998. Extension of life-span by introduction of telomerase into normal human cells. *Science* **279**: 349–352. doi:10.1126/science.279.5349.349

Bruedigam C, Porter AH, Song A, Vroeg In de Wei G, Stoll T, Straube J, Cooper L, Cheng G, Kahl VFS, Sobinoff AP, et al. 2024. Imetelstat-mediated alterations in fatty acid metabolism to induce ferroptosis as a therapeutic strategy for acute myeloid leukemia. *Nat Cancer* **5**: 47–65. doi:10.1038/s43018-023-00653-5

Bryan C, Rice C, Hoffman H, Harkisheimer M, Sweeney M, Skordalakes E. 2015. Structural basis of telomerase inhibition by the highly specific BIBR1532. *Structure* **23**: 1934–1942. doi:10.1016/j.str.2015.08.006

Cazzola M, Malcovati L. 2005. Myelodysplastic syndromes—coping with ineffective hematopoiesis. *N Engl J Med* **352**: 536–538. doi:10.1056/nejmp048266

*Chen L, Zirnberger Batista LF. 2024. Biogenesis and regulation of telomerase during development and cancer. *Cold Spring Harb Perspect Biol* doi:10.1101/cshperspect.a041692

Chiappori AA, Kolevska T, Spigel DR, Hager S, Rarick M, Gadgeel S, Blais N, Von Pawel J, Hart L, Reck M, et al. 2015. A randomized phase II study of the telomerase inhibitor imetelstat as maintenance therapy for advanced non-small-cell lung cancer. *Ann Oncol* **26**: 354–362. doi:10.1093/annonc/mdu550

Ciuleanu T-E, Joshi H, Gryaznov S, Vitoc V, Obrocea M. 2022. 1193TiP a phase II, multicenter, open-label, dose-finding study evaluating THIO sequenced with cemiplimab in patients with advanced NSCLC. *Ann Oncol* **33**: S1094. doi:10.1016/j.annonc.2022.07.1316

Damm K, Hemmann U, Garin-Chesa P, Hauel N, Kauffmann I, Priepke H, Niestroj C, Daiber C, Enenkel B, Guilliard B, et al. 2001. A highly selective telomerase inhibitor limiting human cancer cell proliferation. *EMBO J* **20**: 6958–6968. doi:10.1093/emboj/20.24.6958

Di Antonio M, Ponjavic A, Radzevičius A, Ranasinghe RT, Catalano M, Zhang X, Shen J, Needham LM, Lee SF, Klenerman D, et al. 2020. Single-molecule visualization of DNA G-quadruplex formation in live cells. *Nat Chem* **12**: 832–837. doi:10.1038/s41557-020-0506-4

Dikmen ZG, Gellert GC, Jackson S, Gryaznov S, Tressler R, Dogan P, Wright WE, Shay JW. 2005. In vivo inhibition of lung cancer by GRN163L: a novel human telomerase inhibitor. *Cancer Res* **65**: 7866–7873. doi:10.1158/0008-5472.CAN-05-1215

Ding X, Cheng J, Pang Q, Wei X, Zhang X, Wang P, Yuan Z, Qian D. 2019. BIBR1532, a selective telomerase inhibitor, enhances radiosensitivity of non-small-cell lung cancer through increasing telomere dysfunction and ATM/CHK1 inhibition. *Int J Radiat Oncol Biol Phys* **105**: 861–874. doi:10.1016/j.ijrobp.2019.08.009

Doğan F, Özateş NP, Bağca BG, Abbaszadeh Z, Söğütlü F, Gasımlı R, Gündüz C, Biray Avcı Ç. 2019. Investigation of the effect of telomerase inhibitor BIBR1532 on breast cancer and breast cancer stem cells. *J Cell Biochem* **120**: 1282–1293. doi:10.1002/jcb.27089

Douglass HO Jr, Lavin PT, Woll J, Conroy JF, Carbone P. 1978. Chemotherapy of advanced measurable colon and rectal carcinoma with oral 5-fluorouracil, alone or in combination with cyclophosphamide or 6-thioguanine, with intravenous 5-fluorouracil or β-2′-deoxythioguanosine or with oral 3(4-methyl-cyclohexyl)-1(2-chlorethyl)-1-nitrosourea. A phase II-III study of the eastern cooperative oncology group (EST 4273). *Cancer* **42**: 2538–2545. doi:10.1002/1097-0142(197812)42:6<2538::AID-CNCR2820420606>3.0.CO;2-A

Eglenen-Polat B, Kowash RR, Huang HC, Siteni S, Zhu M, Chen K, Bender ME, Mender I, Stastny V, Drapkin BJ, et al. 2024. A telomere-targeting drug depletes cancer initiating cells and promotes anti-tumor immunity in small cell lung cancer. *Nat Commun* **15**: 672. doi:10.1038/s41467-024-44861-8

El-Daly H, Kull M, Zimmermann S, Pantic M, Waller CF, Martens UM. 2005. Selective cytotoxicity and telomere damage in leukemia cells using the telomerase inhibitor BIBR1532. *Blood* **105**: 1742–1749. doi:10.1182/blood-2003-12-4322

Cite this article as *Cold Spring Harb Perspect Biol* doi: 10.1101/cshperspect.a041703

Ellingsen EB, Mangsbo SM, Hovig E, Gaudernack G. 2021. Telomerase as a target for therapeutic cancer vaccines and considerations for optimizing their clinical potential. *Front Immunol* **12:** 682492. doi:10.3389/fimmu.2021.682492

Ellingsen EB, O'Day S, Mezheyeuski A, Gromadka A, Clancy T, Kristedja TS, Milhem M, Zakharia Y. 2023. Clinical activity of combined telomerase vaccination and pembrolizumab in advanced melanoma: results from a phase I trial. *Clin Cancer Res* **29:** 3026–3036. doi:10.1158/1078-0432.CCR-23-0416

Esnault C, Magat T, Zine El Aabidine A, Garcia-Oliver E, Cucchiarini A, Bouchouika S, Lleres D, Goerke L, Luo Y, Verga D, et al. 2023. G4access identifies G-quadruplexes and their associations with open chromatin and imprinting control regions. *Nat Genet* **55:** 1359–1369. doi:10.1038/s41588-023-01437-4

* Feigon J. 2024. Insights into telomerase mechanism from structure. *Cold Spring Harb Perspect Biol* doi:10.1101/cshperspect.a041698

Ferrandon S, Malleval C, El Hamdani B, Battiston-Montagne P, Bolbos R, Langlois JB, Manas P, Gryaznov SM, Alphonse G, Honnorat J, et al. 2015. Telomerase inhibition improves tumor response to radiotherapy in a murine orthotopic model of human glioblastoma. *Mol Cancer* **14:** 3–7. doi:10.1186/s12943-015-0376-3

Fischer-Mertens J, Otte F, Roderwieser A, Rosswog C, Kahlert Y, Werr L, Hellmann AM, Berding M, Chiu B, Bartenhagen C, et al. 2022. Telomerase-targeting compounds imetelstat and 6-thio-dG act synergistically with chemotherapy in high-risk neuroblastoma models. *Cell Oncol (Dordr)* **45:** 991–1003. doi:10.1007/s13402-022-00702-8

Frink RE, Peyton M, Schiller JH, Gazdar AF, Shay JW, Minna JD. 2016. Telomerase inhibitor imetelstat has preclinical activity across the spectrum of non-small cell lung cancer oncogenotypes in a telomere length dependent manner. *Oncotarget* **7:** 31639–31651. doi:10.18632/oncotarget.9335

Gao J, Pickett HA. 2022. Targeting telomeres: advances in telomere maintenance mechanism-specific cancer therapies. *Nat Rev Cancer* **22:** 515–532. doi:10.1038/s41568-022-00490-1

Goldblatt EM, Erickson PA, Gentry ER, Gryaznov SM, Herbert BS. 2009. Lipid-conjugated telomerase template antagonists sensitize resistant HER2-positive breast cancer cells to trastuzumab. *Breast Cancer Res Treat* **118:** 21–32. doi:10.1007/s10549-008-0201-4

Gomez D, Guédin A, Mergny JL, Salles B, Riou JF, Teulade-Fichou MP, Calsou P. 2010. A G-quadruplex structure within the 5′-UTR of TRF2 mRNA represses translation in human cells. *Nucleic Acids Res* **38:** 7187–7198. doi:10.1093/nar/gkq563

Greider CW, Blackburn EH. 1985. Identification of a specific telomere terminal transferase activity in tetrahymena extracts. *Cell* **43:** 405–413. doi:10.1016/0092-8674(85)90170-9

Gryaznov SM, Jackson S, Dikmen G, Harley C, Herbert BS, Wright WE, Shay JW. 2007. Oligonucleotide conjugate GRN163L targeting human telomerase as potential anticancer and antimetastatic agent. *Nucleosides Nucleotides Nucleic Acids* **26:** 1577–1579. doi:10.1080/15257770701547271

Haddley K. 2012. Imetelstat sodium. *Drugs Future* **37:** 0111. doi:10.1358/dof.2012.37.2.1779019

Hemann MT, Strong MA, Hao LY, Greider CW. 2001. The shortest telomere, not average telomere length, is critical for cell viability and chromosome stability. *Cell* **107:** 67–77. doi:10.1016/S0092-8674(01)00504-9

Herbert BS, Gellert GC, Hochreiter A, Pongracz K, Wright WE, Zielinska D, Chin AC, Harley CB, Shay JW, Gryaznov SM. 2005. Lipid modification of GRN163, an N3′→P5′ thio-phosphoramidate oligonucleotide, enhances the potency of telomerase inhibition. *Oncogene* **24:** 5262–5268. doi:10.1038/sj.onc.1208760

Higgins GR, Jamin DC, Shore NA, Momparler R, Hartman G, Siegel SE. 1985. Phase I evaluation of β-2′-deoxythioguanine in pediatric patients with leukemia. *Cancer Treatment Rep* **69:** 699–701.

Hochreiter AE, Xiao H, Goldblatt EM, Gryaznov SM, Miller KD, Badve S, Sledge GW, Herbert BS. 2006. Telomerase template antagonist GRN163L disrupts telomere maintenance, tumor growth, and metastasis of breast cancer. *Clin Cancer Res* **12:** 3184–3192. doi:10.1158/1078-0432.CCR-05-2760

Hu Y, Bobb D, Lu Y, He J, Dome JS. 2014. Effect of telomerase inhibition on preclinical models of malignant rhabdoid tumor. *Cancer Genet* **207:** 403–411. doi:10.1016/j.cancergen.2014.09.002

Huppert JL, Balasubramanian S. 2007. G-quadruplexes in promoters throughout the human genome. *Nucleic Acids Res* **35:** 406–413. doi:10.1093/nar/gkl1057

Hussain M, Yellapragada S, Al Hadidi S. 2023. Differential diagnosis and therapeutic advances in multiple myeloma: a review article. *Blood Lymphat Cancer* **13:** 33–57. doi:10.2147/BLCTT.S272703

Iachettini S, Stevens MF, Frigerio M, Hummersone MG, Hutchinson I, Garner TP, Searle MS, Wilson DW, Munde M, Nanjunda R, et al. 2013. On and off-target effects of telomere uncapping G-quadruplex selective ligands based on pentacyclic acridinium salts. *J Exp Clin Cancer Res* **32:** 1–12. doi:10.1186/1756-9966-32-68

Jafri MA, Ansari SA, Alqahtani MH, Shay JW. 2016. Roles of telomeres and telomerase in cancer, and advances in telomerase-targeted therapies. *Genome Med* **8:** 69. doi:10.1186/s13073-016-0324-x

Jin M, Hurley LH, Xu H. 2023. A synthetic lethal approach to drug targeting of G-quadruplexes based on CX-5461. *Bioorg Med Chem Lett* **91:** 129384. doi:10.1016/j.bmcl.2023.129384

Kawashima T, Kagawa S, Kobayashi N, Shirakiya Y, Umeoka T, Teraishi F, Taki M, Kyo S, Tanaka N, Fujiwara T. 2004. Telomerase-specific replication-selective virotherapy for human cancer. *Clin Cancer Res* **10:** 285–292. doi:10.1158/1078-0432.CCR-1075-3

Keith WN, Thomson CM, Howcroft J, Maitland NJ, Shay JW. 2007. Seeding drug discovery: integrating telomerase cancer biology and cellular senescence to uncover new therapeutic opportunities in targeting cancer stem cells. *Drug Discov Today* **12:** 611–621. doi:10.1016/j.drudis.2007.06.009

Khoury HJ, Collins RH, Blum W, Maness L, Stiff P, Kelsey SM, Reddy A, Smith JA, DiPersio JF. 2010. Prolonged administration of the telomerase vaccine GTNVAC1 is well tolerated and appears to be associated with favorable outcomes in high-risk acute myeloid leukemia (AML). *Blood* **116**: 2190. doi:10.1182/blood.V116.21.2190.2190

Kim NW, Piatyszek MA, Prowse KR, Harley CB, West MD, Ho PL, Coviello GM, Wright WE, Weinrich SL, Shay JW. 1994. Specific association of human telomerase activity with immortal cells and cancer. *Science* **266**: 2011–2015. doi:10.1126/science.7605428

LePage GA, Junga IG, Bowman B. 1964. Biochemical and carcinostatic effects of 2′-deoxythioguanosine. *Cancer Res* **24**: 835–840.

Lim KW, Lacroix L, Yue DJ, Lim JK, Lim JM, Phan AT. 2010. Coexistence of two distinct G-quadruplex conformations in the hTERT promoter. *J Am Chem Soc* **132**: 12331–12342. doi:10.1021/ja101252n

Marian CO, Cho SK, McEllin BM, Maher EA, Hatanpaa KJ, Madden CJ, Mickey BE, Wright WE, Shay JW, Bachoo RM. 2010a. The telomerase antagonist, imetelstat, efficiently targets glioblastoma tumor-initiating cells leading to decreased proliferation and tumor growth. *Clin Cancer Res* **16**: 154–163. doi:10.1158/1078-0432.CCR-09-2850

Marian CO, Wright WE, Shay JW. 2010b. The effects of telomerase inhibition on prostate tumor-initiating cells. *Int J Cancer* **127**: 321–331. doi:10.1002/ijc.25043

* Martin A, Hockemeyer D. 2024. Regulation of human telomerase: from molecular interactions to population genetics. *Cold Spring Harb Perspect Biol* doi:10.1101/cshperspect.a041693

Mascarenhas J, Komrokji RS, Palandri F, Martino B, Niederwieser D, Reiter A, Scott BL, Baer MR, Hoffman R, Odenike O, et al. 2021. Randomized, single-blind, multicenter phase II study of two doses of imetelstat in relapsed or refractory myelofibrosis. *J Clin Oncol* **39**: 2881–2892. doi:10.1200/JCO.20.02864

Mender I, Gryaznov S, Dikmen ZG, Wright WE, Shay JW. 2015. Induction of telomere dysfunction mediated by the telomerase substrate precursor 6-thio-2′-deoxyguanosine. *Cancer Discov* **5**: 82–95. doi:10.1158/2159-8290.CD-14-0609

Mender I, LaRanger R, Luitel K, Peyton M, Girard L, Lai TP, Batten K, Cornelius C, Dalvi MP, Ramirez M, et al. 2018. Telomerase-mediated strategy for overcoming non-small cell lung cancer targeted therapy and chemotherapy resistance. *Neoplasia* **20**: 826–837. doi:10.1016/j.neo.2018.06.002

Mender I, Batten K, Peyton M, Vemula A, Cornelius C, Girard L, Gao B, Minna JD, Shay JW. 2020a. *SLC43A3* is a biomarker of sensitivity to the telomeric DNA damage mediator 6-thio-2′-deoxyguanosine. *Cancer Res* **80**: 929–936. doi:10.1158/0008-5472

Mender I, Zhang A, Ren Z, Han C, Deng Y, Siteni S, Li H, Zhu J, Vemula A, Shay JW, et al. 2020b. Telomere stress potentiates STING-dependent anti-tumor immunity. *Cancer Cell* **38**: 400–411.e6. doi:10.1016/j.ccell.2020.05.020

Mender I, Siteni S, Barron S, Flusche AM, Kubota N, Yu C, Cornelius C, Tedone E, Maziveyi M, Grichuk A, et al. 2023. Activating an adaptive immune response with a telomerase-mediated telomere targeting therapeutic in hepatocellular carcinoma. *Mol Cancer Ther* **22**: 737–750. doi:10.1158/1535-7163

Meyerson M, Counter CM, Eaton EN, Ellisen LW, Steiner P, Caddle SD, Ziaugra L, Beijersbergen RL, Davidoff MJ, Liu

Q, et al. 1997. hEST2, the putative human telomerase catalytic subunit gene, is up-regulated in tumor cells and during immortalization. *Cell* **90**: 785–795. doi:10.1016/S0092-8674(00)80538-3

Monsen RC, Chua EYD, Hopkins JB, Chaires JB, Trent JO. 2023. Structure of a 28.5 kDa duplex-embedded resolution with G-quadruplex system resolved to 7.4 Å resolution with cryo-EM. *Nucleic Acids Res* **51**: 1943–1959. doi:10.1093/nar/gkad014

Nasrollahzadeh A, Bashash D, Kabuli M, Zandi Z, Kashani B, Zaghal A, Mousavi SA, Ghaffari SH. 2020. Arsenic trioxide and BIBR1532 synergistically inhibit breast cancer cell proliferation through attenuation of NF-κB signaling pathway. *Life Sci* **257**: 118060. doi:10.1016/j.lfs.2020.118060

Negrini S, De Palma R, Filaci G. 2020. Anti-cancer immunotherapies targeting telomerase. *Cancers (Basel)* **12**: 2260. doi:10.3390/cancers12082260

Ohanian M, Arellano ML, Levy MY, O'Dwyer K, Babiker H, Mahadevan D, Zhang H, Rastgoo N, Jin Y, Marango J, et al. 2021. A phase 1a/b dose escalation study of the MYC repressor Apto-253 in patients with relapsed or refractory AML or high-risk MDS. *Blood* **138**: 3411. doi:10.1182/blood-2021-150049

Omura GA, Vogler WR, Smalley RV, Maldonado N, Broun GO, Knospe WH, Ahn YS, Faguet GB. 1977. Phase II study of β-2′-deoxythioguanosine in adult acute leukemia. *Cancer Treat Rep* **61**: 1379–1381.

* O'Sullivan RJ, Greenberg RA. 2024. Mechanisms of alternative lengthening of telomeres. *Cold Spring Harb Perspect Biol* doi:10.1101/cshperspect.a041690

Pascolo E, Wenz C, Lingner J, Hauel N, Priepke H, Kauffmann I, Garin-Chesa P, Rettig WJ, Damm K, Schnapp A. 2002. Mechanism of human telomerase inhibition by BIBR1532, a synthetic, non-nucleosidic drug candidate. *J Biol Chem* **277**: 15566–15572. doi:10.1074/jbc.M201266200

* Pickett HA. 2024. Alternate therapeutics. *Cold Spring Harb Perspect Biol* doi:10.1101/cshperspect.a041691

Piñeiro-Hermida S, Bosso G, Sánchez-Vázquez R, Martínez P, Blasco MA. 2023. Telomerase deficiency and dysfunctional telomeres in the lung tumor microenvironment impair tumor progression in NSCLC mouse models and patient-derived xenografts. *Cell Death Differ* **30**: 1585–1600. doi:10.1038/s41418-023-01149-6

Platzbecker U, Santini V, Fenaux P, Sekeres MA, Savona MR, Madanat YF, Díez-Campelo M, Valcárcel D, Illmer T, Jonášová A, et al. 2024. Imetelstat in patients with lower-risk myelodysplastic syndromes who have relapsed or are refractory to erythropoiesis-stimulating agents (IMerge): a multinational, randomised, double-blind, placebo-controlled, phase 3 trial. *Lancet* **403**: 249–260. doi:10.1016/S0140-6736(23)01724-5

* Reddel R. 2024. Alternate cancer perspectives. *Cold Spring Harb Perspect Biol* doi:10.1101/cshperspect.a041689

Riou JF, Guittat L, Mailliet P, Laoui A, Renou E, Petitgenet O, Mégnin-Chanet F, Hélène C, Mergny JL. 2002. Cell senescence and telomere shortening induced by a new series of specific G-quadruplex DNA ligands. *Proc Natl Acad Sci* **99**: 2672–2677. doi:10.1073/pnas.052698099

Cite this article as *Cold Spring Harb Perspect Biol* doi: 10.1101/cshperspect.a041703

Santini V. 2021. Advances in myelodysplastic syndrome. *Curr Opin Oncol* 33: 681–686. doi:10.1097/CCO.0000 00000000790

Seimiya H. 2020. Crossroads of telomere biology and anticancer drug discovery. *Cancer Sci* 111: 3089–3099. doi:10.1111/cas.14540

Seimiya H, Oh-hara T, Suzuki T, Naasani I, Shimazaki T, Tsuchiya K, Tsuruo T. 2002. Telomere shortening and growth inhibition of human cancer cells by novel synthetic telomerase inhibitors MST-312, MST-295, and MST-1991. *Mol Cancer Ther* 1: 657–665.

* Sekne Z, Ludzia P, Balch S, Nguyen THD. 2024. Structural biology of telomerase and associated factors. *Cold Spring Harb Perspect Biol* doi:10.1101/cshperspect.a041697

Shah MA, Eads JR, Sarkar S, Khan S, Sharaiha R, Carr-Locke D, Chang L, Ginsberg G, DiCicco L, Garcia-Marcano L, et al. 2023. Phase II study of telomelysin (OBP-301) in combination with pembrolizumab in gastroesophageal (GEA) adenocarcinoma. *J Clin Oncol* 41: 4052–4052. doi:10.1200/jco.2023.41.16_suppl.4052

Shammas MA, Koley H, Bertheau RC, Neri P, Fulciniti M, Tassone P, Blotta S, Protopopov A, Mitsiades C, Batchu RB, et al. 2008. Telomerase inhibitor GRN163L inhibits myeloma cell growth in vitro and in vivo. *Leukemia* 22: 1410–1418. doi:10.1038/leu.2008.81

Shay JW, Bacchetti S. 1997. A survey of telomerase activity in human cancer. *Eur J Cancer* 33: 787–791. doi:10.1016/S0959-8049(97)00062-2

Shirakawa Y, Tazawa H, Tanabe S, Kanaya N, Noma K, Koujima T, Kashima H, Kato T, Kuroda S, Kikuchi S, et al. 2021. Phase I dose-escalation study of endoscopic intratumoral injection of OBP-301 (telomelysin) with radiotherapy in oesophageal cancer patients unfit for standard treatments. *Eur J Cancer* 153: 98–108. doi:10.1016/j.ejca.2021.04.043

Sugarman ET, Zhang G, Shay JW. 2019. In perspective: an update on telomere targeting in cancer. *Mol Carcinog* 58: 1581–1588. doi:10.1002/mc.23035

Summers PA, Lewis BW, Gonzalez-Garcia J, Porreca RM, Lim AHM, Cadinu P, Martin-Pintado N, Mann DJ, Edel JB, Vannier JB, et al. 2021. Visualising G-quadruplex DNA dynamics in live cells by fluorescence lifetime imaging microscopy. *Nat Commun* 12: 1–11. doi:10.1038/s41467-020-20414-7

Tamakawa RA, Fleisig HB, Wong JMY. 2010. Telomerase inhibition potentiates the effects of genotoxic agents in breast and colorectal cancer cells in a cell cycle-specific manner. *Cancer Res* 70: 8684–8694. doi:10.1158/0008-5472.CAN-10-2227

Tanabe S, Tazawa H, Kanaya N, Noma K, Kagawa S, Kuroda S, Urata Y, Shirakawa Y, Fujiwara T. 2019. Endoscopic intratumoral injection of OBP-301 (telomelysin) with radiotherapy in esophageal cancer patients unfit for standard treatments. *J Clin Oncol* 37: 130. doi:10.1200/jco.2019.37.4_suppl.130

Tefferi A, Lasho TL, Begna KH, Patnaik MM, Zblewski DL, Finke CM, Laborde RR, Wassie E, Schimek L, Hanson CA, et al. 2015. A pilot study of the telomerase inhibitor imetelstat for myelofibrosis. *N Engl J Med* 373: 908–919. doi:10.1056/nejmoa1310523

Tian T, Chen YQ, Wang SR, Zhou X. 2018. G-Quadruplex: a regulator of gene expression and its chemical targeting. *Chem* 4: 1314–1344. doi:10.1016/j.chempr.2018.02.014

Tomita K, Collopy LC. 2018. Telomeres, telomerase, and cancer. In *Encyclopedia of cancer*, 3rd ed. (ed. Boffetta P, Hainaut P), pp. 437–454. Elsevier, Amsterdam.

Topalian SL, Hodi FS, Brahmer JR, Gettinger SN, Smith DC, McDermott DF, Powderly JD, Sosman JA, Atkins MB, Leming PD, et al. 2019. Five-year survival and correlates among patients with advanced melanoma, renal cell carcinoma, or non-small cell lung cancer treated with nivolumab. *JAMA Oncol* 5: 1411–1420. doi:10.1001/jamaoncol.2019.2187

Waksal JA, Bruedigam C, Komrokji RS, Jamieson CHM, Mascarenhas JO. 2023. Telomerase-targeted therapies in myeloid malignancies. *Blood Adv* 7: 4302–4314. doi:10.1182/bloodadvances.2023009903

Wang X, Hu CS, Petersen B, Qiu J, Ye F, Houldsworth J, Eng K, Huang F, Hoffman R. 2018. Imetelstat, a telomerase inhibitor, is capable of depleting myelofibrosis stem and progenitor cells. *Blood Adv* 2: 2378–2388. doi:10.1182/bloodadvances.2018022012

Wright WE, Piatyszek MA, Rainey WE, Byrd W, Shay JW. 1996. Telomerase activity in human germline and embryonic tissues and cells. *Dev Genet* 18: 173–179. doi:10.1002/(SICI)1520-6408(1996)18:2<173::AID-DVG10>3.0.CO;2-3

Wu X, Smavadati S, Nordfjäll K, Karlsson K, Qvarnström F, Simonsson M, Bergqvist M, Gryaznov S, Ekman S, Paulsson-Karlsson Y. 2012. Telomerase antagonist imetelstat inhibits esophageal cancer cell growth and increases radiation-induced DNA breaks. *Biochim Biophys Acta* 1823: 2130–2135. doi:10.1016/j.bbamcr.2012.08.003

Wu X, Zhang J, Yang S, Kuang Z, Tan G, Yang G, Wei Q, Guo Z. 2017. Telomerase antagonist imetelstat increases radiation sensitivity in esophageal squamous cell carcinoma. *Oncotarget* 8: 13600–13619. doi:10.18632/oncotarget.14618

Yu S, Wei S, Savani M, Lin X, Du K, Mender I, Siteni S, Vasilopoulos T, Reitman ZJ, Ku Y, et al. 2021. A modified nucleoside 6-thio-2′-deoxyguanosine exhibits antitumor activity in gliomas. *Clin Cancer Res* 27: 6800–6814. doi:10.1158/1078-0432.CCR-21-0374

Zhong Z, Shiue L, Kaplan S, de Lange T. 1992. A mammalian factor that binds telomeric TTAGGG repeats in vitro. *Mol Cell Biol* 12: 4834–4843. doi:10.1128/mcb.12.11.4834

Zhou J, Tateishi-Karimata H, Mergny JL, Cheng M, Feng Z, Miyoshi D, Sugimoto N, Li C. 2016. Reevaluation of the stability of G-quadruplex structures under crowding conditions. *Biochimie* 121: 204–208. doi:10.1016/j.biochi.2015.12.012

Zou Y, Sfeir A, Gryaznov SM, Shay JW, Wright WE. 2004. Does a sentinel or a subset of short telomeres determine replicative senescence? *Mol Biol Cell* 15: 3709–3718. doi:10.1091/mbc.E04-0

# Mechanisms of Alternative Lengthening of Telomeres

Roderick J. O'Sullivan[1] and Roger A. Greenberg[2]

[1]Department of Pharmacology and Chemical Biology, UPMC Hillman Cancer Center, University of Pittsburgh, Pittsburgh, Pennsylvania 15261, USA

[2]Department of Cancer Biology, Penn Center for Genome Integrity, Basser Center for BRCA, Perelman School of Medicine, University of Pennsylvania, Philadelphia, Pennsylvania 19104, USA

*Correspondence:* rjo@pitt.edu; rogergr@pennmedicine.upenn.edu

In recent years, significant advances have been made in understanding the intricate details of the mechanisms underlying alternative lengthening of telomeres (ALT). Studies of a specialized DNA strand break repair mechanism, known as break-induced replication, and the advent of telomere-specific DNA damaging strategies and proteomic methodologies to profile the ribonucleoprotein composition of telomeres enabled the discovery of networks of proteins that coordinate the stepwise homology-directed DNA repair and DNA synthesis processes of ALT. These networks couple mediators of homologous recombination, DNA template-switching, long-range template-directed DNA synthesis, and DNA strand resolution with SUMO-dependent liquid condensate formation to create discrete nuclear bodies where telomere extension occurs. This review will discuss the recent findings of how these networks may cooperate to mediate telomere extension by the ALT mechanism and their impact on telomere function and integrity in ALT cancer cells.

## LESSONS FROM YEAST GENETICS

In 1993, Lundblad and Blackburn described the first evidence of an alternative lengthening of telomeres (ALT) in the budding yeast *Saccharomyces cerevisiae* (Lundblad and Blackburn 1993). They observed surviving colonies emerging from a mutant yeast strain, est1⁻, that was defective in telomerase recruitment, displayed successive telomere shortening, and entered senescence. Two categories of "survivors," designated type I and type II, were characterized. Type I survivors had short telomeres but stabilized their chromosome ends by amplifying arrays of subtelomeric tandem repeats (Y′ elements). On the other hand, type II survivors acquired long arrays of telomere repeat ($C_{1-3}$/$TG_{1-3}$) sequences extending up to 12 kilobases (kb). The viability of both types of survivors relied on genes involved in homology-directed DNA break repair (Fig. 1A).

The acquisition and amplification of subtelomeric and telomeric sequences depended on RAD51 and RAD52. Although necessary in type I survivors, RAD51, the principal mediator of strand invasion that primes DNA repair by homologous recombination (HR), was dispensable for type II survivor viability. On the other

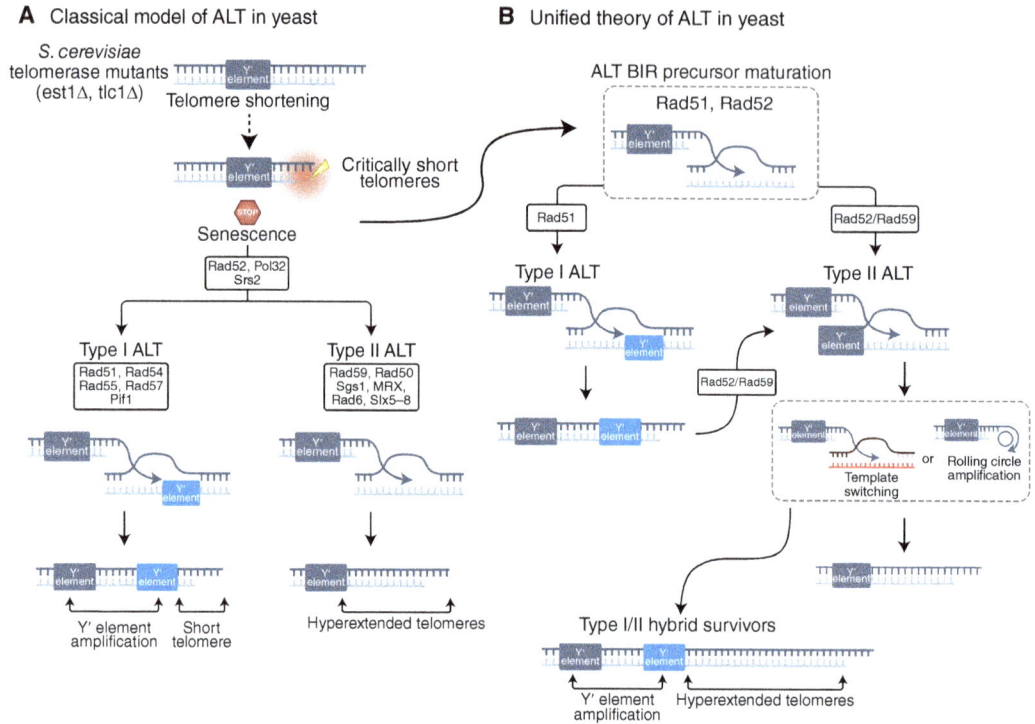

**Figure 1.** The classical and unified theories of alternative lengthening of telomeres (ALT) in yeast. (*A*) Deletion of telomerase (EST1 or TLC1) leads to critically short telomeres and senescence. Yeast cells that bypass or escape senescence engage in homologous recombination (HR)-mediated telomere extension by break-induced replication (BIR). Type I ALT survivors rely on RAD51-mediated HR that amplifies Y′-elements. Type II ALT survivors require RAD52/RAD59 for BIR-mediated hyperextension of telomeres. (*B*) Like the classical model, telomerase deficiency leads to critically short telomeres and senescence. RAD51- and/or RAD52-dependent HR ALT "precursor" development precedes HR pathway choice. The maturation of precursors is chiefly mediated by RAD59/RAD52 and RAD51′ amplifies Y′-elements. The intermediates derived here can also engage RAD52/RAD59-mediated HR and BIR. Sequential rounds of template switching and reinvasion at regions of homology on the same or different telomeres, and possibly rolling-circle amplification using extrachromosomal circular (ecc) DNA templates, lead to hyperextension of telomere repeats. The mature type I/II "hybrid" ALT telomeres harboring amplified Y′-elements and long telomere tracts will emerge.

hand, RAD52, which possesses strand invasion and annealing activities, was essential for both type I and II survivors. POL32, encoding the yeast paralog of the human POLD3 DNA polymerase δ subunit, was also necessary for type I and II survival emergence (Lydeard et al. 2007). This strict genetic dependence of type I and II ALT on RAD52 and POL32 implicated break-induced replication (BIR) as a critical pathway of ALT. BIR is a specialized one-ended DNA break repair mechanism (Kramara et al. 2018). BIR differs from DNA replication as it is largely restricted to the $G_2$-M phase of the cell cycle and

occurs conservatively (Donnianni and Symington 2013). BIR is achieved with a minimal replisome containing PCNA and DNA polymerase δ (PolD) while lacking the replicative CMG helicase and DNA Polε (Lydeard et al. 2007, 2010; Donnianni et al. 2019). BIR DNA synthesis is also discontinuous due to uncoupled leading and lagging strand synthesis and cycles of dissociation and reinvasion of the extending DNA end to reinitiate (Smith et al. 2007). Due to this unstable DNA synthesis program, BIR is error-prone, leading to a higher rate of mutagenesis and chromosomal rearrangements (Deem et al.

2010). Two distinct BIR mechanisms appear to operate in ALT. In one, Rad51 catalyzes strand invasion of the single broken end and initiates template-directed DNA synthesis by DNA Polδ within a dynamic displacement loop (D-loop), also referred to as a migrating bubble, driven by PIF1 helicase (Saini et al. 2013; Wilson et al. 2013). However, a RAD51-independent mechanism involving the annealing of complementary strands by RAD59 can also stimulate DNA Polδ-dependent BIR in type II survivors. Template-switching, where the partially extended DNA end reinvades alternative templates such as the adjacent complementary nascent strand, distal interstitial sequences, or perhaps excised extrachromosomal circular arrays of homologous telomeric sequences, can also reinitiate DNA synthesis in BIR (Fig. 1A; Smith et al. 2007; Sakofsky et al. 2015).

Subsequent genetic studies further refined the genetic relationship between BIR and ALT. RAD54, an ATP-dependent DNA translocase, and the strand exchange mediating protein RAD57, with crucial roles in HR, were only necessary for type I survivors. Factors essential for type II survivors included members of the RAD52 epistasis group like RAD59, MRX (MRE11-RAD50-XRS2), which mediates DNA-end resection analogously to the mammalian Mre11-Rad50-Nbs1 (MRN) complex, and the yeast orthologs of human Bloom helicase (BLM), SGS1 (Cohen and Sinclair 2001; Huang et al. 2001; Johnson et al. 2001), that coordinate DNA end resection and the dissolution of recombination intermediates like D-loops and Holliday junctions (HJs). Type II survivors also depended on ubiquitin and SUMO-modifying enzymes, RAD6 and SLX5-8, implicating these post-translational pathways in ALT (Hu et al. 2013; Churikov et al. 2016). Notably, deleting type I factors favored the emergence of type II survivors and vice versa. For example, deleting RAD51 expedited the emergence of type II yeast survivors. Similarly, deleting the type II factor RAD59 yielded type I survivors only. However, codeletion of a type I and II essential factor completely eradicated survivors. These studies provided invaluable models for the genetic requirements for the survival of ALT, strongly indicating two mutually exclusive and competitive pathways: RAD51-dependent type I and a RAD51-independent type II ALT mechanism that likely involves single-stranded annealing telomeric DNA for homology capture before BIR initiation.

## A UNIFIED THEORY OF ALT IN YEAST

While undeniably powerful, genetic studies of ALT have some limitations. Deleting type I or II genes could, by default, enable dominance of the competing pathway. This could yield survivors through the atavistic engagement of alternative survival mechanisms that confer stochastic survivorship and thus misrepresent a factor's contribution to ALT. The genetic models also predicted each pathway contributes equally to ALT survivorship (Wellinger and Zakian 2012). The expectation was that the cumulative frequency of ALT survivorship after disrupting type I or II ALT pathways would match that observed in telomerase-deficient yeast. However, by applying principles of population genetics (Poisson-based fluctuation analysis), researchers determined that deleting RAD51 or RAD59 yielded survivors at a substantially lower combined frequency (Kockler et al. 2021). This implied that both type I and II pathways likely act simultaneously to promote ALT. Accordingly, survivors harboring coexisting type I Y′ tandem and type II long telomere repeat sequence tracts were identified by long-range sequencing. This observation indicated that the type I and II pathways may not be entirely independent but may converge to generate ALT survivors. This led to a provocative unified model of ALT (Kockler et al. 2021). Central to this unified model is the premise that RAD51 generates ALT precursors, which RAD52/RAD59 later mature into stable ALT survivors.

The unified model postulates that once telomere shortening reaches criticality, checkpoint factors (RIF1, RAD9, TEL1, MEC1) enforce transient cell-cycle arrest before senescence. This is consistent with observations from the tracing of individual telomerase-deficient yeast cells that briefly paused and then reinitiated proliferation before committing to senescence (Teng and Zakian 1999; Teixeira et al. 2004; Xu et al. 2015). End-processing of eroded telomeres by DNA resection factors (SRS2, MRX) primes RAD51-mediated precursor formation by strand invasion and telomere extension by DNA Polδ, conferring sufficient telomere length to bypass growth arrest (Kockler et al. 2021). Maturation of these precursors by RAD59/RAD52-mediated BIR will generate type II ALT survivors with hyperextended telomeres. However, telomeres containing hybrid sequences will be generated should RAD59/RAD52-dependent BIR initiate at telomeres where amplification of Y′ elements by RAD51 has taken place. Template switching and amplification of repeats from circular DNA substrates could also contribute to the structural arrangement of telomeres in hybrid survivors (Fig. 1B).

The yeast "unified model of ALT" remains to be interrogated more broadly. However, the influence that yeast genetic models of ALT have already had on our understanding of ALT in human cancers is hugely significant. The unified model parallels the complex coordination of DNA repair activities reported to mediate ALT in human cancer cells in several aspects. Thus, as before, it seems reasonable to expect that powerful yeast genetics, genome editing, and sequencing will continue to provide vital insights into the ALT mechanism in human cancer.

## ALT IN HUMAN CANCER CELLS

### The First Molecular Details of ALT in Human Cancer Cells

Reddel and colleagues provided the first documentation of telomerase-independent telomere length maintenance in human cancer cell lines (Bryan et al. 1997). They subsequently established that this phenomenon involved recombination between telomere DNA sequences by visualizing the copying and transfer of a neomycin resistance cassette that was experimentally inserted into one telomere and subsequently appeared on several other telomeres located on distinct chromosomes (Bryan et al. 1997; Dunham et al. 2000). Several ALT cancer cell lines examined also exhibited high rates of telomere sister chromatid exchange (t-SCE) (Londoño-Vallejo et al. 2004) and telomere length heterogeneity. These features are not typically observed in primary and telomerase-expressing cancer cells. Their presence was inferred to reflect persistent telomere–telomere recombination in ALT cancer cells. Furthermore, the discovery that mediators of homology-directed repair (HDR), including RAD51, RAD52, BLM, and the MRN, colocalize with telomeres substantiated the evidence that ALT in human cells involves HR mechanisms analogous to those in type I and II yeast survivors. These colocalizations were confined mainly within subnuclear promyelocytic leukemia (PML) bodies, later designated ALT-associated PML bodies (APBs) due to their unique presence in ALT cancer cell lines and tumors (Yeager et al. 1999). APBs are now known to harbor biophysical properties consistent with liquid-phase condensates (Banani et al. 2016). Specifically, the formation of APBs requires noncovalent interactions between SUMO (small ubiquitin-like modifier) modifications and SIM (SUMO-interaction motifs) within PML protein, forming a shell-like structure where as many as 5–10 telomeres and DNA repair factors become sequestered (Draskovic et al. 2009; Zhang et al. 2020). Thus, APBs are widely believed to provide a secure compartment to scaffold homology-directed telomere extension and other vital steps of ALT (Zhang et al. 2021). Although present in only ~5% of ALT cancer cell cultures, the modulation of APB frequency through inhibiting or disrupting protein expression enabled candidate-based screening of factors and molecular pathways contributing to ALT regulation (Jiang et al. 2007, 2009). Complex, looped, circular, and linear extrachromosomal telomeric repeat (ECTR) arrays were also identified in ALT cancer cells (Cesare and Griffith 2004; Henson et al. 2009; Nabetani and Ishikawa 2009). Of these, the presence of partially single-stranded telomeric (CCCTAA)n DNA circles has become a robust marker of ALT. Considered in conjunction with genetic data (i.e., ATRX/DAXX mutations) and APBs, the detection and quantification of C-circles by a PCR-based C-circle assay confirms the ALT status in tumor specimens and quantitatively evaluates ALT in cell lines (Henson et al. 2005, 2009; Loe et al. 2020; de

Cite this article as *Cold Spring Harb Perspect Biol* doi: 10.1101/cshperspect.a041690

Nonneville and Reddel 2021). How C-circles are formed remains unclear, even though mounting evidence has linked C-circle formation with processing of the lagging telomere strand and shown a clear genetic dependence on BLM helicase (Jiang et al. 2024; Lee et al. 2024). ECTRs, such as C-circles, have long been speculated to provide templates for telomere DNA synthesis. Determining the precise molecular function(s) of APBs and the exact origin and functional contribution of ECTRs like C-circles is an ongoing research pursuit.

## Genetic Drivers of ALT Initiation in Human Cancer

While ALT seems to be active only when telomerase reactivation fails to occur, understanding how ALT is activated has been challenging. Clear evidence to support the assumption that ALT emerges due to a gain-of-function in DNA recombination has not materialized. However, ALT telomeres display markers of replication stress, consistent with the idea that persistent telomere damage might stimulate the hyper-recombination phenotype. With the discovery of recurrent mutations in the genes encoding ATRX and DAXX in ALT tumors (Heaphy et al. 2011), several genetic and molecular studies revealed that telomeric chromatin dysregulation is a major factor in ALT emergence. Clinical genomics of large numbers of tumors revealed that ATRX or DAXX loss are frequent concurrent genetic events in cancers that phenotypically exhibit ALT. These clinical studies also uncovered that ATRX or DAXX loss and the initial detection of ALT are more often associated with metastatic disease onset (de Wilde et al. 2012; Kim et al. 2016; Dyer et al. 2017; Singhi et al. 2017). Yet, ALT has been observed in cell lines and tumors that express wild-type ATRX or DAXX, implicating other drivers of ALT (de Nonneville and Reddel 2021). Intriguingly, those tumors often harbor genetic alterations in chromatin modifiers (SETD2, MEN1, ARID1A, CHD8, IDH1) (Chan et al. 2018; Roy et al. 2018), histone variants (histone H3.3) (Schwartzentruber et al. 2012), and factors that maintain chromosomal topology

(SMARCAL1, TOP3A) (Fig. 2A; Diplas et al. 2018; Brosnan-Cashman et al. 2021; de Nonneville et al. 2022). Nonetheless, because of its elevated frequency, the genetic or epigenetic inactivation of ATRX or DAXX leading to loss of the corresponding protein expression has become a robust clinical predictive indicator of ALT activation.

ATRX and DAXX proteins form a multi-functional chromatin remodeling and histone H3.3 deposition complex (Clynes et al. 2013) that manages chromatin assembly at telomeres (Goldberg et al. 2010; Lewis et al. 2010), pericentromeric heterochromatin (McDowell et al. 1999; Teng et al. 2021), zinc finger gene loci (Valle-García et al. 2016), imprinted loci (Voon et al. 2015), and tandem repeat sequences (Law et al. 2010). Several studies have reported various deleterious impacts of ATRX-DAXX inactivation on telomeres, ranging from fundamental alterations in chromatin assembly dynamics and DNA replication (Juhász et al. 2018; Li et al. 2019) to elevated transcription of the telomere repeat-containing lincRNA TERRA that can invade telomeric DNA to form RNA:DNA hybrids (R-loops) that cause stochastic DNA strand break formation (Fig. 2B; Chu et al. 2017; Nguyen et al. 2017). Reintroducing ATRX into ATRX-deficient ALT cancer cells dampens replicative stress at telomeres while also suppressing APB formation and ECTR accumulation, providing evidence that ATRX is a repressor of ALT (Clynes et al. 2015). However, depleting ATRX or DAXX alone does not guarantee ALT activation in cell culture models (Lovejoy et al. 2012). Several studies have shown that exposure to additional stressing agents in conjunction with ATRX-DAXX inactivation can manifest ALT induction (Lovejoy et al. 2012; Li et al. 2019; Clynes et al. 2021).

Related to this theme of chromatin dysfunction-induced replicative stress as a driving factor for ALT initiation, depleting the ASF1a and ASF1b histone H3 chaperones elicited systemic replicative stress followed by the rapid induction of ALT in transformed fibroblasts and provoked a switch from telomerase to ALT-mediated telomere extension in HeLa cells (O'Sullivan et al.

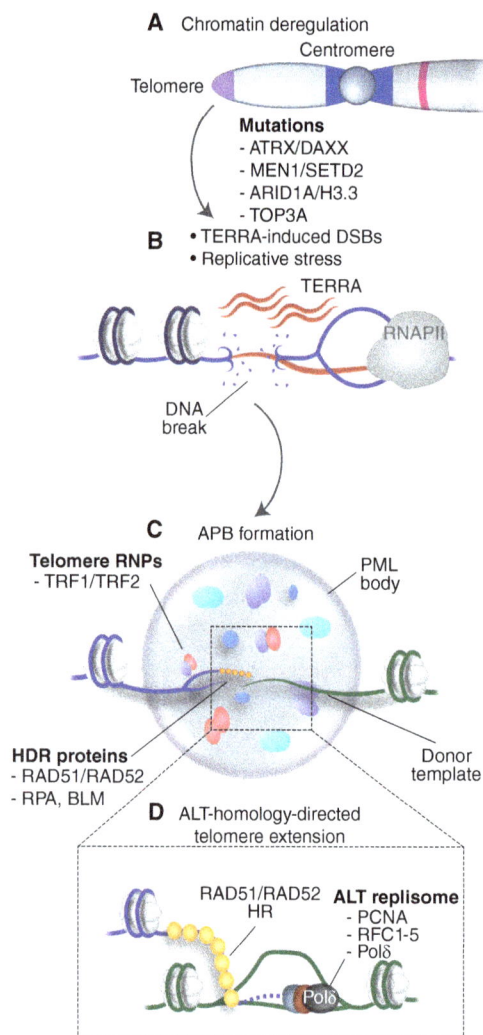

**Figure 2.** The pathway of alternative lengthening of telomeres (ALT) initiation. (*A*) The preponderance of evidence implicates chromatin deregulation due to mutations in chromatin modifiers and remodelers as a principal driver of ALT initiation in human cancer cells. Amplification of TOP3A without chromatin modifier loss can also lead to ALT. (*B*) Chromatin destabilization may license elevated telomeric transcription and TERRA associations, leading to DNA breaks and/or heightened replicative stress within telomeres. (*C*) Damaged telomeres are sequestered to specialized PML nuclear condensates (APBs), harboring mediators of homology-directed repair (HDR). (*D*) RAD51 or RAD52-dependent HR precedes assembly of the ALT replisome (PCNA-RFC-Polδ) that mediates template-directed telomere extension by BIR.

2014). Likewise, mutations in other chromatin modifiers, such as SMARCAL1, can also exert replicative stress at telomeres (Cox et al. 2016). Therefore, replicative stress due to the deregulation of telomeric chromatin is a prerequisite, along with telomerase suppression, for driving the emergence of ALT (Fig. 2B). The destabilization of telomeric chromatin then favors HDR within the APBs, where the sequestration of telomeres occurs through a SUMOylation (Potts and Yu 2007; Chung et al. 2011; Zhang et al. 2021) and BLM-dependent mechanism (Min et al. 2019; Loe et al. 2020), possibly involving liquid–liquid phase separation (Fig. 2C; Zhang et al. 2020). These events may be coordinated with HR that brings distal telomeres into proximity for strand capture and telomere extension by HDR (Fig. 2D). Such mechanisms likely underlie the interchromosomal telomere to telomere transfer of a neomycin resistance marker originally described by Reddel and colleagues (Dunham et al. 2000) and those initially characterized in type I and II ALT yeast survivors.

## A Stepwise Model of ALT in Human Cancer

Building on this knowledge, recent studies have underscored the role of replicative stress and DNA damage as initiating the complex mechanisms underpinning ALT telomere extension. Advances in locus-specific proteomics, such as PiCH (Déjardin and Kingston 2008; Zhang et al. 2023) and bio-ID (Garcia-Exposito et al. 2016; Kaminski et al. 2022), have cataloged the proteome of telomeres in ALT cancer cells, identifying many proteins with specialized roles during ALT. These include mediators of HDR, nascent telomere DNA synthesis, and other processes during ALT. Here, we summarize these recent advances within a sequential, step-by-step model illustrating our current understanding of the ALT mechanism in human cancer cells (Fig. 3).

## Step 1. Priming ALT through Replicative Stress and DNA Breaks

Stalled or damaged replication forks, unresolved TERRA R-loops, G-quadruplexes, and DNA double-strand breaks (DSBs) can destabilize telomeres (Fig. 3). ATRX-DAXX loss, or the inactivation of other chromatin modifiers, likely contributes to the enhanced susceptibility to telomere instability. As mentioned, ATRX-DAXX organizes telomeric chromatin by depositing histone H3.3 (Goldberg et al. 2010; Lewis et al. 2010). With the loss of ATRX-DAXX, the failure to rechromatinize telomeric DNA due to the loss of DAXX-dependent H3.3 deposition causes replicative stress (Clynes and Gibbons 2013; Juhász et al. 2018). This means that nucleosome-free telomeric single-stranded DNA be exposed and unshielded, possibly leading to ATR-kinase signaling activation or nucleolytic cleavage. Indeed, the trimeric single-stranded binding protein RPA is constitutively present at ALT telomeres (Yeager et al. 1999). FANCM, a DNA-dependent ATPase/translocase subunit of the Fanconi anemia (FA) core complex (Pan et al. 2017, 2019; Lu et al. 2019; Silva et al. 2019) and SMARCAL1, an ATP-dependent strand annealing helicase and chromatin remodeler (Cox et al. 2016), can salvage these replication forks by catalyzing fork reversal and restart (Fig. 3; also see Step 4).

ATRX-DAXX loss has also been implicated in causing defective Okazaki fragment maturation on lagging DNA strands (Jiang et al. 2024). Here, BLM can promote the assembly of the ALT damage response and damage-induced replisome by unwinding Okazaki fragments to create long $5'$- flaps on the nascent C-rich telomere strand (Fig. 3; Jiang et al. 2024). Similarly, excessive strand displacements were proposed to be responsible for extrachromosomal telomere DNA formation during ALT (Lee et al. 2024). These observations may relate to original descriptions of $5'$ C-rich overhang telomeric DNA as a hallmark of recombination-dependent telomere maintenance (Oganesian and Karlseder 2011). Deleting BLM expression had the remarkable effect of abolishing the entire downstream assembly of the ALT replisome (Jiang et al. 2024). The preferential link between this BLM-dependent mechanism and lagging strand replication is also notable, given evidence that lagging telomeres appear predisposed for extension in ALT cancer cells (Min et al. 2017) and that BLM appears indispensable for ALT-

HDR (O'Sullivan et al. 2014; Sobinoff et al. 2017; Min et al. 2019; Loe et al. 2020; Jiang et al. 2024).

Telomeres in ALT cancer cells are inherently unstable, exhibiting a propensity for stochastic DNA damage signaling activation whose origin was poorly understood (Cesare et al. 2009). Several recent studies have identified TERRA as a major instigator of this telomere damage (Silva et al. 2019, 2021, 2022). TERRA transcripts are more abundant in ALT cancer cells due to loss of ATRX-DAXX-established heterochromatin and repressive epigenetic marks (Nergadze et al. 2009). Nascent TERRA can hybridize with the complementary telomeric DNA sequence to form cotranscriptional or *cis*-acting TERRA-R-loops (Nguyen et al. 2017). *Cis*-acting TERRA R-loops could block RNA polymerase II, pausing it within telomeres and increasing the probability of DNA breaks. Additionally, mature TERRA molecules can associate and invade telomeric DNA on the same or different chromosomes. The assembly of these *trans*-acting TERRA into R-loops has been implicated in HR during ALT (Feretzaki et al. 2019; Kaminski et al. 2022; Yadav et al. 2022; see Step 2). The direct modulation of TERRA expression that either elevated or reduced the abundance of TERRA R-loops affected ALT initiation differentially (Silva et al. 2019, 2021, 2022). Insufficient TERRA R-loops seemingly impaired ALT. In contrast, excess TERRA R-loops hyperactivated ALT while eliciting cytotoxic DNA damage and replicative stress. As a result, the regulation of TERRA and TERRA R-loops is now viewed as an important determinant of ALT (Fig. 3).

Detecting the events that trigger ALT proved challenging. Given that telomere damage stimulates the ALT phenotypes (i.e., APBs), anchoring the FokI endonuclease directly at the telomere by fusing it to TRF1 provided a way to generate localized DNA breaks at telomeres to stimulate ALT (Cho et al. 2014; Dilley et al. 2016). Implementing this experimental strategy allowed for the synchronous initiation of ALT-associated processes that are typically rare and difficult to assess (Fig. 3). Notably, TRF1-FokI induction bypasses requirements for BLM helicase and RAD52, which appear to initiate the

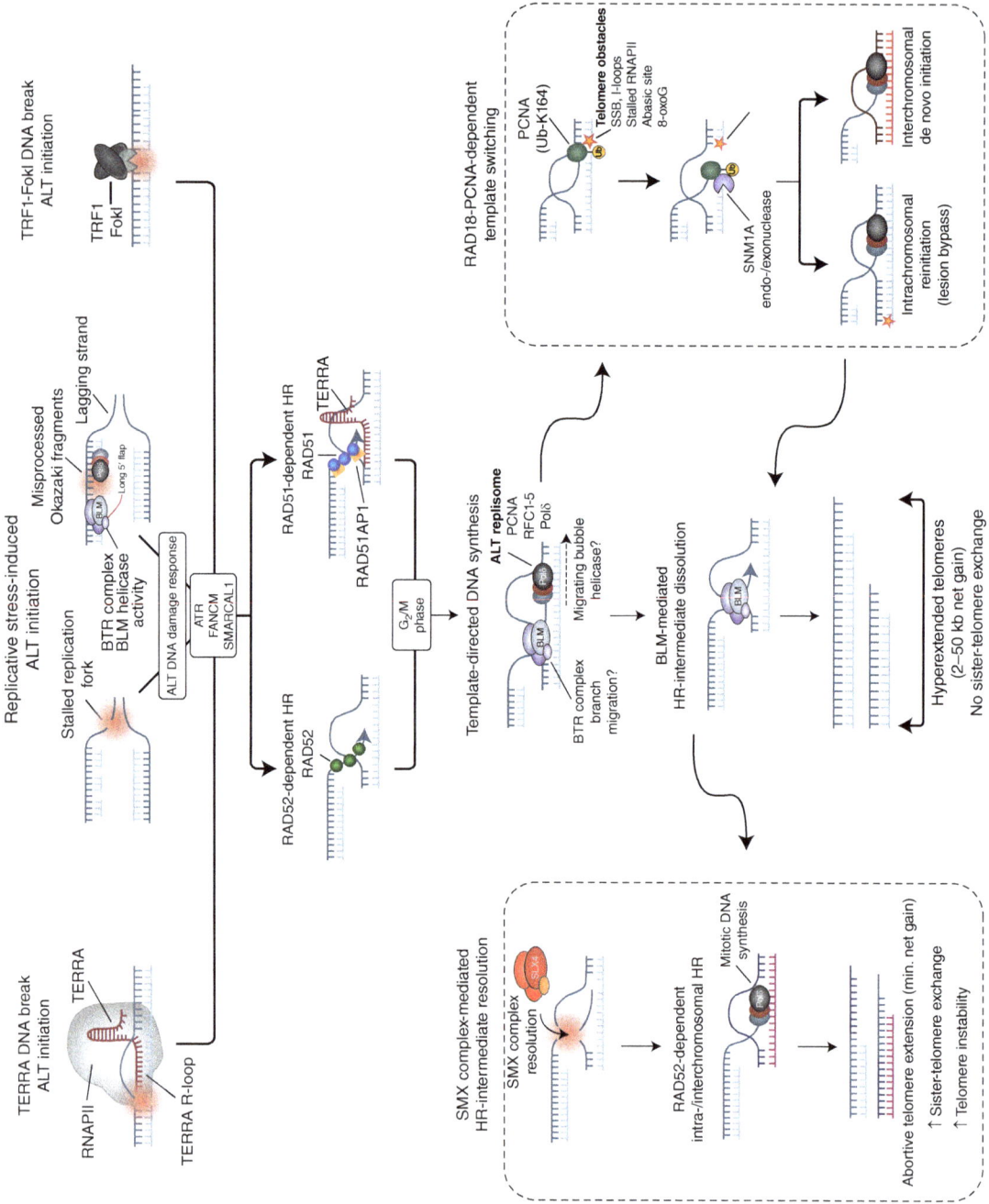

Figure 3. (*See following page for legend.*)

Cite this article as *Cold Spring Harb Perspect Biol* doi: 10.1101/cshperspect.a041690

ALT telomere damage response during replication stress and mediate homology capture by annealing, respectively (Verma et al. 2019; Zhang et al. 2019, 2021; Jiang et al. 2024). TRF1-FokI thus isolates the process of BIR, allowing it to be systematically studied. The application of TRF1-FokI demonstrated the unidirectional nature of processive homology-directed telomere synthesis over many kilobases during ALT (see Step 2).

## Step 2: Establishing Telomere Contacts by Homologous Recombination

Even though it was apparent that telomere damage stimulates HR between telomere sequences on different chromosomes, limitations in real-time tracking of telomeres in living cells meant that visualizing telomere-to-telomere contacts was not feasible. The first direct evidence of interchromosomal telomere contacts was described following the optical visualization of telomere dynamics after telomere DNA breaks were generated by TRF1-FokI (Cho et al. 2014). Upon DNA break formation, damaged telomeres exhibited remarkable dynamic changes in mobility, moving through several microns of nuclear space to contact and cluster with other telomeres. Importantly, these dynamic alterations in telomere trajectory were delayed, albeit not eliminated, in the absence of DNA recombinases RAD51 and the HOP2-MND1 HR-mediator complex that facilitates capture, alignment, and synapsis of homologous telomere sequences (Cho et al. 2014). Interestingly, deletion of the meiotic recombination mediators HOP2 and MND1 also caused PARP inhibitor hypersensitivity and radiosensitivity in a non-ALT setting, indicative of a broader role in the DNA damage response beyond ALT (Koob et al. 2023; Zelceski et al. 2023). The precise mechanism of telomere homology search remains to be determined, although polymerization of F-actin filaments and other motor proteins might be involved (Schrank et al. 2018). Nonetheless, by monitoring the incorporation of the nucleoside analog, EdU, these steps preceded and facilitated new telomere DNA synthesis.

Once homologous telomeres are captured, the RAD51-coated telomeric ssDNA invades

---

Figure 3. A step-by-step alternative lengthening of telomeres (ALT) mechanism in human cancer cells. From the top (left to right) TERRA R-loop-induced replicative stress leads to DNA breaks and the activation of ALT. Replicative stress caused by DNA lesions that stall replication fork progression triggers ATR-kinase activation and the DNA damage response that leads to ALT. In addition, errors in strand displacement synthesis and Okazaki fragment maturation during lagging strand DNA synthesis generate substrates for BLM helicase activity. Aberrant 5'-ssDNA flaps cause further replicative stress and telomeric DNA damage that stimulates the downstream ALT-associated processes. FANCM and SMARCAL1 likely participate in overcoming or mitigating excessive telomere damage here. FANCMs translocase activity could remove TERRA R-loops and other misprocessed replication intermediates. SMARCAL1 catalyzes fork remodeling and/or fork reversal, and diverse DNA resection activities generate the ssDNA substrates necessary for homologous recombination. (Middle) Strand invasion is catalyzed by either RAD52 or RAD51, the latter facilitated by interactions with RAD51AP1 and TERRA R-loop formation. (Bottom center) The assembly of the ALT replisome (PCNA-RFC-Polδ) is followed by long-tract DNA synthesis. BTR (BLM-TOP3A-RMI1/2) complex-dependent branch migration or bubble migration by an undetermined helicase facilitates nascent telomere DNA synthesis. Upon completion of telomere extension and unloading of the ALT replisome, BTR catalyzes the dissolution of HR intermediates to preserve telomere integrity and prevent crossovers. (Bottom right) If the ALT replisome encounters DNA lesions or physical obstacles, RAD18-mediated monoubiquitination of lysine (K) 164 in PCNA recruits SNM1A, which executes endonuclease-mediated DNA nicking and 5'-3' exonuclease resection to bypass the lesion by template switching. The new ssDNA overhang can invade and reprime ALT replisome assembly on the same (intra) or another (inter) telomeric template, eventually concluding with BTR-mediated dissolution. (Bottom left) The failure to remove HR intermediates or bypass obstacles by BTR dissolution or template switching mechanisms creates an opportunity for the SMX nucleolytic complex to resolve HR intermediates. This can lead to RAD52 establishment of mitotic DNA synthesis (MiDaS) and aberrant recombination that yields telomere crossovers (telomere sister chromatid exchanges [tSCE]) and unstable telomeres. These pose a threat to genome stability, potentially by promoting mitotic catastrophe.

---

the donor template dsDNA, forming the D-loop. RAD52 also catalyzes strand invasion and D-loop formation (Fig. 3; Zhang et al. 2019). RAD51-mediated HR may favor interchromosomal HR between distally located telomeres, in contrast with RAD52, which preferentially drives intrachromosomal HR or HR between proximal and/or adjacent chromosomes. RAD51 and its related HR accessory factor, RAD51AP1, exhibit RNA-assisted D-loop formation (Feretzaki et al. 2019; Ouyang et al. 2021). RAD51 and RAD51AP1 directly bind to TERRA RNA, typically *trans*-acting TERRA, using it to invade donor dsDNA on the captured homologous telomere. The RNA:DNA hybrid R-loop generated by the invading TERRA RNA acts as a precursor for subsequent D-loop formation (Fig. 3; Kaminski et al. 2022; Yadav et al. 2022). G-quadruplexes that form on the displaced strand, or binding of RAD51AP1 to TERRA RNA:DNA hybrids are then proposed to stabilize the nascent D-loops, thereby enabling assembly of the ALT replisome and downstream telomere DNA synthesis (see Step 3). Notably, RAD52 does not display such TERRA-related activity during ALT. This suggests that TERRA might dictate whether the RAD51 or RAD52-dependent HR pathway is used in ALT (Fig. 3).

## Step 3: ALT Replisome Assembly and Template-Directed DNA Synthesis

Several studies proved that ALT telomere DNA synthesis involves a mechanism analogous to BIR-mediated DNA DSB repair and telomere extension in ALT yeast survivors (Lydeard et al. 2007, 2010). The generation of localized telomere DNA breaks with TRF1-FokI stimulated the loading of PCNA by the RFC1-5 complex and nascent telomere DNA synthesis by DNA Polδ (Fig. 3; Dilley et al. 2016). Disrupting POLD3, the human ortholog of yeast Pol32, led to extensive telomere shortening. Even though the assembly of the ALT replisome appeared to follow the prior events in HR, the rapid assembly of PCNA-RFC1-5 directly at broken telomeres (i.e., independently of RAD51-mediated HR) was also visualized (Dil-

ley et al. 2016). Importantly, the assembly of the specialized ALT replisome and subsequent telomere-specific DNA synthesis were observed during the $G_2/M$ phase in ALT cancer cells (Dilley et al. 2016; Verma et al. 2019). Pinpointing that telomere extension by ALT takes place in $G_2/M$ phases provided other crucial insights into its unique nature. Unlike normal DNA replication that occurs throughout the S phase and into the early $G_2$ phase, ALT telomere DNA synthesis during $G_2/M$ does not seem subject to replicative checkpoints (Dilley et al. 2016). This allows for continuous long-tract DNA synthesis to the very terminus of the telomere (Dilley et al. 2016). Furthermore, telomeres in ALT cancer cells were also shown to harbor tracts of conservatively replicated telomere DNA, which is again consistent with BIR (Roumelioti et al. 2016).

The precise details of the dynamics and rate of DNA synthesis during telomere extension by ALT remain unanswered. Likewise, whether telomere DNA synthesis requires a migrating bubble, like BIR in yeast (Saini et al. 2013; Wilson et al. 2013), remains to be determined. Similarly, the helicase(s) involved remain to be identified. Other open questions relate to what factors influence the extent of telomere extension. For instance, how the equilibrium between RNAPII passage through telomeric chromatin, stochastic TERRA-binding/R-loop formation (Silva et al. 2022), and the progression of the ALT replisome is controlled will likely influence DNA synthesis dynamics, as demonstrated during BIR in yeast (Liu et al. 2021). Even with that, the rate of telomere DNA synthesis is likely to be discontinuous and variable, interrupted by these and other genomic roadblocks that prevent the ALT replisome from proceeding and must be overcome.

## Step 4: Bypassing Obstacles at Telomeres

To pass through telomeres securely, the active ALT replisome faces several genomic obstacles that must be circumvented (Fig. 3). These include ssDNA gaps (Cesare and Griffith 2004), non-base-paired internal ssDNA bubbles (I-loops) (Mazzucco et al. 2020), R-loops, DNA

Cite this article as *Cold Spring Harb Perspect Biol* doi: 10.1101/cshperspect.a041690

base lesions like abasic sites and 8-oxoguanine (Thosar et al. 2024), and other aberrant and unusual non-B-form secondary structures (Nabetani and Ishikawa 2009). As discussed in Step 1, proteins such as FANCM and SMARCAL1 have roles in stabilizing and restoring stalled replication forks within telomeres at the outset of ALT. FANCM and SMARCAL1 may also assist the recovery of the stalled ALT replisome. Depleting FANCM has catastrophic repercussions for telomere integrity and ALT cancer cell viability (Pan et al. 2017, 2019; Lu et al. 2019; Silva et al. 2019). This may be due to an inability to remove R-loops or DNA-damage-induced DNA intermediates, whose resolution necessitates FANCMs translocase activity. SMARCAL1 might assist in alleviating replicative stress by remodeling the stalled replication fork to enable safe reinitiation of DNA synthesis (Cox et al. 2016). As with FANCM loss, SMARCAL-deficient cells exhibit highly unstable telomeres and compromised viability. Interestingly, the absence of BLM-driven telomere recombination obviates the need for FANCM (Jiang et al. 2024), accounting for the observed suppression of lethality in cells lacking FANCM and BLM (Silva et al. 2019). BLM helicase activity may create damage-dependent telomere intermediates whose resolution requires both SMARCAL1 and FANCM. The nature of these intermediates and whether they occur at replication forks or within telomere recombination intermediates is unknown. The profound impact of FANCM and SMARCAL1 loss on ALT cancer cell viability underscores the importance of securely repairing or bypassing the blocks that the ALT replisome encounters.

Comparative proteomics of telomere composition revealed that, in contrast with telomerase-expressing cancer cells, ALT telomeres are enriched in protein networks that coordinate the cells' diverse DNA repair pathways (Déjardin and Kingston 2008; Garcia-Exposito et al. 2016). PCNA, the sliding clamp that promotes the processive migration of the ALT replisome through telomeres, is an important hub for recruiting several protein complexes that maintain nascent telomere DNA synthesis. This includes recruiting the DNA mismatch repair complexes

(Barroso-González et al. 2021; Sakellariou et al. 2022), whose yeast homolog MSH2 is essential for ALT survivors (Rizki and Lundblad 2001). Interestingly, the noncanonical antirecombination activities of MutSα in heteroduplex rejection that prevents HR between nonidentical DNA sequences and MutSb's G-quadruplex resolution appear to have crucial regulatory roles in ALT (Barroso-González et al. 2021; Sakellariou et al. 2022).

Post-translational modification of PCNA is an important determinant of the efficiency of telomere DNA synthesis. The monoubiquitination of PCNA at lysine (K) 164 by the RAD18 ubiquitin ligase has emerged as a pivotal event during ALT. In response to replicative stress, monoubiquitination of PCNA-K164 (Ub-PCNA) activates DNA damage tolerance pathways that are executed by translesion DNA synthesis (TLS) polymerases, including DNA polymerase η (Polη), to bypass DNA lesions and reinitiate telomere DNA synthesis (Garcia-Exposito et al. 2016). However, by further assessing telomere composition in cells lacking RAD18 (and thus monoubiquitination of PCNA), a specialized nuclease, SNM1A, typically involved in intrastrand cross-link repair, was identified as the principal effector of Ub-PCNAs downstream functions in ALT (Fig. 3). At regions where the ALT replisome encounters blockades, SNM1As endonucleolytic DNA nicking precedes exonucleolytic resection of the template strand, thereby enabling recombination-dependent template switching (Zhang et al. 2023). Its exonucleolytic DNA resection activity generates a new 3' ssDNA strand that can be used in HR to invade DNA past the blockage (Zhang et al. 2023). This invasion can happen on the same or other available templates, such as adjacent chromatids or homologous chromosomes (Fig. 3; Smith et al. 2007; Zhang et al. 2023). Template switching might occur several times as obstacles are encountered, so active telomere DNA synthesis can be restored to achieve maximal telomere extension. Although processes that regulate RAD18-dependent monoubiquitination, including deubiquitinating enzymes (DUBs) and other inhibitory PTMs, as well as negative regulators of PCNA, likely have some roles, our knowledge of the mechanisms that limit and disassemble the ALT replisome remains limited.

## Step 5: Unraveling HR Intermediates to Complete ALT

Dismantling HR intermediates marks a pivotal step in securing the efficient conclusion of telomere extension. The BTR complex, comprised of BLM, TOP3A, RMI1, and RMI2, is implicated in multiple steps of ALT, including APB formation, replisome assembly, $G_4$ resolution, and branch migration (Lu et al. 2019; Min et al. 2019; Loe et al. 2020; Zhang et al. 2021). The BTR complex is also chiefly responsible for the timely dissolution of HJ and D-loops to ensure recombining telomeres are disentangled before mitosis (Wyatt and West 2014). BTR complex dissolution activity also rescues stalled telomere DNA synthesis as part of a recombination-mediated replication fork restart mechanism (Fig. 3). The absence of BTRs resolvase activity and the persistence of unresolved HR structures creates an opportunity for a specialized endonucleolytic complex known as SMX (SLX1-4, MUS81-EME1, XPF-ERCC1) (Sobinoff et al. 2017). SMX acts like an all-purpose nuclease that promiscuously cleaves HR intermediates in a manner that can lead to telomere exchange events (i.e., transfer of telomeres between mitotic chromatids), or telomere rapid deletions where larger segments of telomeric DNA are excised and mitotic DNA synthesis (MiDaS) that is driven by RAD52 and DNA Polδ (Fig. 3; Wyatt et al. 2013; Sobinoff et al. 2017). MiDaS is typically unproductive, failing to manifestly lengthen telomeres, although a recent study suggested the contrary, implicating translesion DNA polymerase-mediated MiDaS in telomere extension (Lu et al. 2024). The reasons for this are unknown. Such uncontrolled DNA synthesis in mitosis may originate from transcription-replication conflicts and thus can be highly detrimental to genomic stability (Groelly et al. 2022). In contrast, depleting SLX4 in ALT cancer cells can cause telomere hyperextension. Although an enhanced dissolution activity of BLM was implicated, increased DNA Polδ localization and more frequent telomere extension

events were also observed following SLX4 loss (Sobinoff et al. 2017). The other actions of BLM in ALT, such as facilitating the assembly of the ALT replisome at telomere intermediates, may also contribute to the hyperextension observed. Of note, while SLX4 loss leads to telomere extension, depleting other SMX constituents such as XPF seems to compromise telomere DNA synthesis (Guh et al. 2022). Thus, much remains to be ascertained concerning the contribution of each SMX member and their coordinated activities during ALT.

A key question was how the opposing forces of BTR and SMX are coordinated. An auxiliary protein, SLX4IP, was found to physically bridge the BTR and SMX complexes by directly interacting with BLM and the XPF subunit of SMX (Panier et al. 2019). Importantly, by binding to and negatively regulating BLM, SLX4IP prevented premature or unwarranted HR-intermediate resolution. How exactly this is achieved is still unclear. However, the disrupting SLX4IP together with SLX4 elicited catastrophic levels of telomere instability, stemming from the out-of-control dissolution of telomeric HR intermediates by BLM (Panier et al. 2019). Thus, SLX4IP has been implicated in regulating the delicate balance in maintaining an efficient and productive ALT telomere extension mechanism, preventing it from veering into a nonproductive and deleterious one.

## CONCLUSIONS AND OUTLOOK

In this review, we have sought to distill the complexity of the ALT mechanism into four primary steps, each being highly coordinated and integrated to achieve telomere length extension as efficiently as possible. We sought to emphasize the crucial lessons learned in the yeast model system that provided a foundation for studies in human cells. It is now well understood that the ALT mechanism is delicately orchestrated. Realizing this to be the case allowed for new paradigms to be established. Whereas it might seem more appealing to therapeutically inhibit ALT, interventions that cause deviates or interfere with any step in the mechanism can have dire effects on ALT cancer cell survival. Thus, delineating the steps of the ALT mechanism also revealed putative targets for ALT therapies. To date, FANCM seems to be a leading candidate. In the coming years, it will be exciting to witness the next advances in ALT, which might include the development of the first dedicated anti-ALT inhibitors. However, deeper mechanistic studies will certainly remain essential for unraveling the complexities of ALT.

## ACKNOWLEDGMENTS

We are indebted to the community of researchers and clinicians whose contributions have advanced our understanding of ALT's complexity. We thank Anna Malkova and Raymund Wellinger for their helpful discussions.

## REFERENCES

Banani SF, Rice AM, Peeples WB, Lin Y, Jain S, Parker R, Rosen MK. 2016. Compositional control of phase-separated cellular bodies. *Cell* **166**: 651–663. doi:10.1016/j.cell.2016.06.010

Barroso-González J, García-Expósito L, Galaviz P, Lynskey ML, Allen JAM, Hoang S, Watkins SC, Pickett HA, O'Sullivan RJ. 2021. Anti-recombination function of MutSα restricts telomere extension by ALT-associated homology-directed repair. *Cell Rep* **37**: 110088. doi:10.1016/j.celrep.2021.110088

Brosnan-Cashman JA, Davis CM, Diplas BH, Meeker AK, Rodriguez FJ, Heaphy CM. 2021. SMARCAL1 loss and alternative lengthening of telomeres (ALT) are enriched in giant cell glioblastoma. *Mod Pathol* **34**: 1810–1819. doi:10.1038/s41379-021-00841-7

Bryan TM, Englezou A, Dalla-Pozza L, Dunham MA, Reddel RR. 1997. Evidence for an alternative mechanism for maintaining telomere length in human tumors and tumor-derived cell lines. *Nat Med* **3**: 1271–1274. doi:10.1038/nm1197-1271

Cesare AJ, Griffith JD. 2004. Telomeric DNA in ALT cells is characterized by free telomeric circles and heterogeneous t-loops. *Mol Cell Biol* **24**: 9948–9957. doi:10.1128/MCB.24.22.9948-9957.2004

Cesare AJ, Kaul Z, Cohen SB, Napier CE, Pickett HA, Neumann AA, Reddel RR. 2009. Spontaneous occurrence of telomeric DNA damage response in the absence of chromosome fusions. *Nat Struct Mol Biol* **16**: 1244–1251. doi:10.1038/nsmb.1725

Chan CS, Laddha SV, Lewis PW, Koletsky MS, Robzyk K, Da Silva E, Torres PJ, Untch BR, Li J, Bose P, et al. 2018. ATRX, DAXX or MEN1 mutant pancreatic neuroendocrine tumors are a distinct α-cell signature subgroup. *Nat Commun* **9**: 4158. doi:10.1038/s41467-018-06498-2

Cho NW, Dilley RL, Lampson MA, Greenberg RA. 2014. Interchromosomal homology searches drive directional

ALT telomere movement and synapsis. *Cell* **159**: 108–121. doi:10.1016/j.cell.2014.08.030

Chu H-P, Cifuentes-Rojas C, Kesner B, Aeby E, Lee H, Wei C, Oh HJ, Boukhali M, Haas W, Lee JT. 2017. TERRA RNA antagonizes ATRX and protects telomeres. *Cell* **170**: 86–101.e16. doi:10.1016/j.cell.2017.06.017

Chung I, Leonhardt H, Rippe K. 2011. De novo assembly of a PML nuclear subcompartment occurs through multiple pathways and induces telomere elongation. *J Cell Sci* **124**: 3603–3618. doi:10.1242/jcs.084681

Churikov D, Charifi F, Eckert-Boulet N, Silva S, Simon M-N, Lisby M, Géli V. 2016. SUMO-dependent relocalization of eroded telomeres to nuclear pore complexes controls telomere recombination. *Cell Rep* **15**: 1242–1253. doi:10.1016/j.celrep.2016.04.008

Clynes D, Gibbons RJ. 2013. ATRX and the replication of structured DNA. *Curr Opin Genet Dev* **23**: 289–294. doi:10.1016/j.gde.2013.01.005

Clynes D, Higgs DR, Gibbons RJ. 2013. The chromatin remodeller ATRX: a repeat offender in human disease. *Trends Biochem Sci* **38**: 461–466. doi:10.1016/j.tibs.2013.06.011

Clynes D, Jelinska C, Xella B, Ayyub H, Scott C, Mitson M, Taylor S, Higgs DR, Gibbons RJ. 2015. Suppression of the alternative lengthening of telomere pathway by the chromatin remodelling factor ATRX. *Nat Commun* **6**: 7538. doi:10.1038/ncomms8538

Clynes D, Goncalves T, Kent T, Shepherd S, Cunniffe S, Kim S, Humphrey T, GIbbons R. 2021. Induction of the ALT pathway requires loss of ATRX-DAXX in concert with genotoxic lesions at telomeres.

Cohen H, Sinclair DA. 2001. Recombination-mediated lengthening of terminal telomeric repeats requires the Sgs1 DNA helicase. *Proc Natl Acad Sci* **98**: 3174–3179. doi:10.1073/pnas.061579598

Cox KE, Maréchal A, Flynn RL. 2016. SMARCAL1 resolves replication stress at ALT telomeres. *Cell Rep* **14**: 1032–1040. doi:10.1016/j.celrep.2016.01.011

Deem A, Keszthelyi A, Blackgrove T, Vayl A, Coffey B, Mathur R, Chabes A, Malkova A. 2010. Break-induced replication is highly inaccurate. *PLoS Biol* **9**: e1000594. doi:10.1371/journal.pbio.1000594

Déjardin J, Kingston RE. 2008. Purification of proteins associated with specific genomic loci. *Cell* **136**: 175–186. doi:10.1016/j.cell.2008.11.045

de Nonneville A, Reddel RR. 2021. Alternative lengthening of telomeres is not synonymous with mutations in ATRX/DAXX. *Nat Commun* **12**: 1552. doi:10.1038/s41467-021-21794-0

de Nonneville A, Salas S, Bertucci F, Sobinoff AP, Adélaïde J, Guille A, Finetti P, Noble JR, Churikov D, Chaffanet M, et al. 2022. TOP3A amplification and ATRX inactivation are mutually exclusive events in pediatric osteosarcomas using ALT. *EMBO Mol Med* **14**: e15859. doi:10.15252/emmm.202215859

de Wilde RF, Heaphy CM, Maitra A, Meeker AK, Edil BH, Wolfgang CL, Ellison TA, Schulick RD, Molenaar IQ, Valk GD, et al. 2012. Loss of ATRX or DAXX expression and concomitant acquisition of the alternative lengthening of telomeres phenotype are late events in a small subset of MEN-1 syndrome pancreatic neuroendocrine tumors. *Mod Pathol* **25:** 1033–1039. doi:10.1038/modpa thol.2012.53

Dilley RL, Verma P, Cho NW, Winters HD, Wondisford AR, Greenberg RA. 2016. Break-induced telomere synthesis underlies alternative telomere maintenance. *Nature* **539:** 54–58. doi:10.1038/nature20099

Diplas BH, He X, Brosnan-Cashman JA, Liu H, Chen LH, Wang Z, Moure CJ, Killela PJ, Loriaux DB, Lipp ES, et al. 2018. The genomic landscape of TERT promoter wild-type-IDH wildtype glioblastoma. *Nat Commun* **9:** 2087. doi:10.1038/s41467-018-04448-6

Donnianni RA, Symington LS. 2013. Break-induced replication occurs by conservative DNA synthesis. *Proc Natl Acad Sci* **110:** 13475–13480. doi:10.1073/pnas.1309 800110

Donnianni RA, Zhou Z-X, Lujan SA, Al-Zain A, Garcia V, Glancy E, Burkholder AB, Kunkel TA, Symington LS. 2019. DNA polymerase δ synthesizes both strands during break-induced replication. *Mol Cell* **76:** 371–381.e4. doi:10.1016/j.molcel.2019.07.033

Draskovic I, Arnoult N, Steiner V, Bacchetti S, Lomonte P, Londoño-Vallejo A. 2009. Probing PML body function in ALT cells reveals spatiotemporal requirements for telomere recombination. *Proc Natl Acad Sci* **106:** 15726–15731. doi:10.1073/pnas.0907689106

Dunham MA, Neumann AA, Fasching CL, Reddel RR. 2000. Telomere maintenance by recombination in human cells. *Nat Genet* **26:** 447–450. doi:10.1038/82586

Dyer MA, Qadeer ZA, Valle-Garcia D, Bernstein E. 2017. ATRX and DAXX: mechanisms and mutations. *Cold Spring Harb Perspect Med* **7:** a026567. doi:10.1101/cshper spect.a026567

Feretzaki M, Pospisilova M, Fernandes RV, Lunardi T, Krejci L, Lingner J. 2019. RAD51-dependent recruitment of TERRA lncRNA to telomeres through R-loops. *Nature* **587:** 303–308. doi:10.1038/s41586-020-2815-6

Garcia-Exposito L, Bournique E, Bergoglio V, Bose A, Barroso-Gonzalez J, Zhang S, Roncaioli JL, Lee M, Wallace CT, Watkins SC, et al. 2016. Proteomic profiling reveals a specific role for translesion DNA polymerase η in the alternative lengthening of telomeres. *Cell Rep* **17:** 1858–1871. doi:10.1016/j.celrep.2016.10.048

Goldberg AD, Banaszynski LA, Noh K-M, Lewis PW, Elsaesser SJ, Stadler S, Dewell S, Law M, Guo X, Li X, et al. 2010. Distinct factors control histone variant H3.3 localization at specific genomic regions. *Cell* **140:** 678–691. doi:10.1016/j.cell.2010.01.003

Groelly FJ, Dagg RA, Petropoulos M, Rossetti GG, Prasad B, Panagopoulos A, Paulsen T, Karamichali A, Jones SE, Ochs F, et al. 2022. Mitotic DNA synthesis is caused by transcription-replication conflicts in BRCA2-deficient cells. *Mol Cell* **82:** 3382–3397.e7. doi:10.1016/j.molcel .2022.07.011

Guh C-Y, Shen H-J, Chen LW, Chiu P-C, Liao I-H, Lo C-C, Chen Y, Hsieh Y-H, Chang T-C, Yen C-P, et al. 2022. XPF activates break-induced telomere synthesis. *Nat Commun* **13:** 5781. doi:10.1038/s41467-022-33428-0

Heaphy CM, de Wilde RF, Jiao Y, Klein AP, Edil BH, Shi C, Bettegowda C, Rodriguez FJ, Eberhart CG, Hebbar S, et al. 2011. Altered telomeres in tumors with *ATRX* and *DAXX* mutations. *Science* **333:** 425–425. doi:10.1126/science.120 7313

Henson JD, Hannay JA, McCarthy SW, Royds JA, Yeager TR, Robinson RA, Wharton SB, Jellinek DA, Arbuckle SM, Yoo J, et al. 2005. A robust assay for alternative lengthening of telomeres in tumors shows the significance of alternative lengthening of telomeres in sarcomas and astrocytomas. *Clin Cancer Res* **11:** 217–225. doi:10.1158/ 1078-0432.217.11.1

Henson JD, Cao Y, Huschtscha LI, Chang AC, Au AYM, Pickett HA, Reddel RR. 2009. DNA C-circles are specific and quantifiable markers of alternative-lengthening-of-telomeres activity. *Nat Biotechnol* **27:** 1181–1185. doi:10 .1038/nbt.1587

Hu Y, Tang H-B, Liu N-N, Tong X-J, Dang W, Duan Y-M, Fu X-H, Zhang Y, Peng J, Meng F-L, et al. 2013. Telomerase-null survivor screening identifies novel telomere recombination regulators. *PLoS Genet* **9:** e1003208. doi:10 .1371/journal.pgen.1003208

Huang P-H, Pryde FE, Lester D, Maddison RL, Borts RH, Hickson ID, Louis EJ. 2001. SGS1 is required for telomere elongation in the absence of telomerase. *Curr Biol* **11:** 125–129. doi:10.1016/S0960-9822(01)00021-5

Jiang W-Q, Zhong Z-H, Henson JD, Reddel RR. 2007. Identification of candidate alternative lengthening of telomeres genes by methionine restriction and RNA interference. *Oncogene* **26:** 4635–4647. doi:10.1038/sj.onc .1210260

Jiang W-Q, Zhong Z-H, Nguyen A, Henson JD, Toouli CD, Braithwaite AW, Reddel RR. 2009. Induction of alternative lengthening of telomeres-associated PML bodies by p53/p21 requires HP1 proteins. *J Cell Biol* **185:** 797–810. doi:10.1083/jcb.200810084

Jiang H, Zhang T, Kaur H, Shi T, Krishnan A, Kwon Y, Sung P, Greenberg RA. 2024. BLM helicase unwinds lagging strand substrates to assemble the ALT telomere damage response. *Mol Cell* **84:** 1684–1698.e9. doi:10.1016/j .molcel.2024.03.011

Johnson FB, Marciniak RA, McVey M, Stewart SA, Hahn WC, Guarente L. 2001. The *Saccharomyces cerevisiae* WRN homolog Sgs1p participates in telomere maintenance in cells lacking telomerase. *EMBO J* **20:** 905–913. doi:10.1093/emboj/20.4.905

Juhász S, Elbakry A, Mathes A, Löbrich M. 2018. ATRX promotes DNA repair synthesis and sister chromatid exchange during homologous recombination. *Mol Cell* **71:** 11–24.e7. doi:10.1016/j.molcel.2018.05.014

Kaminski N, Wondisford AR, Kwon Y, Lynskey ML, Bhargava R, Barroso-González J, García-Expósito L, He B, Xu M, Mellacheruvu D, et al. 2022. RAD51AP1 regulates ALT-HDR through chromatin-directed homeostasis of TERRA. *Mol Cell* **82:** 4001–4017.e7. doi:10.1016/j.mol cel.2022.09.025

Kim JY, Brosnan-Cashman JA, An S, Kim SJ, Song K-B, Kim M-S, Kim M-J, Hwang DW, Meeker AK, Yu E, et al. 2016. Alternative lengthening of telomeres in primary pancreatic neuroendocrine tumors is associated with aggressive clinical behavior and poor survival. *Clin Cancer Res* **23:** 1598–1606. doi:10.1158/1078-0432.CCR-16-1147

Kockler ZW, Comeron JM, Malkova A. 2021. A unified alternative telomere-lengthening pathway in yeast survivor cells. *Mol Cell* **81:** 1816–1829.e5. doi:10.1016/j.molcel .2021.02.004

Cite this article as *Cold Spring Harb Perspect Biol* doi: 10.1101/cshperspect.a041690

Koob L, Friskes A, van Bergen L, Feringa FM, van den Broek B, Koeleman ES, van Beek E, Schubert M, Blomen VA, Brummelkamp TR, et al. 2023. MND1 enables homologous recombination in somatic cells primarily outside the context of replication. *Mol Oncol* **17**: 1192–1211. doi:10.1002/1878-0261.13448

Kramara J, Osia B, Malkova A. 2018. Break-induced replication: the where, the why, and the how. *Trends Genet* **34**: 518–531. doi:10.1016/j.tig.2018.04.002

Law MJ, Lower KM, Voon HPJ, Hughes JR, Garrick D, Viprakasit V, Mitson M, Gobbi MD, Marra M, Morris A, et al. 2010. ATR-X syndrome protein targets tandem repeats and influences allele-specific expression in a size-dependent manner. *Cell* **143**: 367–378. doi:10.1016/j.cell.2010.09.023

Lee J, Lee J, Sohn EJ, Taglialatela A, O'Sullivan RJ, Ciccia A, Min J. 2024. Extrachromosomal telomere DNA derived from excessive strand displacements. *Proc Natl Acad Sci* **121**: e2318438121. doi:10.1073/pnas.2318438121

Lewis PW, Elsaesser SJ, Noh K-M, Stadler SC, Allis CD. 2010. Daxx is an H3.3-specific histone chaperone and cooperates with ATRX in replication-independent chromatin assembly at telomeres. *Proc Natl Acad Sci* **107**: 14075–14080. doi:10.1073/pnas.1008850107

Li F, Deng Z, Zhang L, Wu C, Jin Y, Hwang I, Vladimirova O, Xu L, Yang L, Lu B, et al. 2019. ATRX loss induces telomere dysfunction and necessitates induction of alternative lengthening of telomeres during human cell immortalization. *EMBO J* **38**: e96659. doi:10.15252/embj.201796659

Liu L, Yan Z, Osia BA, Twarowski J, Sun L, Kramara J, Lee R, Kumar S, Elango R, Li H, et al. 2021. Tracking break induced replication reveals its stalling at roadblocks. *Nature* **590**: 655–659. doi:10.1038/s41586-020-03172-w

Loe TK, Li JSZ, Zhang Y, Azeroglu B, Boddy MN, Denchi EL. 2020. Telomere length heterogeneity in ALT cells is maintained by PML-dependent localization of the BTR complex to telomeres. *Genes Dev* **34**: 650–662. doi:10.1101/gad.333963.119

Londoño-Vallejo JA, Der-Sarkissian H, Cazes L, Bacchetti S, Reddel RR. 2004. Alternative lengthening of telomeres is characterized by high rates of telomeric exchange. *Cancer Res* **64**: 2324–2327. doi:10.1158/0008-5472.CAN-03-4035

Lovejoy CA, Li W, Reisenweber S, Thongthip S, Bruno J, de Lange T, De S, Petrini JHJ, Sung PA, Jasin M, et al. 2012. Loss of ATRX, genome instability, and an altered DNA damage response are hallmarks of the alternative lengthening of telomeres pathway. *PLoS Genet* **8**: e1002772. doi:10.1371/journal.pgen.1002772

Lu R, O'Rourke JJ, Sobinoff AP, Allen JAM, Nelson CB, Tomlinson CG, Lee M, Reddel RR, Deans AJ, Pickett HA. 2019. The FANCM-BLM-TOP3A-RMI complex suppresses alternative lengthening of telomeres (ALT). *Nat Commun* **10**: 2252. doi:10.1038/s41467-019-10180-6

Lu R, Nelson CB, Rogers S, Cesare AJ, Sobinoff AP, Pickett HA. 2024. Distinct modes of telomere synthesis and extension contribute to alternative lengthening of telomeres. *iScience* **27**: 108655. doi:10.1016/j.isci.2023.108655

Lundblad V, Blackburn EH. 1993. An alternative pathway for yeast telomere maintenance rescues est1⁻ senescence. *Cell* **73**: 347–360. doi:10.1016/0092-8674(93)90234-H

Lydeard JR, Jain S, Yamaguchi M, Haber JE. 2007. Break-induced replication and telomerase-independent telomere maintenance require Pol32. *Nature* **448**: 820–823. doi:10.1038/nature06047

Lydeard JR, Lipkin-Moore Z, Sheu Y-J, Stillman B, Burgers PM, Haber JE. 2010. Break-induced replication requires all essential DNA replication factors except those specific for pre-RC assembly. *Genes Dev* **24**: 1133–1144. doi:10.1101/gad.1922610

Mazzucco G, Huda A, Galli M, Piccini D, Giannattasio M, Pessina F, Doksani Y. 2020. Telomere damage induces internal loops that generate telomeric circles. *Nat Commun* **11**: 5297. doi:10.1038/s41467-020-19139-4

McDowell TL, Gibbons RJ, Sutherland H, O'Rourke DM, Bickmore WA, Pombo A, Turley H, Gatter K, Picketts DJ, Buckle VJ, et al. 1999. Localization of a putative transcriptional regulator (ATRX) at pericentromeric heterochromatin and the short arms of acrocentric chromosomes. *Proc Natl Acad Sci* **96**: 13983–13988. doi:10.1073/pnas.96.24.13983

Min J, Wright WE, Shay JW. 2017. Alternative lengthening of telomeres can be maintained by preferential elongation of lagging strands. *Nucleic Acids Res* **45**: gkw1295. doi:10.1093/nar/gkw1295

Min J, Wright WE, Shay JW. 2019. Clustered telomeres in phase-separated nuclear condensates engage mitotic DNA synthesis through BLM and RAD52. *Genes Dev* **33**: 814–827. doi:10.1101/gad.324905.119

Nabetani A, Ishikawa F. 2009. Unusual telomeric DNAs in human telomerase-negative immortalized cells. *Mol Cell Biol* **29**: 703–713. doi:10.1128/MCB.00603-08

Nergadze SG, Farnung BO, Wischnewski H, Khoriauli L, Vitelli V, Chawla R, Giulotto E, Azzalin CM. 2009. CpG-island promoters drive transcription of human telomeres. *RNA* **15**: 2186–2194. doi:10.1261/rna.1748309

Nguyen DT, Voon HPJ, Xella B, Scott C, Clynes D, Babbs C, Ayyub H, Kerry J, Sharpe JA, Sloane-Stanley JA, et al. 2017. The chromatin remodelling factor ATRX suppresses R-loops in transcribed telomeric repeats. *EMBO Rep* **18**: 914–928. doi:10.15252/embr.201643078

Oganesian L, Karlseder J. 2011. Mammalian 5′ C-rich telomeric overhangs are a mark of recombination-dependent telomere maintenance. *Mol Cell* **42**: 224–236. doi:10.1016/j.molcel.2011.03.015

O'Sullivan RJ, Arnoult N, Lackner DH, Oganesian L, Haggblom C, Corpet A, Almouzni G, Karlseder J. 2014. Rapid induction of alternative lengthening of telomeres by depletion of the histone chaperone ASF1. *Nat Struct Mol Biol* **21**: 167–174. doi:10.1038/nsmb.2754

Ouyang J, Yadav T, Zhang J-M, Yang H, Rheinbay E, Guo H, Haber DA, Lan L, Zou L. 2021. RNA transcripts stimulate homologous recombination by forming DR-loops. *Nature* **594**: 283–288. doi:10.1038/s41586-021-03538-8

Pan X, Drosopoulos WC, Sethi L, Madireddy A, Schildkraut CL, Zhang D. 2017. FANCM, BRCA1, and BLM cooperatively resolve the replication stress at the ALT telomeres. *Proc Natl Acad Sci* **114**: E5940–E5949. doi:10.1073/pnas.1708065114

Pan X, Chen Y, Biju B, Ahmed N, Kong J, Goldenberg M, Huang J, Mohan N, Klosek S, Parsa K, et al. 2019. FANCM suppresses DNA replication stress at ALT telomeres by disrupting TERRA R-loops. *Sci Rep* **9:** 19110. doi:10.1038/s41598-019-55537-5

Panier S, Maric M, Hewitt G, Mason-Osann E, Gali H, Dai A, Labadorf A, Guervilly J-H, Ruis P, Segura-Bayona S, et al. 2019. SLX4IP antagonizes promiscuous BLM activity during ALT maintenance. *Mol Cell* **76:** 27–43.e11. doi:10.1016/j.molcel.2019.07.010

Potts PR, Yu H. 2007. The SMC5/6 complex maintains telomere length in ALT cancer cells through SUMOylation of telomere-binding proteins. *Nat Struct Mol Biol* **14:** 581–590. doi:10.1038/nsmb1259

Rizki A, Lundblad V. 2001. Defects in mismatch repair promote telomerase-independent proliferation. *Nature* **411:** 713–716. doi:10.1038/35079641

Roumelioti F, Sotiriou SK, Katsini V, Chiourea M, Halazonetis TD, Gagos S. 2016. Alternative lengthening of human telomeres is a conservative DNA replication process with features of break-induced replication. *EMBO Rep* **17:** 1731–1737. doi:10.15252/embr.201643169

Roy S, LaFramboise WA, Liu T-C, Cao D, Luvison A, Miller C, Lyons MA, O'Sullivan RJ, Zureikat AH, Hogg ME, et al. 2018. Loss of chromatin-remodeling proteins and/or CDKN2A associates with metastasis of pancreatic neuroendocrine tumors and reduced patient survival times. *Gastroenterology* **154:** 2060–2063.e8.

Saini N, Ramakrishnan S, Elango R, Ayyar S, Zhang Y, Deem A, Ira G, Haber JE, Lobachev KS, Malkova A. 2013. Migrating bubble during break-induced replication drives conservative DNA synthesis. *Nature* **502:** 389–392. doi:10.1038/nature12584

Sakellariou D, Bak ST, Isik E, Barroso SI, Porro A, Aguilera A, Bartek J, Janscak P, Peña-Diaz J. 2022. MutSβ regulates G4-associated telomeric R-loops to maintain telomere integrity in ALT cancer cells. *Cell Rep* **39:** 110602. doi:10.1016/j.celrep.2022.110602

Sakofsky CJ, Ayyar S, Deem AK, Chung W-H, Ira G, Malkova A. 2015. Translesion polymerases drive microhomology-mediated break-induced replication leading to complex chromosomal rearrangements. *Mol Cell* **60:** 860–872. doi:10.1016/j.molcel.2015.10.041

Schrank BR, Aparicio T, Li Y, Chang W, Chait BT, Gundersen GG, Gottesman ME, Gautier J. 2018. Nuclear ARP2/3 drives DNA break clustering for homology-directed repair. *Nature* **559:** 61–66. doi:10.1038/s41586-018-0237-5

Schwartzentruber J, Korshunov A, Liu X-Y, Jones DTW, Pfaff E, Jacob K, Sturm D, Fontebasso AM, Quang D-AK, Tönjes M, et al. 2012. Driver mutations in histone H3.3 and chromatin remodelling genes in paediatric glioblastoma. *Nature* **482:** 226–231. doi:10.1038/nature10833

Silva B, Pentz R, Figueira AM, Arora R, Lee YW, Hodson C, Wischnewski H, Deans AJ, Azzalin CM. 2019. FANCM limits ALT activity by restricting telomeric replication stress induced by deregulated BLM and R-loops. *Nat Commun* **10:** 2253. doi:10.1038/s41467-019-10179-z

Silva B, Arora R, Bione S, Azzalin CM. 2021. TERRA transcription destabilizes telomere integrity to initiate break-induced replication in human ALT cells. *Nat Commun* **12:** 3760. doi:10.1038/s41467-021-24097-6

Silva B, Arora R, Azzalin CM. 2022. The alternative lengthening of telomeres mechanism jeopardizes telomere integrity if not properly restricted. *Proc Natl Acad Sci* **119:** e2208669119. doi:10.1073/pnas.2208669119

Singhi AD, Liu T-C, Roncaioli JL, Cao D, Zeh HJ, Zureikat AH, Tsung A, Marsh JW, Lee KK, Hogg ME, et al. 2017. Alternative lengthening of telomeres and loss of DAXX/ATRX expression predicts metastatic disease and poor survival in patients with pancreatic neuroendocrine tumors. *Clin Cancer Res* **23:** 600–609. doi:10.1158/1078-0432.CCR-16-1113

Smith CE, Llorente B, Symington LS. 2007. Template switching during break-induced replication. *Nature* **447:** 102–105. doi:10.1038/nature05723

Sobinoff AP, Allen JA, Neumann AA, Yang SF, Walsh ME, Henson JD, Reddel RR, Pickett HA. 2017. BLM and SLX4 play opposing roles in recombination-dependent replication at human telomeres. *EMBO J* **36:** 2907–2919. doi:10.15252/embj.201796889

Teixeira MT, Arneric M, Sperisen P, Lingner J. 2004. Telomere length homeostasis is achieved via a switch between telomerase-extendible and -nonextendible states. *Cell* **117:** 323–335. doi:10.1016/S0092-8674(04)00334-4

Teng S-C, Zakian VA. 1999. Telomere-telomere recombination is an efficient bypass pathway for telomere maintenance in *Saccharomyces cerevisiae*. *Mol Cell Biol* **19:** 8083–8093. doi:10.1128/MCB.19.12.8083

Teng Y-C, Sundaresan A, O'Hara R, Gant VU, Li M, Martire S, Warshaw JN, Basu A, Banaszynski LA. 2021. ATRX promotes heterochromatin formation to protect cells from G-quadruplex DNA-mediated stress. *Nat Commun* **12:** 3887. doi:10.1038/s41467-021-24206-5

Thosar SA, Barnes RP, Detwiler A, Bhargava R, Wondisford A, O'Sullivan RJ, Opresko PL. 2024. Oxidative guanine base damage plays a dual role in regulating productive ALT-associated homology-directed repair. *Cell Rep* **43:** 113656. doi:10.1016/j.celrep.2023.113656

Valle-García D, Qadeer ZA, McHugh DS, Ghiraldini FG, Chowdhury AH, Hasson D, Dyer MA, Recillas-Targa F, Bernstein E. 2016. ATRX binds to atypical chromatin domains at the 3′ exons of zinc finger genes to preserve H3K9me3 enrichment. *Epigenetics* **11:** 398–414. doi:10.1080/15592294.2016.1169351

Verma P, Dilley RL, Zhang T, Gyparaki MT, Li Y, Greenberg RA. 2019. RAD52 and SLX4 act nonepistatically to ensure telomere stability during alternative telomere lengthening. *Genes Dev* **33:** 221–235. doi:10.1101/gad.319723.118

Voon HPJ, Hughes JR, Rode C, De La Rosa-Velázquez IA, Jenuwein T, Feil R, Higgs DR, Gibbons RJ. 2015. ATRX plays a key role in maintaining silencing at interstitial heterochromatic loci and imprinted genes. *Cell Rep* **11:** 405–418. doi:10.1016/j.celrep.2015.03.036

Wellinger RJ, Zakian VA. 2012. Everything you ever wanted to know about *Saccharomyces cerevisiae* telomeres: beginning to end. *Genetics* **191:** 1073–1105. doi:10.1534/genetics.111.137851

Wilson MA, Kwon Y, Xu Y, Chung W-H, Chi P, Niu H, Mayle R, Chen X, Malkova A, Sung P, et al. 2013. Pif1 helicase and Polδ promote recombination-coupled DNA synthesis via bubble migration. *Nature* **502:** 393–396. doi:10.1038/nature12585

Cite this article as *Cold Spring Harb Perspect Biol* doi: 10.1101/cshperspect.a041690

Wyatt HDM, West SC. 2014. Holliday junction resolvases. *Cold Spring Harb Perspect Biol* **6:** a023192. doi:10.1101/cshperspect.a023192

Wyatt HDM, Sarbajna S, Matos J, West SC. 2013. Coordinated actions of SLX1-SLX4 and MUS81-EME1 for Holliday junction resolution in human cells. *Mol Cell* **52:** 234–247. doi:10.1016/j.molcel.2013.08.035

Xu Z, Fallet E, Paoletti C, Fehrmann S, Charvin G, Teixeira MT. 2015. Two routes to senescence revealed by real-time analysis of telomerase-negative single lineages. *Nat Commun* **6:** 7680. doi:10.1038/ncomms8680

Yadav T, Zhang J-M, Ouyang J, Leung W, Simoneau A, Zou L. 2022. TERRA and RAD51AP1 promote alternative lengthening of telomeres through an R- to D-loop switch. *Mol Cell* **82:** 3985–4000.e4. doi:10.1016/j.molcel.2022.09.026

Yeager TR, Neumann AA, Englezou A, Huschtscha LI, Noble JR, Reddel RR. 1999. Telomerase-negative immortalized human cells contain a novel type of promyelocytic leukemia (PML) body. *Cancer Res* **59:** 4175–4179.

Zelceski A, Francica P, Lingg L, Mutlu M, Stok C, Liptay M, Alexander J, Baxter JS, Brough R, Gulati A, et al. 2023. MND1 and PSMC3IP control PARP inhibitor sensitivity in mitotic cells. *Cell Rep* **42:** 112484. doi:10.1016/j.celrep.2023.112484

Zhang J-M, Yadav T, Ouyang J, Lan L, Zou L. 2019. Alternative lengthening of telomeres through two distinct break-induced replication pathways. *Cell Rep* **26:** 955–968.e3. doi:10.1016/j.celrep.2018.12.102

Zhang H, Zhao R, Tones J, Liu M, Dilley RL, Chenoweth DM, Greenberg RA, Lampson MA. 2020. Nuclear body phase separation drives telomere clustering in ALT cancer cells. *Mol Biol cell* **31:** 2048–2056. doi:10.1091/mbc.E19-10-0589

Zhang J-M, Genois M-M, Ouyang J, Lan L, Zou L. 2021. Alternative lengthening of telomeres is a self-perpetuating process in ALT-associated PML bodies. *Mol Cell* **81:** 1027–1042.e4. doi:10.1016/j.molcel.2020.12.030

Zhang T, Rawal Y, Jiang H, Kwon Y, Sung P, Greenberg RA. 2023. Break-induced replication orchestrates resection-dependent template switching. *Nature* **619:** 201–208. doi:10.1038/s41586-023-06177-3

# Therapeutic Opportunities for Alternative Lengthening of Telomeres (ALT) Cancers

Jixuan Gao and Hilda A. Pickett

Telomere Length Regulation Unit, Children's Medical Research Institute, Faculty of Medicine and Health, University of Sydney, Westmead NSW 2145, Australia

*Correspondence:* hpickett@cmri.org.au

Cancers that rely on activation of the alternative lengthening of telomeres (ALT) pathway predominantly affect children and adolescents, and are associated with catastrophic outcomes due to a lack of clinically effective, targeted therapeutics. The exponential rise in our understanding of the ALT mechanism in recent years has led to the identification of many therapeutic targets and strategies for patients suffering from these cancers. These include targeting replication fork remodelers and DNA damage response pathways to exacerbate telomere-specific replication stress, inhibiting ALT-mediated telomere synthesis to induce telomere dysfunction, and using oncolytic viruses to selectively kill ALT cancer cells. Herein we will evaluate the advantages and shortfalls of these therapeutic strategies, and discuss current diagnostic opportunities that are a necessary accompaniment to direct ALT therapeutics to patients.

Alternative lengthening of telomeres (ALT) is one of two characterized mechanisms of telomere length maintenance adopted by cancer cells to establish replicative immortality. Importantly, there is no evidence of ALT activity in normal human cells. Clinical surveys indicate that ALT is active in 10%–15% of all cancers, but this may be challenged by improved clinical diagnostics for ALT detection (Dilley and Greenberg 2015). The prevalence of ALT is substantially elevated in tumors of mesenchymal or neuroepithelial cell origins, such as bone and soft tissue sarcomas and neuroendocrine tumors of the pancreas. ALT is found in more than 60% of osteosarcoma, leiomyosarcoma, undifferentiated pleiomorphic sarcoma, and astrocytoma (MacKenzie et al. 2021). While these tumor types are typically defined as rare cancers, which command limited commercial interest, many disproportionately affect young people, are aggressive with poor outcomes, and have treatment regimens that have remained stagnant over several decades, all of which contribute to an exceptionally high burden of disease. Furthermore, there is potential for ALT to be active in subpopulations of cells within a tumor or in metastatic cancers treated with antitelomerase therapies, emphasizing the need for ALT-targeted therapeutics (Henson et al. 2002; Gocha et al. 2013; Recagni et al. 2020).

There are currently no ALT-specific targets or therapeutic options for ALT-positive cancers, and, consequently, ALT activity is not considered in cancer diagnosis and prognosis. However, we are at the cusp of a change, and new telomere maintenance mechanism-specific targets are emerging and being rapidly exploited for future precision treatment approaches.

## THE ALT MECHANISM

ALT is a homology-directed repair (HDR) pathway in which a variety of homologous recombination (HR) mechanisms become deployed at telomeres with the cumulative effect of telomere extension to achieve telomere length maintenance. ALT telomeres are structurally distinct. This is attributed to frequent loss-of-function mutations in the α-thalassemia/mental retardation X-linked (ATRX)/death-domain-associated (DAXX) histone chaperone complex, the interspersion of telomere variant repeats throughout the telomere repeat array, altered nucleoprotein stoichiometry caused by the recruitment of proteins such as nuclear receptors and chromatin remodelers that displace the shelterin telomere-binding complex, and modification of telomeric chromatin through changes in histone constituents and nucleosome density (Baird et al. 1995; Conomos et al. 2012; de Wilde et al. 2012; Lee et al. 2014; Li et al. 2019; Zhang and Zou 2020; Lu and Pickett 2022). In addition, telomeric DNA is transcribed from subtelomeric regions of the chromosomes into telomeric repeat-containing RNA (TERRA) that can invade the telomere of origin in *cis*, or other telomeres in *trans* (Chu et al. 2017). Invasion of TERRA results in the formation of RNA–DNA hybrid structures including R-loops and displacement-loops (D-loops), which are a source of replication stress. TERRA levels are elevated in ALT-positive cancer cells in comparison to telomerase-positive cell lines (Azzalin et al. 2007; Ng et al. 2009; Arora et al. 2014), and triggering TERRA transcription increases the formation of RNA–DNA hybrids specifically at ALT telomeres (Arora et al. 2014). These properties have a combined impact on ALT cells by conferring a heightened level of telomere-specific replication stress. This is an important and fundamental distinction of ALT telomeres.

Despite the negative connotations associated with heightened levels of replication stress,

it is this inherent damage that propagates ALT activity as a mechanism of DNA repair through homology-directed telomere extension. Specifically, failure to alleviate telomere replication stress through fork regression and replication restart results in the deterioration of stalled forks to form DNA double-strand breaks (DSBs) (Cox Kelli et al. 2016; Lu et al. 2019; Brenner and Nandakumar 2022). Resected DSBs then facilitate homology-directed searches, strand invasion, and telomere synthesis by a mechanism analogous to break-induced replication (BIR), known as break-induced telomere synthesis (BITS) (Dilley et al. 2016; Sobinoff and Pickett 2020). This presents a catch-22, whereby telomeric replication stress is the cause of debilitating DNA damage, but is also the essential instigator of ALT activity. How these two outcomes are managed is pertinent to ALT cell survival. Despite enduring telomere dysfunction and replication stress, ALT telomeres are not prone to end-to-end fusions and the associated lethal consequences (Cesare et al. 2009).

Other key components of the ALT mechanism that are relevant for therapeutic targeting include the presence of a telomere-associated subset of promyelocytic leukemia (PML) nuclear bodies called ALT-associated PML bodies (APBs), telomere length heterogeneity attributed to stochastic BITS events, prevalent telomere sister-chromatid exchange events (T-SCEs) that form following crossover resolution of HR-intermediates, and abundant extrachromosomal telomeric repeat (ECTR) DNA that is generated as a consequence of the mechanism. As our understanding of the ALT mechanism becomes more sophisticated, it is clear that the defined phenotypic characteristics of ALT cells and the inherent structural aberrations that promote recombination-mediated repair at ALT telomeres make direct pharmacological targeting of ALT cancers a tangible possibility.

## ALT DIAGNOSTICS

For ALT-targeted therapeutics to succeed, a reliable companion diagnostic is required to identify ALT activity, and match patient to treatment. A variety of potential assays exist. Loss of ATRX/DAXX protein function is a common and widespread feature of ALT cancers (Clatterbuck Soper and Meltzer 2023). DAXX is a histone chaperone that deposits histone variant H3.3 into chromatin. When acting in concert with ATRX, an SWI/SNF-like chromatin remodeling protein, this function is targeted to telomeres (Lewis et al. 2010). Loss of ATRX/DAXX reduces the deposition of H3.3, thereby destabilizing telomeric chromatin, and promoting ALT activity (Zhang and Zou 2020). Immunohistochemistry (IHC) to detect loss of nuclear localization of ATRX or DAXX can be used as a reliable diagnostic tool for the detection of ALT-positive cancers (Heaphy and Singhi 2022). Loss of nuclear ATRX/DAXX is tightly correlated with ALT activity in pancreatic neuroendocrine tumors (PanNETs) (Heaphy et al. 2011), and in isocitrate dehydrogenase (IDH) 1-mutant astrocytomas, a form of glioma in which ALT is used as the dominant TMM (Mellai et al. 2017; Ferreira et al. 2020; Gritsch et al. 2022).

Loss-of-function mutations in the ATRX/DAXX genes can also abolish protein function, making genetic testing for ATRX/DAXX mutations another potential method to identify ALT status. Mutations include frameshift insertions/deletions, introduction of premature stop codons, and missense single-nucleotide variants (SNVs) (Schwartzentruber et al. 2012). ATRX and DAXX are included in several multigene cancer panel sequencing tests, including the UW-OncoPlex Cancer Gene Panel and MSK-IMPACT, making this option immediately clinically applicable. The clinical significance of ATRX/DAXX mutations is, again, well-established in PanNETS, with a strong correlation between ATRX, and particularly DAXX, mutations, and ALT status, reduced patient survival, and metastatic disease (Hong et al. 2020; Gisder et al. 2023). Other malignancies with strong associations between ATRX mutations and ALT activity include glioblastoma multiforme (GBM) and neuroblastoma (Kurihara et al. 2014). Furthermore, ATRX/DAXX mutations have been used to categorize ALT status in several clinical studies, including phase I trials of PARP and ATR inhibitors in advanced solid tumors, multiomics characterization of tumor

subgroups, and a phase II trial of the WEE1 inhibitor adavosertib with irinotecan for relapsed neuroblastoma, medulloblastoma, and rhabdomyosarcoma (NCT04170153 [clinicaltrials.gov/study/NCT04170153]; NCT05076513 [clinicaltrials.gov/study/NCT05076513]; NCT05234450 [clinicaltrials.gov/study/NCT05234450]; NCT05687136 [clinicaltrials.gov/study/NCT05687136]) (Cole et al. 2023).

Despite the correlation between ATRX/DAXX mutations and ALT activity in some tumor types, this is not universal across all malignancies (Heaphy et al. 2011; Ferreira et al. 2020; Gritsch et al. 2022; Mori et al. 2024). Other genetic mutations that have been found to be associated with ALT activity include loss of the replication fork remodeler SMARCAL1 that is particularly associated with ALT activity in glioblastomas (Diplas et al. 2018; Brosnan-Cashman et al. 2021), and loss of the Holliday junction (HJ) cleavage protein SLX4IP (Panier et al. 2019). While only a proportion of ALT-positive tumors exhibit inactivating mutations in ATRX/DAXX, SMARCAL1, and SLX4IP, mutations in these genes may contribute to a future genetic signature of ALT. Machine learning approaches have had relative success in categorizing ALT activity based on genomic readouts, with major features including telomere repeat content and the presence and positioning of telomere variant repeats, tumor-associated genetic mutations including ATRX/DAXX and hTERT promoter mutations, and the presence of ALT-associated telomere fusions (ALT-TFs) (Lee et al. 2018; Sieverling et al. 2020; de Nonneville and Reddel 2021; Muyas et al. 2024). ALT-TFs involve the fusion of small DNA fragments with microhomology at their ends, appear to be well tolerated in ALT cells, and are distinct from conventional end-to-end fusion events (Cesare et al. 2009; Muyas et al. 2024).

As sequencing technologies develop and long-read sequencing becomes increasingly used in the telomere field, diagnostic tests that specifically focus on telomere length heterogeneity are likely to hold promise. Oxford Nanopore Technology (ONT) and Pacific Biosciences (PacBio) sequencing platforms can accurately quantify telomere length at the end of each individual chromosome, and improved sequencing error rates are providing further insight into the heterogeneity of telomere sequences throughout the telomere repeat array (Kim et al. 2021; Sholes et al. 2022; Tham et al. 2023; Karimian et al. 2024; Sanchez et al. 2024; Schmidt et al. 2024). Comparing telomere length distribution patterns using telomere profiling has been used successfully to identify ALT activity, and other comparisons of chromosome end-specific telomere length variations, for instance, p:q arm length ratios, have utility in the detection of ALT (Perrem et al. 2001; Schmidt et al. 2024).

Native or nondenaturing telomere fluorescence in situ hybridization (ALT-FISH) uses a fluorescent telomere probe that, in the absence of heat-induced denaturation, detects single-stranded telomeric DNA, including ECTRs (Loe et al. 2020; Claude et al. 2021; Frank et al. 2022; Azeroglu et al. 2024). ALT-FISH offers several advantages as a clinical diagnostic over other techniques in that it does not require DNA extraction, can be performed in a high-throughput 384-well plate format, and provides single-cell data that inform about tumor heterogeneity (Claude et al. 2021; Frank et al. 2022; Azeroglu et al. 2024). Also clinically relevant is the C-circle assay, which relies on the presence of extrachromosomal C-circles to serve as templates for rolling circle amplification, catalyzed by Phi polymerase (Henson et al. 2009). The products of this isothermic polymerase reaction are then quantified using a $^{32}$P- or DIG-labeled telomeric probe or by telomere quantitative PCR (Henson et al. 2009; Lau et al. 2013; Idilli et al. 2021). The C-circle assay is highly sensitive, and amenable to high-throughput adaptations. In addition, C-circles are secreted, making them a viable blood-based biomarker (Henson et al. 2009).

Robust single-molecule strategies and techniques that can be applied to liquid biopsies for accurate ALT detection are a priority, and a necessity in the progress and application of ALT therapeutics.

## SYNTHETIC LETHAL INTERACTIONS WITH ALT

One of the major vulnerabilities that can be exploited for the treatment of ALT-positive cancers is the elevated levels of replication stress that are inherent to ALT telomeres (Fig. 1). Synthetic lethal approaches that exacerbate telomere-specific replication stress by targeting replication fork remodelers, TERRA, or the DNA damage response (DDR), represent exciting therapeutic opportunities, and drug development in this area is gaining traction.

## Targeting Replication Fork Remodelers

The Fanconi anemia complementation group M (FANCM) protein is currently the most promising therapeutic target for ALT-positive cancers. FANCM is an ATP-dependent fork remodeling enzyme that is recruited to stalled replication forks by FAAP24 and MHF1/2 to trigger fork regression, while promoting branch migration and the displacement of R-loops and D-loops (Ciccia et al. 2007; Gari et al. 2008; Yan et al. 2010; Coulthard et al. 2013; Schwab et al. 2015; Ling et al. 2016; Huang et al. 2019; O'Rourke et al. 2019; Silva et al. 2019). FANCM plays a critical role in the management of replication stress at ALT telomeres, and FANCM depletion causes catastrophic levels of telomere damage specifically in ALT cells (Lu et al. 2019; Pan et al. 2019; Silva et al. 2019). ALT cell dependency on FANCM is further evident by FANCM being the standout essential gene in ALT cell lines from Project Achilles hosted on the Cancer Dependency Map Portal (Lu and Pickett 2022; DepMap 2024). The ability of FANCM to remodel stalled forks is dependent on DNA binding, as well as its ATP-dependent translocase activity and its protein–protein interaction with the RMI1/2 subcomplex of BLM/TOP3A/RMI1/2 (BTR), providing multiple opportunities for drug development (Fig. 1).

The first FANCM–RMI1/2 protein–protein interaction inhibitor identified from a high-throughput small molecule screen was PIP-199 (Fig. 1; Table 1; Voter et al. 2016; Lu et al. 2019; Yang et al. 2021b). More recently, PIP-199 and its analogs were found to be chemically unstable in aqueous solutions, and failed to show any binding to the RMI1/2 complex (Wu et al. 2023). Consequently, the effects of PIP-199 on ALT activity and ALT cell viability are likely to

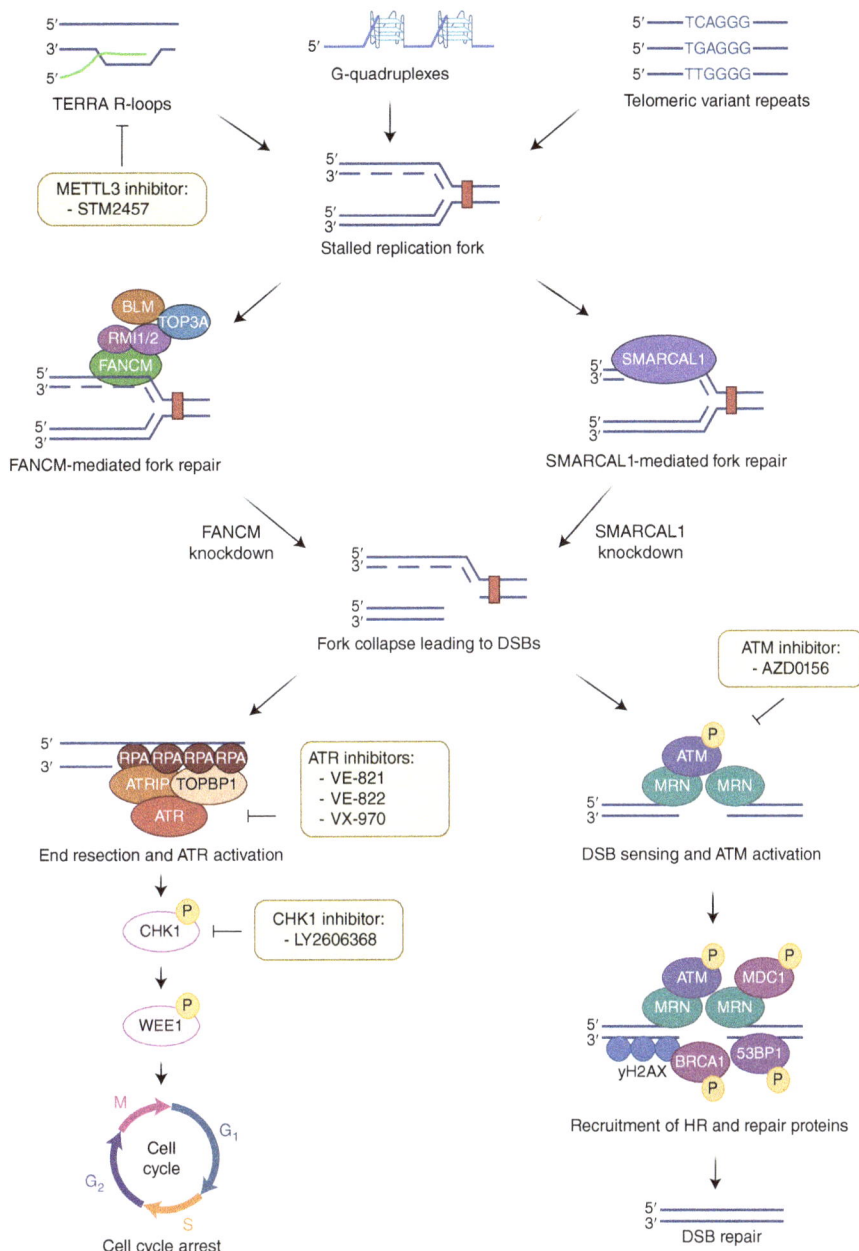

**Figure 1.** Synthetic lethal interactions with alternative lengthening of telomeres (ALT). Variant repeats and secondary structures formed in ALT telomeres enhance replication stress. The STM2457 METTL3 inhibitor impairs TERRA R/D-loop formation to trigger telomere instability and cell death. Replication stress manifests as stalled replication forks, a common occurrence in ALT cancers. Stalled forks can be repaired by the replication fork remodeler SMARCAL1 or by FANCM, with its partner the BLM-TOP3A-RMI1/2 (BTR) complex. In the absence of FANCM or SMARCAL1-mediated fork repair, the lesion deteriorates into a double-strand break (DSB) that can be repaired either by ATR or ATM pathway activation. In the ATR-mediated repair pathway, BLM-EXO1-DNA2 resects the DSB, allowing RPA coating of ssDNA, triggering recruitment of ATRIP, TOPBP1, and activation of ATR. ATR phosphorylates CHK1 and WEE1, which stalls cell cycle progression until repair is complete. ATR inhibition or CHK1 inhibition by molecules such as VE-821 or LY2606368, respectively, can stimulate inappropriate cell cycle progression, accumulation of unrepaired DNA damage, and cell death. In the ATM-mediated repair pathway, the DSB is detected by the MRN complex, which recruits and activates ATM. The ATM kinase then phosphorylates histone H2AX that then recruits MDC1, BRCA1, and 53BP1 to initiate DSB repair. AZD0156 inhibits ATM-mediated DSB repair. (Figure generated with BioRender, https://biorender.com/.)

**Table 1.** Potential agents for the treatment of alternative lengthening of telomeres (ALT)-positive cancers

| Agent | Class of agent | Mechanism of action | Stage of development |
|---|---|---|---|
| PIP-199 | Small molecule inhibitor (SMI) | Disrupts the Fanconi anemia complementation group M (FANCM)–BLM-TOP3A-RMI (BTR) interaction to induce toxic replication stress | Halted in preclinical studies due to stability issues |
| STM2457 | SMI | Inhibits METTL3 to impair telomere repeat-containing RNA (TERRA) R/D-loop formation, telomere instability, and cell death | Preclinical, in vitro |
| VE-821 | SMI | Inhibits ATR to block fork repair and trigger cell cycle progression | Preclinical, in vivo |
| VE-822 | SMI | Inhibits ATR to block fork repair and trigger cell cycle progression | Phase I clinical trial |
| LY2606368 | SMI | Inhibits CHK1, downstream target of ATR | Phase II clinical trial |
| AZD0156 | SMI | ATM inhibitor, blocks double-strand break (DSB) repair | Phase I clinical trial, in vivo efficacy for ALT |
| Trabectedin | SMI | Alkylating agent that binds minor groove of DNA to induce DSBs | FDA-approved, in vitro efficacy for ALT |
| Niraparib | SMI | PARP inhibitor that synergizes with ATRi, RP-3500; exact mechanism unknown | FDA-approved, in vivo efficacy for ALT |
| KU-60019 | SMI | ATM/DNA-PKcs inhibitor that synergizes with DNA synthesis inhibitor, triciribine; exact mechanism unknown | Preclinical, in vivo |
| AICAR | SMI | Inhibits DNA localization and ssDNA annealing functions of RAD52 | Preclinical, in vitro |
| CYT-0851 | SMI | RAD51 inhibitor | Phase II clinical trial |
| JQ1 | SMI | Inhibits BRD4-mediated RAD51AP1 transcription and protein expression | Preclinical, in vivo |
| Zelpolib | SMI | Inhibits Pol δ–mediated homologous recombination (HR) | Preclinical, in vitro |
| ML-792 | SMI | Inhibits small ubiquitin-like modifiers (SUMOs) E1 enzyme | Preclinical, in vitro efficacy for ALT |
| AO/854 | SMI | Inhibits Bloom (BLM)-mediated DNA repair | Preclinical, in vivo |
| 2-D08 | SMI | Inhibits SUMOylation to block the assembly of promyelocytic leukemia (PML) proteins into ALT-associated PML bodies (APBs) | Preclinical, in vitro efficacy for ALT |
| ICP0-null HSV-1 | Therapeutic virus | Replicates in α-thalassemia/mental retardation X-linked (ATRX)-negative cells trigger cell lysis | Preclinical, in vitro efficacy for ALT |

stem from off-target effects rather than from the dissociation of the FANCM-BTR complex (Lu et al. 2019). FANCM is currently the subject of several drug development campaigns to identify small-molecule inhibitors of this essential ALT protein.

Approaches to achieve toxic levels of telomere replication stress have sparked widespread interest across the ALT field, with additional proteins that elicit a replication stress response being identified as promising ALT targets. SMARCAL1 is a replication fork remodeler in the SWI/SNF family of proteins that has both translocase and helicase functions. SMARCAL1 promotes branch migration and fork reversal, and preferentially localizes to APBs in ALT cells, with loss of SMARCAL1 resulting in increased telomeric replication stress and ALT activity (Fig. 1; Bétous et al. 2012; Ciccia et al. 2012; Cox Kelli et al. 2016;

Halder et al. 2022). SMARCAL1 depletion elicits a more subdued induction of telomere replication stress than FANCM depletion (Poole et al. 2015), and loss-of-function mutations in SMARCAL1 have been shown to promote ALT activity in glioblastoma (Brosnan-Cashman et al. 2021; Liu et al. 2023). This is indicative of the levels of replication stress caused by loss of SMARCAL1 being insufficient to cause cell death, but rather having an ALT-promoting effect during tumorigenesis, an important caveat to be taken into account when considering replication stress modulators as ALT therapeutic targets.

## Targeting TERRA

RNA–DNA hybrids formed by transcribed TERRA and telomeric DNA create R-loops, which generate replication stress and sustain ALT activity through the initiation and execution of BIR. TERRA invasion into telomeres is mediated by RAD51 and RAD51AP1, and R-loops are stabilized by SUMO-SIM-mediated interactions between RAD51AP1 and the UAF1 complex (Kaminski et al. 2022). Interestingly, the formation of R-loops generates G-quadruplexes at telomeres, which can further stimulate ALT activity (Yadav et al. 2022). TERRA levels are elevated in ALT-positive cancer cells compared to telomerase-positive cells, making TERRA a potential therapeutic target for ALT cancers (Azzalin et al. 2007; Arora et al. 2014). R-loop repressors include RNaseH1, an endoribonuclease that selectively degrades the RNA component of the R-loop, and FANCM, which unwinds telomeric R-loops using its ATPase/translocase activity (Azzalin 2025). The therapeutic potential of targeting TERRA has been investigated using the laboratory-based transcription activator-like effector (T-TALE) TERRA repressor system. Using this system, inhibition of TERRA was found to decrease telomere replication stress and impair ALT-mediated telomere lengthening, resulting in an increased frequency of telomere signal-free ends (Arora et al. 2014; Silva et al. 2021). This study provides encouraging evidence to prompt further investigations into druggable targets to inhibit TERRA and R-loop formation.

One potential target is the methyltransferase METTL3. METTL3 catalyzes the $m^6A$ modification on TERRA, promoting TERRA stabilization and association with telomeric DNA (Fig. 1; Chen et al. 2022). Depletion of METTL3 reduces the $m^6A$ modification, leading to impaired RAD51 recruitment to telomeres, improper D-loop formation, telomere instability, and cell death (Chen et al. 2022; Vaid et al. 2024). From a clinical perspective, ALT-positive cancer cells had copy number gains of METTL3, and high METTL3 expression predicted poor outcomes in ALT-positive cancer patients (Vaid et al. 2024). Inhibition of METTL3 activity using the methyltransferase inhibitor STM2457 compromised the deposition of $m^6A$ in newly formed TERRA molecules and increased telomeric DNA damage (Table 1), and METTL3 suppression in ALT-positive SK-N-FI xenografts reduced tumor formation (Vaid et al. 2024). However, METTL3 has a range of targets other than TERRA, and plays a wider role in DNA repair. Therefore, it is unclear whether TERRA is the primary target, supporting further exploration of off-target effects and the identification of other TERRA-targeting agents.

## Targeting the DNA Damage Response

Targeting the DDR has been extensively explored as a means of cancer therapy, based on the rationale that cancer cell–specific defects in the DDR result in increased reliance on compensatory pathways that in turn represent vulnerabilities that can be targeted to confer tumor-specific toxicities. ALT cancers display a replication stress defect, and therefore have a heightened reliance on associated DNA repair pathways.

The ATM and ATR kinases play critical roles in orchestrating DNA repair pathways. ATR responds to a wide range of lesions including heavily resected ssDNA, DNA cross-links, base adducts, and stalled DNA polymerases that are common at sites of replication stress in ALT-positive cancer cells (Maréchal and

Zou 2013; Zeman and Cimprich 2014; Barnieh et al. 2021). During replication stress, resected DNA coated by RPA (ssDNA–RPA complexes) triggers activation of ATR. ATR then phosphorylates and activates CHK1 and WEE1 to block cell cycle progression (Fig. 1; Saldivar et al. 2017; Gupta et al. 2022). ATR also suppresses replication origin firing to prevent depletion of the nuclear pool of RPA (Toledo et al. 2013). ATR inhibitors are a rational strategy for the treatment of ALT cancers; however, evidence supporting the universal lethality of ATR inhibitors, including VE-821 and VE-822, across different ALT-positive cancer cell lines has been controversial (Fig. 1; Table 1; Flynn et al. 2015; Deeg et al. 2016; Laroche-Clary et al. 2020). The ATR inhibitor VX-970 is currently in clinical trials in combination with the immune checkpoint inhibitor avelumab for the treatment of DDR-deficient solid tumors. Enrollment in this study requires the mutational status of DDR genes including ATRX, RAD51, and FANCM to be assessed (NCT04266912 [clinicaltrials.gov/study/NCT04266912]), thereby enabling the evaluation of clinical outcome data for patients with ATRX mutations that have a high propensity to use ALT.

CHK1 acts downstream from ATR to block cell cycle progression and prevent the accumulation of unrepaired DNA damage through uncontrolled cell division (da Costa et al. 2023). After recruitment to DSBs by ATRIP and TOPBP1, ATR phosphorylates CHK1 (Fig. 1; Liu et al. 2000; Zhao and Piwnica-Worms 2001). Activated CHK1 then induces CDC25A degradation to prevent the phosphorylation and activation of CDK2, which results in cell cycle arrest to enable DNA repair (Xiao et al. 2003; Karlsson-Rosenthal and Millar 2006). Several CHK1 inhibitors, including the ATP-competitive CHK1 inhibitor prexasertib (LY2606368), have progressed to clinical trials, although their efficacy is yet to be demonstrated (Table 1). Given the presence of elevated telomeric replication stress in ALT cells, CHK1 inhibitors may trigger cell cycle progression before the repair of these stalled forks and elicit selective toxicity to ALT cancer cells, but this remains to be examined.

ATM responds to DSBs generated by collapsed replication forks, common in cells with elevated replication stress, making it an attractive target for ALT-positive cancer cells (Ammazzalorso et al. 2010). After its recruitment to DSBs by the MRN complex (MRE1, RAD50, and NBS1), ATM phosphorylates CHK2, which in turn inhibits CDC25A/C to block cell cycle progression (Fig. 1; Falck et al. 2001). ATM also phosphorylates histone H2AX to recruit a number of proteins such as MDC1, BRCA1, and 53BP1 to initiate DSB repair (Fig. 1; Stewart et al. 2003; Ciccia and Elledge 2010). ALT-positive neuroblastoma cells have been reported to have constitutively high ATM activation attributable to spontaneous telomere dysfunction. ATM activation was further shown to confer chemoresistance to temozolomide and irinotecan, and treatment with the ATM inhibitor AZD0156 was able to reverse this chemoresistance primarily in ALT cells (Table 1; Koneru et al. 2021). The role of ATM in other cancer types with high ALT prevalence remains to be determined.

## INHIBITION OF THE ALT MECHANISM

Inhibition of ALT-mediated telomere extension is predicted to cause gradual telomere erosion to a point at which the telomeres no longer fulfill their capping function. At this point, the cell will succumb to senescence or apoptosis. While it is difficult to predict the number of cell divisions it will take for a specific cell to reach the telomere dysfunction threshold, the heterogeneity of telomere lengths in ALT cells suggests that the very short telomeres may hit this threshold rapidly and that this may be sufficient for growth arrest. This lag phase of telomere shortening represents an inherent therapeutic weakness for both telomerase-positive and ALT-positive cancers, during which cellular resistance mechanisms may manifest. In addition, ALT-mediated telomere lengthening encompasses multiple extension and repair pathways, and it is unclear how compensatory and interchangeable these pathways are. Despite these mechanistic caveats, several major executors of the ALT mechanism may hold promise as ALT therapeutic targets.

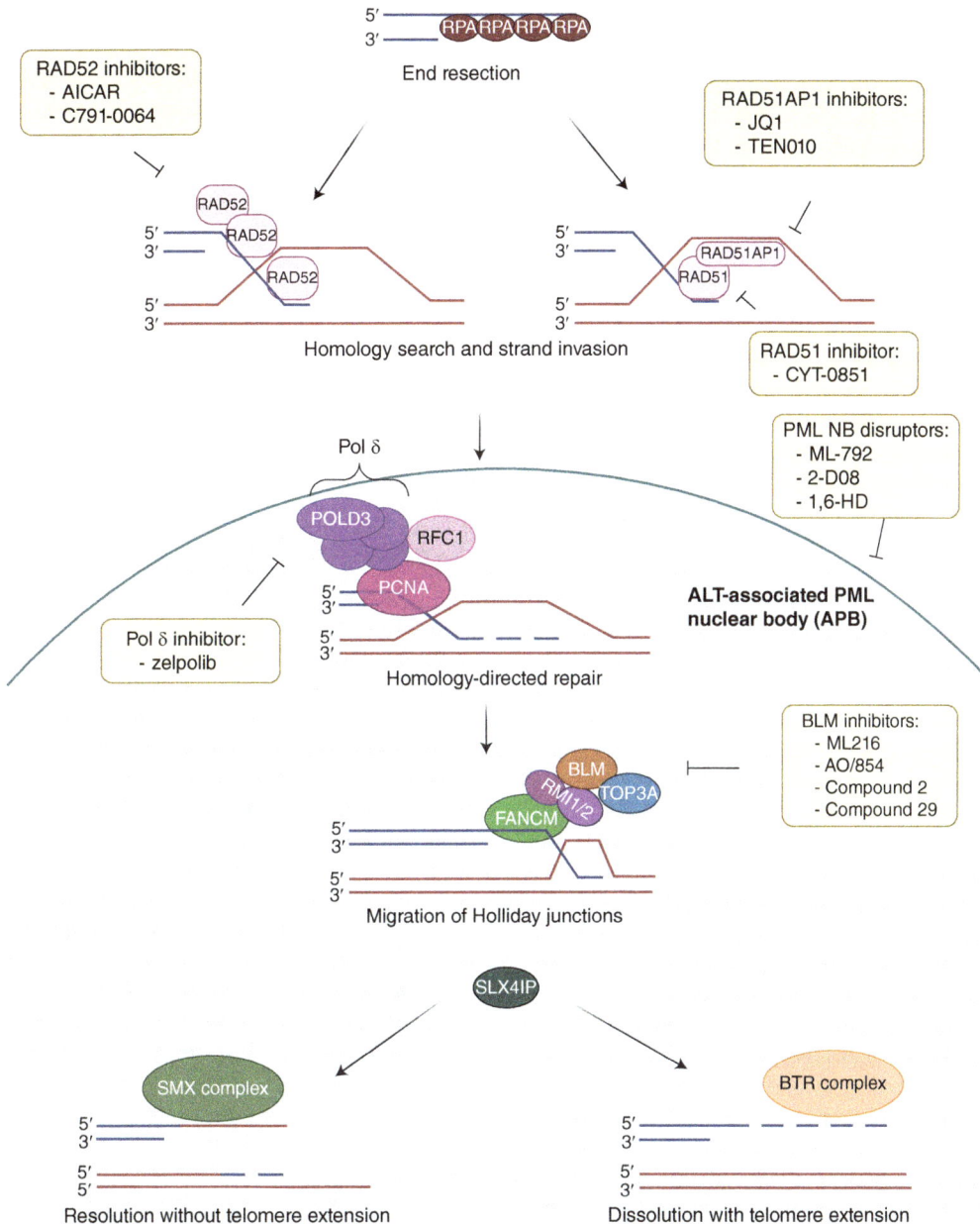

**Figure 2.** Inhibition of the alternative lengthening of telomeres (ALT) mechanism. Double-strand breaks (DSBs) at ALT telomeres can be resected by BLM-EXO1-DNA2, resulting in ssDNA that is coated by RPA. Strand invasion then occurs via either RAD52 (the dominant pathway) or RAD51 with its partner RAD51AP1. Following strand invasion, the RFC-PCNA-Pol δ replisome mediates homology-directed repair at ALT-associated promyelocytic leukemia (PML) nuclear bodies (APBs). The activity of Pol δ can be impaired by zelpolib, while the formation of PML nuclear bodies can be impaired using ML-792, 2-D08, and 1,6-HD. FANCM and its interaction with the BLM-TOP3A-RMI1-RMI2 (BTR) complex mediate the migration of Holliday junctions and telomere synthesis, which can be inhibited by various BLM inhibitors. SLX4IP then regulates BTR-mediated dissolution or SLX1-SLX4, MUS81-EME1, XPF-ERCC1 (SMX)-mediated resolution of recombination intermediates. (Figure generated with BioRender, https://biorender.com/.)

## Targeting Telomere Extension

ALT activity is primarily dependent on RAD52, but can switch to a RAD51-dependent pathway when RAD52 activity is compromised (Min et al. 2017, 2019; Sobinoff and Pickett 2017; Verma et al. 2019; Zhang et al. 2019). ALT-mediated telomere extension relies on the RFC-PCNA-Pol δ replisome and the BLM-TOP3A-RMI (BTR) complex for branch migration and dissolution (Dilley et al. 2016; Sobinoff et al. 2017). Concomitant resolution of recombination intermediates, involving the SLX1-SLX4, MUS81-EME1, XPF-ERCC1 (SMX) complex, also contributes to the ALT pathway (Fig. 2; Sobinoff et al. 2017).

Depletion of RAD52 leads to defective processing of telomere recombination intermediates and telomere loss, but is not lethal to ALT cells (Verma et al. 2019). This suggests that inhibition of either RAD51 or RAD52 alone will not be sufficient to trigger death in ALT cancer cells. Nevertheless, several RAD51 and RAD52 inhibitors exist. These include AMP-activated protein kinase agonist 5-aminoimidazole-4-carboxamide ribonucleotide (AICAR), a RAD52 selective inhibitor that reduces the localization of RAD52 to cisplatin-triggered DSBs and inhibits the single-strand annealing activity of RAD52 in the U-2 OS ALT-positive cell line (Sullivan et al. 2016), the flavonoid, (-)-epicatechin gallate, C791-0064, and small molecule inhibitors 103 and G23 (Fig. 2; Table 1; Hengel et al. 2016; Huang et al. 2016; Yang et al. 2021a). The RAD51 inhibitor CYT-0851 is in phase 1/2 clinical trials and has demonstrated a partial response in 30%–74% of patients with relapsed or refractory B-cell malignancies and advanced solid tumors (Ishida et al. 2009; Huang et al. 2011; Takaku et al. 2011; Balbous et al. 2016; Berte et al. 2016; Lindemann et al. 2021; Lynch et al. 2021; Schürmann et al. 2021; Scott et al. 2021). Investigation of these compounds as single or combination treatments in the context of ALT may provide therapeutic insight and opportunities.

RAD51-associated protein 1 (RAD51AP1) is an essential mediator of telomere length maintenance in ALT-positive cancer cells (Bar-roso-González et al. 2019). RAD51AP1 interacts with RAD51 to stimulate D-loop formation, and functions to repair DSBs that occur spontaneously or following treatment with DNA-damaging agents (Fig. 2; Wiese et al. 2007). ALT-positive cancer cells exhibit higher levels of RAD51AP1 attributable to MMS21-associated SUMOylation of RAD51AP1 (Barroso-González et al. 2019). Depletion of RAD51AP1 caused a reduction in ALT activity, increased telomere damage, and impaired telomere synthesis (Barroso-González et al. 2019). This was accompanied by telomere shortening and an accumulation of cytosolic ECTR fragments that were sensed by cGAS, ultimately targeting RAD51AP1-depleted cells for elimination by the engagement of autophagic cellular programs (Barroso-González et al. 2019). While there are currently no RAD51AP1 inhibitors, the RAD51AP1 promoter was recently found to be positively regulated by the bromodomain and extraterminal motif (BET) protein BRD4 (Ni et al. 2021). JQ1 is a thienotriazolodiazepine-based BET inhibitor that competitively binds to acetyl-lysine recognition motifs to displace BRD4 from superenhancer regions (Fig. 2; Table 1; Filippakopoulos et al. 2010; Wang et al. 2023). Treatment of cervical cancer cells with JQ1 reduced RAD51AP1 transcription and protein expression and caused potent radiosensitization, with RAD51AP1 identified as the major BRD4 target gene involved in radiosensitivity (Ni et al. 2021). This implicates BRD4 disruption as a means to down-regulate RAD51AP1 in ALT cancers. JQ1 has low oral bioavailability and poor pharmacokinetics (PK), hindering its clinical progress (Shorstova et al. 2021); however, its analog TEN010 has improved PK and has progressed to clinical trials, potentially warranting further investigation in the context of ALT (Fig. 2; Table 1).

POLD3 is an accessory subunit of the DNA polymerase enzyme Pol δ that plays a vital role in telomere extension by RAD51 and RAD52, suggesting that its inhibition may broadly block complementary pathways of ALT-associated telomere synthesis (Fig. 2). Consistent with this role, reducing the amount of functional POLD3 increased telomere dysfunction-induced foci (TIF) and reduced the production of telomeric DNA (Dilley et al. 2016; Zhang et al. 2019). There are two potential methods of inhibiting POLD3 activity from a therapeutic perspective. First, POLD3 can be ADP-ribosylated at S422 by PARP1 in response to replication stress (Richards et al. 2023). While wild-type POLD3 can rescue the BIR-impairment caused by depletion of POLD3, the S422A mutant POLD3 cannot, suggesting that ribosylation of this residue enhances the activity of Pol δ in BIR (Richards et al. 2023). This study implicates PARP inhibitors, many of which are in clinical trials or have been U.S. Food and Drug Administration (FDA)-approved, in abating Pol δ-mediated BIR in ALT-positive cancer cells. Second, Pol δ can be directly inhibited. Using structure-based in silico screens, zelpolib was identified as a noncompetitive inhibitor of the catalytic activity of Pol δ that reduces DNA synthesis and fiber lengths and exerts cell growth inhibition (Fig. 2; Table 1; Mishra et al. 2019). Further investigation of these approaches in ALT cells is required.

The Bloom helicase (BLM) plays an integral role in several stages of the ALT mechanism. This includes roles in DNA resection and fork remodeling during the initiation of ALT, branch migration and telomere extension, and in the dissolution of replication intermediates at later stages of the process (Fig. 2; Sobinoff et al. 2017; Lu et al. 2019). Depletion of BLM in ALT-positive cells decreases telomere extension events and other ALT phenotypes, but is not acutely toxic (Sobinoff et al. 2017). Several compounds chemically inhibit the helicase activity of BLM, including ML216, compound 2, compound 29, and AO/854 (Fig. 2; Table 1; Nguyen et al. 2013; Yin et al. 2019; Chen et al. 2021; Ma et al. 2022). The most effective of these is AO/854, which inhibits both the DNA binding and helicase activity of BLM to trigger DNA damage, and has also been shown to reduce tumorigenesis in vivo (Ma et al. 2022). The efficacy of BLM inhibitors may be further potentiated by inhibition of EXD2 activity. EXD2 is a $3'-5'$ exonuclease that localizes to APBs, where it processes replication forks for RAD52-dependent BIR (Broderick et al. 2023). Loss of EXD2 caused ALT-positive cancer cells to become reliant on RAD52-independent BIR, with codepletion of EXD2 and BLM conferring synthetic lethality in ALT cells (Broderick et al. 2023). While the efficacy of BLM inhibitors toward ALT cancers is likely to be overshadowed by the wide-ranging roles of BLM in global DNA repair, developing agents against targets such as EXD2 that are synthetic lethal with BLM inhibition may offer ALT-selective cytotoxicity.

Disrupting the resolution of telomere replication fork intermediates, which occur at later stages of ALT-associated BIR, can also pose toxicity to ALT cancer cells. SLX4 acts as a scaffold for endonucleases SLX1, MUS81-EME1, and XPF-ERCC1 to form the SMX complex, which functions to promote the resolution of recombination intermediates at ALT telomeres (Fig. 2; Sobinoff and Pickett 2017; Hoang and O'Sullivan 2020). The balance between BTR-mediated dissolution and SMX-mediated resolution of recombination intermediates is regulated by SLX4IP (Panier et al. 2019). Loss of SLX4IP increases ALT phenotypes and reduces the growth of ALT-positive cancer cells, and these cytotoxic effects are potentiated when combined with SLX4 knockdown (Panier et al. 2019).

## Targeting ALT-Associated PML Bodies

APBs are a subset of PML nuclear bodies that are specific to ALT cells. APBs are phase-separated clusters of telomeric DNA, telomere-binding proteins, DNA recombination and repair factors, and PML proteins, which are the primary sites of ALT-mediated telomere synthesis. Structurally, APBs consist of a core of telomeric DNA, surrounded by a shell formed by PML and Sp100 proteins (Grobelny et al. 2000). The covalent attachment of small ubiquitin-like modifiers (SUMOs) onto these proteins by the process of SUMOylation triggers APB assembly (Müller et al. 1998; Chung et al. 2011). While several PML protein isoforms exist, the main variant that supports ALT-mediated telomere synthesis is PML-IV (Zhang et al. 2021).

Strategies to target PML proteins or disrupt APB assembly and content have the potential to inhibit the ALT mechanism, although ALT-specific lethality is yet to be demonstrated. There are several options to achieve this, with the first being inhibition of PML protein. CRISPR knockout of PML disrupted the interaction between telomeres and the BTR complex, resulting in reduced C-circle production, loss of telomere length heterogeneity, and progressive telomere shortening (Loe et al. 2020). Targeting SUMOylation also provides a viable means by which to disrupt APB functionality. The small molecule inhibitor 2′,3′,4′-trihydroxy flavone (2-D08) inhibits SUMOylation of PML by the UBC9 SUMO ligase, a process that is required for the assembly of PML proteins into APBs (Fig. 2; Table 1; Kim et al. 2014; Sahin et al. 2014; Zhou et al. 2019). More recently, the SUMO E1 enzyme inhibitor ML-792 has been shown to impair SUMOylation in ALT-positive osteosarcoma cells, leading to reduced telomere synthesis and inhibition of cellular growth specifically in ALT-positive cancer cells (Fig. 2; Table 1; Zhao et al. 2024). It is also notable that the MMS21 SUMO ligase, part of the SMC5/6 complex, localizes to APBs in ALT cells and SUMOylates the shelterin components TRF1 and TRF2, with inhibition of TRF1 and TRF2 SUMOylation inhibiting APB formation and reducing ALT-associated HR, triggering telomere shortening and senescence in ALT cancer cells (Potts and Yu 2007).

Disruption of liquid–liquid phase separation may provide a further opportunity by which to disrupt the APB formation. The organic compound 1,6-hexanediol (1,6-HD) disrupts liquid–liquid phase separation and alters the compartmentalization of chromatin (Ulianov et al. 2021), yet the direct impact of this compound on APBs and ALT activity remains unclear.

## ONCOLYTIC VIRUSES

Research into oncolytic viruses as cancer therapeutics is gaining traction due to their potent ability to induce the lysis of tumor cells. Herpes simplex virus type 1 (HSV-1) is of particular interest because it can infect a wide range of cancer types, be treated using anti-herpetic drugs to control disease from infection, and its glycoproteins can be modified to target specific cancer cells (De Clercq 2004; Liu et al. 2006;

Sokolowski et al. 2015). Various mutations of this virus have entered clinical trials, with the HSV-1-derived oncolytic immunotherapy Talimogene laherparepvec (T-vec) receiving FDA approval for metastatic melanoma in 2015 (Andtbacka et al. 2015; Koch et al. 2020; Khushalani et al. 2023).

The rationale for using HSV-1 as a therapeutic for ALT cancers stems from the frequent ATRX deficiency associated with ALT (Ma et al. 2018; Han et al. 2019; Koch et al. 2020). ATRX and DAXX are constitutive components of PML nuclear bodies, which are the essential first line of defence during viral infection. In ALT cells, APBs are able to form in the absence of ATRX, creating a unique feature that can be exploited therapeutically. Specifically, wild-type HSV-1 degrades PML components including ATRX, but a mutant HSV-1 lacking ICP0 is unable to infect cells with intact PML bodies. This led to the identification of a synthetic lethal opportunity, in which an ICP0-null HSV-1 selectively infected and killed ATRX-deficient ALT cancer cells (Lukashchuk and Everett 2010; Han et al. 2019), with mutant HSV-1 being 10,000- to 1,000-fold more effective at infecting ATRX-deficient cells compared to ATRX-expressing cells (Han et al. 2019). Intriguingly, this study revealed that ATRX also functions to transcriptionally up-regulate PML and reduce PML proteasomal degradation (Han et al. 2019), indicative of the underlying complexity associated with PML body formation and function. While encouraging, further modifications to enhance viral selectivity, combined with extensive in vivo studies, are required for the clinical development of HSV-1 for the treatment of ALT cancers.

## DRUG REPOSITIONING

Many DNA-damaging agents are FDA-approved. Repurposing these drugs for the treatment of ALT cancers can expedite the development and review process, fast-tracking therapies to patients. Trabectedin is an FDA-approved chemotherapeutic used for the treatment of advanced soft tissue sarcoma and ovarian cancer. Mechanistically, trabectedin is an alkylating

Cite this article as *Cold Spring Harb Perspect Biol* doi: 10.1101/cshperspect.a041691

agent that induces the formation of DNA adducts that are stabilized by hydrogen bonds between trabectedin and nucleotides. This results in lesions similar to interstrand cross-links (ICLs), which, if unrepaired, will deteriorate into DSBs (Casado et al. 2008). ALT cell lines have been shown to be exceptionally sensitive to trabectedin (Pompili et al. 2017). However, despite the high prevalence of ALT in malignancies treated with trabectedin in the clinic, efficacy and outcomes remain to be correlated with ALT stratification.

Another class of agents that may exploit the excessive replication stress of ALT-positive cancers are inhibitors of the poly(ADP-ribose) polymerase (PARP) family of ADP-ribosyl transferases. The best-described enzyme of this 18-member family is PARP1, which adds poly (ADP)-ribose (PAR) chains to proteins and nucleic acids as part of the DDR (Ray Chaudhuri and Nussenzweig 2017; Wang et al. 2019; Groslambert et al. 2021). PARP inhibitors, such as olaparib, niraparib, rucaparib, and talazoparib, are FDA-approved for HR-deficient breast, pancreatic, and metastatic prostate cancers, and are thought to function by locking PARP onto DNA, inhibiting replication fork progression and heightening genomic replication stress (Colicchia et al. 2017; Ngoi et al. 2021; Li et al. 2023). PARP inhibitors are, therefore, predicted to exacerbate the inherent replication stress and DNA repair defect in ALT-positive cancer cells. In reality, PARP inhibitors as single agents elicit indiscriminate lethality against both ALT-positive and telomerase-positive cancer cells, presumably due to their diverse roles in DNA replication and repair (Hoang et al. 2020). However, selectivity against ALT-positive cancers was achieved when PARP inhibitors were administered in combination with ATR inhibitors (Zimmermann et al. 2022).

Unlike targeting specific DNA damage repair pathway components to elicit rational synthetic lethal interactions with ALT, high-throughput drug screens allow the unbiased discovery of synthetic lethal interactions. One ALT-targeting drug discovered via this approach is ponatinib, a tyrosine kinase inhibitor that inhibits multiple kinases including BCR-ABL, that is FDA-approved for the treatment of several types of leukemia (Pulte et al. 2022). Ponatinib reduced the viability of ALT-positive cancer cells in vitro and tumors in vivo, and synergized with the DNA synthesis inhibitor triciribine and the ATM/DNA-PKcs inhibitor KU-60019 in ALT cells (Table 1; Kusuma et al. 2023). Despite the absence of a clear mechanism of action, high-throughput screens of FDA-approved drugs have the clear advantage that agents can be fast-tracked through clinical trials, having already passed dose-escalation studies.

## CONCLUDING REMARKS

Understanding the intricacies of the ALT mechanism has advanced substantially over the past decade and through this process, key therapeutic targets have been identified. However, the search for and development of a clinically efficacious ALT-selective agent still eludes. Although there are available inhibitors for many proteins that are integral to the ALT mechanism, most of these agents have not been studied in ALT-positive malignancies. Given the availability of such resources, investigating whether they confer ALT-selective toxicity and hold the same effect as siRNA- or shRNA-mediated depletion of the targets will be a straightforward approach in the search for ALT therapeutics. The most attractive therapeutic strategies include heightening replication stress through the inhibition of replication fork remodelers such as FANCM and SMARCAL1, dual inhibition of PARP and ATR-mediated DNA repair, as well as combined inhibition of DNA synthesis and ATM/DNA-PKcs. Developing an ALT-selective treatment that can enter the clinic will provide the first therapeutic designed specifically to treat this subset of aggressive and hard-to-treat cancers.

## ACKNOWLEDGMENTS

The Pickett laboratory is supported by the National Health and Medical Research Council of Australia and the Medical Research Future Fund. H.A.P. is a cofounder and shareholder of Tessellate Bio.

# REFERENCES

Ammazzalorso F, Pirzio LM, Bignami M, Franchitto A, Pichierri P. 2010. ATR and ATM differently regulate WRN to prevent DSBs at stalled replication forks and promote replication fork recovery. *EMBO J* **29:** 3156–3169. doi:10.1038/emboj.2010.205

Andtbacka RHI, Kaufman HL, Collichio F, Amatruda T, Senzer N, Chesney J, Delman KA, Spitler LE, Puzanov I, Agarwala SS, et al. 2015. Talimogene laherparepvec improves durable response rate in patients with advanced melanoma. *J Clin Oncol* **33:** 2780–2788. doi:10.1200/JCO.2014.58.3377

Arora R, Lee Y, Wischnewski H, Brun CM, Schwarz T, Azzalin CM. 2014. RNaseh1 regulates TERRA-telomeric DNA hybrids and telomere maintenance in ALT tumour cells. *Nat Commun* **5:** 5220. doi:10.1038/ncomms6220

Azeroglu B, Ozbun L, Pegoraro G, Lazzerini Denchi E. 2024. Native FISH: a low- and high-throughput assay to analyze the alternative lengthening of telomere (ALT) pathway. *Methods Cell Biol* **182:** 265–284. doi:10.1016/bs.mcb.2022.10.010

Azzalin CM. 2025. TERRA and the alternative lengthening of telomeres: a dangerous affair. *FEBS Lett* **599:** 157–165. doi:10.1002/1873-3468.14844

Azzalin CM, Reichenbach P, Khoriauli L, Giulotto E, Lingner J. 2007. Telomeric repeat containing RNA and RNA surveillance factors at mammalian chromosome ends. *Science* **318:** 798–801. doi:10.1126/science.1147182

Baird DM, Jeffreys AJ, Royle NJ. 1995. Mechanisms underlying telomere repeat turnover, revealed by hypervariable variant repeat distribution patterns in the human Xp/Yp telomere. *EMBO J* **14:** 5433–5443. doi:10.1002/j.1460-2075.1995.tb00227.x

Balbous A, Cortes U, Guilloteau K, Rivet P, Pinel B, Duchesne M, Godet J, Boissonnade O, Wager M, Bensadoun RJ, et al. 2016. A radiosensitizing effect of RAD51 inhibition in glioblastoma stem-like cells. *BMC Cancer* **16:** 604. doi:10.1186/s12885-016-2647-9

Barnieh FM, Loadman PM, Falconer RA. 2021. Progress towards a clinically-successful ATR inhibitor for cancer therapy. *Curr Res Pharmacol Drug Discov* **2:** 100017. doi:10.1016/j.crphar.2021.100017

Barroso-González J, García-Expósito L, Hoang SM, Lynskey ML, Roncaioli JL, Ghosh A, Wallace CT, de Vitis M, Modesti M, Bernstein KA, et al. 2019. RAD51AP1 is an essential mediator of alternative lengthening of telomeres. *Mol Cell* **76:** 11–26.e7. doi:10.1016/j.molcel.2019.06.043

Berte N, Piée-Staffa A, Piecha N, Wang M, Borgmann K, Kaina B, Nikolova T. 2016. Targeting homologous recombination by pharmacological inhibitors enhances the killing response of glioblastoma cells treated with alkylating drugs. *Mol Cancer Ther* **15:** 2665–2678. doi:10.1158/1535-7163.MCT-16-0176

Bétous R, Mason AC, Rambo RP, Bansbach CE, Badu-Nkansah A, Sirbu BM, Eichman BF, Cortez D. 2012. SMARCAL1 catalyzes fork regression and Holliday junction migration to maintain genome stability during DNA replication. *Genes Dev* **26:** 151–162. doi:10.1101/gad.178459.111

Brenner KA, Nandakumar J. 2022. Consequences of telomere replication failure: the other end-replication problem. *Trends Biochem Sci* **47:** 506–517. doi:10.1016/j.tibs.2022.03.013

Broderick R, Cherdyntseva V, Nieminuszczy J, Dragona E, Kyriakaki M, Evmorfopoulou T, Gagos S, Niedzwiedz W. 2023. Pathway choice in the alternative telomere lengthening in neoplasia is dictated by replication fork processing mediated by EXD2's nuclease activity. *Nat Commun* **14:** 2428. doi:10.1038/s41467-023-38029-z

Brosnan-Cashman JA, Davis CM, Diplas BH, Meeker AK, Rodriguez FJ, Heaphy CM. 2021. SMARCAL1 loss and alternative lengthening of telomeres (ALT) are enriched in giant cell glioblastoma. *Mod Pathol* **34:** 1810–1819. doi:10.1038/s41379-021-00841-7

Casado JA, Río P, Marco E, García-Hernández V, Domingo A, Pérez L, Tercero JC, Vaquero JJ, Albella B, Gago F, et al. 2008. Relevance of the Fanconi anemia pathway in the response of human cells to trabectedin. *Mol Cancer Ther* **7:** 1309–1318. doi:10.1158/1535-7163.MCT-07-2432

Cesare AJ, Kaul Z, Cohen SB, Napier CE, Pickett HA, Neumann AA, Reddel RR. 2009. Spontaneous occurrence of telomeric DNA damage response in the absence of chromosome fusions. *Nat Struct Mol Biol* **16:** 1244–1251. doi:10.1038/nsmb.1725

Chen X, Ali YI, Fisher CE, Arribas-Bosacoma R, Rajasekaran MB, Williams G, Walker S, Booth JR, Hudson JJ, Roe SM, et al. 2021. Uncovering an allosteric mode of action for a selective inhibitor of human Bloom syndrome protein. *eLife* **10:** e65339. doi:10.7554/eLife.65339

Chen L, Zhang C, Ma W, Huang J, Zhao Y, Liu H. 2022. METTL3-mediated m$^6$A modification stabilizes TERRA and maintains telomere stability. *Nucleic Acids Res* **50:** 11619–11634. doi:10.1093/nar/gkac1027

Chu HP, Cifuentes-Rojas C, Kesner B, Aeby E, Lee HG, Wei C, Oh HJ, Boukhali M, Haas W, Lee JT. 2017. TERRA RNA antagonizes ATRX and protects telomeres. *Cell* **170:** 86–101.e16. doi:10.1016/j.cell.2017.06.017

Chung I, Leonhardt H, Rippe K. 2011. De novo assembly of a PML nuclear subcompartment occurs through multiple pathways and induces telomere elongation. *J Cell Sci* **124:** 3603–3618. doi:10.1242/jcs.084681

Ciccia A, Elledge SJ. 2010. The DNA damage response: making it safe to play with knives. *Mol Cell* **40:** 179–204. doi:10.1016/j.molcel.2010.09.019

Ciccia A, Ling C, Coulthard R, Yan Z, Xue Y, Meetei AR, Laghmani el H, Joenje H, McDonald N, de Winter JP, et al. 2007. Identification of FAAP24, a Fanconi anemia core complex protein that interacts with FANCM. *Mol Cell* **25:** 331–343. doi:10.1016/j.molcel.2007.01.003

Ciccia A, Nimonkar AV, Hu Y, Hajdu I, Achar YJ, Izhar L, Petit SA, Adamson B, Yoon JC, Kowalczykowski SC, et al. 2012. Polyubiquitinated PCNA recruits the ZRANB3 translocase to maintain genomic integrity after replication stress. *Mol Cell* **47:** 396–409. doi:10.1016/j.molcel.2012.05.024

Clatterbuck Soper SF, Meltzer PS. 2023. ATRX/DAXX: guarding the genome against the hazards of ALT. *Genes (Basel)* **14:** 790. doi:10.3390/genes14040790

Claude E, de Lhoneux G, Pierreux CE, Marbaix E, de Ville de Goyet M, Boulanger C, Van Damme A, Brichard B, Decottignies A. 2021. Detection of alternative lengthening of telomeres mechanism on tumor sections. *Mol Biomed* **2:** 32. doi:10.1186/s43556-021-00055-y

Cite this article as *Cold Spring Harb Perspect Biol* doi: 10.1101/cshperspect.a041691

Cole KA, Ijaz H, Surrey LF, Santi M, Liu X, Minard CG, Maris JM, Voss S, Reid JM, Fox E, et al. 2023. Pediatric phase 2 trial of a WEE1 inhibitor, adavosertib (AZD1775), and irinotecan for relapsed neuroblastoma, medulloblastoma, and rhabdomyosarcoma. *Cancer* 129: 2245–2255. doi:10.1002/cncr.34786

Colicchia V, Petroni M, Guarguaglini G, Sardina F, Sahún-Roncero M, Carbonari M, Ricci B, Heil C, Capalbo C, Belardinilli F, et al. 2017. PARP inhibitors enhance replication stress and cause mitotic catastrophe in MYCN-dependent neuroblastoma. *Oncogene* 36: 4682–4691. doi:10.1038/onc.2017.40

Conomos D, Stutz MD, Hills M, Neumann AA, Bryan TM, Reddel RR, Pickett HA. 2012. Variant repeats are interspersed throughout the telomeres and recruit nuclear receptors in ALT cells. *J Cell Biol* 199: 893–906. doi:10.1083/jcb.201207189

Coulthard R, Deans AJ, Swuec P, Bowles M, Costa A, West SC, McDonald NQ. 2013. Architecture and DNA recognition elements of the Fanconi anemia FANCM-FAAP24 complex. *Structure* 21: 1648–1658. doi:10.1016/j.str.2013.07.006

Cox Kelli E, Maréchal A, Flynn Rachel L. 2016. SMARCAL1 resolves replication stress at ALT telomeres. *Cell Rep* 14: 1032–1040. doi:10.1016/j.celrep.2016.01.011

da Costa AABA, Chowdhury D, Shapiro GI, D'Andrea AD, Konstantinopoulos PA. 2023. Targeting replication stress in cancer therapy. *Nat Rev Drug Discov* 22: 38–58. doi:10.1038/s41573-022-00558-5

De Clercq E. 2004. Antiviral drugs in current clinical use. *J Clin Virol* 30: 115–133. doi:10.1016/j.jcv.2004.02.009

Deeg KI, Chung I, Bauer C, Rippe K. 2016. Cancer cells with alternative lengthening of telomeres do not display a general hypersensitivity to ATR inhibition. *Front Oncol* 6: 186.

de Nonneville A, Reddel RR. 2021. Alternative lengthening of telomeres is not synonymous with mutations in ATRX/DAXX. *Nat Commun* 12: 1552. doi:10.1038/s41467-021-21794-0

DepMap. 2024. DepMap 24Q4 Public. https://plus.figshare.com/articles/dataset/DepMap_24Q4_Public/27993248

de Wilde RF, Heaphy CM, Maitra A, Meeker AK, Edil BH, Wolfgang CL, Ellison TA, Schulick RD, Molenaar IQ, Valk GD, et al. 2012. Loss of ATRX or DAXX expression and concomitant acquisition of the alternative lengthening of telomeres phenotype are late events in a small subset of MEN-1 syndrome pancreatic neuroendocrine tumors. *Mod Pathol* 25: 1033–1039. doi:10.1038/modpathol.2012.53

Dilley RL, Greenberg RA. 2015. ALTernative telomere maintenance and cancer. *Trends Cancer* 1: 145–156. doi:10.1016/j.trecan.2015.07.007

Dilley RL, Verma P, Cho NW, Winters HD, Wondisford AR, Greenberg RA. 2016. Break-induced telomere synthesis underlies alternative telomere maintenance. *Nature* 539: 54–58. doi:10.1038/nature20099

Diplas BH, He X, Brosnan-Cashman JA, Liu H, Chen LH, Wang Z, Moure CJ, Killela PJ, Loriaux DB, Lipp ES, et al. 2018. The genomic landscape of TERT promoter wildtype-IDH wildtype glioblastoma. *Nat Commun* 9: 2087. doi:10.1038/s41467-018-04448-6

Falck J, Mailand N, Syljuåsen RG, Bartek J, Lukas J. 2001. The ATM-Chk2-Cdc25A checkpoint pathway guards against radioresistant DNA synthesis. *Nature* 410: 842–847. doi:10.1038/35071124

Ferreira MSV, Sørensen MD, Pusch S, Beier D, Bouillon AS, Kristensen BW, Brümmendorf TH, Beier CP, Beier F. 2020. Alternative lengthening of telomeres is the major telomere maintenance mechanism in astrocytoma with isocitrate dehydrogenase 1 mutation. *J Neurooncol* 147: 1–14. doi:10.1007/s11060-020-03394-y

Filippakopoulos P, Qi J, Picaud S, Shen Y, Smith WB, Fedorov O, Morse EM, Keates T, Hickman TT, Felletar I, et al. 2010. Selective inhibition of BET bromodomains. *Nature* 468: 1067–1073. doi:10.1038/nature09504

Flynn RL, Cox KE, Jeitany M, Wakimoto H, Bryll AR, Ganem NJ, Bersani F, Pineda JR, Suvà ML, Benes CH, et al. 2015. Alternative lengthening of telomeres renders cancer cells hypersensitive to ATR inhibitors. *Science* 347: 273–277.

Frank L, Rademacher A, Mücke N, Tirier SM, Koeleman E, Knotz C, Schumacher S, Stainczyk Sabine A, Westermann F, Fröhling S, et al. 2022. ALT-FISH quantifies alternative lengthening of telomeres activity by imaging of single-stranded repeats. *Nucleic Acids Res* 50: e61. doi:10.1093/nar/gkac113

Gari K, Décaillet C, Stasiak AZ, Stasiak A, Constantinou A. 2008. The Fanconi anemia protein FANCM can promote branch migration of Holliday junctions and replication forks. *Mol Cell* 29: 141–148. doi:10.1016/j.molcel.2007.11.032

Gisder DM, Overheu O, Keller J, Nöpel-Dünnebacke S, Uhl W, Reinacher-Schick A, Tannapfel A, Tischoff I. 2023. DAXX, ATRX, and MSI in PanNET and their metastases: correlation with histopathological data and prognosis. *Pathobiology* 90: 71–80. doi:10.1159/000524920

Gocha AR, Nuovo G, Iwenofu OH, Groden J. 2013. Human sarcomas are mosaic for telomerase-dependent and telomerase-independent telomere maintenance mechanisms: implications for telomere-based therapies. *Am J Pathol* 182: 41–48. doi:10.1016/j.ajpath.2012.10.001

Gritsch S, Batchelor TT, Gonzalez Castro LN. 2022. Diagnostic, therapeutic, and prognostic implications of the 2021 World Health Organization classification of tumors of the central nervous system. *Cancer* 128: 47–58. doi:10.1002/cncr.33918

Grobelny JV, Godwin AK, Broccoli D. 2000. ALT-associated PML bodies are present in viable cells and are enriched in cells in the $G_2/M$ phase of the cell cycle. *J Cell Sci* 113: 4577–4585. doi:10.1242/jcs.113.24.4577

Groslambert J, Prokhorova E, Ahel I. 2021. ADP-ribosylation of DNA and RNA. *DNA Repair (Amst)* 105: 103144. doi:10.1016/j.dnarep.2021.103144

Gupta N, Huang TT, Horibata S, Lee JM. 2022. Cell cycle checkpoints and beyond: exploiting the ATR/CHK1/WEE1 pathway for the treatment of PARP inhibitor-resistant cancer. *Pharmacol Res* 178: 106162. doi:10.1016/j.phrs.2022.106162

Halder S, Ranjha L, Taglialatela A, Ciccia A, Cejka P. 2022. Strand annealing and motor driven activities of SMARCAL1 and ZRANB3 are stimulated by RAD51 and the paralog complex. *Nucleic Acids Res* 50: 8008–8022. doi:10.1093/nar/gkac583

Han M, Napier CE, Frölich S, Teber E, Wong T, Noble JR, Choi EHY, Everett RD, Cesare AJ, Reddel RR. 2019. Synthetic lethality of cytolytic HSV-1 in cancer cells with ATRX and PML deficiency. *J Cell Sci* **132:** jcs222349. doi:10.1242/jcs.222349

Heaphy CM, Singhi AD. 2022. The diagnostic and prognostic utility of incorporating DAXX, ATRX, and alternative lengthening of telomeres to the evaluation of pancreatic neuroendocrine tumors. *Hum Pathol* **129:** 11–20. doi:10.1016/j.humpath.2022.07.015

Heaphy CM, de Wilde RF, Jiao Y, Klein AP, Edil BH, Shi C, Bettegowda C, Rodriguez FJ, Eberhart CG, Hebbar S, et al. 2011. Altered telomeres in tumors with *ATRX* and *DAXX* mutations. *Science* **333:** 425. doi:10.1126/science.1207313

Hengel SR, Malacaria E, Folly da Silva Constantino L, Bain FE, Diaz A, Koch BG, Yu L, Wu M, Pichierri P, Spies MA, et al. 2016. Small-molecule inhibitors identify the RAD52-ssDNA interaction as critical for recovery from replication stress and for survival of BRCA2 deficient cells. *eLife* **5:** e14740. doi:10.7554/eLife.14740

Henson JD, Neumann AA, Yeager TR, Reddel RR. 2002. Alternative lengthening of telomeres in mammalian cells. *Oncogene* **21:** 598–610. doi:10.1038/sj.onc.1205058

Henson JD, Cao Y, Huschtscha LI, Chang AC, Au AY, Pickett HA, Reddel RR. 2009. DNA C-circles are specific and quantifiable markers of alternative-lengthening-of-telomeres activity. *Nat Biotechnol* **27:** 1181–1185. doi:10.1038/nbt.1587

Hoang SM, O'Sullivan RJ. 2020. Alternative lengthening of telomeres: building bridges to connect chromosome ends. *Trends Cancer* **6:** 247–260. doi:10.1016/j.trecan.2019.12.009

Hoang SM, Kaminski N, Bhargava R, Barroso-González J, Lynskey ML, García-Expósito L, Roncaioli JL, Wondisford AR, Wallace CT, Watkins SC, et al. 2020. Regulation of ALT-associated homology-directed repair by poly-ADP-ribosylation. *Nat Struct Mol Biol* **27:** 1152–1164. doi:10.1038/s41594-020-0512-7

Hong X, Qiao S, Li F, Wang W, Jiang R, Wu H, Chen H, Liu L, Peng J, Wang J, et al. 2020. Whole-genome sequencing reveals distinct genetic bases for insulinomas and nonfunctional pancreatic neuroendocrine tumours: leading to a new classification system. *Gut* **69:** 877–887. doi:10.1136/gutjnl-2018-317233

Huang F, Motlekar NA, Burgwin CM, Napper AD, Diamond SL, Mazin AV. 2011. Identification of specific inhibitors of human RAD51 recombinase using high-throughput screening. *ACS Chem Biol* **6:** 628–635. doi:10.1021/cb100428c

Huang F, Goyal N, Sullivan K, Hanamshet K, Patel M, Mazina OM, Wang CX, An WF, Spoonamore J, Metkar S, et al. 2016. Targeting BRCA1- and BRCA2-deficient cells with RAD52 small molecule inhibitors. *Nucleic Acids Res* **44:** 4189–4199. doi:10.1093/nar/gkw087

Huang J, Zhang J, Bellani MA, Pokharel D, Gichimu J, James RC, Gali H, Ling C, Yan Z, Xu D, et al. 2019. Remodeling of interstrand crosslink proximal replisomes is dependent on ATR, FANCM, and FANCD2. *Cell Rep* **27:** 1794–1808. e5. doi:10.1016/j.celrep.2019.04.032

Idilli AI, Segura-Bayona S, Lippert TP, Boulton SJ. 2021. A C-circle assay for detection of alternative lengthening of telomere activity in FFPE tissue. *STAR Protoc* **2:** 100569. doi:10.1016/j.xpro.2021.100569

Ishida T, Takizawa Y, Kainuma T, Inoue J, Mikawa T, Shibata T, Suzuki H, Tashiro S, Kurumizaka H. 2009. DIDS, a chemical compound that inhibits RAD51-mediated homologous pairing and strand exchange. *Nucleic Acids Res* **37:** 3367–3376. doi:10.1093/nar/gkp200

Kaminski N, Wondisford AR, Kwon Y, Lynskey ML, Bhargava R, Barroso-González J, García-Expósito L, He B, Xu M, Mellacheruvu D, et al. 2022. RAD51AP1 regulates ALT-HDR through chromatin-directed homeostasis of TERRA. *Mol Cell* **82:** 4001–4017.e7. doi:10.1016/j.molcel.2022.09.025

Karimian K, Groot A, Huso V, Kahidi R, Tan KT, Sholes S, Keener R, McDyer JF, Alder JK, Li H, et al. 2024. Human telomere length is chromosome specific and conserved across individuals. bioRxiv doi:10.1101/2023.12.21.572870

Karlsson-Rosenthal C, Millar JBA. 2006. Cdc25: mechanisms of checkpoint inhibition and recovery. *Trends Cell Biol* **16:** 285–292. doi:10.1016/j.tcb.2006.04.002

Khushalani NI, Harrington KJ, Melcher A, Bommareddy PK, Zamarin D. 2023. Breaking the barriers in cancer care: the next generation of herpes simplex virus–based oncolytic immunotherapies for cancer treatment. *Mol Ther Oncolytics* **31:** 100729. doi:10.1016/j.omto.2023.100729

Kim YS, Keyser SGL, Schneekloth JS Jr. 2014. Synthesis of 2′,3′,4′-trihydroxyflavone (2-D08), an inhibitor of protein sumoylation. *Bioorg Med Chem Lett* **24:** 1094–1097. doi:10.1016/j.bmcl.2014.01.010

Kim E, Kim J, Kim C, Lee J. 2021. Long-read sequencing and de novo genome assemblies reveal complex chromosome end structures caused by telomere dysfunction at the single nucleotide level. *Nucleic Acids Res* **49:** 3338–3353. doi:10.1093/nar/gkab141

Koch MS, Lawler SE, Chiocca EA. 2020. HSV-1 oncolytic viruses from bench to bedside: an overview of current clinical trials. *Cancers (Basel)* **12:** 3514. doi:10.3390/cancers12123514

Koneru B, Farooqi A, Nguyen TH, Chen WH, Hindle A, Eslinger C, Makena MR, Burrow TA, Wilson J, Smith A, et al. 2021. ALT neuroblastoma chemoresistance due to telomere dysfunction-induced ATM activation is reversible with ATM inhibitor AZD0156. *Sci Transl Med* **13:** eabd5750. doi:10.1126/scitranslmed.abd5750

Kurihara S, Hiyama E, Onitake Y, Yamaoka E, Hiyama K. 2014. Clinical features of ATRX or DAXX mutated neuroblastoma. *J Pediatr Surg* **49:** 1835–1838. doi:10.1016/j.jpedsurg.2014.09.029

Kusuma FK, Prabhu A, Tieo G, Ahmed SM, Dakle P, Yong WK, Pathak E, Madan V, Jiang YY, Tam WL, et al. 2023. Signalling inhibition by ponatinib disrupts productive alternative lengthening of telomeres (ALT). *Nat Commun* **14:** 1919. doi:10.1038/s41467-023-37633-3

Laroche-Clary A, Chaire V, Verbeke S, Algéo M-P, Malykh A, Le Loarer F, Italiano A. 2020. ATR inhibition broadly sensitizes soft-tissue sarcoma cells to chemotherapy independent of alternative lengthening telomere (ALT) status. *Sci Rep* **10:** 7488.

Lau LM, Dagg RA, Henson JD, Au AY, Royds JA, Reddel RR. 2013. Detection of alternative lengthening of telomeres by telomere quantitative PCR. *Nucleic Acids Res* **41**: e34. doi:10.1093/nar/gks781

Lee M, Hills M, Conomos D, Stutz MD, Dagg RA, Lau LM, Reddel RR, Pickett HA. 2014. Telomere extension by telomerase and ALT generates variant repeats by mechanistically distinct processes. *Nucleic Acids Res* **42**: 1733–1746. doi:10.1093/nar/gkt1117

Lee M, Teber ET, Holmes O, Nones K, Patch AM, Dagg RA, Lau LMS, Lee JH, Napier CE, Arthur JW, et al. 2018. Telomere sequence content can be used to determine ALT activity in tumours. *Nucleic Acids Res* **46**: 4903–4918. doi:10.1093/nar/gky297

Lewis PW, Elsaesser SJ, Noh KM, Stadler SC, Allis CD. 2010. Daxx is an H3.3-specific histone chaperone and cooperates with ATRX in replication-independent chromatin assembly at telomeres. *Proc Natl Acad Sci* **107**: 14075–14080. doi:10.1073/pnas.1008850107

Li F, Deng Z, Zhang L, Wu C, Jin Y, Hwang I, Vladimirova O, Xu L, Yang L, Lu B, et al. 2019. ATRX loss induces telomere dysfunction and necessitates induction of alternative lengthening of telomeres during human cell immortalization. *EMBO J* **38**: e96659. doi:10.15252/embj.2017 96659

Li Q, Qian W, Zhang Y, Hu L, Chen S, Xia Y. 2023. A new wave of innovations within the DNA damage response. *Signal Transduct Target Ther* **8**: 338. doi:10.1038/s41392-023-01548-8

Lindemann A, Patel AA, Tang L, Tanaka N, Gleber-Netto FO, Bartels MD, Wang L, McGrail DJ, Lin SY, Frank SJ, et al. 2021. Combined inhibition of Rad51 and Wee1 enhances cell killing in HNSCC through induction of apoptosis associated with excessive DNA damage and replication stress. *Mol Cancer Ther* **20**: 1257–1269. doi:10 .1158/1535-7163.MCT-20-0252

Ling C, Huang J, Yan Z, Li Y, Ohzeki M, Ishiai M, Xu D, Takata M, Seidman M, Wang W. 2016. Bloom syndrome complex promotes FANCM recruitment to stalled replication forks and facilitates both repair and traverse of DNA interstrand crosslinks. *Cell Discov* **2**: 16047. doi:10 .1038/celldisc.2016.47

Liu Q, Guntuku S, Cui XS, Matsuoka S, Cortez D, Tamai K, Luo G, Carattini-Rivera S, DeMayo F, Bradley A, et al. 2000. Chk1 is an essential kinase that is regulated by Atr and required for the G(2)/M DNA damage checkpoint. *Genes Dev* **14**: 1448–1459. doi:10.1101/gad.14.12 .1448

Liu TC, Zhang T, Fukuhara H, Kuroda T, Todo T, Canron X, Bikfalvi A, Martuza RL, Kurtz A, Rabkin SD. 2006. Dominant-negative fibroblast growth factor receptor expression enhances antitumoral potency of oncolytic herpes simplex virus in neural tumors. *Clin Cancer Res* **12**: 6791–6799. doi:10.1158/1078-0432.CCR-06-0263

Liu H, Xu C, Diplas BH, Brown A, Strickland LM, Yao H, Ling J, McLendon RE, Keir ST, Ashley DM, et al. 2023. Cancer-associated *SMARCAL*1 loss-of-function mutations promote alternative lengthening of telomeres and tumorigenesis in telomerase-negative glioblastoma cells. *Neuro Oncol* **25**: 1563–1575. doi:10.1093/neuonc/noad 022

Loe TK, Li JSZ, Zhang Y, Azeroglu B, Boddy MN, Denchi EL. 2020. Telomere length heterogeneity in ALT cells is maintained by PML-dependent localization of the BTR complex to telomeres. *Genes Dev* **34**: 650–662. doi:10.1101/ gad.333963.119

Lu R, Pickett HA. 2022. Telomeric replication stress: the beginning and the end for alternative lengthening of telomeres cancers. *Open Biol* **12**: 220011. doi:10.1098/rsob .220011

Lu R, O'Rourke JJ, Sobinoff AP, Allen JAM, Nelson CB, Tomlinson CG, Lee M, Reddel RR, Deans AJ, Pickett HA. 2019. The FANCM-BLM-TOP3A-RMI complex suppresses alternative lengthening of telomeres (ALT). *Nat Commun* **10**: 2252. doi:10.1038/s41467-019-10180-6

Lukashchuk V, Everett RD. 2010. Regulation of ICP0-null mutant herpes simplex virus type 1 infection by ND10 components ATRX and hDaxx. *J Virol* **84**: 4026–4040. doi:10.1128/JVI.02597-09

Lynch RC, Bendell JC, Advani RH, Falchook GS, Munster PN, Patel MR, Gutierrez M, Burness ML, Palmisiano N, Hamadani M, et al. 2021. First-in-human phase I/II study of CYT-0851, a first-in-class inhibitor of RAD51-mediated homologous recombination in patients with advanced solid and hematologic cancers. *J Clin Oncol* **39**: 3006. doi:10.1200/JCO.2021.39.15_suppl.3006

Ma W, He H, Wang H. 2018. Oncolytic herpes simplex virus and immunotherapy. *BMC Immunol* **19**: 40. doi:10.1186/ s12865-018-0281-9

Ma XY, Xu HQ, Zhao JF, Ruan Y, Chen B. 2022. Discovery of a novel Bloom's syndrome protein (BLM) inhibitor suppressing growth and metastasis of prostate cancer. *Int J Mol Sci* **23**: 14798. doi:10.3390/ijms232314798

MacKenzie D Jr, Watters AK, To JT, Young MW, Muratori J, Wilkoff MH, Abraham RG, Plummer MM, Zhang D. 2021. ALT positivity in human cancers: prevalence and clinical insights. *Cancers (Basel)* **13**: 2384. doi:10.3390/ cancers13102384

Maréchal A, Zou L. 2013. DNA damage sensing by the ATM and ATR kinases. *Cold Spring Harb Perspect Biol* **5**: a012716. doi:10.1101/cshperspect.a012716

Mellai M, Annovazzi L, Senetta R, Dell'Aglio C, Mazzucco M, Cassoni P, Schiffer D. 2017. Diagnostic revision of 206 adult gliomas (including 40 oligoastrocytomas) based on ATRX, IDH1/2 and 1p/19q status. *J Neurooncol* **131**: 213–222. doi:10.1007/s11060-016-2296-5

Min J, Wright WE, Shay JW. 2017. Alternative lengthening of telomeres mediated by mitotic DNA synthesis engages break-induced replication processes. *Mol Cell Biol* **37**: e00226-17. doi:10.1128/MCB.00226-1

Min J, Wright WE, Shay JW. 2019. Clustered telomeres in phase-separated nuclear condensates engage mitotic DNA synthesis through BLM and RAD52. *Genes Dev* **33**: 814–827. doi:10.1101/gad.324905.119

Mishra B, Zhang S, Zhao H, Darzynkiewicz Z, Lee EYC, Lee M, Zhang Z. 2019. Discovery of a novel DNA polymerase inhibitor and characterization of its antiproliferative properties. *Cancer Biol Ther* **20**: 474–486. doi:10.1080/ 15384047.2018.1529126

Mori JO, Keegan J, Flynn RL, Heaphy CM. 2024. Alternative lengthening of telomeres: mechanism and the pathogenesis of cancer. *J Clin Pathol* **77**: 82–86. doi:10.1136/jcp-2023-209005

Müller S, Matunis MJ, Dejean A. 1998. Conjugation with the ubiquitin-related modifier SUMO-1 regulates the partitioning of PML within the nucleus. *EMBO J* **17**: 61–70. doi:10.1093/emboj/17.1.61

Muyas F, Rodriguez MJG, Cascão R, Afonso A, Sauer CM, Faria CC, Cortés-Ciriano I, Flores I. 2024. The ALT pathway generates telomere fusions that can be detected in the blood of cancer patients. *Nat Commun* **15**: 82. doi:10.1038/s41467-023-44287-8

Ng LJ, Cropley JE, Pickett HA, Reddel RR, Suter CM. 2009. Telomerase activity is associated with an increase in DNA methylation at the proximal subtelomere and a reduction in telomeric transcription. *Nucleic Acids Res* **37**: 1152–1159. doi:10.1093/nar/gkn1030

Ngoi NYL, Pham MM, Tan DSP, Yap TA. 2021. Targeting the replication stress response through synthetic lethal strategies in cancer medicine. *Trends Cancer* **7**: 930–957. doi:10.1016/j.trecan.2021.06.002

Nguyen GH, Dexheimer TS, Rosenthal AS, Chu WK, Singh DK, Mosedale G, Bachrati CZ, Schultz L, Sakurai M, Savitsky P, et al. 2013. A small molecule inhibitor of the BLM helicase modulates chromosome stability in human cells. *Chem Biol* **20**: 55–62. doi:10.1016/j.chembiol.2012.10.016

Ni M, Li J, Zhao H, Xu F, Cheng J, Yu M, Ke G, Wu X. 2021. BRD4 inhibition sensitizes cervical cancer to radiotherapy by attenuating DNA repair. *Oncogene* **40**: 2711–2724. doi:10.1038/s41388-021-01735-3

O'Rourke JJ, Bythell-Douglas R, Dunn EA, Deans AJ. 2019. ALT control, delete: FANCM as an anti-cancer target in alternative lengthening of telomeres. *Nucleus* **10**: 221–230. doi:10.1080/19491034.2019.1685246

Pan X, Chen Y, Biju B, Ahmed N, Kong J, Goldenberg M, Huang J, Mohan N, Klosek S, Parsa K, et al. 2019. FANCM suppresses DNA replication stress at ALT telomeres by disrupting TERRA R-loops. *Sci Rep* **9**: 19110. doi:10.1038/s41598-019-55537-5

Panier S, Maric M, Hewitt G, Mason-Osann E, Gali H, Dai A, Labadorf A, Guervilly JH, Ruis P, Segura-Bayona S, et al. 2019. SLX4IP antagonizes promiscuous BLM activity during ALT maintenance. *Mol Cell* **76**: 27–43.e11. doi:10.1016/j.molcel.2019.07.010

Perrem K, Colgin LM, Neumann AA, Yeager TR, Reddel RR. 2001. Coexistence of alternative lengthening of telomeres and telomerase in hTERT-transfected GM847 cells. *Mol Cell Biol* **21**: 3862–3875. doi:10.1128/MCB.21.12.3862-3875.2001

Pompili L, Leonetti C, Biroccio A, Salvati E. 2017. Diagnosis and treatment of ALT tumors: is Trabectedin a new therapeutic option? *J Exp Clin Cancer Res* **36**: 189. doi:10.1186/s13046-017-0657-3

Poole LA, Zhao R, Glick GG, Lovejoy CA, Eischen CM, Cortez D. 2015. SMARCAL1 maintains telomere integrity during DNA replication. *Proc Natl Acad Sci* **112**: 14864–14869. doi:10.1073/pnas.1510750112

Potts PR, Yu H. 2007. The SMC5/6 complex maintains telomere length in ALT cancer cells through SUMOylation of telomere-binding proteins. *Nat Struct Mol Biol* **14**: 581–590. doi:10.1038/nsmb1259

Pulte ED, Chen H, Price LSL, Gudi R, Li H, Okusanya OO, Ma L, Rodriguez L, Vallejo J, Norsworthy KJ, et al. 2022. FDA approval summary: revised indication and dosing regimen for ponatinib based on the results of the OPTIC trial. *Oncologist* **27**: 149–157. doi:10.1093/oncolo/oyab040

Ray Chaudhuri A, Nussenzweig A. 2017. The multifaceted roles of PARP1 in DNA repair and chromatin remodelling. *Nat Rev Mol Cell Biol* **18**: 610–621. doi:10.1038/nrm.2017.53

Recagni M, Bidzinska J, Zaffaroni N, Folini M. 2020. The role of alternative lengthening of telomeres mechanism in cancer: translational and therapeutic implications. *Cancers (Basel)* **12**: 949. doi:10.3390/cancers12040949

Richards F, Llorca-Cardenosa MJ, Langton J, Buch-Larsen SC, Shamkhi NF, Sharma AB, Nielsen ML, Lakin ND. 2023. Regulation of Rad52-dependent replication fork recovery through serine ADP-ribosylation of PolD3. *Nat Commun* **14**: 4310. doi:10.1038/s41467-023-40071-w

Sahin U, Ferhi O, Jeanne M, Benhenda S, Berthier C, Jollivet F, Niwa-Kawakita M, Faklaris O, Setterblad N, de Thé H, et al. 2014. Oxidative stress-induced assembly of PML nuclear bodies controls sumoylation of partner proteins. *J Cell Biol* **204**: 931–945. doi:10.1083/jcb.201305148

Saldivar JC, Cortez D, Cimprich KA. 2017. The essential kinase ATR: ensuring faithful duplication of a challenging genome. *Nat Rev Mol Cell Biol* **18**: 622–636. doi:10.1038/nrm.2017.67

Sanchez SE, Gu Y, Wang Y, Golla A, Martin A, Shomali W, Hockemeyer D, Savage SA, Artandi SE. 2024. Digital telomere measurement by long-read sequencing distinguishes healthy aging from disease. *Nat Commun* **15**: 5148. doi:10.1038/s41467-024-49007-4

Schmidt TT, Tyer C, Rughani P, Haggblom C, Jones JR, Dai X, Frazer KA, Gage FH, Juul S, Hickey S, et al. 2024. High resolution long-read telomere sequencing reveals dynamic mechanisms in aging and cancer. *Nat Commun* **15**: 5149. doi:10.1038/s41467-024-48917-7

Schürmann L, Schumacher L, Roquette K, Brozovic A, Fritz G. 2021. Inhibition of the DSB repair protein RAD51 potentiates the cytotoxic efficacy of doxorubicin via promoting apoptosis-related death pathways. *Cancer Lett* **520**: 361–373. doi:10.1016/j.canlet.2021.08.006

Schwab RA, Nieminuszczy J, Shah F, Langton J, Lopez Martinez D, Liang CC, Cohn MA, Gibbons RJ, Deans AJ, Niedzwiedz W. 2015. The Fanconi anemia pathway maintains genome stability by coordinating replication and transcription. *Mol Cell* **60**: 351–361. doi:10.1016/j.molcel.2015.09.012

Schwartzentruber J, Korshunov A, Liu XY, Jones DT, Pfaff E, Jacob K, Sturm D, Fontebasso AM, Quang DA, Tönjes M, et al. 2012. Driver mutations in histone H3.3 and chromatin remodelling genes in paediatric glioblastoma. *Nature* **482**: 226–231. doi:10.1038/nature10833

Scott DE, Francis-Newton NJ, Marsh ME, Coyne AG, Fischer G, Moschetti T, Bayly AR, Sharpe TD, Haas KT, Barber L, et al. 2021. A small-molecule inhibitor of the BRCA2-RAD51 interaction modulates RAD51 assembly and potentiates DNA damage-induced cell death. *Cell Chem Biol* **28**: 835–847.e5. doi:10.1016/j.chembiol.2021.02.006

Sholes SL, Karimian K, Gershman A, Kelly TJ, Timp W, Greider CW. 2022. Chromosome-specific telomere lengths and the minimal functional telomere revealed by nanopore sequencing. *Genome Res* **32**: 616–628. doi:10.1101/gr.275868.121

Cite this article as *Cold Spring Harb Perspect Biol* doi: 10.1101/cshperspect.a041691

Shorstova T, Foulkes WD, Witcher M. 2021. Achieving clinical success with BET inhibitors as anti-cancer agents. *Br J Cancer* **124:** 1478–1490. doi:10.1038/s41416-021-01321-0

Sieverling L, Hong C, Koser SD, Ginsbach P, Kleinheinz K, Hutter B, Braun DM, Cortés-Ciriano I, Xi R, Kabbe R, et al. 2020. Genomic footprints of activated telomere maintenance mechanisms in cancer. *Nat Commun* **11:** 733. doi:10.1038/s41467-019-13824-9

Silva B, Pentz R, Figueira AM, Arora R, Lee YW, Hodson C, Wischnewski H, Deans AJ, Azzalin CM. 2019. FANCM limits ALT activity by restricting telomeric replication stress induced by deregulated BLM and R-loops. *Nat Commun* **10:** 2253. doi:10.1038/s41467-019-10179-z

Silva B, Arora R, Bione S, Azzalin CM. 2021. TERRA transcription destabilizes telomere integrity to initiate break-induced replication in human ALT cells. *Nat Commun* **12:** 3760. doi:10.1038/s41467-021-24097-6

Sobinoff AP, Pickett HA. 2017. Alternative lengthening of telomeres: DNA repair pathways converge. *Trends Genet* **33:** 921–932. doi:10.1016/j.tig.2017.09.003

Sobinoff AP, Pickett HA. 2020. Mechanisms that drive telomere maintenance and recombination in human cancers. *Curr Opin Genet Dev* **60:** 25–30. doi:10.1016/j.gde.2020.02.006

Sobinoff AP, Allen JA, Neumann AA, Yang SF, Walsh ME, Henson JD, Reddel RR, Pickett HA. 2017. BLM and SLX4 play opposing roles in recombination-dependent replication at human telomeres. *EMBO J* **36:** 2907–2919. doi:10.15252/embj.201796889

Sokolowski NA, Rizos H, Diefenbach RJ. 2015. Oncolytic virotherapy using herpes simplex virus: how far have we come? *Oncolytic Virother* **4:** 207–219. doi:10.2147/OV.S66086

Stewart GS, Wang B, Bignell CR, Taylor AM, Elledge SJ. 2003. MDC1 is a mediator of the mammalian DNA damage checkpoint. *Nature* **421:** 961–966. doi:10.1038/nature01446

Sullivan K, Cramer-Morales K, McElroy DL, Ostrov DA, Haas K, Childers W, Hromas R, Skorski T. 2016. Identification of a small molecule inhibitor of RAD52 by structure-based selection. *PLoS ONE* **11:** e0147230. doi:10.1371/journal.pone.0147230

Takaku M, Kainuma T, Ishida-Takaku T, Ishigami S, Suzuki H, Tashiro S, van Soest RW, Nakao Y, Kurumizaka H. 2011. Halenaquinone, a chemical compound that specifically inhibits the secondary DNA binding of RAD51. *Genes Cells* **16:** 427–436. doi:10.1111/j.1365-2443.2011.01494.x

Tham CY, Poon L, Yan T, Koh JYP, Ramlee MK, Teoh VSI, Zhang S, Cai Y, Hong Z, Lee GS, et al. 2023. High-throughput telomere length measurement at nucleotide resolution using the PacBio high fidelity sequencing platform. *Nat Commun* **14:** 281. doi:10.1038/s41467-023-35823-7

Toledo LI, Altmeyer M, Rask MB, Lukas C, Larsen DH, Povlsen LK, Bekker-Jensen S, Mailand N, Bartek J, Lukas J. 2013. ATR prohibits replication catastrophe by preventing global exhaustion of RPA. *Cell* **155:** 1088–1103. doi:10.1016/j.cell.2013.10.043

Ulianov SV, Velichko AK, Magnitov MD, Luzhin AV, Golov AK, Ovsyannikova N, Kireev II, Gavrikov AS, Mishin AS,

Garaev AK, et al. 2021. Suppression of liquid–liquid phase separation by 1,6-hexanediol partially compromises the 3D genome organization in living cells. *Nucleic Acids Res* **49:** 10524–10541. doi:10.1093/nar/gkab249

Vaid R, Thombare K, Mendez A, Burgos-Panadero R, Djos A, Jachimowicz D, Lundberg KI, Bartenhagen C, Kumar N, Tümmler C, et al. 2024. METTL3 drives telomere targeting of TERRA lncRNA through m⁶A-dependent R-loop formation: a therapeutic target for ALT-positive neuroblastoma. *Nucleic Acids Res* **52:** 2648–2671. doi:10.1093/nar/gkad1242

Verma P, Dilley RL, Zhang T, Gyparaki MT, Li Y, Greenberg RA. 2019. RAD52 and SLX4 act nonepistatically to ensure telomere stability during alternative telomere lengthening. *Genes Dev* **33:** 221–235. doi:10.1101/gad.319723.118

Voter AF, Manthei KA, Keck JL. 2016. A high-throughput screening strategy to identify protein–protein interaction inhibitors that block the fanconi anemia DNA repair pathway. *J Biomol Screen* **21:** 626–633. doi:10.1177/1087057116635503

Wang Y, Luo W, Wang Y. 2019. PARP-1 and its associated nucleases in DNA damage response. *DNA Repair (Amst)* **81:** 102651. doi:10.1016/j.dnarep.2019.102651

Wang ZQ, Zhang ZC, Wu YY, Pi YN, Lou SH, Liu TB, Lou G, Yang C. 2023. Bromodomain and extraterminal (BET) proteins: biological functions, diseases, and targeted therapy. *Signal Transduct Target Ther* **8:** 420. doi:10.1038/s41392-023-01647-6

Wiese C, Dray E, Groesser T, San Filippo J, Shi I, Collins DW, Tsai MS, Williams GJ, Rydberg B, Sung P, et al. 2007. Promotion of homologous recombination and genomic stability by RAD51AP1 via RAD51 recombinase enhancement. *Mol Cell* **28:** 482–490. doi:10.1016/j.molcel.2007.08.027

Wu X, Krishna Sudhakar H, Alcock LJ, Lau YH. 2023. Mannich base PIP-199 is a chemically unstable pan-assay interference compound. *J Med Chem* **66:** 11271–11281. doi:10.1021/acs.jmedchem.3c00674

Xiao Z, Chen Z, Gunasekera AH, Sowin TJ, Rosenberg SH, Fesik S, Zhang H. 2003. Chk1 mediates S and $G_2$ arrests through Cdc25A degradation in response to DNA-damaging agents. *J Biol Chem* **278:** 21767–21773. doi:10.1074/jbc.M300229200

Yadav T, Zhang JM, Ouyang J, Leung W, Simoneau A, Zou L. 2022. TERRA and RAD51AP1 promote alternative lengthening of telomeres through an R- to D-loop switch. *Mol Cell* **82:** 3985–4000.e4. doi:10.1016/j.molcel.2022.09.026

Yan Z, Delannoy M, Ling C, Daee D, Osman F, Muniandy PA, Shen X, Oostra AB, Du H, Steltenpool J, et al. 2010. A histone-fold complex and FANCM form a conserved DNA-remodeling complex to maintain genome stability. *Mol Cell* **37:** 865–878. doi:10.1016/j.molcel.2010.01.039

Yang Q, Li Y, Sun R, Li J. 2021a. Identification of a RAD52 inhibitor inducing synthetic lethality in BRCA2-deficient cancer cells. *Front Pharmacol* **12:** 637825. doi:10.3389/fphar.2021.637825

Yang SY, Chang EYC, Lim J, Kwan HH, Monchaud D, Yip S, Stirling Peter C, Wong JMY. 2021b. G-quadruplexes mark alternative lengthening of telomeres. *NAR Cancer* **3:** zcab031. doi:10.1093/narcan/zcab031

Yin QK, Wang CX, Wang YQ, Guo QL, Zhang ZL, Ou TM, Huang SL, Li D, Wang HG, Tan JH, et al. 2019. Discovery of isaindigotone derivatives as novel Bloom's syndrome protein (BLM) helicase inhibitors that disrupt the BLM/DNA interactions and regulate the homologous recombination repair. *J Med Chem* **62:** 3147–3162. doi:10.1021/acs.jmedchem.9b00083

Zeman MK, Cimprich KA. 2014. Causes and consequences of replication stress. *Nat Cell Biol* **16:** 2–9. doi:10.1038/ncb2897

Zhang JM, Zou L. 2020. Alternative lengthening of telomeres: from molecular mechanisms to therapeutic outlooks. *Cell Biosci* **10:** 30. doi:10.1186/s13578-020-00391-6

Zhang JM, Yadav T, Ouyang J, Lan L, Zou L. 2019. Alternative lengthening of telomeres through two distinct break-induced replication pathways. *Cell Rep* **26:** 955–968.e3. doi:10.1016/j.celrep.2018.12.102

Zhang JM, Genois MM, Ouyang J, Lan L, Zou L. 2021. Alternative lengthening of telomeres is a self-perpetuating process in ALT-associated PML bodies. *Mol Cell* **81:** 1027–1042.e4. doi:10.1016/j.molcel.2020.12.030

Zhao H, Piwnica-Worms H. 2001. ATR-mediated checkpoint pathways regulate phosphorylation and activation of human Chk1. *Mol Cell Biol* **21:** 4129–4139. doi:10.1128/MCB.21.13.4129-4139.2001

Zhao R, Xu M, Yu X, Wondisford AR, Lackner RM, Salsman J, Dellaire G, Chenoweth DM, O'Sullivan RJ, Zhao X, et al. 2024. SUMO promotes DNA repair protein collaboration to support alternative telomere lengthening in the absence of PML. *Genes Dev* **38:** 614–630. doi:10.1101/gad.351667.124

Zhou P, Chen X, Li M, Tan J, Zhang Y, Yuan W, Zhou J, Wang G. 2019. 2-D08 as a SUMOylation inhibitor induced ROS accumulation mediates apoptosis of acute myeloid leukemia cells possibly through the deSUMOylation of NOX2. *Biochem Biophys Res Commun* **513:** 1063–1069. doi:10.1016/j.bbrc.2019.04.079

Zimmermann M, Bernier C, Kaiser B, Fournier S, Li L, Desjardins J, Skeldon A, Rimkunas V, Veloso A, Young JTF, et al. 2022. Guiding ATR and PARP inhibitor combinations with chemogenomic screens. *Cell Rep* **40:** 111081. doi:10.1016/j.celrep.2022.111081

# Telomere Dynamics in Zebrafish Aging and Disease

Miguel Godinho Ferreira

Institute for Research on Cancer and Aging of Nice (IRCAN), CNRS UMR7284, INSERM U1081, Université Cote d'Azur, 06107 Nice, France

*Correspondence:* miguel-godinho.ferreira@unice.fr

Fish telomere lengths vary significantly across the numerous species, implicating diverse life strategies and environmental adaptations. Zebrafish have telomere dynamics that are comparable to humans and are emerging as a key model in which to unravel the systemic effects of telomere shortening on aging and interorgan communication. Here, we discuss zebrafish telomere biology, focusing on the organismal impact of telomere attrition beyond cellular senescence, with particular emphasis on how telomeric shortening in specific tissues can unleash widespread organ dysfunction and disease. This highlights a novel aspect of tissue communication, whereby telomere shortening in one organ can propagate through biological networks, influencing the aging process systemically. These discoveries position zebrafish as a valuable model for studying the complex interactions between telomeres, aging, and tissue cross talk, providing important insights with direct relevance to human health and longevity.

## FISH TELOMERES

Telomeres in fish are composed of the canonical vertebrate repeat sequence (TTAGGG)n that varies in length from 2 to 25 kb across different species. This variation reflects a broad taxonomic diversity. Techniques such as fluorescence in situ hybridization (FISH) have identified telomeric sequences on the chromosomes of ~80 fish species (Ocalewicz 2013), illustrating widespread distribution across distinct taxonomic groups. Notably, in addition to typical terminal locations, fish chromosomes also exhibit interstitial telomeric sequences (ITSs). These ITSs, found in various chromosomal regions, including pericentromeric areas and along nucleolar organizer regions, are often considered remnants of past chromosomal fusion events (Meyne et al. 1990; Vicari et al. 2022).

The enzyme telomerase (*tert*), the reverse transcriptase that adds telomeric DNA to the ends of chromosomes, shows varied expression levels in different tissues and across species (Wai et al. 2005; Wu et al. 2006; Yu et al. 2006). In most fish, low levels of telomerase activity are present in somatic tissues, with high levels detected in reproductive organs (Elmore et al. 2008; Lau et al. 2008; Pfennig et al. 2008; Anchelin et al. 2011). Certain fish species exhibit an "indeterminate" growth pattern and continue to grow throughout their lives. However, the growth rate significantly decreases upon the onset of sexual maturity and with advancing age (Dutta 1994; McDowall 1994). Despite continuous growth and the presence of telomerase activity, telomere length nevertheless declines with age (Hatakeyama et al. 2008; Hartmann et al. 2009; Anchelin et al. 2011; Henriques et al. 2013). This suggests that the continuous turnover of cells may exceed the capacity of telomerase to maintain telomere length, leading to gradual telomere shortening.

Diversity in telomere length is likely to reflect different life strategies, as seen in species such as the turquoise killifish (*Nothobranchius furzeri*), in which longer-lived strains show telomere attrition with age, in contrast to their short-lived counterparts. Specifically, *Nothobranchius* sp. populations captured across regions with a strong gradient rainfall, showed that strains from wetter climates had shorter telomeres and longer life spans than closely related strains that lived in drier regions (Reichard et al. 2021). As observed in other vertebrates, this correlation is reversed within species, as individuals with shorter telomeres are prone to disease and early death (Wilbourn et al. 2018). Male killifish, which are typically larger, have shorter life span than females and possess shorter telomeres (Reichard et al. 2021). This effect may be driven by an increased number of cell divisions in males as compared to females. Moreover, faster growing juvenile fish had shorter telomeres at the onset of adulthood, indicating that environmental selection for fast growth may underlie the telomere length—life span dynamics. Under laboratory conditions, the longer-lived killifish MZM strain (maximum life span of 8 mo) shows telomere shortening with age (Hartmann et al. 2009). However, the GRZ strain, the shortest living organism in laboratory conditions (maximum life span of 3–4 mo), does not show appreciable telomere shortening during its lifetime (Hartmann et al. 2009). It is possible that the fast-lived GRZ killifish simply lacks sufficient cell divisions to undergo critical telomere shortening during its lifetime. Despite this, telomerase-deficient GRZ killifish display age-associated defects but fail to undergo premature death, denoting the complex role telomere shortening plays in this extremely fast-lived

animal (Harel et al. 2015). Telomere dynamics are influenced by developmental and regenerative processes. For example, telomerase expression is transiently up-regulated upon organ regeneration, such as the fin and heart, providing an additional layer of telomere length regulation (Elmore et al. 2008; Anchelin et al. 2011; Bednarek et al. 2015).

## TELOMERES IN ZEBRAFISH

Zebrafish (*Danio rerio*) have an average life span of 3–4 years in the laboratory, and have emerged as a valuable model for studying telomere biology, due to their human-like telomere length that shortens during lifetime, the evolutionary conservation of shelterin, and their current utility in human disease modeling (Carneiro et al. 2016a). Further evaluation of the role telomeres and telomerase play in zebrafish will contribute to our understanding of vertebrate telomere evolution and the critical role telomere shortening plays in aging and age-associated disease.

Telomere length in zebrafish ranges from ~10 to 22 kb, with variability observed between different outbred strains (Anchelin et al. 2011; Henriques et al. 2013). Telomerase expression increases from embryonic stages to early adulthood but then declines with age (Anchelin et al. 2011, 2013). Terminal restriction fragment (TRF) analysis showed that zebrafish larvae exhibit long telomeres (~12 kb), which shorten as the organisms grow and age (Henriques et al. 2013; Carneiro et al. 2016b). In sexually mature males, the longest telomeres are observed in the testis (~9.9 kb), followed by the muscle (~8.7 kb), with the shortest length recorded in the gut (~7.9 kb). Analysis in different tissues across various ages revealed significant telomere shortening in the caudal fin (decline of 45–90 bp/mo), a less proliferative tissue, whereas the more proliferative kidney marrow exhibited telomere elongation from larval to adult stages and subsequent shortening post–18 months (Fig. 1; Carneiro et al. 2016b). This pattern correlates with higher telomerase activity in the kidney, which may contribute to less telomere shortening early in life in the kidney marrow compared to the fin. Telomere length in the

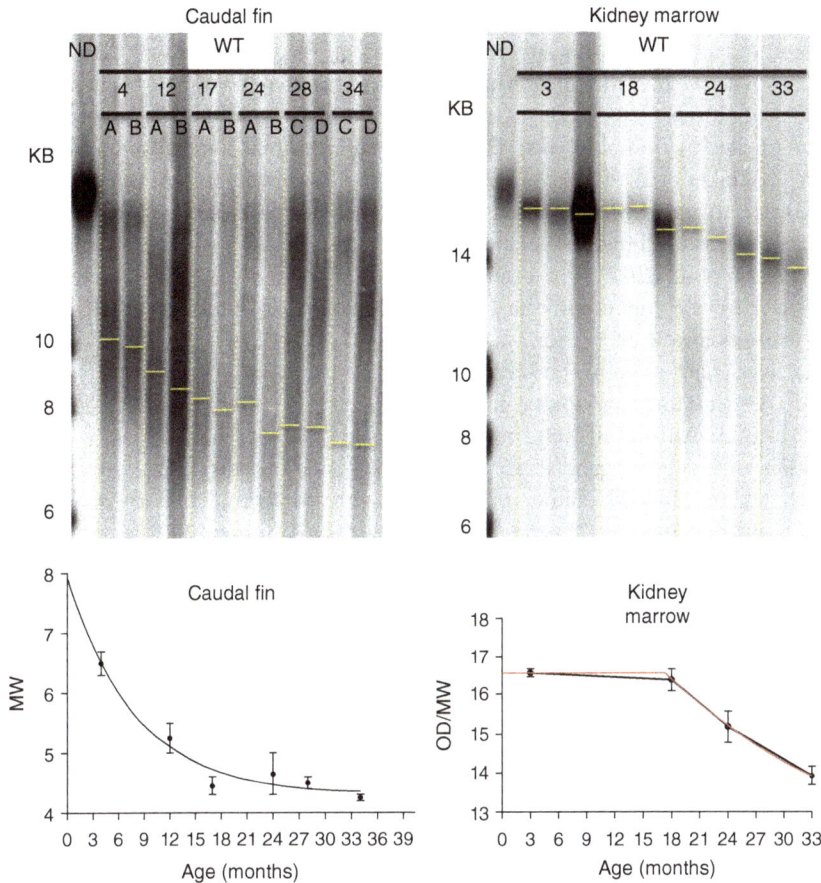

Figure 1. Telomere length in zebrafish is tissue-specific and declines both in fast and slow proliferative tissues. Southern blotting terminal restriction fragments (TRFs) (*top*) and mean telomere length analysis (*bottom*) of caudal fin and kidney marrow of wild-type (WT) zebrafish. (kb) Molecular weight markers, (ND) nondigested genomic DNA, (OD) optical density, (MW) molecular weight. Numbers designate age of the fish in months. Fin—*A, B, C,* and *D* refer to different fins from the same individual. Kidney marrow—three siblings were used for each age. (Figure based on data in Carneiro et al. 2016b.)

gut and muscle of wild-type (WT) zebrafish significantly declines by the age of 24 months (Carneiro et al. 2016b). In contrast, telomeres in the testis do not show notable shortening, highlighting tissue-specific dynamics in telomere length regulation. The gut exhibits a linear decline in telomere length up to 24 months, followed by stabilization (Carneiro et al. 2016b). This stabilization may be due to cellular selection processes eliminating cells with extremely short telomeres, or possibly recombination mechanisms that maintain shorter telomere lengths in older age.

## SHELTERIN IN ZEBRAFISH

Research into telomere-binding proteins and their interactions in zebrafish has revealed conserved evolutionary aspects in vertebrate telomere biology. The identification of zebrafish homologs of human shelterin complex proteins TRF1, TRF2 (termed *terfa* in zebrafish), RAP1, TPP1, POT1, and TIN2 indicates that these proteins retain functional domain structures (Xie et al. 2011; Myler et al. 2021). Two notable exceptions can be assigned to Terfa. Unlike other vertebrates, zebrafish Terfa lacks both an amino-terminal basic-domain, implicated in telomeric D-loop stability, and a Rap1-binding site (Myler et al. 2021). However, the presence of alternative evolutionarily conserved Rap1-Terfa binding sites has been inferred by homology with invertebrate species (Gaullier et al. 2016).

Similar to mice, the expression levels of shelterin genes in zebrafish exhibit tissue-specific variations (Wagner et al. 2017). For example, *terfa* mRNA levels were highest in the brain and muscle, whereas lowest expression was observed in the liver. *tpp1* mRNA showed peak expression in the heart but was lowest in the intestine and ovaries. An overall trend toward down-regulation of shelterin gene expression was apparent during zebrafish aging, particularly in the brain and ovaries (Wagner et al. 2017). Spatiotemporal expression patterns of shelterin genes may play a role in various biological processes, including development, tissue homeostasis, and aging.

Studies using CRISPR-Cas9 showed the effects of targeted gene mutations of zebrafish shelterin genes. Surprisingly, while mutations in some genes are lethal in early development stages, others do not appear to exhibit clear phenotypes. For example, *tin2* and *terf1* homozygous mutants can grow to adulthood, whereas *pot1* homozygous mutant larvae die between 12 and 15 days postfertilization (dpf) (Ma et al. 2022). As in mammals, Terfa protects against telomere DNA damage responses (DDRs) and ATM activation (Ying et al. 2022). Knockdown of Terfa using siRNA leads to a DDR, evident by γH2AX foci that are not restricted to telomeres but also include pericentromeric regions (Ying et al. 2022). Homozygous *terfa* mutants are embryonic lethal, while heterozygous fish are viable but show accelerated aging phenotypes, including lipofuscin accumulation, decreased neurogenesis, and shorter life span (Kishi et al. 2008; Ying et al. 2022). These defects can be rescued by mutations in either *tp53* or *atm* (Ying et al. 2022). Tpp1, another shelterin component identified in zebrafish was shown to localize to telomeres and interact with both Pot1 and Tin2 (Xie et al. 2011); however, it is currently unknown whether it recruits telomerase as in mammals. Knocking down *tpp1* in zebrafish leads to em-

bryonic abnormalities, highlighting the critical role these proteins play in development. Future studies will reveal the essential functions of these proteins in telomere maintenance and overall cellular viability. Moreover, differences in viability between zebrafish and mammalian shelterin may determine their changing roles in telomere protection during development and cell differentiation, as recently observed in mammals (Markiewicz-Potoczny et al. 2021; Ruis and Boulton 2021).

## TELOMERASE ACTIVITY IN ZEBRAFISH

Sequence analysis revealed strong homology between zebrafish telomerase reverse transcriptase (*tert*) and human TERT, particularly at functional domains such as the amino terminus, TR-binding site, and RT motifs (Lau et al. 2008). Despite showing ~50% identity within these domains, *tert* exhibits only 22% identity outside these regions. Zebrafish telomerase is catalytically active and dependent on the TR domain (Imamura et al. 2008). Moreover, the metal-binding motifs A and C in the RT domain, which are essential for the catalytic activity of human TERT, are conserved in zebrafish. Conclusively, mutation of RT motif A results in a catalytically inactive *tert* mutant that is unable to rescue the defects of *tert*-deficient larvae, emphasizing the functional similarities (Imamura et al. 2008).

Analysis of the zebrafish *tert* promoter identified transcription factor-binding sites for Sp1, c-Myc, NF-κB, and estrogen receptor (ER), similar to those in the human TERT promoter, although their relative position to the initiation codon is not conserved (Anchelin et al. 2011). In addition, EGFPLuc reporter assays confirmed activation of the zebrafish telomerase promoter by c-Myc and NF-κB (Anchelin et al. 2011), mirroring the regulation observed in humans. Interestingly, the *tert* promoter of WT zebrafish contains two ETS sites (−120 and −81), similar to de novo mutations often associated with human tumors known as telomerase promoter mutations (TPMs) (Lopes-Bastos et al. 2023). While telomerase expression levels increase from embryo to adult stages in zebrafish, a drastic decline is observed in aged fish (Anchelin et al. 2011). Furthermore, weakened up-regulation of telomerase expression in regenerating fins of old fish correlates with impaired regeneration capacity, highlighting the complex regulation of telomerase expression during aging and tissue regeneration (see below).

The telomerase RNA component (*terc*) in teleost fish, including zebrafish, that provides the catalytic site and serves as a scaffold for the telomerase holoenzyme, displays a significant reduction in length. In contrast to *tert*, *terc* shows considerable variation in its size, sequence, and structure. Teleost fish, which possess smaller genomes, have the shortest *terc* (312–348 nt) among vertebrates, whereas cartilaginous fishes, with larger genomes, exhibit the longest vertebrate *terc* (478–559 nt). This variability reflects the structural plasticity that has occurred throughout fish evolution (Xie et al. 2008). While genetic depletion of *terc* in zebrafish results in impaired myelopoiesis, *terc* overexpression increases the number of neutrophils and macrophages (Alcaraz-Pérez et al. 2014; García-Castillo et al. 2021). These results underscore the critical role of *terc* in hematopoiesis.

## *TERC* KO AND NONCANONICAL ROLES OF TELOMERASE

Apart from its catalytic role in telomere maintenance, work spearheaded by the ML Cayuela laboratory identified *terc* noncatalytic functions (Alcaraz-Pérez et al. 2014). Specifically, *terc*, independently of *tert*, was reported to bind directly to DNA sequences of master myeloid genes, controlling their expression by recruiting RNA polymerase II (RNA Pol II) (García-Castillo et al. 2021). Genetic knockout (KO) of *terc* resulted in embryos with depleted myeloid gene expression and a reduction in the number of neutrophils in the caudal hematopoietic tissue (CHT), which was independent of telomere shortening and *tert* expression (García-Castillo et al. 2021). This study revealed that *terc* exerts its function directly by regulating *spi1a* and *csf3b* expression, the master regulators of zebrafish myelopoiesis. Moreover, short single-stranded oligonucleotides based on the CR4/CR5 domain of *terc* were able

to enhance myelopoiesis through direct interaction with promoter regions and RNA Pol II (Martínez-Balsalobre et al. 2023). Consistent with the zebrafish results, decreased levels of human TERC led to reduced myeloid gene expression in human neutrophil and monocyte progenitor cell lines, without impacting telomerase activity or telomere length (García-Castillo et al. 2021). These studies suggest that TERC governs the expression of gene networks essential for myelopoiesis in both zebrafish and humans, beyond its canonical role in telomere biology.

Similarly, *tert* appears to promote the development of hematopoietic cells in zebrafish through a mechanism unrelated to telomerase activity (Imamura et al. 2008). Embryos of zebrafish *tert* morpholino (phosphorodiamidate antisense oligomers or morpholino oligomer [MO]) knockdowns displayed abnormal differentiation and apoptosis of hematopoietic stem and/or progenitor cells, which led to the circulation of immature blood cells and anemia without any apparent telomere shortening (Imamura et al. 2008). This parallels the reduced number of circulating blood cells, including red blood cells, white blood cells, and platelets, observed in human telomere biology disorders, such as dyskeratosis congenita.

## TELOMERASE MUTANT ZEBRAFISH

Zebrafish have emerged as a complementary vertebrate model for investigating the consequences of telomerase deficiency and telomere shortening in aging, cancer, and regeneration. Three noteworthy aspects of telomere shortening in zebrafish underscore its importance as a model for human telomere biology.

Firstly, telomere shortening occurs in both high-turnover tissues (e.g., gut) and low-turnover organs (e.g., fin), regardless of differences in proliferation rates, similar to observations in humans (Carneiro et al. 2016b). Beyond cell proliferation, the link between higher reactive oxygen species (ROS) levels and telomere shortening remains to be explored in zebrafish. Considering the known genotoxic effects of ROS, especially in G-rich DNA regions like telomeres, tissues with high metabolic rates, such as the brain and heart, are expected to be particularly vulnerable.

Secondly, pronounced telomere shortening in zebrafish gut and fin within the first year parallels the accentuated telomere shortening observed in early human life (Carneiro et al. 2016b), followed by length stabilization at later ages that may reflect the elimination of cells with extremely short telomeres via apoptosis.

Lastly, the accumulation of short telomeres and telomere damage in zebrafish over time anticipates the onset of tissue-specific aging phenotypes and associated diseases, including intestinal inflammation, cachexia, and, surprisingly, cancer.

These findings support the hypothesis that telomere shortening contributes significantly to DNA damage, tissue dysfunction, and age-related diseases in zebrafish aging, resembling observations in humans. Consequently, various larval and adult zebrafish models are emerging to further explore the mechanisms underlying these events.

## HAPLOINSUFFICIENCY AND GENETIC ANTICIPATION

Telomerase expression in zebrafish, akin to mice and humans, displays haploinsufficiency. Offspring from older heterozygous $tert^{+/-}$ fish exhibit shorter telomeres and more pronounced signs of cachexia and fertility issues compared to offspring from younger $tert^{+/-}$ parents (Anchelin et al. 2013; Scahill et al. 2017). This suggests that $tert$ haploinsufficiency affects gametogenesis and worsens with parental aging due to ongoing telomere shortening. Notably, in humans, oogenesis occurs only during embryonic development, while spermatogenesis continues throughout life. In contrast, gametogenesis occurs in both sexes throughout fish lives, having a greater impact on spermatogenesis given the number of cell divisions. Thus, male zebrafish become infertile earlier than females (Carneiro et al. 2016b), impacting the translation of these findings from zebrafish to humans. Moreover, with each successive generation of $tert^{+/-}$ incrossing, phenotypes become increasingly severe. First-generation $tert^{-/-}$ progeny (G1) dis-

play body wasting and early infertility, while this is visible in $tert^{+/-}$ and $tert^{+/+}$ zebrafish only after multiple $tert^{+/-}$ incrosses, rendering the line infertile (Henriques and Ferreira 2024). This genetic anticipation, as seen in heterozygous deficiencies, has been characterized in human telomere biology disorders and CAST/Ei wild-derived short telomere mice (Hathcock et al. 2002; Hao et al. 2005).

In contrast to humans and zebrafish, where telomerase deficiency results in immediate severe phenotypes, most telomerase mutant organisms lack overt dysfunction in the first mutant generation. Several model systems, including ciliates (Yu et al. 1990), yeast (Lundblad and Szostak 1989), plants (Riha et al. 2001), nematodes (Meier et al. 2006), and inbred laboratory mice (Blasco et al. 1997; Rudolph et al. 1999), exhibit organismal functional decline in the absence of telomerase only across multiple generations. Similar to laboratory mice, $tert$ KOs in *Caenorhabditis elegans* and *Arabidopsis thaliana* do not manifest observable impacts on survival and reproduction in the first-generation telomerase KO. However, defects arise when homozygous telomerase KOs are incrossed for several generations, leading to telomere functional decline and genetic anticipation, as seen in heterozygous deficiencies in humans and zebrafish, due to critically short telomeres that interfere with cell proliferation.

## CELLULAR DEFECTS OF TELOMERASE-DEFICIENT ZEBRAFISH

During aging, short telomeres undergo deprotection and are recognized as DNA damage (Cayuela et al. 2019). Consequently, both WT and $tert$ mutant zebrafish accumulate γH2AX foci at telomeres as they age, particularly in the gut epithelia (Carneiro et al. 2016b). The presence of γH2AX telomeric foci correlates with telomere shortening, supporting the notion that shortened telomeres are perceived as DNA damage, activating the DDR. Damaged DNA activates the ATM kinase, which mediates H2AX phosphorylation. Besides γH2AX foci, aged WT and $tert^{-/-}$ zebrafish also exhibit reduced staining of proliferating cell nuclear antigen (PCNA) and increased levels of p53, as well as markers of cell senescence, including p21 (*cdkn1a*), p16/p15 (*cdkn2a/b*), and senescence-associated β-galactosidase (SA-β-gal) staining (Henriques et al. 2013; Carneiro et al. 2016b; El Maï et al. 2020, 2023; Marzullo et al. 2022). This indicates that telomere shortening not only reduces cell proliferation but also induces senescence, contributing to tissue homeostasis disruption during aging. These effects are mediated by p53 activation, as the combination of *tert*$^{-/-}$ and *tp53*$^{-/-}$ mutations rescues both the replication rate and the senescence phenotype (Anchelin et al. 2013; Henriques et al. 2013; El Maï et al. 2020; Şerifoğlu et al. 2024a, b). Cells with critically short telomeres use one of two mechanisms to evade proliferation: (1) activation of a cell death program through p53-dependent expression of proapoptotic proteins like PUMA, or (2) cell-cycle arrest through up-regulation of the CDK inhibitor p16/p15 (*cdkn2a/b*), resulting in cellular senescence. The determinants of whether a cell with short telomeres undergoes senescence or apoptosis remain largely unclear, although evidence suggests that most cells are capable of both outcomes (El Maï et al. 2020). This raises questions about how damaged cells in an organism decide whether to persist in a dysfunctional state or undergo programmed cell death.

p53 stabilization induces cell-cycle arrest, apoptosis, senescence, and autophagy. p53 activation in young *tert*$^{-/-}$ zebrafish does not affect mitochondrial function. However, older *tert*$^{-/-}$ zebrafish exhibit mitochondrial dysfunction in gut and testis, accompanied by disrupted membrane integrity, reduced ATP levels, and accumulation of ROS (Henriques et al. 2013; Carneiro et al. 2016b; El Maï et al. 2020). These alterations are concurrent with the onset of senescence. In contrast to what was previously observed in *tert* KO mice (Sahin et al. 2011), p53 stabilization and mitochondrial dysfunction were not accompanied by a down-regulation of PGC1a, a regulator of mitochondrial biogenesis (Henriques et al. 2013). Instead, DDR and p53 response activation resulted in the reduction in mitochondrial OxPhos defenses with age, with decreased SOD2 expression in response to AKT-dependent activation and FoxO1 and FoxO4 phosphorylation

(El Maï et al. 2020). The interaction between the contradictory antiproliferative p53 and prosurvival mTOR/AKT pathways plays a crucial role in modulating cell fate decisions between apoptosis and senescence. Thus, activation of the mTOR/AKT pathway in response to short telomeres acts as a negative regulator of apoptosis and induces senescence by increasing intracellular ROS levels and activating p16/p15 (*cdkn2a/b*) transcription.

The precise mechanism responsible for inducing cellular senescence remains unclear. Indeed, beyond cell-intrinsic responses, zebrafish studies suggest that there may be a tissue-dependent component that dictates cell fate decision. Despite lacking observable tissue defects, young *tert* mutant zebrafish display elevated levels of p53, apoptosis, and reduced proliferative capacity (El Maï et al. 2020). This heightened apoptosis leads to an increased demand for cell proliferation from neighboring cells, a phenomenon known as apoptosis-induced compensatory proliferation (Fogarty and Bergmann 2017). Consequently, tissue degeneration becomes evident in aging *tert*$^{-/-}$ zebrafish (El Maï et al. 2020), particularly in tissues lacking readily available stem cells or constrained by tissue-intrinsic genetic programs inhibiting cell division. In such cases, cellular hypertrophy (a feature of cell senescence) serves as an alternative strategy for maintaining tissue homeostasis.

The combined effect of telomere shortening and p53 activation leads to loss of tissue integrity, triggering the AKT-dependent pro-proliferative pathway. The simultaneous action of these opposing forces within the cell culminates in cellular senescence (Fig. 2). This is supported by the loss of *tp53* function, which results in the rescue of tissue integrity, and prevents activation of AKT, increased levels of ROS, and induction of senescence. This model is also supported by inhibition of the mTOR/AKT pathway upon telomere shortening, either by genetically modulating the *ztor* pathway or chemically by directly inhibiting AKT. In both scenarios, there is a reduction of cell senescence and rescued tissue integrity. Overall, these findings demonstrate the importance of in vivo tissue damage and the interplay between telomere

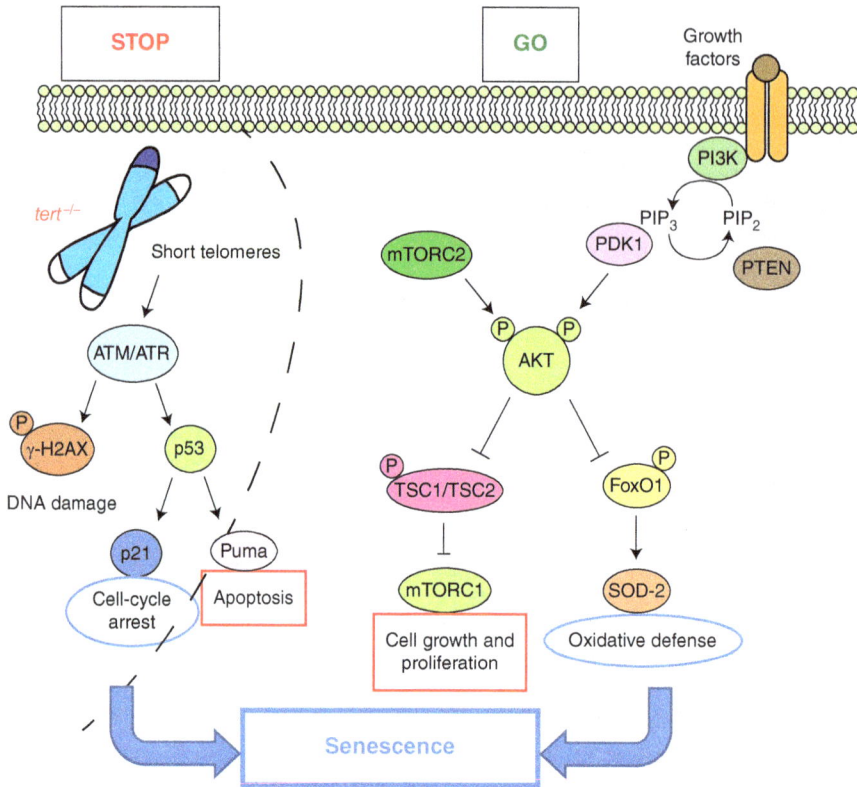

**Figure 2.** Senescence results from the integration of antiproliferation (DDR/p53) and pro-proliferative (mTOR/AKT) pathways. Inhibition of cell proliferation by short telomeres results in a progressive loss of tissue cellularity, eventually leading to tissue damage. As age progresses, loss of tissue homeostasis triggers the pro-proliferative mTOR/AKT pathway. Akt phosphorylates FoxO, inducing its translocation from the nucleus to the cytoplasm. Loss of FoxO transcriptional activity reduces mitochondrial SOD2 expression generating mitochondrial oxidative stress through increased reactive oxygen species (ROS) levels. Mitochondrial dysfunction eventually triggers p15/16 expression and cell senescence. (Figure based on data in El Maï et al. 2020.)

shortening/p53 and AKT/FoxO signaling pathways. It also highlights the necessity for integrating in vitro cell studies with in vivo tissue experiments, which are uniquely amenable in zebrafish. Growth factors present in cultured media are a source of pro-proliferative signaling and continuous cell divisions that differ from the normal homeostatic cell turnover of most tissues.

## ORGAN DEFECTS OF TELOMERASE-DEFICIENT ZEBRAFISH

Zebrafish exhibit various age-associated phenotypes akin to those observed in human aging. These include osteoporosis, spine curvature (ky-

phosis), loss of ciliated cells and hearing, retinal atrophy, cataracts, susceptibility to infections, loss of body mass (wasting), neurodegeneration, altered behavior, and cancer (Gerhard and Cheng 2002; Kishi 2004; Kishi et al. 2009). The onset of many of these age-related changes is accelerated in the absence of telomerase, allowing telomerase-dependent and -independent phenotypes to be differentiated (Fig. 3; Carneiro et al. 2016a). In adult $tert^{-/-}$ zebrafish, there is a progressive shortening of telomeres, along with the accumulation of senescence and systemic inflammation, occurring over a relatively short period of typically between 6 and 9 months, as opposed to the 18–36 months ob-

Cite this article as *Cold Spring Harb Perspect Biol* doi: 10.1101/cshperspect.a041696

| | | $tert^{-/-}$ | | | | | $tert^{+/+}$ | | | | | | | |
|---|---|---|---|---|---|---|---|---|---|---|---|---|---|---|
| | | 3 m | 6 m | 9 m | 12 m | TOD | 3 m | 6 m | 9 m | 12 m | 18 m | 24 m | 36 m | TOD |
| EARLY | Impaired reproduction | + | ++ | +++ | +++ | +++ | – | – | – | – | + | ++ | +++ | +++ |
| EARLY | Gut inflammation | –/+ | ++ | ++ | +++ | +++ | – | – | – | – | + | ++ | ++ | +++ |
| INTERMED. | Fat loss | – | – | + | + | +++ | – | – | – | – | – | ++ | +++ | +++ |
| INTERMED. | Anemia | – | – | + | ++ | ++ | – | – | – | – | – | + | ++ | ++ |
| INTERMED. | Swim bladder infection | – | – | + | + | ++ | – | – | – | – | – | + | ++ | ++ |
| LATE | Decreased gut villi | – | – | | – | +++ | – | – | – | – | – | – | – | +++ |
| LATE | Cachexia | – | – | | – | +++ | – | – | – | – | – | – | – | +++ |

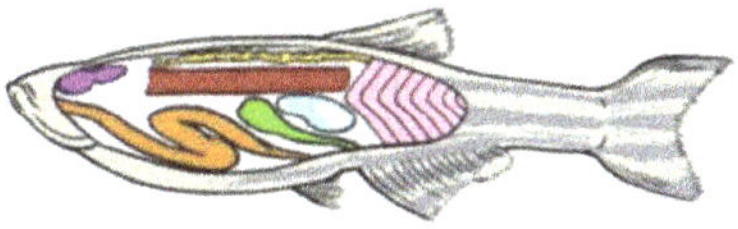

Figure 3. *tert* mutant zebrafish recapitulate old age phenotypes. Time-dependent organ dysfunction in aging $tert^{-/-}$ and wild-type (WT) zebrafish. Semiquantitative analysis was performed using a score ranging from (–) to (+++), depending on the severity and extent of the defects: (–) none, (+) minimal to mild, (++) moderate, (+++) severe. (TOD) Time of death. (*Below*) Schematic representation of the major zebrafish organs highlighting the brain, gut, kidney, reproductive organs, gas bladder, fat, skin, and muscle. (Figure based on data in Henriques et al. 2013 and Carneiro et al. 2016b.)

served in WT zebrafish. This enables an exceptional temporal analysis from young to old $tert^{-/-}$ animals (Carneiro et al. 2016b). Tissue degeneration in aging zebrafish follows a time- and tissue-dependent pattern, with highly proliferative tissues such as the intestine, testis, and blood being among the first to be affected (Fig. 3). This pattern mirrors the human scenario, where mutations affecting telomerase or telomere stability result in telomere biology disorders, particularly impacting highly proliferative tissues like the gut and blood.

Telomere shortening and the accumulation of a DDR in specific tissues precede aging-associated organ dysfunction. Accumulation of short telomeres and persistent DDR in one tissue, such as the gut, may be sufficient to dictate the rate of aging in other tissues via non-cell-autonomous mechanisms. Remarkably, telomerase absence triggers premature manifestation of aging phenotypes in most tissues, even where telomeres do not reach the shortened lengths characteristic of *tert* mutants during aging. Significantly, using percentage of life instead of absolute age to gauge the progression of aging phenotypes reveals surprisingly similar kinetics between the dynamics of WT and $tert^{-/-}$ phenotypes over time, locally and systemically (Carneiro et al. 2016b).

## FUNCTIONAL DEFECTS OF TELOMERASE-DEFICIENT ZEBRAFISH

Telomere shortening is associated with tissue-specific aging phenotypes in zebrafish, accompanied by non-tissue-specific diseases like cachexia, infections, cancer, and behavior impairment (Carneiro et al. 2016b; Espigares et al. 2021). The emergence of age-associated pathology correlates with the appearance of short telomeres in various tissues, such as gut and blood. $tert^{-/-}$ mutants recapitulate these diseases prematurely.

### Cachexia

Cachexia, characterized by weight loss, reduction of fat, and sarcopenia, increases frailty and mortality. In aging zebrafish, cachexia and spine deformation become increasingly prevalent after 24 months, with 38% of older WT fish displaying cachexia (Carneiro et al. 2016b). First-generation $tert^{-/-}$ mutants had significantly shorter life spans and exhibited cachexia from 12 months onward, affecting all mutants at the time of death. Histopathological analysis revealed tissue alterations in fat, gut, muscle, and testis associated with cachexia.

## Bacterial Infections

The incidence of swim bladder infection (aerocystitis) increases with age, affecting up to 30% of WT zebrafish by 36 months, often concurrent with cachexia (Carneiro et al. 2016b). $tert^{-/-}$ zebrafish showed a lower incidence of swim bladder infection but also in association with cachexia. In addition, postmortem bacteriological studies identified *Vibrio alginolyticus* and *Shewanella putrefaciens*, suggestive of aerocystitis as a likely cause of death in aged zebrafish.

## Cancer

Spontaneous tumor incidence was observed in both WT and $tert^{-/-}$ zebrafish, with *tert* mutants showing an earlier onset at 6–9 months compared to 18 months in WT (Carneiro et al. 2016b). Tumors were mainly invasive, affecting reproductive tissues in males and showing a heterogenous pattern, mostly of epithelial origin, in females (Carneiro et al. 2016b). Thus, the absence of telomerase did not significantly change incidence/malignancy rates but, like other age-associated phenotypes, it accelerated the onset of zebrafish cancer to earlier life.

In general, an increase in inflammation preceded cancer development in aging zebrafish (Carneiro et al. 2016b; Lex et al. 2020). Telomere shortening facilitates cancer development in a non-cell-autonomous manner. Studies using zebrafish chimeras revealed an increased incidence of invasive melanoma when tumors were generated in short telomere *tert* mutants (Lex et al. 2020). Adjacent tissues such as the skin and distant organs like the intestine in these mutants exhibited heightened levels of senescence and inflammation. Thus, in addition to the cell-autonomous role of short telomeres in contributing to genome instability, telomere shortening with age causes systemic chronic inflammation, which in turn leads to an increased incidence of tumors.

Approximately 90% of human tumors reactivate telomerase to elongate telomeres and protect chromosome ends. In the absence of telomerase, cancer cells can maintain telomeres through a recombination-based mechanism known as alternative lengthening of telomeres (ALT). Studies on zebrafish brain tumors observed reduced telomerase expression, which was often associated with ALT and the expression of TERRA (Idilli et al. 2020). Reexpression of telomerase in brain tumors not only prevented ALT but also reduced proliferation and malignancy, extending zebrafish survival. In contrast, zebrafish melanoma acquires the expression of telomerase upon tumor progression (Lopes-Bastos et al. 2023). The absence of telomerase in melanoma does not show reduced cancer incidence or improved survival, nor do they develop ALT (Idilli et al. 2020; Lopes-Bastos et al. 2023). Instead, telomerase deficiency increases tumor immunogenicity, leading to slower growth and even melanoma regression (Lopes-Bastos et al. 2023).

## Behavioral Impairment

Aging zebrafish exhibited decreased exploratory behavior, perturbed anxiety-like behaviors (bottom dwelling and resistance to stress stimuli) along with reduced locomotion (Espigares et al. 2021; Vasconcelos et al. 2023; Hudock and Kenney 2024). $tert^{-/-}$ fish displayed similar behavioral changes at younger ages, suggestive of accelerated cognitive decline (Espigares et al. 2021). A judgment bias experimental paradigm revealed that *tert* mutants were more pessimistic in response to ambiguous stimuli compared to WT zebrafish (Espigares et al. 2021), which is potentially linked to increased inflammation caused by telomere shortening. These results suggest that decreased physical state and life span associated with telomerase deficiency are linked to anxiety perception and impaired judgment bias.

## ORGAN REGENERATION IN TELOMERASE-DEFICIENT ZEBRAFISH

Tissue regeneration is a critical process that has been extensively studied in zebrafish due to their remarkable ability to completely regenerate various organs, such as the brain, spinal cord, retina, heart, and fins, even in mature adult stages. However, this regenerative capacity diminishes with age and is influenced by factors such as telomerase activity (Anchelin et al. 2011). The significant study by Bednarek and colleagues has shed light on the role of telomerase activity in tissue regeneration, particularly in response to heart damage (Bednarek et al. 2015). In this study, the authors found that ventricular cryoinjury induces telomerase hyperactivation in cardiomyocytes, leading to a sharp increase in proliferation and a transient elongation of telomeres. Additionally, cryoinjury triggers an initial accumulation of senescent cells, which are limited to the injured region and cleared upon wound closure. These findings suggest that cellular senescence may contribute to tissue remodeling in vivo, as seen in other organisms, including mammals (Antelo-Iglesias et al. 2021). However, in chronic pathological conditions such as aging (modeled by $tert^{-/-}$ fish), the regenerative capacity of tissues, especially the heart, is impaired. Persistent inhibition of the proliferative response and accumulation of senescent cells beyond the injured area contribute to tissue dysfunction and prevent heart regeneration. This persistent accumulation of senescent cells may overload the tissue with senescence-associated secretory phenotype (SASP), leading to a chronic inflammatory microenvironment and formation of fibrotic tissue (scarring) that further exacerbates tissue dysfunction.

The accumulation of senescent cells in aged tissues raises questions about the mechanisms underlying their persistence. Impairment in the clearance of senescent cells by the immune system and the impact of senescence on stem and progenitor cells are potential factors limiting not only organ regeneration but also homeostatic tissue renewal. Efforts to extend human healthy life span have led to pharmacological tests and biological therapies aimed at preventing telomere shortening and eliminating senescent cells during aging. However, the long-term effects of these interventions on human health and disease remain unclear. Preventing telomere shortening and eliminating senescent cells may delay aging, but it may also give rise to pathological conditions, such as cancer. Studies, such as these, highlight the potential for zebrafish in telomere research, since they combine phenotypes observed in humans with the power of genetic manipulation and their ability to regenerate vital organs, such as the brain and the heart.

## GENETIC RESCUE OF TELOMERASE-DEFICIENT ZEBRAFISH

Even though telomere decline in the absence of telomerase (or ALT) cannot be compensated, there have been several instances of genetic rescue of the premature aging phenotypes of *tert*-deficient zebrafish. This presupposes that shortened telomeres are detected as DNA damage by the cell before they are completely lost. Such was previously proposed for human replicative senescence studies in which loss of p53/Rb function would afford further rounds of cell division (and telomere shortening) before genomic instability (crisis) would initiate (Shay and Wright 2005). The same was observed at the organismal level in telomerase KO mice. Late-generation (G4) *tert* KO mice lacking *tp53* function increased longevity and fertility up to G6 (Chin et al. 1999).

### p53

*TP53*, often referred to as the "guardian" of the genome, is a tumor suppressor gene crucial for regulating cell-cycle progression and genome stability. Mutations in *TP53* are prevalent in human tumors. Zebrafish *tp53-M214K* mutants, equivalent to human M246 mutations, lack transcription of key genes involved in cell-cycle regulation and DNA repair, and develop soft-tissue tumors by 9 months of age resulting in reduced longevity (Berghmans et al. 2005; Şerifoğlu et al. 2024a). Aging *tert* mutants possess elevated p53 levels and increased transcription of its target genes, contributing to cellular senescence, apoptosis, and mitochondrial dysfunction. Loss of *tp53* function in *tert tp53-M214K* double mutants increased life span and delayed the onset of male infertility and other aging-related disorders (Şerifoğlu et al. 2024a), highlighting the role of p53 in regulating aging phenotypes. However, chronic inflammation was not rescued in *tert tp53-M214K* fish. Allowing for cell proliferation in the presence of inflammation may result in protumorigenic effects, that ultimately impair the health and longevity of the organism.

### cGAS-STING and IFN Inflammation

The cGAS-STING pathway, involved in DDRs and cell senescence, is activated by cytosolic DNA sensing, and triggers the production of type I interferons (IFNs) and inflammatory cytokines (Zierhut 2024). Human cell culture studies showed that the cGAS-STING response is triggered by telomere shortening (Nassour et al. 2019, 2023). While the specific activators of cGAS in *tert* mutant zebrafish remain unidentified, potential triggers include the presence of micronuclei, disrupted mitochondria, increased expression of transposable elements, and TERRA (Nassour et al. 2023; Şerifoğlu et al. 2024b). The absence of *sting* function in telomerase-deficient zebrafish, eliminates cell senescence, inflammation and increases cell proliferation, thus reverting tissue damage imposed to proliferative tissues. Unlike *tert tp53-M214K* double mutants, *tert sting* zebrafish are resistant to accelerated tumorigenesis typical of *tert* mutants (Şerifoğlu et al. 2024a,b). Thus, the absence of cGAS-STING reduces IFN systemic inflammation characteristic of *tert* mutants, resulting in increased longevity, improved male fertility, and recovery of age-associated phenotypes, including reduced spontaneous cancer incidence (Şerifoğlu et al. 2024b).

### Tissue-Specific Expression of Telomerase

The gut, a central organ in aging, plays a crucial role in nutrient uptake, immune-modulation, and interaction with gut microbiota. Enterocyte-specific telomerase expression prolongs gut homeostasis with age (El Maï et al. 2023), demonstrating the importance of telomere maintenance in this organ. As observed in aging mice (Okumura et al. 2021), senescent cells accumulate in the gut of *tert* mutants, secreting inflammatory molecules that induce DNA damage in distant organs, leading to systemic aging (El Maï et al. 2023). Delaying gut aging counters gut microbiota dysbiosis, suggesting a link between gut health, microbiota, and systemic aging. Therefore, restoring telomere length in critical tissues like the gut is sufficient to temporarily rescue local and systemic homeostasis in aging animals. This is confirmed by an increase of 40% in longevity of gut-rescued telomerase mutants (El Maï et al. 2023). Similarly, telomere shortening in mice triggers gut inflammation through the YAP pathway (Chakravarti et al. 2020). Mosaic expression of telomerase in the LGR5 cells of late-generation *tert* mice improved intestinal function and inflammation; however, no significant systemic effects were reported.

Ultimately, whether systemic aging signals will dominate over locally rejuvenated tissues remains to be examined. Nonetheless, such manipulations are likely to be effective only before telomeres reach a critical length. Mouse studies have indicated that telomerase reactivation in tissues with extensive telomere erosion strongly promotes tumorigenesis and malignancy (Ding et al. 2012). Time- and tissue-dependent telomerase reactivation will enable the definition and testing of such a beneficial period for telomerase therapies.

### ORGAN COMMUNICATION IN PREMATURE AGING OF TELOMERASE-DEFICIENT ZEBRAFISH

The discovery that tissue-specific rescue of telomerase promotes organismal recovery of *tert* mutants suggests the existence of communication between the initially impacted tissue and distal organs (Fig. 4). As a consequence, dysfunction in one tissue is sufficient to trigger aging in the whole organism.

A number of nonmutually exclusive hypotheses can be envisaged to explain this phenomenon.

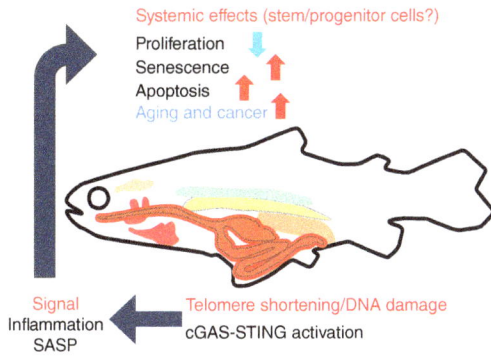

**Figure 4.** Short telomeres trigger chronic inflammation resulting in tissue damage and aging. Telomere shortening begins in specific organs during aging, such as the gut. Critically short telomeres disrupt local homeostasis, causing cellular senescence and type I inflammation through activation of the cGAS-STING pathway. Systemic inflammation and senescence-associated secretory phenotype (SASP) interferes with the proliferation of stem and progenitor cells in distant tissues leading to tissue damage, increased senescence, and apoptosis, resulting in remote organ dysfunction and fueling tumorigenesis with age.

## Humoral Hypothesis

*tert* mutant zebrafish exhibit signs of early cell senescence and SASP linked to gut dysfunction. Inflammatory cytokines from the affected gut may diffuse throughout the organism, influencing tissue homeostasis. Chimera assays demonstrated that <1% of injected blastula cells with very short telomeres can elicit a systemic response in WT embryos. This response consists of SASP molecules, inflammatory cytokines, among others (Lex et al. 2020). Using models such as these, targeting specific pathways, including senescence and the DDR, may offer insights into the mechanisms underlying tissue communication. Drugs such as senolytics and senomorphs, inhibitors of DDR pathways, and anti-inflammatory agents are being investigated for their effects on aging phenotypes.

## Immunosenescence Hypothesis

Telomere shortening in immune cells, particularly lymphoid cells, may reduce the clearance of senescent cells and contribute to organ dysfunction locally and systemically. Investigating whether immune cells from old fish and telomerase mutants are sufficient to cause systemic inflammation independent of other tissues is a priority. Distinguishing between the direct effects of an immune system with critically short telomeres and its role in amplifying inflammatory processes in other organs, such as the gut, will provide insights into the role of telomere shortening in aging.

## Microbiota and Associated Metabolites Hypothesis

Dysfunctional gut homeostasis and microbiota dysbiosis are linked to various age-associated phenotypes. The primary causes that trigger the feedforward loop between gut dysfunction and microbiota dysbiosis remain unclear. Telomere shortening in the gut may represent a good candidate to initiate these events, as seen in telomerase mutant zebrafish (El Maï et al. 2023). Microbiota metabolites, such as short-chain fatty acids, can impact immune system maturation and homeostasis, contributing to systemic effects (Liu et al. 2023). A decline in short-chain fatty acid levels may lead to increased gut inflammation, highlighting the potential role of microbiota and associated metabolites in age-related systemic inflammation.

## PERSPECTIVES

Telomere shortening is a well-established contributor to human aging. However, the multiple mechanisms operating across various biological scales—from molecules and cells to tissues and organisms—limit our understanding of how telomere damage affects tissue homeostasis and contributes to organism decline. A decade of research has demonstrated the relevance of zebrafish, which display human-like telomeres, and exhibit an acceleration of aging-associated diseases and tissue dysfunction upon deletion of telomerase. Importantly, these phenotypes manifest in the first mutant generation, mimicking what is observed in human aging and reflecting the typical "anticipation phenomenon" observed in individuals with TBDs such as dyskeratosis congenita. Both phenotypes and molecular mechanisms associated with telomere dysfunction are highly conserved from mammals to zebrafish, making zebrafish a powerful model for telomere research.

Several challenges and opportunities lie ahead in utilizing zebrafish for telomere research. These include testing the impact of tissue-specific telomerase therapeutics and identifying the molecular mechanisms underlying tissue dysfunction induced by short telomeres during aging. Zebrafish have already facilitated the identification of crucial links between tissue repair, telomerase activation, and cellular senescence, paving the way for a deeper understanding of how telomere dysfunction intersects with other hallmarks of aging.

## ACKNOWLEDGMENTS

Although this review aims at providing an overview of zebrafish telomere biology, it is not a complete compilation of data and I apologize to colleagues whose work I was unable to cite. I am extremely grateful for current and past laboratory members and the zebrafish community for the vital support and encouragement in helping me make the transition from a single-cell model system to a whole organism that ages as we do. I am thankful to Rita Araujo, Luis Batista, Herve Techer, and Eirini Trompouki for critically reading the manuscript.

## REFERENCES

Alcaraz-Pérez F, García-Castillo J, García-Moreno D, López-Muñoz A, Anchelin M, Angosto D, Zon LI, Mulero V, Cayuela ML. 2014. A non-canonical function of telomerase RNA in the regulation of developmental myelopoiesis in zebrafish. *Nat Commun* 5: 3228. doi:10.1038/ncomms4228

Anchelin M, Murcia L, Alcaraz-Pérez F, García-Navarro EM, Cayuela ML. 2011. Behaviour of telomere and telomerase during aging and regeneration in zebrafish. *PLoS ONE* 6: e16955. doi:10.1371/journal.pone.0016955

Anchelin M, Alcaraz-Perez F, Martinez CM, Bernabe-Garcia M, Mulero V, Cayuela ML. 2013. Premature aging in telomerase-deficient zebrafish. *Dis Model Mech* 6: 1101–1112. doi:10.1242/dmm.011635

Antelo-Iglesias L, Picallos-Rabina P, Estévez-Souto V, Da Silva-Álvarez S, Collado M. 2021. The role of cellular senescence in tissue repair and regeneration. *Mech Ageing Dev* 198: 111528. doi:10.1016/j.mad.2021.111528

Bednarek D, González-Rosa JM, Guzmán-Martínez G, Gutiérrez-Gutiérrez Ó, Aguado T, Sánchez-Ferrer C, Marques IJ, Galardi-Castilla M, de Diego I, Gómez MJ, et al. 2015. Telomerase is essential for zebrafish heart regeneration. *Cell Rep* 12: 1691–1703. doi:10.1016/j.celrep.2015.07.064

Berghmans S, Murphey RD, Wienholds E, Neuberg D, Kutok JL, Fletcher CDM, Morris JP, Liu TX, Schulte-Merker S, Kanki JP, et al. 2005. *Tp53* mutant zebrafish develop malignant peripheral nerve sheath tumors. *Proc Natl Acad Sci* 102: 407–412. doi:10.1073/pnas.0406252102

Blasco MA, Lee HW, Hande MP, Samper E, Lansdorp PM, DePinho RA, Greider CW. 1997. Telomere shortening and tumor formation by mouse cells lacking telomerase RNA. *Cell* 91: 25–34. doi:10.1016/S0092-8674(01)80006-4

Carneiro MC, de Castro IP, Ferreira MG. 2016a. Telomeres in aging and disease: lessons from zebrafish. *Dis Model Mech* 9: 737–748. doi:10.1242/dmm.025130

Carneiro MC, Henriques CM, Nabais J, Ferreira T, Carvalho T, Ferreira MG. 2016b. Short telomeres in key tissues initiate local and systemic aging in zebrafish. *PLoS Genet* 12: e1005798. doi:10.1371/journal.pgen.1005798

Cayuela ML, Claes KBM, Ferreira MG, Henriques CM, van Eeden F, Varga M, Vierstraete J, Mione MC. 2019. The zebrafish as an emerging model to study DNA damage in aging, cancer and other diseases. *Front Cell Dev Biol* 6: 178. doi:10.3389/fcell.2018.00178

Chakravarti D, Hu B, Mao X, Rashid A, Li J, Li J, Liao W, Whitley EM, Dey P, Hou P, et al. 2020. Telomere dysfunction activates YAP1 to drive tissue inflammation. *Nat Commun* 11: 4766. doi:10.1038/s41467-020-18420-w

Chin L, Artandi S, Shen Q, Tam A, Lee SL, Gottlieb GJ, Greider CW, DePinho RA. 1999. P53 deficiency rescues the adverse effects of telomere loss and cooperates with telomere dysfunction to accelerate carcinogenesis. *Cell* 97: 527–538. doi:10.1016/S0092-8674(00)80762-X

Ding Z, Wu CJ, Jaskelioff M, Ivanova E, Kost-Alimova M, Protopopov A, Chu GC, Wang G, Lu X, Labrot ES, et al. 2012. Telomerase reactivation following telomere dysfunction yields murine prostate tumors with bone metastases. *Cell* 148: 896–907. doi:10.1016/j.cell.2012.01.039

Dutta H. 1994. Growth in fishes. *Gerontology* 40: 97–112.

El Maï M, Marzullo M, de Castro IP, Ferreira MG. 2020. Opposing p53 and mTOR/AKT promote an in vivo switch from apoptosis to senescence upon telomere shortening in zebrafish. *eLife* 9: 1–26. doi:10.7554/eLife.54935

El Maï M, Bird M, Allouche A, Targen S, Şerifoğlu N, Lopes-Bastos B, Guigonis J, Kang D, Pourcher T, Yue J, et al. 2023. Gut-specific telomerase expression counteracts systemic aging in telomerase-deficient zebrafish. *Nat Aging* 3: 567–584. doi:10.1038/s43587-023-00401-5

Elmore LW, Norris MW, Sircar S, Bright AT, McChesney PA, Winn RN, Holt SE. 2008. Upregulation of telomerase function during tissue regeneration. *Exp Biol Med* 233: 958–967. doi:10.3181/0712-RM-345

Espigares F, Abad-Tortosa D, Varela SAM, Ferreira MG, Oliveira RF. 2021. Short telomeres drive pessimistic judgement bias in zebrafish. *Biol Lett* 17: rsbl.2020.0745. doi:10.1098/rsbl.2020.0745

Fogarty CE, Bergmann A. 2017. Killers creating new life: caspases drive apoptosis-induced proliferation in tissue repair and disease. *Cell Death Differ* **24**: 1390–1400. doi:10.1038/cdd.2017.47

García-Castillo J, Alcaraz-Pérez F, Martínez-Balsalobre E, García-Moreno D, Rossmann MP, Fernández-Lajarín M, Bernabé-García M, Pérez-Oliva AB, Rodríguez-Cortez VC, Bueno C, et al. 2021. Telomerase RNA recruits RNA polymerase II to target gene promoters to enhance myelopoiesis. *Proc Natl Acad Sci* **118**: e2015528118. doi:10.1073/pnas.2015528118

Gaullier G, Miron S, Pisano S, Buisson R, Le Bihan YV, Tellier-Lebègue C, Messaoud W, Roblin P, Guimarães BG, Thai R, et al. 2016. A higher-order entity formed by the flexible assembly of RAP1 with TRF2. *Nucleic Acids Res* **44**: 1962–1976. doi:10.1093/nar/gkv1531

Gerhard GS, Cheng KC. 2002. A call to fins! Zebrafish as a gerontological model. *Aging Cell* **1**: 104–111. doi:10.1046/j.1474-9728.2002.00012.x

Hao LY, Armanios M, Strong MA, Karim B, Feldser DM, Huso D, Greider CW. 2005. Short telomeres, even in the presence of telomerase, limit tissue renewal capacity. *Cell* **123**: 1121–1131. doi:10.1016/j.cell.2005.11.020

Harel I, Benayoun BA, Machado B, Singh PP, Hu CKK, Pech MF, Valenzano DR, Zhang E, Sharp SC, Artandi SE, et al. 2015. A platform for rapid exploration of aging and diseases in a naturally short-lived vertebrate. *Cell* **160**: 1013–1026. doi:10.1016/j.cell.2015.01.038

Hartmann N, Rudolph KL, Reichwald K, Lechel AA, Graf M, Kirschner J, Dorn A, Terzibasi E, Wellner J, Platzer M, et al. 2009. Telomeres shorten while Tert expression increases during ageing of the short-lived fish *Nothobranchius furzeri*. *Mech Ageing Dev* **130**: 290–296. doi:10.1016/j.mad.2009.01.003

Hatakeyama H, Nakamura K, Izumiyamashimomura N, Ishii A, Tsuchida S, Takubo K, Ishikawa N. 2008. The teleost *Oryzias latipes* shows telomere shortening with age despite considerable telomerase activity throughout life. *Mech Ageing Dev* **129**: 550–557. doi:10.1016/j.mad.2008.05.006

Hathcock KS, Hemann MT, Opperman KK, Strong MA, Greider CW, Hodes RJ. 2002. Haploinsufficiency of mTR results in defects in telomere elongation. *Proc Natl Acad Sci* **99**: 3591–3596. doi:10.1073/pnas.012549799

Henriques CM, Ferreira MG. 2024. Telomere length is an epigenetic trait—implications for the use of telomerase-deficient organisms to model human disease. *Dis Model Mech* **17**: 1–6. doi:10.1242/dmm.050581

Henriques CM, Carneiro MC, Tenente IM, Jacinto A, Ferreira MG. 2013. Telomerase is required for zebrafish lifespan. *PLoS Genet* **9**: e1003214. doi:10.1371/journal.pgen.1003214

Hudock J, Kenney JW. 2024. Aging in zebrafish is associated with reduced locomotor activity and strain dependent changes in bottom dwelling and thigmotaxis. *PLoS ONE* **19**: e0300227. doi:10.1371/journal.pone.0300227

Idilli AI, Cusanelli E, Pagani F, Berardinelli F, Bernabé M, Cayuela ML, Poliani PL, Mione MC. 2020. Expression of tert prevents ALT in zebrafish brain tumors. *Front Cell Dev Biol* **8**: 1–18. doi:10.3389/fcell.2020.00065

Imamura S, Uchiyama J, Koshimizu E, Hanai JI, Raftopoulou C, Murphey RD, Bayliss PE, Imai Y, Burns CE, Masutomi K, et al. 2008. A Non-canonical function of zebrafish telomerase reverse transcriptase is required for

developmental hematopoiesis. *PLoS ONE* **3**: e3364. doi:10.1371/journal.pone.0003364

Kishi S. 2004. Functional aging and gradual senescence in zebrafish. *Ann NY Acad Sci* **1019**: 521–526. doi:10.1196/annals.1297.097

Kishi S, Bayliss PE, Uchiyama J, Koshimizu E, Qi J, Nanjappa P, Imamura S, Islam A, Neuberg D, Amsterdam A, et al. 2008. The identification of zebrafish mutants showing alterations in senescence-associated biomarkers. *PLoS Genet* **4**: e1000152. doi:10.1371/journal.pgen.1000152

Kishi S, Slack BE, Uchiyama J, Zhdanova IV. 2009. Zebrafish as a genetic model in biological and behavioral gerontology: where development meets aging in vertebrates—a mini-review. *Gerontology* **55**: 430–441. doi:10.1159/000228892

Lau BWM, Wong AOL, Tsao GSW, So KF, Yip HKF. 2008. Molecular cloning and characterization of the zebrafish (*Danio rerio*) telomerase catalytic subunit (telomerase reverse transcriptase, TERT). *J Mol Neurosci* **34**: 63–75. doi:10.1007/s12031-007-0072-x

Lex K, Maia Gil M, Lopes-Bastos B, Figueira M, Marzullo M, Giannetti K, Carvalho T, Ferreira MG. 2020. Telomere shortening produces an inflammatory environment that increases tumor incidence in zebrafish. *Proc Natl Acad Sci* **117**: 15066–15074. doi:10.1073/pnas.1920049117

Liu X, Shao J, Liao Y, Wang L, Jia Y, Dong P, Liu Z, He D, Li C, Zhang X. 2023. Regulation of short-chain fatty acids in the immune system. *Front Immunol* **14**: 1186892. doi:10.3389/fimmu.2023.118689

Lopes-Bastos B, Nabais J, Ferreira T, El Maï M, Bird M, Targen S, Kang D, Yue JX, Carvalho T, Ferreira MG. 2023. Absence of telomerase leads to immune response and tumor regression in zebrafish melanoma. bioRxiv doi:10.1101/2023.03.24.534079

Lundblad V, Szostak JW. 1989. A mutant with a defect in telomere elongation leads to senescence in yeast. *Cell* **57**: 633–643. doi:10.1016/0092-8674(89)90132-3

Ma J, Tang D, Gao P, Liang S, Zhang R. 2022. Knockout of Shelterin subunit genes in zebrafish results in distinct outcomes. *Biochem Biophys Res Commun* **617**: 22–29. doi:10.1016/j.bbrc.2022.05.079

Markiewicz-Potoczny M, Lobanova A, Loeb AM, Kirak O, Olbrich T, Ruiz S, Lazzerini Denchi E. 2021. TRF2-mediated telomere protection is dispensable in pluripotent stem cells. *Nature* **589**: 110–115. doi:10.1038/s41586-020-2959-4

Martínez-Balsalobre E, García-Castillo J, García-Moreno D, Naranjo-Sánchez E, Fernández-Lajarín M, Blasco MA, Alcaraz-Pérez F, Mulero V, Cayuela ML. 2023. Telomerase RNA-based aptamers restore defective myelopoiesis in congenital neutropenic syndromes. *Nat Commun* **14**: 5912. doi:10.1038/s41467-023-41472-7

Marzullo M, El Maï M, Ferreira M. 2022. Whole-mount senescence-associated β-galactosidase (SA-β-GAL) activity detection protocol for adult zebrafish. *Bio Protocol* **12**: 1–26. doi:10.21769/BioProtoc.4457

McDowall RM. 1994. On size and growth in freshwater fish. *Ecol Freshw Fish* **3**: 67–79. doi:10.1111/j.1600-0633.1994.tb00108.x

Meier B, Clejan I, Liu Y, Lowden M, Gartner A, Hodgkin J, Ahmed S. 2006. trt-1 is the *Caenorhabditis elegans* cata-

lytic subunit of telomerase. *PLoS Genet* **2**: e18. doi:10
.1371/journal.pgen.0020018

Meyne J, Baker RJ, Hobart HH, Hsu TC, Ryder OA, Ward
OG, Wiley JE, Wurster-Hill DH, Yates TL, Moyzis RK.
1990. Distribution of non-telomeric sites of the
(TTAGGG)n telomeric sequence in vertebrate chromosomes. *Chromosoma* **99**: 3–10. doi:10.1007/BF01737283

Myler LR, Kinzig CG, Sasi NK, Zakusilo G, Cai SW, De
Lange T. 2021. The evolution of metazoan shelterin.
*Genes Dev* **35**: 1625–1641. doi:10.1101/gad.348835.121

Nassour J, Radford R, Correia A, Fusté JM, Schoell B, Jauch
A, Shaw RJ, Karlseder J. 2019. Autophagic cell death re-
stricts chromosomal instability during replicative crisis.
*Nature* **565**: 659–663. doi:10.1038/s41586-019-0885-0

Nassour J, Aguiar LG, Correia A, Schmidt TT, Mainz L,
Przetocka S, Haggblom C, Tadepalle N, Williams A,
Shokhirev MN, et al. 2023. Telomere-to-mitochondria
signalling by ZBP1 mediates replicative crisis. *Nature*
**614**: 767–773. doi:10.1038/s41586-023-05710-8

Ocalewicz K. 2013. Telomeres in fishes. *Cytogenet Genome
Res* **141**: 114–125. doi:10.1159/000354278

Okumura S, Konishi Y, Narukawa M, Sugiura Y, Yoshimoto
S, Arai Y, Sato S, Yoshida Y, Tsuji S, Uemura K, et al. 2021.
Gut bacteria identified in colorectal cancer patients promote tumourigenesis via butyrate secretion. *Nat Commun* **12**: 1–14. doi:10.1038/s41467-021-25965-x

Pfennig F, Kind B, Zieschang F, Busch M, Gutzeit HO. 2008.
*Tert* expression and telomerase activity in gonads and
somatic cells of the Japanese medaka (*Oryzias latipes*).
*Dev Growth Differ* **50**: 131–141. doi:10.1111/j.1440-
169X.2008.00986.x

Reichard M, Giannetti K, Ferreira T, Maouche A, Vrtílek M,
Polačik M, Blažek R, Ferreira MG. 2021. Lifespan and
telomere length variation across populations of wild-derived African killifish. *Mol Ecol* **31**: 5979–5992. doi:10
.1111/mec.16287

Riha K, McKnight TD, Griffing LR, Shippen DE. 2001. Living with genome instability: plant responses to telomere
dysfunction. *Science* **291**: 1797–1800. doi:10.1126/science
.1057110

Rudolph KL, Chang S, Lee HW, Blasco M, Gottlieb GJ, Greider C, DePinho RA. 1999. Longevity, stress response, and
cancer in aging telomerase-deficient mice. *Cell* **96**: 701–
712. doi:10.1016/S0092-8674(00)80580-2

Ruis P, Boulton SJ. 2021. The end protection problem—an
unexpected twist in the tail. *Genes Dev* **35**: 1–21. doi:10
.1101/gad.344044.120

Sahin E, Colla S, Liesa M, Moslehi J, Müller FL, Guo M,
Cooper M, Kotton D, Fabian AJ, Walkey C, et al. 2011.
Telomere dysfunction induces metabolic and mitochondrial compromise. *Nature* **470**: 359–365. doi:10.1038/na
ture09787

Scahill CM, Digby Z, Sealy IM, White RJ, Collins JE, Busch-
Nentwich EM. 2017. The age of heterozygous telomerase
mutant parents influences the adult phenotype of their
offspring irrespective of genotype in zebrafish. *Wellcome
Open Res* **2**: 77. doi:10.12688/wellcomeopenres.12530.1

Şerifoğlu N, Lopes-Bastos B, Ferreira MG. 2024a. Lack of
telomerase reduces cancer incidence and increases lifespan of zebrafish tp53M214K mutants. *Sci Rep* **14**: 5382.
doi:10.1038/s41598-024-56153-8

Şerifoğlu N, Allavena G, Lopes-Bastos B, Marzullo M, Bou-
sounis P, Trompouki E, Ferreira MG. 2024b. cGAS-
STING is responsible for aging of telomerase deficient
zebrafish. bioRxiv doi:10.1101/2024.03.11.584360

Shay JW, Wright WE. 2005. Senescence and immortalization: role of telomeres and telomerase. *Carcinogenesis* **26**:
867–874. doi:10.1093/carcin/bgh296

Vasconcelos RO, Gordillo-Martinez F, Ramos A, Lau IH.
2023. Effects of noise exposure and ageing on anxiety
and social behaviour in zebrafish. *Biology (Basel)* **12**:
1165. doi:10.3390/biology12091165

Vicari MR, Bruschi DP, Cabral-De-mello DC, Nogaroto V.
2022. Telomere organization and the interstitial telomeric
sites involvement in insects and vertebrates chromosome
evolution. *Genet Mol Biol* **45**: 1–22. doi:10.1590/1678-
4685-gmb-2022-0071

Wagner KDD, Ying Y, Leong W, Jiang J, Hu X, Chen Y,
Michiels JFF, Lu Y, Gilson E, Wagner N, et al. 2017.
The differential spatiotemporal expression pattern of
shelterin genes throughout lifespan. *Aging* **9**: 1219–
1232. doi:10.18632/aging.101223

Wai HY, Yeoh E, Brenner S, Venkatesh B. 2005. Cloning and
expression of the reverse transcriptase component of puf-
ferfish (*Fugu rubripes*) telomerase. *Gene* **353**: 207–217.
doi:10.1016/j.gene.2005.04.038

Wilbourn RV, Moatt JP, Froy H, Walling CA, Boonekamp JJ,
Nussey DH. 2018. The relationship between telomere
length and mortality risk in non-model vertebrate systems: a meta-analysis. *Phil Trans R Soc B* **373**:
20160447. doi:10.1098/rstb.2016.0447

Wu L, Multani AS, He H, Cosme-Blanco W, Deng YY, Deng
JM, Bachilo O, Pathak S, Tahara H, Bailey SM, et al. 2006.
Pot1 deficiency initiates DNA damage checkpoint activation and aberrant homologous recombination at telomeres. *Cell* **126**: 49–62. doi:10.1016/j.cell.2006.05.037

Xie M, Mosig A, Qi X, Li Y, Stadler PF, Chen JLJL. 2008.
Structure and function of the smallest vertebrate telomerase RNA from teleost fish. *J Biol Chem* **283**: 2049–2059.
doi:10.1074/jbc.M708032200

Xie Y, Yang D, He Q, Songyang Z. 2011. Zebrafish as a model
system to study the physiological function of telomeric
protein TPP1. *PLoS ONE* **6**: e16440. doi:10.1371/journal
.pone.0016440

Ying Y, Hu X, Han P, Mendez-Bermudez A, Bauwens S Y, Eid
R, Tan L, Gilson E, Ye J, Lu, et al. 2022. The non-telomeric
evolutionary trajectory of TRF2 in zebrafish reveals its specific roles in neurodevelopment and aging. *Nucleic Acids
Res* **50**: 2081–2095. doi:10.1093/nar/gkac065

Yu GL, Bradley JD, Attardi LD, Blackburn EH. 1990. In vivo
alteration of telomere sequences and senescence caused
by mutated *Tetrahymena* telomerase RNAs. *Nature* **344**:
126–132. doi:10.1038/344126a0

Yu RMK, Chen EXH, Kong RYC, Ng PKS, Mok HOL, Au
DWT. 2006. Hypoxia induces telomerase reverse transcriptase (TERT) gene expression in non-tumor fish tissues in vivo: the marine medaka (*Oryzias melastigma*)
model. *BMC Mol Biol* **7**: 27. doi:10.1186/1471-2199-7-27

Zierhut C. 2024. Potential cGAS-STING pathway functions
in DNA damage responses, DNA replication and DNA
repair. *DNA Repair (Amst)* **133**: 103608. doi:10.1016/j
.dnarep.2023.103608

Cite this article as *Cold Spring Harb Perspect Biol* doi: 10.1101/cshperspect.a041696

# Life and Death without Telomerase: The *Saccharomyces cerevisiae* Model

Veronica Martinez-Fernandez, Aurélia Barascu, and Maria Teresa Teixeira

Sorbonne Université, CNRS, Institut de Biologie Physico-Chimique, Laboratoire de Biologie Moléculaire et Cellulaire des Eucaryotes, LBMCE, F-75005 Paris, France

*Correspondence:* teresa.teixeira@cnrs.fr

*Saccharomyces cerevisiae*, a model organism in telomere biology, has been instrumental in pioneering a comprehensive understanding of the molecular processes that occur in the absence of telomerase across eukaryotes. This exploration spans investigations into telomere dynamics, intracellular signaling cascades, and organelle-mediated responses, elucidating their impact on proliferative capacity, genome stability, and cellular variability. Through the lens of budding yeast, numerous sources of cellular heterogeneity have been identified, dissected, and modeled, shedding light on the risks associated with telomeric state transitions, including the evasion of senescence. Moreover, the unraveling of the intricate interplay between the nucleus and other organelles upon telomerase inactivation has provided insights into eukaryotic evolution and cellular communication networks. These contributions, akin to milestones achieved using budding yeast, such as the discovery of the cell cycle, DNA damage checkpoint mechanisms, and DNA replication and repair processes, have been of paramount significance for the telomere field. Particularly, these insights extend to understanding replicative senescence as an anticancer mechanism in humans and enhancing our understanding of eukaryotes' evolution.

Telomeres are located at the termini of linear chromosomes in eukaryotes. They consist of G-rich repetitive DNA sequences that extend from the 5′ to the 3′ protruding end and are associated with specialized proteins and noncoding RNA. The primary role of telomeres is to protect chromosome ends from being mistakenly recognized as DNA double-strand breaks (DSBs), thereby preventing the inappropriate activation of DNA repair processes that could otherwise compromise genome stability (Jain and Cooper 2010; de Lange 2018). During DNA replication, telomeres undergo a natural shortening process attributed to the DNA end-replication problem (Lingner et al. 1995; Soudet et al. 2014; Takai et al. 2024). To counteract this shortening, telomerase, a specialized ribonucleoprotein complex, synthesizes telomeric repeats de novo using its reverse transcriptase subunit and an RNA template (Vasianovich and Wellinger 2017; Ha et al. 2022; Lue and Autexier 2023). Telomerase plays a pivotal role in enabling the unlimited proliferation of eukaryotic cells by maintaining telomere length. Nevertheless, telomerase activity is repressed in many somatic cells of humans and other vertebrates.

This leads to telomere erosion, ultimately triggering replicative senescence—a permanent cell-cycle arrest due to the activation of the DNA damage checkpoint (DDC) (Enomoto et al. 2002; d'Adda di Fagagna et al. 2003; Ijpma and Greider 2003). Some rare telomerase-negative cells, known as "postsenescent survivors," bypass the massive cell proliferation arrest associated with telomere shortening and maintain their telomeres through homology-directed repair (HDR) within telomeric sequences (Lundblad and Blackburn 1993; Dilley et al. 2016). Persistently short telomeres cause a spectrum of early and fatal degenerative syndromes, such as dyskeratosis congenita, pulmonary fibrosis, and bone marrow failure (Stanley and Armanios 2015). Conversely, cancer cells bypass telomere-initiated senescence by both suppressing the signaling pathways that control replicative senescence and pathological lengthening of their telomeres (Artandi and DePinho 2000). Thus, understanding the processes of telomere shortening and assessing their impact on cells is crucial for unraveling the molecular mechanisms underlying human diseases. Moreover, this knowledge enhances our understanding of how evolution has exploited the advent of linear chromosomes in eukaryotes to drive innovative biological applications.

From the initial stages of the molecular characterization of telomeres during the 1970s and 1980s, it has become evident that telomeres' structural and dynamic aspects are remarkably preserved throughout eukaryotic evolution (Teixeira and Gilson 2005). Telomeric DNA at the extremities of chromosomes is often composed of repeated short motifs rich in "TGs" in the sequence running from the 5′ to the 3′ end (degenerated $TG_{1-3}$ in *Saccharomyces cerevisiae*,

Fig. 1). Early experiments in telomere research used ciliates as models, drawn by the abundant presence of mini-chromosomes encoding rRNA in this species, which provided an exceptional source of DNA extremities for biochemical studies (Blackburn and Gall 1978). However, a transformative shift occurred with the introduction of *S. cerevisiae* as a model organism. This yeast species offered additional experimental tools, including a broad range of molecular biology techniques, gene-editing capabilities, and robust genetic resources, accelerating telomere research. In a groundbreaking experiment, the termini of a *Tetrahymena* mini-chromosome were ligated to a yeast linearized plasmid (Szostak and Blackburn 1982). The resulting recombinant molecule exhibited remarkable stability within yeast cells following transformation. This result underscored the capacity of the yeast telomere maintenance system to recognize and maintain the DNA ends of a distantly related organism, establishing the evolutionary conservation of the telomere maintenance system across two distinct eukaryotic domains. Building on the cloning of yeast telomeres and the maintenance of such a linear plasmid, a genetic screen was devised to identify yeast mutants exhibiting a gradual reduction in telomere length, designated as *est* (ever shorter telomeres). Mutant alleles of *EST1*, *EST2*, *EST3*, and *EST4*, later identified as *CDC13*, did not lead to immediate cellular lethality; instead, yeast mutants gradually lost their proliferation ability over time, a phenotype termed replicative senescence (Lundblad and Szostak 1989; Lendvay et al. 1996). The synergy between ciliates and budding yeast models continued with the discovery of the telomerase catalytic subunit. First purified from ciliates, where it is very abundant, protein

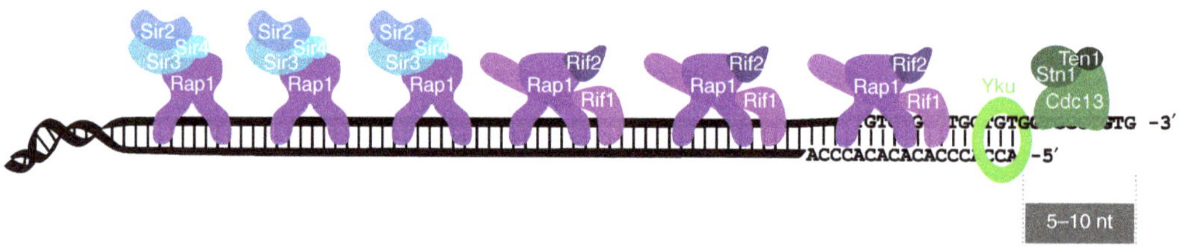

**Figure 1.** Scheme of major telomeric proteins in *Saccharomyces cerevisiae*.

Cite this article as *Cold Spring Harb Perspect Biol* doi: 10.1101/cshperspect.a041699

sequencing revealed *EST2* as the budding yeast ortholog encoding the catalytic subunit of telomerase. Specific amino acid mutations in the Est2 putative catalytic site displayed telomere shortening and senescence, establishing telomerase as a reverse transcriptase with origins common to reverse transcriptases of retroelements (Lingner et al. 1997; Nakamura and Cech 1998).

The cloning of yeast telomeres enabled the purification and characterization of the binding protein Rap1, an abundant nuclear protein also involved in transcription regulation at many gene promoters genome-wide (Berman et al. 1986; Shore and Nasmyth 1987; Longtine et al. 1989). In addition, Cdc13, which binds to the ssDNA telomeric G-strand, was discovered in a seminal yeast genetic screen for cell-cycle components and later demonstrated to be a major telomeric protein (Hartwell et al. 1973; Garvik et al. 1995; Lin and Zakian 1996). Telomeric proteins also comprise Rif1 and Rif2, which are multifunctional proteins that bind telomeric DNA either directly or through their interaction with Rap1; the Sir complex (Sir2, Sir3, and Sir4), which associates with telomeres through Rap1 and mediates the heterochromatinization of telomeric regions by deacetylating histones; and, finally, Stn1 and Ten1, which form a tripartite complex with Cdc13—the CST complex—with similarities to RPA, at single-stranded telomeric DNA (Fig. 1; Gao et al. 2007; Wellinger and Zakian 2012).

In this review, we will describe the consequences of experimental inactivation of telomerase in the model organism *S. cerevisiae*, which has been credited as a model for the first steps of replicative senescence, a phenomenon first described as the Hayflick limit (Hayflick 1965). Many mammalian species with short life spans and low body masses, such as mouse laboratory strains, do not repress telomerase, and their telomeres do not shorten with age (Gomes et al. 2011), which precludes these species from serving as direct models for the study of telomere length-dependent replicative aging. We will explore how *S. cerevisiae* can serve as a vanguard model of replicative senescence, leveraging advanced experimental approaches encompassing modern genetics, genomics, quantitative biology, and mathematical modeling.

## BEFORE ALL: WHY DO TELOMERES SHORTEN?

The discovery of cellular DNA semi-conservative replication mechanisms prompted Watson (1972) and Olovnikov (1973) to realize that if replication of telomeres required RNA primers, removal of the 5′-terminal primer would lead to one of the daughter DNA strands being shorter than its parental counterpart. This was consistent with the discovery that telomeres end with a 3′-overhang in ciliates (Zakian 1989). However, evidence that both ends of each linear DNA molecule possess 3′-overhangs, combined with the discovery of a transient increase of single-stranded G-rich tails late in the S-phase in budding yeast, indicated that the DNA end-replication problem was more complex than expected (Wellinger et al. 1993b, 1996). In particular, it was hypothesized that the shortening of DNA extremities at the passage of the replisome occurs primarily at the newly synthesized leading strand, which stops prematurely due to the absence of a template from the leading parental molecule—the equivalent of the 3′-overhang. In contrast, the lagging strand parental molecule would appear unaffected (Lingner et al. 1995). Hence, while the 3′-overhang at the lagging strand is preserved by the passage of the replication fork, leading-strand synthesis results in a blunt end that is further processed to regenerate the 3′-tail (Fig. 2; Faure et al. 2010; Soudet et al. 2014). As a consequence, the shortening rate of telomeres is theoretically strictly dependent on the overhang length. Thus, the overhang length and its regulation at both the leading and lagging strands are key parameters in replicative senescence.

A significant breakthrough has recently brought attention to the molecular intricacies underlying the reestablishment of the 3′-overhang on the lagging strand. Experiments employing an in vitro reconstituted DNA replication system from *S. cerevisiae* demonstrated that the RNA primer of the final Okazaki fragment cannot be placed once the replisome approaches

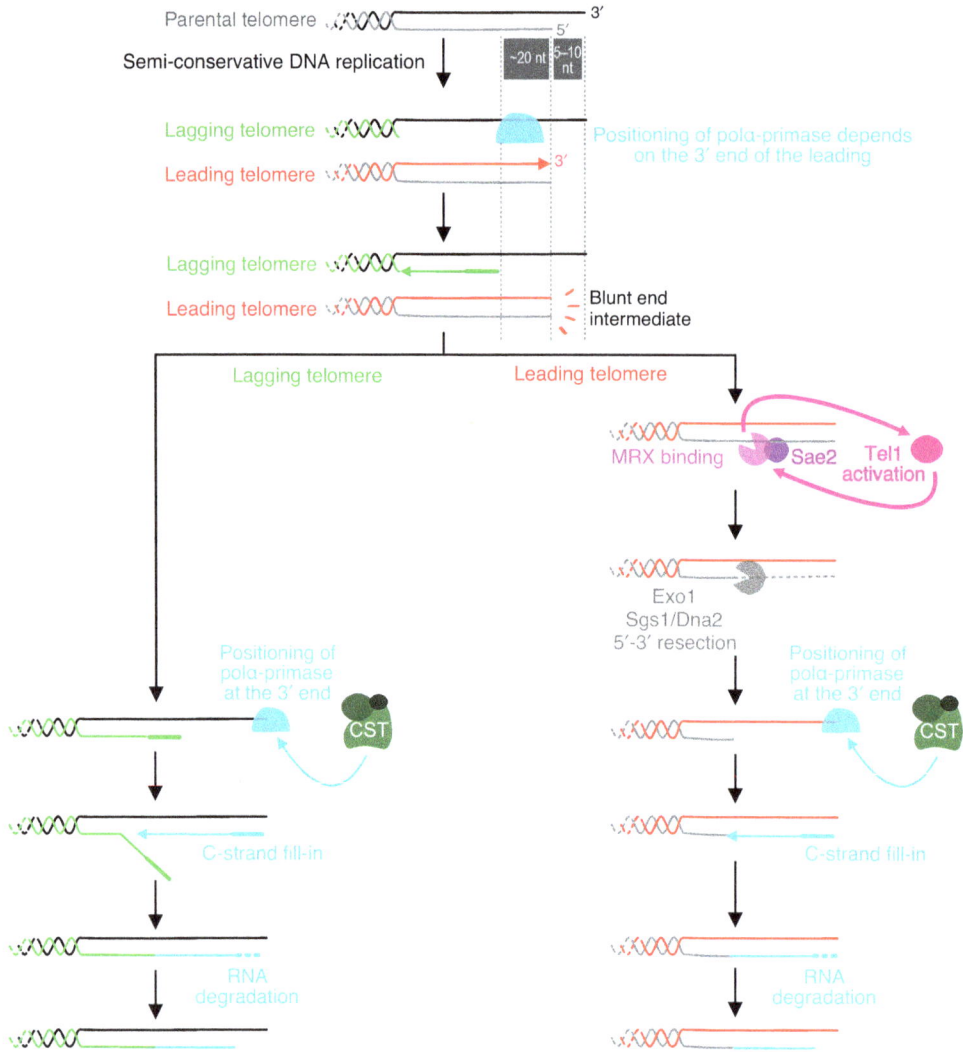

**Figure 2.** The current model of the DNA end-replication problem. When the replisome reaches the telomere, synthesis of the leading strand may proceed to the 5′ end of the template, creating a blunt intermediate. This intermediate is processed to form a 3′-overhang of ~40 nt by the action of Mre11 assisted by Sae2 and Tel1 and Exo1, Sgs1/Dna2. The last Okazaki fragment laid in the context of the replisome during the lagging-strand synthesis leaves an overhang. The telomeric complex Cdc13-Stn1-Ten1 (CST) ensures that a last telomeric Okazaki fragment is initiated near or at the 3′ end, both at lagging and processed leading strands.

~20 nt from the end of the leading-strand template (Takai et al. 2024). This is consistent with cryo-electron microscopy structures of the yeast replisome containing Polα-primase, where the primase active site is positioned ~75 Å from the point of template unwinding, corresponding to 20–30 nt of ssDNA (Jones et al. 2023). As a result, the replisome is unable to support the conventional lagging-strand synthesis along a protruding single-stranded lagging-strand template. This feature of the semi-conservative DNA replication machinery appears to be conserved between yeast and humans, as is the molecular machinery that evolved to limit the overhang length and consequently the telomere shortening rate: the CST complex. Accordingly,

Cite this article as *Cold Spring Harb Perspect Biol* doi: 10.1101/cshperspect.a041699

CST proteins from *Candida glabrata*, a species closely related to *S. cerevisiae*, were found to enable the priming reaction by Polα-primase on an ssDNA template and to promote the primase-to-polymerase switch, reducing the length of the RNA primer (Lue et al. 2014). Since the overhang length was measured to be about 5–10 nt in budding yeast (Soudet et al. 2014), which is much shorter than that in humans, the yeast CST may enable positioning the Polα-primase right at the 3′ extremity of the parental lagging molecule. Yet, whether and how the RNA primer is then removed remains unsolved.

The processing of the leading-strand synthesis product to reconstitute the 3′-tail telomeric structure was first discovered and best dissected in budding yeast. Although there are some differences between humans and yeast, the fundamental principle is conserved: The blunt end of the leading strand undergoes processing similarly to DNA DSBs, albeit with restricted and regulated action by telomeric proteins through convergent evolution (Myler et al. 2023; Pizzul et al. 2024). First, in *S. cerevisiae*, in vivo data suggest that the synthesis of the new molecule at the leading strand proceeds up to the end of the template (Soudet et al. 2014). Second, processing follows the passage of the replication fork, is cell-cycle-restricted and is independent of telomerase activity (Wellinger et al. 1993a,b 1996; Dionne and Wellinger 1996, 1998). Third, maturation of the blunt end requires limited activation of the DDC and the restricted action of nucleases well known for their roles in processing DSBs and stalled replication forks, thereby generating ssDNA (Larrivée et al. 2004; Soudet et al. 2014). In brief, the current model proposes that the MRX complex composed of the Mre11 nuclease, Rad50 and Xrs2, is recruited to telomeres along with Tel1, and joined by Sae2. Next, while Mre11 endo- or exonuclease activities may be dispensable in this context, Exo1 and Sgs1/Dna2 collaborate to 5′-3′ resect the C-strand (Krogh et al. 2005; Bonetti et al. 2009). Subsequently, Rap1-dependent binding of Rif2 at telomeres limits the resection of the C-strand (Bonetti et al. 2010; Cejka and Symington 2021).

Importantly, relatively longer telomeres destabilize MRX binding and Tel1 activation to initiate C-strand resection via Rif2 inhibition. This was demonstrated using two systems to generate a single shorter telomere in cells, which could be compared with other telomeres in the same cells or control strains. In the first system, a set of strains was constructed in which a unique site for the HO endonuclease was introduced in a subtelomeric locus, with short (80-bp) or long (250-bp) telomeric repeats flanking the HO site. Conditional expression of the HO endonuclease thus generated a unique short or long telomere, followed by a few nontelomeric base pairs (Diede and Gottschling 1999, 2001). In the second system, telomeric repeats were modified to contain two site-specific recombinase Flp1 recognition target sites, resulting in native telomeric ends with variable short lengths upon in vivo recombination (Marcand et al. 1999). Although some differences exist between the two systems, particularly an additional abridged Mec1 DDC activation in the HO system possibly due to the presence of the nontelomeric DNA extremity, both systems suggested differential processing of short and long telomeres, the short telomeres being subjected to longer 5′–3′ resection (Michelson et al. 2005; Negrini et al. 2007; McGee et al. 2010; Fallet et al. 2014).

In budding yeast, the telomere shortening rate is constant, independent of telomere length (Marcand et al. 1999), suggesting that 5- to 10-nt-long G-overhangs are reconstituted after each telomere replication. Therefore, while the extent of resection is higher at short telomeres, C-strand synthesis might be stimulated by a putative higher recruitment of Cdc13 (Negrini et al. 2007; McGee et al. 2010). Consequently, natural chromosome ends serve a dual role: They both initiate and restrict DDC and DNA repair activities typically observed at DSBs. The mechanism coordinating this delicate balance remains unclear but is crucial for maintaining genomic integrity.

## PROCESSING AND SIGNALING OF THE SHORTEST TELOMERES IN CELLS

The finding that *est* mutants in budding yeast display progressive shortening of telomeres and simultaneous gradual increase of cell mortality largely contributed to set the stage for the paradigmatic association of telomere length with cell proliferation capacity in eukaryotes (Lundblad and Szostak 1989; Greider 1990; Lendvay et al. 1996). Accordingly, in human primary fibroblasts, the original cells in which the Hayflick limit was first described, telomeres shorten upon serial passage and cells gradually lose viability. This process is reversed whenever telomerase is reexpressed (Harley et al. 1990; Counter et al. 1992). In the absence of certain DDC factors, cells can survive the replicative senescence limit, and telomeres shorten further until a crisis step marked by high cellular mortality and high genome instability is reached, suggesting a worsening of telomere function (Artandi and DePinho 2000).

The detailed analysis of the phenotype of yeast cells lacking telomerase activity has offered invaluable insights mirroring the Hayflick limit (Lundblad and Szostak 1989). Initially, these cells exhibited viability spanning several generations before mortality became apparent, suggesting a delayed phenotype possibly linked to an accumulation of short telomeres preceding its onset. Additionally, cell cultures displayed significant cell-to-cell heterogeneity in terms of cell proliferation capacity, underscoring complex mechanisms at work (Xu and Teixeira 2019). Consequently, this phenomenon observed in yeast was also termed replicative senescence.

The early finding that telomerase activity relies on the major DDC kinases Tel1 and Mec1 (Ritchie et al. 1999) highlighted the intimate link between DSBs, which generate accidental linear DNA ends, and natural chromosomal ends, and inspired the first model of replicative senescence (Blackburn 2000). In this model:

1. Telomeres would switch from "capped" states, which do not activate the DDC, to "uncapped," which do activate the DDC.

2. Shorter telomeres would more frequently adopt an uncapped state compared to longer telomeres. Therefore, the proportion of uncapped telomeres would increase with the culturing of cells, along with telomere shortening.

3. Cells accumulating uncapped telomeres stop growing due to increasing DDC activation surpassing a threshold.

An important facet of this model was rapidly verified at the experimental level: The full activation of the DDC in senescent cells was first confirmed in yeast and then in mammalian cells (Enomoto et al. 2002; d'Adda di Fagagna et al. 2003; Ijpma and Greider 2003). Cultures of telomerase-negative budding yeast cells indeed accumulate large dumbbell-shaped cells, indicating a $G_2/M$ cell-cycle arrest, similar to cells exposed to irreparable chromosomal DSBs. This cell-cycle arrest is suppressed in mutants of the DDC pathway, such as Ddc2-Mec1-[Rad17-Mec3-Ddc1 (9-1-1)]-Rad24, and is accompanied by the activation of DCC mediators Rad9 and Mrc1, as well as the Rad53 DDC effector (Teixeira 2013). An important prediction of this model is that a few uncapped telomeres should be sufficient to trigger a DDC and the subsequent lack of cell viability, despite the presence of longer telomeres, as shown in telomerase-negative mice (Hemann et al. 2001).

Subsequently, the molecular structure of capped versus uncapped telomeres started to be defined, along with the concept of telomere protection, the anti-checkpoint function of telomeres, and the discovery of the shelterin complex. When placed near to an accidental chromosomal break, telomere sequences inhibit DDC (Michelson et al. 2005). Yet, a limited DDC activity and a restricted action of DNA repair factors are key for telomere maintenance by telomerase and regeneration of the G-overhang as described above. Mutations in essential telomeric proteins were shown to unleash the activation of DDC at telomeres, leading to cell-cycle arrest and unscheduled DNA repair activities such as telomere fusions or degradations. Cdc13 in budding yeast constituted the archetype of a protein binding to telomeric ssDNA, which protects telomeres from Mec1-Rad9 DDC activation (Garvik et al. 1995). Mutations in Cdc13 led to rapid accumulation of telomeric ssDNA, caused by Exo1-mediated 5′-to-3′ resection of the telomeric C-strand (Maringele and Lydall 2002, 2009; Dewar and Lydall 2010).

Cite this article as *Cold Spring Harb Perspect Biol* doi: 10.1101/cshperspect.a041699

Similarly, telomere fusions were observed when Rap1 was mutated in certain conditions favoring nonhomologous end joining in budding yeast (Pardo and Marcand 2005). In mammalian cells, dissection of the roles of individual telomeric proteins in preventing DDC activation led to the concept of telomeres acting as "shelters" for natural chromosomal ends to protect them from uncontrolled DDC activation. In line with this concept, the name shelterin was chosen to describe the protein complexes at telomeres (de Lange 2005).

Although short telomeres are expected to bind fewer telomeric proteins and therefore provide a compromised shelter, it remained unclear whether telomeres deficient in a specific telomeric protein would mimic telomere signaling at the onset of senescence. Experiments performed in the absence of telomerase in *S. cerevisiae* demonstrated that senescent cultures accumulate cells with more ssDNA, dependent on Exo1 activity, and more bound RPA, Ddc2, Mre11, and Rad52 proteins, especially at their shortest telomeres (Khadaroo et al. 2009; Fallet et al. 2014). This supports a model in which short telomeres become uncapped due to a lack of sufficient telomeric-binding proteins. Consequently, short telomeres would accumulate subtelomeric ssDNA such that cells cannot distinguish uncapped telomeres from processed accidental DSBs, leading to a permanent activation of the Mec1 DDC. The lack of Rap1 and Rif2, and perhaps insufficient Cdc13 binding to compensate for C-strand higher resection, would therefore underlie the more frequent uncapping of short telomeres. However, the exact structure of a short telomere at the point of dysfunction has not been established. For instance, the contribution of processes such as replication difficulties and fork stalling at shorter telomeres cannot be excluded, as they would result in a similar activation of Mec1-dependent DDC (Teixeira 2013; Fallet et al. 2014).

## FROM SINGLE TELOMERES TO SINGLE CELLS

Important open questions include whether a specific length threshold exists at which an in-

dividual telomere no longer restricts the DDC, and how many uncapped or unprotected telomeres are required to stop cell division. To address this, the Flp1 system was used to generate a single critically short telomere in telomerase-negative cells, revealing an accelerated onset of replicative senescence and preferential binding of Mec1 to the shortest telomere (Abdallah et al. 2009). This result supported a model in which a single critically short telomere is sufficient to induce senescence. However, it could not rule out the possibility that several short telomeres or a series of subsequent events starting with the first telomere reaching a critical length would be required to shift the fate of cells into cell division arrest. The huge interclonal and intercellular variability in the replicative senescence phenotype, particularly the variations in cell proliferation potential among cells in a senescent culture, poses significant challenges to addressing this fundamental threshold question.

However, it is through the formal analysis of the heterogeneity of the senescence phenotype that some questions were answered. While mitosis of telomerase-negative cells generates two cells with identical senescence onset, meiosis can segregate a factor influencing senescence onset among telomerase-negative spores. Moreover, the frequency of segregation correlates with the length of the shortest telomere being the determining factor (Xu et al. 2013). In other words, in contrast to mitosis, meiosis segregates two distinct telomere lengths, which can be considered alleles of each chromosome end. This could explain the greater heterogeneity among cells resulting from meiosis compared to mitosis. Importantly, it was observed that senescence onset is predominantly genetically determined, although not through classical Mendelian inheritance. Instead, it follows a pattern where the shortest telomere exerts dominance over the others (1 out of 32 telomeres for *S. cerevisiae*). Additionally, this work establishes a mathematical model of telomere-length homeostasis based on experimental data (Teixeira et al. 2004), with a specific focus on the shortest telomere, which is expected to be under selective pressure in a culture of telomerase-negative cells, as later modeled (Rat et al. 2023). This model corroborates earlier findings regarding the variability of telomere lengths (Shampay and Blackburn 1988), and, importantly, it highlights a potential practical implication of these variances.

To dissect the heterogeneity of replicative senescence, single-cell studies were undertaken using microfluidics-based live-cell imaging, an emerging technology easily applicable to *S. cerevisiae*. This technology enables the continuous recording of consecutive cell divisions over several days. In one study, the effects of telomerase inactivation were examined specifically in mother cells, which are subjected to a phenomenon known as mother cell aging (Xie et al. 2015). In budding yeast, asymmetrical cell division produces two distinct cells: the mother and daughter. Mother cell aging is characterized by the retention of a significant portion of the cytoplasm and the accumulation of markers of aging until cell death (Denoth et al. 2014). As a result, mother cells can undergo ~20–30 rounds of budding before ceasing division. Spores with a deleted telomerase gene were germinated, before being introduced in a microfluidics circuit. As only mother cells are retained in this circuit and buds are washed away, consecutive cell divisions of mothers are recorded up to cell death. In this context, it was demonstrated that yeast telomerase-negative mother cells experience transient cell-cycle arrests dependent on DDC activation, which seemed stochastic and independent of telomere length (Xie et al. 2015).

In another study aimed at independently assessing replicative senescence, cells were placed in microcavities allowing for indefinite growth, facilitated by a microfluidics circuit designed to remove excess cells (Fig. 3A; Xu et al. 2015). Initially employed to investigate mother cell aging (Fehrmann et al. 2013), a modification to a strain that buds in opposite directions resulted in frequent replacement of the cell at the tip of the microcavity by daughter cells (Fig. 3B). Consequently, in contrast to the experimental setting above, the lineage being tracked here largely excludes old mother cells. Another notable difference lies in the method of telomerase inactivation. In the first study, heterozygous diploids for telomerase activity are tetrad-dissected, and phenotypic changes of telomerase-negative

spore-derived clones examined numerous generations following telomerase inactivation. Conversely, the second study employs an artificial conditional promoter to repress the expression of an essential telomerase subunit by the addition of a drug to yeast media passing through the microfluidics circuit. This approach results in cells initially possessing shorter yet stable telomeres due to the artificial promoter but allows for immediate examination of the consequences upon telomerase gene repression. This latter method reveals that telomerase-deficient cells undergo a highly variable number of cell divisions (10–70) before undergoing an abrupt and irreversible transition to two to four markedly prolonged cell cycles, ultimately leading to cell death (Xu et al. 2015; Coutelier et al. 2018). These lineages were dubbed type A lineages (Fig. 3C). Thus, these data suggest that despite enormous heterogeneity in the number of cell divisions, cells enter senescence with similar abrupt kinetics. However, it was also found that shortly after telomerase inactivation, nearly half of the lineages undergo frequent and reversible DDC-dependent cell-cycle arrests, which may occur consecutively. This novel phenotype resembles the stochastic and reversible cell-cycle arrests detected in the first study. These lineages were labeled as type B (Fig. 3C). In addition, the large majority of consecutive abnormally long cell cycles, characteristic of telomerase-negative lineages, followed or not by normal cell cycles, rely on adaptation to the DNA damage pathway (Coutelier et al. 2018). Adaptation allows mitosis to occur despite persistent and unrepaired DNA damage following a dampening of the DDC, with consequences for genome integrity (see below).

Quantitative data from this work helped to develop two complementary mathematical models of senescence. One model describes the behavior of cells that abruptly enter replicative senescence (type A), while the other describes the behavior of cells experiencing reversible stochastic cell-cycle arrests (type B) (Bourgeron et al. 2015; Eugène et al. 2017; Martin et al. 2021). First, the abrupt switch observed in type A cells is consistent with a scenario in which senescence onset is triggered by the first telo-

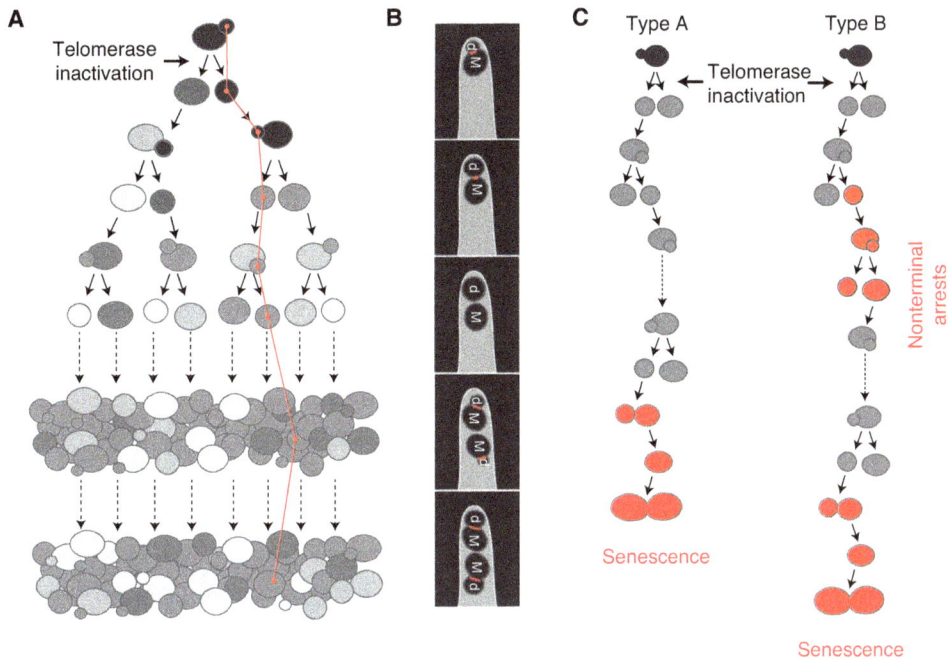

**Figure 3.** Replicative senescence in single-cell lineages. (*A*) Microfluidics enabled monitoring of the consecutive cell divisions from telomerase inactivation to cell death in single-cell lineages (red line) circumventing population study biases. (*B*) Scheme of microfluidics circuit to analyze single-cell lineages. Cells invade microcavities. The cell placed at the tip of the microcavity is often replaced by its daughter ([M] mother cell; [d] daughter cell). The duration of each cell division is scored for consecutive cell divisions from telomerase inactivation to cell death. (*C*) Schematic of the two types of cell lineages detected in microfluidics experiments using telomerase-negative cells. Although the overall number of cell divisions is highly variable in the absence of telomerase, nearly half of the type A lineage proliferated as rapidly as wild-type cells before abruptly arresting cell divisions. Type B lineages exhibited intermittent periods of cell-cycle arrest at the DNA damage checkpoint (DDC), called nonterminal arrests, followed by resumption of normal cell cycles and subsequent arrest and terminal senescence. Consequences of nonterminal arrests are the possible accumulation of genomic instability in cells with substantial proliferation potential.

mere reaching a critically short length, representing a deterministic threshold. Heterogeneity among type A lineages can be fully explained by (1) the initial telomere length distribution; and (2) the asymmetry of the telomere replication mechanism (Soudet et al. 2014). Second, numerical simulations suggest that nonterminal arrests typical of type B lineages are compatible with an age-dependent process distinct from terminal senescence. In this case, the frequency of the first reversible cell-cycle arrest gradually increases with telomere shortening. Thus, these two replicative senescence processes depict distinct cellular states, each governed by its own mechanisms. While the uncapping of telomeres

at a predetermined length can theoretically be associated with the collapse of telomeric protein structures below a defined sequence length threshold, the gradual escalation in the occurrence of nonfatal telomere events remains challenging to define. We speculate that these reversible arrests may be attributed to stalling of replication forks, which are known to encounter difficulties when traversing telomere repeat motifs (Ivessa et al. 2002; Makovets et al. 2004; Teixeira 2013). In the absence of telomerase, which typically acts at problematic replication forks through telomeres (Miller et al. 2006; Matmati et al. 2020; Huda et al. 2023), the frequent resumption of the cell cycle could be associated with facilitated repair of stalled replication forks, assisted at telomeres by numerous replisome-associated factors. An alternative, nonexclusive hypothesis suggests the involvement of reactive oxygen species (ROS) (see below).

In conclusion, single-cell data in yeast support the concept of a telomere length threshold, where a single telomere below this threshold triggers an irreversible cell-cycle arrest, thus replicative senescence. However, it is important to note that there is also a stochastic component to replicative senescence that is less dependent on telomere length. Further mathematical simulations show that these cells are expected to accumulate in late senescent cultures (Rat et al. 2023). In human cultured cells, five dysfunctional telomeres are required to trigger senescence (Kaul et al. 2012), and telomere dysfunction in these cells often appears independent of telomere length. While the deterministic activation threshold of the DDC may differ from that in S. cerevisiae, we speculate that late-passage cultures of human cells may consist of cells that follow the stochastic principles governing senescence described in S. cerevisiae, expected to accumulate in late senescent cultures (Rat et al. 2023). Overall, dissecting the structures of the shortest telomeres in different contexts will be crucial for defining the molecular determinants of replicative senescence.

One important discovery from single-cell analysis of senescence is the impact of mortality rates on senescence rates. When a strain is affected by a telomere-independent comorbidity that influences its survival, senescence is dramatically accelerated, and telomere shortening increases with each population doubling (Rat et al. 2023). Notably, this phenomenon can occur independently of the comorbidity specific to telomerase-negative cells, and its impact on experiments in which senescence is investigated in various mutant contexts prompts a reevaluation and reinterpretation of the results. This effect has often been overlooked because even a 10-fold increase in mortality rate per cell division, from 0.3%–0.5% to 5%, has an almost negligible effect on colony size or cultures grown to saturation, as in senescence assays. In conclusion, analyzing senescence at different scales, whether examining single cells or populations, offers unique and complementary perspectives.

## POSTSENESCENT SURVIVAL

Shortly after replicative senescence was defined in budding yeast cells, it was observed that a minor population could survive telomerase inactivation. These cells were called survivors, and the mechanism(s) enabling their telomere reelongation are primarily based on HDR within telomeric sequences (Lundblad and Blackburn 1993). These mechanisms, known as alternative lengthening of telomeres (ALT), are remarkably conserved across eukaryotic evolution, as many human cancers rely on ALT for their survival (Cesare and Reddel 2010; Pickett and Reddel 2015). Similar to budding yeast, ALT cancers exhibit chromosome ends with very long telomeres displaying significant length variations, accumulate circular DNA containing telomeric repeats, and have telomeres located at specific subnuclear sites—ALT-associated promyelocytic leukemia bodies (APBs) in human cells or near nuclear pore complexes in S. cerevisiae (Aguilera et al. 2023)—which seem to favor their specialized HDR maintenance mechanisms. In recent years, there has been considerable research interest in the initial mechanistic transition to this epigenetic state of telomere maintenance and the pathways involved in ALT. This interest is particularly fueled by the crucial role of ALT in cancer therapies, including ALT-dependent relapses after telomerase-inactivation-based cancer treatments.

In the context of budding yeast, the transition to survival within a senescent cell population is an infrequent event, estimated at a frequency of $2 \times 10^{-5}$ (Kockler et al. 2021), and is not genetically determined (Makovets et al. 2008). Initially, senescent cultures were thought to give rise to two distinct survivor types, type I or type II, which depend on distinct genetic pathways based on Rad52 and Pol32-dependent break-induced replication (BIR) mechanisms and/or possibly rolling circle replication of telomeric circles (Larrivée and Wellinger 2006; Aguilera et al. 2022), as elegantly demonstrated in S. cerevisiae's related yeast Kluveromyces lactis

(Topcu et al. 2005). Type I survivors acquire multiple tandem $Y'$ subtelomeric elements through replication-dependent recombination, which, even on X-only telomeres, exhibit short $TG_{1-3}$ terminal repeats, and depend on Rad51 for their formation. In contrast, type II survivors display heterogeneous $TG_{1-3}$ repeats, sometimes at remarkably elevated levels at their chromosome ends, and rely on Rad59, Sgs1, and Rad50 for their emergence. However, as stated earlier, many mutants of these genes exhibit increased mortality rates independent of telomerase inactivation, which may impact senescence rates, and therefore the opportunity to form type I or type II survivors. Additionally, some mutations affect telomere length prior to telomerase inactivation, which could influence survivors' outcomes (Lebel et al. 2009). These considerations have prompted a reevaluation of the effects of these mutations on ALT activation.

Recent careful quantification of survivor frequency using a mathematical method based on the stochastic nature of ALT activation demonstrated that Rad51, Rad59, Rad52, and Pol32, and a functional DDC are all required for survivor emergence (Kockler et al. 2021). Both type I and type II survivors are mechanistically linked. The first step involves the swapping and amplification of $Y'$ subtelomeric elements among chromosome ends, followed by the addition of long and heterogeneous telomeres, both mechanisms depending on BIR factors, although to different extents. These steps follow a "maturation" phase, during which the shortest telomere in cells would trigger a series of DDC activation and limited recombination events to prevent the onset of senescence. This would correspond to a probation period where opportunities to mature into stable survivors can arise. This could also reflect the behavior of type B cells observed in microfluidics, which undergo adaptation-dependent consecutive prolonged cell cycles followed by normal cell cycles as described above (Xie et al. 2015; Xu et al. 2015; Kockler et al. 2021). The fact that type B lineages enter senescence in a manner decoupled from the initial telomere length distribution prior to telomerase activity further supports this model (Martin et al. 2021; Rat et al. 2023). Importantly,

survivors in budding yeast are themselves subjected to senescence when subcloned, confirming that the survival state does not result from the acquisition of a mutation or an irreversible event (Misino et al. 2022). Accordingly, the reintroduction of telomerase in these cells enables a switch back to normal telomerase-mediated telomere maintenance. In budding yeast, the ALT mechanisms might be no more than an opportunistic chance to survive following the stochastic re-elongation of the shortest telomere, then the second shortest, and so on. Whether a rarer event, characterized by the generation of a longer telomeric donor template through the rolling circle replication of telomeric circular DNA—a potential byproduct of preceding transactions—serves as the driving force behind sudden and significant elongation(s), is an appealing hypothesis for understanding the transition into a matured survivor.

An important aspect of ALT in both budding yeast and mammalian cells is its relationship with telomeric noncoding transcription, specifically transcription comprising subtelomeric and telomeric regions (Azzalin et al. 2007; Luke et al. 2008; Zeinoun et al. 2023; Azzalin 2024). The most studied telomeric transcript is TERRA, which is found to be abundant at short telomeres at the onset of senescence and in survivors, along with its derivative secondary structures known as telR-loops (Cusanelli et al. 2013; Graf et al. 2017; Misino et al. 2018, 2022; Zeinoun et al. 2023; Azzalin 2024). TelR-loops are RNA:DNA hybrids composed of telomeric RNA hybridizing with complementary DNA at telomeres. These structures can theoretically stem from TERRA RNA invading double-stranded telomeric DNA in *cis* or *trans*, and accumulate at short telomeres due to a reduced presence of Rif2, which recruits RNAseH2, the enzyme that digests the RNA component of telR-loops (Graf et al. 2017).

While RNA:DNA hybrids are known to impair replication fork progression at many genomic sites, promoting or inhibiting the accumulation of telR-loops has been shown to foster or suppress the emergence of survivors, respectively, in both budding yeast and mammalian cells (Balk et al. 2013; Azzalin 2024). This seemingly counterintuitive finding can be interpreted by recognizing that, although telR-loops pose a threat to cell viability for the vast majority of cells, they may also represent an opportunity for long-term survival, due to their ability to trigger repair pathways that favor the onset of ALT.

Taken together, a possible scenario for the emergence of survivors in yeast senescent cultures is that their precursors may stem from the recurrent reversible cell-cycle arrests observed in type B cells during single-cell analyses, prior to telomeres reaching the critically short lengths that trigger irreversible DDC activation. However, the precise telomeric structure at the origin of these events remains elusive. As telR-loops accumulate at short telomeres and potentially cause replication stress, they may initiate the first steps of ALT.

## GENOMIC INSTABILITY TRIGGERED BY TELOMERE EROSION

In 1938, Barbara McClintock described, based on her studies of broken chromosome end stability in *Zea mays*, how the natural ends of chromosomes protect chromosome termini from breakage-fusion-bridge cycles and therefore from genomic instability (McClintock 1941). Subsequently, chromosomal aberrations and aneuploidy were detected in late cultures of senescent cells since the 1960s, later defined as the crisis stage, a phenotype reverted by telomerase reexpression (discussed in Artandi and DePinho 2000; Maciejowski and de Lange 2017). These observations gave rise to the paradox of telomere shortening being both tumor suppressive because it stops indefinite proliferation through replicative senescence and pro-oncogenic since it causes genomic instability at the crisis stage (Blasco et al. 1997; Hackett et al. 2001; Hackett and Greider 2003; Maciejowski and de Lange 2017; Dewhurst et al. 2021).

In budding yeast, the initial study by Lundblad and Szostak in 1989 investigating the effects of telomerase inactivation (Lundblad and Szostak 1989) included a careful assessment of genomic instability. Similarly to human cells approaching senescence and/or crisis, the authors detected an increase in the frequency of loss of an artificial nonessential chromosome as telomeres shortened in *S. cerevisiae*. Subsequently, Greider and colleagues started to elucidate the mechanisms by which chromosome ends endured the loss of telomerase. They discovered that telomere shortening leads to a gradient of genomic instability from the end to the interior of the chromosome (Hackett et al. 2001; Hackett and Greider 2003). These studies employed an elegant experimental design, estimating the mutation rate per cell division through fluctuation assays. They tracked the loss of function of reporter genes located at different positions on a chromosome, ranging from the telomere to the centromere, in telomerase-negative cells. During the peak of replicative senescence, mutagenesis near chromosome ends exhibited ~10-fold increase compared to other regions adjacent to the centromere. Most observed mutations were categorized as gross chromosomal rearrangements, likely resulting from end resection by Exo1 at dysfunctional telomeres combined with events compatible with BIR healing at potentially critically short telomeres (Hackett and Greider 2003). Therefore, similar to mammalian cells in late senescence and/or crisis, telomerase-negative cells in budding yeast also undergo a genetic catastrophe stage as they approach senescence, with genomic instability being higher in regions near telomeres, implying a role of short telomeres in initiating the process.

Telomere-associated chromatin structure has also been discussed as a determinant of genomic instability during senescence (Prado et al. 2017; Henninger and Teixeira 2020). Rap1 was already discussed in this work as an essential protein protecting telomeres from fusion (Pardo and Marcand 2005) and a transcriptional regulator of numerous nontelomeric genes by promoting nucleosome displacement (Platt et al. 2013; Song et al. 2019; Mivelaz et al. 2020). During senescence, Rap1 occupancy decreases from shortened telomeres and subtelomeric regions and relocates instead to the promoters of a new set of genes, known as NRTS (new Rap1 targets at senescence) in a Mec1-dependent manner. Within NRTS, Rap1 negatively regulates genes encoding histone proteins, potential-

Cite this article as *Cold Spring Harb Perspect Biol* doi: 10.1101/cshperspect.a041699

ly explaining the widespread decrease in histone levels observed in senescent cells (Platt et al. 2013). Building on this finding, experiments conducted by Barrientos-Moreno and colleagues (Barrientos-Moreno et al. 2018) in pre-senescence cells aimed to elucidate the synergistic role of Mec1/ATR and Tel1/ATM kinases, along with histone occupancy, in preventing telomere fusions. The authors demonstrated that *tlc1 mec1* mutants effectively suppress telomere-to-telomere fusion by reducing the available pool of histones. Furthermore, Mec1-dependent depletion of histones during telomere shortening could prevent telomere-to-telomere fusion by favoring homologous recombination through a Rad51-independent mechanism.

Taken together, both short telomeres and chromatin modifications during senescence contribute to the increase in genomic instability characterized by gross chromosomal rearrangements. Knowing that senescent cultures are composed of a highly heterogeneous population of cells at variable "ages" (Rat et al. 2023), some remaining questions are (1) When does genomic instability occur along individual routes to senescence—during senescence or crisis? (2) What is the threshold of telomere shortening required to become dysfunctional—is there a limit to telomere length? and (3) What precisely distinguishes the structure of a functional telomere from a dysfunctional one?

To address the first question, microfluidics technology provided important insights yet again. Introducing a live fluorescent biosensor for the early stage of DDC activation revealed that during the most prolonged terminal cell cycles, as well as during the extended nonterminal arrests characteristic of type B lineages (see above), the early stage of DDC was activated and activation persisted throughout the subsequent mitotic events (Coutelier et al. 2018). This suggested that these mitoses occurred under the condition of "adaptation to DDC." In *S. cerevisiae*, "adaptation to DDC" was first defined by Sandell and Zakian (1993) who engineered telomere loss from the end of a disomic supernumerary chromosome. After many hours of DDC-dependent cell-cycle arrest, they observed cells recovering without repairing the

damaged chromosome. Later on, factors involved in this process were characterized, and mutants were isolated (Toczyski et al. 1997). Finally, as expected for a process that enables mitosis despite the presence of DNA damage, "adaptation to DDC" was shown to be mutagenic (Galgoczy and Toczyski 2001). Introducing some of these mutants into telomerase-inactivated cells suppressed most prolonged consecutive cell cycles, as detected in microfluidics. Similarly, genomic instability associated with replicative senescence was reduced. These findings suggest that genomic instability in telomerase-negative budding yeast cells is triggered by telomeric damage, which activates the DDC followed by "adaptation to DDC." Typically, this occurs right before senescence, having minimal effects on progeny. However, it can also occur during nonterminal arrests in type B cells. In these cases, a significant progeny with genome rearrangements can proliferate, contributing to the spread of genomic variants within the population (Coutelier et al. 2018).

Budding yeast has been instrumental in delineating the pathways that lead to increased genome instability in senescent cell populations. Essentially, this process is likely associated with telomeres in cells that have experienced substantial telomere erosion. However, the precise mechanism triggering DNA repair and the consequent genomic rearrangements remains unclear. It is yet to be determined whether it is an extremely eroded, dysfunctional telomere, or a more stochastic replication error that initiates this cascade of events. Such information will help pinpoint the stage at which genomic instability increases in human cells and clarify the pro-oncogenic role of telomeres.

## EFFECTS OF TELOMERASE INACTIVATION ON CELL METABOLISM AND CELLULAR ORGANELLES

Upon telomerase inactivation, senescent cells remain metabolically active and undergo significant alterations in morphology, metabolism, and subcellular organelles, in addition to profound remodeling of chromatin and telomere folding modification, as discussed earlier (Nautiyal et al. 2002; Platt et al. 2013; Wagner et al. 2020). In this context, budding yeast senescent cells activate genes involved in DNA damage, environmental stresses, and oxidative stress, as well as show significant changes in mitochondrial morphology, characterized by increased mitochondrial proliferation (Nautiyal et al. 2002; Xie et al. 2015). Since mitochondrial defects lead to reduced mitophagy and subsequent accumulation of mitochondrial mass, it is plausible that the increase in mitochondrial proliferation may be a compensatory response to the loss of mitochondrial function (Dalle et al. 2014; Rizza et al. 2018; Nassour et al. 2019; Picca et al. 2020). However, recent data indicate that inhibiting autophagy or mitophagy does not significantly affect the rate of replicative senescence in *S. cerevisiae* (Zeinoun et al. 2024). This suggests that either mitophagy is not down-regulated in budding yeast senescent cells or that the mitochondria effects may not drive the kinetics of senescence at the population level. Nevertheless, long-term effects on type B nonterminal arrests, rare genomic instability, and postsenescence survival warrant further investigation. More broadly, the relationship between telomere erosion and cell metabolism is far from being understood, and budding yeast could contribute as a model to delineate the temporal and mechanistic connections among these events at single-cell resolution.

Mitochondria serve as the main sites for ATP production and are at the heart of a complex network that includes various organelles such as the endoplasmic reticulum (ER), Golgi apparatus, peroxisomes, lipid droplets, endosomes, and lysosomes. This network is crucial for maintaining cellular homeostasis and supporting various biological processes including macromolecule synthesis, bioenergetics, and stress sensing. Mitochondrial calcium homeostasis is vital in maintaining mitochondrial function and can be regulated through ER–mitochondria membrane contact sites. Additionally, mitochondria play a central role in the production of ROS (Balaban et al. 2005; Passos et al. 2010; Stenberg et al. 2022).

Mitochondrial dysfunction is a key feature of the senescence phenotype across many or-ganisms. In mammalian cells, the removal of mitochondria is sufficient to reverse many aspects of the senescence phenotype (Correia-Melo et al. 2016). It has been proposed that replicative senescence results in the repression of master regulators of mitochondrial biogenesis and metabolism, PGC-1α and PGC-1β, by the activation of p53, a major DDC factor in mammalian cells (Sahin et al. 2011; Schank et al. 2020). Overall, data obtained in mammalian cells suggest that mitochondrial dysfunction and the resultant increase in ROS may contribute to a positive feedback loop that locks senescence into an irreversible state. Moreover, mitochondria could act as potential sources of additional cell-to-cell heterogeneity and contribute to increasing genome instability during the senescence process (Passos et al. 2007; Sahin et al. 2011; Sahin and DePinho 2012; Ahmed and Lingner 2018).

Dysfunctional mitochondria are often associated with elevated oxygen consumption, reduced efficiency of oxidative phosphorylation, a decline in membrane potential, increased mitochondrial DNA damage, and higher levels of ROS (Passos et al. 2007, 2010; Moiseeva et al. 2009; Korolchuk et al. 2017; El Maï et al. 2020). Accordingly, both mammalian and budding yeast senescent cells exhibit enhanced levels of ROS and oxidative damage (Passos and von Zglinicki 2005; Passos et al. 2010; Nelson et al. 2018; Ahmed and Lingner 2020; Zeinoun et al. 2024). As ROS are highly reactive molecules, ROS generated by mitochondria or other cytoplasmic enzymes can freely diffuse into the nuclei and cause DNA damage (Cadet and Wagner 2013). Because of their G-rich nature, telomeres are particularly susceptible to oxidative damage as they accumulate the oxidized base 8-oxoguanine (8-oxoG) (Hewitt et al. 2012; Qian et al. 2019). In budding yeast, the absence of Ogg1, the DNA glycosylase central to base excision repair (BER) that is responsible for specific excision of 8-oxoG, results in increased levels of 8-oxoG at telomeres. This is accompanied by a simultaneous decrease of Rap1 binding to telomeres and subsequent alteration of telomere length homeostasis (Lu and Liu 2010). Further studies in mammalian cells have established that

8-oxoG alters telomeric G-quadruplex stability and telomerase activity, and accelerates telomere shortening (Aeby et al. 2016; Fouquerel et al. 2016, 2019; Ahmed and Lingner 2020). As DNA repair mechanisms are less efficient at telomeres (Zhou et al. 2013), exposure to severe oxidative stress can lead to persistent 8-oxoG, resulting in telomere shortening and eventual telomeric fusion. This, in turn, could contribute to genome instability (Chen et al. 2005; Passos et al. 2007; Coluzzi et al. 2014), a situation that might be recapitulated in senescent cells, turning ROS increase into a source of cell-to-cell variability in senescence. This idea could be readily tested in telomerase-negative *OGG1*-deleted budding yeast cells, where genomic instability can be assessed using established fluctuation assays.

Mitochondria also contribute to the biosynthesis of nucleotides, and their dysfunction could lead to reduced nucleotide levels, potentially exacerbating telomere shortening and genome stability (Desler et al. 2007; Gupta et al. 2013; Coluzzi et al. 2019). Furthermore, in mammalian cells, an increase in mitochondrial $Ca^{2+}$ levels leads to a decrease in mitochondrial membrane potential, resulting in an increase in ROS production, which further accelerates replicative senescence (Wiel et al. 2014).

As a sensor of ROS increases, the mammalian MAPK p38 pathway appears pivotal in replicative senescence (Iwasa et al. 2003). Activation of p38 in response to various stressors, including oxidative stress, can induce premature senescence (Wang et al. 2002; Barascu et al. 2012; Martínez-Limón et al. 2020), and it is important for maintaining high ROS levels following DNA damage triggered by dysfunctional telomeres (Passos et al. 2010). Similarly, in budding yeast, the MAPK Hog1, which is the ortholog of mammalian p38, was shown to be essential for protecting against a range of stressors, including osmotic (Brewster et al. 1993) and oxidative stress (Haghnazari and Heyer 2004). Hog1 is also involved in replicative senescence in budding yeast, but its effects differ from the effects of MAPK p38 in mammalian cells. In telomerase-negative budding yeast cells, Hog1 is activated by its canonical activation pathway,

through the MAPKK Pbs2, but not in response to DDC activation, since this activation is Mec1-independent (Zeinoun et al. 2024). Notably, both Hog1 and Mec1 appear to contribute to ROS control in senescence.

Considering the evolutionary conservation of senescence, we can assume that using budding yeast, paired with sophisticated genetic tools, will continue to provide a deeper understanding of the relationship and chronological sequence of events involving telomere shortening, mitochondria dysfunction, and metabolic changes.

## PERSPECTIVES

As *S. cerevisiae* has consistently led the way in scientific discovery, pioneering many foundational paradigms, we anticipate it will be the first among eukaryotes to provide a comprehensive understanding of all molecular processes occurring in the absence of telomerase. This encompasses the exploration of telomere structures, signaling cascades within the cell, cellular responses originating from various organelles, and the resultant implications on proliferation capacity, genome stability, and insights into the sources of cell-to-cell variability.

Using budding yeast as a model, the priorities will be to elucidate the precise alterations in molecular composition and structure of telomeres as they shorten due to the DNA end-replication problem throughout the cell cycle. These molecular transformations potentially trigger senescence at a critically short length or activate ALT, or ultimately increase the risk of initiating genomic instability—conditions that remain largely enigmatic to date. Unraveling the sources of heterogeneity during this process and delineating the risks associated with transitions between telomeric states, such as the potential for cells to evade senescence, are paramount.

Furthermore, achieving a comprehensive understanding of the intricate cross talk between the nucleus and other organelles necessitates probing the precise choreography of events occurring in cells upon telomerase inactivation. This exploration promises insights into the evolution of eukaryotes, shedding light on the origins of complex cellular communication systems and the remarkable adaptability of eukaryotic organisms.

Replicative senescence, exemplified by its role as an anticancer mechanism in humans, underscores the potential of these discoveries to inform therapeutic strategies and deepen our understanding of cellular evolution.

## ACKNOWLEDGMENTS

We thank all Telomere Biology team members for fruitful discussions. We wish to acknowledge Elisabetta Citterio from Life Science Editors for English editing of the manuscript. During the preparation of this work, we occasionally used ChatGPT 3.5/4.0 and DeepL to improve the readability and language of the manuscript. After using this tool, the authors reviewed and edited the content as needed and take full responsibility for the content of the published article. Work in MTT's laboratory is supported by the CNRS, Sorbonne University, the "Fondation de la Recherche Medicale" (Equipe FRM EQU202003010428), by the French National Research Agency (ANR) as part of the "Investissements d'Avenir" Program (LabEx Dynamo ANR-11-LABX-0011-01) and the French National Cancer Institute (INCa_15192).

## REFERENCES

Abdallah P, Luciano P, Runge KW, Lisby M, Géli V, Gilson E, Teixeira MT. 2009. A two-step model for senescence triggered by a single critically short telomere. *Nat Cell Biol* **11**: 988–993. doi:10.1038/ncb1911

Aeby E, Ahmed W, Redon S, Simanis V, Lingner J. 2016. Peroxiredoxin 1 protects telomeres from oxidative damage and preserves telomeric DNA for extension by telomerase. *Cell Rep* **17**: 3107–3114. doi:10.1016/j.celrep.2016.11.071

Aguilera P, Dubarry M, Hardy J, Lisby M, Simon MN, Géli V. 2022. Telomeric C-circles localize at nuclear pore complexes in *Saccharomyces cerevisiae*. *EMBO J* **41**: e108736. doi:10.15252/embj.2021108736

Aguilera P, Dubarry M, Géli V, Simon MN. 2023. NPCs and APBs: two HUBs of non-canonical homology-based recombination at telomeres? *Cell Cycle* **22**: 1163–1168. doi:10.1080/15384101.2023.2206350

Ahmed W, Lingner J. 2018. PRDX1 and MTH1 cooperate to prevent ROS-mediated inhibition of telomerase. *Genes Dev* **32**: 658–669. doi:10.1101/gad.313460.118

Ahmed W, Lingner J. 2020. PRDX1 counteracts catastrophic telomeric cleavage events that are triggered by DNA repair activities post oxidative damage. *Cell Rep* **33**: 108347. doi:10.1016/j.celrep.2020.108347

Artandi SE, DePinho RA. 2000. A critical role for telomeres in suppressing and facilitating carcinogenesis. *Curr Opin Genet Dev* **10**: 39–46. doi:10.1016/S0959-437X(99)00047-7

Azzalin CM. 2024. TERRA and the alternative lengthening of telomeres: a dangerous affair. *FEBS Lett* doi:10.1002/1873-3468.14844

Azzalin CM, Reichenbach P, Khoriauli L, Giulotto E, Lingner J. 2007. Telomeric repeat containing RNA and RNA surveillance factors at mammalian chromosome ends. *Science* **318**: 798–801. doi:10.1126/science.1147182

Balaban RS, Nemoto S, Finkel T. 2005. Mitochondria, oxidants, and aging. *Cell* **120**: 483–495. doi:10.1016/j.cell.2005.02.001

Balk B, Maicher A, Dees M, Klermund J, Luke-Glaser S, Bender K, Luke B. 2013. Telomeric RNA-DNA hybrids affect telomere-length dynamics and senescence. *Nat Struct Mol Biol* **20**: 1199–1205. doi:10.1038/nsmb.2662

Barascu A, Le CC, Pennarun G, Genet D, Imam N, Lopez B, Bertrand P. 2012. Oxidative stress induces an ATM-independent senescence pathway through p38 MAPK-mediated lamin B1 accumulation. *EMBO J* **31**: 1080–1094. doi:10.1038/emboj.2011.492

Barrientos-Moreno M, Murillo-Pineda M, Muñoz-Cabello AM, Prado F. 2018. Histone depletion prevents telomere fusions in pre-senescent cells. *PLoS Genet* **14**: e1007407. doi:10.1371/journal.pgen.1007407

Berman J, Tachibana CY, Tye BK. 1986. Identification of a telomere-binding activity from yeast. *Proc Natl Acad Sci* **83**: 3713–3717. doi:10.1073/pnas.83.11.3713

Blackburn EH. 2000. Telomere states and cell fates. *Nature* **408**: 53–56. doi:10.1038/35040500

Blackburn EH, Gall JG. 1978. A tandemly repeated sequence at the termini of the extrachromosomal ribosomal RNA genes in *Tetrahymena*. *J Mol Biol* **120**: 33–53. doi:10.1016/0022-2836(78)90294-2

Blasco MA, Lee HW, Hande MP, Samper E, Lansdorp PM, DePinho RA, Greider CW. 1997. Telomere shortening and tumor formation by mouse cells lacking telomerase RNA. *Cell* **91**: 25–34. doi:10.1016/S0092-8674(01)80006-4

Bonetti D, Martina M, Clerici M, Lucchini G, Longhese MP. 2009. Multiple pathways regulate 3' overhang generation at *S. cerevisiae* telomeres. *Mol Cell* **35**: 70–81. doi:10.1016/j.molcel.2009.05.015

Bonetti D, Clerici M, Anbalagan S, Martina M, Lucchini G, Longhese MP. 2010. Shelterin-like proteins and Yku inhibit nucleolytic processing of *Saccharomyces cerevisiae* telomeres. *PLoS Genet* **6**: e1000966. doi:10.1371/journal.pgen.1000966

Bourgeron T, Xu Z, Doumic M, Teixeira MT. 2015. The asymmetry of telomere replication contributes to replicative senescence heterogeneity. *Sci Rep* **5**: 15326. doi:10.1038/srep15326

Brewster JL, de VT, Dwyer ND, Winter E, Gustin MC. 1993. An osmosensing signal transduction pathway in yeast. *Science* **259**: 1760–1763. doi:10.1126/science.7681220

Cite this article as *Cold Spring Harb Perspect Biol* doi: 10.1101/cshperspect.a041699

Cadet J, Wagner JR. 2013. DNA base damage by reactive oxygen species, oxidizing agents, and UV radiation. *Cold Spring Harb Perspect Biol* **5:** a012559. doi:10.1101/cshperspect.a012559

Cejka P, Symington LS. 2021. DNA end resection: mechanism and control. *Annu Rev Genet* **55:** 285–307. doi:10.1146/annurev-genet-071719-020312

Cesare AJ, Reddel RR. 2010. Alternative lengthening of telomeres: models, mechanisms and implications. *Nat Rev Genet* **11:** 319–330. doi:10.1038/nrg2763

Chen JH, Ozanne SE, Hales CN. 2005. Heterogeneity in premature senescence by oxidative stress correlates with differential DNA damage during the cell cycle. *DNA Repair (Amst)* **4:** 1140–1148. doi:10.1016/j.dnarep.2005.06.003

Coluzzi E, Colamartino M, Cozzi R, Leone S, Meneghini C, O'Callaghan N, Sgura A. 2014. Oxidative stress induces persistent telomeric DNA damage responsible for nuclear morphology change in mammalian cells. *PLoS ONE* **9:** e110963. doi:10.1371/journal.pone.0110963

Coluzzi E, Leone S, Sgura A. 2019. Oxidative stress induces telomere dysfunction and senescence by replication fork arrest. *Cells* **8:** 19. doi:10.3390/cells8010019

Correia-Melo C, Marques FD, Anderson R, Hewitt G, Hewitt R, Cole J, Carroll BM, Miwa S, Birch J, Merz A, et al. 2016. Mitochondria are required for pro-ageing features of the senescent phenotype. *EMBO J* **35:** 724–742. doi:10.15252/embj.201592862

Counter CM, Avilion AA, LeFeuvre CE, Stewart NG, Greider CW, Harley CB, Bacchetti S. 1992. Telomere shortening associated with chromosome instability is arrested in immortal cells which express telomerase activity. *EMBO J* **11:** 1921–1929. doi:10.1002/j.1460-2075.1992.tb05245.x

Coutelier H, Xu Z, Morisse MC, Lhuillier-Akakpo M, Pelet S, Charvin G, Dubrana K, Teixeira MT. 2018. Adaptation to DNA damage checkpoint in senescent telomerase-negative cells promotes genome instability. *Genes Dev* **32:** 1499–1513. doi:10.1101/gad.318485.118

Cusanelli E, Romero CA, Chartrand P. 2013. Telomeric noncoding RNA TERRA is induced by telomere shortening to nucleate telomerase molecules at short telomeres. *Mol Cell* **51:** 780–791. doi:10.1016/j.molcel.2013.08.029

d'Adda di Fagagna F, Reaper PM, Clay-Farrace L, Fiegler H, Carr P, Von ZT, Saretzki G, Carter NP, Jackson SP. 2003. A DNA damage checkpoint response in telomere-initiated senescence. *Nature* **426:** 194–198. doi:10.1038/nature02118

Dalle PP, Nelson G, Otten EG, Korolchuk VI, Kirkwood TB, von ZT, Shanley DP. 2014. Dynamic modelling of pathways to cellular senescence reveals strategies for targeted interventions. *PLoS Comp Biol* **10:** e1003728. doi:10.1371/journal.pcbi.1003728

de Lange T. 2005. Shelterin: the protein complex that shapes and safeguards human telomeres. *Genes Dev* **19:** 2100–2110. doi:10.1101/gad.1346005

de Lange T. 2018. Shelterin-mediated telomere protection. *Annu Rev Genet* **52:** 223–247. doi:10.1146/annurev-genet-032918-021921

Denoth LA, Julou T, Barral Y. 2014. Budding yeast as a model organism to study the effects of age. *FEMS Microbiol Rev* **38:** 300–325. doi:10.1111/1574-6976.12060

Desler C, Munch-Petersen B, Stevnsner T, Matsui S, Kulawiec M, Singh KK, Rasmussen LJ. 2007. Mitochondria as determinant of nucleotide pools and chromosomal stability. *Mutat Res* **625:** 112–124. doi:10.1016/j.mrfmmm.2007.06.002

Dewar JM, Lydall D. 2010. Pif1- and Exo1-dependent nucleases coordinate checkpoint activation following telomere uncapping. *EMBO J* **29:** 4020–4034. doi:10.1038/emboj.2010.267

Dewhurst SM, Yao X, Rosiene J, Tian H, Behr J, Bosco N, Takai KK, de Lange T, Imieliński M. 2021. Structural variant evolution after telomere crisis. *Nat Commun* **12:** 2093. doi:10.1038/s41467-021-21933-7

Diede SJ, Gottschling DE. 1999. Telomerase-mediated telomere addition in vivo requires DNA primase and DNA polymerases α and δ. *Cell* **99:** 723–733. doi:10.1016/S0092-8674(00)81670-0

Diede SJ, Gottschling DE. 2001. Exonuclease activity is required for sequence addition and Cdc13p loading at a de novo telomere. *Curr Biol* **11:** 1336–1340. doi:10.1016/S0960-9822(01)00400-6

Dilley RL, Verma P, Cho NW, Winters HD, Wondisford AR, Greenberg RA. 2016. Break-induced telomere synthesis underlies alternative telomere maintenance. *Nature* **539:** 54–58. doi:10.1038/nature20099

Dionne I, Wellinger RJ. 1996. Cell cycle-regulated generation of single-stranded G-rich DNA in the absence of telomerase. *Proc Natl Acad Sci* **93:** 13902–13907. doi:10.1073/pnas.93.24.13902

Dionne I, Wellinger RJ. 1998. Processing of telomeric DNA ends requires the passage of a replication fork. *Nucleic Acids Res* **26:** 5365–5371. doi:10.1093/nar/26.23.5365

El Maï M, Marzullo M, de Castro IP, Ferreira MG. 2020. Opposing p53 and mTOR/AKT promote an in vivo switch from apoptosis to senescence upon telomere shortening in zebrafish. *eLife* **9:** e54935. doi:10.7554/eLife.54935

Enomoto S, Glowczewski L, Berman J. 2002. MEC3, MEC1, and DDC2 are essential components of a telomere checkpoint pathway required for cell cycle arrest during senescence in *Saccharomyces cerevisiae*. *Mol Biol Cell* **13:** 2626–2638. doi:10.1091/mbc.02-02-0012

Eugène S, Bourgeron T, Xu Z. 2017. Effects of initial telomere length distribution on senescence onset and heterogeneity. *J Theor Biol* **413:** 58–65. doi:10.1016/j.jtbi.2016.11.010

Fallet E, Jolivet P, Soudet J, Lisby M, Gilson E, Teixeira MT. 2014. Length-dependent processing of telomeres in the absence of telomerase. *Nucleic Acids Res* **42:** 3648–3665. doi:10.1093/nar/gkt1328

Faure V, Coulon S, Hardy J, Géli V. 2010. Cdc13 and telomerase bind through different mechanisms at the lagging- and leading-strand telomeres. *Mol Cell* **38:** 842–852. doi:10.1016/j.molcel.2010.05.016

Fehrmann S, Paoletti C, Goulev Y, Ungureanu A, Aguilaniu H, Charvin G. 2013. Aging yeast cells undergo a sharp entry into senescence unrelated to the loss of mitochondrial membrane potential. *Cell Rep* **5:** 1589–1599. doi:10.1016/j.celrep.2013.11.013

Fouquerel E, Lormand J, Bose A, Lee HT, Kim GS, Li J, Sobol RW, Freudenthal BD, Myong S, Opresko PL. 2016. Oxidative guanine base damage regulates human telomerase activity. *Nat Struct Mol Biol* **23:** 1092–1100. doi:10.1038/nsmb.3319

Fouquerel E, Barnes RP, Uttam S, Watkins SC, Bruchez MP, Opresko PL. 2019. Targeted and persistent 8-oxoguanine base damage at telomeres promotes telomere loss and crisis. *Mol Cell* **75:** 117–130.e6. doi:10.1016/j.molcel.2019.04.024

Galgoczy DJ, Toczyski DP. 2001. Checkpoint adaptation precedes spontaneous and damage-induced genomic instability in yeast. *Mol Cell Biol* **21:** 1710–1718. doi:10.1128/MCB.21.5.1710-1718.2001

Gao H, Cervantes RB, Mandell EK, Otero JH, Lundblad V. 2007. RPA-like proteins mediate yeast telomere function. *Nat Struct Mol Biol* **14:** 208–214. doi:10.1038/nsmb1205

Garvik B, Carson M, Hartwell L. 1995. Single-stranded DNA arising at telomeres in cdc13 mutants may constitute a specific signal for the RAD9 checkpoint. *Mol Cell Biol* **15:** 6128–6138. doi:10.1128/MCB.15.11.6128

Gomes NM, Ryder OA, Houck ML, Charter SJ, Walker W, Forsyth NR, Austad SN, Venditti C, Pagel M, Shay JW, et al. 2011. Comparative biology of mammalian telomeres: hypotheses on ancestral states and the roles of telomeres in longevity determination. *Aging Cell* **10:** 761–768. doi:10.1111/j.1474-9726.2011.00718.x

Graf M, Bonetti D, Lockhart A, Serhal K, Kellner V, Maicher A, Jolivet P, Teixeira MT, Luke B. 2017. Telomere length determines TERRA and R-loop regulation through the cell cycle. *Cell* **170:** 72–85.e14. doi:10.1016/j.cell.2017.06.006

Greider CW. 1990. Telomeres, telomerase and senescence. *Bioessays* **12:** 363–369. doi:10.1002/bies.950120803

Gupta A, Sharma S, Reichenbach P, Marjavaara L, Nilsson AK, Lingner J, Chabes A, Rothstein R, Chang M. 2013. Telomere length homeostasis responds to changes in intracellular dNTP pools. *Genetics* **193:** 1095–1105. doi:10.1534/genetics.112.149120

Ha T, Kaiser C, Myong S, Wu B, Xiao J. 2022. Next generation single-molecule techniques: imaging, labeling, and manipulation in vitro and in cellulo. *Mol Cell* **82:** 304–314. doi:10.1016/j.molcel.2021.12.019

Hackett JA, Greider CW. 2003. End resection initiates genomic instability in the absence of telomerase. *Mol Cell Biol* **23:** 8450–8461. doi:10.1128/MCB.23.23.8450-8461.2003

Hackett JA, Feldser DM, Greider CW. 2001. Telomere dysfunction increases mutation rate and genomic instability. *Cell* **106:** 275–286. doi:10.1016/S0092-8674(01)00457-3

Haghnazari E, Heyer WD. 2004. The Hog1 MAP kinase pathway and the Mec1 DNA damage checkpoint pathway independently control the cellular responses to hydrogen peroxide. *DNA Repair (Amst)* **3:** 769–776. doi:10.1016/j.dnarep.2004.03.043

Harley CB, Futcher AB, Greider CW. 1990. Telomeres shorten during ageing of human fibroblasts. *Nature* **345:** 458–460. doi:10.1038/345458a0

Hartwell LH, Mortimer RK, Culotti J, Culotti M. 1973. Genetic control of the cell division cycle in yeast. V: Genetic analysis of cdc mutants. *Genetics* **74:** 267–286. doi:10.1093/genetics/74.2.267

Hayflick L. 1965. The limited in vitro lifetime of human diploid cell strains. *Exp Cell Res* **37:** 614–636. doi:10.1016/0014-4827(65)90211-9

Hemann MT, Strong MA, Hao LY, Greider CW. 2001. The shortest telomere, not average telomere length, is critical for cell viability and chromosome stability. *Cell* **107:** 67–77. doi:10.1016/S0092-8674(01)00504-9

Henninger E, Teixeira MT. 2020. Telomere-driven mutational processes in yeast. *Curr Opin Genet Dev* **60:** 99–106. doi:10.1016/j.gde.2020.02.018

Hewitt G, Jurk D, Marques FD, Correia-Melo C, Hardy T, Gackowska A, Anderson R, Taschuk M, Mann J, Passos JF. 2012. Telomeres are favoured targets of a persistent DNA damage response in ageing and stress-induced senescence. *Nat Commun* **3:** 708. doi:10.1038/ncomms1708

Huda A, Arakawa H, Mazzucco G, Galli M, Petrocelli V, Casola S, Chen L, Doksani Y. 2023. The telomerase reverse transcriptase elongates reversed replication forks at telomeric repeats. *Sci Adv* **9:** eadf2011. doi:10.1126/sciadv.adf2011

Ijpma AS, Greider CW. 2003. Short telomeres induce a DNA damage response in *Saccharomyces cerevisiae*. *Mol Biol Cell* **14:** 987–1001. doi:10.1091/mbc.02-04-0057

Ivessa AS, Zhou JQ, Schulz VP, Monson EK, Zakian VA. 2002. *Saccharomyces* Rrm3p, a 5′ to 3′ DNA helicase that promotes replication fork progression through telomeric and subtelomeric DNA. *Genes Dev* **16:** 1383–1396. doi:10.1101/gad.982902

Iwasa H, Han J, Ishikawa F. 2003. Mitogen-activated protein kinase p38 defines the common senescence-signalling pathway. *Genes Cells* **8:** 131–144. doi:10.1046/j.1365-2443.2003.00620.x

Jain D, Cooper JP. 2010. Telomeric strategies: means to an end. *Annu Rev Genet* **44:** 243–269. doi:10.1146/annurev-genet-102108-134841

Jones ML, Aria V, Baris Y, Yeeles JTP. 2023. How Pol α-primase is targeted to replisomes to prime eukaryotic DNA replication. *Mol Cell* **83:** 2911–2924.e16. doi:10.1016/j.molcel.2023.06.035

Kaul Z, Cesare AJ, Huschtscha LI, Neumann AA, Reddel RR. 2012. Five dysfunctional telomeres predict onset of senescence in human cells. *EMBO Rep* **13:** 52–59. doi:10.1038/embor.2011.227

Khadaroo B, Teixeira MT, Luciano P, Eckert-Boulet N, Germann SM, Simon MN, Gallina I, Abdallah P, Gilson E, Géli V, et al. 2009. The DNA damage response at eroded telomeres and tethering to the nuclear pore complex. *Nat Cell Biol* **11:** 980–987. doi:10.1038/ncb1910

Kockler ZW, Comeron JM, Malkova A. 2021. A unified alternative telomere-lengthening pathway in yeast survivor cells. *Mol Cell* **81:** 1816–1829.e5. doi:10.1016/j.molcel.2021.02.004

Korolchuk VI, Miwa S, Carroll B, von ZT. 2017. Mitochondria in cell senescence: is mitophagy the weakest link? *EBioMed* **21:** 7–13. doi:10.1016/j.ebiom.2017.03.020

Krogh BO, Llorente B, Lam A, Symington LS. 2005. Mutations in Mre11 phosphoesterase motif I that impair *Saccharomyces cerevisiae* Mre11-Rad50-Xrs2 complex stability in addition to nuclease activity. *Genetics* **171:** 1561–1570. doi:10.1534/genetics.105.049478

Larrivée M, Wellinger RJ. 2006. Telomerase- and capping-independent yeast survivors with alternate telomere states. *Nat Cell Biol* **8:** 741–747. doi:10.1038/ncb1429

Larrivée M, LeBel C, Wellinger RJ. 2004. The generation of proper constitutive G-tails on yeast telomeres is dependent on the MRX complex. *Genes Dev* **18:** 1391–1396. doi:10.1101/gad.1199404

Lebel C, Rosonina E, Sealey DC, Pryde F, Lydall D, Maringele L, Harrington LA. 2009. Telomere maintenance and survival in *Saccharomyces cerevisiae* in the absence of telomerase and RAD52. *Genetics* **182:** 671–684. doi:10.1534/genetics.109.102939

Lendvay TS, Morris DK, Sah J, Balasubramanian B, Lundblad V. 1996. Senescence mutants of *Saccharomyces cerevisiae* with a defect in telomere replication identify three additional EST genes. *Genetics* **144:** 1399–1412. doi:10.1093/genetics/144.4.1399

Lin JJ, Zakian VA. 1996. The *Saccharomyces* CDC13 protein is a single-strand TG1–3 telomeric DNA-binding protein in vitro that affects telomere behavior in vivo. *Proc Natl Acad Sci* **93:** 13760–13765. doi:10.1073/pnas.93.24.13760

Lingner J, Cooper JP, Cech TR. 1995. Telomerase and DNA end replication: no longer a lagging strand problem? *Science* **269:** 1533–1534. doi:10.1126/science.7545310

Lingner J, Hughes TR, Shevchenko A, Mann M, Lundblad V, Cech TR. 1997. Reverse transcriptase motifs in the catalytic subunit of telomerase. *Science* **276:** 561–567. doi:10.1126/science.276.5312.561

Longtine MS, Wilson NM, Petracek ME, Berman J. 1989. A yeast telomere binding activity binds to two related telomere sequence motifs and is indistinguishable from RAP1. *Curr Genet* **16:** 225–239. doi:10.1007/BF00422108

Lu J, Liu Y. 2010. Deletion of Ogg1 DNA glycosylase results in telomere base damage and length alteration in yeast. *EMBO J* **29:** 398–409. doi:10.1038/emboj.2009.355

Lue NF, Autexier C. 2023. Orchestrating nucleic acid–protein interactions at chromosome ends: telomerase mechanisms come into focus. *Nat Struct Mol Biol* **30:** 878–890. doi:10.1038/s41594-023-01022-7

Lue NF, Chan J, Wright WE, Hurwitz J. 2014. The CDC13-STN1-TEN1 complex stimulates Pol α activity by promoting RNA priming and primase-to-polymerase switch. *Nat Commun* **5:** 5762. doi:10.1038/ncomms6762

Luke B, Panza A, Redon S, Iglesias N, Li Z, Lingner J. 2008. The Rat1p 5′ to 3′ exonuclease degrades telomeric repeat-containing RNA and promotes telomere elongation in *Saccharomyces cerevisiae*. *Mol Cell* **32:** 465–477. doi:10.1016/j.molcel.2008.10.019

Lundblad V, Blackburn EH. 1993. An alternative pathway for yeast telomere maintenance rescues est1⁻ senescence. *Cell* **73:** 347–360. doi:10.1016/0092-8674(93)90234-H

Lundblad V, Szostak JW. 1989. A mutant with a defect in telomere elongation leads to senescence in yeast. *Cell* **57:** 633–643. doi:10.1016/0092-8674(89)90132-3

Lydall D. 2009. Taming the tiger by the tail: modulation of DNA damage responses by telomeres. *EMBO J* **28:** 2174–2187. doi:10.1038/emboj.2009.176

Maciejowski J, de Lange T. 2017. Telomeres in cancer: tumour suppression and genome instability. *Nat Rev Mol Cell Biol* **18:** 175–186. doi:10.1038/nrm.2016.171

Makovets S, Herskowitz I, Blackburn EH. 2004. Anatomy and dynamics of DNA replication fork movement in yeast telomeric regions. *Mol Cell Biol* **24:** 4019–4031. doi:10.1128/MCB.24.9.4019-4031.2004

Makovets S, Williams TL, Blackburn EH. 2008. The telotype defines the telomere state in *Saccharomyces cerevisiae* and is inherited as a dominant non-Mendelian characteristic in cells lacking telomerase. *Genetics* **178:** 245–257. doi:10.1534/genetics.107.083030

Marcand S, Brevet V, Gilson E. 1999. Progressive *cis*-inhibition of telomerase upon telomere elongation. *EMBO J* **18:** 3509–3519. doi:10.1093/emboj/18.12.3509

Maringele L, Lydall D. 2002. EXO1-dependent single-stranded DNA at telomeres activates subsets of DNA damage and spindle checkpoint pathways in budding yeast *yku70Δ* mutants. *Genes Dev* **16:** 1919–1933. doi:10.1101/gad.225102

Martin H, Doumic M, Teixeira MT, Xu Z. 2021. Telomere shortening causes distinct cell division regimes during replicative senescence in *Saccharomyces cerevisiae*. *Cell Biosci* **11:** 180. doi:10.1186/s13578-021-00693-3

Martínez-Limón A, Joaquin M, Caballero M, Posas F, de Nadal E. 2020. The p38 pathway: from biology to cancer therapy. *Int J Mol Sci* **21:** 1913. doi:10.3390/ijms21061913

Matmati S, Lambert S, Géli V, Coulon S. 2020. Telomerase repairs collapsed replication forks at telomeres. *Cell Rep* **30:** 3312–3322.e3. doi:10.1016/j.celrep.2020.02.065

McClintock B. 1941. The stability of broken ends of chromosomes in *Zea mays*. *Genetics* **26:** 234–282. doi:10.1093/genetics/26.2.234

McGee JS, Phillips JA, Chan A, Sabourin M, Paeschke K, Zakian VA. 2010. Reduced Rif2 and lack of Mec1 target short telomeres for elongation rather than double-strand break repair. *Nat Struct Mol Biol* **17:** 1438–1445. doi:10.1038/nsmb.1947

Michelson RJ, Rosenstein S, Weinert T. 2005. A telomeric repeat sequence adjacent to a DNA double-stranded break produces an anticheckpoint. *Genes Dev* **19:** 2546–2559. doi:10.1101/gad.1293805

Miller KM, Rog O, Cooper JP. 2006. Semi-conservative DNA replication through telomeres requires Taz1. *Nature* **440:** 824–828. doi:10.1038/nature04638

Misino S, Bonetti D, Luke-Glaser S, Luke B. 2018. Increased TERRA levels and RNase H sensitivity are conserved hallmarks of post-senescent survivors in budding yeast. *Differentiation* **100:** 37–45. doi:10.1016/j.diff.2018.02.002

Misino S, Busch A, Wagner CB, Bento F, Luke B. 2022. TERRA increases at short telomeres in yeast survivors and regulates survivor associated senescence (SAS). *Nucleic Acids Res* **50:** 12829–12843. doi:10.1093/nar/gkac1125

Mivelaz M, Cao AM, Kubik S, Zencir S, Hovius R, Boichenko I, Stachowicz AM, Kurat CF, Shore D, Fierz B. 2020. Chromatin fiber invasion and nucleosome displacement by the Rap1 transcription factor. *Mol Cell* **77:** 488–500.e9. doi:10.1016/j.molcel.2019.10.025

Moiseeva O, Bourdeau V, Roux A, Deschênes-Simard X, Ferbeyre G. 2009. Mitochondrial dysfunction contributes to oncogene-induced senescence. *Mol Cell Biol* **29:** 4495–4507. doi:10.1128/MCB.01868-08

Myler LR, Toia B, Vaughan CK, Takai K, Matei AM, Wu P, Paull TT, de Lange T, Lottersberger F. 2023. DNA-PK and the TRF2 iDDR inhibit MRN-initiated resection at leading-end telomeres. *Nat Struct Mol Biol* **30:** 1346–1356. doi:10.1038/s41594-023-01072-x

Nakamura TM, Cech TR. 1998. Reversing time: origin of telomerase. *Cell* **92:** 587–590. doi:10.1016/S0092-8674(00)81123-X

Nassour J, Radford R, Correia A, Fusté JM, Schoell B, Jauch A, Shaw RJ, Karlseder J. 2019. Autophagic cell death restricts chromosomal instability during replicative crisis. *Nature* **565:** 659–663. doi:10.1038/s41586-019-0885-0

Nautiyal S, DeRisi JL, Blackburn EH. 2002. The genome-wide expression response to telomerase deletion in *Saccharomyces cerevisiae*. *Proc Natl Acad Sci* **99:** 9316–9321. doi:10.1073/pnas.142162499

Negrini S, Ribaud V, Bianchi A, Shore D. 2007. DNA breaks are masked by multiple Rap1 binding in yeast: implications for telomere capping and telomerase regulation. *Genes Dev* **21:** 292–302. doi:10.1101/gad.400907

Nelson G, Kucheryavenko O, Wordsworth J, von ZT. 2018. The senescent bystander effect is caused by ROS-activated NF-κB signalling. *Mech Ageing Dev* **170:** 30–36. doi:10.1016/j.mad.2017.08.005

Olovnikov AM. 1973. A theory of marginotomy. The incomplete copying of template margin in enzymic synthesis of polynucleotides and biological significance of the phenomenon. *J Theor Biol* **41:** 181–190. doi:10.1016/0022-5193(73)90198-7

Pardo B, Marcand S. 2005. Rap1 prevents telomere fusions by nonhomologous end joining. *EMBO J* **24:** 3117–3127. doi:10.1038/sj.emboj.7600778

Passos JF, von Zglinicki T. 2005. Mitochondria, telomeres and cell senescence. *Exp Gerontol* **40:** 466–472. doi:10.1016/j.exger.2005.04.006

Passos JF, Saretzki G, Ahmed S, Nelson G, Richter T, Peters H, Wappler I, Birket MJ, Harold G, Schaeuble K, et al. 2007. Mitochondrial dysfunction accounts for the stochastic heterogeneity in telomere-dependent senescence. *PLoS Biol* **5:** e110. doi:10.1371/journal.pbio.0050110

Passos JF, Nelson G, Wang C, Richter T, Simillion C, Proctor CJ, Miwa S, Olijslagers S, Hallinan J, Wipat A, et al. 2010. Feedback between p21 and reactive oxygen production is necessary for cell senescence. *Mol Syst Biol* **6:** 347. doi:10.1038/msb.2010.5

Picca A, Calvani R, Coelho-Junior HJ, Landi F, Bernabei R, Marzetti E. 2020. Inter-organelle membrane contact sites and mitochondrial quality control during aging: a geroscience view. *Cells* **9:** 598. doi:10.3390/cells9030598

Pickett HA, Reddel RR. 2015. Molecular mechanisms of activity and derepression of alternative lengthening of telomeres. *Nat Struct Mol Biol* **22:** 875–880. doi:10.1038/nsmb.3106

Pizzul P, Casari E, Rinaldi C, Gnugnoli M, Mangiagalli M, Tisi R, Longhese MP. 2024. Rif2 interaction with Rad50 counteracts Tel1 functions in checkpoint signalling and DNA tethering by releasing Tel1 from MRX binding. *Nucleic Acids Res* **52:** 2355–2371. doi:10.1093/nar/gkad1246

Platt JM, Ryvkin P, Wanat JJ, Donahue G, Ricketts MD, Barrett SP, Waters HJ, Song S, Chavez A, Abdallah KO, et al. 2013. Rap1 relocalization contributes to the chromatin-mediated gene expression profile and pace of cell senescence. *Genes Dev* **27:** 1406–1420. doi:10.1101/gad.218776.113

Prado F, Jimeno-González S, Reyes JC. 2017. Histone availability as a strategy to control gene expression. *RNA Biol* **14:** 281–286. doi:10.1080/15476286.2016.1189071

Qian W, Kumar N, Roginskaya V, Fouquerel E, Opresko PL, Shiva S, Watkins SC, Kolodieznyi D, Bruchez MP, Van Houten B. 2019. Chemoptogenetic damage to mitochondria causes rapid telomere dysfunction. *Proc Natl Acad Sci* **116:** 18435–18444. doi:10.1073/pnas.1910574116

Rat A, Doumic M, Teixeira MT, Xu Z. 2023. Individual cell fate and population dynamics revealed by a mathematical model linking telomere length and replicative senescence. bioRxiv doi:10.1101/2023.11.22.568287

Ritchie KB, Mallory JC, Petes TD. 1999. Interactions of TLC1 (which encodes the RNA subunit of telomerase), TEL1, and MEC1 in regulating telomere length in the yeast *Saccharomyces cerevisiae*. *Mol Cell Biol* **19:** 6065–6075. doi:10.1128/MCB.19.9.6065

Rizza S, Cardaci S, Montagna C, Di GG, De ZD, Bordi M, Maiani E, Campello S, Borreca A, Puca AA, et al. 2018. S-nitrosylation drives cell senescence and aging in mammals by controlling mitochondrial dynamics and mitophagy. *Proc Natl Acad Sci* **115:** E3388–E3397. doi:10.1073/pnas.1722452115

Sahin E, DePinho RA. 2012. Axis of ageing: telomeres, p53 and mitochondria. *Nat Rev Mol Cell Biol* **13:** 397–404. doi:10.1038/nrm3352

Sahin E, Colla S, Liesa M, Moslehi J, Müller FL, Guo M, Cooper M, Kotton D, Fabian AJ, Walkey C, et al. 2011. Telomere dysfunction induces metabolic and mitochondrial compromise. *Nature* **470:** 359–365. doi:10.1038/nature09787

Sandell LL, Zakian VA. 1993. Loss of a yeast telomere: arrest, recovery, and chromosome loss. *Cell* **75:** 729–739. doi:10.1016/0092-8674(93)90493-A

Schank M, Zhao J, Wang L, Li Z, Cao D, Nguyen LN, Dang X, Khanal S, Nguyen LNT, Thakuri BKC, et al. 2020. Telomeric injury by KML001 in human T cells induces mitochondrial dysfunction through the p53-PGC-1α pathway. *Cell Death Dis* **11:** 1030. doi:10.1038/s41419-020-03238-7

Shampay J, Blackburn EH. 1988. Generation of telomere-length heterogeneity in *Saccharomyces cerevisiae*. *Proc Natl Acad Sci* **85:** 534–538. doi:10.1073/pnas.85.2.534

Shore D, Nasmyth K. 1987. Purification and cloning of a DNA binding protein from yeast that binds to both silencer and activator elements. *Cell* **51:** 721–732. doi:10.1016/0092-8674(87)90095-X

Song S, Perez JV, Svitko W, Ricketts MD, Dean E, Schultz D, Marmorstein R, Johnson FB. 2019. Rap1-mediated nucleosome displacement can regulate gene expression in senescent cells without impacting the pace of senescence. *Aging Cell* **19:** e13061. doi:10.1111/acel.13061

Soudet J, Jolivet P, Teixeira MT. 2014. Elucidation of the DNA end-replication problem in *Saccharomyces cerevisiae*. *Mol Cell* **53:** 954–964. doi:10.1016/j.molcel.2014.02.030

Stanley SE, Armanios M. 2015. The short and long telomere syndromes: paired paradigms for molecular medicine. *Curr Opin Genet Dev* **33:** 1–9. doi:10.1016/j.gde.2015.06.004

Stenberg S, Li J, Gjuvsland AB, Persson K, Demitz-Helin E, González Peña C, Yue JX, Gilchrist C, Ärengård T, Ghiaci

P, et al. 2022. Genetically controlled mtDNA deletions prevent ROS damage by arresting oxidative phosphorylation. *eLife* **11**: e76095. doi:10.7554/eLife.76095

Szostak JW, Blackburn EH. 1982. Cloning yeast telomeres on linear plasmid vectors. *Cell* **29**: 245–255. doi:10.1016/0092-8674(82)90109-X

Takai H, Aria V, Borges P, Yeeles JTP, de Lange T. 2024. CST-polymerase α-primase solves a second telomere end-replication problem. *Nature* **627**: 664–670. doi:10.1038/s41586-024-07137-1

Teixeira MT. 2013. Saccharomyces cerevisiae as a model to study replicative senescence triggered by telomere shortening. *Front Oncol* **3**: 101. doi:10.3389/fonc.2013.00101

Teixeira MT, Gilson E. 2005. Telomere maintenance, function and evolution: the yeast paradigm. *Chromosome Res* **13**: 535–548. doi:10.1007/s10577-005-0999-0

Teixeira MT, Arneric M, Sperisen P, Lingner J. 2004. Telomere length homeostasis is achieved via a switch between telomerase-extendible and -nonextendible states. *Cell* **117**: 323–335. doi:10.1016/S0092-8674(04)00334-4

Toczyski DP, Galgoczy DJ, Hartwell LH. 1997. CDC5 and CKII control adaptation to the yeast DNA damage checkpoint. *Cell* **90**: 1097–1106. doi:10.1016/S0092-8674(00)80375-X

Topcu Z, Nickles K, Davis C, McEachern MJ. 2005. Abrupt disruption of capping and a single source for recombinationally elongated telomeres in *Kluyveromyces lactis*. *Proc Natl Acad Sci* **102**: 3348–3353. doi:10.1073/pnas.0408770102

Vasianovich Y, Wellinger RJ. 2017. Life and death of yeast telomerase RNA. *J Mol Biol* **429**: 3242–3254. doi:10.1016/j.jmb.2017.01.013

Wagner T, Pérez-Martínez L, Schellhaas R, Barrientos-Moreno M, Öztürk M, Prado F, Butter F, Luke B. 2020. Chromatin modifiers and recombination factors promote a telomere fold-back structure, that is lost during replicative senescence. *PLoS Genet* **16**: e1008603. doi:10.1371/journal.pgen.1008603

Wang W, Chen JX, Liao R, Deng Q, Zhou JJ, Huang S, Sun P. 2002. Sequential activation of the MEK-extracellular signal-regulated kinase and MKK3/6-p38 mitogen-activated protein kinase pathways mediates oncogenic ras-induced premature senescence. *Mol Cell Biol* **22**: 3389–3403. doi:10.1128/MCB.22.10.3389-3403.2002

Watson JD. 1972. Origin of concatemeric T7 DNA. *Nat New Biol* **239**: 197–201. doi:10.1038/newbio239197a0

Wellinger RJ, Zakian VA. 2012. Everything you ever wanted to know about *Saccharomyces cerevisiae* telomeres: beginning to end. *Genetics* **191**: 1073–1105. doi:10.1534/genetics.111.137851

Wellinger RJ, Wolf AJ, Zakian VA. 1993a. Origin activation and formation of single-strand TG1-3 tails occur sequentially in late S phase on a yeast linear plasmid. *Mol Cell Biol* **13**: 4057–4065.

Wellinger RJ, Wolf AJ, Zakian VA. 1993b. Saccharomyces telomeres acquire single-strand TG1-3 tails late in S phase. *Cell* **72**: 51–60. doi:10.1016/0092-8674(93)90049-V

Wellinger RJ, Ethier K, Labrecque P, Zakian VA. 1996. Evidence for a new step in telomere maintenance. *Cell* **85**: 423–433. doi:10.1016/S0092-8674(00)81120-4

Wiel C, Lallet-Daher H, Gitenay D, Gras B, Le CB, Augert A, Ferrand M, Prevarskaya N, Simonnet H, Vindrieux D, et al. 2014. Endoplasmic reticulum calcium release through ITPR2 channels leads to mitochondrial calcium accumulation and senescence. *Nat Commun* **5**: 3792. doi:10.1038/ncomms4792

Xie Z, Jay KA, Smith DL, Zhang Y, Liu Z, Zheng J, Tian R, Li H, Blackburn EH. 2015. Early telomerase inactivation accelerates aging independently of telomere length. *Cell* **160**: 928–939. doi:10.1016/j.cell.2015.02.002

Xu Z, Teixeira MT. 2019. The many types of heterogeneity in replicative senescence. *Yeast* **36**: 637–648. doi:10.1002/yea.3433

Xu Z, Duc KD, Holcman D, Teixeira MT. 2013. The length of the shortest telomere as the major determinant of the onset of replicative senescence. *Genetics* **194**: 847–857. doi:10.1534/genetics.113.152322

Xu Z, Fallet E, Paoletti C, Fehrmann S, Charvin G, Teixeira MT. 2015. Two routes to senescence revealed by real-time analysis of telomerase-negative single lineages. *Nat Commun* **6**: 7680. doi:10.1038/ncomms8680

Zakian VA. 1989. Structure and function of telomeres. *Annu Rev Genet* **23**: 579–604. doi:10.1146/annurev.ge.23.120189.003051

Zeinoun B, Teixeira MT, Barascu A. 2023. TERRA and telomere maintenance in the yeast *Saccharomyces cerevisiae*. *Genes (Basel)* **14**: 618. doi:10.3390/genes14030618

Zeinoun B, Teixeira MT, Barascu A. 2024. Hog1 acts in a Mec1-independent manner to counteract oxidative stress following telomerase inactivation in *Saccharomyces cerevisiae*. *Commun Biol* **7**: 761. doi:10.1038/s42003-024-06464-3

Zhou J, Liu M, Fleming AM, Burrows CJ, Wallace SS. 2013. Neil3 and NEIL1 DNA glycosylases remove oxidative damages from quadruplex DNA and exhibit preferences for lesions in the telomeric sequence context. *J Biol Chem* **288**: 27263–27272. doi:10.1074/jbc.M113.479055

# Shedding Light on Telomere Replication, Insights from the Fission Yeast *Schizosaccharomyces pombe*

Stéphane Coulon

CNRS, INSERM, Aix Marseille Université, Institut Paoli-Calmettes, CRCM, Equipe labellisée par la Ligue Nationale contre le Cancer, Marseille F-13009, France

*Correspondence:* stephane.coulon@inserm.fr

Over the years, the fission yeast has become a reference model for telomere biology studies as this organism shares with mammals a highly conserved telomere composition. Here, we highlight the latest discoveries in telomere replication in fission yeast and show how this research brings new insights into the understanding of the replication and maintenance of mammalian telomeres.

Telomeres are nucleoprotein structures that protect chromosome ends from degradation. In most organisms, telomeric DNA consists of G-rich repetitive sequences, ending in a single-stranded overhang (G-tail). These sequences are bound by telomeric proteins that form a complex named shelterin. Chromosome ends resemble DNA double-strand breaks (DSBs), and the function of shelterin is to repress undesired DNA repair activities that would lead to end-to-end chromosome fusions (de Lange 2004, 2018). Another important function of shelterin is to recruit telomerase, an enzyme that replenishes telomeric DNA, which is lost during the replication of the chromosome ends.

Unlike the rest of the genome, telomeres encounter the "end-replication problem" due to their location at the termini of the linear chromosomes. The incomplete replication of the linear DNA molecules by the conventional DNA polymerases stems from their inability to lay down the last Okazaki fragment at the very end of linear DNA, which provokes the progressive shortening of telomeres at each cell division (Takai et al. 2024). This problem is further aggravated by the resection and fill-in reaction during the synthesis of the C-rich telomere leading strand (Lingner et al. 1995; Gilson and Géli 2007; Soudet et al. 2014; Takai et al. 2024). To circumvent telomeric loss, shelterin recruits a reverse transcriptase called telomerase, which is able to elongate the 3′ overhang by the addition of telomeric repeats on the leading strand thanks to its intrinsic RNA template (de Lange 2004; Nandakumar and Cech 2013; Armstrong and Tomita 2017; Roake and Artandi 2020; Takai et al. 2024). On the other hand, the G-rich lagging strand is maintained through the fill-in synthesis carried out by the CTC1-STN1-TEN1 (CST)–Polα-primase complex. In addition, among the entire genome, telomeres are one of the most difficult regions to replicate because they encompass multiple difficulties for replication fork progression (Higa et al. 2017; Maestroni et al. 2017b; Bonnell et al. 2021). Indeed, G-rich DNA that is prone to form secondary structures such as G-quadruplex (G4), transcription including R-loop and RNA:DNA hybrids, the telomeric loop (in human cells), heterochromatin and telomere compaction are diverse sources of endogenous blocks. This high replication stress at terminal DNA is a challenge to replication and thus telomeres are considered as fragile sites (Sfeir et al. 2009). Any defect in the replication process may cause abrupt loss of telomeric DNA that could be deleterious for telomere maintenance and the overall genome stability. The consequence of high replication stress at telomeres is that incoming replication forks that progress toward the end will slow down, pause, and may occasionally collapse. This phenomenon is exacerbated by the absence of a natural converging fork, so that cells have to complete replication at all costs.

The fission yeast *Schizosaccharomyces pombe* is a useful model for fundamental research, such as genome organization, differential gene regulation, cell-cycle control, signal transduction, or cellular morphogenesis (Hayles and Nurse 2018). This is also the case for the field of the telomere biology since the fission yeast shelterin-like complex highlights evolutionarily conserved elements of telomere length regulation between fission yeast and mammalian cells (Nakamura et al. 1997; Miyoshi et al. 2008). In this work, I review the previous and latest discoveries in telomere replication in fission yeast and show how these discoveries can benefit the understanding of the replication of mammalian telomeres. Before then, it is essential to describe the structure of chromosome extremities, as well as to detail the function of all known players to grasp the full complexity of telomere replication in *S. pombe*.

## TELOMERE BIOLOGY IN FISSION YEAST

### The Telomeric DNA

*S. pombe* cells have a small haploid genome consisting of three chromosomes I, II, and III of 5.6, 4.6, and 3.5 Mb, respectively (Fig. 1; Wood et al. 2002). The telomeres in fission yeast are ~300 bp

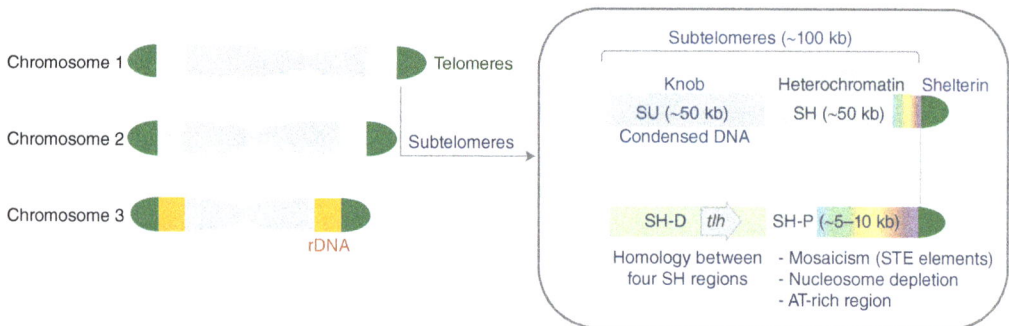

**Figure 1.** Subtelomere structure in fission yeast. *Schizosaccharomyces pombe* cells have three chromosomes terminating in ~300 bp of telomeric repeats at each end. The sequences adjacent to telomeric DNA are subtelomeres (~100 kb) that are divided in two distinct regions: SH (homologous sequences) and SU (unique sequence). The SH region possesses common DNA sequences that are homologous among subtelomeres, and they form heterochromatin. The SU regions do not show high homology among each other but they share a condensed chromatin structure, a knob. The SH region can be further subdivided into the telomere-proximal (SH-P) and the telomere-distal (SH-D) regions. The SH-D region displays a high homology among the four ends of chromosomes and contains the open reading frames of putative telomere-linked helicases (*tlh*) of the RecQ family. The proximal region is a highly complex mosaic structure of segments that encompass the subtelomeric sequences STE, and it spans ~5–10 kb adjacent to the telomeric repeats. The SH-P region exhibits a high percentage of exclusive poly[dA:dT] tracts, correlates with nucleosome depletion, and may coincide with DNA replication origins. Ribosomal (rDNA) repeats are located adjacent to the telomeres in chromosome 3.

Cite this article as *Cold Spring Harb Perspect Biol* doi: 10.1101/cshperspect.a041704

long and are composed of irregular repeats ($G_{2-6}$TTAC[A]) terminating in a single-stranded $3'$ overhang. Immediately adjacent to telomeric DNA are the subtelomeric sequences. The number of subtelomeres varies from four to six per haploid genome, depending on the presence or absence of subtelomeric sequences between the telomeres and the ribosomal DNA (rDNA) repeats at the ends of Chr III.

The subtelomeres span ~100 kb separating the telomeres from the euchromatic arms of the chromosomes. They consist of two distinct regions, SH (homologous sequences) and SU (unique sequence) (Fig. 1; for review, see Yadav et al. 2021). The SH region (~50 kb) includes common DNA sequences in which homologs are found on different subtelomeres, and they all form heterochromatin, while the SU regions (~50 kb) do not show high homology among subtelomeres. SU regions do not have the mark of heterochromatin (H3K9me and HP1$^{Swi6}$) but they share a condensed chromatin structure called knob (Matsuda et al. 2015; Tashiro et al. 2016).

The SH region can be further divided into a telomere-proximal (SH-P) region and a telomere-distal (SH-D) region (Oizumi et al. 2021). The proximal region is a highly complex mosaic of segments that encompass the subtelomeric sequences STE1, STE2, and STE3 and spans 5–10 kb adjacent to telomeric repeats. The SH-P region exhibits a high percentage of poly[dA: dT] tracks correlated with nucleosome depletion, and may coincide with DNA replication origins (Segurado et al. 2003; van Emden et al. 2019; Barnes and Korber 2021). The SH-D region displays a high homology between the four ends of chromosomes and contains the open reading frames of putative telomere-linked helicases (tlh) of the RecQ family (Oizumi et al. 2021).

The heterochromatic SH regions of subtelomeres (distal and proximal) are highly enriched in histone 3 methylation at lysine 9 (H3K9me) and associated heterochromatin protein 1 (HP1) homologs. The methyltransferase Clr4-complex (CLRC) methylates H3K9, which serves as a binding site for the heterochromatin proteins (Swi6, Chp1, and Chp2; HP1 family). Chp2 plays an important role in the recruitment of SHREC,

a histone deacetylase complex (HDAC), that participates in heterochromatin establishment by ensuring a low level of acetylation of histones H3 and H4 (a transcriptional activation histone mark) (Bjerling et al. 2002; Gómez et al. 2005; Sugiyama et al. 2007; Motamedi et al. 2008). Thus, hypermethylation and hypoacetylation of histone tails at subtelomeres contribute to telomere silencing. Two mechanisms are responsible for establishing heterochromatin at subtelomeres (Kanoh et al. 2005). On one hand, shelterin mainly promotes heterochromatin assembly through its association with CLRC, which methylates histone H3K9. This is mediated by the Ccq1 protein (Wang et al. 2016), which also cooperates with Taz1 to recruit SHREC (Sugiyama et al. 2007). On the other hand, the RNA interference (RNAi) machinery uses cenH sequence embedded within the subtelomeric tlh genes as a template for the production of double-stranded RNA (dsRNA) that is processed into small interfering RNA (siRNA) by Dicer. siRNAs are loaded into RITS complex (Ago1), guiding it back to the complementary nascent RNA (transcribed by RNA polymerase II), where it directs the deposition of H3K9me2 through the CLRC (Kanoh et al. 2005).

In spite of their silent heterochromatin status, telomeres are transcribed in long noncoding RNA named TERRA (Bah et al. 2012). Transcription of TERRA starts within the adjacent subtelomeric sequences and contains a variable number of telomeric G-rich sequences. In addition to TERRA, a variety of telomeric transcripts are produced (Bah et al. 2012; Greenwood and Cooper 2012; Moravec et al. 2016). Telomeric transcripts play key roles in telomere maintenance, and their physiological levels are essential for the integrity of telomeric DNA. However, telomere transcription also generates conflicts with replication, which is another source of stress for the cell.

## Telomeric Proteins of Fission Yeast
### Shelterin

The fission yeast shelterin complex consists of Taz1 (TRF1/TRF2 ortholog) that specifically binds to duplex telomeric DNA, the G-tail-binding protein Pot1, and the four shelterin proteins Tpz1 (TPP1 ortholog), Rap1, Poz1, and Ccq1 that link Taz1 and Pot1 through a network of protein–protein interactions (Fig. 2A; Moser and Nakamura 2009; Dehé and Cooper 2010). Poz1 is necessary to bridge the Pot1-Tpz1 complex to Taz1-Rap1 complex, and thus likely fulfills the similar functional roles as mammalian TIN2, which connects TRF1 and TRF2 to the POT1-TPP1 complex (Miyoshi et al. 2008). This is a remarkably conserved structural feature between the shelterin of *S. pombe* and the human shelterin.

The shelterin complex is essential for the spatiotemporal regulation of telomerase and telomere elongation. The group of F. Ishikawa was the first to propose that shelterin status may exist in two distinguishable modes (Miyoshi et al. 2008): a "closed" configuration when all the components of the shelterin are assembled together that protects and prevents telomerase action, and a "semi-open" configuration that allows telomerase recruitment (Fig. 2). Biochemical studies and structure analysis from the Qiao laboratory also supported this model, further showing that a stable shelterin bridge linking telomeric double-stranded DNA (dsDNA) to single-stranded DNA (ssDNA) is formed via a hierarchical assembly process enforcing a negative regulation of telomerase in a nonextendible mode (Jun et al. 2013; Liu et al. 2015; Kim et al. 2017; Liu et al. 2021). Structural insights from the group of M. Lei further demonstrated that a Taz1, Poz1, and Tpz1-Ccq1 subcomplex can dimerize leading to a higher complex organization of the shelterin (Sun et al. 2022). This potential dimeric conformation of shelterin also supports the two-state model (Fig. 2B).

## Telomerase

In fission yeast, the core components of telomerase comprise the telomerase RNA (TER1) and the catalytic reverse transcriptase protein Trt1 (Nakamura et al. 1997; Webb and Zakian 2007). Est1 is a regulatory subunit of Trt1 that directly binds TER1 and directs telomerase to telomeres through an interaction with Ccq1 (Beernink et al. 2003; Webb and Zakian 2012).

Deleting either of the protein subunits of telomerase or its RNA component leads to replicative senescence: telomeres gradually shorten until the cells either cease dividing or die (crisis). This cell-cycle arrest is caused by a DNA damage checkpoint that is activated as a result of unprotected short telomeres being recognized as irreparable DSBs. A small subset of cells eventually recovers from growth arrest as survivors emerge (Nakamura et al. 1998; Jain et al. 2010). Recent observations also indicate that the role of telomerase is not restricted to telomere elongation, as telomerase also participates in telomere replication (see below).

## The Stn1-Ten1 Complex

In humans, the CST complex is an ssDNA-binding protein that shares homology with replication protein A (RPA) (Sun et al. 2009). CST interacts with DNA polymerase α-primase for the synthesis of the complementary C-rich strand (Casteel et al. 2008; Wang et al. 2012) and also acts as a terminator of telomere elongation by promoting C-strand fill-in synthesis (Miyake et al. 2009; Chen et al. 2012; Stewart et al. 2018). In fission yeast, STN1 and TEN1 orthologs have been identified, but the third component of CST complex, CTC1, is lacking (Martín et al. 2007; Moser and Nakamura 2009). Despite that, the Stn1-Ten1 (ST) complex fulfills identical function in fission yeast with the help of the Pot1-Tpz-Ccq1 (PTP) subcomplex that recruits and stabilizes this accessory element (Garg et al. 2014; Miyagawa et al. 2014; Carvalho Borges et al. 2023).

## Rif1

RIF1 is a highly conserved and multifunctional protein. In humans, it has emerged as a key regulator of DNA processing during DSB repair inhibiting resection of DNA ends, as well as a key factor protecting stalled forks during DNA replication stress, and also as a master controller of replication timing (for review, see Blasiak et al. 2021). In budding and fission yeasts, Rif1 controls the timing of replication by regulating the firing of late origins (Hayano et al. 2012;

Cite this article as *Cold Spring Harb Perspect Biol* doi: 10.1101/cshperspect.a041704

Davé et al. 2014; Mattarocci et al. 2014; Hiraga et al. 2018). Indeed, Rif1 associates with protein phosphatase 1 (PP1) and counteracts MCM helicase phosphorylation by the Dbf4-dependent kinase (DDK), the major regulator of origin firing. It has been shown in fission yeast that Rif1 is able to bind specific sites that adopt G4-like structures and prevents the firing of late or dormant origins on chromosome arms (Kanoh et al. 2015). At subtelomeres, Rif1 is also recruited through its interaction with Taz1 (Tazumi et al. 2012; Ogawa et al. 2018). Models propose that Rif1 generates local chromatin structures that may exert long-range suppressive effects on origin firing over the 50–100 kb segment in fission yeast.

In budding and fission yeasts, Rif1 was first identified as a negative regulator of telomere length (Hardy et al. 1992; Kanoh and Ishikawa 2001; Miller et al. 2005). Long telomeres observed in $rif1\Delta$ cells evade the replication timing control, likely due to the diminished PP1-phosphatase activity (Kedziora et al. 2018). Along the same line in *S. pombe*, Rif1-PP1 is thought to dephosphorylate Ccq1, a key step for telomerase inhibition (Vaurs et al. 2022). Although the role of Rif1 in replication fork protection has been shown in budding yeast (Mattarocci et al. 2017; Hiraga et al. 2018), so far there is no evidence that this function is conserved in fission yeast. Rif1 is also known to suppress the resolution of telomere entanglements of mitotic chromosomes in fission yeast (Zaaijer et al. 2016; Kanoh et al. 2023; Nageshan et al. 2024).

## Functions of Shelterin

### DNA End Protection

Shelterin fulfills several essential functions. First, it protects telomeric DNA from degradation and end-to-end chromosome fusions by inhibiting DNA repair pathways (Ferreira and Cooper 2001, 2004; Carneiro et al. 2010; Pan et al. 2015; Audry et al. 2024). Although the mechanisms involved are still unknown, Taz1 and Rap1 play a specific role in non-homologous end-joining (NHEJ) inhibition (Khayat et al. 2021). Second, shelterin prevents check-

point activation by defining a chromatin-privileged region that excludes the H4K20me2 histone mark, specific to DSBs (Carneiro et al. 2010). This prevents stable association of the checkpoint protein Cbr2[53BP1] and therefore prevents cell-cycle arrest.

### Telomerase Regulation

Telomerase is recruited to telomeres in late S/$G_2$ phase concomitantly with telomere replication (Moser et al. 2009). This coincides with the semi-open state of shelterin that allows the recruitment and activation of telomerase (Fig. 2B), which is both temporally and spatially regulated via the state of the Pot1-Tpz1-Ccq1 subcomplex. Rad3[ATR]- and Tel1[ATM]-dependent phosphorylation of Ccq1 at Thr93 promotes the late S/$G_2$ phase recruitment of telomerase to telomeres (Moser et al. 2011; Yamazaki et al. 2012; Chang et al. 2013) by fostering Ccq1-Est1 interaction (Moser et al. 2011; Webb and Zakian 2012). In a collaborative manner, Tpz1 interaction with Trt1 further engages the telomerase core complex (Trt1-TER1) for the 3′ end extension (Armstrong et al. 2014; Hu et al. 2016). Phosphorylation of Thr93 of Ccq1 is thought to be removed by the action of phosphatase associated with Rif1 (Vaurs et al. 2022), while at the same time, SUMOylation of Tpz1 at Lys242, by the SUMO ligase Pli1, becomes detectable (Garg et al. 2014; Miyagawa et al. 2014). The sumoylation of Tpz1 promotes recruitment of the ST complex via the SUMO-interacting motif (SIM) of Stn1 (Matmati et al. 2018; Mennie et al. 2019). Stn1 dephosphorylation at Ser74 by the Ssu72 phosphatase is also required for efficient Stn1 recruitment to telomeres (Escandell et al. 2019). In its turn, the ST complex negatively regulates the association of telomerase with telomeres and promotes the recruitment of the polymerase α-primase complex for the synthesis of the complementary strand (C-strand). Reformation of the telomere into its closed configuration by shelterin seals the end of the telomere elongation phase. Recent studies have demonstrated that telomeric proteins such as Taz1, the Pot1-Tpz1-Poz1 (PTP) subcomplex, the ST complex, and the telomerase itself are also involved in chromosome end replication. They

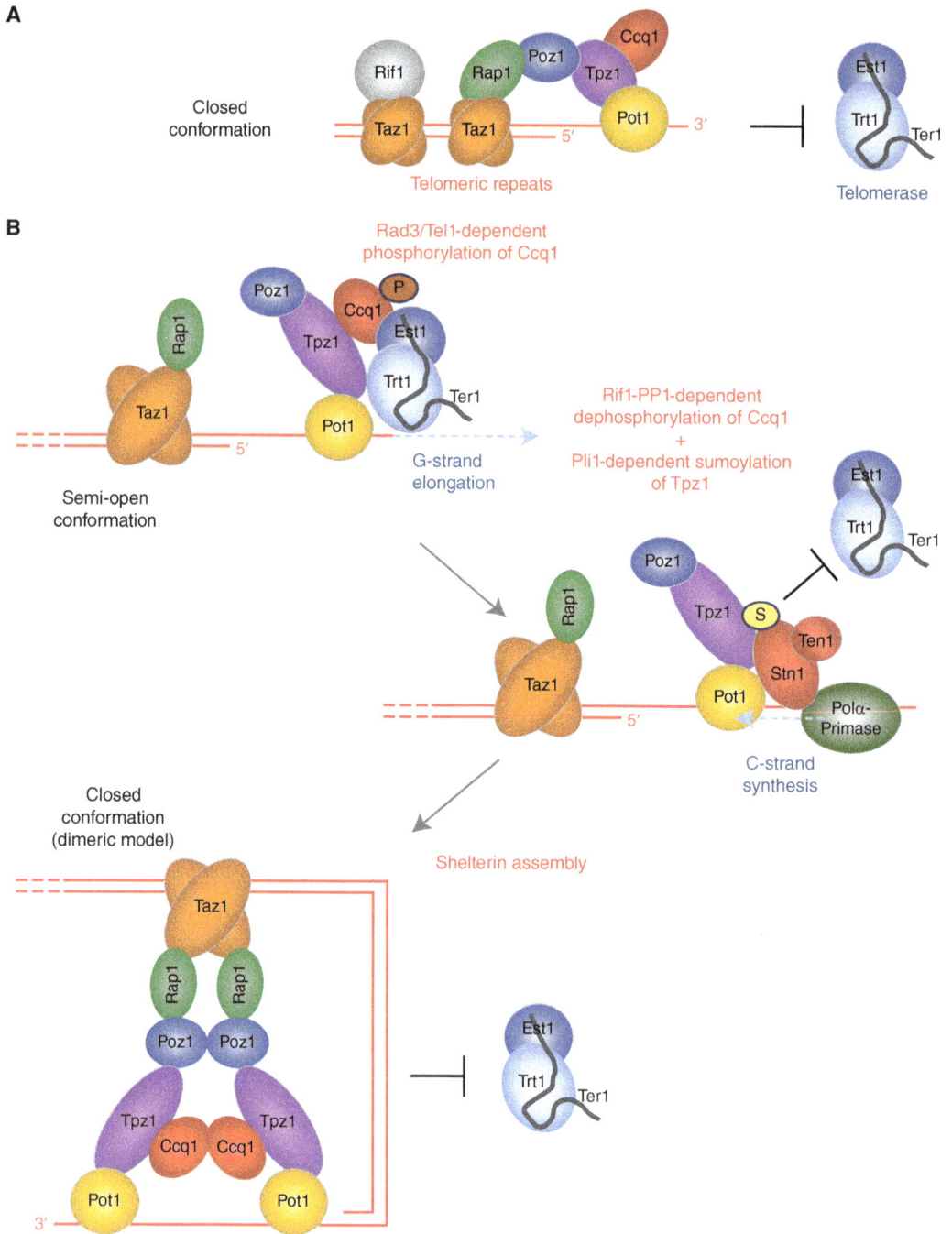

Figure 2. (*See following page for legend.*)

can now be considered as essential factors that participate in telomere replication (see below).

## Telomere Replication

As mentioned above, telomeres are among the most difficult regions to replicate, in yeast and in mammals, because of the many obstacles that impede replication fork progression (Higa et al. 2017; Maestroni et al. 2017b; Bonnell et al. 2021). Although DNA-bound telomeric proteins, such as Rap1 and TRF1, may represent barriers to the progression of the fork (Ohki and Ishikawa 2004; Douglas and Diffley 2021), it is well established, for more than one decade now, that they also facilitate the replication of DNA ends in both humans and fission yeast (Miller et al. 2006; Sfeir et al. 2009). In humans, for example, TRF1 and TRF2 act as scaffold proteins to recruit accessory factors to resolve DNA structures that form at telomeres. Along the same line, recent studies in fission yeast also highlighted the involvement of telomeric proteins in the replication of telomeric and subtelomeric regions (see below).

Having defined in the preceding paragraphs the role of each telomeric protein in telomere maintenance in *S. pombe*, we will now focus on how the cells use these factors to ensure an efficient replication of chromosome ends.

## TELOMERE REPLICATION IN FISSION YEAST

### Replication Dynamic of Telomeres

In fission yeast, telomeres are replicated in the late $S/G_2$ phase (Moser et al. 2009). Indeed, Rif1

and Taz1 control the timing of replication by regulating the firing of late origins (Hayano et al. 2012; Tazumi et al. 2012; Davé et al. 2014; Ogawa et al. 2018). Rif1 is recruited through its interaction with Taz1 at subtelomeres and associates with PP1 phosphatase to counteract MCM phosphorylation, which is required for helicase activation. The outcome is that subtelomeric replication origins are strongly suppressed by PP1, the current view being that PP1 is enriched around telomeres, forming a "PP1-zone" that restrains the firing efficiency and, ultimately, the global timing of DNA replication. A consequence of this is, first, that telomere replication mostly proceeds unidirectionally from a centromere-proximal origin, and, second, that a converging fork from the opposite direction cannot rescue a stalled replication fork. For the right arm of the chromosome II (C2-R), the most active origin is located 25 kb away from telomeric sequences and in the majority of cells, the outward-moving fork from this origin is responsible for the replication of the entire subtelomeric region of C2-R (Fig. 3; Vaurs et al. 2023). The C-rich leading strand is replicated by Polε while the G-rich lagging strand is replicated by Polδ. The two-dimensional gel (2D-gel) electrophoresis analysis of replicating DNA has shown that replication forks naturally slow and eventually stall as they approach telomeric chromatin (Miller et al. 2006). Current models suggest that forks may eventually reverse, generating a four-way DNA junction known as a chicken-foot (Carr and Lambert 2013; Berti et al. 2020). In the absence of an upcoming converging fork, the stalling

---

**Figure 2.** Control of telomerase recruitment by the shelterin in fission yeast. (*A*) Schematic representation of the telomeric structure in fission yeast. In the closed configuration, telomerase is inhibited by the shelterin. (*B*) Regulation of telomerase recruitment to telomeres. When telomeres are replicated in the late $S/G_2$ phase, shelterin folds into a semi-open configuration that allows the recruitment of telomerase. Rad3[ATR]- and Tel1[ATM]-dependent phosphorylation of Ccq1 promotes Ccq1-Est1 interaction. In a collaborative manner, Tpz1 interaction with Trt1 further engages the telomerase core complex (Trt1-TER1) for the 3′end extension. Ccq1 phosphorylation is thought to disappear through the action of the phosphatase-associated function of Rif1, while at the same time, SUMOylation of Tpz1 by the SUMO ligase Pli1 promotes the recruitment of the Stn1-Ten1 (ST). The function of ST complex is to negatively regulate the association of telomerase at telomeres and to promote the recruitment of polymerase α-primase complex to initiate the synthesis of the complementary strand (C-rich strand). The reformation of shelterin into the closed configuration seals the end of the telomere elongation phase. Structural insights indicate that telomeric proteins can dimerize, suggesting that closed conformation exists in a dimeric form as depicted.

---

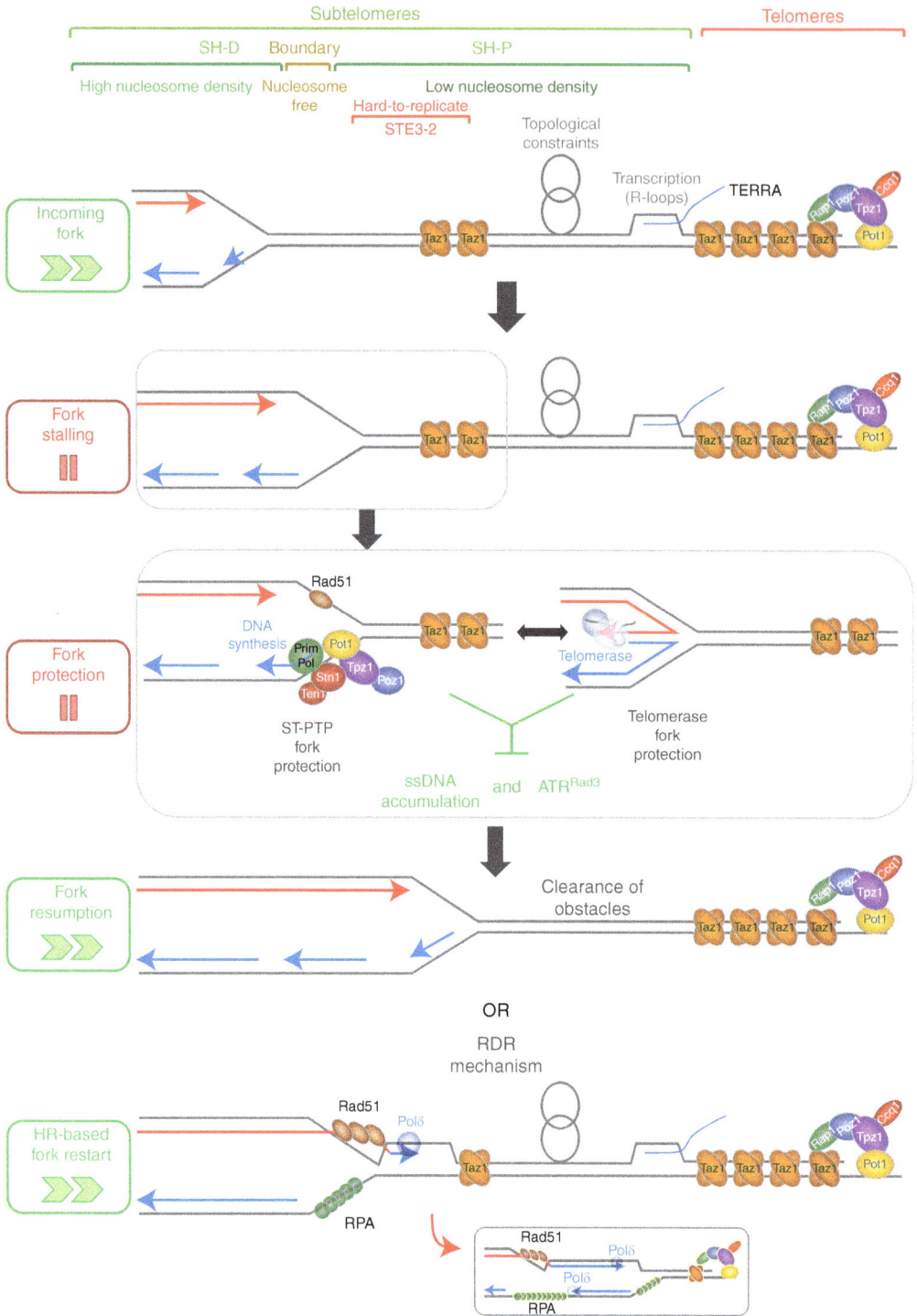

**Figure 3.** (*See following page for legend.*)

Cite this article as *Cold Spring Harb Perspect Biol* doi: 10.1101/cshperspect.a041704

may persist, occasionally resulting in the fork collapse. To avoid the deleterious consequences of fork collapse, the pathways that rely on homologous recombination (HR) evolved for the restart of DNA replication after fork collapse.

## Homologous Recombination and Replication

Due to the absence of converging forks, the restart of replication through an HR-based mechanism might be of great importance at DNA ends more than anywhere else in the genome. There are several HR-based mechanisms that are involved in the restart of replication forks. Break-induced replication (BIR), well documented in budding yeast, is one way to resume and finish DNA replication at chromosome ends (Sakofsky and Malkova 2017; Liu and Malkova 2022). In BIR, the resection of the DSB allows the 3′ end to invade the sister chromatid so that replication can proceed via a migrating D-loop until the end of chromosome. In this noncanonical conservative DNA replication process (as opposed to canonical semi-conservative where polymerase Polε synthesizes the leading and Polδ the lagging strand), leading and lagging strand synthesis are not coupled

and the DNA is synthesized by Polδ on both strands. An alternative scenario is that restart occurs at dysfunctional fork without a DSB intermediate (Fig. 3). In fission yeast, studies of engineered replication fork barriers (RFBs) highlighted an HR-dependent restart. It involves an initial fork reversal step that allows regulation of resection via Ku binding and subsequent displacement by the Mre11-Rad50-Nbs1 complex (MRN) (Lambert et al. 2010; Ait Saada et al. 2018; Audoynaud et al. 2023). In this process, named recombination-dependent replication (RDR), both the leading and lagging strands are synthesized by Polδ similar to BIR but, in contrast, replication remains semi-conservative. Whether the restart of the dysfunctional fork is performed by BIR or by RDR is currently unknown. In both cases, the fork restart is associated with various types of genetic instability, since ectopic recombination events can restart the fork at the wrong locus thus causing rearrangements or loss of genetic material. In addition, subtelomeric regions of *S. pombe* are composed of a mosaic of repeated elements that favors and promotes these recombination events (Maestroni et al. 2017a; Kanoh 2023), and whether carried out by BIR or RDR, the restarted DNA replication is intrinsically

---

Figure 3. Model of replication of terminal sequences in fission yeast. Subtelomere-telomere replication mostly proceeds unidirectionally from a centromere-proximal origin. Replication barriers such as positively supercoiled DNA, R-loops, and DNA-bound proteins are obstacles that prevent fork progression. Therefore, replication forks naturally slow down and eventually stall as they approach telomeric chromatin. We have identified a hard-to-replicate region of the C2-R, located within the STE3 and STE2 elements (named STE3-2). STE3-2 region is located at the boundary between the high- and low-density nucleosome regions highlighting a change in chromatin structure that can further impede fork progression. Our current model is that fork stalling at the STE3-2 is a crucial step for chromosomal end replication. The role of Taz1 at STE3-2 is unclear, but Taz1 could act as a barrier to the progression of the fork. This natural pausing of replication at STE3-2 may provide time for removing obstacles in front of the fork. In the meantime, forks need to be stabilized and protected. We propose that Pot1-Tpz1-Poz1 recruits ST and Polα to initiate lagging strand synthesis to counteract active resection, thus avoiding the accumulation of single-stranded DNA (ssDNA) and activation of Rad3 at the fork. Taz1 could also participate in the protection or stabilization of the fork at this stage. We also proposed that the four-way branched DNA structures arising from reversal of the stalled forks may produce a substrate for telomerase. The recruitment of telomerase may also protect the reversed fork and prevent recombination by acting as a repair enzyme. When the replication stress is resolved, the replication resumes. In this speculative scenario, the action of telomerase and ST-PTP-Polα are two anti-recombinogenic processes that prevent the homologous recombination (HR)-dependent fork restart. If the fork collapses or breaks, the restart can occur through the HR-based mechanism, either recombination-dependent replication (represented here) or break-induced replication. In both cases, the leading and lagging strands may use Polδ for their synthesis in contrast to canonical replication.

---

error-prone (Lambert et al. 2005; Iraqui et al. 2012).

In budding yeast, collapsed forks within telomere repeats are known to relocate to the nuclear periphery for nuclear pore complex (NPC) anchorage and SUMO-based mechanisms are central for the regulation of HR activity at NPCs (Churikov et al. 2016; Aguilera et al. 2020). There is no evidence so far that a such a mechanism exists in fission yeast. Nevertheless, nontelomeric dysfunctional forks do relocate and anchor to NPCs through a SUMO- and SUMO-targeted ubiquitin ligase (STUbL)-dependent pathways in *S. pombe* (Kramarz et al. 2020). In this setting, clearance of SUMO conjugates is then necessary to allow RDR. It has been further observed that in *taz1Δ* cells, dysfunctional telomeres are subjected to increased SUMO chain conjugation, and that STUbL activity likely promotes telomere entanglements (Nie et al. 2017). Along the same line, the sumoylation of the RecQ helicase homolog Rqh1 has been shown to promote the processing of dysfunctional telomeres (Rog et al. 2010). These data support the model that collapsed telomeric forks in fission yeast, like in budding yeast, might be targeted to the NPC through a SUMO- and STUbL-dependent mechanism to promote RDR. Further work will be necessary to confirm these hypotheses.

## STE3-2—a Hard-to-Replicate Region

Consistent with the observation that replication forks slow down and stall when approaching chromosome ends, we have recently identified a hard-to-replicate region of the C2-R, located within STE3 and STE2 elements (named STE3-2) (Vaurs et al. 2023). We anticipate that in addition to the expected constraints such as positively supercoiled DNA, R-loops, DNA:RNA hybrids, etc., this region possesses specific properties responsible for the replication stress that remain to be identified. Notably, the subtelomeric region (from *tlh2* to the telomeric repeats of C2-R) comprises a unique chromatin structure displaying very low nucleosome density in contrast to the *tlh2* region, where nucleosome density is high (van Emden et al. 2019). Thus,

the STE3-2 region resembles an insulator element located at the boundary between the high- and low-density nucleosome regions, the last being also prone to bind other factors (Fig. 3). This change in chromatin structure may also impede fork progression. Therefore, when a replication fork progresses unidirectionally from the origin through subtelomeres and toward the telomeric repeats, it is likely that it slows down and/or pauses when approaching the STE3-2 region. Thus, it is tempting to speculate that the function of STE3-2 region is to naturally stall incoming replication forks providing time for removing obstacles in front of the forks. This will ensure a temporal regulation of the replication of terminal sequences. Nevertheless, the underlying mechanisms are currently unknown. In the following paragraphs, we will examine the involvement of telomeric proteins in the replication of the STE3-2 region and will highlight the importance of HR as a rescue pathway when telomeric proteins are deficient.

## ROLE OF TELOMERIC PROTEINS IN TELOMERE REPLICATION

### Stn1-Ten1 Complex and Pot1-Tpz1-Poz1 Subcomplex

Both *stn1* and *ten1* are essential genes, deletion mutants of which eventually survive by circularizing their chromosomes (Martín et al. 2007). This suggests that both proteins fulfill crucial functions during telomere maintenance. As described above, the ST complex is recruited to telomeres through the interaction between the SIM domain of Stn1 and the SUMO-modified Tpz1. ST complex inhibits telomerase action and promotes the synthesis of the C-strand at telomeres. The isolation of thermosensitive alleles of *stn1* (*stn1-1* and *stn1-226*) by the group of F. Ishikawa and our laboratory also allowed the conclusion that the ST complex fulfils essential functions in replicating terminal DNA (Takikawa et al. 2016; Matmati et al. 2018). Indeed, at restrictive temperatures, these mutants lose their telomeres and subtelomeres, exhibit replication fork collapse at subtelomeres (observed by 2D gel analysis), and accumulate ssDNA.

Recently, we further investigated the role of the ST complex in the replication dynamic of chromosome ends. To this end, we employed the polymerase usage-sequencing (PuSeq) approach that relies on mutant replicative polymerases that insert ribonucleotides at increased frequency during DNA synthesis (Daigaku et al. 2015; Keszthelyi et al. 2015). Thereby, PuSeq allows the tracking of polymerase usage across the genome by mapping the strand-specific location of ribonucleotides. Noticeably, we have established that Stn1 is crucial for the replication of STE3-2 (Vaurs et al. 2023). Indeed, when Stn1 function was compromised, Polδ synthesized the lagging strand and also participated in the synthesis of the leading strand over the STE3 and STE2 regions. The use of Polδ on both strands reflects the occurrence of HR-mediated replication fork restart (Fig. 3). We also confirmed the recruitment of Stn1 to subtelomeres and found that its binding relies on the integrity of the PTP subcomplex but is independent of Taz1 (Carvalho Borges et al. 2023; Vaurs et al. 2023). In full agreement with these results, Pot1 and Tpz1 are also recruited to the STE3-2 region (Carvalho Borges et al. 2023). Studies with the thermosensitive allele of *pot1* (*pot1-1*) further revealed that Pot1 is crucial for subtelomere replication as *pot1-1* cells accumulate ssDNA at restrictive temperatures, similarly to the *stn1* thermosensitive mutants (Carvalho Borges et al. 2023). Remarkably, the overexpression of Stn1 restored the viability and telomere defects of the *pot1-1* mutant, connecting Pot1 function directly to the recruitment of Stn1. Thus, the PTP complex recruits the ST complex to subtelomeres to allow their efficient replication. In addition, the observation that overexpression of the catalytic subunit of DNA polymerase α (Pol1) was able to restore viability and telomere defects of *pot1-1* and *stn1-226* further demonstrated that ST and PTP complexes promote DNA synthesis (Matmati et al. 2018; Carvalho Borges et al. 2023). Along the same lines, inactivation of Exo1 significantly limited the accumulation of ssDNA. Notably, the deletion of *exo1* also suppressed the growth defect of *stn1-226* cells. In summary, our current model is that when forks are stalled at STE3-2, PTP recruits ST and Polα to initiate lagging strand synthesis, which counteracts active resection, thus avoiding accumulation of ssDNA. This anti-recombinogenic process prevents HR-based restart, giving cells time to remove obstacles ahead of the fork and resume replication.

Remarkably, genetic studies underscored the importance of the HR pathway when the ST complex is dysfunctional, as exemplified by the synthetic lethality of the *stn1-226* and *rad51Δ* (Matmati et al. 2018). Along the same lines, recruitment of RPA, Rad51, and Rad52 was observed by ChIP in *stn1* mutants (Takikawa et al. 2016; Vaurs et al. 2023). These observations confirmed the need for HR as a backup pathway to restart stalled forks (Fig. 3). Whether BIR or RDR restarts the forks is not established but these mechanisms are not exclusive. Genetic analysis also revealed the involvement of MRN, Ctp1, and Mus81 in the restart process, although their exact roles remain undetermined (Vaurs et al. 2023).

How can the recruitment of telomeric protein at subtelomeres be explained? Pot1 is the subunit of the PTP complex that binds the G-rich telomeric strand in different modes, thanks to its DNA-binding domains (Baumann and Cech 2001; Dickey and Wuttke 2014). Throughout the subtelomeric region and particularly within the STE3 and STE2, a multitude of single telomeric sequences (GGTTAC) are disseminated in the duplex DNA. Passage of the replication fork exposes ssDNA, and GGTTAC sites become potential binding sites for Pot1 on the lagging strand. Thus, it is very likely that the recruitment of PTP and ST complexes rely on the binding of Pot1 to internal single-stranded telomeric sequences that are present in subtelomeres.

## Taz1

Taz1 is the main component of shelterin that binds telomeric DNA duplex as a dimer, each monomer recognizing a telomeric GGTTA sequence (Cooper et al. 1997; Spink et al. 2000; Deng et al. 2015). Tandem repeats of this sequence were identified at internal locations in chromosome arms and are known as nontelomeric-binding sites for Taz1 that promote the assembly of heterochromatin (Zofall et al. 2016; Toteva et al. 2017). Same tandem repeats were also found in STE3 and likely account for the detection of Taz1 binding by ChIP at subtelomeres (Kanoh et al. 2005; Vaurs et al. 2023). Loss of Taz1 results in several telomere defects including dramatic elongation of the telomeric tracts due to hyperactivity of telomerase through the different phases of the cell cycle (Miller et al. 2005; Dehé et al. 2012). The lack of Taz1 prevents shelterin from folding telomeres into the nonextendible state, which likely results in uncontrollable telomerase action. In this setting, long telomeres are prone to form entanglements that may account for the cold sensitivity of *taz1Δ* cells (Miller and Cooper 2003). Importantly, Taz1 also plays a prominent role in the efficient replication fork progression through terminal DNA (Miller et al. 2006; Vaurs et al. 2023). Indeed, loss of Taz1 leads to the accumulation of stalled replication forks at telomeres and subtelomeres. Genetic analysis further established the necessity of factors such as Rad51 (Matmati et al. 2020) or ST complex (Matmati et al. 2018) to sustain growth of the *taz1Δ* cells and for the maintenance of their telomeres. This underscores the importance of the HR pathway and the requirement of the ST-dependent lagging strand synthesis to circumvent the absence of Taz1. Nevertheless, the function of Taz1 in the replication of telomeres remains enigmatic. To date, the putative Taz1-interacting factors, such as helicases, RNases, or any other activities capable of facilitating replication, have not been identified. It seems clear that Taz1 like TRF1 promotes efficient telomere replication (Sfeir et al. 2009), but it is also possible that DNA-bound Taz1 impedes, to some extent, fork progression, acting as a barrier like its *Saccharomyces cerevisiae* and human counterparts Rap1 and TRF1, respectively (Ohki and Ishikawa 2004; Douglas and Diffley 2021). How Taz1 controls the progression of replication forks needs further investigations.

## Telomerase

Deleting either the protein subunits of telomerase or its RNA component leads to the gradual short-ening of telomeres until definitive arrest of growth. The occurrence of stochastic events at the DNA level during replicative senescence is probably responsible for the heterogeneity of senescence kinetics observed from clone to clone. This led us to investigate replication dynamics in *S. pombe* cells lacking telomerase to unmask the replication stress occurring at telomeres. 2D gel analysis revealed that replication of telomeres is severely impaired without telomerase and correlates with an accumulation of replication intermediates, indicating that telomerase per se plays a role in telomere replication (Matmati et al. 2020). How could telomerase participate in telomere replication? In-depth analysis allowed us to establish that reversed replication forks are transient structures at telomeres and that telomerase can bind to reversed forks to protect them, thereby allowing the completion of their replication. It was also established by the group of S. Lambert that the Ku heterodimer can load onto the terminally arrested forks that undergo reversal at artificial RFB (Teixeira-Silva et al. 2017). Its association with and removal from reversed forks controls end resection and replication fork restart mediated by HR. Thus, we proposed that four-way branched DNA structures emanating from stalled forks may produce a substrate for either Ku or telomerase, consistent with the observation that the absence of telomerase was deleterious for the viability of the *pku70Δ* cells (Baumann and Cech 2000). While telomerase provides the means to resume telomere replication either by de novo telomere synthesis or by fork protection, Ku-binding controls end resection and HR-mediated replication fork restart. Consistent with this model, Rad51, Mre11, and Ctp1 are required to sustain the viability of *S. pombe* cells lacking telomerase activity. This indicates that telomerase may protect against telomere recombination by acting as a repair enzyme of the reversed forks. These genetic interactions can also explain the necessity for HR-dependent maintenance of telomeres in the absence of telomerase.

How can telomerase bind to the reversed fork? This is a question that remains unanswered. However, the observation that the *tpz1*[K75A] mutation, which lies in the TEL-patch-like domain of Tpz1 that affects its interaction with Trt1 and telomerase processivity (Armstrong et al. 2014), also impairs the replication of telomeres, suggesting that Tpz1, and thus the PTP complex, is necessary for the telomerase activity at reversed replication forks (Matmati et al. 2020). A nonmutually exclusive possibility is that telomerase is recruited independently of PTP. One possibility is that telomerase can recognize directly DNA structures that emanate from replication intermediates and/or through its interaction with RPA (Dehé et al. 2012; Luciano et al. 2012).

## Rif1

Among the telomeric proteins, Rif1 remains among the most enigmatic. As mentioned above, Rif1 physically interacts with Taz1 and is also able to bind directly to chromatin, recognizing secondary DNA structures (Xu et al. 2010; Masai et al. 2019). Rif1 is well known to suppress origin firing by recruiting PP1 phosphatases, which counteracts the phosphorylation of MCM, making Rif1 a crucial regulator of replication timing (Hayano et al. 2012; Davé et al. 2014; Mattarocci et al. 2014; Hiraga et al. 2018). Indeed, current models propose that Rif1 generates local chromatin structures that may exert long-range suppressive effects on origin firing.

In contrast to what was observed in mammalian cells, there is no evidence so far for a role of SpRif1 in replication fork protection; however, deletion of the *rif1* gene restores growth defect and telomere/subtelomere loss in the *stn1-1*, *stn1-226*, and *pot1-1* mutants at the restrictive temperature (Takikawa et al. 2016; Matmati et al. 2018; Carvalho Borges et al. 2023). The analysis of replication dynamics by the PuSeq method in *stn1-226* mutant cells lacking Rif1 allowed us to show that firing of cryptic origins within the subtelomeres reestablishes a conventional replication profile (abrogation of the DNA polymerase δ/δ configuration, indicating conservative DNA replication, observed at the STE3-2 region) (Vaurs et al. 2023). This indicates that the firing of a subtelomeric origin, usually repressed by Rif1, allows efficient replication of the terminal sequences. Thus, in an indirect manner, Rif1 participates in the replication of subtelomeres

and telomeres. Nevertheless, a direct role of Rif1 in replication fork protection is not excluded and would require further investigations.

## ROLE OF NONTELOMERIC PROTEINS IN TELOMERE REPLICATION

Of course, it is well understood that replication factors that are essential for the replication of the entire genome are also crucial for the replication of the telomeres. Nevertheless, telomeres present multiple additional difficulties for replication and therefore some additional adaptations are required for their replication. We have detailed the role of telomeric proteins in the replication of telomeres, but certain specific needs at chromosome ends are also supported by general replication factors.

### Replication Protein A

RPA is a highly conserved heterotrimeric ssDNA-binding protein involved in DNA repair, DNA recombination, and replication (Wold 1997). RPA becomes enriched at telomeres in late S phase and is thought to be brought to telomeres by the incoming replication fork (Moser et al. 2009; Luciano et al. 2012). Interestingly, several alleles of RPA result in a telomere-shortening phenotype in both budding and fission yeasts (Smith and Rothstein 1995, 1999; Ono et al. 2003; Schramke et al. 2004). In particular, in *S. pombe*, a mutation in the second OB-fold domain of the largest RPA subunit (Rpa1-D223Y), leads to substantial telomere shortening (Ono et al. 2003; Luciano et al. 2012). In addition, the Rpa1-D223Y mutation provokes dramatic replication defects at telomeres and is synthetically lethal with *taz1* deletion (Kibe et al. 2007; Audry et al. 2015). The aberrant telomeric DNA structures arising in Rpa1-D223Y mutant originate during lagging telomere replication and require the HR pathway to be processed. Remarkably, the ectopic expression of a potent G4 unwinder, the *S. cerevisiae* Pif1 helicase, restored telomere length and replication defect of *rpa1-D223Y* cells. Thus, it appears that RPA prevents the formation of G-rich structures, such as G4s, during the transient accumulation of telomeric ssDNA that occurs at the lagging telomere. In addition, we uncovered that RPA interacts with the telomerase RNA subunit TER1 in a cell-cycle-dependent manner (Luciano et al. 2012). Intriguingly, this interaction occurs earlier than the recruitment of telomerase to telomeres (Moser et al. 2009). Although the significance of this interaction is not clearly established, it is feasible that RPA facilitates telomerase action at telomeres during the early stage of their replication, perhaps at reversed replication forks. Further work will be necessary to determine the intricate role of RPA during the replication of telomeres.

## Swi1

Swi1 is the fission yeast ortholog of Timeless, a member of the fork protection complex (FPC) that is a part of the replisome and travels with the fork. FPC ensures proper replication fork pausing and the smooth passage of the replication forks at hard-to-replicate regions, including telomeres (reviewed by Leman and Noguchi 2012). Remarkably, both the down-regulation of Timeless and deletion of *swi1* lead to shortened telomeres in human cells and in fission yeast, respectively (Leman and Noguchi 2012). It has been established that Swi1 ensures the proper replication of telomeres, presumably by maintaining the fork in a stable conformation that prevents its collapse (Gadaleta et al. 2016). Interestingly, TRF1 and TRF2 are associated with the Timeless protein, suggesting that a connection between shelterin elements and FPC can modulate replication fork progression (Leman and Noguchi 2012). This observation raises a question of whether Swi1 could also fulfill a similar function of stabilizing the fork at STE3-2 region by interacting, for example, with Taz1.

## Pfh1

Pfh1 DNA helicase is the sole member of the Pif1 family DNA helicases in the fission yeast (Pinter et al. 2008; Sabouri 2017). In vitro, Pfh1 recognizes and unwinds G4 structures (Wallgren et al. 2016), and in vivo Pfh1 is essential for the replication of regions that are difficult to replicate and promotes fork movement through G4 (Sabouri

et al. 2012, 2014). At telomeres, Pfh1 facilitates telomeric DNA replication (McDonald et al. 2014; Wallgren et al. 2016), suggesting that unresolved G4 structures cause replication fork pausing and that one of Pfh1's functions is to unwind G4 structures ahead of the replication fork. Another nonmutually exclusive possibility is that Pfh1 removes stably DNA-bound protein complexes or RNA:DNA hybrids that may block DNA replication (Sabouri 2017). Pfh1 interacts with several core proteins of the replisome, such as MCM helicase, replicative DNA polymerases, RPA, and the processivity clamp PCNA in an S phase–dependent manner (McDonald et al. 2016), suggesting that Pfh1 travels with the replisome and promotes replication at hard-to-replicate sites such as telomeres.

In addition to RPA, Swi1, and Pfh1, it is likely that other accessory factors facilitate the replication of chromosome ends and remain to be discovered in fission yeast.

## MECHANISMS CONSERVED IN FISSION YEAST AND HUMANS

Telomeric repeats in yeasts are only 300 bp long, but we have seen that hard-to-replicate regions extend to ~7 kb, and that telomeric proteins are required for replication of these sequences. As in mammalian cells, these subtelomeric heterochromatinized regions are the binding site for shelterin subunits, nucleosomes, and probably other protein factors. In the end, they share many more similarities than expected with human telomeric sequences, and some peculiarities of their replication may also be conserved across species. The aim of this section is not to provide a complete review of the known players and mechanisms involved in telomere replication in humans, but to highlight certain discoveries made in fission yeast that are or could be relevant to humans.

### POT1-TPP1 and CST–Polα/Primase in Humans

Although human CST is able to bind ssDNA on its own, the CST–Polα/primase complex is likely brought to telomeres owing to its interaction

 Cite this article as *Cold Spring Harb Perspect Biol* doi: 10.1101/cshperspect.a041704

with the POT1/TPP1 to complete telomere replication by synthesis of the C-rich strands at both the lagging and leading telomeres (Cai and de Lange 2023). In fission yeast, it is highly unlikely that ST complex binds directly to ssDNA because it lacks the CTC1 homolog. Despite this, ST fulfils functions similar to those in mammalian cells thanks to its interaction with Tpz1. Importantly, we have highlighted in *S. pombe* the role of Pot1 and the ST complex during the replication of telomeres and subtelomeres. Our current model suggests that PTP-Stn1-Ten1-Polα/primase protects the fork by curbing DNA end resection and promoting lagging-strand synthesis. Human CST is also known to promote replication restart at telomeres and at nontelomeric sites such as repetitive sequences under stress conditions (Stewart et al. 2018). POT1 is unlikely to contribute to the genome-wide DNA replication function of CST, but POT1 and CST appear to participate in telomere replication (Pinzaru et al. 2016). Indeed, POT1 and TPP1 could act at stalled replication forks to promote lagging strand synthesis by recruiting CST–Polα/primase, and it emerges as a key regulator of telomere replication in human cells as well as in fission yeast. If so, the function of POT1-TPP1 at stalled replication forks could be reminiscent of the role of 53BP1-RIF1-shieldin at DSBs in recruiting CST-Polα/primase to counteract resection (Mirman et al. 2018).

## Telomerase

In fission yeast, there is now mounting evidence that telomerase can recognize replication intermediates that originate from collapsed replication forks thereby shielding telomeres from HR (Dehé et al. 2012; Matmati et al. 2020). Observation that a catalytically dead version of telomerase aggravates replicative senescence compared to the cells lacking telomerase indicates that the activity of telomerase is necessary to protect stalled replication fork (Matmati et al. 2020). In line with these key findings in fission yeast, the group of Y. Doksani recently demonstrated that telomeric repeats are hotspots for replication fork reversal and that telomerase is able to elongate the G-rich nascent strand of the reversed replication forks in human cells (Huda et al. 2023). In mouse, the binding of telomerase to reversed replication forks has been also observed in RTEL1-deficient cell lines (Margalef et al. 2018). In this genetic context, telomerase seems to inappropriately bind to and to stabilize reversed replication forks, thereby preventing replication fork restart. What emerges from these studies is that fork reversal occurs at telomeric DNA and this intermediate structure is recognized by telomerase that acts as an accessory factor of replication at telomeres. In both fission yeast and humans, telomerase recruitment and activation is a complex, regulated process orchestrated by shelterin. Is telomerase recruitment to the reversed forks also controlled by shelterin? This remains to be determined.

## RPA

RPA acts as the first sensor of replication stress by binding and protecting ssDNA at the fork. It also acts as a platform to recruit several partners in the DNA damage response pathway, including the ATR kinase (Zou and Elledge 2003; Maréchal and Zou 2015). At telomeres, the RPA-to-POT1 switch is thought to occur on ssDNA to prevent RPA-dependent ATR activation (Flynn et al. 2011). Our laboratory has been pioneering the studies of the role of RPA in telomere maintenance in yeast models, further showing that it can prevent the formation of G-rich structures and facilitate the action of telomerase (Luciano et al. 2012; Audry et al. 2015). In line with these observations, we recently identified mutations in *RPA1* and *RPA2* genes that cause telomere biology disorders, confirming the functions and importance of RPA for telomere stability in humans (Sharma et al. 2021; Kochman et al. 2024). Indeed, RPA might recruit several factors that are necessary to limit the replication stress before being replaced by POT1.

## Replication Fork Restart by HR

In this review, we have repeatedly highlighted the important role of the HR pathway in several fission yeast mutants. HR-dependent fork restart is an evolutionarily conserved mechanism that is also engaged at arrested forks in human cells (Ait Saada et al. 2018; Berti et al. 2020). Factors such as BRCA2 and RAD51 perform important functions during telomere replication and in telomere protection by promoting HR, although the exact mechanisms remain unknown (Verdun and Karlseder 2006; Badie et al. 2010). Indeed, HR could promote telomere elongation by inter- or intratelomere recombination and could facilitate the formation of the protective T-loop (Tarsounas and West 2005). HR is also used by cancer cells that do not reactivate telomerase and instead maintain their telomeres through the ALT mechanism (Cesare and Reddel 2010). Thus, in human cells, HR seems to play a predominant role in telomere replication; however, the manner in which HR facilitates replication remains to be determined. It is possible that the subunits of shelterin are involved in this process. Indeed, TRF1 and TRF2 may modulate RAD51 telomeric functions, based at least on observations in vitro (Bower and Griffith 2014). This might represent an avenue for future research.

## CONCLUDING REMARKS

During the last decade, our knowledge of the mechanisms and players involved in telomere replication in fission yeast has greatly expanded. Nevertheless, some gray areas remain such as the precise function of the major telomeric factor Taz1 in telomere replication, the spatiotemporal regulation of replication events at chromosomal ends, and the precise mechanisms used for fork protection and restart. Recent studies underscored that replication stress at the chromosome ends is not restricted to telomeric tracts but extends to the SH-P region.

Indeed, telomeric chromatin (encompassing the telomeric DNA repeats and over half a kilobase of subtelomeric DNA) is organized in a unique protection pattern in which telomere-binding proteins and RNA impose a distinct nucleosome arrangement (Greenwood et al. 2018). This structure, called the "telosome," may impede the progression of the fork and Taz1 is thought to facilitate this process. The

SH-P subtelomeric regions (5–10 kb from repeats) are enriched in H3K9me/HP1 and exhibit a low density of nucleosomes that may allow extension of the binding of shelterin proteins (van Emden et al. 2019). Firing of origins that are normally repressed by Rif1 in the SH-P region suggests that replication factors are also bound to these regions. Therefore, the telomeric and subtelomeric regions of fission yeast create complex heterochromatin structures that resemble in many aspects the human telomeres that stretch over 10–15 kb and where shelterin proteins, nucleosomes, and heterochromatin coexist (Nishibuchi and Déjardin 2017). On the basis of these similarities, uncovered mechanisms in fission yeast are undoubtedly relevant for mammalian cells too.

Human telomere replication necessitates more factors than telomere replication in *S. pombe*. One reason for that might be related to the fact that telomerase is constitutively expressed in yeast while it is repressed in most human cells. Moreover, fission yeast telomerase appears to protect collapsed replication forks (Matmati et al. 2020; Huda et al. 2023). Hence, human somatic cells, which are deprived of telomerase activity, may have evolved numerous additional mechanisms to allow efficient telomeric fork progression. This highlights the importance of understanding fork stabilization and restart in fission yeast, and of understanding the role of telomerase in these processes. In-depth comprehension of these mechanisms in yeast will allow us to better elucidate many aspects of the telomere biology in humans.

## COMPETING INTEREST STATEMENT

The authors declare no competing interests.

## ACKNOWLEDGMENTS

I am very grateful to Dmitri Churikov and Karel Naiman for many suggestions and for helpful comments on the manuscript. S.C. is supported by the Agence Nationale de la Recherche (ANR22-CE12 TeloSTAB and ANR-20-CE12 TeloRPA), by la Ligue Nationale Contre le Cancer (LNCC-Equipe labélisée) and by the Can-céropôle PACA.

## REFERENCES

Aguilera P, Whalen J, Minguet C, Churikov D, Freudenreich C, Simon MN, Géli V. 2020. The nuclear pore complex prevents sister chromatid recombination during replicative senescence. *Nat Commun* **11:** 160. doi:10.1038/s41467-019-13979-5

Ait Saada AA, Lambert SAE, Carr AM. 2018. Preserving replication fork integrity and competence via the homologous recombination pathway. *DNA Repair (Amst)* **71:** 135–147. doi:10.1016/j.dnarep.2018.08.017

Armstrong CA, Tomita K. 2017. Fundamental mechanisms of telomerase action in yeasts and mammals: understanding telomeres and telomerase in cancer cells. *Open Biol* **7:** 160338. doi:10.1098/rsob.160338

Armstrong CA, Pearson SR, Amelina H, Moiseeva V, Tomita K. 2014. Telomerase activation after recruitment in fission yeast. *Curr Biol* **24:** 2006–2011. doi:10.1016/j.cub.2014.07.035

Audoynaud C, Schirmeisen K, Ait Saada A, Gesnik A, Fernández-Varela P, Boucherit V, Ropars V, Chaudhuri A, Fréon K, Charbonnier JB, et al. 2023. RNA: DNA hybrids from Okazaki fragments contribute to establish the Ku-mediated barrier to replication-fork degradation. *Mol Cell* **83:** 1061–1074.e6. doi:10.1016/j.molcel.2023.02.008

Audry J, Maestroni L, Delagoutte E, Gauthier T, Nakamura TM, Gachet Y, Saintomé C, Géli V, Coulon S. 2015. RPA prevents G-rich structure formation at lagging-strand telomeres to allow maintenance of chromosome ends. *EMBO J* **34:** 1942–1958. doi:10.15252/embj.201490773

Audry J, Zhang H, Kerr C, Berkner KL, Runge KW. 2024. Ccq1 restrains Mre11-mediated degradation to distinguish short telomeres from double-strand breaks. *Nucleic Acids Res* **52:** 3722–3739. doi:10.1093/nar/gkae044

Badie S, Escandell JM, Bouwman P, Carlos AR, Thanasoula M, Gallardo MM, Suram A, Jaco I, Benitez J, Herbig U, et al. 2010. BRCA2 acts as a RAD51 loader to facilitate telomere replication and capping. *Nat Struct Mol Biol* **17:** 1461–1469. doi:10.1038/nsmb.1943

Bah A, Wischnewski H, Shchepachev V, Azzalin CM. 2012. The telomeric transcriptome of *Schizosaccharomyces pombe*. *Nucleic Acids Res* **40:** 2995–3005. doi:10.1093/nar/gkr1153

Barnes T, Korber P. 2021. The active mechanism of nucleosome depletion by Poly(dA:dT) tracts in vivo. *Int J Mol Sci* **22:** 8233. doi:10.3390/ijms22158233

Baumann P, Cech TR. 2000. Protection of telomeres by the Ku protein in fission yeast. *Mol Biol Cell* **11:** 3265–3275. doi:10.1091/mbc.11.10.3265

Baumann P, Cech TR. 2001. Pot1, the putative telomere end-binding protein in fission yeast and humans. *Science* **292:** 1171–1175. doi:10.1126/science.1060036

Beernink HTH, Miller K, Deshpande A, Bucher P, Cooper JP. 2003. Telomere maintenance in fission yeast requires an est1 ortholog. *Curr Biol* **13:** 575–580. doi:10.1016/S0960-9822(03)00169-6

Berti M, Cortez D, Lopes M. 2020. The plasticity of DNA replication forks in response to clinically relevant genotoxic stress. *Nat Rev Mol Cell Biol* **21:** 633–651. doi:10.1038/s41580-020-0257-5

Bjerling P, Silverstein RA, Thon G, Caudy A, Grewal SIS, Ekwall K. 2002. Functional divergence between histone deacetylases in fission yeast by distinct cellular localization and in vivo specificity. *Mol Cell Biol* **22:** 2170–2181. doi:10.1128/MCB.22.7.2170-2181.2002

Blasiak J, Szczepańska J, Sobczuk A, Fila M, Pawlowska E. 2021. RIF1 links replication timing with fork reactivation and DNA double-strand break repair. *Int J Mol Sci* **22:** 11440. doi:10.3390/ijms222111440

Bonnell E, Pasquier E, Wellinger RJ. 2021. Telomere replication: solving multiple end replication problems. *Front Cell Dev Biol* **9:** 668171. doi:10.3389/fcell.2021.668171

Bower BD, Griffith JD. 2014. TRF1 and TRF2 differentially modulate Rad51-mediated telomeric and nontelomeric displacement loop formation in vitro. *Biochemistry* **53:** 5485–5495. doi:10.1021/bi5006249

Cai SW, de Lange T. 2023. CST–polα/primase: the second telomere maintenance machine. *Genes Dev* **37:** 555–569. doi:10.1101/gad.350479.123

Carneiro T, Khair L, Reis CC, Borges V, Moser BA, Nakamura TM, Ferreira MG. 2010. Telomeres avoid end detection by severing the checkpoint signal transduction pathway. *Nature* **467:** 228–232. doi:10.1038/nature09353

Carr AM, Lambert SAE. 2013. Replication stress-induced genome instability: the dark side of replication maintenance by homologous recombination. *J Mol Biol* **425:** 4733–4744. doi:10.1016/j.jmb.2013.04.023

Carvalho Borges PC, Bouabboune C, Escandell JM, Matmati S, Coulon S, Ferreira MG. 2023. Pot1 promotes telomere DNA replication via the Stn1-Ten1 complex in fission yeast. *Nucleic Acids Res* **51:** 12325–12336. doi:10.1093/nar/gkad1036

Casteel DE, Zhuang S, Zeng Y, Perrino FW, Boss GR, Goulian M, Pilz RB. 2008. A DNA polymerase-α primase cofactor with homology to replication protein A-32 regulates DNA replication in mammalian cells. *J Biol Chem* **284:** 5807–5818. doi:10.1074/jbc.M807593200

Cesare AJ, Reddel RR. 2010. Alternative lengthening of telomeres: models, mechanisms and implications. *Nat Rev Genet* **11:** 319–330. doi:10.1038/nrg2763

Chang YT, Moser BA, Nakamura TM. 2013. Fission yeast shelterin regulates DNA polymerases and Rad3(ATR) kinase to limit telomere extension. *PLoS Genet* **9:** e1003936. doi:10.1371/journal.pgen.1003936

Chen LY, Redon S, Lingner J. 2012. The human CST complex is a terminator of telomerase activity. *Nature* **488:** 540–544. doi:10.1038/nature11269

Churikov D, Charifi F, Eckert-Boulet N, Silva S, Simon MN, Lisby M, Géli V. 2016. SUMO-dependent relocalization of eroded telomeres to nuclear pore complexes controls telomere recombination. *Cell Rep* **15:** 1242–1253. doi:10.1016/j.celrep.2016.04.008

Cooper JP, Nimmo ER, Allshire RC, Cech TR. 1997. Regulation of telomere length and function by a Myb-domain protein in fission yeast. *Nature* **385:** 744–747. doi:10.1038/385744a0

Daigaku Y, Keszthelyi A, Müller CA, Miyabe I, Brooks T, Retkute R, Hubank M, Nieduszynski CA, Carr AM. 2015. A global profile of replicative polymerase usage. *Nat Struct Mol Biol* **22:** 192–198. doi:10.1038/nsmb.2962

Davé A, Cooley C, Garg M, Bianchi A. 2014. Protein phosphatase 1 recruitment by Rif1 regulates DNA replication origin firing by counteracting DDK activity. *Cell Rep* **7:** 53–61. doi:10.1016/j.celrep.2014.02.019

Dehé PM, Cooper JP. 2010. Fission yeast telomeres forecast the end of the crisis. *FEBS Lett* **584:** 3725–3733. doi:10.1016/j.febslet.2010.07.045

Dehé PM, Rog O, Ferreira MG, Greenwood J, Cooper JP. 2012. Taz1 enforces cell-cycle regulation of telomere synthesis. *Mol Cell* **46:** 797–808. doi:10.1016/j.molcel.2012.04.022

de Lange T. 2004. Opinion: T-loops and the origin of telomeres. *Nat Rev Mol Cell Biol* **5:** 323–329. doi:10.1038/nrm1359

de Lange T. 2018. Shelterin-mediated telomere protection. *Annu Rev Genet* **52:** 223–247. doi:10.1146/annurev-genet-032918-021921

Deng W, Wu J, Wang F, Kanoh J, Dehe P-M, Inoue H, Chen J, Lei M. 2015. Fission yeast telomere-binding protein Taz1 is a functional but not a structural counterpart of human TRF1 and TRF2. *Cell Res* **25:** 881–884. doi:10.1038/cr.2015.76

Dickey TH, Wuttke DS. 2014. The telomeric protein Pot1 from *Schizosaccharomyces pombe* binds ssDNA in two modes with differing 3′ end availability. *Nucleic Acids Res* **42:** 9656–9665. doi:10.1093/nar/gku680

Douglas ME, Diffley JFX. 2021. Budding yeast Rap1, but not telomeric DNA, is inhibitory for multiple stages of DNA replication in vitro. *Nucleic Acids Res* **49:** 5671–5683. doi:10.1093/nar/gkab416

Escandell JM, Carvalho ES, Gallo-Fernández M, Reis CC, Matmati S, Luís IM, Abreu IA, Coulon S, Ferreira MG. 2019. Ssu72 phosphatase is a conserved telomere replication terminator. *EMBO J* **38:** e100476. doi:10.15252/embj.2018100476

Ferreira MG, Cooper JP. 2001. The fission yeast Taz1 protein protects chromosomes from Ku-dependent end-to-end fusions. *Mol Cell* **7:** 55–63. doi:10.1016/S1097-2765(01)00154-X

Ferreira MG, Cooper JP. 2004. Two modes of DNA double-strand break repair are reciprocally regulated through the fission yeast cell cycle. *Genes Dev* **18:** 2249–2254. doi:10.1101/gad.315804

Flynn RL, Centore RC, O'sullivan RJ, Rai R, Tse A, Songyang Z, Chang S, Karlseder J, Zou L. 2011. TERRA and hnRNPA1 orchestrate an RPA-to-POT1 switch on telomeric single-stranded DNA. *Nature* **471:** 532–536. doi:10.1038/nature09772

Gadaleta MC, Das MM, Tanizawa H, Chang YT, Noma K, Nakamura TM, Noguchi E. 2016. Swi1Timeless prevents repeat instability at fission yeast telomeres. *PLoS Genet* **12:** e1005943. doi:10.1371/journal.pgen.1005943

Garg M, Gurung RL, Mansoubi S, Ahmed JO, Davé A, Watts FZ, Bianchi A. 2014. Tpz1TPP1 SUMOylation reveals evolutionary conservation of SUMO-dependent Stn1 telomere association. *EMBO Rep* **15:** 871–877. doi:10.15252/embr.201438919

Gilson E, Géli V. 2007. How telomeres are replicated. *Nat Rev Mol Cell Biol* **8:** 825–838. doi:10.1038/nrm2259

Gómez EB, Espinosa JM, Forsburg SL. 2005. *Schizosaccharomyces pombe mst2⁺* encodes a MYST family histone acetyltransferase that negatively regulates telomere silencing. *Mol Cell Biol* **25:** 8887–8903. doi:10.1128/MCB.25.20.8887-8903.2005

Greenwood J, Cooper JP. 2012. Non-coding telomeric and subtelomeric transcripts are differentially regulated by telomeric and heterochromatin assembly factors in fission yeast. *Nucleic Acids Res* **40:** 2956–2963. doi:10.1093/nar/gkr1155

Greenwood J, Patel H, Cech TR, Cooper JP. 2018. Fission yeast telosomes: non-canonical histone-containing chromatin structures dependent on shelterin and RNA. *Nucleic Acids Res* **46:** 8865–8875. doi:10.1093/nar/gky605

Hardy CF, Sussel L, Shore D. 1992. A RAP1-interacting protein involved in transcriptional silencing and telomere length regulation. *Genes Dev* **6:** 801–814. doi:10.1101/gad.6.5.801

Hayano M, Kanoh Y, Matsumoto S, Renard-Guillet C, Shirahige K, Masai H. 2012. Rif1 is a global regulator of timing of replication origin firing in fission yeast. *Genes Dev* **26:** 137–150. doi:10.1101/gad.178491.111

Hayles J, Nurse P. 2018. Introduction to fission yeast as a model system. *Cold Spring Harb Protoc* **2018:** 323–333. doi:10.1101/pdb.top079749

Higa M, Fujita M, Yoshida K. 2017. DNA replication origins and fork progression at mammalian telomeres. *Genes (Basel)* **8:** 112. doi:10.3390/genes8040112

Hiraga S, Monerawela C, Katou Y, Shaw S, Clark KR, Shirahige K, Donaldson AD. 2018. Budding yeast Rif1 binds to replication origins and protects DNA at blocked replication forks. *EMBO Rep* **19:** e46222. doi:10.15252/embr.201846222

Hu X, Liu J, Jun HI, Kim JK, Qiao F. 2016. Multi-step coordination of telomerase recruitment in fission yeast through two coupled telomere-telomerase interfaces. *eLife* **5:** e15470. doi:10.7554/eLife.15470

Huda A, Arakawa H, Mazzucco G, Galli M, Petrocelli V, Casola S, Chen L, Doksani Y. 2023. The telomerase reverse transcriptase elongates reversed replication forks at telomeric repeats. *Sci Adv* **9:** eadf2011. doi:10.1126/sciadv.adf2011

Iraqui I, Chekkal Y, Jmari N, Pietrobon V, Fréon K, Costes A, Lambert SAE. 2012. Recovery of arrested replication forks by homologous recombination is error-prone. *PLoS Genet* **8:** e1002976. doi:10.1371/journal.pgen.1002976

Jain D, Hebden AK, Nakamura TM, Miller KM, Cooper JP. 2010. HAATI survivors replace canonical telomeres with blocks of generic heterochromatin. *Nature* **467:** 223–227. doi:10.1038/nature09374

Jun HI, Liu J, Jeong H, Kim JK, Qiao F. 2013. Tpz1 controls a telomerase-nonextendible telomeric state and coordinates switching to an extendible state via Ccq1. *Genes Dev* **27:** 1917–1931. doi:10.1101/gad.219485.113

Kanoh J. 2023. Roles of specialized chromatin and DNA structures at subtelomeres in *Schizosaccharomyces pombe*. *Biomolecules* **13:** 810. doi:10.3390/biom13050810

Kanoh J, Ishikawa F. 2001. Sprap1 and spRif1, recruited to telomeres by Taz1, are essential for telomere function

Cite this article as *Cold Spring Harb Perspect Biol* doi: 10.1101/cshperspect.a041704

in fission yeast. *Curr Biol* **11**: 1624–1630. doi:10.1016/ S0960-9822(01)00503-6

Kanoh J, Sadaie M, Urano T, Ishikawa F. 2005. Telomere binding protein Taz1 establishes Swi6 heterochromatin independently of RNAi at telomeres. *Curr Biol* **15**: 1808–1819. doi:10.1016/j.cub.2005.09.041

Kanoh Y, Matsumoto S, Fukatsu R, Kakusho N, Kono N, Renard-Guillet C, Masuda K, Iida K, Nagasawa K, Shirahige K, et al. 2015. Rif1 binds to G quadruplexes and suppresses replication over long distances. *Nat Struct Mol Biol* **22**: 889–897. doi:10.1038/nsmb.3102

Kanoh Y, Ueno M, Hayano M, Kudo S, Masai H. 2023. Aberrant association of chromatin with nuclear periphery induced by Rif1 leads to mitotic defect. *Life Sci Alliance* **6**: e202201603. doi:10.26508/lsa.202201603

Kedziora S, Gali VK, Wilson RHC, Clark KRM, Nieduszyn-ski CA, Hiraga S, Donaldson AD. 2018. Rif1 acts through protein phosphatase 1 but independent of replication timing to suppress telomere extension in budding yeast. *Nucleic Acids Res* **46**: 3993–4003. doi:10.1093/nar/gky132

Keszthelyi A, Daigaku Y, Ptasińska K, Miyabe I, Carr AM. 2015. Mapping ribonucleotides in genomic DNA and exploring replication dynamics by polymerase usage se-quencing (Pu-seq). *Nat Protoc* **10**: 1786–1801. doi:10 .1038/nprot.2015.116

Khayat F, Cannavo E, Alshmery M, Foster WR, Chahwan C, Maddalena M, Smith C, Oliver AW, Watson AT, Carr AM, et al. 2021. Inhibition of MRN activity by a telomere protein motif. *Nat Commun* **12**: 3856. doi:10.1038/ s41467-021-24047-2

Kibe T, Ono Y, Sato K, Ueno M. 2007. Fission yeast Taz1 and RPA are synergistically required to prevent rapid telomere loss. *Mol Biol Cell* **18**: 2378–2387. doi:10.1091/mbc.e06-12-1084

Kim JK, Liu J, Hu X, Yu C, Roskamp K, Sankaran B, Huang L, Komives EA, Qiao F. 2017. Structural basis for shelterin bridge assembly. *Mol Cell* **68**: 698–713.e6. doi:10.1016/j .molcel.2017.10.032

Kochman R, Ba I, Yates M, Pirabakaran V, Gourmelon F, Churikov D, Lafaille M, Kermasson L, Hamelin C, Marois I, et al. 2024. Heterozygous RPA2 variant as a novel ge-netic cause of telomere biology disorders. *Genes Dev* **38**: 755–771.

Kramarz K, Schirmeisen K, Boucherit V, Ait Saada A, Lovo C, Palancade B, Freudenreich C, Lambert SAE. 2020. The nuclear pore primes recombination-dependent DNA synthesis at arrested forks by promoting SUMO removal. *Nat Commun* **11**: 5643–5615. doi:10.1038/s41467-020-19516-z

Lambert S, Watson A, Sheedy DM, Martin B, Carr AM. 2005. Gross chromosomal rearrangements and elevated recombination at an inducible site-specific replication fork barrier. *Cell* **121**: 689–702. doi:10.1016/j.cell.2005 .03.022

Lambert SAE, Mizuno K, Blaisonneau J, Martineau S, Cha-net R, Fréon K, Murray JM, Carr AM, Baldacci G. 2010. Homologous recombination restarts blocked replication forks at the expense of genome rearrangements by tem-plate exchange. *Mol Cell* **39**: 346–359. doi:10.1016/j .molcel.2010.07.015

Leman AR, Noguchi E. 2012. Local and global functions of Timeless and Tipin in replication fork protection. *Cell Cycle* **11**: 3945–3955. doi:10.4161/cc.21989

Lingner J, Cooper J, Cech T. 1995. Telomerase and DNA end replication: no longer a lagging strand problem? *Science* **269**: 1533–1534. doi:10.1126/science.7545310

Liu L, Malkova A. 2022. Break-induced replication: unrav-eling each step. *Trends Genet* **38**: 752–765. doi:10.1016/j .tig.2022.03.011

Liu J, Yu C, Hu X, Kim JK, Bierma JC, Jun HI, Rychnovsky SD, Huang L, Qiao F. 2015. Dissecting fission yeast shel-terin interactions via MICro-MS links disruption of shel-terin bridge to tumorigenesis. *Cell Rep* **12**: 2169–2180. doi:10.1016/j.celrep.2015.08.043

Liu J, Hu X, Bao K, Kim JK, Zhang C, Jia S, Qiao F. 2021. The cooperative assembly of shelterin bridge provides a kinet-ic gateway that controls telomere length homeostasis. *Nu-cleic Acids Res* **49**: 8110–8119. doi:10.1093/nar/gkab550

Luciano P, Coulon S, Faure V, Corda Y, Bos J, Brill SJ, Gilson E, Simon MN, Géli V. 2012. RPA facilitates telomerase activity at chromosome ends in budding and fis-sion yeasts. *EMBO J* **31**: 2034–2046. doi:10.1038/emboj .2012.40

Maestroni L, Audry J, Matmati S, Arcangioli B, Géli V, Cou-lon S. 2017a. Eroded telomeres are rearranged in quies-cent fission yeast cells through duplications of subtelo-meric sequences. *Nat Commun* **8**: 1684. doi:10.1038/ s41467-017-01894-6

Maestroni L, Matmati S, Coulon S. 2017b. Solving the telo-mere replication problem. *Genes (Basel)* **8**: 55. doi:10 .3390/genes8020055

Maréchal A, Zou L. 2015. RPA-coated single-stranded DNA as a platform for post-translational modifications in the DNA damage response. *Cell Res* **25**: 9–23. doi:10.1038/cr .2014.147

Margalef P, Kotsantis P, Borel V, Bellelli R, Panier S, Boulton SJ. 2018. Stabilization of reversed replication forks by telomerase drives telomere catastrophe. *Cell* **172**: 439–453.e14. doi:10.1016/j.cell.2017.11.047

Martín V, Du LL, Rozenzhak S, Russell P. 2007. Protection of telomeres by a conserved Stn1 Ten1 complex. *Proc Natl Acad Sci* **104**: 14038–14043. doi:10.1073/pnas .0705497104

Masai H, Fukatsu R, Kakusho N, Kanoh Y, Moriyama K, Ma Y, Iida K, Nagasawa K. 2019. Rif1 promotes association of G-quadruplex (G4) by its specific G4 binding and oligo-merization activities. *Sci Rep* **9**: 8618–8615. doi:10.1038/ s41598-019-44736-9

Matmati S, Vaurs M, Escandell JM, Maestroni L, Nakamura TM, Ferreira MG, Géli V, Coulon S. 2018. The fission yeast Stn1-Ten1 complex limits telomerase activity via its SUMO-interacting motif and promotes telomeres rep-lication. *Sci Adv* **4**: eaar2740. doi:10.1126/sciadv.aar2740

Matmati S, Lambert SAE, Géli V, Coulon S. 2020. Telome-rase repairs collapsed replication forks at telomeres. *Cell Rep* **30**: 3312–3322.e3. doi:10.1016/j.celrep.2020.02.065

Matsuda A, Chikashige Y, Ding DQ, Ohtsuki C, Mori C, Asakawa H, Kimura H, Haraguchi T, Hiraoka Y. 2015. Highly condensed chromatins are formed adjacent to subtelomeric and decondensed silent chromatin in fission yeast. *Nat Commun* **6**: 7753. doi:10.1038/ncomms8753

Mattarocci S, Shyian M, Lemmens L, Damay P, Altintas DM, Shi T, Bartholomew CR, Thomä NH, Hardy CFJ, Shore D. 2014. Rif1 controls DNA replication timing in yeast through the PP1 phosphatase Glc7. *Cell Rep* **7**: 62–69. doi:10.1016/j.celrep.2014.03.010

Mattarocci S, Reinert JK, Bunker RD, Fontana GA, Shi T, Klein D, Cavadini S, Faty M, Shyian M, Hafner L, et al. 2017. Rif1 maintains telomeres and mediates DNA repair by encasing DNA ends. *Nat Struct Mol Biol* **24**: 588–595. doi:10.1038/nsmb.3420

McDonald KR, Sabouri N, Webb CJ, Zakian VA. 2014. The Pif1 family helicase Pfh1 facilitates telomere replication and has an RPA-dependent role during telomere lengthening. *DNA Repair (Amst)* **24**: 80–86. doi:10.1016/j.dnarep.2014.09.008

McDonald KR, Guise AJ, Pourbozorgi-Langroudi P, Cristea IM, Zakian VA, Capra JA, Sabouri N. 2016. Pfh1 is an accessory replicative helicase that interacts with the replisome to facilitate fork progression and preserve genome integrity. *PLoS Genet* **12**: e1006238. doi:10.1371/journal.pgen.1006238

Mennie AK, Moser BA, Hoyle A, Low RS, Tanaka K, Nakamura TM. 2019. Tpz1TPP1 prevents telomerase activation and protects telomeres by modulating the Stn1-Ten1 complex in fission yeast. *Commun Biol* **2**: 297–215. doi:10.1038/s42003-019-0546-8

Miller KM, Cooper JP. 2003. The telomere protein Taz1 is required to prevent and repair genomic DNA breaks. *Mol Cell* **11**: 303–313. doi:10.1016/S1097-2765(03)00041-8

Miller KM, Ferreira MG, Cooper JP. 2005. Taz1, Rap1 and Rif1 act both interdependently and independently to maintain telomeres. *EMBO J* **24**: 3128–3135. doi:10.1038/sj.emboj.7600779

Miller KM, Rog O, Cooper JP. 2006. Semi-conservative DNA replication through telomeres requires Taz1. *Nature* **440**: 824–828. doi:10.1038/nature04638

Mirman Z, Lottersberger F, Takai H, Kibe T, Gong Y, Takai K, Bianchi A, Zimmermann M, Durocher D, de Lange T. 2018. 53BP1-RIF1-shieldin counteracts DSB resection through CST- and Polα-dependent fill-in. *Nature* **560**: 112–116. doi:10.1038/s41586-018-0324-7

Miyagawa K, Low RS, Santosa V, Tsuji H, Moser BA, Fujisawa S, Harland JL, Raguimova ON, Go A, Ueno M, et al. 2014. SUMOylation regulates telomere length by targeting the shelterin subunit Tpz1(Tpp1) to modulate shelterin-Stn1 interaction in fission yeast. *Proc Natl Acad Sci* **111**: 5950–5955. doi:10.1073/pnas.1401359111

Miyake Y, Nakamura M, Nabetani A, Shimamura S, Tamura M, Yonehara S, Saito M, Ishikawa F. 2009. RPA-like mammalian Ctc1-Stn1-Ten1 complex binds to single-stranded DNA and protects telomeres independently of the Pot1 pathway. *Mol Cell* **36**: 193–206. doi:10.1016/j.molcel.2009.08.009

Miyoshi T, Kanoh J, Saito M, Ishikawa F. 2008. Fission yeast Pot1-Tpp1 protects telomeres and regulates telomere length. *Science* **320**: 1341–1344. doi:10.1126/science.1154819

Moravec M, Wischnewski H, Bah A, Hu Y, Liu N, Lafranchi L, King MC, Azzalin CM. 2016. TERRA promotes telomerase-mediated telomere elongation in *Schizosaccharomyces pombe*. *EMBO Rep* **17**: 999–1012. doi:10.15252/embr.201541708

Moser BA, Nakamura TM. 2009. Protection and replication of telomeres in fission yeast. *Biochem Cell Biol* **87**: 747–758. doi:10.1139/O09-037

Moser BA, Subramanian L, Chang YT, Noguchi C, Noguchi E, Nakamura TM. 2009. Differential arrival of leading and lagging strand DNA polymerases at fission yeast telomeres. *EMBO J* **28**: 810–820. doi:10.1038/emboj.2009.31

Moser BA, Chang YT, Kosti J, Nakamura TM. 2011. Tel1^ATM and Rad3ATR kinases promote Ccq1-Est1 interaction to maintain telomeres in fission yeast. *Nat Struct Mol Biol* **18**: 1408–1413. doi:10.1038/nsmb.2187

Motamedi MR, Hong EJE, Li X, Gerber S, Denison C, Gygi S, Moazed D. 2008. HP1 proteins form distinct complexes and mediate heterochromatic gene silencing by nonoverlapping mechanisms. *Mol Cell* **32**: 778–790. doi:10.1016/j.molcel.2008.10.026

Nageshan RK, Ortega R, Krogan N, Cooper JP. 2024. Fate of telomere entanglements is dictated by the timing of anaphase midregion nuclear envelope breakdown. *Nat Commun* **15**: 4707. doi:10.1038/s41467-024-48382-2

Nakamura TM, Morin GB, Chapman KB, Weinrich SL, Andrews WH, Lingner J, Harley CB, Cech TR. 1997. Telomerase catalytic subunit homologs from fission yeast and human. *Science* **277**: 955–959. doi:10.1126/science.277.5328.955

Nakamura TM, Cooper JP, Cech TR. 1998. Two modes of survival of fission yeast without telomerase. *Science* **282**: 493–496. doi:10.1126/science.282.5388.493

Nandakumar J, Cech TR. 2013. Finding the end: recruitment of telomerase to telomeres. *Nat Rev Mol Cell Biol* **14**: 69–82. doi:10.1038/nrm3505

Nie M, Moser BA, Nakamura TM, Boddy MN. 2017. SUMO-targeted ubiquitin ligase activity can either suppress or promote genome instability, depending on the nature of the DNA lesion. *PLoS Genet* **13**: e1006776-25. doi:10.1371/journal.pgen.1006776

Nishibuchi G, Déjardin J. 2017. The molecular basis of the organization of repetitive DNA-containing constitutive heterochromatin in mammals. *Chromosome Res* **25**: 77–87. doi:10.1007/s10577-016-9547-3

Ogawa S, Kido S, Handa T, Ogawa H, Asakawa H, Takahashi TS, Nakagawa T, Hiraoka Y, Masukata H. 2018. Shelterin promotes tethering of late replication origins to telomeres for replication-timing control. *EMBO J* **37**: e98997. doi:10.15252/embj.201898997

Ohki R, Ishikawa F. 2004. Telomere-bound TRF1 and TRF2 stall the replication fork at telomeric repeats. *Nucleic Acids Res* **32**: 1627–1637. doi:10.1093/nar/gkh309

Oizumi Y, Kaji T, Tashiro S, Takeshita Y, Date Y, Kanoh J. 2021. Complete sequences of *Schizosaccharomyces pombe* subtelomeres reveal multiple patterns of genome variation. *Nat Commun* **12**: 611–616. doi:10.1038/s41467-020-20595-1

Ono Y, Tomita K, Matsuura A, Nakagawa T, Masukata H, Uritani M, Ushimaru T, Ueno M. 2003. A novel allele of fission yeast rad11 that causes defects in DNA repair and telomere length regulation. *Nucleic Acids Res* **31**: 7141–7149. doi:10.1093/nar/gkg917

Pan L, Hildebrand K, Stutz C, Thomä N, Baumann TRCP. 2015. Minishelterins separate telomere length regulation and end protection in fission yeast. *Genes Dev* **29**: 1164–1174. doi:10.1101/gad.261123.115

Pinter SF, Aubert SD, Zakian VA. 2008. The *Schizosaccharomyces pombe* Pfh1p DNA helicase is essential for the maintenance of nuclear and mitochondrial DNA. *Mol Cell Biol* **28:** 6594–6608. doi:10.1128/MCB.00191-08

Pinzaru AM, Hom RA, Beal A, Phillips AF, Ni E, Cardozo T, Nair N, Choi J, Wuttke DS, Sfeir A, et al. 2016. Telomere replication stress induced by POT1 inactivation accelerates tumorigenesis. *Cell Rep* **15:** 2170–2184. doi:10.1016/j.celrep.2016.05.008

Roake CM, Artandi SE. 2020. Regulation of human telomerase in homeostasis and disease. *Nat Rev Mol Cell Biol* **21:** 384–397. doi:10.1038/s41580-020-0234-z

Rog O, Miller KM, Ferreira MG, Cooper JP. 2010. Sumoylation of RecQ helicase controls the fate of dysfunctional telomeres. *Mol Cell* **33:** 559–569. doi:10.1016/j.molcel.2009.01.027

Sabouri N. 2017. The functions of the multi-tasking Pfh1Pif1 helicase. *Curr Genet* **63:** 621–626. doi:10.1007/s00294-016-0675-2

Sabouri N, McDonald KR, Webb CJ, Cristea IM, Zakian VA. 2012. DNA replication through hard-to-replicate sites, including both highly transcribed RNA Pol II and Pol III genes, requires the *S. pombe* Pfh1 helicase. *Genes Dev* **26:** 581–593. doi:10.1101/gad.184697.111

Sabouri N, Capra JA, Zakian VA. 2014. The essential *Schizosaccharomyces pombe* Pfh1 DNA helicase promotes fork movement past G-quadruplex motifs to prevent DNA damage. *BMC Biol* **12:** 101. doi:10.1186/s12915-014-0101-5

Sakofsky CJ, Malkova A. 2017. Break induced replication in eukaryotes: mechanisms, functions, and consequences. *Crit Rev Biochem Mol Biol* **52:** 395–413. doi:10.1080/10409238.2017.1314444

Schramke V, Luciano P, Brevet V, Guillot S, Corda Y, Longhese MP, Gilson E, Géli V. 2004. RPA regulates telomerase action by providing Est1p access to chromosome ends. *Nat Genet* **36:** 46–54. doi:10.1038/ng1284

Segurado M, de Luis A, Antequera F. 2003. Genome-wide distribution of DNA replication origins at A+T-rich islands in *Schizosaccharomyces pombe*. *EMBO Rep* **4:** 1048–1053. doi:10.1038/sj.embor.7400008

Sfeir A, Kosiyatrakul ST, Hockemeyer D, MacRae SL, Karlseder J, Schildkraut CL, de Lange T. 2009. Mammalian telomeres resemble fragile sites and require TRF1 for efficient replication. *Cell* **138:** 90–103. doi:10.1016/j.cell.2009.06.021

Sharma R, Sahoo SS, Honda M, Granger SL, Goodings C, Sanchez L, Künstner A, Busch H, Beier F, Pruett-Miller SM, et al. 2021. Gain-of-function mutations in RPA1 cause a syndrome with short telomeres and somatic genetic rescue. *Blood* **139:** 1039–1051. doi:10.1182/blood.2021011980

Smith J, Rothstein R. 1995. A mutation in the gene encoding the *Saccharomyces cerevisiae* single-stranded DNA-binding protein Rfa1 stimulates a *RAD52*-independent pathway for direct-repeat recombination. *Mol Cell Biol* **15:** 1632–1641. doi:10.1128/MCB.15.3.1632

Smith J, Rothstein R. 1999. An allele of RFA1 suppresses RAD52-dependent double-strand break repair in *Saccharomyces cerevisiae*. *Genetics* **151:** 447–458. doi:10.1093/genetics/151.2.447

Soudet J, Jolivet P, Teixeira MT. 2014. Elucidation of the DNA end-replication problem in *Saccharomyces cerevisiae*. *Mol Cell* **53:** 954–964. doi:10.1016/j.molcel.2014.02.030

Spink KG, Evans RJ, Chambers A. 2000. Sequence-specific binding of Taz1p dimers to fission yeast telomeric DNA. *Nucleic Acids Res* **28:** 527–533. doi:10.1093/nar/28.2.527

Stewart JA, Wang Y, Ackerson SM, Schuck PL. 2018. Emerging roles of CST in maintaining genome stability and human disease. *Front Biosci* **23:** 1564–1586. doi:10.2741/4661

Sugiyama T, Cam HP, Sugiyama R, Noma K, Zofall M, Kobayashi R, Grewal SIS. 2007. SHREC, an effector complex for heterochromatic transcriptional silencing. *Cell* **128:** 491–504. doi:10.1016/j.cell.2006.12.035

Sun J, Yu EY, Yang Y, Confer LA, Sun SH, Wan K, Lue NF, Lei M. 2009. Stn1–Ten1 is an Rpa2–Rpa3-like complex at telomeres. *Genes Dev* **23:** 2900–2914. doi:10.1101/gad.1851909

Sun H, Wu Z, Zhou Y, Lu Y, Lu H, Chen H, Shi S, Zeng Z, Wu J, Lei M. 2022. Structural insights into Pot1-ssDNA, Pot1-Tpz1 and Tpz1-Ccq1 Interactions within fission yeast shelterin complex. *PLoS Genet* **18:** e1010308. doi:10.1371/journal.pgen.1010308

Takai H, Aria V, Borges P, Yeeles JTP, de Lange T. 2024. CST–polymerase α-primase solves a second telomere end-replication problem. *Nature* **627:** 664–670. doi:10.1038/s41586-024-07137-1

Takikawa M, Tarumoto Y, Ishikawa F. 2016. Fission yeast Stn1 is crucial for semi-conservative replication at telomeres and subtelomeres. *Nucleic Acids Res* **45:** 1255–1269. doi:10.1093/nar/gkw1176

Tarsounas M, West SC. 2005. Recombination at mammalian telomeres: an alternative mechanism for telomere protection and elongation. *Cell Cycle* **4:** 672–674. doi:10.4161/cc.4.5.1689

Tashiro S, Handa T, Matsuda A, Ban T, Takigawa T, Miyasato K, Ishii K, Kugou K, Ohta K, Hiraoka Y, et al. 2016. Shugoshin forms a specialized chromatin domain at subtelomeres that regulates transcription and replication timing. *Nat Commun* **7:** 10393. doi:10.1038/ncomms10393

Tazumi A, Fukuura M, Nakato R, Kishimoto A, Takenaka T, Ogawa S, Song J, Takahashi TS, Nakagawa T, Shirahige K, et al. 2012. Telomere-binding protein Taz1 controls global replication timing through its localization near late replication origins in fission yeast. *Gene Dev* **26:** 2050–2062. doi:10.1101/gad.194282.112

Teixeira-Silva A, Ait Saada AA, Hardy J, Iraqui I, Nocente MC, Fréon K, Lambert SAE. 2017. The end-joining factor Ku acts in the end-resection of double strand break-free arrested replication forks. *Nat Commun* **8:** 1982. doi:10.1038/s41467-017-02144-5

Toteva T, Mason B, Kanoh Y, Brøgger P, Green D, Verhein-Hansen J, Masai H, Thon G. 2017. Establishment of expression-state boundaries by Rif1 and Taz1 in fission yeast. *Proc Natl Acad Sci* **114:** 1093–1098. doi:10.1073/pnas.1614837114

van Emden TS, Forn M, Forné I, Sarkadi Z, Capella M, Martín Caballero L, Fischer-Burkart S, Brönner C, Simonetta M, Toczyski D, et al. 2019. Shelterin and subtelomeric DNA sequences control nucleosome maintenance and genome stability. *EMBO Rep* **20**: e47181. doi:10 .15252/embr.201847181

Vaurs M, Audry J, Runge KW, Géli V, Coulon S. 2022. A proto-telomere is elongated by telomerase in a shelterin-dependent manner in quiescent fission yeast cells. *Nucleic Acids Res* **50**: 11682–11695. doi:10.1093/nar/gkac986

Vaurs M, Naiman K, Bouabboune C, Rai S, Ptasińska K, Rives M, Matmati S, Carr AM, Géli V, Coulon S. 2023. Stn1-Ten1 and Taz1 independently promote replication of subtelomeric fragile sequences in fission yeast. *Cell Rep* **42**: 112537. doi:10.1016/j.celrep.2023.112537

Verdun RE, Karlseder J. 2006. The DNA damage machinery and homologous recombination pathway act consecutively to protect human telomeres. *Cell* **127**: 709–720. doi:10 .1016/j.cell.2006.09.034

Wallgren M, Mohammad JB, Yan KP, Pourbozorgi-Langroudi P, Ebrahimi M, Sabouri N. 2016. G-rich telomeric and ribosomal DNA sequences from the fission yeast genome form stable G-quadruplex DNA structures in vitro and are unwound by the Pfh1 DNA helicase. *Nucleic Acids Res* **44**: 6213–6231. doi:10.1093/nar/gkw349

Wang F, Stewart JA, Kasbek C, Zhao Y, Wright WE, Price CM. 2012. Human CST has independent functions during telomere duplex replication and C-strand fill-in. *Cell Rep* **2**: 1096–1103. doi:10.1016/j.celrep.2012.10.007

Wang J, Cohen AL, Letian A, Tadeo X, Moresco JJ, Liu J, Yates JR, Qiao F, Jia S. 2016. The proper connection between shelterin components is required for telomeric heterochromatin assembly. *Genes Dev* **30**: 827–839. doi:10 .1101/gad.266718.115

Webb CJ, Zakian VA. 2007. Identification and characterization of the *Schizosaccharomyces pombe* TER1 telomerase RNA. *Nat Struct Mol Biol* **15**: 34–42. doi:10.1038/ nsmb1354

Webb CJ, Zakian VA. 2012. *Schizosaccharomyces pombe* Ccq1 and TER1 bind the 14-3-3-like domain of Est1,

which promotes and stabilizes telomerase-telomere association. *Genes Dev* **26**: 82–91. doi:10.1101/gad.181826 .111

Wold MS. 1997. Replication protein A: a heterotrimeric, single-stranded DNA-binding protein required for eukaryotic DNA metabolism. *Annu Rev Biochem* **66**: 61– 92. doi:10.1146/annurev.biochem.66.1.61

Wood V, Gwilliam R, Rajandream MA, Lyne M, Lyne R, Stewart A, Sgouros J, Peat N, Hayles J, Baker S, et al. 2002. The genome sequence of *Schizosaccharomyces pombe*. *Nature* **415**: 871–880. doi:10.1038/nature724

Xu D, Muniandy P, Leo E, Yin J, Thangavel S, Shen X, Ii M, Agama K, Guo R, Fox D, et al. 2010. Rif1 provides a new DNA-binding interface for the Bloom syndrome complex to maintain normal replication. *EMBO J* **29**: 3140–3155. doi:10.1038/emboj.2010.186

Yadav RK, Matsuda A, Lowe BR, Hiraoka Y, Partridge JF. 2021. Subtelomeric chromatin in the fission yeast *S. pombe*. *Microorganisms* **9**: 1977. doi:10.3390/microorgan isms9091977

Yamazaki H, Tarumoto Y, Ishikawa F. 2012. Tel1(ATM) and Rad3(ATR) phosphorylate the telomere protein Ccq1 to recruit telomerase and elongate telomeres in fission yeast. *Genes Dev* **26**: 241–246. doi:10.1101/gad.177873.111

Zaaijer S, Shaikh N, Nageshan RK, Cooper JP. 2016. Rif1 regulates the fate of DNA entanglements during mitosis. *Cell Rep* **16**: 148–160. doi:10.1016/j.celrep.2016.05.077

Zofall M, Smith DR, Mizuguchi T, Dhakshnamoorthy J, Grewal SIS. 2016. Taz1-Shelterin promotes facultative heterochromatin assembly at chromosome-internal sites containing late replication origins. *Mol Cell* **62**: 862–874. doi:10.1016/j.molcel.2016.04.034

Zou L, Elledge SJ. 2003. Sensing DNA damage through ATRIP recognition of RPA-ssDNA complexes. *Science* **300**: 1542–1548. doi:10.1126/science.1083430

# Maintaining Telomeres without Telomerase in *Drosophila*: Novel Mechanisms and Rapid Evolution to Save a Genus

Stefano Cacchione,[1,5] Giovanni Cenci,[1,2,5] Anne-Marie Dion-Côté,[3,5] Daniel A. Barbash,[4,5] and Grazia Daniela Raffa[1]

[1]Department of Biology and Biotechnology, Sapienza University of Rome, 00185 Rome, Italy

[2]Fondazione Cenci Bolognetti, Istituto Pasteur, 00161 Roma, Italy

[3]Département de Biologie, Université de Moncton, Moncton, New Brunswick E1A 3E9, Canada

[4]Department of Molecular Biology and Genetics, Cornell University, Ithaca, New York 14853, USA

*Correspondence:* graziadaniela.raffa@uniroma1.it

Telomere maintenance is crucial for preventing the linear eukaryotic chromosome ends from being mistaken for DNA double-strand breaks, thereby avoiding chromosome fusions and the loss of genetic material. Unlike most eukaryotes that use telomerase for telomere maintenance, *Drosophila* relies on retrotransposable elements—specifically *HeT-A*, *TAHRE*, and *TART* (collectively referred to as HTT)—which are regulated and precisely targeted to chromosome ends. *Drosophila* telomere protection is mediated by a set of fast-evolving proteins, termed terminin, which bind to chromosome termini without sequence specificity, balancing DNA damage response factors to avoid erroneous repair mechanisms. This unique telomere capping mechanism highlights an alternative evolutionary strategy to compensate for telomerase loss. The modulation of recombination and transcription at *Drosophila* telomeres offers insights into the diverse mechanisms of telomere maintenance. Recent studies at the population level have begun to reveal the architecture of telomere arrays, the diversity among the HTT subfamilies, and their relative frequencies, aiming to understand whether and how these elements have evolved to reach an equilibrium with the host and to resolve genetic conflicts. Further studies may shed light on the complex relationships between telomere transcription, recombination, and maintenance, underscoring the adaptive plasticity of telomeric complexes across eukaryotes.

## TRANSPOSABLE ELEMENTS PROTECT CHROMOSOME ENDS IN *DROSOPHILA*

Telomeres are nucleoprotein structures that protect chromosome ends from being recognized as DNA double-strand breaks and act as buffers for coding sequences against erosion due to the end replication problem (de Lange 2009). While in most eukaryotes, telomere sequence repeats are synthesized by the telomerase reverse transcriptase (RT), in the *Drosophila* genus, telomerase components are absent, and telomeres

---

[5]These authors equally contributed to this work.

are maintained by telomere-specific retrotransposable elements (Cenci et al. 2005; Arkhipova 2012; Raffa et al. 2013). In *Drosophila melanogaster*, three families of long terminal repeat (LTR) retrotransposons from the *jockey* clade named *HeT-A*, *TAHRE*, and *TART* (collectively referred to herein as HTTs) are assembled in tandem head-to-tail arrays at chromosome ends (Levis et al. 1993; Abad et al. 2004) and precisely target telomeres via regulated transposition events (Fig. 1). The sequence heterogeneity of *Drosophila* telomeres implies that their protection relies on factors that specifically bind the chromosome termini, independently of their sequence. A set of fast-evolving proteins (referred to as terminin, see below) emerged in *Drosophila*, with the ability to recognize without sequence specificity natural chromosome ends and to shield them from being detected as double-strand breaks by the DNA damage response (DDR) apparatus, while avoiding unintended binding to genuine double-strand breaks that require proper repair. This sequence-independent telomere capping in *Drosophila* requires a delicate balance between DDR factors and terminin proteins, suppression of recombination and nonhomologous end-joining (NHEJ) pathways, and regulation of the expression and retrotransposition of telomere elements, which have evolved to limit their propagation solely at telomeres.

## *DROSOPHILA* TELOMERES: ORIGIN AND STRUCTURE

The sequence and structure of HTTs vary considerably across species (Villasante et al. 2007;

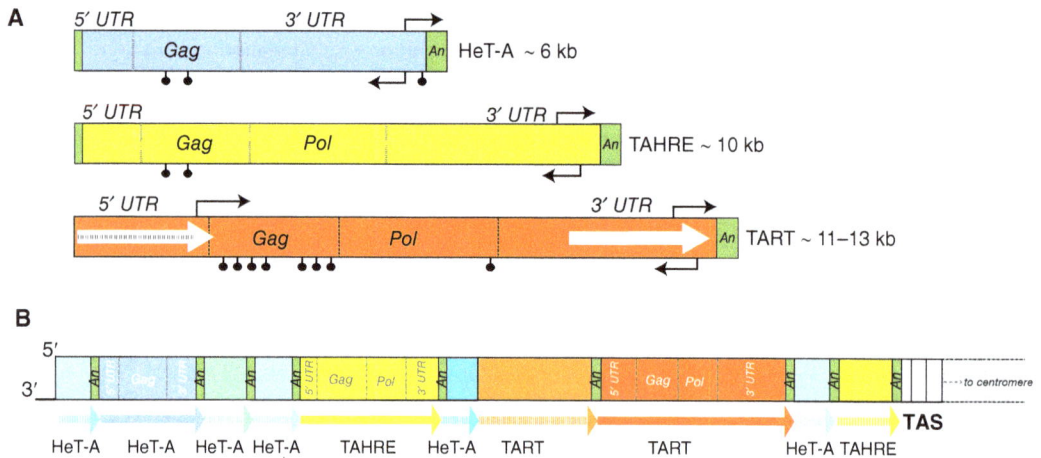

**Figure 1.** Organization of *Drosophila* telomeres. (*A*) Schematic representation of the telomeric *Drosophila* retrotransposons *HeT-A*, *TAHRE*, and *TART*. Black arrows denote transcription start sites of major sense and antisense transcripts. Green boxes: poly(A) regions (An). *HeT-A* elements are found in tandem and the sense promoter in the distal part of the 3′ untranslated region drives the transcription of a downstream element. White arrows in *TART* element denote "perfect nonterminal repeats (PNTR)" (Pardue and DeBaryshe 2011). The dotted line in the 5′ PNTR denotes sequences generated during a new transposition event, through reverse transcription of the sequence of the 3′ PNTR. The 3′ PNTR contains termination sequences. Black dots denote regions prone to the formation of G4 structures (Jedlicka et al. 2023). (*B*) Schematic representation of a hypothetical "average" *Drosophila* telomere, composed of copies of the HTT elements, *HeT-A* (64%), *TART* (19%), and *TAHRE* (17%), arrayed with their 3′ poly(A) ends (An, green rectangles) pointing toward the centromere (Abad et al. 2004; George et al. 2006; McGurk et al. 2021). Progressive telomere erosion shortens the elements at their 5′ ends and it is estimated that about 80% and 90% of the *HeT-A* and *TART* elements, respectively, are incomplete. Full-length copies are labeled by solid arrows; the boxes within the rectangles denote the coding and noncoding regions of the elements (see text for further details). (TAS) Telomere-associated subtelomeric sequences.

Pardue and DeBaryshe 2008; Saint-Leandre et al. 2019). *TART*-like and *HeT-A*-like elements were discovered in *Drosophila yakuba* and *Drosophila virilis* and localize to telomeres, demonstrating that they have a conserved function in telomere structure across at least 45 million years of evolution (Danilevskaya et al. 1998; Casacuberta and Pardue 2003a,b). Villasante and colleagues extended the search to several additional species, including ones from the *Sophophora, Drosophila,* and Hawaiian *Drosophila* subgenera (Villasante et al. 2007). All species examined have elements with both an ORF1 containing zinc knuckles (GAG) and an ORF2 containing an RT domain, and phylogenetic analysis suggests that these elements have evolved vertically from a common ancestor (Fig. 1A). However, the ORFs and the untranslated regions (UTRs) show extensive structural differences. Furthermore, most species contain "half telomeric-retrotransposons" (HTRs) that, like *D. melanogaster HeT-A*, lack the RT (ORF2) and most likely rely on RTs provided in *trans* to mobilize. These HTRs have recurrently evolved from full-length *TART-* or *TAHRE*-like elements, but telomere localization was not demonstrated for most of them (Casacuberta and Pardue 2005).

In the case of *HeT-A*, productive telomere elongation events by target-primed reverse transcription (TPRT) depend on the formation of a ribonucleoprotein assembly at the 3′ OH that includes the *HeT-A*-encoded template RNA bound to the GAG-like proteins encoded by ORF1 that provide specific targeting at telomeres, plus the RT that is encoded by a different element, likely *TAHRE*. In the case of these nonautonomous telomeric elements, the RNA template and the enzymatic activity required for telomere elongation are encoded by separate loci, as also occurs for telomerase ribonucleoproteins. The RTs of telomerase, non-LTR retrotransposons, and group II introns, which may have provided the structural basis for chromosome linearization (de Lange 2015), share some similar structural and functional features, suggesting a common ancestry (Nakamura and Cech 1998 and references in Belfort et al. 2011). Therefore, the acquisition of a retrotrans-

poson-based mechanism of telomere maintenance, following telomerase loss in the Dipteran ancestor, estimated to have occurred around 260 million years ago (Mason et al. 2015), may have recapitulated some of the ancient events that occurred in the early eukaryotic progenitors during the transition from circular to linear genomes: repurposing extant RTs and a suitable RNA template to extend the DNA ends. RTs have also been coopted as an end-extension mechanism in other animals, including *Bombyx mori* and Bdelloid rotifers (Arkhipova et al. 2017; also for review, see Servant and Deininger 2015).

Interestingly, Jedlička et al. (2023) identified G-quadruplex-forming sequences in *Drosophila* HTT and at terminal sequences from other species with noncanonical telomeres. This finding suggests that noncanonical telomeric sequences might be under selective pressure to maintain a G-rich bias that could provide advantages similar to those conferred by G-quadruplexes at canonical telomeres (Abad and Villasante 1999; Bryan 2020).

The sequence of the retroelements and the structures of telomere arrays can vary and turnover considerably even within the same genus (Fig. 1B). Using a combination of long-read assembly and cytogenetic mapping, Saint-Leandre et al. (2019) explored telomere structures in species diverging up to ~15 MY within the *melanogaster* subgroup. They observed a high level of structural variation among validated telomeric elements and proposed recurrent evolution of new telomeric families. Strikingly, they also discovered that *Drosophila biarmipes* has completely lost telomeric retrotransposons and instead appears to form its telomeres from satellite DNA sequences and fragments of Helitron DNA transposons, likely relying on recombination for telomere maintenance.

Retrotransposon-based telomeres are complex mosaic structures that vary in the number and proportion of full-length and partial copies present at each chromosome end (George et al. 2006; Villasante et al. 2007). They can also display high interindividual diversity within species. Analyses of short-read sequencing data from 84 *D. melanogaster* strains isolated from five continents revealed that total telomere length varies widely, from 143 kb to 1.2 Mb, with a median of 400 kb (McGurk et al. 2021). This corresponds roughly to a median of 12 insertions at each of the eight telomeres, the majority of which are *HeT-A* insertions from five subfamilies (64%) and the rest being distributed between *TART* (19%, three subfamilies) and *TAHRE* (17%), consistent with earlier estimates (Abad et al. 2004; George et al. 2006). Most of the insertions are incomplete, from about 80% for *HeT-A* and *TAHRE* to 93% for *TART*. HTT composition did not significantly differ between short- and long-telomere strains: short-telomere strains displayed evidence of unhealed terminal deficiencies, whereas long-telomere strains also tended to accumulate other transposable elements (TEs). This is consistent with HTTs being codependent for telomere elongation and suggests that telomere length is under host genome control when telomeres shorten, rather than being subjected to the mobilization of a specific HTT. Moreover, these analyses show that HTTs are not randomly interspersed, and that every HTT tends to neighbor itself more than expected, indicating that HTTs tend to be reverse transcribed more than once at a given location. This pattern was much stronger for *TART* (see also George et al. 2006), and could also result from recombination between termini as previously observed for *HeT-A* (Kahn et al. 2000). Altogether, these analyses suggest that *HeT-A* relies on *TAHRE* for reverse transcription, and that *TAHRE* and *TART* are not mobilized at the same time. Whether these results are a complete depiction of wild populations remains unclear, as the lines analyzed were inbred for 12 generations and thus affected by genetic drift and possibly laboratory selection; gaining an accurate representation of telomere length and composition in natural populations will require the sequencing of recently captured wild flies.

Further advances will come from the increased use of long-read sequencing techniques, such as Oxford Nanopore Technologies (ONT) sequencing (Deamer et al. 2016) and real-time SMRT sequencing technology (Eid et al. 2009). Long-read sequencing is already transforming the field of human genomics, allowing the completion of telomere-to-telomere (T2T) assembly (Nurk et al. 2022) and promising to routinely obtain phased assemblies from diploid genomes (Miga and Eichler 2023; Rautiainen et al. 2023). For example, several groups have used long-read sequencing to address telomere length variation across chromosomes and individuals (Karimian et al. 2024), telomere attrition in aging and disease (Sanchez et al. 2024; Schmidt et al. 2024), and telomere repeat variation in cancer (Schmidt et al. 2024; Tan et al. 2024).

## EVOLUTION OF HTTS: ARE *DROSOPHILA* TELOMERIC TEs IN CONFLICT WITH THEIR HOST?

HTTs have often been considered as domesticated elements or adaptive symbionts that serve an essential host function of *Drosophila* due to loss of telomerase (Kidwell and Lisch 2001; Servant and Deininger 2015). But are HTTs really different from other TEs by being devoid of the potential to cause genetic conflict?

Genetic conflicts occur when selfish DNA proliferates in a manner that reduces the fitness of the host in which it is replicating (Burt and Trivers 2008). Ongoing genetic conflicts result in recurrent cycles of evolution of both the selfish DNA and host defense systems. Undomesticated TEs are widely considered to be selfish DNAs, as they can evolve rapidly in structure and sequence and invade new species after horizontal transfer. Likewise, many host genes involved in TE defense also evolve rapidly with evolutionary signatures of adaptive evolution (Emerson and Thomas 2009; Obbard et al. 2009; Kolaczkowski et al. 2011; Wells et al. 2023). Active TEs also often show population-genetic patterns that are consistent with them having host fitness effects that lead to their removal from the population by negative selection (for review, see Lee and Langley 2010). However, selfish elements can be either neutral or deleterious to their hosts, and concluding that genetic conflict occurs between telomeric TEs and their host requires demonstrating that the overproliferation of HTTs, at the telomere or elsewhere, reduces host fitness (Werren 2011).

Cite this article as *Cold Spring Harb Perspect Biol* doi: 10.1101/cshperspect.a041708

Telomere length varies widely among *D. melanogaster* strains and some have exceptionally long telomeres (Wei et al. 2017). One study found that a long-telomere strain has reduced fitness compared to a short-telomere strain (Walter et al. 2007). This study warrants further investigation, for example, to determine whether fitness shows a continuous distribution relative to telomere lengths across many strains. Another prediction of genetic conflict between telomeric TEs and their hosts is that HTT elements would at least occasionally "escape" from the telomere and insert elsewhere in the genome. At least in *D. melanogaster*, this does not seem to occur, as nontelomeric insertions were not found in an extensive population sample (McGurk et al. 2021). Although there are fragments of HTTs found in heterochromatin, they may originate from gene conversion or rearrangements rather than retrotransposition (Danilevskaya et al. 1993; Losada et al. 1997; Berloco et al. 2005).

On one hand, the absence of transposition of HTTs at nontelomeric sites and the rarity of strains with unusual expansions of specific HTTs, may suggest that genetic conflict is not a major feature of HTTs or that conflicts are quickly resolved as lineages evolve. On the other hand, the rapid evolution of HTTs and the adaptive evolution of telomere capping proteins do not seem consistent with stable and complete domestication.

It has been argued that for domestication to be complete, protein-coding sequences from the TE (retrotransposase in particular) must be separated from other sequences to prevent future replication and "resurrection" (Jangam et al. 2017; Blumenstiel 2019). It has also been suggested that HTTs should be considered to be in a mutualistic relationship: HTTs provide benefit to their host by protecting chromosome ends, whereas their host provides them with a "safe haven" to propagate that minimizes fitness cost. However, mutualism/cooperation is often unstable and may evolve to either complete co-option/domestication or return to genetic conflict (Cosby et al. 2019). Further arguing against complete domestication is the high turnover of HTTs across species and their retaining trans-

positional activity as whole TEs (Markova et al. 2020). Another perspective is that telomere-specific localization may reflect a niche strategy of HTTs: landing in a safe haven minimizes deleterious consequences to the host. Analogous to mutualism/cooperation, niche strategies may also not be fully stable and thus be susceptible to episodic conflict. McGurk et al. (2021) emphasized the instability of this niche, suggesting that continuous end erosion and high rates of terminal deletions may drive rapid turnover of families without necessarily involving genetic conflict.

In a different interpretation, the rapid turnover of telomeric TEs across *Drosophila* species may represent a direct consequence of cycles of genetic conflict, as host species periodically suppress overactive HTTs followed by the evolution of new HTT families (Saint-Leandre et al. 2019). HTT rapid evolution is also paralleled by the concomitant rapid evolution of the capping proteins that protect telomeres (see below) (Raffa et al. 2011); these dynamic changes in the sequences of both telomeric DNA and telomeric proteins are reminiscent of an evolutionary arms race resulting from a genetic conflict (Lee et al. 2017).

A fascinating case of host sequence capture by a TE also may be indicative of a history of genetic conflict between HTTs and their host. *TART-A* elements in *D. melanogaster* have captured a segment of the gene *nuclear export factor 2* (*nxf2*), which is required for piRNA-dependent silencing of TEs (Ellison et al. 2020). These *TART-A* elements now produce piRNAs homologous to *nxf2*, and *nxf2* expression is lower in *D. melanogaster* than in its sister species *Drosophila simulans*. The authors suggest that *TART-A* is selfishly suppressing a host defense system by controlling *nxf2* expression to increase its copy number.

It is reasonable to conclude that no single mechanistic or evolutionary interpretation fully explains all the observed patterns of HTT and telomere cap evolution. Some of the seemingly contradictory results may reflect different timescales. For example, the absence of nontelomeric HTT insertions in contemporary *D. melanogaster* populations does not preclude that some HTT families may have caused genetic conflicts in the past. A deeper understanding of *Drosophila* telomere evolution will benefit from extensive sampling from the wild as well as carefully controlled genetic experiments that manipulate telomere length and determine consequent fitness effects.

## REGULATION OF TELOMERE PROTECTION

The characterization of mutations in at least 12 loci of *D. melanogaster* that lead to telomere fusions in mitotic cells has allowed the identification of capping proteins that are enriched at, or exclusive to, telomeres (Cenci et al. 2005; Raffa et al. 2013; Cacchione et al. 2020). The fast-evolving proteins HOAP, HipHop, Moi, Ver, and Tea that bind to DNA in a sequence-independent manner constitute the terminin complex, which, analogous to the shelterin complex in many other eukaryotes, localizes only at telomeres and serves telomere-specific functions (Cacchione et al. 2020). Moreover, in vitro and in vivo results indicate that Moi, Tea, and Ver form the MTV subcomplex with single-strand binding properties that interacts with DNA overhangs (Zhang et al. 2016; Cicconi et al. 2017; Cheng et al. 2018; Vedelek et al. 2021; Chen et al. 2022). Telomere fusions are also prevented by several nonterminin proteins, which are evolutionarily conserved and play roles not restricted to telomeres. These proteins include Heterochromatin Protein 1a (HP1a), Without children (Woc), the Mre-11-Rad50-Nbs1 (MRN) complex, ATM, Eff/UbcD1, Peo/AKTIP, and Separase (Cacchione et al. 2020).

However, it remains unclear how the DDR is inhibited at telomeres, and whether fusions of deprotected telomeres depend on end resection, telomeric transcription, chromatin state, and/or the cell cycle. NHEJ rather than homologous recombination (HR) is thought to be the main mechanism by which fusions are formed, because head-to-head attachment of *HeT-A* elements with NHEJ signatures at the junctions were detected in *mre11*, *nbs*, and *atm* mutants that are defective for HR (Gao et al. 2009; Morciano et al. 2013). Depletion of ligase IV, which is required for NHEJ, also suppresses fusions of

unprotected telomeres in *aubergine (aub)* and *armitage (armi)* mutants, further supporting this view (Khurana et al. 2010). In addition, telomere fusions in mutants lacking the kinases ATM and ATR or the MRN complex can be suppressed by additional loss of H2A.Z, suggesting that in the absence of this histone variant, telomeres adopt alternative structures that either restore capping or affect damage detection (Rong 2008). However, in checkpoint kinase mutant (*mec1Δtel1Δ*) pre-senescent budding yeast cells, depletion of histones protects yeast telomeres from fusion by favoring Rad51-independent HR instead of inhibiting NHEJ (Barrientos-Moreno et al. 2018). It thus cannot be ruled out that the same mechanism can occur in *Drosophila mre11*, *nbs*, or *atm* mutants upon loss of H2A.Z. Finally, suppression of telomere fusions can also occur through epigenetic mechanisms. Loqs is a *Drosophila* dsRNA-binding protein that is homologous to the mammalian TRBP/PACT protein required for siRNA and micro-RNA biogenesis. Mutations in the *loquacious (loqs)* gene reduce the frequency of telomeric fusions in *cav* and *ver* mutants but not in *nbs1* (Porrazzo et al. 2022). As low levels of Loqs are associated with more efficient DNA repair (Porrazzo et al. 2022), repression of endo-siRNA biogenesis in flies could result in the accumulation of *cis* Natural Antisense Transcripts (*cis*-NATs), from which they derive (Okamura et al. 2008). These transcripts are known to promote chromatin remodeling and favor DNA repair activities and the resolution of R-loops at telomeres (Zhao et al. 2020).

## THE ESTABLISHMENT OF TELOMERE PROTECTION COMPLEXES DURING DEVELOPMENT

A poorly understood aspect is how the capping complex is assembled during development, and whether it is established de novo or is instead dependent on an inherited condition previously set in either the female or male germline. Unfortunately, loss-of-function mutations in genes coding for telomere capping proteins cause lethality in flies due to extensive telomere fusions in somatic cells, thus hampering the analysis of the capping function in the germline. Insights come from the identification of mutations with either maternal or paternal effects, resulting in the failure of telomere protection in early embryos.

Female flies bearing hypomorphic mutations in either *mre11* or *nbs* develop normally but produce embryos with arrested development due to telomere covalent fusions and mitotic failure, likely caused by the exclusion of Mre11 and Rad50 from chromatin (Gao et al. 2009). Somatic cells deficient for MRN complex function exhibit defective HOAP and terminin recruitment at telomeres, but this has not been addressed with *mre11* or *nbs* maternal effect mutations at embryonic telomeres. *Drosophila* maternal effect mutations in ATM kinase, which regulates MRN, induce covalent chromosome fusions during early embryonic cell division, similarly to the above-mentioned mutations in *nbs* and *mre11*. These defects are independent of the MRN complex activity, as *atm* mutant embryos show neither obvious reduction in the total level of Mre11/Nbs proteins nor a mislocalization of Rad50 (Morciano et al. 2013). However, mutant embryos exhibit a significant reduction of HipHop, but not of other terminin components, indicating that telomere destabilization could result from an inefficient loading of HipHop onto chromosome ends (Morciano et al. 2013). Consistently, females hemizygous for a hypomorphic *hiphop* mutation were also sterile and laid embryos that died as a consequence of rampant telomere fusions (Cui et al. 2021). These findings indicate that HipHop is required for the reestablishment of a capped status of chromosome ends that derives from the mother. Interestingly, the same hypomorph *hiphop* mutation specifically derepressed telomeric elements in the germline by a mechanism that is not associated with the piRNA-mediated pathway, although it could depend on HP1a activities. Yet, this derepression did not result in telomere lengthening (Cui et al. 2021).

Insights on the mechanisms of telomere maintenance in developing embryos can also be obtained through the investigation of telomere dynamics during male gametogenesis, where telomeres undergo distinct changes. Spermio-genesis is particularly intriguing, as telomere maintenance during this postmeiotic phase is prolonged compared to other spermatogenic stages and provides a plentiful supply of synchronized, differentiated nuclei that are particularly suitable for imaging of telomeres. Among the rare paternal-effect mutants, *ms(3)K81* has provided relevant insights into the mechanisms of telomere maintenance in the male germline. Most embryos from K81-deficient fathers arrest during mid-cleavage, with a few surviving to late embryogenesis as gynogenetic haploids containing only the maternal genome (Yasuda et al. 1995; Dubruille et al. 2010; Gao et al. 2011). *K81* encodes a male germline-specific paralog of HipHop that replaces HipHop in spermatid telomeres and remains associated with paternal telomeres until zygote formation (Dubruille et al. 2010; Gao et al. 2011). Although the loss of capping proteins in *k81* mutant spermatids does not impede sperm maturation, it triggers telomere fusions upon fertilization, impacting the segregation of the paternal genome and resulting in haploid embryos with maternal chromosomes (Dubruille et al. 2010; Gao et al. 2011). Thus, K81 emerges as a key factor in determining a paternal imprint on sperm chromatin that is crucial for the functional reset of the paternal chromosomes after fertilization. However, whether K81 is also required for the maintenance of the capping complex on postmeiotic telomeres is still debated (Gao et al. 2011). Additional insights into the maintenance of telomere capping during genome-wide chromatin remodeling, which occurs when spermatids mature into sperm, derive from the characterization of the *deadbeat (ddbt)* male sterile mutation (Yamaki et al. 2016). *ddbt* mutant males show normal development of sperm but no viable offspring as a consequence of an early arrest during embryo development and a selective elimination of paternal chromosomes. *ddbt* encodes a specialized, telomere-enriched sperm nuclear basic protein (SNBP) that localizes to spermatid chromosome ends after the meiotic replacement of HipHop with K81. Its recruitment depends on K81 and its loss produces mature spermatids with uncapped telomeres that are devoid of HP1a, HOAP, and K81, thus indicating that it functions to retain a proper protection of sperm telomeres (Yamaki et al. 2016).

Why K81 replaces HipHop during spermiogenesis remains an open question. During the spermiogenesis of most animals, including *Drosophila*, sperm chromatin undergoes a massive reorganization to transform sperm DNA into a mitotically competent nucleus. This mainly occurs through the replacement of nucleosomal histones with SNBPs, such as the protamines in mammals (Lewis et al. 2003) as well as the activity of testis-specific chromatin factors, such as HP1E (Levine et al. 2015). Improper sperm-packaged DNA results in an unsuccessful post-fertilization process and nonviable embryos undergoing premature division (Dubruille et al. 2023). From this perspective, K81, which specifically regulates telomeric chromatin organization, might represent a further level of safeguarding a competent male pronucleus from genome instability that could have escaped from the regulation of other sperm chromatin remodeling mechanisms. The observation that telomere dynamics also change in mammalian sperm nuclei supports the view that specialized telomere factors are required for sperm telomere maintenance (Reig-Viader et al. 2016).

## REGULATION OF TELOMERE LENGTH

HTT transposition must be a regulated process, whereby telomere shortening signals new transpositions to chromosome ends. Regulation is also a likely process to evolve if telomere length affects host fitness. While telomere elongation is controlled by both the accessibility of the chromosome end for transpositions and the regulation of HTT expression, increased transcription of retrotransposons may not necessarily correlate with higher transposition rates and vice versa (Siudeja et al. 2021). For example, flies deficient for the NHEJ repair proteins Ku70/Ku80 have increased rates of HTT telomere attachments but do not appear to increase *HeT-A* transcript levels (Melnikova et al. 2005). Telomere length control has been extensively reviewed elsewhere (Shpiz and Kalmykova 2011); here, we briefly summarize the key findings and then focus on studies published subsequently.

## piRNA Regulation of HTT Transcription in the Female Germline and in Somatic Cells

Retrotransposon repression in the germline is regulated by a defense pathway based on the production of Piwi-interacting small RNAs (piRNAs) that are transmitted from one generation to the next through the maternal cytoplasm (Le Thomas et al. 2014). piRNAs are 23–30 nt RNAs with sequences complementary to their target element and are processed from precursors transcribed from discrete piRNA clusters (Brennecke et al. 2007), loci spanning up to 100 kb that contain remnants of old TE insertions and preferentially produce piRNAs targeting active elements (Said et al. 2022). Single recently integrated elements and repetitive transgenic sequences can also generate piRNAs (discussed in Komarov et al. 2020; Chen and Aravin 2021; Luo et al. 2023) and the telomeres are a remarkable example as they act both as piRNA production loci and piRNA targets (Maxwell et al. 2006; Shpiz and Kalmykova 2009; Radion et al. 2017). The modulation of canonical (promoter-driven) and noncanonical (promoter-independent) transcription at HTT retrotransposons drives the generation of forward and reverse transcripts with different fates, and implicates specific mechanisms and transcription factors (Danilevskaya et al. 1999; George et al. 2010; Huang et al. 2017; Radion et al. 2018; ElMaghraby et al. 2019; Kordyukova and Kalmykova 2019; Zhao et al. 2019; Kordyukova et al. 2020; Sato and Siomi 2020; Kalmykova and Sokolova 2023). The germline-specific complex RDC is formed by Rhino (paralog of the HP1a heterochromatic protein), Deadlock, and Cutoff (Mohn et al. 2014; Gamez et al. 2020), and is recruited to retrotransposons following PIWI-dependent deposition of H3K9 methylation marks by the SETDB1/Egg methyltransferase. RDC fuels noncanonical transcription through Moonshiner, TFIIA-S, and TRF2, represses splicing of piRNA precursor transcripts through Cutoff and UAP56, and suppresses their polyadenylation and termination (Chen et al. 2016). While Maelstrom suppresses canonical transcription, the nonspecific lethal (NLS) complex binds to the promoters in

Cite this article as *Cold Spring Harb Perspect Biol* doi: 10.1101/cshperspect.a041708

the 3′ UTRs of the telomeric transposons and may promote canonical transcription of piRNA precursors (Iyer et al. 2023). As a result, different types of transcripts are then produced, with diverse fates (Fig. 2).

Sense transcripts can be translated or become RNA templates for reverse transcription. They can also serve as piRNA precursors when bound by Aub in the nuage, a granular perinuclear region enriched in piRNA processing factors (Brennecke et al. 2007; Lim and Kai 2007; Webster et al. 2015). Antisense transcripts can be converted into piRNAs either through the ping-pong cycle or through phased piRNA biogenesis. piRNAs assemble with PIWI proteins to form RISC complexes, and display a specific nucleotide bias at defined positions: piRNAs with a uridine at the 5′ end (1U bias) assemble with Piwi or Aub, while Ago3 binds preferentially piRNAs with an adenine at position 10 (10A bias) (Brennecke et al. 2007). Nascent transcripts are targeted by Piwi RISC complexes that mainly contain—but are not limited to—piRNA products of phased biosynthesis, largely targeting sense transcripts (Senti et al. 2015; Sato and Siomi 2020). Piwi complexes initiate cotranscriptional silencing by base pairing to nascent transcripts and recruiting Eggless/SetDB1 (Ninova et al. 2020), which promotes the recruitment of HP1a and RDC complexes. The piRNA pathway is connected to the heterochromatin machinery through the Sphinx complex (Schnabl et al. 2021; Andreev et al. 2022). Inefficient cotranscriptional silencing leads to the deregulation of TE transcripts and the prominent accumulation of TE nuclear foci, detectable by FISH. Inefficient posttranscriptional silencing in the cytoplasm results in cytoplasmic foci enriched in TE transcripts and the GAG proteins they encode. *HeT-A*, *TART*, and *TAHRE* are robustly dependent on the secondary piRNA biogenesis (ping-pong) pathway, which pro-

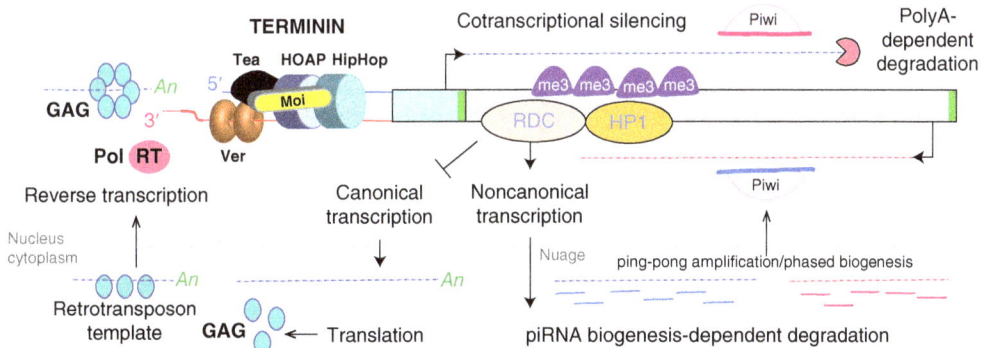

**Figure 2.** Telomere homeostasis in the *Drosophila* female germline. The Terminin telomere capping complex binds DNA in a sequence-independent fashion: HOAP/HipHop bind duplex DNA, Moi/TEA/Ver bind single-strand DNA. Transcription of the HTT telomere retroelements produces both sense and antisense transcripts (blue and red dotted lines, respectively), which undergo diverse fates: in the cytoplasm, a fraction of sense transcripts is translated, can be assembled with their cognate GAG proteins, and can be recruited at the 3′ end of the chromosome to elongate telomeres by target-primed reverse transcription (TPRT). TPRT is mediated by a reverse transcriptase (Pol) that is either encoded by an element of the same family (in the case of *TART* and *TAHRE*) or by a different family (for *HeT-A*). Both sense and antisense transcripts are subjected to degradation either in the nucleus or in the nuage, where they become precursors of piRNAs, via the ping-pong and phased biogenesis pathways. piRNAs assemble with a member of the PIWI protein family to form RISC complexes, reenter the nucleus, and target their complementary HTT nascent transcripts. Then they initiate cotranscriptional silencing through recruitment of the RDC complex, the establishment of H3K9 methylated chromatin compartments, and the propagation of a complex series of epigenetic marks. Altogether, these changes set the balance between canonical and noncanonical transcription, which ultimately regulate the expression levels of the HTT elements (see text for details).

vides the major contribution to generating Piwi-bound piRNAs in the germline (Senti et al. 2015).

The regulation of HTTs in somatic cells and their mobilization rate are still poorly understood, but several studies have explored their expression. For example, Perrat et al. (2013) found that *HeT-A* transcripts are expressed at higher levels in αβ neurons, where Aub or Ago3 abundance is low, compared to other adult brain cells. Interestingly, *HeT-A* levels increased dramatically in both guts and heads of *ago2* and *dicer-2* mutants, which control the endo-siRNA pathway, and in *ago3* and *armi* mutants, but are less strictly dependent on Aub and Piwi (Siudeja et al. 2021). This greater impact of Ago3 compared to Aub and Piwi on *HeT-A* repression in the soma, contrasts with the ovarian germline, where *HeT-A*, *TART*, and *TAHRE* expression is most strongly affected by Piwi. This heterogeneity highlights the different architectures of piRNA production pathways in somatic versus germline cells, and the different repertoires of piRNAs that control transcriptional silencing in different contexts (Senti et al. 2015). Past studies have overestimated the rate of de novo somatic insertions in the genome for several families of elements (Perrat et al. 2013; Treiber and Waddell 2017). However, the recent implementation of long-read sequencing techniques has allowed the assessment of bona fide de novo transposition events of several families of mobile elements in somatic cell genomes, and suggests that the abundance of small RNAs that control TE repression is not always a reliable predictor of transposition activity (Siudeja et al. 2021).

## Other Factors that Regulate HTT Expression

HP1a, encoded by *Su(var)205*, controls the silencing of telomeric sequences, and flies heterozygous for *Su(var)205* mutations show increased telomere length. These roles are due to the interaction between *Su(var)205* and H3K9me3 (Perrini et al. 2004) and are independent of HP1 function in capping. It has been proposed that HP1a functions both upstream and downstream from piRNA processing (Teo et al. 2018; Ilyin et al. 2021).

Besides the piRNA pathway, the expression of HTT elements is also regulated by specific factors that bind enhancers or insulators to block transcription in the oocyte or in somatic cells. For example, in ovarian follicle cells, *HeT-A* repression is independent of the Piwi pathway and is controlled by specific isoforms of the Mod (mdg4) protein, which binds to enhancers located in the subtelomeric TAS-R repeats and blocks their activity (Takeuchi et al. 2022). In the germline, *TART* expression is restricted to a specific stage during oogenesis, within a permissive region in the germarium characterized by reduced levels of the Piwi protein (Dufourt et al. 2014; Théron et al. 2018) and by a transient reduction in the level of the BEAF32 insulator protein, which normally impedes enhancer–promoter interactions (Sokolova et al. 2023). The strong derepression of *HeT-A*, *TART*, and *TAHRE* has also been observed in larval brains and ovaries from flies deficient in transcription cofactors Mediator, Scalloped, and E2F1-Dp, but the rate of de novo insertion in somatic cells has not been determined (Liu et al. 2023).

It has been proposed that the epigenetic status of the HTT domains affects the positioning of telomeres within the nucleus. Although limited by the sensitivity of DNA FISH techniques, extant data suggest that *HeT-A*-enriched DNA is detectable as discrete foci in nurse cells close to the nuclear envelope, many of them overlapping with HOAP and likely reflecting nonrandom telomere positioning. In contrast, *TART*-enriched DNA does not appear to always colocalize with *HeT-A* enriched foci. The peripheric *HeT-A* localization is lost in piRNA-biogenesis mutants, upon loss of Rhino binding and H3K9me marks (Radion et al. 2018).

The interplay between protection and elongation mechanisms is also relevant. The mechanism by which the HTTs achieve accurate end-targeting remains unclear. Notably, both ectopically expressed and endogenous transposon proteins are equally targeted to chromosome ends (Rashkova et al. 2002), indicating that transposon-encoded factors may interact with host-encoded factors with a controlled stoichiometry, and that while *TART* may supply the RT, *HeT-A* Gag aids nuclear targeting of

*TART*. In somatic cells, *HeT-A* is expressed predominantly during S-phase, suggesting a connection between HTT telomere insertions and DNA replication (Zhang et al. 2014). Ver interacts with virus-like spheres formed by *HeT-A* Gag assembled with the *HeT-A* RNA template, likely helping the recruitment of the RNPs at the telomere. However, it is unknown whether Ver or other terminin components participate in the recruitment of the RT (encoded by an element of the *TART* or *TAHRE* subfamilies) for the assembly and positioning of productive elongation complexes at the terminal 3′ OH, to extend telomeres through TPRT and second strand synthesis. It is also unknown whether the assembly of GAG proteins with the sense transcripts during sphere formation occurs independently from translation, and what fraction of Gag-associated transcripts will actually be used as templates for reverse transcription, thus contributing to net telomere lengthening in somatic cells.

Complex relationships may exist between capping and retrotransposon regulation in both the soma and the germline, and capping proteins may influence the transcription, stability, or trafficking of telomere transcripts. For example, the HOAP protein encoded by *caravaggio (cav)* is among the most rapidly evolving proteins in *Drosophila* (Schmid and Tautz 1997; Shareef et al. 2001; Cenci et al. 2003). Yet, *D. yakuba* HOAP properly localizes to telomeres when expressed in *D. melanogaster cav* mutants and prevents telomere fusions (Saint-Leandre et al. 2020). Interestingly, however, *D. melanogaster* HTTs are derepressed in these experiments and significant telomere lengthening occurs within 10 generations. These results suggest that adaptive evolution of HOAP has occurred recurrently to restrict cycles of telomere overgrowth, and challenge the view that telomere capping and length regulation in flies are uncoupled events.

Notably, the viable allele of *hiphop* that derepresses HTTs (discussed above) does not lead to telomere overgrowth (Cui et al. 2021), suggesting that a functional telomere cap may be needed for the recruitment of the HTT RNA–protein complexes and for their insertion at chromosome ends.

Sequencing of terminal DNA provides evidence that telomere elongation may also be achieved by recombination-based mechanisms, including gene conversion that either involves sequences on the homologous chromosome or tandemly arranged sequences on the same chromosome (Cenci et al. 2005; Cacchione et al. 2020). These mechanisms are primarily responsible for attaching *HeT-A* elements to terminally deleted chromosomes (Mikhailovsky et al. 1999; Kahn et al. 2000). Moreover, X chromosome telomeres harbor tandem *TART-A* elements that share terminal repeats, consistent with the occurrence of recombination events between the 3′ and 5′ terminal repeats (George et al. 2006; McGurk et al. 2021).

Recombination-based mechanisms have been postulated as an alternative event to *HeT-A* attachments in determining *HeT-A* abundance in long-telomere strains from natural populations (Wei et al. 2017). Large differences in *HeT-A* abundance among strains could potentially result from unequal crossing-over that can increase and decrease large blocks of *HeT-A* arrays, thus generating a large spectrum of different sizes. The probability of crossing-over presumably increases when telomeres become longer (Wei et al. 2017).

A moderate increase in *HeT-A* expression is observed in the ovaries of both young females deficient in nuclear lamin components and aged females. This increase is accompanied by the appearance of DNA damage foci and the accumulation of the recombinase Rad51 at telomeres, suggesting that increased recombination rates may be a hallmark of physiological aging (Morgunova et al. 2022).

*Telomere elongator (Tel)* and *Enhancer of terminal gene conversion (E(tc))* dominant mutations increase the abundance of *HeT-A* and *TART* sequences at telomeres and regulate HTT recombination/gene conversion (Melnikova and Georgiev 2002; Siriaco et al. 2002). Characterization of the mutation *Telomere elongator 1 (Tel[1])* revealed that deletions within intron 8 of the gene *Ino80* result in a long-telomere dominant phenotype (Reddy et al. 2022). Yet, the molecular mechanism that underlies this misregulation remains unclear. The long telomeres in salivary-gland polytene chromosomes from *Tel* mutants recruit telomere capping proteins and tend to associate with each other, at increasing frequency as the telomeres become longer. It is thus likely that these associations between long telomeres are mediated by recombination events (Siriaco et al. 2002; Reddy et al. 2022).

Hence, understanding *Tel* function will help decipher how *Drosophila* regulates telomere length independently of telomere capping. Furthermore, a detailed characterization of *Tel* could provide fundamental insights into recombination-mediated telomere maintenance mechanisms, such as the ALT pathway of some human cancers and normal mammalian somatic cells (Neumann et al. 2013; Hoang and O'Sullivan 2020) and the recombination/gene conversion-based pathways of rare telomerase-deficient yeast survivors (Kass-Eisler and Greider 2000).

## CONCLUSIONS AND PERSPECTIVES

Diverse strategies have evolved in eukaryotes to maintain linear chromosomes, possibly including protective complexes and structures such as the T-loop, as well as telomerase-mediated maintenance (de Lange 2004). Simultaneously, various protein complexes related to CST have evolved to fill crucial roles, including telomere lengthening of the 3′ strand, complementary strand fill-in, and the binding of ssDNA overhangs at telomeres (Myler et al. 2021). These proteins use diverse OB-fold-binding modes that confer a spectrum of affinities, ranging from high to very poor or no sequence-specificity (Barbour and Wuttke 2023). The MTV subcomplex of Terminin may have evolved from an ancestral CST complex (Cicconi et al. 2017; Cheng et al. 2018; Lue 2018).

*Drosophila* telomeres that depend on retrotransposon targeting of telomeres and sequence-independent capping complexes that bind telomeres and protect ssDNA, may represent just one possible evolutionary path that allowed for telomerase loss. With ongoing advancements in genome characterization across species, we speculate that other creative strategies to compensate for telomerase loss will be discovered. Noteworthy instances, such as a *Drosophila* species in which recombination rather than retrotransposons became the pivotal mechanism for telomere maintenance, indicate that suppression of recombination at telomeres can be loosened (Saint-Leandre et al. 2019).

Studies on ALT telomeres in human cells have shed light on the significance of recombination for chromosome maintenance, the complex relationships of telomere transcripts with chromatin and the mitigation of replication stress (Glousker and Lingner 2021; Loe et al. 2023; Zeinoun et al. 2023; Azzalin 2024). *Drosophila* telomeres also offer an intriguing model for examining the modulation of recombination and transcription. The turnover of the HTT transcripts and their targeting at telomeres in the soma and germline may be different. Specifically, germline retrotransposons are provided in *trans* and retrotransposon RNA recruitment at telomeres might occur through protein-mediated mechanisms or association with ssDNA, forming R-loops, akin to mechanisms seen at ALT telomeres. Advanced sequencing techniques may uncover higher-than-estimated recombination levels at fly telomeres, potentially reflecting recombination-driven telomere maintenance mechanisms as observed in yeast cells that survive telomerase loss (Henninger and Teixeira 2020; Apte et al. 2021). The complex relationships between telomere transcription and telomere maintenance, may be a general hallmark of telomere evolution, driving the plasticity and functional adaptation of telomeric complexes in diverse species under varying conditions in the soma and the germline.

## ACKNOWLEDGMENTS

This work was supported by grants from NSERC discovery grants RGPIN-2019-05744 (to A.M.D.C.); NIGMS R35GM153275 (to D.A.B.); AIRC IG 26496 (to G.D.R.).

## REFERENCES

Abad JP, Villasante A. 1999. The 3′ non-coding region of the *Drosophila melanogaster* HeT-A telomeric retrotransposon contains sequences with propensity to form G-quad-

Cite this article as *Cold Spring Harb Perspect Biol* doi: 10.1101/cshperspect.a041708

ruplex DNA. *FEBS Lett* **453:** 59–62. doi:10.1016/S0014-5793(99)00695-X

Abad JP, De Pablos B, Osoegawa K, De Jong PJ, Martín-Gallardo A, Villasante A. 2004. TAHRE, a novel telomeric retrotransposon from *Drosophila melanogaster*, reveals the origin of *Drosophila* telomeres. *Mol Biol Evol* **21:** 1620–1624. doi:10.1093/molbev/msh180

Andreev VI, Yu C, Wang J, Schnabl J, Tirian L, Gehre M, Handler D, Duchek P, Novatchkova M, Baumgartner L, et al. 2022. Panoramix SUMOylation on chromatin connects the piRNA pathway to the cellular heterochromatin machinery. *Nat Struct Mol Biol* **29:** 130–142. doi:10.1038/s41594-022-00721-x

Apte MS, Masuda H, Wheeler DL, Cooper JP. 2021. RNAi and Ino80 complex control rate limiting translocation step that moves rDNA to eroding telomeres. *Nucleic Acids Res* **49:** 8161–8176. doi:10.1093/nar/gkab586

Arkhipova IR. 2012. Telomerase, retrotransposons, and evolution. In *Telomerases: chemistry, biology, and clinical applications* (ed. Lue NF, Autexier C). Wiley, Hoboken, NJ.

Arkhipova IR, Yushenova IA, Rodriguez F. 2017. Giant reverse transcriptase-encoding transposable elements at telomeres. *Mol Biol Evol* **34:** 2245–2257. doi:10.1093/molbev/msx159

Azzalin CM. 2024. TERRA and the alternative lengthening of telomeres: a dangerous affair. *FEBS Lett* doi:10.1002/1873-3468.14844

Barbour AT, Wuttke DS. 2023. RPA-like single-stranded DNA-binding protein complexes including CST serve as specialized processivity factors for polymerases. *Curr Opin Struct Biol* **81:** 102611. doi:10.1016/j.sbi.2023.102611

Barrientos-Moreno M, Murillo-Pineda M, Muñoz-Cabello AM, Prado F. 2018. Histone depletion prevents telomere fusions in pre-senescent cells. *PLoS Genet* **14:** e1007407. doi:10.1371/journal.pgen.1007407

Belfort M, Curcio MJ, Lue NF. 2011. Telomerase and retrotransposons: reverse transcriptases that shaped genomes. *Proc Natl Acad Sci* **108:** 20304–20310. doi:10.1073/pnas.1100269109

Berloco M, Fanti L, Sheen F, Levis RW, Pimpinelli S. 2005. Heterochromatic distribution of HeT-A- and TART-like sequences in several *Drosophila* species. *Cytogenet Genome Res* **110:** 124–133. doi:10.1159/000084944

Blumenstiel JP. 2019. Birth, school, work, death, and resurrection: the life stages and dynamics of transposable element proliferation. *Genes (Basel)* **10:** 336. doi:10.3390/genes10050336

Brennecke J, Aravin AA, Stark A, Dus M, Kellis M, Sachidanandam R, Hannon GJ. 2007. Discrete small RNA-generating loci as master regulators of transposon activity in *Drosophila*. *Cell* **128:** 1089–1103. doi:10.1016/j.cell.2007.01.043

Bryan TM. 2020. G-Quadruplexes at telomeres: friend or foe? *Molecules* **25:** 3686. doi:10.3390/molecules25163686

Burt A, Trivers R. 2008. *Genes in conflict: the biology of selfish genetic elements.* Harvard University Press, Cambridge, MA.

Cacchione S, Cenci G, Raffa GD. 2020. Silence at the end: how *Drosophila* regulates expression and transposition of telomeric retroelements. *J Mol Biol* **432:** 4305–4321. doi:10.1016/j.jmb.2020.06.004

Casacuberta E, Pardue ML. 2003a. *HeT-A* elements in *Drosophila virilis*: retrotransposon telomeres are conserved across the *Drosophila* genus. *Proc Natl Acad Sci* **100:** 14091–14096. doi:10.1073/pnas.1936193100

Casacuberta E, Pardue ML. 2003b. Transposon telomeres are widely distributed in the *Drosophila* genus: *TART* elements in the *virilis* group. *Proc Natl Acad Sci* **100:** 3363–3368. doi:10.1073/pnas.0230353100

Casacuberta E, Pardue ML. 2005. HeT-A and TART, two *Drosophila* retrotransposons with a bona fide role in chromosome structure for more than 60 million years. *Cytogenet Genome Res* **110:** 152–159. doi:10.1159/000084947

Cenci G, Siriaco G, Raffa GD, Kellum R, Gatti M. 2003. The *Drosophila* HOAP protein is required for telomere capping. *Nat Cell Biol* **5:** 82–84. doi:10.1038/ncb902

Cenci G, Ciapponi L, Gatti M. 2005. The mechanism of telomere protection: a comparison between *Drosophila* and humans. *Chromosoma* **114:** 135–145. doi:10.1007/s00412-005-0005-9

Chen P, Aravin AA. 2021. Transposon-taming piRNAs in the germline: where do they come from? *Mol Cell* **81:** 3884–3885. doi:10.1016/j.molcel.2021.09.017

Chen YA, Stuwe E, Luo Y, Ninova M, Le Thomas A, Rozhavskaya E, Li S, Vempati S, Laver JD, Patel DJ, et al. 2016. Cutoff suppresses RNA polymerase II termination to ensure expression of piRNA precursors. *Mol Cell* **63:** 97–109. doi:10.1016/j.molcel.2016.05.010

Chen T, Wei X, Courret C, Cui M, Cheng L, Wu J, Ahmad K, Larracuente AM, Rong YS. 2022. The nanoCUT&RUN technique visualizes telomeric chromatin in *Drosophila*. *PLoS Genet* **18:** e1010351. doi:10.1371/journal.pgen.1010351

Cheng L, Cui M, Rong YS. 2018. MTV sings jubilation for telomere biology in *Drosophila*. *Fly (Austin)* **12:** 41–45. doi:10.1080/19336934.2017.1325979

Cicconi A, Micheli E, Vernì F, Jackson A, Gradilla AC, Cipressa F, Raimondo D, Bosso G, Wakefield JG, Ciapponi L, et al. 2017. The *Drosophila* telomere-capping protein Verrocchio binds single-stranded DNA and protects telomeres from DNA damage response. *Nucleic Acids Res* **45:** 3068–3085. doi:10.1093/nar/gkw1244

Cosby RL, Chang NC, Feschotte C. 2019. Host-transposon interactions: conflict, cooperation, and cooption. *Genes Dev* **33:** 1098–1116. doi:10.1101/gad.327312.119

Cui M, Bai Y, Li K, Rong YS. 2021. Taming active transposons at *Drosophila* telomeres: the interconnection between HipHop's roles in capping and transcriptional silencing. *PLoS Genet* **17:** e1009925. doi:10.1371/journal.pgen.1009925

Danilevskaya O, Lofsky A, Kurenova EV, Pardue ML. 1993. The Y chromosome of *Drosophila melanogaster* contains a distinctive subclass of Het-A-related repeats. *Genetics* **134:** 531–543. doi:10.1093/genetics/134.2.531

Danilevskaya ON, Tan C, Wong J, Alibhai M, Pardue ML. 1998. Unusual features of the *Drosophila melanogaster* telomere transposable element *HeT-A* are conserved in *Drosophila yakuba* telomere elements. *Proc Natl Acad Sci* **95:** 3770–3775. doi:10.1073/pnas.95.7.3770

Danilevskaya ON, Traverse KL, Hogan NC, DeBaryshe PG, Pardue ML. 1999. The two *Drosophila* telomeric transposable elements have very different patterns of transcription. *Mol Cell Biol* **19:** 873–881. doi:10.1128/MCB.19.1.873

Deamer D, Akeson M, Branton D. 2016. Three decades of nanopore sequencing. *Nat Biotechnol* **34:** 518–524. doi:10.1038/nbt.3423

de Lange T. 2004. T-loops and the origin of telomeres. *Nat Rev Mol Cell Biol* **5:** 323–329. doi:10.1038/nrm1359

de Lange T. 2009. How telomeres solve the end-protection problem. *Science* **326:** 948–952. doi:10.1126/science.1170633

de Lange T. 2015. A loopy view of telomere evolution. *Front Genet* **6:** 321. doi:10.3389/fgene.2015.00321

Dubruille R, Orsi GA, Delabaere L, Cortier E, Couble P, Marais GA, Loppin B. 2010. Specialization of a *Drosophila* capping protein essential for the protection of sperm telomeres. *Curr Biol* **20:** 2090–2099. doi:10.1016/j.cub.2010.11.013

Dubruille R, Herbette M, Revel M, Horard B, Chang CH, Loppin B. 2023. Histone removal in sperm protects paternal chromosomes from premature division at fertilization. *Science* **382:** 725–731. doi:10.1126/science.adh0037

Dufourt J, Dennis C, Boivin A, Gueguen N, Théron E, Goriaux C, Pouchin P, Ronsseray S, Brasset E, Vaury C. 2014. Spatio-temporal requirements for transposable element piRNA-mediated silencing during *Drosophila* oogenesis. *Nucleic Acids Res* **42:** 2512–2524. doi:10.1093/nar/gkt1184

Eid J, Fehr A, Gray J, Luong K, Lyle J, Otto G, Peluso P, Rank D, Baybayan P, Bettman B, et al. 2009. Real-time DNA sequencing from single polymerase molecules. *Science* **323:** 133–138. doi:10.1126/science.1162986

Ellison CE, Kagda MS, Cao W. 2020. Telomeric TART elements target the piRNA machinery in *Drosophila*. *PLoS Biol* **18:** e3000689. doi:10.1371/journal.pbio.3000689

ElMaghraby MF, Andersen PR, Pühringer F, Hohmann U, Meixner K, Lendl T, Tirian L, Brennecke J. 2019. A heterochromatin-specific RNA export pathway facilitates piRNA production. *Cell* **178:** 964–979.e20. doi:10.1016/j.cell.2019.07.007

Emerson RO, Thomas JH. 2009. Adaptive evolution in zinc finger transcription factors. *PLoS Genet* **5:** e1000325. doi:10.1371/journal.pgen.1000325

Gamez S, Srivastav S, Akbari OS, Lau NC. 2020. Diverse defenses: a perspective comparing Dipteran Piwi-piRNA pathways. *Cells* **9:** 2180. doi:10.3390/cells9102180

Gao G, Bi X, Chen J, Srikanta D, Rong YS. 2009. Mre11-Rad50-Nbs complex is required to cap telomeres during *Drosophila* embryogenesis. *Proc Natl Acad Sci* **106:** 10728–10733. doi:10.1073/pnas.0902707106

Gao G, Cheng Y, Wesolowska N, Rong YS. 2011. Paternal imprint essential for the inheritance of telomere identity in *Drosophila*. *Proc Natl Acad Sci* **108:** 4932–4937. doi:10.1073/pnas.1016792108

George JA, DeBaryshe PG, Traverse KL, Celniker SE, Pardue ML. 2006. Genomic organization of the *Drosophila* telomere retrotransposable elements. *Genome Res* **16:** 1231–1240. doi:10.1101/gr.5348806

George JA, Traverse KL, DeBaryshe PG, Kelley KJ, Pardue ML. 2010. Evolution of diverse mechanisms for protecting chromosome ends by *Drosophila* TART telomere retrotransposons. *Proc Natl Acad Sci* **107:** 21052–21057. doi:10.1073/pnas.1015926107

Glousker G, Lingner J. 2021. Challenging endings: how telomeres prevent fragility. *Bioessays* **43:** e2100157. doi:10.1002/bies.202100157

Henninger E, Teixeira MT. 2020. Telomere-driven mutational processes in yeast. *Curr Opin Genet Dev* **60:** 99–106. doi:10.1016/j.gde.2020.02.018

Hoang SM, O'Sullivan RJ. 2020. Alternative lengthening of telomeres: building bridges to connect chromosome ends. *Trends Cancer* **6:** 247–260. doi:10.1016/j.trecan.2019.12.009

Huang X, Fejes Tóth K, Aravin AA. 2017. piRNA biogenesis in *Drosophila melanogaster*. *Trends Genet* **33:** 882–894. doi:10.1016/j.tig.2017.09.002

Ilyin AA, Stolyarenko AD, Zenkin N, Klenov MS. 2021. Complex genetic interactions between Piwi and HP1a in the repression of transposable elements and tissue-specific genes in the ovarian germline. *Int J Mol Sci* **22:** 13430. doi:10.3390/ijms222413430

Iyer SS, Sun Y, Seyfferth J, Manjunath V, Samata M, Alexiadis A, Kulkarni T, Gutierrez N, Georgiev P, Shvedunova M, et al. 2023. The NSL complex is required for piRNA production from telomeric clusters. *Life Sci Alliance* **6:** e202302194. doi:10.26508/lsa.202302194

Jangam D, Feschotte C, Betrán E. 2017. Transposable element domestication as an adaptation to evolutionary conflicts. *Trends Genet* **33:** 817–831. doi:10.1016/j.tig.2017.07.011

Jedlička P, Tokan V, Kejnovská I, Hobza R, Kejnovský E. 2023. Telomeric retrotransposons show propensity to form G-quadruplexes in various eukaryotic species. *Mob DNA* **14:** 3. doi:10.1186/s13100-023-00291-9

Kahn T, Savitsky M, Georgiev P. 2000. Attachment of *HeT-A* sequences to chromosomal termini in *Drosophila melanogaster* may occur by different mechanisms. *Mol Cell Biol* **20:** 7634–7642. doi:10.1128/MCB.20.20.7634-7642.2000

Kalmykova AI, Sokolova OA. 2023. Retrotransposons and telomeres. *Biochemistry (Mosc)* **88:** 1739–1753. doi:10.1134/S0006297923110068

Karimian K, Groot A, Huso V, Kahidi R, Tan KT, Sholes S, Keener R, McDyer JF, Alder JK, Li H, et al. 2024. Human telomere length is chromosome end-specific and conserved across individuals. *Science* **384:** 533–539. doi:10.1126/science.ado0431

Kass-Eisler A, Greider CW. 2000. Recombination in telomere-length maintenance. *Trends Biochem Sci* **25:** 200–204. doi:10.1016/S0968-0004(00)01557-7

Khurana JS, Xu J, Weng Z, Theurkauf WE. 2010. Distinct functions for the *Drosophila* piRNA pathway in genome maintenance and telomere protection. *PLoS Genet* **6:** e1001246. doi:10.1371/journal.pgen.1001246

Kidwell MG, Lisch DR. 2001. Perspective: transposable elements, parasitic DNA, and genome evolution. *Evolution (N Y)* **55:** 1–24. doi:10.1111/j.0014-3820.2001.tb01268.x

Kolaczkowski B, Hupalo DN, Kern AD. 2011. Recurrent adaptation in RNA interference genes across the *Drosoph*-

*ila* phylogeny. *Mol Biol Evol* **28:** 1033–1042. doi:10.1093/molbev/msq284

Komarov PA, Sokolova O, Akulenko N, Brasset E, Jensen S, Kalmykova A. 2020. Epigenetic requirements for triggering heterochromatinization and Piwi-interacting RNA production from transgenes in the *Drosophila* germline. *Cells* **9:** 922. doi:10.3390/cells9040922

Kordyukova MY, Kalmykova AI. 2019. Nature and functions of telomeric transcripts. *Biochemistry (Mosc)* **84:** 137–146. doi:10.1134/S0006297919020044

Kordyukova M, Sokolova O, Morgunova V, Ryazansky S, Akulenko N, Glukhov S, Kalmykova A. 2020. Nuclear Ccr4-Not mediates the degradation of telomeric and transposon transcripts at chromatin in the *Drosophila* germline. *Nucleic Acids Res* **48:** 141–156. doi:10.1093/nar/gkz1072

Lee YC, Langley CH. 2010. Transposable elements in natural populations of *Drosophila melanogaster*. *Phil Trans R Soc Lond B Biol Sci* **365:** 1219–1228. doi:10.1098/rstb.2009.0318

Lee YC, Leek C, Levine MT. 2017. Recurrent innovation at genes required for telomere integrity in *Drosophila*. *Mol Biol Evol* **34:** 467–482. doi:10.1093/molbev/msw248

Le Thomas A, Stuwe E, Li S, Du J, Marinov G, Rozhkov N, Chen YC, Luo Y, Sachidanandam R, Toth KF, et al. 2014. Transgenerationally inherited piRNAs trigger piRNA biogenesis by changing the chromatin of piRNA clusters and inducing precursor processing. *Genes Dev* **28:** 1667–1680. doi:10.1101/gad.245514.114

Levine MT, Vander Wende HM, Malik HS. 2015. Mitotic fidelity requires transgenerational action of a testis-restricted HP1. *eLife* **4:** e07378. doi:10.7554/eLife.07378

Levis RW, Ganesan R, Houtchens K, Tolar LA, Sheen FM. 1993. Transposons in place of telomeric repeats at a *Drosophila* telomere. *Cell* **75:** 1083–1093. doi:10.1016/0092-8674(93)90318-K

Lewis JD, Song Y, de Jong ME, Bagha SM, Ausió J. 2003. A walk though vertebrate and invertebrate protamines. *Chromosoma* **111:** 473–482. doi:10.1007/s00412-002-0226-0

Lim AK, Kai T. 2007. Unique germ-line organelle, nuage, functions to repress selfish genetic elements in *Drosophila melanogaster*. *Proc Natl Acad Sci* **104:** 6714–6719. doi:10.1073/pnas.0701920104

Liu M, Xie XJ, Li X, Ren X, Sun J, Lin Z, Hemba-Waduge RU, Ji JY. 2023. Transcriptional coupling of telomeric retrotransposons with the cell cycle. bioRxiv doi:10.1101/2023.09.30.560321

Loe TK, Lazzerini Denchi E, Tricola GM, Azeroglu B. 2023. ALTercations at telomeres: stress, recombination and extrachromosomal affairs. *Biochem Soc Trans* **51:** 1935–1946. doi:10.1042/BST20230265

Losada A, Abad JP, Villasante A. 1997. Organization of DNA sequences near the centromere of the *Drosophila melanogaster* Y chromosome. *Chromosoma* **106:** 503–512. doi:10.1007/s004120050272

Lue NF. 2018. Evolving linear chromosomes and telomeres: a C-strand-centric view. *Trends Biochem Sci* **43:** 314–326. doi:10.1016/j.tibs.2018.02.008

Luo Y, He P, Kanrar N, Fejes Toth K, Aravin AA. 2023. Maternally inherited siRNAs initiate piRNA cluster formation. *Mol Cell* **83:** 3835–3851.e7. doi:10.1016/j.molcel.2023.09.033

Markova DN, Christensen SM, Betrán E. 2020. Telomere-specialized retroelements in *Drosophila*: adaptive symbionts of the genome, neutral, or in conflict? *Bioessays* **42:** e1900154. doi:10.1002/bies.201900154

Mason JM, Randall TA, Capkova Frydrychova R. 2015. Telomerase lost? *Chromosoma* **125:** 65–73. doi:10.1007/s00412-015-0528-7

Maxwell PH, Belote JM, Levis RW. 2006. Identification of multiple transcription initiation, polyadenylation, and splice sites in the *Drosophila melanogaster* TART family of telomeric retrotransposons. *Nucleic Acids Res* **34:** 5498–5507. doi:10.1093/nar/gkl709

McGurk MP, Dion-Côté AM, Barbash DA. 2021. Rapid evolution at the *Drosophila* telomere: transposable element dynamics at an intrinsically unstable locus. *Genetics* **217:** iyaa027. doi:10.1093/genetics/iyaa027

Melnikova L, Georgiev P. 2002. *Enhancer of terminal gene conversion*, a new mutation in *Drosophila melanogaster* that induces telomere elongation by gene conversion. *Genetics* **162:** 1301–1312. doi:10.1093/genetics/162.3.1301

Melnikova L, Biessmann H, Georgiev P. 2005. The Ku protein complex is involved in length regulation of *Drosophila* telomeres. *Genetics* **170:** 221–235. doi:10.1534/genetics.104.034538

Miga KH, Eichler EE. 2023. Envisioning a new era: complete genetic information from routine, telomere-to-telomere genomes. *Am J Hum Genet* **110:** 1832–1840. doi:10.1016/j.ajhg.2023.09.011

Mikhailovsky S, Belenkaya T, Georgiev P. 1999. Broken chromosomal ends can be elongated by conversion in *Drosophila melanogaster*. *Chromosoma* **108:** 114–120. doi:10.1007/s004120050358

Mohn F, Sienski G, Handler D, Brennecke J. 2014. The rhino-deadlock-cutoff complex licenses noncanonical transcription of dual-strand piRNA clusters in *Drosophila*. *Cell* **157:** 1364–1379. doi:10.1016/j.cell.2014.04.031

Morciano P, Zhang Y, Cenci G, Rong YS. 2013. A hypomorphic mutation reveals a stringent requirement for the ATM checkpoint protein in telomere protection during early cell division in *Drosophila*. *G3 (Bethesda)* **3:** 1043–1048. doi:10.1534/g3.113.006312

Morgunova VV, Sokolova OA, Sizova TV, Malaev LG, Babaev DS, Kwon DA, Kalmykova AI. 2022. Dysfunction of Lamin B and physiological aging cause telomere instability in *Drosophila* germline. *Biochemistry (Mosc)* **87:** 1600–1610. doi:10.1134/S000629792212015X

Myler LR, Kinzig CG, Sasi NK, Zakusilo G, Cai SW, de Lange T. 2021. The evolution of metazoan shelterin. *Genes Dev* **35:** 1625–1641. doi:10.1101/gad.348835.121

Nakamura TM, Cech TR. 1998. Reversing time: origin of telomerase. *Cell* **92:** 587–590. doi:10.1016/S0092-8674(00)81123-X

Neumann AA, Watson CM, Noble JR, Pickett HA, Tam PP, Reddel RR. 2013. Alternative lengthening of telomeres in normal mammalian somatic cells. *Genes Dev* **27:** 18–23. doi:10.1101/gad.205062.112

Ninova M, Godneeva B, Chen YA, Luo Y, Prakash SJ, Jankovics F, Erdélyi M, Aravin AA, Fejes Tóth K. 2020. The SUMO ligase Su(var)2-10 controls hetero- and euchromatic gene expression via establishing H3K9 trimethylation and negative feedback regulation. *Mol Cell* **77**: 571–585.e4. doi:10.1016/j.molcel.2019.09.033

Nurk S, Koren S, Rhie A, Rautiainen M, Bzikadze AV, Mikheenko A, Vollger MR, Altemose N, Uralsky L, Gershman A, et al. 2022. The complete sequence of a human genome. *Science* **376**: 44–53. doi:10.1126/science.abj6987

Obbard DJ, Gordon KH, Buck AH, Jiggins FM. 2009. The evolution of RNAi as a defence against viruses and transposable elements. *Philos Trans R Soc Lond B Biol Sci* **364**: 99–115. doi:10.1098/rstb.2008.0168

Okamura K, Balla S, Martin R, Liu N, Lai EC. 2008. Two distinct mechanisms generate endogenous siRNAs from bidirectional transcription in *Drosophila melanogaster*. *Nat Struct Mol Biol* **15**: 581–590. doi:10.1038/nsmb.1438

Pardue ML, DeBaryshe PG. 2008. *Drosophila* telomeres: a variation on the telomerase theme. *Fly (Austin)* **2**: 101–110. doi:10.4161/fly.6393

Pardue ML, DeBaryshe PG. 2011. Retrotransposons that maintain chromosome ends. *Proc Natl Acad Sci* **108**: 20317–20324. doi:10.1073/pnas.1100278108

Perrat PN, DasGupta S, Wang J, Theurkauf W, Weng Z, Rosbash M, Waddell S. 2013. Transposition-driven genomic heterogeneity in the *Drosophila* brain. *Science* **340**: 91–95. doi:10.1126/science.1231965

Perrini B, Piacentini L, Fanti L, Altieri F, Chichiarelli S, Berloco M, Turano C, Ferraro A, Pimpinelli S. 2004. HP1 controls telomere capping, telomere elongation, and telomere silencing by two different mechanisms in *Drosophila*. *Mol Cell* **15**: 467–476. doi:10.1016/j.molcel.2004.06.036

Porrazzo A, Cipressa F, De Gregorio A, De Pittà C, Sales G, Ciapponi L, Morciano P, Esposito G, Tabocchini MA, Cenci G. 2022. Low dose rate γ-irradiation protects fruit fly chromosomes from double strand breaks and telomere fusions by reducing the esi-RNA biogenesis factor loquacious. *Commun Biol* **5**: 905. doi:10.1038/s42003-022-03885-w

Radion E, Ryazansky S, Akulenko N, Rozovsky Y, Kwon D, Morgunova V, Olovnikov I, Kalmykova A. 2017. Telomeric retrotransposon HeT-A contains a bidirectional promoter that initiates divergent transcription of piRNA precursors in *Drosophila* germline. *J Mol Biol* **429**: 3280–3289. doi:10.1016/j.jmb.2016.12.002

Radion E, Morgunova V, Ryazansky S, Akulenko N, Lavrov S, Abramov Y, Komarov PA, Glukhov SI, Olovnikov I, Kalmykova A. 2018. Key role of piRNAs in telomeric chromatin maintenance and telomere nuclear positioning in *Drosophila* germline. *Epigenetics Chromatin* **11**: 40. doi:10.1186/s13072-018-0210-4

Raffa GD, Ciapponi L, Cenci G, Gatti M. 2011. Terminin: a protein complex that mediates epigenetic maintenance of *Drosophila* telomeres. *Nucleus* **2**: 383–391. doi:10.4161/nucl.2.5.17873

Raffa GD, Cenci G, Ciapponi L, Gatti M. 2013. Organization and evolution of *Drosophila* terminin: similarities and differences between *Drosophila* and human telomeres. *Front Oncol* **3**: 112. doi:10.3389/fonc.2013.00112

Rashkova S, Karam SE, Kellum R, Pardue ML. 2002. Gag proteins of the two *Drosophila* telomeric retrotransposons are targeted to chromosome ends. *J Cell Biol* **159**: 397–402. doi:10.1083/jcb.200205039

Rautiainen M, Nurk S, Walenz BP, Logsdon GA, Porubsky D, Rhie A, Eichler EE, Phillippy AM, Koren S. 2023. Telomere-to-telomere assembly of diploid chromosomes with Verkko. *Nat Biotechnol* **41**: 1474–1482. doi:10.1038/s41587-023-01662-6

Reddy HM, Randall TA, Cipressa F, Porrazzo A, Cenci G, Frydrychova RC, Mason JM. 2022. Identification of the *telomere elongation* mutation in *Drosophila*. *Cells* **11**: 3484. doi:10.3390/cells11213484

Reig-Viader R, Garcia-Caldés M, Ruiz-Herrera A. 2016. Telomere homeostasis in mammalian germ cells: a review. *Chromosoma* **125**: 337–351. doi:10.1007/s00412-015-0555-4

Rong YS. 2008. Loss of the histone variant H2A.Z restores capping to checkpoint-defective telomeres in *Drosophila*. *Genetics* **180**: 1869–1875. doi:10.1534/genetics.108.095547

Said I, McGurk MP, Clark AG, Barbash DA. 2022. Patterns of piRNA regulation in *Drosophila* revealed through transposable element clade inference. *Mol Biol Evol* **39**: msab336. doi:10.1093/molbev/msab336

Saint-Leandre B, Nguyen SC, Levine MT. 2019. Diversification and collapse of a telomere elongation mechanism. *Genome Res* **29**: 920–931. doi:10.1101/gr.245001.118

Saint-Leandre B, Christopher C, Levine MT. 2020. Adaptive evolution of an essential telomere protein restricts telomeric retrotransposons. *eLife* **9**: e60987. doi:10.7554/eLife.60987

Sanchez SE, Gu Y, Wang Y, Golla A, Martin A, Shomali W, Hockemeyer D, Savage SA, Artandi SE. 2024. Digital telomere measurement by long-read sequencing distinguishes healthy aging from disease. *Nat Commun* **15**: 5148. doi:10.1038/s41467-024-49007-4

Sato K, Siomi MC. 2020. The piRNA pathway in *Drosophila* ovarian germ and somatic cells. *Proc Jpn Acad Ser B Phys Biol Sci* **96**: 32–42. doi:10.2183/pjab.96.003

Schmid KJ, Tautz D. 1997. A screen for fast evolving genes from *Drosophila*. *Proc Natl Acad Sci* **94**: 9746–9750. doi:10.1073/pnas.94.18.9746

Schmidt TT, Tyer C, Rughani P, Haggblom C, Jones JR, Dai X, Frazer KA, Gage FH, Juul S, Hickey S, et al. 2024. High resolution long-read telomere sequencing reveals dynamic mechanisms in aging and cancer. *Nat Commun* **15**: 5149. doi:10.1038/s41467-024-48917-7

Schnabl J, Wang J, Hohmann U, Gehre M, Batki J, Andreev VI, Purkhauser K, Fasching N, Duchek P, Novatchkova M, et al. 2021. Molecular principles of Piwi-mediated cotranscriptional silencing through the dimeric SFiNX complex. *Genes Dev* **35**: 392–409. doi:10.1101/gad.347989.120

Senti KA, Jurczak D, Sachidanandam R, Brennecke J. 2015. piRNA-guided slicing of transposon transcripts enforces their transcriptional silencing via specifying the nuclear piRNA repertoire. *Genes Dev* **29**: 1747–1762. doi:10.1101/gad.267252.115

Servant G, Deininger PL. 2015. Insertion of retrotransposons at chromosome ends: adaptive response to chromosome maintenance. *Front Genet* **6**: 358. doi:10.3389/fgene.2015.00358

Shareef MM, King C, Damaj M, Badagu R, Huang DW, Kellum R. 2001. *Drosophila* heterochromatin protein 1 (HP1)/origin recognition complex (ORC) protein is associated with HP1 and ORC and functions in heterochromatin-induced silencing. *Mol Biol Cell* **12:** 1671–1685. doi:10.1091/mbc.12.6.1671

Shpiz S, Kalmykova A. 2009. Epigenetic transmission of piRNAs through the female germline. *Genome Biol* **10:** 208. doi:10.1186/gb-2009-10-2-208

Shpiz S, Kalmykova A. 2011. Role of piRNAs in the *Drosophila* telomere homeostasis. *Mob Genet Elements* **1:** 274–278. doi:10.4161/mge.18301

Siriaco GM, Cenci G, Haoudi A, Champion LE, Zhou C, Gatti M, Mason JM. 2002. *Telomere elongation (Tel)*, a new mutation in *Drosophila melanogaster* that produces long telomeres. *Genetics* **160:** 235–245. doi:10.1093/genetics/160.1.235

Siudeja K, van den Beek M, Riddiford N, Boumard B, Wurmser A, Stefanutti M, Lameiras S, Bardin AJ. 2021. Unraveling the features of somatic transposition in the *Drosophila* intestine. *EMBO J* **40:** e106388. doi:10.15252/embj.2020106388

Sokolova O, Morgunova V, Sizova TV, Komarov PA, Olenkina OM, Babaev DS, Mikhaleva EA, Kwon DA, Erokhin M, Kalmykova A. 2023. The insulator BEAF32 controls the spatial-temporal expression profile of the telomeric retrotransposon *TART* in the *Drosophila* germline. *Development* **150:** dev201678. doi:10.1242/dev.201678

Takeuchi C, Yokoshi M, Kondo S, Shibuya A, Saito K, Fukaya T, Siomi H, Iwasaki YW. 2022. Mod(mdg4) variants repress telomeric retrotransposon *HeT-A* by blocking subtelomeric enhancers. *Nucleic Acids Res* **50:** 11580–11599. doi:10.1093/nar/gkac1034

Tan KT, Slevin MK, Leibowitz ML, Garrity-Janger M, Shan J, Li H, Meyerson M. 2024. Neotelomeres and telomere-spanning chromosomal arm fusions in cancer genomes revealed by long-read sequencing. *Cell Genom* **4:** 100588. doi:10.1016/j.xgen.2024.100588

Teo RYW, Anand A, Sridhar V, Okamura K, Kai T. 2018. Heterochromatin protein 1a functions for piRNA biogenesis predominantly from pericentric and telomeric regions in *Drosophila*. *Nat Commun* **9:** 1735. doi:10.1038/s41467-018-03908-3

Théron E, Maupetit-Mehouas S, Pouchin P, Baudet L, Brasset E, Vaury C. 2018. The interplay between the Argonaute proteins Piwi and Aub within *Drosophila* germarium is critical for oogenesis, piRNA biogenesis and TE silencing. *Nucleic Acids Res* **46:** 10052–10065. doi:10.1093/nar/gky695

Treiber CD, Waddell S. 2017. Resolving the prevalence of somatic transposition in *Drosophila*. *eLife* **6:** e28297. doi:10.7554/eLife.28297

Vedelek B, Kovács A, Boros IM. 2021. Evolutionary mode for the functional preservation of fast-evolving *Drosophila* telomere capping proteins. *Open Biol* **11:** 210261. doi:10.1098/rsob.210261

Villasante A, Abad JP, Planelló R, Méndez-Lago M, Celniker SE, de Pablos B. 2007. *Drosophila* telomeric retrotranspo-

sons derived from an ancestral element that was recruited to replace telomerase. *Genome Res* **17:** 1909–1918. doi:10.1101/gr.6365107

Walter MF, Biessmann MR, Benitez C, Török T, Mason JM, Biessmann H. 2007. Effects of telomere length in *Drosophila melanogaster* on life span, fecundity, and fertility. *Chromosoma* **116:** 41–51. doi:10.1007/s00412-006-0081-5

Webster A, Li S, Hur JK, Wachsmuth M, Bois JS, Perkins EM, Patel DJ, Aravin AA. 2015. Aub and Ago3 are recruited to Nuage through two mechanisms to form a ping-pong complex assembled by Krimper. *Mol Cell* **59:** 564–575. doi:10.1016/j.molcel.2015.07.017

Wei KH, Reddy HM, Rathnam C, Lee J, Lin D, Ji S, Mason JM, Clark AG, Barbash DA. 2017. A pooled sequencing approach identifies a candidate meiotic driver in *Drosophila*. *Genetics* **206:** 461–465. doi:10.1534/genetics.116.197335

Wells JN, Chang NC, McCormick J, Coleman C, Ramos N, Jin B, Feschotte C. 2023. Transposable elements drive the evolution of metazoan zinc finger genes. *Genome Res* **33:** 1325–1339. doi:10.1101/gr.277966.123

Werren JH. 2011. Selfish genetic elements, genetic conflict, and evolutionary innovation. *Proc Natl Acad Sci* **108:** 10863–10870. doi:10.1073/pnas.1102343108

Yamaki T, Yasuda GK, Wakimoto BT. 2016. The Deadbeat paternal effect of uncapped sperm telomeres on cell cycle progression and chromosome behavior in *Drosophila melanogaster*. *Genetics* **203:** 799–816. doi:10.1534/genetics.115.182436

Yasuda GK, Schubiger G, Wakimoto BT. 1995. Genetic characterization of ms(3)K81, a paternal effect gene of *Drosophila melanogaster*. *Genetics* **140:** 219–229. doi:10.1093/genetics/140.1.219

Zeinoun B, Teixeira MT, Barascu A. 2023. TERRA and telomere maintenance in the yeast *Saccharomyces cerevisiae*. *Genes (Basel)* **14:** 618. doi:10.3390/genes14030618

Zhang L, Beaucher M, Cheng Y, Rong YS. 2014. Coordination of transposon expression with DNA replication in the targeting of telomeric retrotransposons in *Drosophila*. *EMBO J* **33:** 1148–1158.

Zhang Y, Zhang L, Tang X, Bhardwaj SR, Ji J, Rong YS. 2016. MTV, an ssDNA protecting complex essential for transposon-based telomere maintenance in *Drosophila*. *PLoS Genet* **12:** e1006435. doi:10.1371/journal.pgen.1006435

Zhao K, Cheng S, Miao N, Xu P, Lu X, Zhang Y, Wang M, Ouyang X, Yuan X, Liu W, et al. 2019. A Pandas complex adapted for piRNA-guided transcriptional silencing and heterochromatin formation. *Nat Cell Biol* **21:** 1261–1272. doi:10.1038/s41556-019-0396-0

Zhao S, Zhang X, Chen S, Zhang S. 2020. Natural antisense transcripts in the biological hallmarks of cancer: powerful regulators hidden in the dark. *J Exp Clin Cancer Res* **39:** 187. doi:10.1186/s13046-020-01700-0

# Index

www.ingramcontent.com/pod-product-compliance
Lightning Source LLC
Chambersburg PA
CBHW061929190326
41458CB00009B/2698